WileyPLUS

ALL THE HELP, **RESOURCES,** AND PERSONAL SUPPORT YOU AND YOUR STUDENTS NEED!

www.wileyplus.com/resources

1st DAY OF CLASS *... AND BEYOND!*

2-Minute Tutorials and all of the resources you and your students need to get started

WileyPLUS
Student Partner Program

Student support from an experienced student user

Wiley Faculty Network

Collaborate with your colleagues, find a mentor, attend virtual and live events, and view resources
www.WhereFacultyConnect.com

WileyPLUS

Quick Start

Pre-loaded, ready-to-use assignments and presentations created by subject matter experts

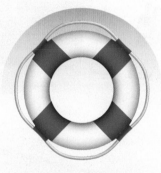

Technical Support 24/7
FAQs, online chat, and phone support
www.wileyplus.com/support

© Courtney Keating/iStockphoto

Your *WileyPLUS* Account Manager, providing personal training and support

9th Edition

Materials Science and Engineering

SI Version

WILLIAM D. CALLISTER, JR.
Department of Metallurgical Engineering
The University of Utah

DAVID G. RETHWISCH
Department of Chemical and Biochemical Engineering
The University of Iowa

WILEY

ISBN: 978-1-118-31922-2

Printed in Asia

10 9 8 7 6 5 4

Dedicated to
Bill Stenquist, editor and friend

Preface

In this ninth edition we have retained the objectives and approaches for teaching materials science and engineering that were presented in previous editions. **The first, and primary, objective** is to present the basic fundamentals on a level appropriate for university/college students who have completed their freshmen calculus, chemistry, and physics courses.

The **second objective** is to present the subject matter in a logical order, from the simple to the more complex. Each chapter builds on the content of previous ones.

The third objective, or philosophy, that we strive to maintain throughout the text is that if a topic or concept is worth treating, then it is worth treating in sufficient detail and to the extent that students have the opportunity to fully understand it without having to consult other sources; in addition, in most cases, some practical relevance is provided.

The fourth objective is to include features in the book that will expedite the learning process. These learning aids include the following:

- Numerous illustrations, now presented in full color, and photographs to help visualize what is being presented

- Learning objectives, to focus students' attention on what they should be getting from each chapter

- "Why Study . . ." and "Materials of Importance" items as well as case studies that provide relevance to topic discussions

- "Concept Check" questions that test whether a student understands the subject matter on a conceptual level

- Key terms, and descriptions of key equations, highlighted in the margins for quick reference

- End-of-chapter questions and problems designed to progressively develop students' understanding of concepts and facility with skills

- Answers to selected problems, so students can check their work

- A glossary, a global list of symbols, and references to facilitate understanding of the subject matter

- End-of-chapter summary tables of important equations and symbols used in these equations

- Processing/Structure/Properties/Performance correlations and summary concept maps for four materials (steels, glass-ceramics, polymer fibers, and silicon semiconductors), which integrate important concepts from chapter to chapter

- Materials of Importance sections that lend relevance to topical coverage by discussing familiar and interesting materials and their applications

The fifth objective is to enhance the teaching and learning process by using the newer technologies that are available to most instructors and today's engineering students.

New/Revised Content

Several important changes have been made with this ninth edition. One of the most significant is the incorporation of several new sections, as well as revisions/amplifications of other sections. These include the following:

- Reorganization in the sequencing and content of several chapters. These changes were made in response to suggestions from adopters of previous editions.
- Numerous new and revised example problems
- Revised, expanded, and updated tables
- Two new case studies: "Liberty Ship Failures" (Chapter 1) and "Use of Composites in the Boeing 787 Dreamliner" (Chapter 16)
- Bond hybridization in carbon (Chapter 2)
- Revision of discussions on crystallographic planes and directions to include the use of equations for the determination of planar and directional indices (Chapter 3)
- Revised discussion on determination of grain size (Chapter 6)
- New section on the structure of carbon fibers (Chapter 14)
- Revised/expanded discussions on structures, properties, and applications of the nanocarbons: fullerenes, carbon nanotubes, and graphene (Chapter 14)
- Revised/expanded discussion on structural composites: laminar composites and sandwich panels (Chapter 16)
- New section on structure, properties, and applications of nanocomposite materials (Chapter 16)
- Tutorial videos. In *WileyPLUS*, Tutorial Videos help students with their "muddiest points" in conceptual understanding and problem-solving.
- Exponents and logarithms. In *WileyPLUS*, the exponential functions and natural logarithms have been added to the Exponents and Logarithms section of the Math Skills Review.

Online Learning Resources—Student Companion Site at www.wiley.com/college/callister.

Also found on the book's website is a Students' Companion page on which are posted several important instructional elements for the student that complement the text; these include the following:

- **Answers to Concept Check questions**, questions which are found in the print book.
- **Library of Case Studies.** One way to demonstrate principles of *design* in an engineering curriculum is via case studies: analyses of problem-solving strategies applied to real-world examples of applications/devices/failures encountered by engineers. Five case studies are provided as follows: (1) Materials Selection for a Torsionally Stressed Cylindrical Shaft; (2) Automobile Valve Spring; (3) Failure of an Automobile Rear Axle; (4) Artificial Total Hip Replacement; and (5) Chemical Protective Clothing.
- **Mechanical Engineering (ME) Module.** This module treats materials science/ engineering topics not covered in the printed text that are relevant to mechanical engineering.
- **Extended Learning Objectives.** This is a more extensive list of learning objectives than is provided at the beginning of each chapter. These direct the student to study the subject material to a greater depth.

- **Student Lecture PowerPoint® Slides**. These slides (in both Adobe Acrobat® PDF and PowerPoint® formats) are virtually identical to the lecture slides provided to an instructor for use in the classroom. The student set has been designed to allow for note taking on printouts.

- **Index of Learning Styles**. Upon answering a 44-item questionnaire, a user's learning-style preference (i.e., the manner in which information is assimilated and processed) is assessed.

Online Resources for Instructors—Instructors Companion Site at www.wiley.com/college/callister.

The Instructor Companion Site is available for instructors who have adopted this text. Please visit the website to register for access. Resources that are available include the following:

- **All resources found on the Student Companion Site.** (Except for the Student Lecture PowerPoint® Slides.)

- **Instructor Solutions Manual.** Detailed solutions for all end-of-chapter questions and problems (in both Word® and Adobe Acrobat® PDF formats).

- *Virtual Materials Science and Engineering (VMSE).* This web-based software package consists of interactive simulations and animations that enhance the learning of key concepts in materials science and engineering. Included in VMSE are eight modules and a materials properties/cost database. Titles of these modules are as follows: (1) Metallic Crystal Structures and Crystallography; (2) Ceramic Crystal Structures; (3) Repeat Unit and Polymer Structures; (4) Dislocations; (5) Phase Diagrams; (6) Diffusion; (7) Tensile Tests; and (8) Solid-Solution Strengthening.

- **Image Gallery.** Illustrations from the book. Instructors can use them in assignments, tests, or other exercises they create for students.

- **Art PowerPoint Slides.** Book art loaded into PowerPoints, so instructors can more easily use them to create their own PowerPoint Slides.

- **Lecture Note PowerPoints.** These slides, developed by the authors and Peter M. Anderson (The Ohio State University), follow the flow of topics in the text, and include materials taken from the text as well as other sources. Slides are available in both Adobe Acrobat® PDF and PowerPoint® formats. [*Note:* If an instructor doesn't have available all fonts used by the developer, special characters may not be displayed correctly in the PowerPoint version (i.e., it is not possible to embed fonts in PowerPoints); however, in the PDF version, these characters will appear correctly.]

- **Solutions to Case Study Problems.**

- **Solutions to Problems in the Mechanical Engineering Web Module.**

- **Suggested Course Syllabi for the Various Engineering Disciplines.** Instructors may consult these syllabi for guidance in course/lecture organization and planning.

- **Experiments and Classroom Demonstrations.** Instructions and outlines for experiments and classroom demonstrations that portray phenomena and/or illustrate principles that are discussed in the book; references are also provided that give more detailed accounts of these demonstrations.

WileyPLUS

WileyPLUS is a research-based online environment for effective teaching and learning.

WileyPLUS builds students' confidence by taking the guesswork out of studying by providing them with a clear roadmap: what is assigned, what is required for each assignment, and whether assignments are done correctly. Independent research has shown that students using *WileyPLUS* will take more initiative so the instructor has a greater impact on their achievement in the classroom and beyond. *WileyPLUS* also helps students study and progress at a pace that's right for them. Our integrated resources–available 24/7–function like a personal tutor, directly addressing each student's demonstrated needs by providing specific problem-solving techniques.

What do students receive with *WileyPLUS*?

- The complete digital textbook.
- Navigation assistance, including links to relevant sections in the online textbook.
- Immediate feedback on performance and progress, 24/7.
- Integrated, multi-media resources—to include *VMSE* (*Virtual Materials Science & Engineering*), tutorial videos, a Math Skills Review, flashcards, and much more; these resources provide multiple study paths and encourage more active learning.

What do instructors receive with *WileyPLUS*?

- The ability to effectively and efficiently personalize and manage their course.
- The ability to track student performance and progress, and easily identify those who are falling behind.
- Media-rich course materials and assessment resources including a complete Solutions Manual, PowerPoint® Lecture Slides, Extended Learning Objectives, and much more. www.WileyPLUS.com

Feedback

We have a sincere interest in meeting the needs of educators and students in the materials science and engineering community, and therefore we solicit feedback on this edition. Comments, suggestions, and criticisms may be submitted to the authors via email at the following address: billcallister@comcast.net.

Acknowledgments

Since we undertook the task of writing this and previous editions, instructors and students, too numerous to mention, have shared their input and contributions on how to make this work more effective as a teaching and learning tool. To all those who have helped, we express our sincere thanks.

We express our appreciation to those who have made contributions to this edition. We are especially indebted to the following:

Audrey Butler of The University of Iowa, and Bethany Smith and Stephen Krause of Arizona State University, for helping to develop material in the *WileyPLUS* course.

Grant Head for his expert programming skills, which he used in developing the *Virtual Materials Science and Engineering* software.

Eric Hellstrom and Theo Siegrist of Florida State University for their feedback and suggestions for this edition.

In addition, we thank the many instructors who participated in the fall 2011 marketing survey; their valuable contributions were driving forces for many of the changes and additions to this ninth edition.

We are also indebted to Dan Sayre, Executive Editor; Jennifer Welter, Senior Product Designer; and Jessica Knecht, Editorial Program Assistant, for their guidance and assistance on this revision.

Last, but certainly not least, we deeply and sincerely appreciate the continual encouragement and support of our families and friends.

William D. Callister, Jr.
David G. Rethwisch
January 2014

Contents

List of Symbols

The number of the section in which a symbol is introduced or explained is given in parentheses.

A = area

Å = angstrom unit

A_i = atomic weight of element i (2.2)

APF = atomic packing factor (4.2)

a = lattice parameter: unit cell x-axial length (4.2)

a = crack length of a surface crack (10.5)

at% = atom percent (6.6)

B = magnetic flux density (induction) (21.2)

B_r = magnetic remanence (21.7)

BCC = body-centered cubic crystal structure (4.3)

b = lattice parameter: unit cell y-axial length (3.4)

$\mathbf{b}$ = Burgers vector (6.7)

C = capacitance (19.18)

C_i = concentration (composition) of component i in wt% (6.6)

C_i' = concentration (composition) of component i in at% (6.6)

C_v, C_p = heat capacity at constant volume, pressure (20.2)

CPR = corrosion penetration rate (18.3)

CVN = Charpy V-notch (10.6)

%CW = percent cold work (9.10)

c = lattice parameter: unit cell z-axial length (3.4)

c = velocity of electromagnetic radiation in a vacuum (22.2)

D = diffusion coefficient (7.3)

D = dielectric displacement (19.19)

DP = degree of polymerization (5.5)

d = diameter

d = average grain diameter (9.8)

d_{hkl} = interplanar spacing for planes of Miller indices h, k, and l (4.19)

E = energy (2.5)

E = modulus of elasticity or Young's modulus (8.3)

$\mathscr{E}$ = electric field intensity (19.3)

E_f = Fermi energy (19.5)

E_g = band gap energy (19.6)

$E_r(t)$ = relaxation modulus (15.4)

%EL = ductility, in percent elongation (8.4)

e = electric charge per electron (19.7)

e^- = electron (18.2)

erf = Gaussian error function (7.4)

exp = e, the base for natural logarithms

F = force, interatomic or mechanical (2.5, 8.2)

$\mathscr{F}$ = Faraday constant (18.2)

FCC = face-centered cubic crystal structure (4.2)

G = shear modulus (8.3)

H = magnetic field strength (21.2)

H_c = magnetic coercivity (21.7)

HB = Brinell hardness (8.5)

HCP = hexagonal close-packed crystal structure (4.4)

HK = Knoop hardness (8.5)

HRB, HRF = Rockwell hardness: B and F scales (8.5)

HR15N, HR45W = superficial Rockwell hardness: 15N and 45W scales (8.5)

HV = Vickers hardness (8.5)

h = Planck's constant (22.2)

(hkl) = Miller indices for a crystallographic plane (3.7)

$(hkil)$ = Miller indices for a crystallographic plane, hexagonal crystals (3.7)

I = electric current (19.2)

I = intensity of electromagnetic radiation (22.3)

i = current density (18.3)

i_C = corrosion current density (18.4)

J = diffusion flux (7.3)

J = electric current density (19.3)

K_c = fracture toughness (10.5)

K_{Ic} = plane strain fracture toughness for mode I crack surface displacement (10.5)

k = Boltzmann's constant (6.2)

k = thermal conductivity (20.4)

l = length

l_c = critical fiber length (16.4)

ln = natural logarithm

log = logarithm taken to base 10

M = magnetization (21.2)

$\overline{M}_n$ = polymer number-average molecular weight (5.5)

$\overline{M}_w$ = polymer weight-average molecular weight (5.5)

mol% = mole percent

N = number of fatigue cycles (10.8)

N_A = Avogadro's number (4.5)

N_f = fatigue life (10.8)

n = principal quantum number (2.3)

n = number of atoms per unit cell (4.5)

n = strain-hardening exponent (8.4)

n = number of electrons in an electrochemical reaction (18.2)

n = number of conducting electrons per cubic meter (19.7)

n = index of refraction (22.5)

n' = for ceramics, the number of formula units per unit cell (4.10)

n_i = intrinsic carrier (electron and hole) concentration (19.10)

P = dielectric polarization (19.19)

P–B ratio = Pilling–Bedworth ratio (18.10)

p = number of holes per cubic meter (19.10)

Q = activation energy

Q = magnitude of charge stored (19.18)

R = atomic radius (4.2)

R = gas constant

%RA = ductility, in percent reduction in area (8.4)

r = interatomic distance (2.5)

r = reaction rate (18.3)

r_A, r_C = anion and cation ionic radii (4.6)

S = fatigue stress amplitude (10.8)

SEM = scanning electron microscopy or microscope

T = temperature

T_c = Curie temperature (21.6)

T_C = superconducting critical temperature (21.12)

T_g = glass transition temperature (15.12, 17.8)

T_m = melting temperature

TEM = transmission electron microscopy or microscope

TS = tensile strength (8.4)

t = time

t_r = rupture lifetime (10.12)

U_r = modulus of resilience (8.4)

$[uvw]$ = indices for a crystallographic direction (3.6)

$[uvtw], [UVW]$ = indices for a crystallographic direction, hexagonal crystals (3.6)

V = electrical potential difference (voltage) (18.2, 19.2)

V_C = unit cell volume (4.4)

V_C = corrosion potential (18.4)

V_H = Hall voltage (19.14)

V_i = volume fraction of phase i (11.8)

v = velocity

vol% = volume percent

W_i = mass fraction of phase i (11.8)

wt% = weight percent (6.6)

x = length

x = space coordinate

Y = dimensionless parameter or function in fracture toughness expression (10.5)

y = space coordinate

z = space coordinate

α = lattice parameter: unit cell y–z interaxial angle (3.4)

α, β, γ = phase designations

α_l = linear coefficient of thermal expansion (20.3)

β = lattice parameter: unit cell x–z interaxial angle (3.4)

γ = lattice parameter: unit cell x–y interaxial angle (3.4)

γ = shear strain (8.2)

Δ = precedes the symbol of a parameter to denote finite change

ϵ = engineering strain (8.2)

ϵ = dielectric permittivity (19.18)

ϵ_r = dielectric constant or relative permittivity (19.18)

$\dot{\epsilon}_s$ = steady-state creep rate (10.12)

ϵ_T = true strain (8.4)

η = viscosity (14.8)

η = overvoltage (18.4)

2θ = Bragg diffraction angle (4.20)

θ_D = Debye temperature (20.2)

λ = wavelength of electromagnetic radiation (4.19)

μ = magnetic permeability (21.2)

μ_B = Bohr magneton (21.2)

μ_r = relative magnetic permeability (21.2)

μ_e = electron mobility (19.7)

μ_h = hole mobility (19.10)

ν = Poisson's ratio (8.3)

ν = frequency of electromagnetic radiation (22.2)

ρ = density (4.5)

ρ = electrical resistivity (19.2)

ρ_t = radius of curvature at the tip of a crack (10.5)

σ = engineering stress, tensile or compressive (8.2)

σ = electrical conductivity (19.3)

σ^* = longitudinal strength (composite) (16.5)

σ_c = critical stress for crack propagation (10.5)

σ_{fs} = flexural strength (14.7)

σ_m = maximum stress (10.5)

σ_m = mean stress (10.7)

σ'_m = stress in matrix at composite failure (16.5)

σ_T = true stress (8.4)

σ_w = safe or working stress (8.7)

σ_y = yield strength (8.4)

τ = shear stress (8.2)

τ_c = fiber–matrix bond strength/ matrix shear yield strength (16.4)

τ_{crss} = critical resolved shear stress (9.5)

χ_m = magnetic susceptibility (21.2)

Subscripts

c = composite

cd = discontinuous fibrous composite

cl = longitudinal direction (aligned fibrous composite)

ct = transverse direction (aligned fibrous composite)

f = final

f = at fracture

f = fiber

i = instantaneous

m = matrix

$m, \max$ = maximum

$\min$ = minimum

0 = original

0 = at equilibrium

0 = in a vacuum

© iStockphoto/Mark Oleksiy

© blickwinkel/Alamy

© iStockphoto/Jill Chen

A familiar item fabricated from three different material types is the beverage container. Beverages are marketed in aluminum (metal) cans (top), glass (ceramic) bottles (center), and plastic (polymer) bottles (bottom).

© iStockphoto/Mark Oleksiy

© blickwinkel/Alamy

1.1 HISTORICAL PERSPECTIVE

Materials are probably more deep seated in our culture than most of us realize. Transportation, housing, clothing, communication, recreation, and food production—virtually every segment of our everyday lives is influenced to one degree or another by materials. Historically, the development and advancement of societies have been intimately tied to the members' ability to produce and manipulate materials to fill their needs. In fact, early civilizations have been designated by the level of their materials development (Stone Age, Bronze Age, Iron Age).[1]

The earliest humans had access to only a very limited number of materials, those that occur naturally: stone, wood, clay, skins, and so on. With time, they discovered techniques for producing materials that had properties superior to those of the natural ones; these new materials included pottery and various metals. Furthermore, it was discovered that the properties of a material could be altered by heat treatments and by the addition of other substances. At this point, materials utilization became a selection process that involved deciding from a given, rather limited set of materials, the one best suited for an application by virtue of its characteristics. It was not until relatively recent times that scientists came to understand the relationships between the structural elements of materials and their properties. This knowledge, acquired over approximately the past 100 years, has empowered them to fashion, to a large degree, the characteristics of materials. Thus, tens of thousands of different materials have evolved with rather specialized characteristics that meet the needs of our modern and complex society, including metals, plastics, glasses, and fibers.

The development of many technologies that make our existence so comfortable has been intimately associated with the accessibility of suitable materials. An advancement in the understanding of a material type is often the forerunner to the stepwise progression of a technology. For example, automobiles would not have been possible without the availability of inexpensive steel or some other comparable substitute. In the contemporary era, sophisticated electronic devices rely on components that are made from what are called *semiconducting materials*.

1.2 MATERIALS SCIENCE AND ENGINEERING

Sometimes it is useful to subdivide the discipline of materials science and engineering into *materials science* and *materials engineering* subdisciplines. Strictly speaking, materials science involves investigating the relationships that exist between the structures and

---

[1]The approximate dates for the beginnings of the Stone, Bronze, and Iron ages are 2.5 million BC, 3500 BC, and 1000 BC, respectively.

properties of materials. In contrast, materials engineering involves, on the basis of these structure–property correlations, designing or engineering the structure of a material to produce a predetermined set of properties.[2] From a functional perspective, the role of a materials scientist is to develop or synthesize new materials, whereas a materials engineer is called upon to create new products or systems using existing materials and/or to develop techniques for processing materials. Most graduates in materials programs are trained to be both materials scientists and materials engineers.

Structure is, at this point, a nebulous term that deserves some explanation. In brief, the structure of a material usually relates to the arrangement of its internal components. *Subatomic structure* involves electrons within the individual atoms and interactions with their nuclei. On an atomic level, structure encompasses the organization of atoms or molecules relative to one another. The next larger structural realm, which contains large groups of atoms that are normally agglomerated together, is termed *microscopic,* meaning that which is subject to direct observation using some type of microscope. Finally, structural elements that can be viewed with the naked eye are termed *macroscopic*.

The notion of *property* deserves elaboration. While in service use, all materials are exposed to external stimuli that evoke some type of response. For example, a specimen subjected to forces experiences deformation, while a polished metal surface reflects light. A property is a material trait in terms of the kind and magnitude of response to a specific imposed stimulus. Generally, definitions of properties are made independent of material shape and size.

Virtually all important properties of solid materials may be grouped into six different categories: mechanical, electrical, thermal, magnetic, optical, and deteriorative. For each, there is a characteristic type of stimulus capable of provoking different responses. Mechanical properties relate deformation to an applied load or force; examples include elastic modulus (stiffness), strength, and toughness. For electrical properties, such as electrical conductivity and dielectric constant, the stimulus is an electric field. The thermal behavior of solids can be represented in terms of heat capacity and thermal conductivity. Magnetic properties demonstrate the response of a material to the application of a magnetic field. For optical properties, the stimulus is electromagnetic or light radiation; index of refraction and reflectivity are representative optical properties. Finally, deteriorative characteristics relate to the chemical reactivity of materials. The chapters that follow discuss properties that fall within each of these six classifications.

In addition to structure and properties, two other important components are involved in the science and engineering of materials—namely, *processing* and *performance*. With regard to the relationships of these four components, the structure of a material depends on how it is processed. Furthermore, a material's performance is a function of its properties. Thus, the interrelationship among processing, structure, properties, and performance is as depicted in the schematic illustration shown in Figure 1.1. Throughout this text, we draw attention to the relationships among these four components in terms of the design, production, and utilization of materials.

We present an example of these processing-structure-properties-performance principles in Figure 1.2, a photograph showing three thin disk specimens placed over some printed matter. It is obvious that the optical properties (i.e., the light transmittance) of each of the three materials are different; the one on the left is transparent (i.e., virtually all of the

Figure 1.1   The four components of the discipline of materials science and engineering, and their interrelationship.

---

[2]Throughout this text, we draw attention to the relationships between material properties and structural elements.

Figure 1.2 Three thin disk specimens of aluminum oxide that have been placed over a printed page in order to demonstrate their differences in light-transmittance characteristics. The disk on the left is *transparent* (i.e., virtually all light that is reflected from the page passes through it), whereas the one in the center is *translucent* (meaning that some of this reflected light is transmitted through the disk). The disk on the right is *opaque*—that is, none of the light passes through it. These differences in optical properties are a consequence of differences in structure of these materials, which have resulted from the way the materials were processed.

Specimen preparation, P. A. Lessing

reflected light passes through it), whereas the disks in the center and on the right are, respectively, translucent and opaque. All of these specimens are of the same material, aluminum oxide, but the leftmost one is what we call a *single crystal*—that is, has a high degree of perfection—which gives rise to its transparency. The center one is composed of numerous and very small single crystals that are all connected; the boundaries between these small crystals scatter a portion of the light reflected from the printed page, which makes this material optically translucent. Finally, the specimen on the right is composed not only of many small, interconnected crystals, but also of a large number of very small pores or void spaces. These pores also effectively scatter the reflected light and render this material opaque.

Thus, the structures of these three specimens are different in terms of crystal boundaries and pores, which affect the optical transmittance properties. Furthermore, each material was produced using a different processing technique. If optical transmittance is an important parameter relative to the ultimate in-service application, the performance of each material will be different.

1.3 WHY STUDY MATERIALS SCIENCE AND ENGINEERING?

Why do we study materials? Many an applied scientist or engineer, whether mechanical, civil, chemical, or electrical, is at one time or another exposed to a design problem involving materials, such as a transmission gear, the superstructure for a building, an oil refinery component, or an integrated circuit chip. Of course, materials scientists and engineers are specialists who are totally involved in the investigation and design of materials.

Many times, a materials problem is one of selecting the right material from the thousands available. The final decision is normally based on several criteria. First, the in-service conditions must be characterized, for these dictate the properties required of the material. Only on rare occasions does a material possess the maximum or ideal combination of properties. Thus, it may be necessary to trade one characteristic for another. The classic example involves strength and ductility; normally, a material having a high strength has only a limited ductility. In such cases, a reasonable compromise between two or more properties may be necessary.

A second selection consideration is any deterioration of material properties that may occur during service operation. For example, significant reductions in mechanical strength may result from exposure to elevated temperatures or corrosive environments.

Finally, probably the overriding consideration is that of economics: What will the finished product cost? A material may be found that has the ideal set of properties but is prohibitively expensive. Here again, some compromise is inevitable. The cost of a finished piece also includes any expense incurred during fabrication to produce the desired shape.

The more familiar an engineer or scientist is with the various characteristics and structure–property relationships, as well as the processing techniques of materials, the more proficient and confident he or she will be in making judicious materials choices based on these criteria.

C A S E   S T U D Y

Liberty Ship Failures

The following case study illustrates one role that materials scientists and engineers are called upon to assume in the area of materials performance: analyze mechanical failures, determine their causes, and then propose appropriate measures to guard against future incidents.

The failure of many of the World War II Liberty ships[3] is a well-known and dramatic example of the brittle fracture of steel that was thought to be ductile.[4] Some of the early ships experienced structural damage when cracks developed in their decks and hulls. Three of them catastrophically split in half when cracks formed, grew to critical lengths, and then rapidly propagated completely around the ships' girths. Figure 1.3 shows one of the ships that fractured the day after it was launched.

Subsequent investigations concluded one or more of the following factors contributed to each failure[5]:

- When some normally ductile metal alloys are cooled to relatively low temperatures, they become susceptible to brittle fracture—that is, they experience a ductile-to-brittle transition upon cooling through a critical range of temperatures. These Liberty ships were constructed of steel that experienced a ductile-to-brittle transition. Some of them were deployed to the frigid North Atlantic, where the once ductile metal experienced brittle fracture when temperatures dropped to below the transition temperature.[6]

- The corner of each hatch (i.e., door) was square; these corners acted as points of stress concentration where cracks can form.

- German U-boats were sinking cargo ships faster than they could be replaced using existing construction techniques. Consequently, it became necessary to revolutionize construction methods to build cargo ships faster and in greater numbers. This was accomplished using prefabricated steel sheets that were assembled by welding rather than by the traditional time-consuming riveting. Unfortunately, cracks in welded structures may propagate unimpeded for large distances, which can lead to catastrophic failure. However, when structures are riveted, a crack ceases to propagate once it reaches the edge of a steel sheet.

- Weld defects and *discontinuities* (i.e., sites where cracks can form) were introduced by inexperienced operators.

---

[3]During World War II, 2,710 Liberty cargo ships were mass-produced by the United States to supply food and materials to the combatants in Europe.

[4]Ductile metals fail after relatively large degrees of permanent deformation; however, very little if any permanent deformation accompanies the fracture of brittle materials. Brittle fractures can occur very suddenly as cracks spread rapidly; crack propagation is normally much slower in ductile materials, and the eventual fracture takes longer. For these reasons, the ductile mode of fracture is usually preferred. Ductile and brittle fractures are discussed in Sections 10.3 and 10.4.

[5]Sections 10.2 through 10.6 discuss various aspects of failure.

[6]This ductile-to-brittle transition phenomenon, as well as techniques that are used to measure and raise the critical temperature range, are discussed in Section 10.6.

(*continued*)

Figure 1.3 The Liberty ship *S.S. Schenectady,* which, in 1943, failed before leaving the shipyard.
(Reprinted with permission of Earl R. Parker, *Brittle Behavior of Engineering Structures,* National Academy of Sciences, National Research Council, John Wiley & Sons, New York, 1957.)

Remedial measures taken to correct these problems included the following:

- Lowering the ductile-to-brittle temperature of the steel to an acceptable level by improving steel quality (e.g., reducing sulfur and phosphorus impurity contents).
- Rounding off hatch corners by welding a curved reinforcement strip on each corner.[7]
- Installing crack-arresting devices such as riveted straps and strong weld seams to stop propagating cracks.

- Improving welding practices and establishing welding codes.

In spite of these failures, the Liberty ship program was considered a success for several reasons, the primary reason being that ships that survived failure were able to supply Allied Forces in the theater of operations and in all likelihood shortened the war. In addition, structural steels were developed with vastly improved resistances to catastrophic brittle fractures. Detailed analyses of these failures advanced the understanding of crack formation and growth, which ultimately evolved into the discipline of fracture mechanics.

---

[7]The reader may note that corners of windows and doors for all of today's marine and aircraft structures are rounded.

1.4 CLASSIFICATION OF MATERIALS

WileyPLUS

Tutorial Video:
What Are the
Different Classes
of Materials?

Solid materials have been conveniently grouped into three basic categories: metals, ceramics, and polymers, a scheme based primarily on chemical makeup and atomic structure. Most materials fall into one distinct grouping or another. In addition, there are the composites that are engineered combinations of two or more different materials. A brief explanation of these material classifications and representative characteristics is offered next. Another category is advanced materials—those used in high-technology applications, such as semiconductors, biomaterials, smart materials, and nanoengineered materials; these are discussed in Section 1.5.

Figure 1.4
Bar chart of room-temperature density values for various metals, ceramics, polymers, and composite materials.

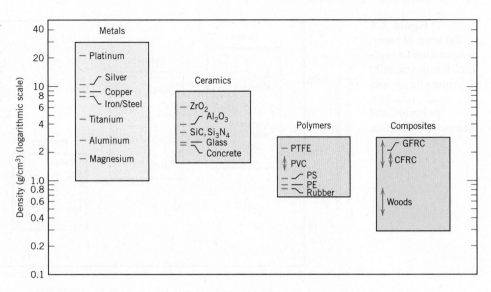

Metals

Metals are composed of one or more metallic elements (e.g., iron, aluminum, copper, titanium, gold, nickel), and often also nonmetallic elements (e.g., carbon, nitrogen, oxygen) in relatively small amounts.[8] Atoms in metals and their alloys are arranged in a very orderly manner (as discussed in Chapter 3) and are relatively dense in comparison to the ceramics and polymers (Figure 1.4). With regard to mechanical characteristics, these materials are relatively stiff (Figure 1.5) and strong (Figure 1.6), yet are ductile (i.e., capable of large amounts of deformation without fracture), and are resistant to fracture (Figure 1.7), which accounts for their widespread use in structural applications. Metallic materials have large numbers of nonlocalized electrons—that is, these electrons are not bound to particular atoms. Many properties of metals are directly attributable to these electrons. For example, metals are extremely good conductors of electricity

WileyPLUS

Tutorial Video:
Metals

Figure 1.5
Bar chart of room-temperature stiffness (i.e., elastic modulus) values for various metals, ceramics, polymers, and composite materials.

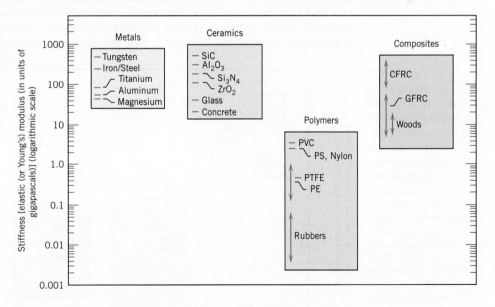

---

[8]The term *metal alloy* refers to a metallic substance that is composed of two or more elements.

Figure 1.6
Bar chart of room-temperature strength (i.e., tensile strength) values for various metals, ceramics, polymers, and composite materials.

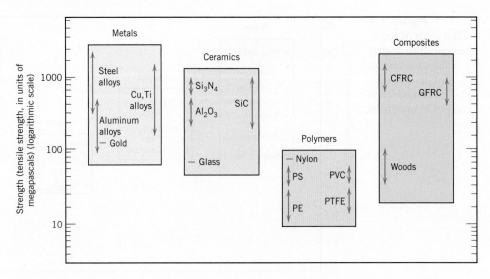

(Figure 1.8) and heat, and are not transparent to visible light; a polished metal surface has a lustrous appearance. In addition, some metals (i.e., Fe, Co, and Ni) have desirable magnetic properties.

Figure 1.9 shows several common and familiar objects that are made of metallic materials. The types and applications of metals and their alloys are discussed further in Chapter 13.

Ceramics

Ceramics are compounds between metallic and nonmetallic elements; they are most frequently oxides, nitrides, and carbides. For example, common ceramic materials include aluminum oxide (or *alumina,* Al_2O_3), silicon dioxide (or *silica,* SiO_2), silicon carbide (SiC), silicon nitride (Si_3N_4), and, in addition, what some refer to as the *traditional ceramics*—those composed of clay minerals (e.g., porcelain), as well as cement and glass. With regard to mechanical behavior, ceramic materials are relatively stiff and strong—stiffnesses and strengths are comparable to those of the metals (Figures 1.5 and 1.6). In addition, they are typically very hard. Historically, ceramics have exhibited extreme brittleness (lack of ductility) and are highly susceptible to fracture (Figure 1.7). However, newer ceramics are being engineered to have improved resistance to fracture; these materials are used for cookware, cutlery, and

WileyPLUS

Tutorial Video:
Ceramics

Figure 1.7
Bar chart of room-temperature resistance to fracture (i.e., fracture toughness) for various metals, ceramics, polymers, and composite materials. (Reprinted from *Engineering Materials 1: An Introduction to Properties, Applications and Design,* third edition, M. F. Ashby and D. R. H. Jones, pages 177 and 178, Copyright 2005, with permission from Elsevier.)

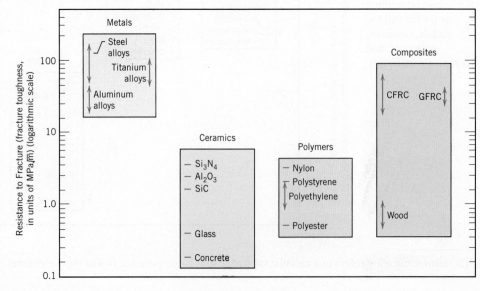

Figure 1.8
Bar chart of room-temperature electrical conductivity ranges for metals, ceramics, polymers, and semiconducting materials.

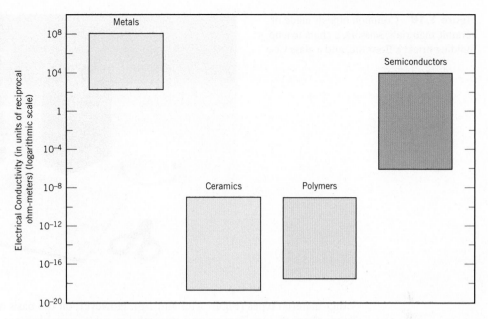

even automobile engine parts. Furthermore, ceramic materials are typically insulative to the passage of heat and electricity (i.e., have low electrical conductivities, Figure 1.8) and are more resistant to high temperatures and harsh environments than are metals and polymers. With regard to optical characteristics, ceramics may be transparent, translucent, or opaque (Figure 1.2), and some of the oxide ceramics (e.g., Fe_3O_4) exhibit magnetic behavior.

Several common ceramic objects are shown in Figure 1.10. The characteristics, types, and applications of this class of materials are also discussed in Chapter 14.

Polymers

Polymers include the familiar plastic and rubber materials. Many of them are organic compounds that are chemically based on carbon, hydrogen, and other nonmetallic elements (i.e., O, N, and Si). Furthermore, they have very large molecular structures, often chainlike in nature, that usually have a backbone of carbon atoms. Some common and familiar polymers are polyethylene (PE), nylon, poly(vinyl chloride) (PVC), polycarbonate (PC), polystyrene (PS), and silicone rubber. These materials typically have low densities (Figure 1.4), whereas their mechanical characteristics are generally dissimilar to those of the metallic and ceramic materials—they are not as stiff or strong as these

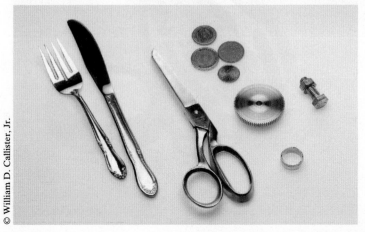

Figure 1.9 Familiar objects made of metals and metal alloys (from left to right): silverware (fork and knife), scissors, coins, a gear, a wedding ring, and a nut and bolt.

Figure 1.10 Common objects made of ceramic materials: scissors, a china teacup, a building brick, a floor tile, and a glass vase.

© William D. Callister, Jr.

other material types (Figures 1.5 and 1.6). However, on the basis of their low densities, many times their stiffnesses and strengths on a per-mass basis are comparable to those of the metals and ceramics. In addition, many of the polymers are extremely ductile and pliable (i.e., plastic), which means they are easily formed into complex shapes. In general, they are relatively inert chemically and unreactive in a large number of environments. One major drawback to the polymers is their tendency to soften and/or decompose at modest temperatures, which, in some instances, limits their use. Furthermore, they have low electrical conductivities (Figure 1.8) and are nonmagnetic.

Figure 1.11 shows several articles made of polymers that are familiar to the reader. Chapters 5 and 15 are devoted to discussions of the structures, properties, applications, and processing of polymeric materials.

WileyPLUS

Tutorial Video:
Polymers

Figure 1.11 Several common objects made of polymeric materials: plastic tableware (spoon, fork, and knife), billiard balls, a bicycle helmet, two dice, a lawn mower wheel (plastic hub and rubber tire), and a plastic milk carton.

© William D. Callister, Jr.

CASE STUDY

Carbonated Beverage Containers

One common item that presents some interesting material property requirements is the container for carbonated beverages. The material used for this application must satisfy the following constraints: (1) provide a barrier to the passage of carbon dioxide, which is under pressure in the container; (2) be nontoxic, unreactive with the beverage, and, preferably, recyclable; (3) be relatively strong and capable of surviving a drop from a height of several feet when containing the beverage; (4) be inexpensive, including the cost to fabricate the final shape; (5) if optically transparent, retain its optical clarity; and (6) be capable of being produced in different colors and/or adorned with decorative labels.

All three of the basic material types—metal (aluminum), ceramic (glass), and polymer (polyester plastic)—are used for carbonated beverage containers (per the chapter-opening photographs). All of these materials are nontoxic and unreactive with beverages. In addition, each material has its pros and cons. For example, the aluminum alloy is relatively strong (but easily dented), is a very good barrier to the diffusion of carbon dioxide, is easily recycled, cools beverages rapidly, and allows labels to be painted onto its surface. However, the cans are optically opaque and relatively expensive to produce. Glass is impervious to the passage of carbon dioxide, is a relatively inexpensive material, and may be recycled, but it cracks and fractures easily, and glass bottles are relatively heavy. Whereas plastic is relatively strong, may be made optically transparent, is inexpensive and lightweight, and is recyclable, it is not as impervious to the passage of carbon dioxide as aluminum and glass. For example, you may have noticed that beverages in aluminum and glass containers retain their carbonization (i.e., "fizz") for several years, whereas those in two-liter plastic bottles "go flat" within a few months.

Composites

A *composite* is composed of two (or more) individual materials that come from the categories previously discussed—metals, ceramics, and polymers. The design goal of a composite is to achieve a combination of properties that is not displayed by any single material and also to incorporate the best characteristics of each of the component materials. A large number of composite types are represented by different combinations of metals, ceramics, and polymers. Furthermore, some naturally occurring materials are composites—for example, wood and bone. However, most of those we consider in our discussions are synthetic (or human-made) composites.

One of the most common and familiar composites is fiberglass, in which small glass fibers are embedded within a polymeric material (normally an epoxy or polyester).[9] The glass fibers are relatively strong and stiff (but also brittle), whereas the polymer is more flexible. Thus, fiberglass is relatively stiff, strong (Figures 1.5 and 1.6), and flexible. In addition, it has a low density (Figure 1.4).

WileyPLUS

Tutorial Video:
Composites

Another technologically important material is the carbon fiber–reinforced polymer (CFRP) composite—carbon fibers that are embedded within a polymer. These materials are stiffer and stronger than glass fiber–reinforced materials (Figures 1.5 and 1.6) but more expensive. CFRP composites are used in some aircraft and aerospace applications, as well as in high-tech sporting equipment (e.g., bicycles, golf clubs, tennis rackets, skis/snowboards) and recently in automobile bumpers. The new Boeing 787 fuselage is primarily made from such CFRP composites.

Chapter 16 is devoted to a discussion of these interesting composite materials.

---

[9]Fiberglass is sometimes also termed a *glass fiber–reinforced polymer* composite (GFRP).

1.5 ADVANCED MATERIALS

Materials utilized in high-technology (or high-tech) applications are sometimes termed *advanced materials*. By *high technology*, we mean a device or product that operates or functions using relatively intricate and sophisticated principles, including electronic equipment (camcorders, CD/DVD players), computers, fiber-optic systems, spacecraft, aircraft, and military rocketry. These advanced materials are typically traditional materials whose properties have been enhanced, but also include newly developed, high-performance materials. Furthermore, they may be of all material types (e.g., metals, ceramics, polymers) and are normally expensive. Advanced materials include semiconductors, biomaterials, and what we may term *materials of the future* (i.e., smart materials and nanoengineered materials), which we discuss next. The properties and applications of a number of these advanced materials—for example, materials that are used for lasers, integrated circuits, magnetic information storage, liquid crystal displays (LCDs), and fiber optics—are also discussed in subsequent chapters.

Semiconductors

Semiconductors have electrical properties that are intermediate between those of electrical conductors (i.e., metals and metal alloys) and insulators (i.e., ceramics and polymers)—see Figure 1.8. Furthermore, the electrical characteristics of these materials are extremely sensitive to the presence of minute concentrations of impurity atoms, for which the concentrations may be controlled over very small spatial regions. Semiconductors have made possible the advent of integrated circuitry that has totally revolutionized the electronics and computer industries (not to mention our lives) over the past three decades.

Biomaterials

Biomaterials are employed in components implanted into the human body to replace diseased or damaged body parts. These materials must not produce toxic substances and must be compatible with body tissues (i.e., must not cause adverse biological reactions). All of the preceding materials—metals, ceramics, polymers, composites, and semiconductors—may be used as biomaterials.

Smart Materials

Smart (or *intelligent*) *materials* are a group of new and state-of-the-art materials now being developed that will have a significant influence on many of our technologies. The adjective *smart* implies that these materials are able to sense changes in their environment and then respond to these changes in predetermined manners—traits that are also found in living organisms. In addition, this *smart* concept is being extended to rather sophisticated systems that consist of both smart and traditional materials.

Components of a smart material (or system) include some type of sensor (which detects an input signal) and an actuator (which performs a responsive and adaptive function). Actuators may be called upon to change shape, position, natural frequency, or mechanical characteristics in response to changes in temperature, electric fields, and/or magnetic fields.

Four types of materials are commonly used for actuators: shape-memory alloys, piezoelectric ceramics, magnetostrictive materials, and electrorheological/magnetorheological fluids. *Shape-memory alloys* are metals that, after having been deformed, revert to their original shape when temperature is changed (see the Materials of Importance box following Section 12.9). *Piezoelectric ceramics* expand and contract in response to an applied electric field (or voltage); conversely, they also generate an electric field when their dimensions are altered (see Section 19.25). The behavior of *magnetostrictive materials* is analogous to that of the piezoelectrics, except that they are responsive to magnetic fields. Also, *electrorheological* and *magnetorheological fluids* are liquids that experience dramatic changes in viscosity upon the application of electric and magnetic fields, respectively.

Materials/devices employed as sensors include optical fibers (Section 22.14), piezoelectric materials (including some polymers), and microelectromechanical systems (MEMS; Section 14.17).

For example, one type of smart system is used in helicopters to reduce aerodynamic cockpit noise created by the rotating rotor blades. Piezoelectric sensors inserted into the blades monitor blade stresses and deformations; feedback signals from these sensors are fed into a computer-controlled adaptive device that generates noise-canceling antinoise.

Nanomaterials

One new material class that has fascinating properties and tremendous technological promise is the *nanomaterials,* which may be any one of the four basic types—metals, ceramics, polymers, or composites. However, unlike these other materials, they are not distinguished on the basis of their chemistry but rather their size; the *nano* prefix denotes that the dimensions of these structural entities are on the order of a nanometer (10^{-9} m)—as a rule, less than 100 nanometers (nm; equivalent to the diameter of approximately 500 atoms).

Prior to the advent of nanomaterials, the general procedure scientists used to understand the chemistry and physics of materials was to begin by studying large and complex structures and then investigate the fundamental building blocks of these structures that are smaller and simpler. This approach is sometimes termed *top-down science.* However, with the development of scanning probe microscopes (Section 6.12), which permit observation of individual atoms and molecules, it has become possible to design and build new structures from their atomic-level constituents, one atom or molecule at a time (i.e., "materials by design"). This ability to arrange atoms carefully provides opportunities to develop mechanical, electrical, magnetic, and other properties that are not otherwise possible. We call this the *bottom-up approach*, and the study of the properties of these materials is termed *nanotechnology.*[10]

Some of the physical and chemical characteristics exhibited by matter may experience dramatic changes as particle size approaches atomic dimensions. For example, materials that are opaque in the macroscopic domain may become transparent on the nanoscale; some solids become liquids, chemically stable materials become combustible, and electrical insulators become conductors. Furthermore, properties may depend on size in this nanoscale domain. Some of these effects are quantum mechanical in origin, whereas others are related to *surface phenomena*—the proportion of atoms located on surface sites of a particle increases dramatically as its size decreases.

Because of these unique and unusual properties, nanomaterials are finding niches in electronic, biomedical, sporting, energy production, and other industrial applications. Some are discussed in this text, including the following:

- Catalytic converters for automobiles (Materials of Importance box, Chapter 6)
- Nanocarbons—Fullerenes, carbon nanotubes, and graphene (Section 14.17)
- Particles of carbon black as reinforcement for automobile tires (Section 16.2)
- Nanocomposites (Section 16.16)
- Magnetic nanosize grains that are used for hard disk drives (Section 21.11)
- Magnetic particles that store data on magnetic tapes (Section 21.11)

---

[10]One legendary and prophetic suggestion as to the possibility of nanoengineered materials was offered by Richard Feynman in his 1959 American Physical Society lecture titled "There's Plenty of Room at the Bottom."

Whenever a new material is developed, its potential for harmful and toxicological interactions with humans and animals must be considered. Small nanoparticles have exceedingly large surface area–to–volume ratios, which can lead to high chemical reactivities. Although the safety of nanomaterials is relatively unexplored, there are concerns that they may be absorbed into the body through the skin, lungs, and digestive tract at relatively high rates, and that some, if present in sufficient concentrations, will pose health risks—such as damage to DNA or promotion of lung cancer.

1.6 MODERN MATERIALS' NEEDS

In spite of the tremendous progress that has been made in the discipline of materials science and engineering within the past few years, technological challenges remain, including the development of even more sophisticated and specialized materials, as well as consideration of the environmental impact of materials production. Some comment is appropriate relative to these issues so as to round out this perspective.

Nuclear energy holds some promise, but the solutions to the many problems that remain necessarily involve materials, such as fuels, containment structures, and facilities for the disposal of radioactive waste.

Significant quantities of energy are involved in transportation. Reducing the weight of transportation vehicles (automobiles, aircraft, trains, etc.), as well as increasing engine operating temperatures, will enhance fuel efficiency. New high-strength, low-density structural materials remain to be developed, as well as materials that have higher-temperature capabilities, for use in engine components.

Furthermore, there is a recognized need to find new and economical sources of energy and to use present resources more efficiently. Materials will undoubtedly play a significant role in these developments. For example, the direct conversion of solar power into electrical energy has been demonstrated. Solar cells employ some rather complex and expensive materials. To ensure a viable technology, materials that are highly efficient in this conversion process yet less costly must be developed.

The hydrogen fuel cell is another very attractive and feasible energy-conversion technology that has the advantage of being nonpolluting. It is just beginning to be implemented in batteries for electronic devices and holds promise as a power plant for automobiles. New materials still need to be developed for more efficient fuel cells and also for better catalysts to be used in the production of hydrogen.

Furthermore, environmental quality depends on our ability to control air and water pollution. Pollution control techniques employ various materials. In addition, materials processing and refinement methods need to be improved so that they produce less environmental degradation—that is, less pollution and less despoilage of the landscape from the mining of raw materials. Also, in some materials manufacturing processes, toxic substances are produced, and the ecological impact of their disposal must be considered.

Many materials that we use are derived from resources that are nonrenewable—that is, not capable of being regenerated, including most polymers, for which the prime raw material is oil, and some metals. These nonrenewable resources are gradually becoming depleted, which necessitates (1) the discovery of additional reserves, (2) the development of new materials having comparable properties with less adverse environmental impact, and/or (3) increased recycling efforts and the development of new recycling technologies. As a consequence of the economics of not only production but also environmental impact and ecological factors, it is becoming increasingly important to consider the "cradle-to-grave" life cycle of materials relative to the overall manufacturing process.

The roles that materials scientists and engineers play relative to these, as well as other environmental and societal issues, are discussed in more detail in Chapter 23.

SUMMARY

Materials Science and Engineering

- There are six different property classifications of materials that determine their applicability: mechanical, electrical, thermal, magnetic, optical, and deteriorative.

- One aspect of materials science is the investigation of relationships that exist between the structures and properties of materials. By *structure*, we mean how some internal component(s) of the material is (are) arranged. In terms of (and with increasing) dimensionality, structural elements include subatomic, atomic, microscopic, and macroscopic.

- With regard to the design, production, and utilization of materials, there are four elements to consider—processing, structure, properties, and performance. The performance of a material depends on its properties, which in turn are a function of its structure(s); furthermore, structure(s) is (are) determined by how the material was processed.

- Three important criteria in materials selection are in-service conditions to which the material will be subjected, any deterioration of material properties during operation, and economics or cost of the fabricated piece.

Classification of Materials

- On the basis of chemistry and atomic structure, materials are classified into three general categories: metals (metallic elements), ceramics (compounds between metallic and nonmetallic elements), and polymers (compounds composed of carbon, hydrogen, and other nonmetallic elements). In addition, composites are composed of at least two different material types.

Advanced Materials

- Another materials category is the advanced materials that are used in high-tech applications, including semiconductors (having electrical conductivities intermediate between those of conductors and insulators), biomaterials (which must be compatible with body tissues), smart materials (those that sense and respond to changes in their environments in predetermined manners), and nanomaterials (those that have structural features on the order of a nanometer, some of which may be designed on the atomic/molecular level).

REFERENCES

Ashby, M. F., and D. R. H. Jones, *Engineering Materials 1: An Introduction to Their Properties, Applications, and Design,* 4th edition, Butterworth-Heinemann, Oxford, England, 2012.

Ashby, M. F., and D. R. H. Jones, *Engineering Materials 2: An Introduction to Microstructures and Processing,* 4th edition, Butterworth-Heinemann, Oxford, England, 2012.

Ashby, M. F., H. Shercliff, and D. Cebon, *Materials Engineering, Science, Processing and Design,* Butterworth-Heinemann, Oxford, England, 2007.

Askeland, D. R., P. P. Fulay, and W. J. Wright, *The Science and Engineering of Materials,* 6th edition, Cengage Learning, Stamford, CT, 2011.

Baillie, C., and L. Vanasupa, *Navigating the Materials World,* Academic Press, San Diego, CA, 2003.

Douglas, E. P., *Introduction to Materials Science and Engineering: A Guided Inquiry,* Pearson Education, Upper Saddle River, NJ, 2014.

Fischer, T., *Materials Science for Engineering Students,* Academic Press, San Diego, CA, 2009.

Jacobs, J. A., and T. F. Kilduff, *Engineering Materials Technology,* 5th edition, Prentice Hall PTR, Paramus, NJ, 2005.

McMahon, C. J., Jr., *Structural Materials,* Merion Books, Philadelphia, PA, 2004.

Murray, G. T., C. V. White, and W. Weise, *Introduction to Engineering Materials,* 2nd edition, CRC Press, Boca Raton, FL, 2007.

Schaffer, J. P., A. Saxena, S. D. Antolovich, T. H. Sanders, Jr., and S. B. Warner, *The Science and Design of Engineering Materials,* 2nd edition, McGraw-Hill, New York, NY, 1999.

Shackelford, J. F., *Introduction to Materials Science for Engineers,* 7th edition, Prentice Hall PTR, Paramus, NJ, 2009.

Smith, W. F., and J. Hashemi, *Foundations of Materials Science and Engineering,* 5th edition, McGraw-Hill, New York, NY, 2010.

Van Vlack, L. H., *Elements of Materials Science and Engineering,* 6th edition, Addison-Wesley Longman, Boston, MA, 1989.

White, M. A., *Physical Properties of Materials,* 2nd edition, CRC Press, Boca Raton, FL, 2012.

QUESTION

1.1 Select one or more of the following modern items or devices and conduct an Internet search in order to determine what specific material(s) is (are) used and what specific properties this (these) material(s) possess(es) in order for the device/item to function properly. Finally, write a short essay in which you report your findings.

Cell phone/digital camera batteries
Cell phone displays
Solar cells
Wind turbine blades
Fuel cells
Automobile engine blocks (other than cast iron)
Automobile bodies (other than steel alloys)

Space telescope mirrors
Military body armor
Sports equipment
  Soccer balls
  Basketballs
  Ski poles
  Ski boots
  Snowboards
  Surfboards
  Golf clubs
  Golf balls
  Kayaks
  Lightweight bicycle frames

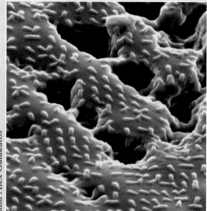

Courtesy Jeffrey Karp, Robert Langer, and Alex Galakatos

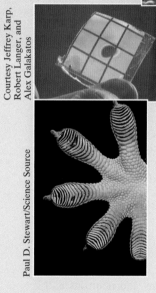

Courtesy Jeffrey Karp, Robert Langer, and Alex Galakatos

Paul D. Stewart/Science Source

The photograph at the bottom of this page is of a gecko.

Geckos, harmless tropical lizards, are extremely fascinating and extraordinary animals. They have very sticky feet (one of which is shown in the third photograph) that cling to virtually any surface. This characteristic makes it possible for them to run rapidly up vertical walls and along the undersides of horizontal surfaces. In fact, a gecko can support its body mass with a single toe! The secret to this remarkable ability is the presence of an extremely large number of microscopically small hairs on each of their toe pads. When these hairs come in contact with a surface, weak forces of attraction (i.e., van der Waals forces) are established between hair molecules and molecules on the surface. The fact that these hairs are so small and so numerous explains why the gecko grips surfaces so tightly. To release its grip, the gecko simply curls up its toes and peels the hairs away from the surface.

Using their knowledge of this mechanism of adhesion, scientists have developed several ultrastrong synthetic adhesives, one of which is an adhesive tape (shown in the second photograph) that is an especially promising tool for use in surgical procedures as a replacement for sutures and staples to close wounds and incisions. This material retains its adhesive nature in wet environments, is biodegradable, and does not release toxic substances as it dissolves during the healing process. Microscopic features of this adhesive tape are shown in the top photograph.

Barbara Peacock/Photodisc/Getty Images, Inc.

An important reason to have an understanding of interatomic bonding in solids is that in some instances, the type of bond allows us to explain a material's properties. For example, consider carbon, which may exist as both graphite and diamond. Whereas graphite is relatively soft and has a "greasy" feel to it, diamond is one of the hardest known materials in nature. In addition, the electrical properties of diamond and graphite are dissimilar: diamond is a poor conductor of electricity, but graphite is a reasonably good conductor. These disparities in properties are directly attributable to a type of interatomic bonding found in graphite that does not exist in diamond (see Section 4.12).

Learning Objectives

After studying this chapter, you should be able to do the following:

1. Name the two atomic models cited, and note the differences between them.

2. Describe the important quantum-mechanical principle that relates to electron energies.

3. (a) Schematically plot attractive, repulsive, and net energies versus interatomic separation for two atoms or ions.

(b) Note on this plot the equilibrium separation and the bonding energy.

4. (a) Briefly describe ionic, covalent, metallic, hydrogen, and van der Waals bonds.

(b) Note which materials exhibit each of these bonding types.

2.1 INTRODUCTION

Some of the important properties of solid materials depend on geometric atomic arrangements and also the interactions that exist among constituent atoms or molecules. This chapter, by way of preparation for subsequent discussions, considers several fundamental and important concepts—namely, atomic structure, electron configurations in atoms and the periodic table, and the various types of primary and secondary interatomic bonds that hold together the atoms that compose a solid. These topics are reviewed briefly, under the assumption that some of the material is familiar to the reader.

Atomic Structure

2.2 FUNDAMENTAL CONCEPTS

Each atom consists of a very small nucleus composed of protons and neutrons and is encircled by moving electrons.[1] Both electrons and protons are electrically charged, the charge magnitude being 1.602×10^{-19}C, which is negative in sign for electrons and positive for protons; neutrons are electrically neutral. Masses for these subatomic particles are extremely small; protons and neutrons have approximately the same mass, 1.67×10^{-27} kg, which is significantly larger than that of an electron, 9.11×10^{-31} kg.

atomic number (Z)

Each chemical element is characterized by the number of protons in the nucleus, or the **atomic number (Z)**.[2] For an electrically neutral or complete atom, the atomic number also equals the number of electrons. This atomic number ranges in integral units from 1 for hydrogen to 92 for uranium, the highest of the naturally occurring elements.

The *atomic mass* (A) of a specific atom may be expressed as the sum of the masses of protons and neutrons within the nucleus. Although the number of protons is the same for

---

[1]Protons, neutrons, and electrons are composed of other subatomic particles such as quarks, neutrinos, and bosons. However, this discussion is concerned only with protons, neutrons, and electrons.

[2]Terms appearing in **boldface** type are defined in the Glossary, which follows Appendix E.

isotope

atomic weight

atomic mass unit
(amu)

all atoms of a given element, the number of neutrons (N) may be variable. Thus atoms of some elements have two or more different atomic masses, which are called **isotopes**. The **atomic weight** of an element corresponds to the weighted average of the atomic masses of the atom's naturally occurring isotopes.[3] The **atomic mass unit (amu)** may be used to compute atomic weight. A scale has been established whereby 1 amu is defined as $\frac{1}{12}$ of the atomic mass of the most common isotope of carbon, carbon 12 (^{12}C) ($A = 12.00000$). Within this scheme, the masses of protons and neutrons are slightly greater than unity, and

$$A \cong Z + N \tag{2.1}$$

mole

The atomic weight of an element or the molecular weight of a compound may be specified on the basis of amu per atom (molecule) or mass per mole of material. In one **mole** of a substance, there are 6.022×10^{23} (Avogadro's number) atoms or molecules. These two atomic weight schemes are related through the following equation:

1 amu/atom (or molecule) = 1 g/mol

For example, the atomic weight of iron is 55.85 amu/atom, or 55.85 g/mol. Sometimes use of amu per atom or molecule is convenient; on other occasions, grams (or kilograms) per mole is preferred. The latter is used in this book.

EXAMPLE PROBLEM 2.1

Average Atomic Weight Computation for Cerium

Cerium has four naturally occurring isotopes: 0.185% of ^{136}Ce, with an atomic weight of 135.907 amu; 0.251% of ^{138}Ce, with an atomic weight of 137.906 amu; 88.450% of ^{140}Ce, with an atomic weight of 139.905 amu; and 11.114% of ^{142}Ce, with an atomic weight of 141.909 amu. Calculate the average atomic weight of Ce.

Solution

The average atomic weight of a hypothetical element M, $\overline{A}_M$, is computed by adding fraction-of-occurrence — atomic weight products for all its isotopes; that is,

$$\overline{A}_M = \sum_i f_{i_M} A_{i_M} \tag{2.2}$$

In this expression, f_{i_M} is the fraction-of-occurrence of isotope i for element M (i.e., the percentage-of-occurrence divided by 100), and A_{i_M} is the atomic weight of the isotope.

For cerium, Equation 2.2 takes the form

$$\overline{A}_{Ce} = f_{^{136}Ce} A_{^{136}Ce} + f_{^{138}Ce} A_{^{138}Ce} + f_{^{140}Ce} A_{^{140}Ce} + f_{^{142}Ce} A_{^{142}Ce}$$

Incorporating values provided in the problem statement for the several parameters leads to

$$\overline{A}_{Ce} = \left(\frac{0.185\%}{100}\right)(135.907 \text{ amu}) + \left(\frac{0.251\%}{100}\right)(137.906 \text{ amu}) + \left(\frac{88.450\%}{100}\right)(139.905 \text{ amu})$$

$$+ \left(\frac{11.114\%}{100}\right)(141.909 \text{ amu})$$

$$= (0.00185)(135.907 \text{ amu}) + (0.00251)(137.906 \text{ amu}) + (0.8845)(139.905 \text{ amu})$$

$$+ (0.11114)(141.909 \text{ amu})$$

$$= 140.115 \text{ amu}$$

---

[3]The term *atomic mass* is really more accurate than *atomic weight* inasmuch as, in this context, we are dealing with masses and not weights. However, atomic weight is, by convention, the preferred terminology and is used throughout this book. The reader should note that it is *not* necessary to divide molecular weight by the gravitational constant.

Concept Check 2.1 Why are the atomic weights of the elements generally not integers? Cite two reasons.

[*The answer may be found at* www.wiley.com/college/callister *(Student Companion Site).*]

2.3 ELECTRONS IN ATOMS

Atomic Models

During the latter part of the nineteenth century, it was realized that many phenomena involving electrons in solids could not be explained in terms of classical mechanics. What followed was the establishment of a set of principles and laws that govern systems of atomic **quantum mechanics** and subatomic entities that came to be known as **quantum mechanics**. An understanding of the behavior of electrons in atoms and crystalline solids necessarily involves the discussion of quantum-mechanical concepts. However, a detailed exploration of these principles is beyond the scope of this text, and only a very superficial and simplified treatment is given.

Bohr atomic model One early outgrowth of quantum mechanics was the simplified **Bohr atomic model**, in which electrons are assumed to revolve around the atomic nucleus in discrete orbitals, and the position of any particular electron is more or less well defined in terms of its orbital. This model of the atom is represented in Figure 2.1.

Another important quantum-mechanical principle stipulates that the energies of electrons are *quantized*—that is, electrons are permitted to have only specific values of energy. An electron may change energy, but in doing so, it must make a quantum jump either to an allowed higher energy (with absorption of energy) or to a lower energy (with emission of energy). Often, it is convenient to think of these allowed electron energies as being associated with *energy levels* or *states*. These states do not vary continuously with energy—that is, adjacent states are separated by finite energies. For example, allowed states for the Bohr hydrogen atom are represented in Figure 2.2*a*. These energies are taken to be negative, whereas the zero reference is the unbound or free electron. Of course, the single electron associated with the hydrogen atom fills only one of these states.

Thus, the Bohr model represents an early attempt to describe electrons in atoms, in terms of both position (electron orbitals) and energy (quantized energy levels).

This Bohr model was eventually found to have some significant limitations because of its inability to explain several phenomena involving electrons. A resolution was reached **wave-mechanical** with a **wave-mechanical model**, in which the electron is considered to exhibit both wave-**model** like and particle-like characteristics. With this model, an electron is no longer treated as a particle moving in a discrete orbital; rather, position is considered to be the probability of an electron's being at various locations around the nucleus. In other words, position is described by a probability distribution or electron cloud. Figure 2.3 compares Bohr and

Figure 2.1 Schematic representation of the Bohr atom.

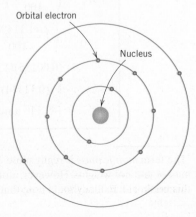

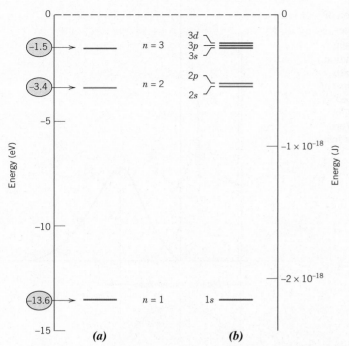

Figure 2.2 (*a*) The first three electron energy states for the Bohr hydrogen atom. (*b*) Electron energy states for the first three shells of the wave-mechanical hydrogen atom. (Adapted from W. G. Moffatt, G. W. Pearsall, and J. Wulff, *The Structure and Properties of Materials,* Vol. I, *Structure,* p. 10. Copyright © 1964 by John Wiley & Sons, New York.)

wave-mechanical models for the hydrogen atom. Both models are used throughout the course of this text; the choice depends on which model allows the simplest explanation.

Quantum Numbers

quantum number

In wave mechanics, every electron in an atom is characterized by four parameters called **quantum numbers**. The size, shape, and spatial orientation of an electron's probability density (or *orbital*) are specified by three of these quantum numbers. Furthermore, Bohr energy levels separate into electron subshells, and quantum numbers dictate the number of states within each subshell. Shells are specified by a *principal quantum number n*, which may take on integral values beginning with unity; sometimes these shells are designated by the letters *K, L, M, N, O*, and so on, which correspond, respectively, to *n* = 1, 2, 3, 4, 5, . . . , as indicated in Table 2.1. Note also that this quantum

Table 2.1 Summary of the Relationships among the Quantum Numbers *n, l, m$_l$*, and Numbers of Orbitals and Electrons

Value of n	Value of l	Values of m$_l$	Subshell	Number of Orbitals	Number of Electrons
1	0	0	1s	1	2
2	0	0	2s	1	2
	1	−1, 0, +1	2p	3	6
3	0	0	3s	1	2
	1	−1, 0, +1	3p	3	6
	2	−2, −1, 0, +1, +2	3d	5	10
4	0	0	4s	1	2
	1	−1, 0, +1	4p	3	6
	2	−2, −1, 0, +1, +2	4d	5	10
	3	−3, −2, −1, 0, +1, +2, +3	4f	7	14

Source: From J. E. Brady and F. Senese, *Chemistry: Matter and Its Changes,* 4th edition. Reprinted with permission of John Wiley & Sons, Inc.

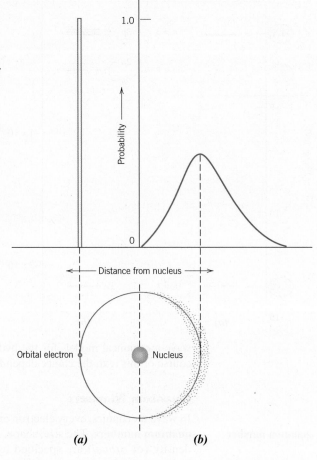

Figure 2.3 Comparison of the (*a*) Bohr and (*b*) wave-mechanical atom models in terms of electron distribution. (Adapted from Z. D. Jastrzeb-ski, *The Nature and Properties of Engineering Materials*, 3rd edition, p. 4. Copyright © 1987 by John Wiley & Sons, New York. Reprinted by permission of John Wiley & Sons, Inc.)

number, and it only, is also associated with the Bohr model. This quantum number is related to the size of an electron's orbital (or its average distance from the nucleus).

The second (or *azimuthal*) quantum number, *l*, designates the subshell. Values of *l* are restricted by the magnitude of *n* and can take on integer values that range from $l = 0$ to $l = (n - 1)$. Each subshell is denoted by a lowercase letter—an *s*, *p*, *d*, or *f*—related to *l* values as follows:

Value of l	Letter Designation
0	*s*
1	*p*
2	*d*
3	*f*

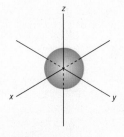

Figure 2.4
Spherical shape of an *s* electron orbital.

Furthermore, electron orbital shapes depend on *l*. For example *s* orbitals are spherical and centered on the nucleus (Figure 2.4). There are three orbitals for a *p* subshell (as explained next); each has a nodal surface in the shape of a dumbbell (Figure 2.5). Axes for these three orbitals are mutually perpendicular to one another like those of an *x-y-z* coordinate system; thus, it is convenient to label these orbitals p_x, p_y, and p_z (see Figure 2.5). Orbital configurations for *d* subshells are more complex and are not discussed here.

Figure 2.5
Orientations and
shapes of (a) p_x,
(b) p_y, and (c) p_z
electron orbitals.

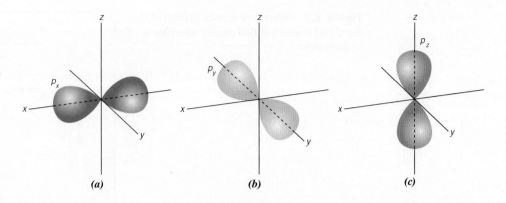

(a)   (b)   (c)

The number of electron orbitals for each subshell is determined by the third (or magnetic) quantum number, m_l; m_l can take on integer values between $-l$ and $+l$, including 0. When $l = 0$, m_l can only have a value of 0 because $+0$ and -0 are the same. This corresponds to an s subshell, which can have only one orbital. Furthermore, for $l = 1$, m_l can take on values of $-1, 0$, and $+1$, and three p orbitals are possible. Similarly, it can be shown that d subshells have five orbitals, and f subshells have seven. In the absence of an external magnetic field, all orbitals within each subshell are identical in energy. However, when a magnetic field is applied, these subshell states split, with each orbital assuming a slightly different energy. Table 2.1 presents a summary of the values and relationships among the n, l, and m_l quantum numbers.

Associated with each electron is a *spin moment*, which must be oriented either up or down. Related to this spin moment is the fourth quantum number, m_s, for which two values are possible: $+\frac{1}{2}$ (for spin up) and $-\frac{1}{2}$ (for spin down).

Thus, the Bohr model was further refined by wave mechanics, in which the introduction of three new quantum numbers gives rise to electron subshells within each shell. A comparison of these two models on this basis is illustrated, for the hydrogen atom, in Figures 2.2a and 2.2b.

A complete energy level diagram for the various shells and subshells using the wave-mechanical model is shown in Figure 2.6. Several features of the diagram are

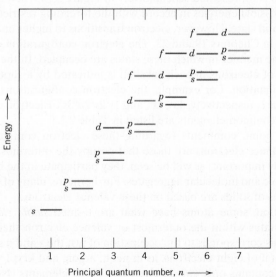

Figure 2.6 Schematic representation of the relative energies of the electrons for the various shells and subshells.
(From K. M. Ralls, T. H. Courtney, and J. Wulff, *Introduction to Materials Science and Engineering*, p. 22. Copyright © 1976 by John Wiley & Sons, New York. Reprinted by permission of John Wiley & Sons, Inc.)

Figure 2.7 Schematic representation of the filled and lowest unfilled energy states for a sodium atom.

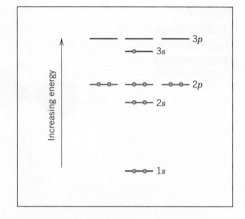

worth noting. First, the smaller the principal quantum number, the lower the energy level; for example, the energy of a 1s state is less than that of a 2s state, which in turn is lower than that of the 3s. Second, within each shell, the energy of a subshell level increases with the value of the l quantum number. For example, the energy of a 3d state is greater than that of a 3p, which is larger than 3s. Finally, there may be overlap in energy of a state in one shell with states in an adjacent shell, which is especially true of d and f states; for example, the energy of a 3d state is generally greater than that of a 4s.

Electron Configurations

electron state

Pauli exclusion principle

The preceding discussion has dealt primarily with **electron states**—values of energy that are permitted for electrons. To determine the manner in which these states are filled with electrons, we use the **Pauli exclusion principle**, another quantum-mechanical concept, which stipulates that each electron state can hold no more than two electrons that must have opposite spins. Thus, s, p, d, and f subshells may each accomodate, respectively, a total of 2, 6, 10, and 14 electrons; the right column of Table 2.1 notes the maximum number of electrons that may occupy each orbital for the first four shells.

Of course, not all possible states in an atom are filled with electrons. For most atoms, the electrons fill up the lowest possible energy states in the electron shells and subshells, two electrons (having opposite spins) per state. The energy structure for a sodium atom is represented schematically in Figure 2.7. When all the electrons occupy the lowest possible energies in accord with the foregoing restrictions, an atom is said to be in its **ground state**. However, electron transitions to higher energy states are possible, as discussed in Chapters 19 and 22. The **electron configuration** or structure of an atom represents the manner in which these states are occupied. In the conventional notation, the number of electrons in each subshell is indicated by a superscript after the shell–subshell designation. For example, the electron configurations for hydrogen, helium, and sodium are, respectively, $1s^1$, $1s^2$, and $1s^2 2s^2 2p^6 3s^1$. Electron configurations for some of the more common elements are listed in Table 2.2.

ground state

electron configuration

At this point, comments regarding these electron configurations are necessary. First, the **valence electrons** are those that occupy the outermost shell. These electrons are extremely important; as will be seen, they participate in the bonding between atoms to form atomic and molecular aggregates. Furthermore, many of the physical and chemical properties of solids are based on these valence electrons.

valence electron

In addition, some atoms have what are termed *stable electron configurations*— that is, the states within the outermost or valence electron shell are completely filled. Normally, this corresponds to the occupation of just the s and p states for the outermost shell by a total of eight electrons, as in neon, argon, and krypton; one exception is helium, which contains only two 1s electrons. These elements (Ne, Ar, Kr, and He) are

Table 2.2

Expected Electron
Configurations for
Some Common
Elements[a]

Element	Symbol	Atomic Number	Electron Configuration
Hydrogen	H	1	$1s^1$
Helium	He	2	$1s^2$
Lithium	Li	3	$1s^2 2s^1$
Beryllium	Be	4	$1s^2 2s^2$
Boron	B	5	$1s^2 2s^2 2p^1$
Carbon	C	6	$1s^2 2s^2 2p^2$
Nitrogen	N	7	$1s^2 2s^2 2p^3$
Oxygen	O	8	$1s^2 2s^2 2p^4$
Fluorine	F	9	$1s^2 2s^2 2p^5$
Neon	Ne	10	$1s^2 2s^2 2p^6$
Sodium	Na	11	$1s^2 2s^2 2p^6 3s^1$
Magnesium	Mg	12	$1s^2 2s^2 2p^6 3s^2$
Aluminum	Al	13	$1s^2 2s^2 2p^6 3s^2 3p^1$
Silicon	Si	14	$1s^2 2s^2 2p^6 3s^2 3p^2$
Phosphorus	P	15	$1s^2 2s^2 2p^6 3s^2 3p^3$
Sulfur	S	16	$1s^2 2s^2 2p^6 3s^2 3p^4$
Chlorine	Cl	17	$1s^2 2s^2 2p^6 3s^2 3p^5$
Argon	Ar	18	$1s^2 2s^2 2p^6 3s^2 3p^6$
Potassium	K	19	$1s^2 2s^2 2p^6 3s^2 3p^6 4s^1$
Calcium	Ca	20	$1s^2 2s^2 2p^6 3s^2 3p^6 4s^2$
Scandium	Sc	21	$1s^2 2s^2 2p^6 3s^2 3p^6 3d^1 4s^2$
Titanium	Ti	22	$1s^2 2s^2 2p^6 3s^2 3p^6 3d^2 4s^2$
Vanadium	V	23	$1s^2 2s^2 2p^6 3s^2 3p^6 3d^3 4s^2$
Chromium	Cr	24	$1s^2 2s^2 2p^6 3s^2 3p^6 3d^5 4s^1$
Manganese	Mn	25	$1s^2 2s^2 2p^6 3s^2 3p^6 3d^5 4s^2$
Iron	Fe	26	$1s^2 2s^2 2p^6 3s^2 3p^6 3d^6 4s^2$
Cobalt	Co	27	$1s^2 2s^2 2p^6 3s^2 3p^6 3d^7 4s^2$
Nickel	Ni	28	$1s^2 2s^2 2p^6 3s^2 3p^6 3d^8 4s^2$
Copper	Cu	29	$1s^2 2s^2 2p^6 3s^2 3p^6 3d^{10} 4s^1$
Zinc	Zn	30	$1s^2 2s^2 2p^6 3s^2 3p^6 3d^{10} 4s^2$
Gallium	Ga	31	$1s^2 2s^2 2p^6 3s^2 3p^6 3d^{10} 4s^2 4p^1$
Germanium	Ge	32	$1s^2 2s^2 2p^6 3s^2 3p^6 3d^{10} 4s^2 4p^2$
Arsenic	As	33	$1s^2 2s^2 2p^6 3s^2 3p^6 3d^{10} 4s^2 4p^3$
Selenium	Se	34	$1s^2 2s^2 2p^6 3s^2 3p^6 3d^{10} 4s^2 4p^4$
Bromine	Br	35	$1s^2 2s^2 2p^6 3s^2 3p^6 3d^{10} 4s^2 4p^5$
Krypton	Kr	36	$1s^2 2s^2 2p^6 3s^2 3p^6 3d^{10} 4s^2 4p^6$

[a]When some elements covalently bond, they form *sp* hybrid bonds. This is especially true
for C, Si, and Ge.

the inert, or noble, gases, which are virtually unreactive chemically. Some atoms of the elements that have unfilled valence shells assume stable electron configurations by gaining or losing electrons to form charged ions or by sharing electrons with other atoms. This is the basis for some chemical reactions and also for atomic bonding in solids, as explained in Section 2.6.

Concept Check 2.2   Give electron configurations for the Fe^{3+} and S^{2-} ions.

[*The answer may be found at* www.wiley.com/college/callister (*Student Companion Site*).]

2.4 THE PERIODIC TABLE

periodic table

All the elements have been classified according to electron configuration in the **periodic table** (Figure 2.8). Here, the elements are situated, with increasing atomic number, in seven horizontal rows called *periods*. The arrangement is such that all elements arrayed in a given column or group have similar valence electron structures, as well as chemical and physical properties. These properties change gradually, moving horizontally across each period and vertically down each column.

The elements positioned in Group 0, the rightmost group, are the *inert gases*, which have filled electron shells and stable electron configurations. Group VIIA and VIA elements are one and two electrons deficient, respectively, from having stable structures. The Group VIIA elements (F, Cl, Br, I, and At) are sometimes termed the *halogens*.

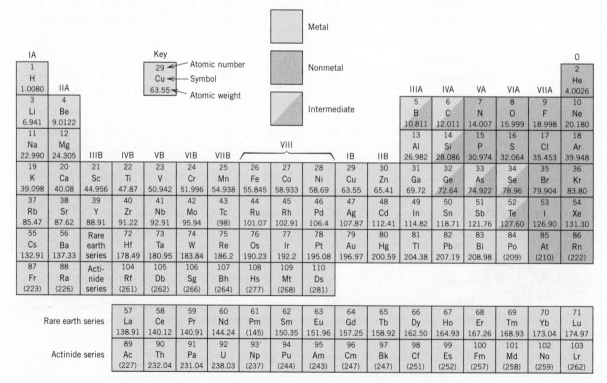

Figure 2.8   The periodic table of the elements. The numbers in parentheses are the atomic weights of the most stable or common isotopes.

The alkali and the alkaline earth metals (Li, Na, K, Be, Mg, Ca, etc.) are labeled as Groups IA and IIA, having, respectively, one and two electrons in excess of stable structures. The elements in the three long periods, Groups IIIB through IIB, are termed the *transition metals*, which have partially filled *d* electron states and in some cases one or two electrons in the next higher energy shell. Groups IIIA, IVA, and VA (B, Si, Ge, As, etc.) display characteristics that are intermediate between the metals and nonmetals by virtue of their valence electron structures.

electropositive

electronegative

As may be noted from the periodic table, most of the elements really come under the metal classification. These are sometimes termed **electropositive** elements, indicating that they are capable of giving up their few valence electrons to become positively charged ions. Furthermore, the elements situated on the right side of the table are **electronegative**—that is, they readily accept electrons to form negatively charged ions, or sometimes they share electrons with other atoms. Figure 2.9 displays electronegativity values that have been assigned to the various elements arranged in the periodic table. As a general rule, electronegativity increases in moving from left to right and from bottom to top. Atoms are more likely to accept electrons if their outer shells are almost full and if they are less "shielded" from (i.e., closer to) the nucleus.

In addition to chemical behavior, physical properties of the elements also tend to vary systematically with position in the periodic table. For example, most metals that reside in the center of the table (Groups IIIB through IIB) are relatively good conductors of electricity and heat; nonmetals are typically electrical and thermal insulators. Mechanically, the metallic elements exhibit varying degrees of *ductility*—the ability to be plastically deformed without fracturing (e.g., the ability to be rolled into thin sheets). Most of the nonmetals are either gases or liquids, or in the solid state are brittle in nature. Furthermore, for the Group IVA elements [C (diamond), Si, Ge, Sn, and Pb], electrical conductivity increases as we move down this column. The Group VB metals (V, Nb, and Ta) have very high melting temperatures, which increase in going down this column.

It should be noted that there is not always this consistency in property variations within the periodic table. Physical properties change in a more or less regular manner; however, there are some rather abrupt changes when one moves across a period or down a group.

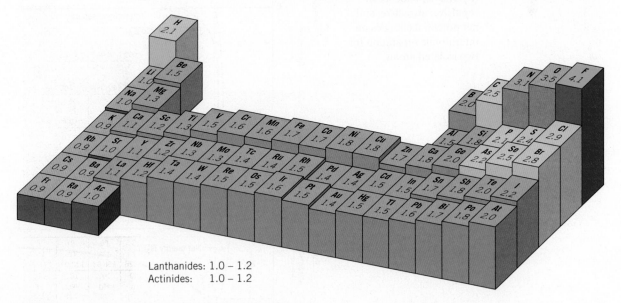

Lanthanides: 1.0 – 1.2
Actinides: 1.0 – 1.2

Figure 2.9 The electronegativity values for the elements.
(From J. E. Brady and F. Senese, *Chemistry: Matter and Its Changes*, 4th edition. This material is reproduced with permission of John Wiley & Sons, Inc.)

Atomic Bonding in Solids

2.5 BONDING FORCES AND ENERGIES

An understanding of many of the physical properties of materials is enhanced by a knowledge of the interatomic forces that bind the atoms together. Perhaps the principles of atomic bonding are best illustrated by considering how two isolated atoms interact as they are brought close together from an infinite separation. At large distances, interactions are negligible because the atoms are too far apart to have an influence on each other; however, at small separation distances, each atom exerts forces on the others. These forces are of two types, attractive (F_A) and repulsive (F_R), and the magnitude of each depends on the separation or interatomic distance (r); Figure 2.10a is a schematic plot of F_A and F_R versus r. The origin of an attractive force F_A depends on the particular type of bonding that exists between the two atoms, as discussed shortly. Repulsive forces arise from interactions between the negatively charged electron clouds for the two atoms and are important only at small values of r as the outer electron shells of the two atoms begin to overlap (Figure 2.10a).

The net force F_N between the two atoms is just the sum of both attractive and repulsive components; that is,

$$F_N = F_A + F_R \tag{2.3}$$

Figure 2.10 (a) The dependence of repulsive, attractive, and net forces on interatomic separation for two isolated atoms. (b) The dependence of repulsive, attractive, and net potential energies on interatomic separation for two isolated atoms.

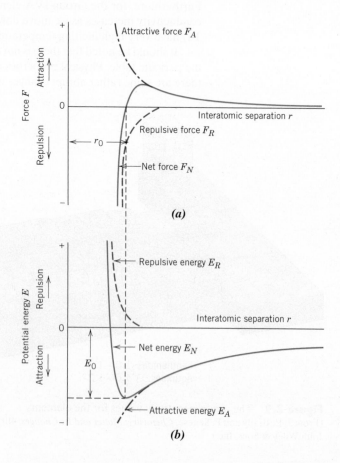

which is also a function of the interatomic separation, as also plotted in Figure 2.10a. When F_A and F_R are equal in magnitude but opposite in sign, there is no net force—that is,

$$F_A + F_R = 0 \tag{2.4}$$

and a state of equilibrium exists. The centers of the two atoms remain separated by the equilibrium spacing r_0, as indicated in Figure 2.10a. For many atoms, r_0 is approximately 0.3 nm. Once in this position, any attempt to move the two atoms farther apart is counteracted by the attractive force, while pushing them closer together is resisted by the increasing repulsive force.

Sometimes it is more convenient to work with the potential energies between two atoms instead of forces. Mathematically, energy (E) and force (F) are related as

Force–potential energy relationship for two atoms

$$E = \int F \, dr \tag{2.5a}$$

And, for atomic systems,

$$E_N = \int_r^\infty F_N \, dr \tag{2.6}$$

$$= \int_r^\infty F_A \, dr + \int_r^\infty F_R \, dr \tag{2.7}$$

$$= E_A + E_R \tag{2.8a}$$

in which E_N, E_A, and E_R are, respectively, the net, attractive, and repulsive energies for two isolated and adjacent atoms.[4]

Figure 2.10b plots attractive, repulsive, and net potential energies as a function of interatomic separation for two atoms. From Equation 2.8a, the net curve is the sum of the attractive and repulsive curves. The minimum in the net energy curve corresponds to the equilibrium spacing, r_0. Furthermore, the **bonding energy** for these two atoms, E_0, corresponds to the energy at this minimum point (also shown in Figure 2.10b); it represents the energy required to separate these two atoms to an infinite separation.

bonding energy

Although the preceding treatment deals with an ideal situation involving only two atoms, a similar yet more complex condition exists for solid materials because force and energy interactions among atoms must be considered. Nevertheless, a bonding energy, analogous to E_0 above, may be associated with each atom. The magnitude of this bonding energy and the shape of the energy–versus–interatomic separation curve vary from material to material, and they both depend on the type of atomic bonding. Furthermore,

---

[4]Force in Equation 2.5a may also be expressed as

$$F = \frac{dE}{dr} \tag{2.5b}$$

Likewise, the force equivalent of Equation 2.8a is as follows:

$$F_N = F_A + F_R \tag{2.3}$$

$$= \frac{dE_A}{dr} + \frac{dE_R}{dr} \tag{2.8b}$$

a number of material properties depend on E_0, the curve shape, and bonding type. For example, materials having large bonding energies typically also have high melting temperatures; at room temperature, solid substances are formed for large bonding energies, whereas for small energies, the gaseous state is favored; liquids prevail when the energies are of intermediate magnitude. In addition, as discussed in Section 8.3, the mechanical stiffness (or modulus of elasticity) of a material is dependent on the shape of its force–versus–interatomic separation curve (Figure 8.4). The slope for a relatively stiff material at the $r = r_0$ position on the curve will be quite steep; slopes are shallower for more flexible materials. Furthermore, how much a material expands upon heating or contracts upon cooling (i.e., its linear coefficient of thermal expansion) is related to the shape of its E–versus–r curve (see Section 20.3). A deep and narrow "trough," which typically occurs for materials having large bonding energies, normally correlates with a low coefficient of thermal expansion and relatively small dimensional alterations for changes in temperature.

primary bond

Three different types of **primary** or chemical **bond** are found in solids—ionic, covalent, and metallic. For each type, the bonding necessarily involves the valence electrons; furthermore, the nature of the bond depends on the electron structures of the constituent atoms. In general, each of these three types of bonding arises from the tendency of the atoms to assume stable electron structures, like those of the inert gases, by completely filling the outermost electron shell.

Secondary or physical forces and energies are also found in many solid materials; they are weaker than the primary ones but nonetheless influence the physical properties of some materials. The sections that follow explain the several kinds of primary and secondary interatomic bonds.

2.6 PRIMARY INTERATOMIC BONDS

Ionic Bonding

ionic bonding

Ionic bonding is perhaps the easiest to describe and visualize. It is always found in compounds composed of both metallic and nonmetallic elements, elements situated at the horizontal extremities of the periodic table. Atoms of a metallic element easily give up their valence electrons to the nonmetallic atoms. In the process, all the atoms acquire stable or inert gas configurations (i.e., completely filled orbital shells) and, in addition, an electrical charge—that is, they become ions. Sodium chloride (NaCl) is the classic ionic material. A sodium atom can assume the electron structure of neon (and a net single positive charge with a reduction in size) by a transfer of its one valence 3s electron to a chlorine atom (Figure 2.11a). After such a transfer, the chlorine ion acquires a net negative charge, an electron configuration identical to that of argon; it is also larger than the chlorine atom. Ionic bonding is illustrated schematically in Figure 2.11b.

coulombic force

The attractive bonding forces are **coulombic**—that is, positive and negative ions, by virtue of their net electrical charge, attract one another. For two isolated ions, the attractive energy E_A is a function of the interatomic distance according to

Attractive energy—interatomic separation relationship

$$E_A = -\frac{A}{r} \tag{2.9}$$

Theoretically, the constant A is equal to

$$A = \frac{1}{4\pi\epsilon_0}(|Z_1|e)(|Z_2|e) \tag{2.10}$$

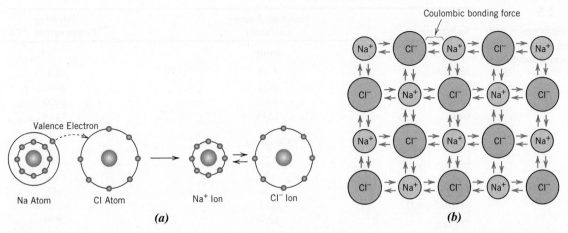

Figure 2.11 Schematic representations of (*a*) the formation of Na$^+$ and Cl$^-$ ions and (*b*) ionic bonding in sodium chloride (NaCl).

Here ϵ_0 is the permittivity of a vacuum (8.85×10^{-12} F/m), $|Z_1|$ and $|Z_2|$ are absolute values of the valences for the two ion types, and e is the electronic charge (1.602×10^{-19}C). The value of A in Equation 2.9 assumes the bond between ions 1 and 2 is totally ionic (see Equation 2.16). Inasmuch as bonds in most of these materials are not 100% ionic, the value of A is normally determined from experimental data rather than computed using Equation 2.10.

An analogous equation for the repulsive energy is[5]

Repulsive energy—interatomic separation relationship

$$E_R = \frac{B}{r^n} \tag{2.11}$$

In this expression, B and n are constants whose values depend on the particular ionic system. The value of n is approximately 8.

Ionic bonding is termed *nondirectional*—that is, the magnitude of the bond is equal in all directions around an ion. It follows that for ionic materials to be stable, all positive ions must have as nearest neighbors negatively charged ions in a three-dimensional scheme, and vice versa. Some of the ion arrangements for these materials are discussed in Chapter 4.

WileyPLUS

Tutorial Video:
Bonding
What Is Ionic Bonding?

Bonding energies, which generally range between 600 and 1500 kJ/mol, are relatively large, as reflected in high melting temperatures.[6] Table 2.3 contains bonding energies and melting temperatures for several ionic materials. Interatomic bonding is typified by ceramic materials, which are characteristically hard and brittle and, furthermore, electrically and thermally insulative. As discussed in subsequent chapters, these properties are a direct consequence of electron configurations and/or the nature of the ionic bond.

---

[5]In Equation 2.11, the value of the constant B is also fit using experimental data.

[6]Sometimes bonding energies are expressed per atom or per ion. Under these circumstances, the electron volt (eV) is a conveniently small unit of energy. It is, by definition, the energy imparted to an electron as it falls through an electric potential of one volt. The joule equivalent of the electron volt is as follows: 1.602×10^{-19} J = 1 eV.

Table 2.3

Bonding Energies and
Melting Temperatures
for Various Substances

Substance	Bonding Energy (kJ/mol)	Melting Temperature (°C)
Ionic		
NaCl	640	801
LiF	850	848
MgO	1000	2800
CaF_2	1548	1418
Covalent		
Cl_2	121	−102
Si	450	1410
InSb	523	942
C (diamond)	713	>3550
SiC	1230	2830
Metallic		
Hg	62	−39
Al	330	660
Ag	285	962
W	850	3414
van der Waals[a]		
Ar	7.7	−189 (@ 69 kPa)
Kr	11.7	−158 (@ 73.2 kPa)
CH_4	18	−182
Cl_2	31	−101
Hydrogen[a]		
HF	29	−83
NH_3	35	−78
H_2O	51	0

[a]Values for van der Waals and hydrogen bonds are energies *between* molecules or atoms (*inter*molecular), not between atoms within a molecule (*intra*molecular).

EXAMPLE PROBLEM 2.2

Computation of Attractive and Repulsive Forces between Two Ions

The atomic radii of K^+ and Br^- ions are 0.138 and 0.196 nm, respectively.

(a) Using Equations 2.9 and 2.10, calculate the force of attraction between these two ions at their equilibrium interionic separation (i.e., when the ions just touch one another).

(b) What is the force of repulsion at this same separation distance?

Solution

(a) From Equation 2.5b, the force of attraction between two ions is

$$F_A = \frac{dE_A}{dr}$$

Whereas, according to Equation 2.9,

$$E_A = -\frac{A}{r}$$

Now, taking the derivation of E_A with respect to r yields the following expression for the force of attraction F_A:

$$F_A = \frac{dE_A}{dr} = \frac{d\left(-\dfrac{A}{r}\right)}{dr} = -\left(\frac{-A}{r^2}\right) = \frac{A}{r^2} \tag{2.12}$$

Now substitution into this equation the expression for A (Eq. 2.10) gives

$$F_A = \frac{1}{4\pi\epsilon_0 r^2}(|Z_1|e)(|Z_2|e) \tag{2.13}$$

Incorporation into this equation values for e and ϵ_0 leads to

$$\begin{aligned} F_A &= \frac{1}{4\pi(8.85 \times 10^{-12}\,\text{F/m})(r^2)}[|Z_1|(1.602 \times 10^{-19}\text{C})][|Z_2|(1.602 \times 10^{-19}\text{C})] \\ &= \frac{(2.31 \times 10^{-28}\,\text{N}\cdot\text{m}^2)(|Z_1|)(|Z_2|)}{r^2} \end{aligned} \tag{2.14}$$

For this problem, r is taken as the interionic separation r_0 for KBr, which is equal to the sum of the K^+ and Br^- ionic radii inasmuch as the ions touch one another—that is,

$$\begin{aligned} r_0 &= r_{K^+} + r_{Br^-} \\ &= 0.138\,\text{nm} + 0.196\,\text{nm} \\ &= 0.334\,\text{nm} \\ &= 0.334 \times 10^{-9}\,\text{m} \end{aligned} \tag{2.15}$$

When we substitute this value for r into Equation 2.14, and taking ion 1 to be K^+ and ion 2 as Br^- (i.e., $Z_1 = +1$ and $Z_2 = -1$), then the force of attraction is equal to

$$F_A = \frac{(2.31 \times 10^{-28}\,\text{N}\cdot\text{m}^2)(|+1|)(|-1|)}{(0.334 \times 10^{-9}\,\text{m})^2} = 2.07 \times 10^{-9}\,\text{N}$$

(b) At the equilibrium separation distance the sum of attractive and repulsive forces is zero according to Equation 2.4. This means that

$$F_R = -F_A = -(2.07 \times 10^{-9}\,\text{N}) = -2.07 \times 10^{-9}\,\text{N}$$

Covalent Bonding

covalent bonding

A second bonding type, **covalent bonding**, is found in materials whose atoms have small differences in electronegativity—that is, that lie near one another in the periodic table. For these materials, stable electron configurations are assumed by the sharing of electrons between adjacent atoms. Two covalently bonded atoms will each contribute at least one electron to the bond, and the shared electrons may be considered to belong to both atoms. Covalent bonding is schematically illustrated in Figure 2.12 for a molecule of hydrogen (H_2). The hydrogen atom has a single $1s$ electron. Each of the atoms can acquire a helium electron configuration (two $1s$ valence electrons) when they share their single electron (right side of Figure 2.12). Furthermore, there is an overlapping of

Figure 2.16 Schematic diagram that shows the formation of sp^2 hybrid orbitals in carbon. (*a*) Promotion of a 2*s* electron to a 2*p* state; (*b*) this promoted electron in a 2*p* state; (*c*) three $2sp^2$ orbitals that form by mixing the single 2*s* orbital with two 2*p* orbitals—the $2p_z$ orbital remains unhybridized.

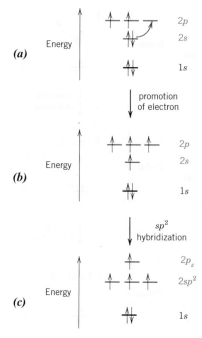

Metallic Bonding

metallic bonding

Metallic bonding, the final primary bonding type, is found in metals and their alloys. A relatively simple model has been proposed that very nearly approximates the bonding scheme. With this model, these valence electrons are not bound to any particular atom in the solid and are more or less free to drift throughout the entire metal. They may be thought of as belonging to the metal as a whole, or forming a "sea of electrons" or an "electron cloud." The remaining nonvalence electrons and atomic nuclei form what are called *ion cores*, which possess a net positive charge equal in magnitude to the total valence electron charge

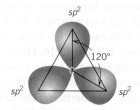

Figure 2.17 Schematic diagram showing three sp^2 orbitals that are coplanar and point to the corners of a triangle; the angle between adjacent orbitals is 120°. (From J. E. Brady and F. Senese, *Chemistry: Matter and Its Changes,* 4th edition. Reprinted with permission of John Wiley & Sons, Inc.)

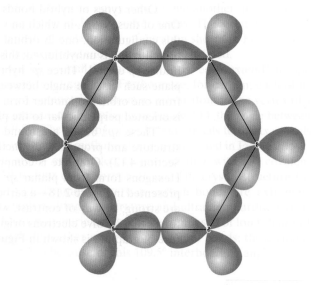

Figure 2.18 The formation of a hexagon by the bonding of six sp^2 triangles to one another.

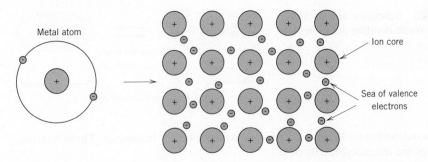

Metal atom

Ion core

Sea of valence electrons

Figure 2.19 Schematic illustration of metallic bonding.

per atom. Figure 2.19 illustrates metallic bonding. The free electrons shield the positively charged ion cores from the mutually repulsive electrostatic forces that they would otherwise exert upon one another; consequently, the metallic bond is nondirectional in character. In addition, these free electrons act as a "glue" to hold the ion cores together. Bonding energies and melting temperatures for several metals are listed in Table 2.3. Bonding may be weak or strong; energies range from 62 kJ/mol for mercury to 850 kJ/mol for tungsten. Their respective melting temperatures are −39°C and 3414°C.

Metallic bonding is found in the periodic table for Group IA and IIA elements and, in fact, for all elemental metals.

Metals are good conductors of both electricity and heat as a consequence of their free electrons (see Sections 19.5, 19.6, and 20.4). Furthermore, in Section 9.4, we note that at room temperature, most metals and their alloys fail in a ductile manner—that is, fracture occurs after the materials have experienced significant degrees of permanent deformation. This behavior is explained in terms of deformation mechanism (Section 9.2), which is implicitly related to the characteristics of the metallic bond.

WileyPLUS

Tutorial Video:
Bonding
What Is Metallic
Bonding?

 Concept Check 2.3 Explain why covalently bonded materials are generally less dense than ionically or metallically bonded ones.

[*The answer may be found at* www.wiley.com/college/callister (*Student Companion Site*).]

2.7 SECONDARY BONDING OR VAN DER WAALS BONDING

secondary bond

van der Waals bond

Secondary bonds, or **van der Waals** (physical) **bonds**, are weak in comparison to the primary or chemical bonds; bonding energies range between about 4 and 30 kJ/mol. Secondary bonding exists between virtually all atoms or molecules, but its presence may be obscured if any of the three primary bonding types is present. Secondary bonding is evidenced for the inert gases, which have stable electron structures. In addition, secondary (or *inter*molecular) bonds are possible between atoms or groups of atoms, which themselves are joined together by primary (or *intra*molecular) ionic or covalent bonds.

dipole

Secondary bonding forces arise from atomic or molecular **dipoles**. In essence, an electric dipole exists whenever there is some separation of positive and negative portions of an atom or molecule. The bonding results from the coulombic attraction between the positive end of one dipole and the negative region of an adjacent one, as indicated in Figure 2.20. Dipole interactions occur between induced dipoles, between induced dipoles and polar molecules (which have permanent dipoles), and between polar molecules. **Hydrogen bonding**, a special type of secondary bonding, is found to exist

hydrogen bonding

Figure 2.20 Schematic illustration of van der Waals bonding between two dipoles.

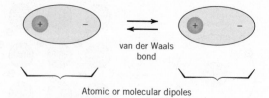

van der Waals bond

Atomic or molecular dipoles

WileyPLUS

Tutorial Video:
Bonding
What Is a Dipole?

WileyPLUS

Tutorial Video:
Bonding
What Is van der
Waals Bonding?

between some molecules that have hydrogen as one of the constituents. These bonding mechanisms are discussed briefly next.

Fluctuating Induced Dipole Bonds

A dipole may be created or induced in an atom or molecule that is normally electrically symmetric—that is, the overall spatial distribution of the electrons is symmetric with respect to the positively charged nucleus, as shown in Figure 2.21a. All atoms experience constant vibrational motion that can cause instantaneous and short-lived distortions of this electrical symmetry for some of the atoms or molecules and the creation of small electric dipoles. One of these dipoles can in turn produce a displacement of the electron distribution of an adjacent molecule or atom, which induces the second one also to become a dipole that is then weakly attracted or bonded to the first (Figure 2.21b); this is one type of van der Waals bonding. These attractive forces, which are temporary and fluctuate with time, may exist between large numbers of atoms or molecules.

The liquefaction and, in some cases, the solidification of the inert gases and other electrically neutral and symmetric molecules such as H_2 and Cl_2 are realized because of this type of bonding. Melting and boiling temperatures are extremely low in materials for which induced dipole bonding predominates; of all possible intermolecular bonds, these are the weakest. Bonding energies and melting temperatures for argon, krypton, methane, and chlorine are also tabulated in Table 2.3.

Polar Molecule–Induced Dipole Bonds

polar molecule

Permanent dipole moments exist in some molecules by virtue of an asymmetrical arrangement of positively and negatively charged regions; such molecules are termed **polar molecules**. Figure 2.22a shows a schematic representation of a hydrogen chloride molecule; a permanent dipole moment arises from net positive and negative charges that are respectively associated with the hydrogen and chlorine ends of the HCl molecule.

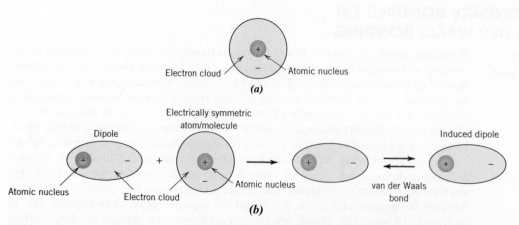

Electron cloud — Atomic nucleus

(a)

Dipole — Electrically symmetric atom/molecule — Induced dipole

Atomic nucleus — Electron cloud — Atomic nucleus — van der Waals bond

(b)

Figure 2.21 Schematic representations of (*a*) an electrically symmetric atom and (*b*) how an electric dipole induces an electrically symmetric atom/molecule to become a dipole—also the van der Waals bond between the dipoles.

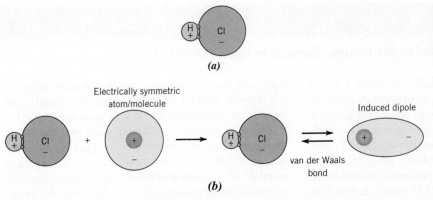

Figure 2.22 Schematic representations of (a) a hydrogen chloride molecule (dipole) and (b) how an HCl molecule induces an electrically symmetric atom/molecule to become a dipole—also the van der Waals bond between these dipoles.

Polar molecules can also induce dipoles in adjacent nonpolar molecules, and a bond forms as a result of attractive forces between the two molecules; this bonding scheme is represented schematically in Figure 2.22b. Furthermore, the magnitude of this bond is greater than for fluctuating induced dipoles.

Permanent Dipole Bonds

Coulombic forces also exist between adjacent polar molecules as in Figure 2.20. The associated bonding energies are significantly greater than for bonds involving induced dipoles.

The strongest secondary bonding type, the hydrogen bond, is a special case of polar molecule bonding. It occurs between molecules in which hydrogen is covalently bonded to fluorine (as in HF), oxygen (as in H_2O), or nitrogen (as in NH_3). For each H—F, H—O, or H—N bond, the single hydrogen electron is shared with the other atom. Thus, the hydrogen end of the bond is essentially a positively charged bare proton unscreened by any electrons. This highly positively charged end of the molecule is capable of a strong attractive force with the negative end of an adjacent molecule, as demonstrated in Figure 2.23 for HF. In essence, this single proton forms a bridge between two negatively charged atoms. The magnitude of the hydrogen bond is generally greater than that of the other types of secondary bonds and may be as high as 51 kJ/mol, as shown in Table 2.3. Melting and boiling temperatures for hydrogen fluoride, ammonia, and water are abnormally high in light of their low molecular weights, as a consequence of hydrogen bonding.

In spite of the small energies associated with secondary bonds, they nevertheless are involved in a number of natural phenomena and many products that we use on a daily basis. Examples of physical phenomena include the solubility of one substance in another, surface tension and capillary action, vapor pressure, volatility, and viscosity. Common applications that make use of these phenomena include *adhesives*—van der Waals bonds form between two surfaces so that they adhere to one another (as discussed in the chapter opener for this chapter); *surfactants*—compounds that lower the surface tension of a liquid, and are found in soaps, detergents, and foaming agents; *emulsifiers*—substances that, when added to two immiscible materials (usually liquids), allow particles of one material to be suspended in another (common emulsions include sunscreens, salad dressings, milk, and mayonnaise); and *desiccants*—materials that form hydrogen bonds with water molecules (and remove moisture from closed containers—e.g., small packets that are often found in cartons of packaged goods); and finally, the strengths, stiffnesses, and softening temperatures of polymers, to some degree, depend on secondary bonds that form between chain molecules.

WileyPLUS

Tutorial Video:
Bonding
What Are the Differences between Ionic, Covalent, Metallic, and van der Waals Types of Bonding?

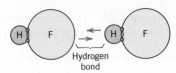

Figure 2.23 Schematic representation of hydrogen bonding in hydrogen fluoride (HF).

MATERIALS OF IMPORTANCE

Water (Its Volume Expansion upon Freezing)

Upon freezing (i.e., transforming from a liquid to a solid upon cooling), most substances experience an increase in density (or, correspondingly, a decrease in volume). One exception is water, which exhibits the anomalous and familiar expansion upon freezing—approximately 9 volume percent expansion. This behavior may be explained on the basis of hydrogen bonding. Each H_2O molecule has two hydrogen atoms that can bond to oxygen atoms; in addition, its single O atom can bond to two hydrogen atoms of other H_2O molecules. Thus, for solid ice, each water molecule participates in four hydrogen bonds, as shown in the three-dimensional schematic of Figure 2.24*a;* here, hydrogen bonds are denoted by dashed lines, and each water molecule has 4 nearest-neighbor molecules. This is a relatively open structure—that is, the molecules are not closely packed together—and as a result, the density is comparatively low. Upon melting, this structure is partially destroyed, such that the water molecules become more closely packed together (Figure 2.24*b*)—at room temperature, the average number of nearest-neighbor water molecules has increased to approximately 4.5; this leads to an increase in density.

Consequences of this anomalous freezing phenomenon are familiar; it explains why icebergs float; why, in cold climates, it is necessary to add antifreeze to an automobile's cooling system (to keep the engine block from cracking); and why freeze–thaw cycles break up the pavement in streets and cause potholes to form.

Photography by S. Tanner

A watering can that ruptured along a side panel–bottom panel seam. Water that was left in the can during a cold late-autumn night expanded as it froze and caused the rupture.

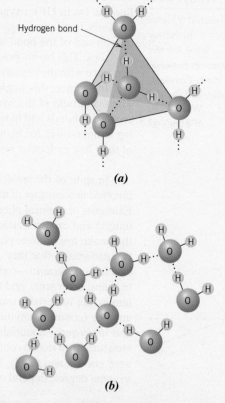

Hydrogen bond

(a)

(b)

Figure 2.24 The arrangement of water (H_2O) molecules in (*a*) solid ice and (*b*) liquid water.

2.8 MIXED BONDING

Sometimes it is illustrative to represent the four bonding types—ionic, covalent, metallic, and van der Waals—on what is called a *bonding tetrahedron*—a three-dimensional tetrahedron with one of these "extreme" types located at each vertex, as shown in Figure 2.25*a*. Furthermore, we should point out that for many real materials, the atomic bonds are mixtures of two or more of these extremes (i.e., *mixed bonds*). Three mixed-bond types—covalent–ionic, covalent–metallic, and metallic–ionic—are also included on edges of this tetrahedron; we now discuss each of them.

For mixed covalent–ionic bonds, there is some ionic character to most covalent bonds and some covalent character to ionic ones. As such, there is a continuum between these two extreme bond types. In Figure 2.25*a*, this type of bond is represented between the ionic and covalent bonding vertices. The degree of either bond type depends on the relative positions of the constituent atoms in the periodic table (see Figure 2.8) or the difference in their electronegativities (see Figure 2.9). The wider the separation (both horizontally—relative to Group IVA—and vertically) from the lower left to the upper right corner (i.e., the greater the difference in electronegativity), the more ionic is the bond. Conversely, the closer the atoms are together (i.e., the smaller the difference in electronegativity), the greater is the degree of covalency. Percent ionic character (%IC) of a bond between elements A and B (A being the most electronegative) may be approximated by the expression

$$\%IC = \left\{1 - \exp[-(0.25)(X_A - X_B)^2]\right\} \times 100 \tag{2.16}$$

where X_A and X_B are the electronegativities for the respective elements.

Another type of mixed bond is found for some elements in Groups IIIA, IVA, and VA of the periodic table (viz., B, Si, Ge, As, Sb, Te, Po, and At). Interatomic bonds for these elements are mixtures of metallic and covalent, as noted on Figure 2.25*a*. These materials are called the *metalloids* or *semi-metals*, and their properties are intermediate between the metals and nonmetals. In addition, for Group IV elements, there is a gradual transition from covalent to metallic bonding as one moves vertically down this column—for example, bonding in carbon (diamond) is purely covalent, whereas for tin and lead, bonding is predominantly metallic.

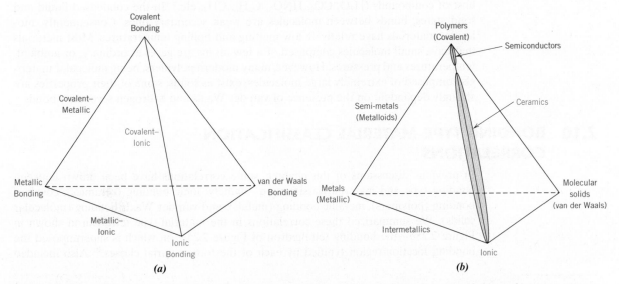

Figure 2.25 (*a*) Bonding tetrahedron: Each of the four extreme (or pure) bonding types is located at one corner of the tetrahedron; three mixed bonding types are included along tetrahedron edges. (*b*) Material-type tetrahedron: correlation of each material classification (metals, ceramics, polymers, etc.) with its type(s) of bonding.

Mixed metallic–ionic bonds are observed for compounds composed of two metals when there is a significant difference between their electronegativities. This means that some electron transfer is associated with the bond inasmuch as it has an ionic component. Furthermore, the larger this electronegativity difference, the greater the degree of ionicity. For example, there is little ionic character to the titanium–aluminum bond for the intermetallic compound $TiAl_3$ because electronegativities of both Al and Ti are the same (1.5; see Figure 2.9). However, a much greater degree of ionic character is present for $AuCu_3$; the electronegativity difference for copper and gold is 0.5.

EXAMPLE PROBLEM 2.3

Calculation of the Percent Ionic Character for the C–H Bond

Compute the percent ionic character (%IC) of the interatomic bond that forms between carbon and hydrogen.

Solution

The %IC of a bond between two atoms/ions, A and B (A being the more electronegative) is a function of their electronegativities X_A and X_B, according to Equation 2.16. The electronegativities for C and H (see Figure 2.9) are $X_C = 2.5$ and $X_H = 2.1$. Therefore, the %IC is

$$\%IC = \left\{1 - \exp[-(0.25)(X_C - X_H)^2]\right\} \times 100$$
$$= \left\{1 - \exp[-(0.25)(2.5 - 2.1)^2]\right\} \times 100$$
$$= 3.9\%$$

Thus the C—H atomic bond is primarily covalent (96.1%).

2.9 MOLECULES

Many common molecules are composed of groups of atoms bound together by strong covalent bonds, including elemental diatomic molecules (F_2, O_2, H_2, etc.), as well as a host of compounds (H_2O, CO_2, HNO_3, C_6H_6, CH_4, etc.). In the condensed liquid and solid states, bonds between molecules are weak secondary ones. Consequently, molecular materials have relatively low melting and boiling temperatures. Most materials that have small molecules composed of a few atoms are gases at ordinary, or ambient, temperatures and pressures. However, many modern polymers, being molecular materials composed of extremely large molecules, exist as solids; some of their properties are strongly dependent on the presence of van der Waals and hydrogen secondary bonds.

2.10 BONDING TYPE-MATERIAL CLASSIFICATION CORRELATIONS

In previous discussions of this chapter, some correlations have been drawn between bonding type and material classification—namely, ionic bonding (ceramics), covalent bonding (polymers), metallic bonding (metals), and van der Waals bonding (molecular solids). We summarized these correlations in the material-type tetrahedron shown in Figure 2.25b—the bonding tetrahedron of Figure 2.25a, on which is superimposed the bonding location/region typified by each of the four material classes.[10] Also included

---

[10]Although most atoms in polymer molecules are covalently bonded, some van der Waals bonding is normally present. We chose not to include van der Waals bonds for polymers because they (van der Waals) are *inter*molecular (i.e., between molecules) as opposed to *intra*molecular (within molecules) and not the principal bonding type.

are those materials having mixed bonding: intermetallics and semi-metals. Mixed ionic–covalent bonding for ceramics is also noted. Furthermore, the predominant bonding type for semiconducting materials is covalent, with the possibility of an ionic contribution.

SUMMARY

Electrons in Atoms
- The two atomic models are Bohr and wave mechanical. Whereas the Bohr model assumes electrons to be particles orbiting the nucleus in discrete paths, in wave mechanics we consider them to be wavelike and treat electron position in terms of a probability distribution.
- The energies of electrons are *quantized*—that is, only specific values of energy are allowed.
- The four electron quantum numbers are n, l, m_l, and m_s. They specify, respectively, electron orbital size, orbital shape, number of electron orbitals, and spin moment.
- According to the Pauli exclusion principle, each electron state can accommodate no more than two electrons, which must have opposite spins.

The Periodic Table
- Elements in each of the columns (or groups) of the periodic table have distinctive electron configurations. For example:
  Group 0 elements (the inert gases) have filled electron shells.
  Group IA elements (the alkali metals) have one electron greater than a filled electron shell.

Bonding Forces and Energies
- *Bonding force* and *bonding energy* are related to one another according to Equations 2.5a and 2.5b.
- Attractive, repulsive, and net energies for two atoms or ions depend on interatomic separation per the schematic plot of Figure 2.10*b*.
- From a plot of interatomic separation versus force for two atoms/ions, the equilibrium separation corresponds to the value at zero force.
- From a plot of interatomic separation versus potential energy for two atoms/ions, the bonding energy corresponds to the energy value at the minimum of the curve.

Primary Interatomic Bonds
- For ionic bonds, electrically charged ions are formed by the transference of valence electrons from one atom type to another.
- The attractive force between two isolated ions that have opposite charges may be computed using Equation 2.13.
- There is a sharing of valence electrons between adjacent atoms when bonding is covalent.
- Electron orbitals for some covalent bonds may overlap or hybridize. Hybridization of s and p orbitals to form sp^3 and sp^2 orbitals in carbon was discussed. Configurations of these hybrid orbitals were also noted.
- With metallic bonding, the valence electrons form a "sea of electrons" that is uniformly dispersed around the metal ion cores and acts as a form of glue for them.

Secondary Bonding or van der Waals Bonding
- Relatively weak van der Waals bonds result from attractive forces between electric dipoles, which may be induced or permanent.
- For hydrogen bonding, highly polar molecules form when hydrogen covalently bonds to a nonmetallic element such as fluorine.

Mixed Bonding
- In addition to van der Waals bonding and the three primary bonding types, covalent–ionic, covalent–metallic, and metallic–ionic mixed bonds exist.
- The percent ionic character (%IC) of a bond between two elements (A and B) depends on their electronegativities (X's) according to Equation 2.16.

Bonding Type-
Material
Classification
Correlations

• Correlations between bonding type and material class were noted:

Polymers—covalent

Metals—metallic

Ceramics—ionic/mixed ionic–covalent

Molecular solids—van der Waals

Semi-metals—mixed covalent–metallic

Intermetallics—mixed metallic–ionic

Important Terms and Concepts

atomic mass unit (amu)
atomic number (Z)
atomic weight (A)
Bohr atomic model
bonding energy
coulombic force
covalent bond
dipole (electric)
electron configuration
electronegative

electron state
electropositive
ground state
hydrogen bond
ionic bond
isotope
metallic bond
mole
Pauli exclusion principle

periodic table
polar molecule
primary bond
quantum mechanics
quantum number
secondary bond
valence electron
van der Waals bond
wave-mechanical model

REFERENCES

Most of the material in this chapter is covered in college-level chemistry textbooks. Two are listed here as references.

Ebbing, D. D., S. D. Gammon, and R. O. Ragsdale, *Essentials of General Chemistry*, 2nd edition, Cengage Learning, Boston, MA, 2006.

Jespersen, N. D, J. E. Brady, and A. Hyslop, *Chemistry: Matter and Its Changes*, 6th edition, Wiley, Hoboken, NJ, 2012.

QUESTIONS AND PROBLEMS

Fundamental Concepts
Electrons in Atoms

2.1 Cite the difference between atomic mass and atomic weight.

2.2 Chromium has four naturally occurring isotopes: 4.34% of ^{50}Cr, with an atomic weight of 49.9460 amu; 83.79% of ^{52}Cr, with an atomic weight of 51.9405 amu; 9.50% of ^{53}Cr, with an atomic weight of 52.9407 amu; and 2.37% of ^{54}Cr, with an atomic weight of 53.9389 amu. On the basis of these data, confirm that the average atomic weight of Cr is 51.9963 amu.

2.3 (a) How many grams are there in one amu of a material?

(b) Mole, in the context of this book, is taken in units of gram-mole. On this basis, how many atoms are there in a gram-mole of a substance?

2.4 (a) Cite two important quantum-mechanical concepts associated with the Bohr model of the atom.

(b) Cite two important additional refinements that resulted from the wave-mechanical atomic model.

Note: In each chapter, most of the terms listed in the Important Terms and Concepts section are defined in the Glossary, which follows Appendix E. The other terms are important enough to warrant treatment in a full section of the text and can be found in the Contents or the Index.

2.5 Relative to electrons and electron states, what does each of the four quantum numbers specify?

2.6 Allowed values for the quantum numbers of electrons are as follows:

$$n = 1, 2, 3, \ldots$$
$$l = 0, 1, 2, 3, \ldots, n - 1$$
$$m_l = 0, \pm 1, \pm 2, \pm 3, \ldots, \pm l$$
$$m_s = \pm \tfrac{1}{2}$$

The relationships between n and the shell designations are noted in Table 2.1. Relative to the subshells,

$l = 0$ corresponds to an s subshell
$l = 1$ corresponds to a p subshell
$l = 2$ corresponds to a d subshell
$l = 3$ corresponds to an f subshell

For the K shell, the four quantum numbers for each of the two electrons in the $1s$ state, in the order of nlm_lm_s, are $100\tfrac{1}{2}$ and $100(-\tfrac{1}{2})$. Write the four quantum numbers for all of the electrons in the L and M shells, and note which correspond to the s, p, and d subshells.

2.7 Give the electron configurations for the following ions: Fe^{2+}, Al^{3+}, Cu^+, Ba^{2+}, Br^-, O^{2-}, Fe^{3+} and S^{2-}.

2.8 Sodium chloride (NaCl) exhibits predominantly ionic bonding. The Na^+ and Cl^- ions have electron structures that are identical to which two inert gases?

The Periodic Table

2.9 With regard to electron configuration, what do all the elements in Group VIIA of the periodic table have in common?

2.10 To what group in the periodic table would an element with atomic number 114 belong?

2.11 Without consulting Figure 2.6 or Table 2.2, determine whether each of the following electron configurations is an inert gas, a halogen, an alkali metal, an alkaline earth metal, or a transition metal. Justify your choices.

(a) $1s^2 2s^2 2p^6 3s^2 3p^6 3d^7 4s^2$

(b) $1s^2 2s^2 2p^6 3s^2 3p^6$

(c) $1s^2 2s^2 2p^5$

(d) $1s^2 2s^2 2p^6 3s^2$

(e) $1s^2 2s^2 2p^6 3s^2 3p^6 3d^2 4s^2$

(f) $1s^2 2s^2 2p^6 3s^2 3p^6 4s^1$

2.12 (a) What electron subshell is being filled for the rare earth series of elements on the periodic table?

(b) What electron subshell is being filled for the actinide series?

Bonding Forces and Energies

2.13 Calculate the force of attraction between a K^+ and an O^{2-} ion whose centers are separated by a distance of 1.6 nm.

2.14 The net potential energy between two adjacent ions, E_N, may be represented by the sum of Equations 2.9 and 2.11; that is,

$$E_N = -\frac{A}{r} + \frac{B}{r^n} \tag{2.17}$$

Calculate the bonding energy E_0 in terms of the parameters A, B, and n using the following procedure:

1. Differentiate E_N with respect to r, and then set the resulting expression equal to zero, because the curve of E_N versus r is a minimum at E_0.

2. Solve for r in terms of A, B, and n, which yields r_0, the equilibrium interionic spacing.

3. Determine the expression for E_0 by substituting r_0 into Equation 2.17.

2.15 For a K^+– Cl^- ion pair, attractive and repulsive energies E_A and E_R, respectively, depend on the distance between the ions r, according to

$$E_A = -\frac{1.436}{r}$$

$$E_R = \frac{5.86 \times 10^{-6}}{r^9}$$

For these expressions, energies are expressed in electron volts per K^+–Cl^- pair, and r is the distance in nanometers. The net energy E_N is just the sum of the preceding two expressions.

(a) Superimpose on a single plot E_N, E_R, and E_A versus r up to 1.2 nm.

(b) On the basis of this plot, determine (i) the equilibrium spacing r_0 between the K^+ and Cl^- ions, and (ii) the magnitude of the bonding energy E_0 between the two ions.

(c) Mathematically determine the r_0 and E_0 values using the solutions to Problem 2.14 and compare these with the graphical results from part (b).

2.16 Consider a hypothetical X^+–Y^- ion pair for which the equilibrium interionic spacing and bonding energy values are 0.38 nm and -6.13 eV, respectively. If it is known that n in Equation 2.17 has a value of 10, using the results of Problem 2.14, determine explicit expressions for attractive and repulsive energies E_A and E_R of Equations 2.9 and 2.11.

2.17 The net potential energy E_N between two adjacent ions is sometimes represented by the expression

$$E_N = -\frac{C}{r} + D \exp\left(-\frac{r}{\rho}\right) \qquad (2.18)$$

in which r is the interionic separation and C, D, and ρ are constants whose values depend on the specific material.

(a) Derive an expression for the bonding energy E_0 in terms of the equilibrium interionic separation r_0 and the constants D and ρ using the following procedure:

1 Differentiate E_N with respect to r, and set the resulting expression equal to zero.

2 Solve for C in terms of D, ρ, and r_0.

3 Determine the expression for E_0 by substitution for C in Equation 2.18.

(b) Derive another expression for E_0 in terms of r_0, C, and ρ using a procedure analogous to the one outlined in part (a).

Primary Interatomic Bonds

2.18 (a) Briefly cite the main differences between ionic, covalent, and metallic bonding.

(b) State the Pauli exclusion principle.

2.19 Make a plot of bonding energy versus melting temperature for the metals listed in Table 2.3. Using this plot, approximate the bonding energy for copper, which has a melting temperature of 1084°C.

2.20 Using Table 2.2, determine the number of covalent bonds that are possible for atoms of the following elements: germanium, phosphorus, selenium, and chlorine.

Secondary Bonding or van der Waals Bonding

2.21 Explain why hydrogen fluoride (HF) has a higher boiling temperature than hydrogen chloride (HCl) (19.4°C vs. −85°C), even though HF has a lower molecular weight.

Mixed Bonding

2.22 Compute the percent ionic character of the interatomic bonds for the following compounds: TiO_2, ZnTe, CsCl, InSb, and $MgCl_2$.

Bonding Type–Material Classification Correlations

2.23 What type(s) of bonding would be expected for each of the following materials: brass (a copper–zinc alloy), rubber, barium sulfide (BaS), solid xenon, bronze, nylon, and aluminum phosphide (AlP)?

Spreadsheet Problems

2.1SS Generate a spreadsheet that allows the user to input values of A, B, and n (Equation 2.17), and then does the following:

(a) Plots on a graph of potential energy versus interatomic separation for two atoms/ions, curves for attractive (E_A), repulsive (E_R), and net (E_N) energies, and

(b) Determines the equilibrium spacing (r_0) and the bonding energy (E_0).

2.2SS Generate a spreadsheet that computes the percent ionic character of a bond between atoms of two elements, once the user has input values for the elements' electronegativities.

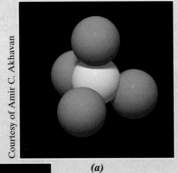

(a)

The illustrations shown present the structure of quartz (SiO_2) from three different dimensional perspectives. White and red balls represent, respectively, silicon and oxygen atoms.

(a) Schematic representation of the most basic structural unit for quartz (as well as for all silicate materials). Each atom of silicon is bonded to and surrounded by four oxygen atoms, whose centers are located at the corners of a tetrahedron. Chemically, this unit is represented as SiO_4^{4-}.

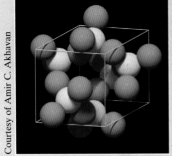

(b)

(b) Sketch of a unit cell for quartz, which is composed of several interconnected SiO_4^{4-} tetrahedra.

(c)

(c) Schematic diagram showing a large number of interconnected SiO_4^{4-} tetrahedra. The shape of this structure is characteristic of that adopted by a single crystal of quartz.

(d)

(d) Photograph of two single crystals of quartz. Note that the shape of the large crystal in the photograph resembles the shape of the structure shown in (c).

The properties of some materials are directly related to their crystal structures. For example, pure and unde-formed magnesium and beryllium, having one crystal structure, are much more brittle (i.e., fracture at lower degrees of deformation) than are pure and undeformed metals such as gold and silver that have yet another crystal structure (see Section 9.4).

Furthermore, significant property differences exist between crystalline and noncrystalline materials hav-ing the same composition. For example, noncrystalline ceramics and polymers normally are optically transparent; the same materials in crystalline (or semicrystalline) form tend to be opaque or, at best, translucent.

Learning Objectives

After studying this chapter, you should be able to do the following:

1. Describe the difference in atomic/molecular structure between crystalline and noncrystalline materials.

2. Given three direction index integers, sketch the direction corresponding to these indices within a unit cell.

3. Specify the Miller indices for a plane that has been drawn within a unit cell.

4. Distinguish between single crystals and poly-crystalline materials.

5. Define *isotropy* and *anisotropy* with respect to material properties.

3.1 INTRODUCTION

Chapter 2 was concerned primarily with the various types of atomic bonding, which are determined by the electron structures of the individual atoms. The present discussion is devoted to the next level of the structure of materials, specifically, to some of the arrangements that may be assumed by atoms in the solid state. Within this framework, concepts of crystallinity and noncrystallinity are introduced. For crystalline solids, the notion of crystal structure is presented, specified in terms of a unit cell. The scheme by which crystallographic points, directions, and planes are expressed is explained. Single crystals, polycrystalline materials, and noncrystalline materials are considered.

Crystal Structures

3.2 FUNDAMENTAL CONCEPTS

crystalline

Solid materials may be classified according to the regularity with which atoms or ions are arranged with respect to one another. A **crystalline** material is one in which the atoms are situated in a repeating or periodic array over large atomic distances—that is, long-range order exists, such that upon solidification, the atoms will position themselves in a repetitive three-dimensional pattern, in which each atom is bonded to its nearest-neighbor atoms. All metals, many ceramic materials, and certain polymers form crystal-line structures under normal solidification conditions. For those that do not crystallize, this long-range atomic order is absent; these *noncrystalline* or *amorphous* materials are discussed briefly at the end of this chapter.

crystal structure

Some of the properties of crystalline solids depend on the **crystal structure** of the material, the manner in which atoms, ions, or molecules are spatially arranged. There is an extremely large number of different crystal structures all having long-range atomic order; these vary from relatively simple structures for metals to exceedingly complex

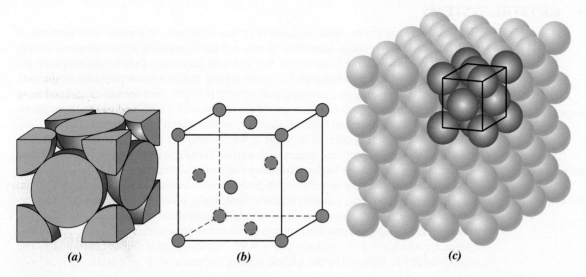

Figure 3.1 For the face-centered cubic crystal structure, (*a*) a hard-sphere unit cell representation, (*b*) a reduced-sphere unit cell, and (*c*) an aggregate of many atoms.
[Figure (*c*) adapted from W. G. Moffatt, G. W. Pearsall, and J. Wulff, *The Structure and Properties of Materials*, Vol. I, *Structure*, p. 51. Copyright © 1964 by John Wiley & Sons, New York.]

ones, as displayed by some of the ceramic and polymeric materials. The present discussion deals with fundamental concepts of crystalline solids. Chapter 4 is devoted to crystal structures for metals, ceramics, and polymers.

When crystalline structures are described, atoms (or ions) are thought of as being solid spheres having well-defined diameters. This is termed the *atomic hard-sphere model* in which spheres representing nearest-neighbor atoms touch one another. An example of the hard-sphere model for the atomic arrangement found in some of the common elemental metals is displayed in Figure 3.1*c*. In this particular case all the atoms are identical. Sometimes the term **lattice** is used in the context of crystal structures; in this sense *lattice* means a three-dimensional array of points coinciding with atom positions (or sphere centers).

lattice

3.3 UNIT CELLS

unit cell

The atomic order in crystalline solids indicates that small groups of atoms form a repetitive pattern. Thus, in describing crystal structures, it is often convenient to subdivide the structure into small repeat entities called **unit cells**. Unit cells for most crystal structures are parallelepipeds or prisms having three sets of parallel faces; one is drawn within the aggregate of spheres (Figure 3.1*c*), which in this case happens to be a cube. A unit cell is chosen to represent the symmetry of the crystal structure, wherein all the atom positions in the crystal may be generated by translations of the unit cell integral distances along each of its edges. Thus, the unit cell is the basic structural unit or building block of the crystal structure and defines the crystal structure by virtue of its geometry and the atom positions within. Convenience usually dictates that parallelepiped corners coincide with centers of the hard-sphere atoms. Furthermore, more than a single unit cell may be chosen for a particular crystal structure; however, we generally use the unit cell having the highest level of geometrical symmetry.

3.4 CRYSTAL SYSTEMS

lattice parameters

crystal system

Because there are many different possible crystal structures, it is sometimes convenient to divide them into groups according to unit cell configurations and/or atomic arrangements. One such scheme is based on the unit cell geometry, that is, the shape of the appropriate unit cell parallelepiped without regard to the atomic positions in the cell. Within this framework, an *xyz* coordinate system is established with its origin at one of the unit cell corners; each of the *x*, *y*, and *z* axes coincides with one of the three parallelepiped edges that extend from this corner, as illustrated in Figure 3.2. The unit cell geometry is completely defined in terms of six parameters: the three edge lengths *a*, *b*, and *c*, and the three interaxial angles α, β, and γ. These are indicated in Figure 3.2, and are sometimes termed the **lattice parameters** of a crystal structure.

On this basis there are seven different possible combinations of *a*, *b*, and *c* and α, β, and γ, each of which represents a distinct **crystal system**. These seven crystal systems are cubic, tetragonal, hexagonal, orthorhombic, rhombohedral,[1] monoclinic, and triclinic. The lattice parameter relationships and unit cell sketches for each are represented in Table 3.1. The cubic system, for which $a = b = c$ and $\alpha = \beta = \gamma = 90°$, has the greatest degree of symmetry. The least symmetry is displayed by the triclinic system, because $a \neq b \neq c$ and $\alpha \neq \beta \neq \gamma$.

Concept Check 3.1   What is the difference between crystal structure and crystal system?

[*The answer may be found at* www.wiley.com/college/callister *(Student Companion Site)*.]

Figure 3.2   A unit cell with *x*, *y*, and *z* coordinate axes, showing axial lengths (*a*, *b*, and *c*) and interaxial angles (α, β, and γ).

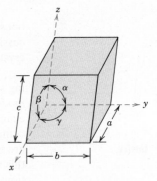

Table 3.1   Lattice Parameter Relationships and Figures Showing Unit Cell Geometries for the Seven Crystal Systems

Crystal System	Axial Relationships	Interaxial Angles	Unit Cell Geometry
Cubic	$a = b = c$	$\alpha = \beta = \gamma = 90°$	
Hexagonal	$a = b \neq c$	$\alpha = \beta = 90°, \gamma = 120°$	

(*continued*)

---

[1]Also called *trigonal*.

Table 3.1 (Continued)

Crystal System	Axial Relationships	Interaxial Angles	Unit Cell Geometry
Tetragonal	$a = b \neq c$	$\alpha = \beta = \gamma = 90°$	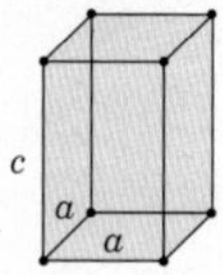
Rhombohedral (Trigonal)	$a = b = c$	$\alpha = \beta = \gamma \neq 90°$	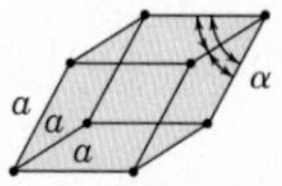
Orthorhombic	$a \neq b \neq c$	$\alpha = \beta = \gamma = 90°$	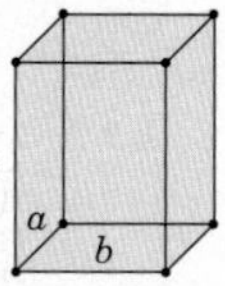
Monoclinic	$a \neq b \neq c$	$\alpha = \gamma = 90° \neq \beta$	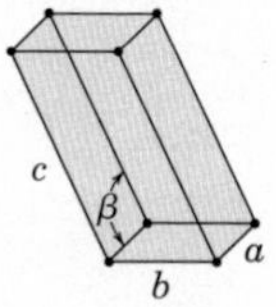
Triclinic	$a \neq b \neq c$	$\alpha \neq \beta \neq \gamma \neq 90°$	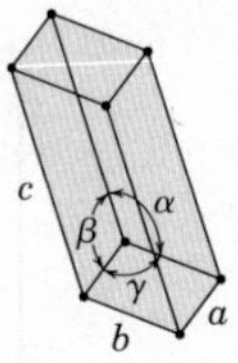

Crystallographic Points, Directions, and Planes

When dealing with crystalline materials, it often becomes necessary to specify a particular point within a unit cell, a crystallographic direction, or some crystallographic plane of atoms. Labeling conventions have been established in which three numbers or indices are used to designate point locations, directions, and planes. The basis for determining index values is the unit cell, with a right-handed coordinate system consisting of three (x, y, and z) axes situated at one of the corners and coinciding with the unit cell edges, as shown in Figure 3.2. For some crystal systems—namely, hexagonal, rhombohedral, monoclinic, and triclinic—the three axes are *not* mutually perpendicular, as in the familiar Cartesian coordinate scheme.

3.5 POINT COORDINATES

Sometimes it is necessary to specify a lattice position within a unit cell. This is possible using three *point coordinate indices*: q, r, and s. These indices are fractional multiples of a, b, and c unit cell edge lengths—that is, q is some fractional length of a along the x axis, r is some fractional length of b along the y axis, and similarly for s; or

$$qa = \text{lattice position referenced to the } x \text{ axis} \qquad (3.1a)$$

$$rb = \text{lattice position referenced to the } y \text{ axis} \qquad (3.1b)$$

$$sc = \text{lattice position referenced to the } z \text{ axis} \qquad (3.1c)$$

To illustrate, consider the unit cell in Figure 3.3, the x-y-z coordinate system with its origin located at a unit cell corner, and the lattice site located at point P. Note how the location of P is related to the products of its q, r, and s coordinate indices and the unit cell edge lengths.[2]

Figure 3.3 The manner in which the q, r, and s coordinates at point P within the unit cell are determined. The q coordinate (which is a fraction) corresponds to the distance qa along the x axis, where a is the unit cell edge length. The respective r and s coordinates for the y and z axes are determined similarly.

EXAMPLE PROBLEM 3.1

Location of Point Having Specified Coordinates

For the unit cell shown in the accompanying sketch (*a*), locate the point having coordinates $\frac{1}{4}$ 1 $\frac{1}{2}$.

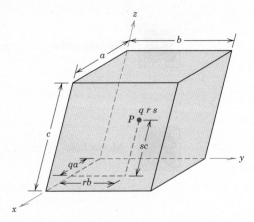

(a)   *(b)*

----

[2]We have chosen not to separate the q, r, and s indices by commas or any other punctuation marks (which is the normal convention).

Solution

From sketch (*a*), edge lengths for this unit cell are as follows: $a = 0.48$ nm, $b = 0.46$ nm, and $c = 0.40$ nm. Furthermore, in light of the preceding discussion, the three point coordinate indices are $q = \frac{1}{4}$, $r = 1$, and $s = \frac{1}{2}$. We use Equations 3.1a through 3.1c to determine lattice positions for this point as follows:

$$\text{lattice position referenced to the } x \text{ axis} = qa$$
$$= \left(\tfrac{1}{4}\right)a = \tfrac{1}{4}(0.48 \text{ nm}) = 0.12 \text{ nm}$$

$$\text{lattice position referenced to the } y \text{ axis} = rb$$
$$= (1)b = (1)(0.46 \text{ nm}) = 0.46 \text{ nm}$$

$$\text{lattice position referenced to the } z \text{ axis} = sc$$
$$= \left(\tfrac{1}{2}\right)c = \left(\tfrac{1}{2}\right)(0.40 \text{ nm}) = 0.20 \text{ nm}$$

To locate the point having these coordinates within the unit cell, first use the x lattice position and move from the origin (point M) 0.12 nm units along the x axis (to point N), as shown in (*b*). Similarly, using the y lattice position, proceed 0.46 nm parallel to the y axis, from point N to point O. Finally, move from this position 0.20 nm units parallel to the z axis to point P (per the z lattice position), as noted again in (*b*). Thus, point P corresponds to the $\frac{1}{4} 1 \frac{1}{2}$ point coordinates.

EXAMPLE PROBLEM 3.2

Specification of Point Coordinate Indices

Specify coordinate indices for all numbered points of the unit cell in the illustration on the next page.

Solution

For this unit cell, coordinate points are located at all eight corners with a single point at the center position.

Point 1 is located at the origin of the coordinate system, and, therefore, its lattice position indices referenced to the x, y, and z axes are $0a$, $0b$, and $0c$, respectively. And from Equations 3.1a through 3.1c,

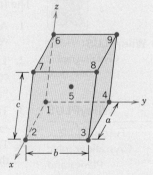

$$\text{lattice position referenced to the } x \text{ axis} = 0a = qa$$
$$\text{lattice position referenced to the } y \text{ axis} = 0b = rb$$
$$\text{lattice position referenced to the } z \text{ axis} = 0c = sc$$

Solving the above three expressions for values of the q, r, and s indices leads to

$$q = \frac{0a}{a} = 0$$

$$r = \frac{0b}{b} = 0$$

$$s = \frac{0c}{c} = 0$$

Therefore this is the 0 0 0 point.

Because point number 2 lies one unit cell edge length along the x axis, its lattice position indices referenced to the x, y, and z axes are a, $0b$, and $0c$, and

$$\text{lattice position index referenced to the } x \text{ axis} = a = qa$$

$$\text{lattice position index referenced to the } y \text{ axis} = 0b = rb$$

$$\text{lattice position index referenced to the } z \text{ axis} = 0c = sc$$

Thus we determine values for the q, r, and s indices as follows:

$$q = 1 \qquad r = 0 \qquad s = 0$$

Hence, point 2 is 1 0 0.

This same procedure is carried out for the remaining seven points in the unit cell. Point indices for all nine points are listed in the following table.

Point Number	q	r	s
1	0	0	0
2	1	0	0
3	1	1	0
4	0	1	0
5	$\frac{1}{2}$	$\frac{1}{2}$	$\frac{1}{2}$
6	0	0	1
7	1	0	1
8	1	1	1
9	0	1	1

3.6 CRYSTALLOGRAPHIC DIRECTIONS

WileyPLUS: VMSE
Crystallographic
Directions

WileyPLUS

Tutorial Video:
Crystallographic
Planes and
Directions

A *crystallographic direction* is defined as a line directed between two points, or a *vector*. The following steps are used to determine the three directional indices:

1. A right-handed x-y-z coordinate system is first constructed. As a matter of convenience, its origin may be located at a unit cell corner.

2. The coordinates of two points that lie on the direction vector (referenced to the coordinate system) are determined—for example, for the vector tail, point 1: x_1, y_1, and z_1; whereas for the vector head, point 2: x_2, y_2, and z_2.

3. Tail point coordinates are subtracted from head point components—that is, $x_2 - x_1$, $y_2 - y_1$, and $z_2 - z_1$.

4. These coordinate differences are then normalized in terms of (i.e., divided by) their respective a, b, and c lattice parameters—that is,

$$\frac{x_2 - x_1}{a} \quad \frac{y_2 - y_1}{b} \quad \frac{z_2 - z_1}{c}$$

which yields a set of three numbers.

5. If necessary, these three numbers are multiplied or divided by a common factor to reduce them to the smallest integer values.

6. The three resulting indices, not separated by commas, are enclosed in square brackets, thus: [uvw]. The u, v, and w integers correspond to the normalized coordinate differences referenced to the x, y, and z axes, respectively.

In summary, the u, v, and w indices may be determined using the following equations:

$$u = n\left(\frac{x_2 - x_1}{a}\right) \qquad (3.2a)$$

$$v = n\left(\frac{y_2 - y_1}{b}\right) \qquad (3.2b)$$

$$w = n\left(\frac{z_2 - z_1}{c}\right) \qquad (3.2c)$$

In these expressions, n is the factor that may be required to reduce u, v, and w to integers.

For each of the three axes, there are both positive and negative coordinates. Thus, negative indices are also possible, which are represented by a bar over the appropriate index. For example, the $[1\bar{1}1]$ direction has a component in the $-y$ direction. Also, changing the signs of all indices produces an antiparallel direction; that is, $[\bar{1}1\bar{1}]$ is directly opposite to $[1\bar{1}1]$. If more than one direction (or plane) is to be specified for a particular crystal structure, it is imperative for maintaining consistency that a positive–negative convention, once established, not be changed.

The [100], [110], and [111] directions are common ones; they are drawn in the unit cell shown in Figure 3.4.

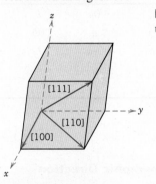

Figure 3.4 The [100], [110], and [111] directions within a unit cell.

EXAMPLE PROBLEM 3.3

Determination of Directional Indices

Determine the indices for the direction shown in the accompanying figure.

Solution

It is first necessary to take note of the vector tail and head coordinates. From the illustration, tail coordinates are as follows:

$$x_1 = a \qquad y_1 = 0b \qquad z_1 = 0c$$

For the head coordinates,

$$x_2 = 0a \qquad y_2 = b \qquad z_2 = c/2$$

Now taking point coordinate differences,

$$x_2 - x_1 = 0a - a = -a$$
$$y_2 - y_1 = b - 0b = b$$
$$z_2 - z_1 = c/2 - 0c = c/2$$

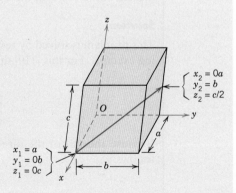

It is now possible to use Equations 3.2a through 3.2c to compute values of u, v, and w. However, because the $z_2 - z_1$ difference is a fraction (i.e., $c/2$), we anticipate that in order to have integer values for the three indices, it is necessary to assign n a value of 2. Thus,

$$u = n\left(\frac{x_2 - x_1}{a}\right) = 2\left(\frac{-a}{a}\right) = -2$$

$$v = n\left(\frac{y_2 - y_1}{b}\right) = 2\left(\frac{b}{b}\right) = 2$$

$$w = n\left(\frac{z_2 - z_1}{c}\right) = 2\left(\frac{c/2}{c}\right) = 1$$

And, finally enclosure of the -2, 2, and 1 indices in brackets leads to $[\bar{2}21]$ as the direction designation.[3]

This procedure is summarized as follows:

	x	y	z
Head coordinates $(x_2, y_2, z_2,)$	$0a$	b	$c/2$
Tail coordinates $(x_1, y_1, z_1,)$	a	$0b$	$0c$
Coordinate differences	$-a$	b	$c/2$
Calculated values of u, v, and w	$u = -2$	$v = 2$	$w = 1$
Enclosure		$[\bar{2}21]$	

EXAMPLE PROBLEM 3.4

Construction of a Specified Crystallographic Direction

Within the following unit cell draw a $[1\bar{1}0]$ direction with its tail located at the origin of the coordinate system, point O.

Solution

This problem is solved by reversing the procedure of the preceding example. For this $[1\bar{1}0]$ direction,

$$u = 1$$
$$v = -1$$
$$w = 0$$

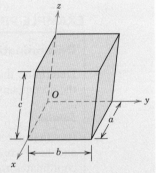

Because the tail of the direction vector is positioned at the origin, its coordinates are as follows:

$$x_1 = 0a$$
$$y_1 = 0b$$
$$z_1 = 0c$$

We now want to solve for the coordinates of the vector head—that is, x_2, y_2, and z_2. This is possible using rearranged forms of Equations 3.2a through 3.2c and incorporating the above values

---

[3]If these u, v, and w values are not integers, it is necessary to choose another value for n.

for the three direction indices (u, v, and w) and vector tail coordinates. Taking the value of n to be 1 because the three direction indices are all integers leads to

$$x_2 = ua + x_1 = (1)(a) + 0a = a$$
$$y_2 = vb + y_1 = (-1)(b) + 0b = -b$$
$$z_2 = wc + z_1 = (0)(c) + 0c = 0c$$

The construction process for this direction vector is shown in the following figure.

Because the tail of the vector is positioned at the origin, we start at the point labeled O and then move in a stepwise manner to locate the vector head. Because the x head coordinate (x_2) is a, we proceed from point O, a units along the x axis to point Q. From point Q, we move b units parallel to the $-y$ axis to point P, because the y head coordinate (y_2) is $-b$. There is no z component to the vector inasmuch as the z head coordinate (z_2) is $0c$. Finally, the vector corresponding to this [$1\bar{1}0$] direction is constructed by drawing a line from point O to point P, as noted in the illustration.

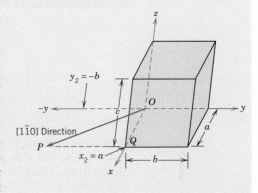

For some crystal structures, several nonparallel directions with different indices are *crystallographically equivalent*, meaning that the spacing of atoms along each direction is the same. For example, in cubic crystals, all the directions represented by the following indices are equivalent: [100], [$\bar{1}00$], [010], [$0\bar{1}0$], [001], and [$00\bar{1}$]. As a convenience, equivalent directions are grouped together into a *family,* which is enclosed in angle brackets, thus: ⟨100⟩. Furthermore, directions in cubic crystals having the same indices without regard to order or sign—for example, [123] and [$\bar{2}1\bar{3}$]—are equivalent. This is, in general, not true for other crystal systems. For example, for crystals of tetragonal symmetry, the [100] and [010] directions are equivalent, whereas the [100] and [001] are not.

Directions in Hexagonal Crystals

A problem arises for crystals having hexagonal symmetry in that some equivalent crystallographic directions do not have the same set of indices. For example, the [111] direction is equivalent to [$\bar{1}01$] rather than to a direction with indices that are combinations of 1s and −1s. This situation is addressed using a four-axis, or *Miller–Bravais,* coordinate system, which is shown in Figure 3.5. The three a_1, a_2, and a_3 axes are all contained

Figure 3.5 Coordinate axis system for a hexagonal unit cell (Miller–Bravais scheme).

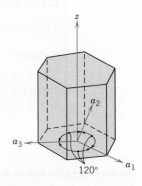

within a single plane (called the *basal plane*) and are at 120° angles to one another. The *z* axis is perpendicular to this basal plane. Directional indices, which are obtained as described earlier, are denoted by four indices, as [*uvtw*]; by convention, the *u*, *v*, and *t* indices relate to vector coordinate differences referenced to the respective a_1, a_2, and a_3 axes in the basal plane; the fourth index pertains to the *z* axis.

Conversion from the three-index system to the four-index system as

$$[UVW] \rightarrow [uvtw]$$

is accomplished using the following formulas[4]:

$$u = \frac{1}{3}(2U - V) \tag{3.3a}$$

$$v = \frac{1}{3}(2V - U) \tag{3.3b}$$

$$t = -(u + v) \tag{3.3c}$$

$$w = W \tag{3.3d}$$

Here, uppercase *U*, *V*, and *W* indices are associated with the three-index scheme (instead of *u*, *v*, and *w* as previously), whereas lowercase *u*, *v*, *t*, and *w* correlate with the Miller–Bravais four-index system. For example, using these equations, the [010] direction becomes [$\bar{1}2\bar{1}0$]. Several directions have been drawn in the hexagonal unit cell of Figure 3.6.

Figure 3.6 For the hexagonal crystal system, the [0001], [$1\bar{1}00$], and [$11\bar{2}0$] directions.

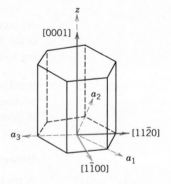

EXAMPLE PROBLEM 3.5

Determination of Directional Indices for a Hexagonal Unit Cell

Determine the indices (four-index system) for the direction shown in the accompanying figure.

Solution

The first thing we need to do is determine *U*, *V*, and *W* indices for the vector referenced to the three-axis scheme represented in the sketch. In this case, we modify Equations 3.2a–3.2c to read as follows:

---

[4]Reduction to the lowest set of integers may be necessary, as discussed earlier.

$$U = n\left(\frac{a_1' - a_1''}{a}\right) \tag{3.4a}$$

$$V = n\left(\frac{a_2' - a_2''}{a}\right) \tag{3.4b}$$

$$W = n\left(\frac{z' - z''}{c}\right) \tag{3.4c}$$

where again, single and double primes for a_1, a_2, and z denote head and tail coordinates, respectively. Because the vector passes through the origin, $a_1'' = a_2'' = 0a$ and $z'' = 0c$. Furthermore, from the sketch, coordinates for the vector head are as follows:

$$a_1' = 0a$$

$$a_2' = -a$$

$$z' = \frac{c}{2}$$

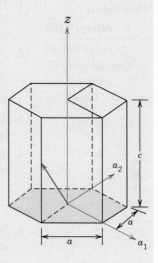

Because the denominator in z' is 2, we assume that $n = 2$. Therefore,

$$U = n\left(\frac{a_1' - a_1''}{a}\right) = 2\left(\frac{0a - 0a}{a}\right) = 0$$

$$V = n\left(\frac{a_2' - a_2''}{a}\right) = 2\left(\frac{-a - 0a}{a}\right) = -2$$

$$W = n\left(\frac{z' - z''}{c}\right) = 2\left(\frac{c/2 - 0c}{c}\right) = 1$$

This direction is represented by enclosing the above indices in brackets—namely, $[0\bar{2}1]$.

Now it becomes necessary to convert these indices into an index set referenced to the four-axis scheme. This requires the use of Equations 3.3a–3.3d. For this $[0\bar{2}1]$ direction,

$$U = 0 \qquad V = -2 \qquad W = 1$$

and

$$u = \frac{1}{3}(2U - V) = \frac{1}{3}[(2)(0) - (-2)] = \frac{2}{3}$$

$$v = \frac{1}{3}(2V - U) = \frac{1}{3}[(2)(-2) - 0] = -\frac{4}{3}$$

$$t = -(u + v) = -\left(\frac{2}{3} - \frac{4}{3}\right) = \frac{2}{3}$$

$$w = W = 1$$

Multiplication of the preceding indices by 3 reduces them to the lowest set, which yields values for u, v, t, and w of 2, −4, 2, and 3, respectively. Hence, the direction vector shown in the figure is $[2\bar{4}23]$.

3.7 CRYSTALLOGRAPHIC PLANES

WileyPLUS: VMSE
Crystallographic
Planes

Miller indices

WileyPLUS

Tutorial Video:
Crystallographic
Planes and
Directions

The orientations of planes for a crystal structure are represented in a similar manner. Again, the unit cell is the basis, with the three-axis coordinate system as represented in Figure 3.2. In all but the hexagonal crystal system, crystallographic planes are specified by three **Miller indices** as (*hkl*). Any two planes parallel to each other are equivalent and have identical indices. The procedure used to determine the *h*, *k*, and *l* index numbers is as follows:

1. If the plane passes through the selected origin, either another parallel plane must be constructed within the unit cell by an appropriate translation, or a new origin must be established at the corner of another unit cell.[5]

2. At this point, the crystallographic plane either intersects or parallels each of the three axes. The coordinate for the intersection of the crystallographic plane with each of the axes is determined (referenced to the origin of the coordinate system). These intercepts for the *x*, *y*, and *z* axes will be designed by *A*, *B*, and *C*, respectively.

3. The reciprocals of these numbers are taken. A plane that parallels an axis is considered to have an infinite intercept and therefore a zero index.

4. The reciprocals of the intercepts are then normalized in terms of (i.e., multiplied by) their respective *a*, *b*, and *c* lattice parameters. That is,

$$\frac{a}{A} \quad \frac{b}{B} \quad \frac{c}{C}$$

5. If necessary, these three numbers are changed to the set of smallest integers by multiplication or by division by a common factor.[6]

6. Finally, the integer indices, not separated by commas, are enclosed within parentheses, thus: (*hkl*). The *h*, *k*, and *l* integers correspond to the normalized intercept reciprocals referenced to the *x*, *y*, and *z* axes, respectively.

In summary, the *h*, *k*, and *l* indices may be determined using the following equations:

$$h = \frac{na}{A} \tag{3.5a}$$

$$k = \frac{nb}{B} \tag{3.5b}$$

$$l = \frac{nc}{C} \tag{3.5c}$$

In these expressions, *n* is the factor that may be required to reduce *h*, *k*, and *l* to integers. An intercept on the negative side of the origin is indicated by a bar or minus sign positioned over the appropriate index. Furthermore, reversing the directions of all indices specifies another plane parallel to, on the opposite side of, and equidistant from the origin. Several low-index planes are represented in Figure 3.7.

---

[5]When selecting a new origin, the following procedure is suggested:

If the crystallographic plane that intersects the origin lies in one of the unit cell faces, move the origin one unit cell distance parallel to the axis that intersects this plane.

If the crystallographic plane that intersects the origin passes through one of the unit cell axes, move the origin one unit cell distance parallel to either of the two other axes.

For all other cases, move the origin one unit cell distance parallel to any of the three unit cell axes.

[6]On occasion, index reduction is not carried out (e.g., for x-ray diffraction studies described in Sections 4.20–4.22); for example, (002) is not reduced to (001). In addition, for ceramic materials, the ionic arrangement for a reduced-index plane may be different from that for a nonreduced one.

Figure 3.7
Representations of a series each of the (a) (001), (b) (110), and (c) (111) crystallographic planes.

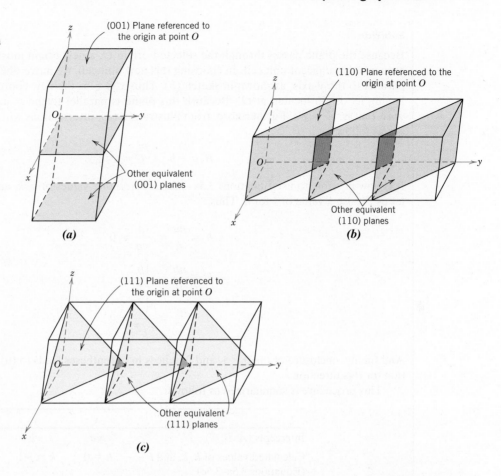

(a)

(b)

(c)

One interesting and unique characteristic of cubic crystals is that planes and directions having the same indices are perpendicular to one another; however, for other crystal systems there are no simple geometrical relationships between planes and directions having the same indices.

EXAMPLE PROBLEM 3.6

Determination of Planar (Miller) Indices

Determine the Miller indices for the plane shown in the accompanying sketch (a).

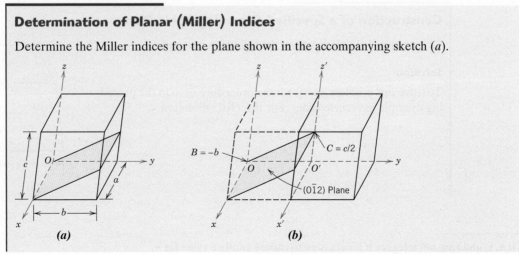

(a)   *(b)*

Solution

Because the plane passes through the selected origin O, a new origin must be chosen at the corner of an adjacent unit cell. In choosing this new unit cell, we move one unit-cell distance parallel to the y-axis, as shown in sketch (b). Thus x'-y-z' is the new coordinate axis system having its origin located at O'. Because this plane is parallel to the x' axis its intercept is ∞a—that is, $A = \infty a$. Furthermore, from illustration (b), intersections with the y and z' axes are as follows:

$$B = -b \qquad C = c/2$$

It is now possible to use Equations 3.5a–3.5c to determine values of h, k, and l. At this point, let us choose a value of 1 for n. Thus,

$$h = \frac{na}{A} = \frac{1a}{\infty a} = 0$$

$$k = \frac{nb}{B} = \frac{1b}{-b} = -1$$

$$l = \frac{nc}{C} = \frac{1c}{c/2} = 2$$

And finally, enclosure of the 0, -1, and 2 indices in parentheses leads to $(0\bar{1}2)$ as the designation for this direction.[7]

This procedure is summarized as follows:

	x	y	z
Intercepts (A, B, C)	∞a	$-b$	$c/2$
Calculated values of h, k, and l (Equations 3.5a–3.5c)	$h = 0$	$k = -1$	$l = 2$
Enclosure		$(0\bar{1}2)$	

EXAMPLE PROBLEM 3.7

Construction of a Specified Crystallographic Plane

Construct a (101) plane within the following unit cell.

Solution

To solve this problem, carry out the procedure used in the preceding example in reverse order. For this (101) direction,

$$h = 1$$
$$k = 0$$
$$l = 1$$

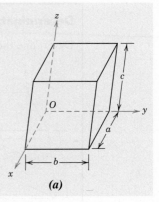

(a)

---

[7]If h, k, and l are not integers, it is necessary to choose another value for n.

Using these h, k, and l indices, we want to solve for the values of A, B, and C using rearranged forms of Equations 3.5a–3.5c. Taking the value of n to be 1—because these three Miller indices are all integers—leads to the following:

$$A = \frac{na}{h} = \frac{(1)(a)}{1} = a$$

$$B = \frac{nb}{k} = \frac{(1)(b)}{0} = \infty b$$

$$C = \frac{nc}{l} = \frac{(1)(c)}{1} = c$$

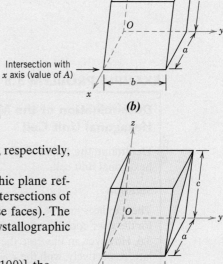

Intersection with
z axis (value of C)

Intersection with
x axis (value of A)

(b)

Thus, this (101) plane intersects the x axis at a (because $A = a$), it parallels the y axis (because $B = \infty b$), and intersects the z axis at c. On the unit cell shown next are noted the locations of the intersections for this plane.

The only plane that parallels the y axis and intersects the x and z axes at axial a and c coordinates, respectively, is shown next.

Note that the representation of a crystallographic plane referenced to a unit cell is by lines drawn to indicate intersections of this plane with unit cell faces (or extensions of these faces). The following guides are helpful with representing crystallographic planes:

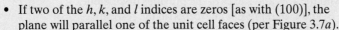
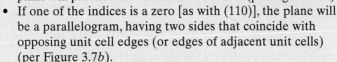

(c)

- If two of the h, k, and l indices are zeros [as with (100)], the plane will parallel one of the unit cell faces (per Figure 3.7a).
- If one of the indices is a zero [as with (110)], the plane will be a parallelogram, having two sides that coincide with opposing unit cell edges (or edges of adjacent unit cells) (per Figure 3.7b).
- If none of the indices is zero [as with (111)], all intersections will pass through unit cell faces (per Figure 3.7c).

Hexagonal Crystals

For crystals having hexagonal symmetry, it is desirable that equivalent planes have the same indices; as with directions, this is accomplished by the Miller–Bravais system shown in Figure 3.5. This convention leads to the four-index ($hkil$) scheme, which is favored in most instances because it more clearly identifies the orientation of a plane in a hexagonal crystal. There is some redundancy in that i is determined by the sum of h and k through

$$i = -(h + k) \tag{3.6}$$

Otherwise, the three h, k, and l indices are identical for both indexing systems.

We determine these indices in a manner analogous to that used for other crystal systems as described previously—that is, taking normalized reciprocals of axial intercepts, as described in the following example problem.

Figure 3.8 presents several of the common planes that are found for crystals having hexagonal symmetry.

Figure 3.8 For the hexagonal crystal system, the (0001), (10$\bar{1}$1), and ($\bar{1}$010) planes.

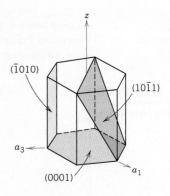

EXAMPLE PROBLEM 3.8

Determination of the Miller–Bravais Indices for a Plane within a Hexagonal Unit Cell

Determine the Miller–Bravais indices for the plane shown in the hexagonal unit cell.

Solution

These indices may be determined in the same manner that was used for the x-y-z coordinate situation and described in Example Problem 3.6. However, in this case the a_1, a_2, and z axes are used and correlate, respectively, with the x, y, and z axes of the previous discussion. If we again take A, B, and C to represent intercepts on the respective a_1, a_2, and z axes, normalized intercept reciprocals may be written as

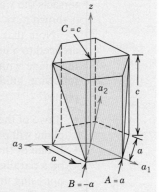

$$\frac{a}{A} \quad \frac{a}{B} \quad \frac{c}{C}$$

Now, because the three intercepts noted on the above unit cell are

$$A = a \qquad B = -a \qquad C = c$$

values of h, k, and l, may be determined using Equations 3.5a–3.5c, as follows (assuming $n = 1$):

$$h = \frac{na}{A} = \frac{(1)(a)}{a} = 1$$

$$k = \frac{na}{B} = \frac{(1)(a)}{-a} = -1$$

$$l = \frac{nc}{C} = \frac{(1)(c)}{c} = 1$$

And, finally, the value of i is found using Equation 3.6, as follows:

$$i = -(h + k) = -[1 + (-1)] = 0$$

Therefore, the ($hkil$) indices are (1$\bar{1}$01).

Notice that the third index is zero (i.e., its reciprocal = ∞), which means this plane parallels the a_3 axis. Inspection of the preceding figure shows that this is indeed the case.

This concludes our discussion on crystallographic points, directions, and planes. A review and summary of these topics is found in Table 3.2.

Table 3.2 Summary of Equations Used to Determine Crystallographic Point, Direction, and Planar Indices

Coordinate Type	Index Symbols	Representative Equation[a]	Equation Symbols
Point	$q\ r\ s$	qa = lattice position referenced to x axis	—
Direction			
Non-hexagonal	$[uvw]$, $[UVW]$	$u = n\left(\dfrac{x_2 - x_1}{a}\right)$	x_1 = tail coordinate—x axis x_2 = head coordinate—x axis
Hexagonal	$[uvtw]$	$u = 3n\left(\dfrac{a_1' - a_1''}{a}\right)$	a_1' = head coordinate—a_1 axis a_1'' = tail coordinate—a_1 axis
		$u = \dfrac{1}{3}(2U - V)$	—
Plane			
Non-hexagonal	(hkl)	$h = \dfrac{na}{A}$	A = plane intercept—x axis
Hexagonal	$(hkil)$	$i = -(h + k)$	—

[a]In these equations a and n denote, respectively, the x-axis lattice parameter, and a reduction-to-integer parameter.

Crystalline and Noncrystalline Materials

3.8 SINGLE CRYSTALS

single crystal

For a crystalline solid, when the periodic and repeated arrangement of atoms is perfect or extends throughout the entirety of the specimen without interruption, the result is a **single crystal**. All unit cells interlock in the same way and have the same orientation. Single crystals exist in nature, but they can also be produced artificially. They are ordinarily difficult to grow because the environment must be carefully controlled.

If the extremities of a single crystal are permitted to grow without any external constraint, the crystal assumes a regular geometric shape having flat faces, as with some of the gemstones; the shape is indicative of the crystal structure. A garnet single crystal is shown in Figure 3.9. Within the past few years, single crystals have become extremely important in many modern technologies, in particular electronic microcircuits, which employ single crystals of silicon and other semiconductors.

3.9 POLYCRYSTALLINE MATERIALS

grain

polycrystalline

Most crystalline solids are composed of a collection of many small crystals or **grains**; such materials are termed **polycrystalline**. Various stages in the solidification of a polycrystalline specimen are represented schematically in Figure 3.10. Initially, small crystals or nuclei form at various positions. These have random crystallographic orientations, as indicated by the square grids. The small grains grow by the successive addition from the surrounding liquid of atoms to the structure of each. The extremities of adjacent grains impinge on one another as the solidification process approaches completion. As indicated in Figure 3.10, the crystallographic orientation varies from grain to grain. Also, there exists some atomic mismatch within the region where two grains meet; this area, called a **grain boundary**, is discussed in more detail in Section 6.8.

grain boundary

Figure 3.9 A garnet single crystal that was found in Tongbei, Fujian Province, China. (Photograph courtesy of Irocks.com, Megan Foreman photo.)

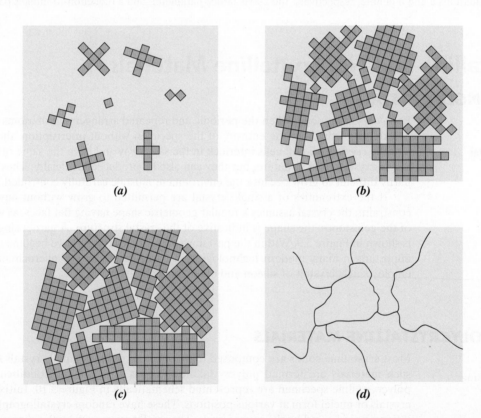

Figure 3.10 Schematic diagrams of the various stages in the solidification of a polycrystalline material; the square grids depict unit cells. (*a*) Small crystallite nuclei. (*b*) Growth of the crystallites; the obstruction of some grains that are adjacent to one another is also shown. (*c*) Upon completion of solidification, grains having irregular shapes have formed. (*d*) The grain structure as it would appear under the microscope; dark lines are the grain boundaries. (Adapted from W. Rosenhain, *An Introduction to the Study of Physical Metallurgy,* 2nd edition, Constable & Company Ltd., London, 1915.)

3.10 ANISOTROPY

anisotropy

isotropic

The physical properties of single crystals of some substances depend on the crystallographic direction in which measurements are taken. For example, the elastic modulus, the electrical conductivity, and the index of refraction may have different values in the [100] and [111] directions. This directionality of properties is termed **anisotropy**, and it is associated with the variance of atomic or ionic spacing with crystallographic direction. Substances in which measured properties are independent of the direction of measurement are **isotropic**. The extent and magnitude of anisotropic effects in crystalline materials are functions of the symmetry of the crystal structure; the degree of anisotropy increases with decreasing structural symmetry—triclinic structures normally are highly anisotropic. The modulus of elasticity values at [100], [110], and [111] orientations for several metals are presented in Table 3.3.

For many polycrystalline materials, the crystallographic orientations of the individual grains are totally random. Under these circumstances, even though each grain may be anisotropic, a specimen composed of the grain aggregate behaves isotropically. Also, the magnitude of a measured property represents some average of the directional values. Sometimes the grains in polycrystalline materials have a preferential crystallographic orientation, in which case the material is said to have a "texture."

The magnetic properties of some iron alloys used in transformer cores are anisotropic—that is, grains (or single crystals) magnetize in a $\langle 100 \rangle$-type direction easier than any other crystallographic direction. Energy losses in transformer cores are minimized by utilizing polycrystalline sheets of these alloys into which have been introduced a *magnetic texture*: most of the grains in each sheet have a $\langle 100 \rangle$-type crystallographic direction that is aligned (or almost aligned) in the same direction, which is oriented parallel to the direction of the applied magnetic field. Magnetic textures for iron alloys are discussed in detail in the Material of Importance box in Chapter 21 following Section 21.9.

Table 3.3
Modulus of Elasticity Values for Several Metals at Various Crystallographic Orientations

	Modulus of Elasticity (GPa)		
Metal	*[100]*	*[110]*	*[111]*
Aluminum	63.7	72.6	76.1
Copper	66.7	130.3	191.1
Iron	125.0	210.5	272.7
Tungsten	384.6	384.6	384.6

Source: R. W. Hertzberg, *Deformation and Fracture Mechanics of Engineering Materials,* 3rd edition. Copyright © 1989 by John Wiley & Sons, New York. Reprinted by permission of John Wiley & Sons, Inc.

3.11 NONCRYSTALLINE SOLIDS

noncrystalline

amorphous

It has been mentioned that **noncrystalline** solids lack a systematic and regular arrangement of atoms over relatively large atomic distances. Sometimes such materials are also called **amorphous** (meaning literally "without form"), or supercooled liquids, inasmuch as their atomic structure resembles that of a liquid.

An amorphous condition may be illustrated by comparison of the crystalline and noncrystalline structures of the ceramic compound silicon dioxide (SiO_2), which may exist in both states. Figures 3.11a and 3.11b present two-dimensional schematic diagrams for both structures of SiO_2. Even though each silicon ion bonds to three oxygen ions for both states, beyond this, the structure is much more disordered and irregular for the noncrystalline structure.

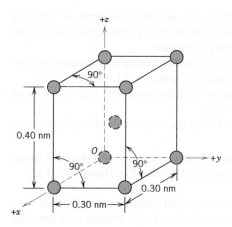

3.3 Sketch a unit cell for the body-centered orthorhombic crystal structure.

Point Coordinates

3.4 List the point coordinates for all atoms that are associated with the FCC unit cell (Figure 3.1).

3.5 List the point coordinates of the titanium, barium, and oxygen ions for a unit cell of the perovskite crystal structure (Figure 4.9).

3.6 List the point coordinates of all atoms that are associated with the diamond cubic unit cell (Figure 4.17).

3.7 Sketch a tetragonal unit cell, and within that cell indicate locations of the $\frac{1}{2}$ 1 $\frac{1}{2}$ and $\frac{1}{4}$ $\frac{1}{2}$ $\frac{3}{4}$ point coordinates.

3.8 Using the Molecule Definition Utility found in the "Metallic Crystal Structures and Crystallography" and "Ceramic Crystal Structures" modules of VMSE, available in WileyPlus, generate a three-dimensional unit cell for the intermetallic compound $AuCu_3$ given the following: (1) the unit cell is cubic with an edge length of 0.370 nm, (2) gold atoms are situated at all cube corners, and (3) copper atoms are positioned at the centers of all unit cell faces.

Crystallographic Directions

3.9 Draw an orthorhombic unit cell, and within that cell a $[12\bar{1}]$ direction.

3.10 Sketch a monoclinic unit cell, and within that cell a $[0\bar{1}1]$ direction.

3.11 What are the indices for the directions indicated by the two vectors in the following sketch?

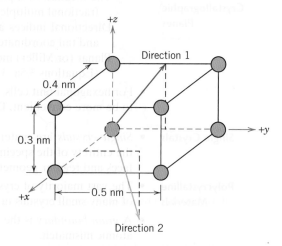

3.12 Within a cubic unit cell, sketch the following directions:

(a) $[\bar{1}10]$ (e) $[\bar{1}\bar{1}1]$

(b) $[\bar{1}\bar{2}1]$ (f) $[\bar{1}22]$

(c) $[0\bar{1}2]$ (g) $[12\bar{3}]$

(d) $[1\bar{3}3]$ (h) $[\bar{1}03]$

3.13 Determine the indices for the directions shown in the following cubic unit cell:

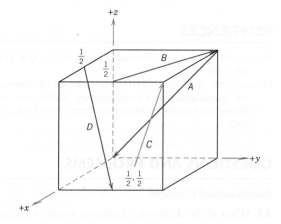

3.14 Determine the indices for the directions shown in the following cubic unit cell:

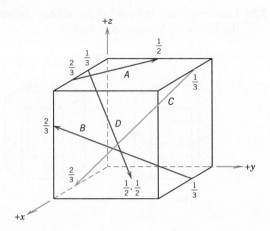

3.15 For tetragonal crystals, cite the indices of directions that are equivalent to each of the following directions:

(a) [001]

(b) [110]

(c) [010]

3.16 Convert the [100] and [111] directions into the four-index Miller–Bravais scheme for hexagonal unit cells.

3.17 Determine indices for the directions shown in the following hexagonal unit cells:

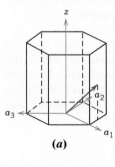

(a)

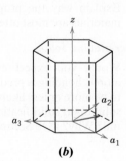

(b)

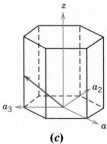

(c)

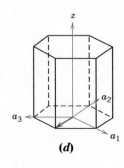

(d)

3.18 Sketch the [$\bar{1}\bar{1}23$] and [$10\bar{1}0$] directions in a hexagonal unit cell.

3.19 Using Equations 3.3a, 3.3b, 3.3c, and 3.3d, derive expressions for each of the three uppercase indices (U, V and W) in terms of the lowercase indices (u, v, t, and w).

Crystallographic Planes

3.20 **(a)** Draw an orthorhombic unit cell, and within that cell a (210) plane.

(b) Draw a monoclinic unit cell, and within that cell a (002) plane.

3.21 What are the indices for the two planes drawn in the following sketch?

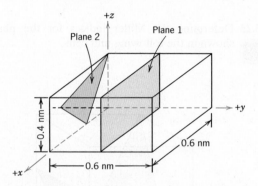

3.22 Sketch within a cubic unit cell the following planes:

(a) ($0\bar{1}\bar{1}$) **(c)** ($10\bar{2}$) **(e)** ($\bar{1}1\bar{1}$) **(g)** ($\bar{1}2\bar{3}$)

(b) ($11\bar{2}$) **(d)** ($1\bar{3}1$) **(f)** ($1\bar{2}\bar{2}$) **(h)** ($0\bar{1}\bar{3}$)

3.23 Determine the Miller indices for the planes shown in the following unit cell:

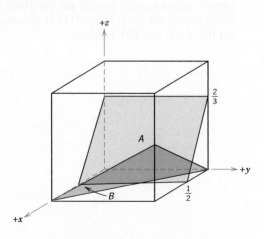

3.24 Determine the Miller indices for the planes shown in the following unit cell:

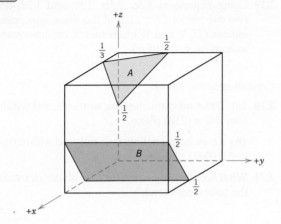

3.25 Determine the Miller indices for the planes shown in the following unit cell:

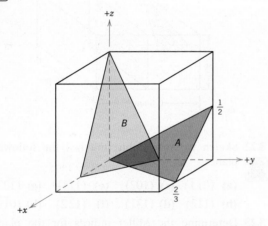

3.26 Cite the indices of the direction that results from the intersection of each of the following pairs of planes within a cubic crystal: **(a)** the (100) and (010) planes, **(b)** the (111) and (11$\bar{1}$) planes, and **(c)** the (10$\bar{1}$) and (001) planes.

3.27 Convert the (010) and (101) planes into the four-index Miller–Bravais scheme for hexagonal unit cells.

3.28 Determine the indices for the planes shown in the following hexagonal unit cells:

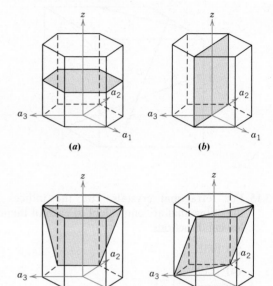

3.29 Sketch the (1$\bar{1}$01) and (11$\bar{2}$0) planes in a hexagonal unit cell.

Polycrystalline Materials

3.30 Explain why the properties of polycrystalline materials are most often isotropic.

Noncrystalline Solids

3.31 Would you expect a material in which the atomic bonding is predominantly ionic in nature to be more or less likely to form a noncrystalline solid upon solidification than a covalent material? Why? (See Section 2.6.)

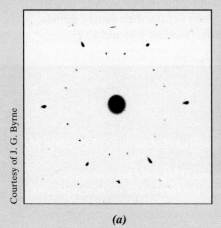

Courtesy of J. G. Byrne

(a)

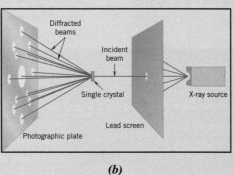

Diffracted beams

Incident beam

Single crystal

X-ray source

Lead screen

Photographic plate

(b)

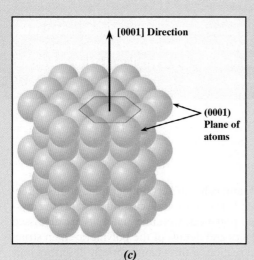

[0001] Direction

(0001) Plane of atoms

(c)

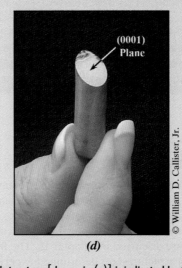

(0001) Plane

© William D. Callister, Jr.

(d)

(a) X-ray diffraction photograph [or Laue photograph (Section 4.20)] for a single crystal of magnesium.

(b) Schematic diagram illustrating how the spots (i.e., the diffraction pattern) in *(a)* are produced. The lead screen blocks out all beams generated from the x-ray source, except for a narrow beam traveling in a single direction. This incident beam is diffracted by individual crystallographic planes in the single crystal (having different orientations), which gives rise to the various diffracted beams that impinge on the photographic plate. Intersections of these beams with the plate appear as spots when the film is developed. The large spot in the center of *(a)* is from the incident beam, which is parallel to a [0001] crystallographic direction. It should be noted that the hexagonal symmetry of magnesium's hexagonal close-packed crystal structure [shown in *(c)*] is indicated by the diffraction spot pattern that was generated.

(d) Photograph of a single crystal of magnesium that was cleaved (or split) along a (0001) plane—the flat surface is a (0001) plane. Also, the direction perpendicular to this plane is a [0001] direction.

(e) Photograph of a *mag wheel*—a lightweight automobile wheel made of magnesium.

iStockphoto

(e)

[Figure *(b)* from J. E. Brady and F. Senese, *Chemistry: Matter and Its Changes*, 4th edition. Copyright © 2004 by John Wiley & Sons, Hoboken, NJ. Reprinted by permission of John Wiley & Sons, Inc.]

Some of the properties of materials may be explained by their crystal structures. For example, (a) pure and un-deformed magnesium and beryllium, having one crystal structure, are much more brittle (i.e., fracture at lower degrees of deformation) than are pure and undeformed metals such as gold and silver that have yet another crystal structure (see Section 9.4); (b) the permanent magnetic and ferroelectric behaviors of some ceramic materials are explained by their crystal structures (Sections 21.5 and 19.24); and (c) the degree of crystallinity of semicrystalline polymers impacts their density, stiffness, strength, and ductility (Sections 4.13 and 15.8).

Learning Objectives

After studying this chapter you should be able to do the following:

1. Draw unit cells for face-centered cubic, body-centered cubic, and hexagonal close-packed crystal structures.
2. Derive the relationships between unit cell edge length and atomic radius for face-centered cubic and body-centered cubic crystal structures.
3. Compute the densities for metals having face-centered cubic and body-centered cubic crystal structures given their unit cell dimensions.
4. Sketch/describe unit cells for sodium chloride, cesium chloride, zinc blende, diamond cubic, fluorite, and perovskite crystal structures. Do

   likewise for the atomic structures of graphite and a silica glass.
5. Given the chemical formula for a ceramic compound and the ionic radii of its component ions, predict the crystal structure.
6. Describe how face-centered cubic and hexagonal close-packed crystal structures may be generated by the stacking of close-packed planes of atoms. Do the same for the sodium chloride crystal structure in terms of close-packed planes of anions.
7. Briefly describe the crystalline state in polymeric materials.

4.1 INTRODUCTION

Fundamental concepts of crystal structures; unit cells; and crystallographic points, directions, and planes were explained in Chapter 3. The present discussion presents the crystal structures found in metals, ceramics, and polymers. Ceramic materials are discussed in greater detail in Chapter 14. Characteristics and details of the polymeric chain structures are discussed in Chapter 15. The final section of this chapter briefly describes how crystal structures are determined experimentally using X-ray diffraction techniques.

Metallic Crystal Structures

The atomic bonding in this group of materials is metallic and thus nondirectional in nature. Consequently, there are minimal restrictions as to the number and position of nearest-neighbor atoms; this leads to relatively large numbers of nearest neighbors and dense atomic packings for most metallic crystal structures. Also, for metals, when we use the hard-sphere model for the crystal structure, each sphere represents an ion core. Table 4.1 presents the atomic radii for a number of metals. Three relatively simple crystal structures are found for most of the common metals: face-centered cubic, body-centered cubic, and hexagonal close-packed.

4.2 THE FACE-CENTERED CUBIC CRYSTAL STRUCTURE

face-centered cubic (FCC)

The crystal structure found for many metals has a unit cell of cubic geometry, with atoms located at each of the corners and the centers of all the cube faces. It is aptly called the **face-centered cubic (FCC)** crystal structure. Some of the familiar metals having this crystal structure are copper, aluminum, silver, and gold (see also Table 4.1). Figure 3.1*a* shows a hard-sphere model for the FCC unit cell, whereas in Figure 3.1*b* the atom centers are represented by small circles to provide a better perspective on atom positions. The aggregate of atoms in Figure 3.1*c* represents a section of crystal consisting of many FCC unit cells. These spheres or ion cores touch one another across a face diagonal; the cube edge length *a* and the atomic radius *R* are related through

Unit cell edge length for face-centered cubic

$$a = 2R\sqrt{2} \tag{4.1}$$

This result is obtained in Example Problem 4.1.

Table 4.1

Atomic Radii and Crystal Structures for 16 Metals

Metal	Crystal Structure[a]	Atomic Radius[b] (nm)	Metal	Crystal Structure	Atomic Radius (nm)
Aluminum	FCC	0.1431	Molybdenum	BCC	0.1363
Cadmium	HCP	0.1490	Nickel	FCC	0.1246
Chromium	BCC	0.1249	Platinum	FCC	0.1387
Cobalt	HCP	0.1253	Silver	FCC	0.1445
Copper	FCC	0.1278	Tantalum	BCC	0.1430
Gold	FCC	0.1442	Titanium (α)	HCP	0.1445
Iron (α)	BCC	0.1241	Tungsten	BCC	0.1371
Lead	FCC	0.1750	Zinc	HCP	0.1332

[a]FCC = face-centered cubic; HCP = hexagonal close-packed; BCC = body-centered cubic.
[b]A nanometer (nm) equals 10^{-9} m; to convert from nanometers to angstrom units (Å), multiply the nanometer value by 10.

WileyPLUS: VMSE
Crystal Systems and Unit Cells for Metals

On occasion, we need to determine the number of atoms associated with each unit cell. Depending on an atom's location, it may be considered to be shared with adjacent unit cells—that is, only some fraction of the atom is assigned to a specific cell. For example, for cubic unit cells, an atom completely within the interior "belongs" to that unit cell, one at a cell face is shared with one other cell, and an atom residing at a corner is shared among eight. The number of atoms per unit cell, *N*, can be computed using the following formula:

$$N = N_i + \frac{N_f}{2} + \frac{N_c}{8} \tag{4.2}$$

where

N_i = the number of interior atoms
N_f = the number of face atoms
N_c = the number of corner atoms

WileyPLUS

Tutorial Video:
FCC Unit Cell
Calculations

For the FCC crystal structure, there are eight corner atoms ($N_c = 8$), six face atoms ($N_f = 6$), and no interior atoms ($N_i = 0$). Thus, from Equation 4.2,

$$N = 0 + \frac{6}{2} + \frac{8}{8} = 4$$

or a total of four whole atoms may be assigned to a given unit cell. This is depicted in Figure 3.1a, where only sphere portions are represented within the confines of the cube. The cell is composed of the volume of the cube that is generated from the centers of the corner atoms, as shown in the figure.

Corner and face positions are really equivalent—that is, translation of the cube corner from an original corner atom to the center of a face atom will not alter the cell structure.

coordination number

atomic packing factor (APF)

Two other important characteristics of a crystal structure are the **coordination number** and the **atomic packing factor (APF)**. For metals, each atom has the same number of nearest-neighbor or touching atoms, which is the coordination number. For face-centered cubics, the coordination number is 12. This may be confirmed by examination of Figure 3.1a; the front face atom has four corner nearest-neighbor atoms surrounding it, four face atoms that are in contact from behind, and four other equivalent face atoms residing in the next unit cell to the front (not shown).

The APF is the sum of the sphere volumes of all atoms within a unit cell (assuming the atomic hard-sphere model) divided by the unit cell volume—that is,

Definition of atomic packing factor

$$\text{APF} = \frac{\text{volume of atoms in a unit cell}}{\text{total unit cell volume}} \tag{4.3}$$

For the FCC structure, the atomic packing factor is 0.74, which is the maximum packing possible for spheres all having the same diameter. Computation of this APF is also included as an example problem. Metals typically have relatively large atomic packing factors to maximize the shielding provided by the free electron cloud.

4.3 THE BODY-CENTERED CUBIC CRYSTAL STRUCTURE

body-centered cubic (BCC)

Another common metallic crystal structure also has a cubic unit cell with atoms located at all eight corners and a single atom at the cube center. This is called a **body-centered cubic (BCC)** crystal structure. A collection of spheres depicting this crystal structure is shown in Figure 4.1c, whereas Figures 4.1a and 4.1b are diagrams of BCC unit cells with

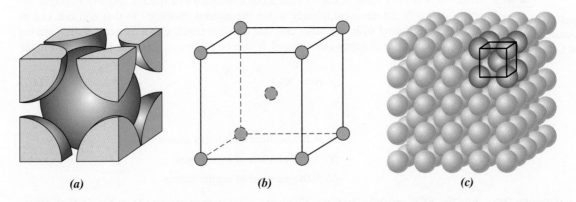

(a)   *(b)*   *(c)*

Figure 4.1 For the body-centered cubic crystal structure, (a) a hard-sphere unit cell representation, (b) a reduced-sphere unit cell, and (c) an aggregate of many atoms.
[Figure (c) from W. G. Moffatt, G. W. Pearsall, and J. Wulff, *The Structure and Properties of Materials*, Vol. I, *Structure*, p. 51. Copyright © 1964 by John Wiley & Sons, New York.]

the atoms represented by hard-sphere and reduced-sphere models, respectively. Center and corner atoms touch one another along cube diagonals, and unit cell length a and atomic radius R are related through

Unit cell edge length for body-centered cubic

$$a = \frac{4R}{\sqrt{3}} \tag{4.4}$$

WileyPLUS: VMSE
Crystal Systems and
Unit Cells for Metals

Chromium, iron, tungsten, and several other metals listed in Table 4.1 exhibit a BCC structure.

Each BCC unit cell has eight corner atoms and a single center atom, which is wholly contained within its cell; therefore, from Equation 4.2, the number of atoms per BCC unit cell is

$$N = N_i + \frac{N_f}{2} + \frac{N_c}{8}$$

$$= 1 + 0 + \frac{8}{8} = 2$$

WileyPLUS

Tutorial Video:
BCC Unit Cell
Calculations

The coordination number for the BCC crystal structure is 8; each center atom has as nearest neighbors its eight corner atoms. Because the coordination number is less for BCC than for FCC, the atomic packing factor is also lower for BCC—0.68 versus 0.74.

It is also possible to have a unit cell that consists of atoms situated only at the corners of a cube. This is called the *simple cubic (SC) crystal structure*; hard-sphere and reduced-sphere models are shown, respectively, in Figures 4.2a and 4.2b. None of the metallic elements have this crystal structure because of its relatively low atomic packing factor (see Concept Check 4.1). The only simple-cubic element is polonium, which is considered to be a metalloid (or semi-metal).

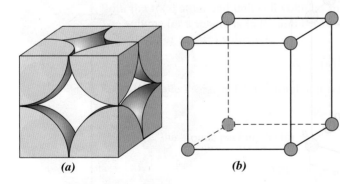

Figure 4.2 For the simple cubic crystal structure, (*a*) a hard-sphere unit cell, and (*b*) a reduced-sphere unit cell.

(a)   *(b)*

4.4 THE HEXAGONAL CLOSE-PACKED CRYSTAL STRUCTURE

hexagonal close-
packed (HCP)

Not all metals have unit cells with cubic symmetry; the final common metallic crystal structure to be discussed has a unit cell that is hexagonal. Figure 4.3a shows a reduced-sphere unit cell for this structure, which is termed **hexagonal close-packed (HCP)**; an assemblage of several HCP unit cells is presented in

EXAMPLE PROBLEM 4.3

Determination of HCP Unit Cell Volume

(a) Calculate the volume of an HCP unit cell in terms of its a and c lattice parameters.

(b) Now provide an expression for this volume in terms of the atomic radius, R, and the c lattice parameter.

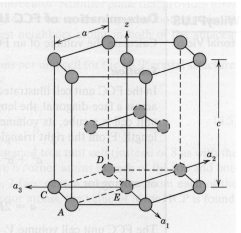

Solution

(a) We use the adjacent reduced-sphere HCP unit cell to solve this problem.

Now, the unit cell volume is just the product of the base area times the cell height, c. This base area is just three times the area of the parallelepiped $ACDE$ shown below. (This $ACDE$ parallelepiped is also labeled in the above unit cell.)

The area of $ACDE$ is just the length of $\overline{CD}$ times the height $\overline{BC}$. But $\overline{CD}$ is just a, and $\overline{BC}$ is equal to

$$\overline{BC} = a\cos(30°) = \frac{a\sqrt{3}}{2}$$

Thus, the base area is just

$$\text{AREA} = (3)(\overline{CD})(\overline{BC}) = (3)(a)\left(\frac{a\sqrt{3}}{2}\right) = \frac{3a^2\sqrt{3}}{2}$$

Again, the unit cell volume V_C is just the product of the AREA and c; thus,

$$
\begin{aligned}
V_C &= \text{AREA}(c) \\
&= \left(\frac{3a^2\sqrt{3}}{2}\right)(c) \\
&= \frac{3a^2c\sqrt{3}}{2}
\end{aligned}
\tag{4.7a}
$$

(b) For this portion of the problem, all we need do is realize that the lattice parameter a is related to the atomic radius R as

$$a = 2R$$

Now making this substitution for a in Equation 4.7a gives

$$
\begin{aligned}
V_C &= \frac{3(2R)^2c\sqrt{3}}{2} \\
&= 6R^2c\sqrt{3}
\end{aligned}
\tag{4.7b}
$$

4.5 DENSITY COMPUTATIONS–METALS

A knowledge of the crystal structure of a metallic solid permits computation of its theoretical density ρ through the relationship

Theoretical density
for metals

$$\rho = \frac{nA}{V_C N_A} \tag{4.8}$$

where

$$n = \text{number of atoms associated with each unit cell}$$
$$A = \text{atomic weight}$$
$$V_C = \text{volume of the unit cell}$$
$$N_A = \text{Avogadro's number } (6.022 \times 10^{23} \text{ atoms/mol})$$

EXAMPLE PROBLEM 4.4

Theoretical Density Computation for Copper

Copper has an atomic radius of 0.128 nm, an FCC crystal structure, and an atomic weight of 63.5 g/mol. Compute its theoretical density, and compare the answer with its measured density.

Solution

Equation 4.8 is employed in the solution of this problem. Because the crystal structure is FCC, n, the number of atoms per unit cell, is 4. Furthermore, the atomic weight A_{Cu} is given as 63.5 g/mol. The unit cell volume V_C for FCC was determined in Example Problem 4.1 as $16R^3\sqrt{2}$, where R, the atomic radius, is 0.128 nm.

Substitution for the various parameters into Equation 4.8 yields

$$\rho = \frac{nA_{Cu}}{V_C N_A} = \frac{nA_{Cu}}{(16R^3\sqrt{2})N_A}$$

$$= \frac{(4 \text{ atoms/unit cell})(63.5 \text{ g/mol})}{[16\sqrt{2}(1.28 \times 10^{-8} \text{ cm})^3/\text{unit cell}](6.022 \times 10^{23} \text{ atoms/mol})}$$

$$= 8.89 \text{ g/cm}^3$$

The literature value for the density of copper is 8.94 g/cm³, which is in very close agreement with the foregoing result.

Ceramic Crystal Structures

Ceramic materials were discussed briefly in Chapter 1, which noted that they are inorganic and nonmetallic materials. Most ceramics are compounds between metallic and nonmetallic elements for which the interatomic bonds are either totally ionic, or predominantly ionic but having some covalent character. The term *ceramic* comes from the Greek word *keramikos,* which means "burnt stuff," indicating that desirable properties of these materials are normally achieved through a high-temperature heat treatment process called firing.

Because ceramics are composed of at least two elements, and often more, their crystal structures are generally more complex than those for metals. The atomic bonding in these materials ranges from purely ionic to totally covalent; many ceramics exhibit a combination of these two bonding types, the degree of ionic character being dependent on the electronegativities of the atoms. Table 4.2 presents the percent ionic character for several common ceramic materials; these values were determined using Equation 2.16 and the electronegativities in Figure 2.9.

Table 4.2

Percent Ionic Character of the Interatomic Bonds for Several Ceramic Materials

Material	Percent Ionic Character
CaF_2	89
MgO	70
NaCl	59
Al_2O_3	63
SiO_2	51
Si_3N_4	34
ZnS	12
SiC	12

4.6 IONIC ARRANGEMENT GEOMETRIES

cation

anion

For those ceramic materials for which the atomic bonding is predominantly ionic, the crystal structures may be thought of as being composed of electrically charged ions instead of atoms. The metallic ions, or **cations**, are positively charged because they have given up their valence electrons to the nonmetallic ions, or **anions**, which are negatively charged. Two characteristics of the component ions in crystalline ceramic materials influence the crystal structure: the magnitude of the electrical charge on each of the component ions, and the relative sizes of the cations and anions. With regard to the first characteristic, the crystal must be electrically neutral; that is, all the cation positive charges must be balanced by an equal number of anion negative charges. The chemical formula of a compound indicates the ratio of cations to anions, or the composition that achieves this charge balance. For example, in calcium fluoride, each calcium ion has a $+2$ charge (Ca^{2+}), and associated with each fluorine ion is a single negative charge (F^-). Thus, there must be twice as many F^- as Ca^{2+} ions, which is reflected in the chemical formula CaF_2.

The second criterion involves the sizes or ionic radii of the cations and anions, r_C and r_A, respectively. Because the metallic elements give up electrons when ionized, cations are ordinarily smaller than anions, and, consequently, the ratio r_C/r_A is less than unity. Each cation prefers to have as many nearest-neighbor anions as possible. The anions also desire a maximum number of cation nearest neighbors.

Stable ceramic crystal structures form when those anions surrounding a cation are all in contact with that cation, as illustrated in Figure 4.4. The coordination number (i.e., number of anion nearest neighbors for a cation) is related to the cation–anion radius ratio. For a specific coordination number, there is a critical or minimum r_C/r_A ratio for which this cation–anion contact is established (Figure 4.4); this ratio may be determined from pure geometrical considerations (see Example Problem 4.5).

The coordination numbers and nearest-neighbor geometries for various r_C/r_A ratios are presented in Table 4.3. For r_C/r_A ratios less than 0.155, the very small cation is bonded to two anions in a linear manner. If r_C/r_A has a value between 0.155 and 0.225, the coordination number for the cation is 3. This means each cation is surrounded by three anions in the form of a planar equilateral triangle, with the cation located in the center. The coordination

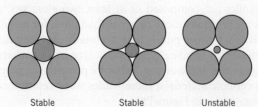

Stable   Stable   Unstable

Figure 4.4 Stable and unstable anion–cation coordination configurations. Red circles represent anions; blue circles denote cations.

Table 4.3

Coordination
Numbers and
Geometries for
Various Cation–Anion
Radius Ratios (r_C/r_A)

Coordination Number	Cation–Anion Radius Ratio	Coordination Geometry
2	<0.155	
3	0.155–0.225	
4	0.225–0.414	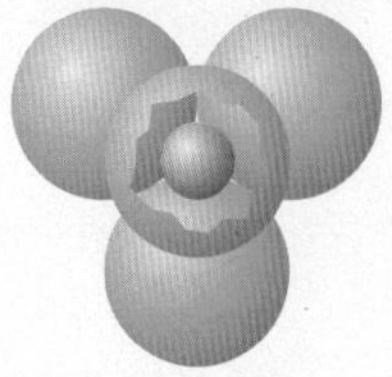
6	0.414–0.732	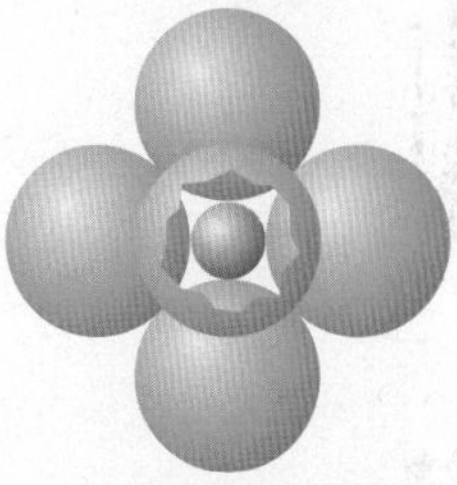
8	0.732–1.0	

Source: W. D. Kingery, H. K. Bowen, and D. R. Uhlmann, *Introduction to Ceramics,* 2nd edition.
Copyright © 1976 by John Wiley & Sons, New York. Reprinted by permission of John Wiley & Sons, Inc.

number is 4 for r_C/r_A between 0.225 and 0.414; the cation is located at the center of a tetrahedron, with anions at each of the four corners. For r_C/r_A between 0.414 and 0.732, the cation may be thought of as being situated at the center of an octahedron surrounded by six anions, one at each corner, as also shown in the table. The coordination number is 8 for r_C/r_A between 0.732 and 1.0, with anions at all corners of a cube and a cation positioned at the center. For a radius ratio greater than unity, the coordination number is 12. The most common coordination numbers for ceramic materials are 4, 6, and 8. Table 4.4 gives the ionic radii for several anions and cations that are common in ceramic materials.

The relationships between coordination number and cation–anion radius ratios (as noted in Table 4.3) are based on geometrical considerations and assuming "hard-sphere" ions; therefore, these relationships are only approximate, and there are exceptions. For example, some ceramic compounds with r_C/r_A ratios greater than 0.414 in which the bonding is highly covalent (and directional) have a coordination number of 4 (instead of 6).

Table 4.4
Ionic Radii for Several Cations and Anions for a Coordination Number of 6

Cation	*Ionic Radius (nm)*	*Anion*	*Ionic Radius (nm)*
Al^{3+}	0.053	Br^-	0.196
Ba^{2+}	0.136	Cl^-	0.181
Ca^{2+}	0.100	F^-	0.133
Cs^+	0.170	I^-	0.220
Fe^{2+}	0.077	O^{2-}	0.140
Fe^{3+}	0.069	S^{2-}	0.184
K^+	0.138		
Mg^{2+}	0.072		
Mn^{2+}	0.067		
Na^+	0.102		
Ni^{2+}	0.069		
Si^{4+}	0.040		
Ti^{4+}	0.061		

The size of an ion depends on several factors. One of these is coordination number: ionic radius tends to increase as the number of nearest-neighbor ions of opposite charge increases. Ionic radii given in Table 4.4 are for a coordination number of 6. Therefore, the radius is greater for a coordination number of 8 and less when the coordination number is 4.

In addition, the charge on an ion will influence its radius. For example, from Table 4.4, the radii for Fe^{2+} and Fe^{3+} are 0.077 and 0.069 nm, respectively, which values may be contrasted to the radius of an iron atom—0.124 nm. When an electron is removed from an atom or ion, the remaining valence electrons become more tightly bound to the nucleus, which results in a decrease in ionic radius. Conversely, ionic size increases when electrons are added to an atom or ion.

EXAMPLE PROBLEM 4.5

Computation of Minimum Cation-to-Anion Radius Ratio for a Coordination Number of 3

Show that the minimum cation-to-anion radius ratio for the coordination number 3 is 0.155.

Solution

For this coordination, the small cation is surrounded by three anions to form an equilateral triangle as shown here, triangle *ABC;* the centers of all four ions are coplanar.

This boils down to a relatively simple plane trigonometry problem. Consideration of the right triangle *APO* makes it clear that the side lengths are related to the anion and cation radii r_A and r_C as

$$\overline{AP} = r_A$$

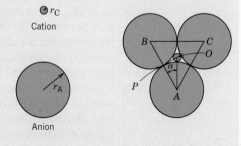

and

$$\overline{AO} = r_A + r_C$$

Furthermore, the side length ratio $\overline{AP}/\overline{AO}$ is a function of the angle α as

$$\frac{\overline{AP}}{\overline{AO}} = \cos\alpha$$

The magnitude of α is 30°, because line $\overline{AO}$ bisects the 60° angle BAC. Thus,

$$\frac{\overline{AP}}{\overline{AO}} = \frac{r_A}{r_A + r_C} = \cos 30° = \frac{\sqrt{3}}{2}$$

Solving for the cation–anion radius ratio,

$$\frac{r_C}{r_A} = \frac{1 - \sqrt{3}/2}{\sqrt{3}/2} = 0.155$$

4.7 AX-TYPE CRYSTAL STRUCTURES

Some of the common ceramic materials are those in which there are equal numbers of cations and anions. These are often referred to as AX compounds, where A denotes the cation and X the anion. There are several different crystal structures for AX compounds; each is typically named after a common material that assumes the particular structure.

Rock Salt Structure

WileyPLUS: VMSE

Perhaps the most common AX crystal structure is the *sodium chloride* (NaCl), or *rock salt,* type. The coordination number for both cations and anions is 6, and therefore the cation–anion radius ratio is between approximately 0.414 and 0.732. A unit cell for this crystal structure (Figure 4.5) is generated from an FCC arrangement of anions with one cation situated at the cube center and one at the center of each of the 12 cube edges. An equivalent crystal structure results from a face-centered arrangement of cations. Thus, the rock salt crystal structure may be thought of as two interpenetrating FCC lattices—one composed of the cations, the other of anions. Some common ceramic materials that form with this crystal structure are NaCl, MgO, MnS, LiF, and FeO.

Cesium Chloride Structure

WileyPLUS: VMSE

Figure 4.6 shows a unit cell for the *cesium chloride* (CsCl) crystal structure; the coordination number is 8 for both ion types. The anions are located at each of the corners of a cube, whereas the cube center is a single cation. Interchange of anions with cations, and vice versa, produces the same crystal structure. This is *not* a BCC crystal structure because ions of two different kinds are involved.

Zinc Blende Structure

WileyPLUS: VMSE

A third AX structure is one in which the coordination number is 4—that is, all ions are tetrahedrally coordinated. This is called the *zinc blende,* or *sphalerite,* structure, after the mineralogical term for zinc sulfide (ZnS). A unit cell is presented in Figure 4.7; all corner and face positions of the cubic cell are occupied by S atoms, whereas the Zn atoms fill interior tetrahedral positions. An equivalent structure results if Zn and S atom

Concept Check 4.2   Table 4.4 gives the ionic radii for K^+ and O^{2-} as 0.138 and 0.140 nm, respectively.

(a) What is the coordination number for each O^{2-} ion?

(b) Briefly describe the resulting crystal structure for K_2O.

(c) Explain why this is called the *antifluorite structure*.

[*The answer may be found at* www.wiley.com/college/callister *(Student Companion Site).*]

4.10 DENSITY COMPUTATIONS-CERAMICS

It is possible to compute the theoretical density of a crystalline ceramic material from unit cell data in a manner similar to that described in Section 4.5 for metals. In this case the density ρ may be determined using a modified form of Equation 4.8, as follows:

Theoretical density
for ceramic materials

$$\rho = \frac{n'(\Sigma A_C + \Sigma A_A)}{V_C N_A} \qquad (4.9)$$

where

n' = the number of formula units within the unit cell[2]

ΣA_C = the sum of the atomic weights of all cations in the formula unit

ΣA_A = the sum of the atomic weights of all anions in the formula unit

V_C = the unit cell volume

N_A = Avogadro's number, 6.022×10^{23} formula units/mol

EXAMPLE PROBLEM 4.7

Theoretical Density Calculation for Sodium Chloride

On the basis of the crystal structure, compute the theoretical density for sodium chloride. How does this compare with its measured density?

Solution

The theoretical density may be determined using Equation 4.9, where n', the number of NaCl units per unit cell, is 4 because both sodium and chloride ions form FCC lattices. Furthermore,

$$\Sigma A_C = A_{Na} = 22.99 \text{ g/mol}$$
$$\Sigma A_A = A_{Cl} = 35.45 \text{ g/mol}$$

---

[2] By *formula unit*, we mean all the ions that are included in the chemical formula unit. For example, for $BaTiO_3$, a formula unit consists of one barium ion, one titanium ion, and three oxygen ions.

Because the unit cell is cubic, $V_C = a^3$, a being the unit cell edge length. For the face of the cubic unit cell shown in the accompanying figure,

$$a = 2r_{Na^+} + 2r_{Cl^-}$$

r_{Na^+} and r_{Cl^-} being the sodium and chlorine ionic radii, respectively, given in Table 4.4 as 0.102 and 0.181 nm.

Thus,

$$V_C = a^3 = (2r_{Na^+} + 2r_{Cl^-})^3$$

Finally,

$$\rho = \frac{n'(A_{Na} + A_{Cl})}{(2r_{Na^+} + 2r_{Cl^-})^3 N_A}$$

$$= \frac{4(22.99 + 35.45)}{[2(0.102 \times 10^{-7}) + 2(0.181 \times 10^{-7})]^3(6.022 \times 10^{23})}$$

$$= 2.14 \text{ g/cm}^3$$

This result compares favorably with the experimental value of 2.16 g/cm³.

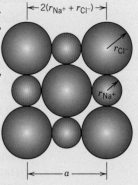

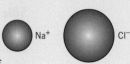

4.11 SILICATE CERAMICS

Silicates are materials composed primarily of silicon and oxygen, the two most abundant elements in Earth's crust; consequently, the bulk of soils, rocks, clays, and sand come under the silicate classification. Rather than characterizing the crystal structures of these materials in terms of unit cells, it is more convenient to use various arrangements of an SiO_4^{4-} tetrahedron (Figure 4.10). Each atom of silicon is bonded to four oxygen atoms, which are situated at the corners of the tetrahedron; the silicon atom is positioned at the center. Because this is the basic unit of the silicates, it is often treated as a negatively charged entity.

Often the silicates are not considered to be ionic because there is a significant covalent character to the interatomic Si–O bonds (Table 4.2), which are directional and relatively strong. Regardless of the character of the Si–O bond, there is a formal charge of −4 associated with every SiO_4^{4-} tetrahedron because each of the four oxygen atoms requires an extra electron to achieve a stable electronic structure. Various silicate structures arise from the different ways in which the SiO_4^{4-} units can be combined into one-, two-, and three-dimensional arrangements.

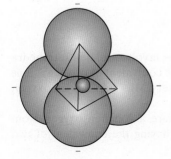

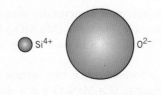

Figure 4.10 A silicon–oxygen (SiO_4^{4-}) tetrahedron.

Silica

Chemically, the most simple silicate material is silicon dioxide, or silica (SiO_2). Structurally, it is a three-dimensional network that is generated when the corner oxygen atoms in each tetrahedron are shared by adjacent tetrahedra. Thus, the material is electrically neutral, and all atoms have stable electronic structures. Under these circumstances the ratio of Si to O atoms is 1:2, as indicated by the chemical formula.

If these tetrahedra are arrayed in a regular and ordered manner, a crystalline structure is formed. There are three primary polymorphic crystalline forms of silica: quartz, cristobalite (Figure 4.11), and tridymite. Their structures are relatively complicated and comparatively open—that is, the atoms are not closely packed together. As a consequence, these crystalline silicas have relatively low densities; for example, at room temperature, quartz has a density of only 2.65 g/cm^3. The strength of the Si–O interatomic bonds is reflected in a relatively high melting temperature, 1710°C.

Silica Glasses

Silica can also be made to exist as a noncrystalline solid or glass having a high degree of atomic randomness, which is characteristic of the liquid; such a material is called *fused silica,* or *vitreous silica.* As with crystalline silica, the SiO_4^{4-} tetrahedron is the basic unit; beyond this structure, considerable disorder exists. The structures for crystalline and noncrystalline silica are compared schematically in Figure 3.11. Other oxides (e.g., B_2O_3, GeO_2) may also form glassy structures (and polyhedral oxide structures similar to that shown in Figure 4.10); these materials, as well as SiO_2, are termed *network formers.*

The common inorganic glasses that are used for containers, windows, and so on are silica glasses to which have been added other oxides such as CaO and Na_2O. These oxides do not form polyhedral networks. Rather, their cations are incorporated within and modify the SiO_4^{4-} network; for this reason, these oxide additives are termed *network modifiers.* For example, Figure 4.12 is a schematic representation of the structure of a sodium–silicate glass. Still other oxides, such as TiO_2 and Al_2O_3, although not network formers, substitute for silicon and become part of and stabilize the network; these are called *intermediates.* From a practical perspective, the addition of these modifiers and intermediates lowers the melting point and viscosity of a glass and makes it easier to form at lower temperatures (Section 17.8).

The Silicates

For the various silicate minerals, one, two, or three of the corner oxygen atoms of the SiO_4^{4-} tetrahedra are shared by other tetrahedra to form some rather complex structures. Some of these, represented in Figure 4.13, have formulas SiO_4^{4-}, $Si_2O_7^{6-}$, $Si_3O_9^{6-}$ and so on; single-chain structures are also possible, as in Figure 4.13e. Positively charged cations such as Ca^{2+}, Mg^{2+}, and Al^{3+} serve two roles: First, they compensate the negative charges from the SiO_4^{4-} units so that charge neutrality is achieved; second, these cations ionically bond the SiO_4^{4-} tetrahedra together.

Simple Silicates

Of these silicates, the most structurally simple ones involve isolated tetrahedra (Figure 4.13a). For example, forsterite (Mg_2SiO_4) has the equivalent of two Mg^{2+} ions associated with each tetrahedron in such a way that every Mg^{2+} ion has six oxygen nearest neighbors.

The $Si_2O_7^{6-}$ ion is formed when two tetrahedra share a common oxygen atom (Figure 4.13b). Akermanite ($Ca_2MgSi_2O_7$) is a mineral having the equivalent of two Ca^{2+} ions and one Mg^{2+} ion bonded to each $Si_2O_7^{6-}$ unit.

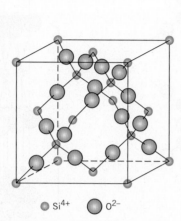

Figure 4.11 The arrangement of silicon and oxygen atoms in a unit cell of cristobalite, a polymorph of SiO_2.

● Si^{4+} ○ O^{2-}

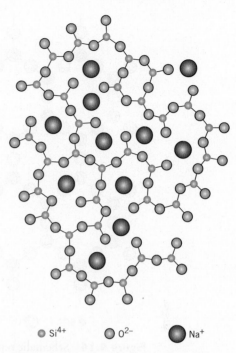

● Si^{4+} ○ O^{2-} ● Na^+

Figure 4.12 Schematic representation of ion positions in a sodium–silicate glass.

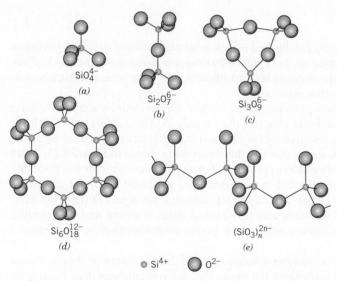

SiO_4^{4-}
(a)

$Si_2O_7^{6-}$
(b)

$Si_3O_9^{6-}$
(c)

$Si_6O_{18}^{12-}$
(d)

$(SiO_3)_n^{2n-}$
(e)

● Si^{4+} ○ O^{2-}

Figure 4.13 Five silicate ion structures formed from SiO_4^{4-} tetrahedra.

Layered Silicates

A two-dimensional sheet or layered structure can also be produced by the sharing of three oxygen ions in each of the tetrahedra (Figure 4.14); for this structure, the repeating unit formula may be represented by $(Si_2O_5)^{2-}$. The net negative charge is associated with the unbonded oxygen atoms projecting out of the plane of the page.

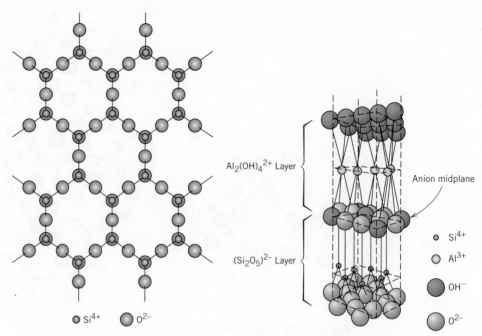

Figure 4.14 Schematic representation of the two-dimensional silicate sheet structure having a repeat unit formula of $(Si_2O_5)^{2-}$.

Figure 4.15 The structure of kaolinite clay. [Adapted from W. E. Hauth, "Crystal Chemistry of Ceramics," *American Ceramic Society Bulletin* 30, no. 4 (1951): 140.]

Electroneutrality is ordinarily established by a second planar sheet structure having an excess of cations, which bond to these unbonded oxygen atoms from the Si_2O_5 sheet. Such materials are called the sheet or layered silicates, and their basic structure is characteristic of the clays and other minerals.

One of the most common clay minerals, kaolinite, has a relatively simple two-layer silicate sheet structure. Kaolinite clay has the formula $Al_2(Si_2O_5)(OH)_4$ in which the silica tetrahedral layer, represented by $(Si_2O_5)^{2-}$, is made electrically neutral by an adjacent $Al_2(OH)_4^{2+}$ layer. A single sheet of this structure is shown in Figure 4.15, which is exploded in the vertical direction to provide a better perspective on the ion positions; the two distinct layers are indicated in the figure. The midplane of anions consists of O^{2-} ions from the $(Si_2O_5)^{2-}$ layer, as well as OH^- ions that are a part of the $Al_2(OH)_4^{2+}$ layer. Whereas the bonding within this two-layered sheet is strong and intermediate ionic-covalent, adjacent sheets are only loosely bound to one another by weak van der Waals forces.

A crystal of kaolinite is made of a series of these double layers or sheets stacked parallel to each other and form small flat plates that are typically less than 1 μm in diameter and nearly hexagonal. Figure 4.16 is an electron micrograph of kaolinite crystals at a high magnification, showing the hexagonal crystal plates, some of which are piled one on top of the other.

These silicate sheet structures are not confined to the clays; other minerals also in this group are talc [$Mg_3(Si_2O_5)_2(OH)_2$] and the micas [e.g., muscovite, $KAl_3Si_3O_{10}(OH)_2$], which are important ceramic raw materials. As might be deduced from the chemical formulas, the structures for some silicates are among the most complex of all the inorganic materials.

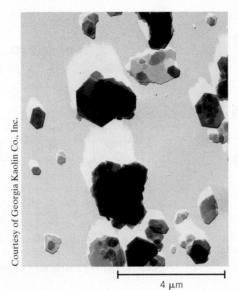

Figure 4.16 Electron micrograph of kaolinite crystals. They are in the form of hexagonal plates, some of which are stacked on top of one another. 7,500×.

Courtesy of Georgia Kaolin Co., Inc.

4 μm

4.12 CARBON

Although not one of the most frequently occurring elements found on Earth, carbon affects our lives in diverse and interesting ways. It exists in the elemental state in nature, and solid carbon has been used by all civilizations since prehistoric times. In today's world, the unique properties (and property combinations) of the several forms of carbon make it extremely important in many commercial sectors, including some cutting-edge technologies.

Carbon exists in two allotropic forms—diamond and graphite—as well as in the amorphous state. The carbon group of materials does not fall within any of the traditional metal, ceramic, or polymer classification schemes. However, we choose to discuss them in this chapter because graphite is sometimes classified as a ceramic. This treatment of the carbons focuses primarily on the structures of diamond and graphite. Discussions on the properties and applications (both current and potential) of diamond and graphite as well as the nanocarbons (i.e., fullerenes, carbon nanotubes, and graphene) are presented in Sections 14.16 and 14.17.

Diamond

WileyPLUS: VMSE

Diamond is a metastable carbon polymorph at room temperature and atmospheric pressure. Its crystal structure is a variant of the zinc blende structure (Figure 4.7) in which carbon atoms occupy all positions (both Zn and S); the unit cell for diamond is shown in Figure 4.17. Each carbon atom has undergone sp^3 hybridization so that it bonds (tetrahedrally) to four other carbons; these are extremely strong covalent bonds discussed in Section 2.6 (and represented in Figure 2.14). The crystal structure of diamond is appropriately called the *diamond cubic* crystal structure, which is also found for other Group IVA elements in the periodic table [e.g., germanium, silicon, and gray tin below 13°C].

Graphite

WileyPLUS: VMSE

Another polymorph of carbon, graphite, has a crystal structure (Figure 4.18) distinctly different from that of diamond; furthermore, it is a stable polymorph at ambient temperature and pressure. For the graphite structure, carbon atoms are located at corners of interlocking regular hexagons that lie in parallel (basal) planes. Within these planes (layers or sheets), sp^2 hybrid orbitals bond each carbon atom to three other adjacent and

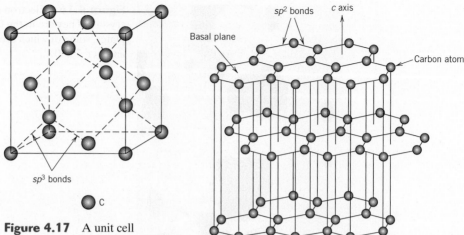

Figure 4.17 A unit cell for the diamond cubic crystal structure.

Figure 4.18 The structure of graphite.

coplanar carbons atoms; these bonds are strong covalent ones.[3] This hexagonal configuration assumed by sp^2 bonded carbon atoms is represented in Figure 2.18. Furthermore, each atom's fourth bonding electron is *delocalized* (i.e., does not belong to a specific atom or bond). Rather, its orbital becomes part of a molecular orbital that extends over adjacent atoms and resides between layers. Furthermore, interlayer bonds are directed perpendicular to these planes (i.e., in the c direction noted in Figure 4.18) and are of the weak van der Waals type.

4.13 POLYMER CRYSTALLINITY

polymer crystallinity

The crystalline state may exist in polymeric materials. However, because it involves molecules instead of just atoms or ions, as with metals and ceramics, the atomic arrangements will be more complex for polymers. We think of **polymer crystallinity** as the packing of molecular chains to produce an ordered atomic array. Crystal structures may be specified in terms of unit cells, which are often quite complex. For example, Figure 4.19 shows the unit cell for polyethylene and its relationship to the molecular chain structure; this unit cell has orthorhombic geometry (Table 3.1). Of course, the chain molecules also extend beyond the unit cell shown in the figure.

Molecular substances having small molecules (e.g., water and methane) are normally either totally crystalline (as solids) or totally amorphous (as liquids). As a consequence of their size and often complexity, polymer molecules are often only partially crystalline (or semicrystalline), having crystalline regions dispersed within the remaining amorphous material. Any chain disorder or misalignment will result in an amorphous region, a condition that is fairly common, because twisting, kinking, and coiling of the chains prevent the strict ordering of every segment of every chain. Other structural effects are also influential in determining the extent of crystallinity, as discussed shortly.

The degree of crystallinity may range from completely amorphous to almost entirely (up to about 95%) crystalline; in contrast, metal specimens are almost always entirely

[3]A single layer of this sp^2 bonded graphite is called *graphene*. Graphene is one of the nanocarbon materials, discussed in Section 13.9.

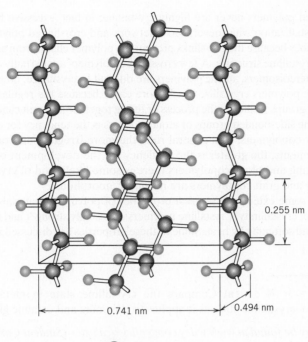

Figure 4.19 Arrangement of molecular chains in a unit cell for polyethylene. (From *Chemical Crystallography: An Introduction to Optical and X-ray Methods,* by Bunn (1945), Fig. 143, p. 233. By permission of Oxford University Press.)

0.255 nm

0.741 nm   0.494 nm

C   H

crystalline, whereas many ceramics are either totally crystalline or totally noncrystalline. Semicrystalline polymers are, in a sense, analogous to two-phase metal alloys, discussed previously.

The density of a crystalline polymer will be greater than an amorphous one of the same material and molecular weight because the chains are more closely packed together for the crystalline structure. The degree of crystallinity by weight may be determined from accurate density measurements, according to

Percent crystallinity (semicrystalline polymer)—dependence on specimen density, and densities of totally crystalline and totally amorphous materials

$$\% \text{ crystallinity} = \frac{\rho_c(\rho_s - \rho_a)}{\rho_s(\rho_c - \rho_a)} \times 100 \qquad (4.10)$$

where ρ_s is the density of a specimen for which the percent crystallinity is to be determined, ρ_a is the density of the totally amorphous polymer, and ρ_c is the density of the perfectly crystalline polymer. The values of ρ_a and ρ_c must be measured by other experimental means.

The degree of crystallinity of a polymer depends on the rate of cooling during solidification as well as on the chain configuration. During crystallization upon cooling through the melting temperature, the chains, which are highly random and entangled in the viscous liquid, must assume an ordered configuration. For this to occur, sufficient time must be allowed for the chains to move and align themselves.

The molecular chemistry as well as chain configuration also influence the ability of a polymer to crystallize. Crystallization is not favored in polymers that are composed of chemically complex repeat units (e.g., polyisoprene). However, crystallization is not easily prevented in chemically simple polymers such as polyethylene and polytetrafluoroethylene, even for very rapid cooling rates.

For linear polymers, crystallization is easily accomplished because there are few restrictions to prevent chain alignment. Any side branches interfere with crystallization, such that branched polymers never are highly crystalline; in fact, excessive branching may prevent any crystallization whatsoever. Most network and crosslinked polymers are almost totally amorphous because the crosslinks prevent the polymer chains from rearranging and aligning into a crystalline structure. A few crosslinked polymers are partially crystalline. With regard to the stereoisomers, atactic polymers are difficult to crystallize; however, isotactic and syndiotactic polymers crystallize much more easily because the regularity of the geometry of the side groups facilitates the process of fitting together adjacent chains. Also, the bulkier or larger the side-bonded groups of atoms, the less is the tendency for crystallization.

For copolymers, as a general rule, the more irregular and random the repeat unit arrangements, the greater is the tendency for the development of noncrystallinity. For alternating and block copolymers there is some likelihood of crystallization. However, random and graft copolymers are normally amorphous.

To some extent, the physical properties of polymeric materials are influenced by the degree of crystallinity. Crystalline polymers are usually stronger and more resistant to dissolution and softening by heat. Some of these properties are discussed in subsequent chapters.

Concept Check 4.3 (a) Compare the crystalline state in metals and polymers. (b) Compare the noncrystalline state as it applies to polymers and ceramic glasses.

[*The answer may be found at* www.wiley.com/college/callister *(Student Companion Site).*]

EXAMPLE PROBLEM 4.8

Computations of the Density and Percent Crystallinity of Polyethylene

(a) Compute the density of totally crystalline polyethylene. The orthorhombic unit cell for polyethylene is shown in Figure 4.19; also, the equivalent of two ethylene repeat units is contained within each unit cell.

(b) Using the answer to part (a), calculate the percent crystallinity of a branched polyethylene that has a density of 0.925 g/cm³. The density for the totally amorphous material is 0.870 g/cm³.

Solution

(a) Equation 4.8, used to determine densities for metals, also applies to polymeric materials and is used to solve this problem. It takes the same form, namely,

$$\rho = \frac{nA}{V_C N_A}$$

where n represents the number of repeat units within the unit cell (for polyethylene $n = 2$), and A is the repeat unit molecular weight, which for polyethylene is

$$A = 2(A_C) + 4(A_H)$$
$$= (2)(12.01 \text{ g/mol}) + (4)(1.008 \text{ g/mol}) = 28.05 \text{ g/mol}$$

Also, V_C is the unit cell volume, which is just the product of the three unit cell edge lengths in Figure 4.19; or

$$V_C = (0.741 \text{ nm})(0.494 \text{ nm})(0.255 \text{ nm})$$
$$= (7.41 \times 10^{-8} \text{ cm})(4.94 \times 10^{-8} \text{ cm})(2.55 \times 10^{-8} \text{ cm})$$
$$= 9.33 \times 10^{-23} \text{ cm}^3/\text{unit cell}$$

Now, substitution into Equation 4.8 of this value, values for n and A cited previously, and the value of N_A leads to

$$\rho = \frac{nA}{V_C N_A}$$

$$= \frac{(2 \text{ repeat units/unit cell})(28.05 \text{ g/mol})}{(9.33 \times 10^{-23} \text{ cm}^3/\text{unit cell})(6.022 \times 10^{23} \text{ repeat units/mol})}$$

$$= 0.998 \text{ g/cm}^3$$

(b) We now use Equation 4.10 to calculate the percent crystallinity of the branched polyethylene with $\rho_c = 0.998 \text{ g/cm}^3$, $\rho_a = 0.870 \text{ g/cm}^3$, and $\rho_s = 0.925 \text{ g/cm}^3$. Thus,

$$\% \text{ crystallinity} = \frac{\rho_c(\rho_s - \rho_a)}{\rho_s(\rho_c - \rho_a)} \times 100$$

$$= \frac{0.998 \text{ g/cm}^3 (0.925 \text{ g/cm}^3 - 0.870 \text{ g/cm}^3)}{0.925 \text{ g/cm}^3 (0.998 \text{ g/cm}^3 - 0.870 \text{ g/cm}^3)} \times 100$$

$$= 46.4\%$$

4.14 POLYMORPHISM AND ALLOTROPY

polymorphism
allotropy

Some metals, as well as nonmetals, may have more than one crystal structure, a phenomenon known as **polymorphism**. When found in elemental solids, the condition is often termed **allotropy**. The prevailing crystal structure depends on both the temperature and the external pressure. One familiar example is found in carbon: graphite is the stable polymorph at ambient conditions, whereas diamond is formed at extremely high pressures. Also, pure iron has a BCC crystal structure at room temperature, which changes to FCC iron at 912°C. Most often a modification of the density and other physical properties accompanies a polymorphic transformation.

4.15 ATOMIC ARRANGEMENTS

WileyPLUS: VMSE
Planar Atomic
Arrangements

The atomic arrangement for a crystallographic plane, which is often of interest, depends on the crystal structure. The (110) atomic planes for FCC and BCC crystal structures are represented in Figures 4.20 and 4.21, respectively. Reduced-sphere unit cells are also included. Note that the atomic packing is different for each case. The circles represent atoms lying in the crystallographic planes as would be obtained from a slice taken through the centers of the full-size hard spheres.

A "family" of planes contains all planes that are *crystallographically equivalent*—that is, having the same atomic packing; a family is designated by indices enclosed in braces—such as {100}. For example, in cubic crystals, the (111), $(\bar{1}\bar{1}\bar{1})$, $(\bar{1}11)$, $(1\bar{1}\bar{1})$, $(11\bar{1})$, $(\bar{1}\bar{1}1)$, $(\bar{1}1\bar{1})$, and $(1\bar{1}1)$ planes all belong to the {111} family. However, for tetragonal crystal structures, the {100} family contains only the (100), $(\bar{1}00)$, (010), and $(0\bar{1}0)$ planes because the (001) and $(00\bar{1})$ planes are not crystallographically equivalent. Also, in the cubic system only, planes having the same indices, irrespective of order and sign, are equivalent. For example, both $(1\bar{2}3)$ and $(3\bar{1}2)$ belong to the {123} family.

4.16 LINEAR AND PLANAR DENSITIES

The two previous sections discussed the equivalency of nonparallel crystallographic directions and planes. Directional equivalency is related to *linear density* in the sense that, for a particular material, equivalent directions have identical linear densities. The

M A T E R I A L   O F   I M P O R T A N C E

Tin (Its Allotropic Transformation)

Another common metal that experiences an allotropic change is tin. White (or β) tin, having a body-centered tetragonal crystal structure at room temperature, transforms, at 13.2°C, to gray (or α) tin, which has a crystal structure similar to that of diamond (i.e., the diamond cubic crystal structure); this transformation is represented schematically as follows:

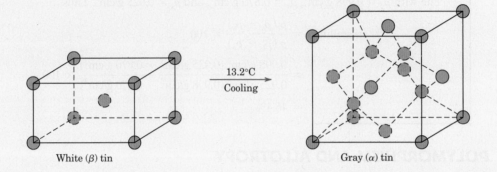

White (β) tin → 13.2°C Cooling → Gray (α) tin

The rate at which this change takes place is extremely slow; however, the lower the temperature (below 13.2°C) the faster the rate. Accompanying this white-to-gray-tin transformation is an increase in volume (27%), and, accordingly, a decrease in density (from 7.30 g/cm^3 to 5.77 g/cm^3). Consequently, this volume expansion results in the disintegration of the white tin metal into a coarse powder of the gray allotrope. For normal subambient temperatures, there is no need to worry about this disintegration process for tin products because of the very slow rate at which the transformation occurs.

This white-to-gray tin transition produced some rather dramatic results in 1850 in Russia. The winter that year was particularly cold, and record low temperatures persisted for extended periods of time. The uniforms of some Russian soldiers had tin buttons, many of which crumbled because of these extreme cold conditions, as did also many of the tin church organ pipes. This problem came to be known as the *tin disease*.

Specimen of white tin (left). Another specimen disintegrated upon transforming to gray tin (right) after it was cooled to and held at a temperature below 13.2°C for an extended period of time.
(Photograph courtesy of Professor Bill Plumbridge, Department of Materials Engineering, The Open University, Milton Keynes, England.)

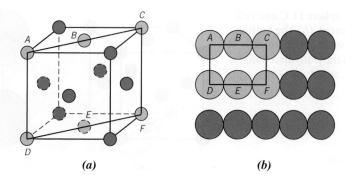

Figure 4.20 (*a*) Reduced-sphere FCC unit cell with the (110) plane. (*b*) Atomic packing of an FCC (110) plane. Corresponding atom positions from (*a*) are indicated.

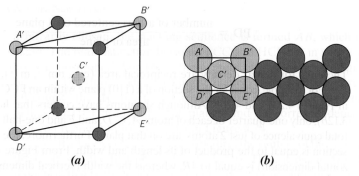

Figure 4.21 (*a*) Reduced-sphere BCC unit cell with the (110) plane. (*b*) Atomic packing of a BCC (110) plane. Corresponding atom positions from (*a*) are indicated.

corresponding parameter for crystallographic planes is *planar density,* and planes having the same planar density values are also equivalent.

Linear density (*LD*) is defined as the number of atoms per unit length whose centers lie on the direction vector for a specific crystallographic direction; that is,

$$LD = \frac{\text{number of atoms centered on direction vector}}{\text{length of direction vector}} \tag{4.11}$$

The units of linear density are reciprocal length (e.g., nm^{-1}, m^{-1}).

For example, let us determine the linear density of the [110] direction for the FCC crystal structure. An FCC unit cell (reduced sphere) and the [110] direction therein are shown in Figure 4.22*a*. Represented in Figure 4.22*b* are the five atoms that lie on the bottom face of this unit cell; here, the [110] direction vector passes from the center of atom *X*, through atom *Y*, and finally to the center of atom *Z*. With regard to the numbers of atoms, it is necessary to take into account the sharing of atoms with adjacent unit cells (as discussed in Section 4.2 relative to atomic packing factor computations). Each of the *X* and *Z* corner atoms is also shared with one other adjacent unit cell along this [110] direction (i.e., one-half of each of these atoms belongs to the unit cell being considered), whereas atom *Y* lies entirely within the unit cell. Thus, there is an equivalence of two atoms along the [110] direction vector in the unit cell. Now, the direction vector length is equal to 4*R* (Figure 4.22*b*); thus, from Equation 4.11, the [110] linear density for FCC is

$$LD_{110} = \frac{2 \text{ atoms}}{4R} = \frac{1}{2R} \tag{4.12}$$

Figure 4.25 (*a*) Close-packed stacking sequence for the face-centered cubic structure. (*b*) A corner has been removed to show the relation between the stacking of close-packed planes of atoms and the FCC crystal structure; the heavy triangle outlines a (111) plane. [Figure (*b*) from W. G. Moffatt, G. W. Pearsall, and J. Wulff, *The Structure and Properties of Materials,* Vol. I, *Structure*, p. 51. Copyright © 1964 by John Wiley & Sons, New York.]

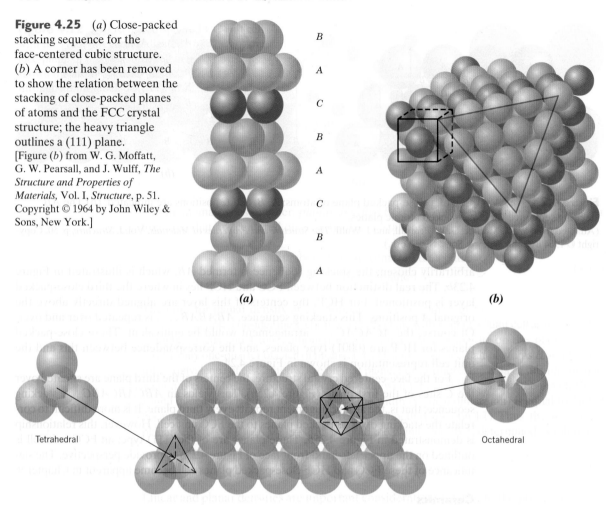

(*a*)

(*b*)

Tetrahedral

Octahedral

Figure 4.26 The stacking of one plane of close-packed (orange) spheres (anions) on top of another (blue spheres); the geometries of tetrahedral and octahedral positions between the planes are noted. (From W. G. Moffatt, G. W. Pearsall, and J. Wulff, *The Structure and Properties of Materials,* Vol. I, *Structure.* Copyright © 1964 by John Wiley & Sons, New York. Reprinted by permission of John Wiley & Sons, Inc.)

tetrahedral position

octahedral position

one type; this is termed a **tetrahedral position**, because straight lines drawn from the centers of the surrounding spheres form a four-sided tetrahedron. The other site type in Figure 4.26 involves six ion spheres, three in each of the two planes. Because an octahedron is produced by joining these six sphere centers, this site is called an **octahedral position**. Thus, the coordination numbers for cations filling tetrahedral and octahedral positions are 4 and 6, respectively. Furthermore, for each of these anion spheres, one octahedral and two tetrahedral positions exist.

Ceramic crystal structures of this type depend on two factors: (1) the stacking of the close-packed anion layers (both FCC and HCP arrangements are possible, which correspond to *ABCABC* . . . and *ABABAB* . . . sequences, respectively), and (2) the manner in which the interstitial sites are filled with cations. For example, consider the rock salt crystal structure discussed earlier. The unit cell has cubic symmetry, and each cation (Na^+ ion) has six Cl^- ion nearest neighbors, as may be verified from Figure 4.5. That is, the Na^+ ion at the center has as nearest neighbors the six Cl^- ions that reside at the centers of each of the cube faces. The crystal structure, having cubic symmetry, may be considered in terms of an FCC array of close-packed planes of anions, and all planes are

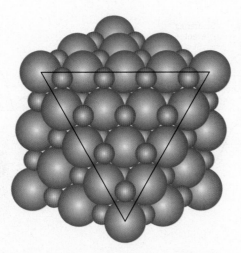

Figure 4.27 A section of the rock salt crystal structure from which a corner has been removed. The exposed plane of anions (green spheres inside the triangle) is a (111)-type plane; the cations (red spheres) occupy the interstitial octahedral positions.

of the {111} type. The cations reside in octahedral positions because they have as nearest neighbors six anions. Furthermore, all octahedral positions are filled, because there is a single octahedral site per anion, and the ratio of anions to cations is 1:1. For this crystal structure, the relationship between the unit cell and close-packed anion plane stacking schemes is illustrated in Figure 4.27.

Other, but not all, ceramic crystal structures may be treated in a similar manner; included are the zinc blende and perovskite structures. The *spinel structure* is one of the $A_mB_nX_p$ types, which is found for magnesium aluminate or spinel ($MgAl_2O_4$). With this structure, the O^{2-} ions form an FCC lattice, whereas Mg^{2+} ions fill tetrahedral sites and Al^{3+} ions reside in octahedral positions. Magnetic ceramics, or ferrites, have a crystal structure that is a slight variant of this spinel structure, and the magnetic characteristics are affected by the occupancy of tetrahedral and octahedral positions (see Section 21.5).

WileyPLUS: VMSE

X-RAY DIFFRACTION: DETERMINATION OF CRYSTAL STRUCTURES

Historically, much of our understanding regarding the atomic and molecular arrangements in solids has resulted from x-ray diffraction investigations; furthermore, x-rays are still very important in developing new materials. We now give a brief overview of the diffraction phenomenon and how, using x-rays, atomic interplanar distances and crystal structures are deduced.

4.18 THE DIFFRACTION PHENOMENON

Diffraction occurs when a wave encounters a series of regularly spaced obstacles that (1) are capable of scattering the wave, and (2) have spacings that are comparable in magnitude to the wavelength. Furthermore, diffraction is a consequence of specific phase relationships established between two or more waves that have been scattered by the obstacles.

Consider waves 1 and 2 in Figure 4.28a, which have the same wavelength (λ) and are in phase at point $O-O'$. Now let us suppose that both waves are scattered in such a way that they traverse different paths. The phase relationship between the scattered waves, which depends upon the difference in path length, is important.

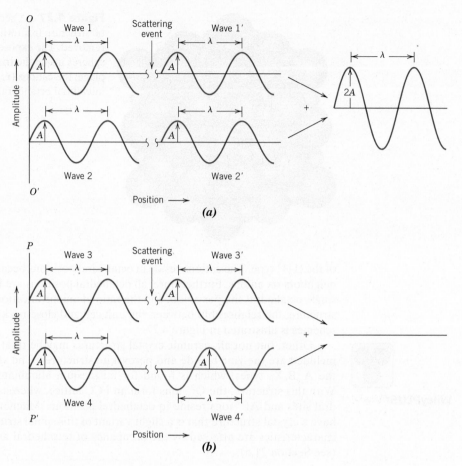

Figure 4.28 (*a*) Demonstration of how two waves (labeled 1 and 2) that have the same wavelength λ and remain in phase after a scattering event (waves 1′ and 2′) constructively interfere with one another. The amplitudes of the scattered waves add together in the resultant wave. (*b*) Demonstration of how two waves (labeled 3 and 4) that have the same wavelength and become out of phase after a scattering event (waves 3′ and 4′) destructively interfere with one another. The amplitudes of the two scattered waves cancel one another.

diffraction

One possibility results when this path length difference is an integral number of wavelengths. As noted in Figure 4.28*a*, these scattered waves (now labeled 1′ and 2′) are still in phase. They are said to mutually reinforce (or constructively interfere with) one another; when amplitudes are added, the wave shown on the right side of the figure results. This is a manifestation of **diffraction**, and we refer to a *diffracted beam* as one composed of a large number of scattered waves that mutually reinforce one another.

Other phase relationships are possible between scattered waves that will not lead to this mutual reinforcement. The other extreme is that demonstrated in Figure 4.28*b*, in which the path length difference after scattering is some integral number of *half*-wavelengths. The scattered waves are out of phase—that is, corresponding amplitudes cancel or annul one another, or destructively interfere (i.e., the resultant wave has zero amplitude), as indicated on the right side of the figure. Of course, phase relationships intermediate between these two extremes exist, resulting in only partial reinforcement.

4.19 X-RAY DIFFRACTION AND BRAGG'S LAW

X-rays are a form of electromagnetic radiation that have high energies and short wavelengths—wavelengths on the order of the atomic spacings for solids. When a beam of x-rays impinges on a solid material, a portion of this beam is scattered in all directions by the electrons associated with each atom or ion that lies within the beam's path. Let us now examine the necessary conditions for diffraction of x-rays by a periodic arrangement of atoms.

Consider the two parallel planes of atoms $A–A'$ and $B–B'$ in Figure 4.29, which have the same h, k, and l Miller indices and are separated by the interplanar spacing d_{hkl}. Now assume that a parallel, monochromatic, and coherent (in-phase) beam of x-rays of wavelength λ is incident on these two planes at an angle θ. Two rays in this beam, labeled 1 and 2, are scattered by atoms P and Q. Constructive interference of the scattered rays $1'$ and $2'$ occurs also at an angle θ to the planes if the path length difference between $1–P–1'$ and $2–Q–2'$ (i.e., $\overline{SQ} + \overline{QT}$) is equal to a whole number, n, of wavelengths—that is, the condition for diffraction is

$$n\lambda = \overline{SQ} + \overline{QT} \tag{4.15}$$

Bragg's law—relationship among x-ray wavelength, interatomic spacing, and angle of diffraction for constructive interference

or

$$n\lambda = d_{hkl} \sin\theta + d_{hkl} \sin\theta$$
$$= 2d_{hkl} \sin\theta \tag{4.16}$$

Bragg's law

Equation 4.16 is known as **Bragg's law**; n is the order of reflection, which may be any integer $(1, 2, 3, \ldots)$ consistent with $\sin\theta$ not exceeding unity. Thus, we have a simple expression relating the x-ray wavelength and interatomic spacing to the angle of the diffracted beam. If Bragg's law is not satisfied, then the interference will be nonconstructive so as to yield a very low-intensity diffracted beam.

The magnitude of the distance between two adjacent and parallel planes of atoms (i.e., the interplanar spacing d_{hkl}) is a function of the Miller indices (h, k, and l) as well as the lattice parameter(s). For example, for crystal structures that have cubic symmetry,

Interplanar spacing for a plane having indices h, k, and l

$$d_{hkl} = \frac{a}{\sqrt{h^2 + k^2 + l^2}} \tag{4.17}$$

Figure 4.29 Diffraction of x-rays by planes of atoms ($A–A'$ and $B–B'$).

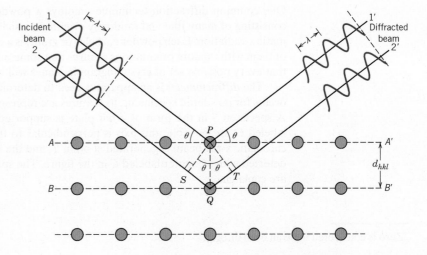

Table 4.6

X-Ray Diffraction Reflection Rules and Reflection Indices for Body-Centered Cubic, Face-Centered Cubic, and Simple Cubic Crystal Structures

Crystal Structure	Reflections Present	Reflection Indices for First Six Planes
BCC	$(h + k + l)$ even	110, 200, 211, 220, 310, 222
FCC	h, k, and l either all odd or all even	111, 200, 220, 311, 222, 400
Simple cubic	All	100, 110, 111, 200, 210, 211

in which a is the lattice parameter (unit cell edge length). Relationships similar to Equation 4.17, but more complex, exist for the other six crystal systems noted in Table 3.1.

Bragg's law, Equation 4.16, is a necessary but not sufficient condition for diffraction by real crystals. It specifies when diffraction will occur for unit cells having atoms positioned only at cell corners. However, atoms situated at other sites (e.g., face and interior unit cell positions as with FCC and BCC) act as extra scattering centers, which can produce out-of-phase scattering at certain Bragg angles. The net result is the absence of some diffracted beams that, according to Equation 4.16, should be present. Specific sets of crystallographic planes that do not give rise to diffracted beams depend on crystal structure. For the BCC crystal structure, $h + k + l$ must be even if diffraction is to occur, whereas for FCC, h, k, and l must all be either odd or even; diffracted beams for all sets of crystallographic planes are present for the simple cubic crystal structure (Figure 4.2). These restrictions, called *reflection rules*, are summarized in Table 4.6.[4]

Concept Check 4.4 For cubic crystals, as values of the planar indices h, k, and l increase, does the distance between adjacent and parallel planes (i.e., the interplanar spacing) increase or decrease? Why?

[*The answer may be found at* www.wiley.com/college/callister (*Student Companion Site*).]

4.20 DIFFRACTION TECHNIQUES

One common diffraction technique employs a powdered or polycrystalline specimen consisting of many fine and randomly oriented particles that are exposed to monochromatic x-radiation. Each powder particle (or grain) is a crystal, and having a large number of them with random orientations ensures that some particles are properly oriented such that every possible set of crystallographic planes will be available for diffraction.

The *diffractometer* is an apparatus used to determine the angles at which diffraction occurs for powdered specimens; its features are represented schematically in Figure 4.30. A specimen S in the form of a flat plate is supported so that rotations about the axis labeled O are possible; this axis is perpendicular to the plane of the page. The monochromatic x-ray beam is generated at point T, and the intensities of diffracted beams are detected with a counter labeled C in the figure. The specimen, x-ray source, and counter are coplanar.

---

[4]Zero is considered to be an even integer.

Figure 4.30 Schematic diagram of an x-ray diffractometer; T = x-ray source, S = specimen, C = detector, and O = the axis around which the specimen and detector rotate.

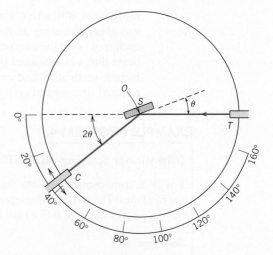

The counter is mounted on a movable carriage that may also be rotated about the O axis; its angular position in terms of 2θ is marked on a graduated scale.[5] Carriage and specimen are mechanically coupled such that a rotation of the specimen through θ is accompanied by a 2θ rotation of the counter; this ensures that the incident and reflection angles are maintained equal to one another (Figure 4.30). Collimators are incorporated within the beam path to produce a well-defined and focused beam. Utilization of a filter provides a near-monochromatic beam.

As the counter moves at constant angular velocity, a recorder automatically plots the diffracted beam intensity (monitored by the counter) as a function of 2θ; 2θ is termed the *diffraction angle*, which is measured experimentally. Figure 4.31 shows a diffraction pattern for a powdered specimen of lead. The high-intensity peaks result when the Bragg diffraction condition is satisfied by some set of crystallographic planes. These peaks are plane-indexed in the figure.

Other powder techniques have been devised in which diffracted beam intensity and position are recorded on a photographic film instead of being measured by a counter.

One of the primary uses of x-ray diffractometry is for the determination of crystal structure. The unit cell size and geometry may be resolved from the angular positions of the diffraction peaks, whereas the arrangement of atoms within the unit cell is associated with the relative intensities of these peaks.

X-rays, as well as electron and neutron beams, are also used in other types of material investigations. For example, crystallographic orientations of single crystals are possible

Figure 4.31
Diffraction pattern for powdered lead. (Courtesy of Wesley L. Holman.)

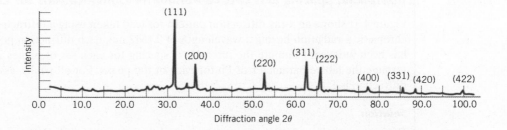

---

[5]Note that the symbol θ has been used in two different contexts for this discussion. Here, θ represents the angular locations of both x-ray source and counter relative to the specimen surface. Previously (e.g., Equation 4.16), it denoted the angle at which the Bragg criterion for diffraction is satisfied.

using x-ray diffraction (or Laue) photographs. The chapter-opening photograph (*a*) was generated using an incident x-ray beam that was directed on a magnesium crystal; each spot (with the exception of the darkest one near the center) resulted from an x-ray beam that was diffracted by a specific set of crystallographic planes. Other uses of x-rays include qualitative and quantitative chemical identifications and the determination of residual stresses and crystal size.

EXAMPLE PROBLEM 4.9

Interplanar Spacing and Diffraction Angle Computations

For BCC iron, compute **(a)** the interplanar spacing and **(b)** the diffraction angle for the (220) set of planes. The lattice parameter for Fe is 0.2866 nm. Assume that monochromatic radiation having a wavelength of 0.1790 nm is used, and the order of reflection is 1.

Solution

(a) The value of the interplanar spacing d_{hkl} is determined using Equation 4.17, with $a = 0.2866$ nm, and $h = 2$, $k = 2$, and $l = 0$ because we are considering the (220) planes. Therefore,

$$d_{hkl} = \frac{a}{\sqrt{h^2 + k^2 + l^2}}$$

$$= \frac{0.2866 \text{ nm}}{\sqrt{(2)^2 + (2)^2 + (0)^2}} = 0.1013 \text{ nm}$$

(b) The value of θ may now be computed using Equation 4.16, with $n = 1$ because this is a first-order reflection:

$$\sin\theta = \frac{n\lambda}{2d_{hkl}} = \frac{(1)(0.1790 \text{ nm})}{(2)(0.1013 \text{ nm})} = 0.884$$

$$\theta = \sin^{-1}(0.884) = 62.13°$$

The diffraction angle is 2θ, or

$$2\theta = (2)(62.13°) = 124.26°$$

EXAMPLE PROBLEM 4.10

Interplanar Spacing and Lattice Parameter Computations for Lead

Figure 4.31 shows an x-ray diffraction pattern for lead taken using a diffractometer and monochromatic x-radiation having a wavelength of 0.1542 nm; each diffraction peak on the pattern has been indexed. Compute the interplanar spacing for each set of planes indexed; also, determine the lattice parameter of Pb for each of the peaks. For all peaks, assume the order of diffraction is 1.

Solution

For each peak, in order to compute the interplanar spacing and the lattice parameter we must employ Equations 4.16 and 4.17, respectively. The first peak of Figure 4.31, which results from diffraction by the (111) set of planes, occurs at $2\theta = 31.3°$; the corresponding interplanar spacing for this set of planes, using Equation 4.16, is equal to

$$d_{111} = \frac{n\lambda}{2\sin\theta} = \frac{(1)(0.1542 \text{ nm})}{(2)\left[\sin\left(\dfrac{31.3°}{2}\right)\right]} = 0.2858 \text{ nm}$$

And, from Equation 4.17, the lattice parameter a is determined as

$$a = d_{hkl}\sqrt{h^2 + k^2 + l^2}$$
$$= d_{111}\sqrt{(1)^2 + (1)^2 + (1)^2}$$
$$= (0.2858 \text{ nm})\sqrt{3} = 0.4950 \text{ nm}$$

Similar computations are made for the next four peaks; the results are tabulated below:

Peak Index	2θ	d_{hkl}(nm)	a(nm)
200	36.6	0.2455	0.4910
220	52.6	0.1740	0.4921
311	62.5	0.1486	0.4929
222	65.5	0.1425	0.4936

SUMMARY

Metallic Crystal Structures

- Most common metals exist in at least one of three relatively simple crystal structures:
  Face-centered cubic (FCC), which has a cubic unit cell (Figure 3.1).
  Body-centered cubic (BCC), which also has a cubic unit cell (Figure 4.1).
  Hexagonal close-packed, which has a unit cell of hexagonal symmetry (Figure 4.2a).

- Unit cell edge length (a) and atomic radius (R) are related according to
  Equation 4.1 for face-centered cubic, and
  Equation 4.4 for body-centered cubic.

- Two features of a crystal structure are
  Coordination number—the number of nearest-neighbor atoms, and
  Atomic packing factor—the fraction of solid-sphere volume in the unit cell.

Density Computations— Metals

- The theoretical density of a metal (ρ) is a function of the number of equivalent atoms per unit cell, the atomic weight, the unit cell volume, and Avogadro's number (Equation 4.8).

Ceramic Crystal Structures

- Interatomic bonding in ceramics ranges from purely ionic to totally covalent.
- For predominantly ionic bonding:
  Metallic cations are positively charged, whereas nonmetallic ions have negative charges.
  Crystal structure is determined by (1) the charge magnitude on each ion and (2) the radius of each type of ion.

- Many of the simpler crystal structures are described in terms of unit cells:
  Rock salt (Figure 4.5)
  Cesium chloride (Figure 4.6)
  Zinc blende (Figure 4.7)
  Fluorite (Figure 4.8)
  Perovskite (Figure 4.9)

Density Computations— Ceramics
- The theoretical density of a ceramic material can be computed using Equation 4.9.

Silicate Ceramics
- For the silicates, structure is more conveniently represented in terms of interconnecting SiO_4^{4-} tetrahedra (Figure 4.10). Relatively complex structures may result when other cations (e.g., Ca^{2+}, Mg^{2+}, Al^{3+}) and anions (e.g., OH^-) are added.
- Silicate ceramics include the following:
    Crystalline silica (SiO_2) (as cristobalite, Figure 4.11)
    Layered silicates (Figures 4.14 and 4.15)
    Noncrystalline silica glasses (Figure 3.11)

Carbon
- Carbon (sometimes also considered a ceramic) can exist in several polymorphic forms:
    Diamond (Figure 4.17)
    Graphite (Figure 4.18)

Polymer Crystallinity
- When the molecular chains are aligned and packed in an ordered atomic arrangement, the condition of crystallinity is said to exist.
- Amorphous polymers are also possible wherein the chains are misaligned and disordered.
- In addition to being entirely amorphous, polymers may also exhibit varying degrees of crystallinity—that is, crystalline regions are interdispersed within amorphous areas.
- Crystallinity is facilitated for polymers that are chemically simple and that have regular and symmetrical chain structures.
- The percent crystallinity of a semicrystalline polymer is dependent on its density, as well as the densities of the totally crystalline and totally amorphous materials, according to Equation 4.10.

Polymorphism and Allotropy
- Polymorphism occurs when a specific material can have more than one crystal structure. Allotropy is polymorphism for elemental solids.

Linear and Planar Densities
- Crystallographic directional and planar equivalencies are related to atomic linear and planar densities, respectively.
    Linear density (for a specific crystallographic direction) is defined as the number of atoms per unit length whose centers lie on the vector for this direction (Equation 4.11).
    Planar density (for a specific crystallographic plane) is taken as the number of atoms per unit area that are centered on the particular plane (Equation 4.13).
- For a given crystal structure, planes having identical atomic packing yet different Miller indices belong to the same family.

Close-Packed Crystal Structures
- Both FCC and HCP crystal structures may be generated by the stacking of close-packed planes of atoms on top of one another. With this scheme A, B, and C denote possible atom positions on a close-packed plane.
    The stacking sequence for HCP is $ABABAB. \ldots$
    The stacking sequence for FCC is $ABCABCABC. \ldots$
- Close-packed planes for FCC and HCP are {111} and {0001}, respectively.
- Some ceramic crystal structures can be generated from the stacking of close-packed planes of anions; cations fill interstitial tetrahedral and/or octahedral positions that exist between adjacent planes.

X-Ray Diffraction: Determination of Crystal Structures	• X-ray diffractometry is used for crystal structure and interplanar spacing determinations. A beam of x-rays directed on a crystalline material may experience diffraction (constructive interference) as a result of its interaction with a series of parallel atomic planes.
	• Bragg's law specifies the condition for diffraction of x-rays—Equation 4.16.

Important Terms and Concepts

allotropy	cation	hexagonal close-packed (HCP)
anion	coordination number	octahedral position
atomic packing factor (APF)	crystallinity (polymer)	polymer crystallinity
body-centered cubic (BCC)	diffraction	polymorphism
Bragg's law	face-centered cubic (FCC)	tetrahedral position

REFERENCES

Buerger, M. J., *Elementary Crystallography,* Wiley, New York, 1956.

Chiang, Y. M., D. P. Birnie, III, and W. D. Kingery, *Physical Ceramics: Principles for Ceramic Science and Engineering,* Wiley, New York, 1997.

Cullity, B. D., and S. R. Stock, *Elements of X-Ray Diffraction,* 3rd edition, Prentice Hall, Upper Saddle River, NJ, 2001.

DeGraef, M., and M. E. McHenry, *Structure of Materials: An Introduction to Crystallography, Diffraction, and Symmetry,* Cambridge University Press, New York, 2007.

Hammond, C., *The Basics of Crystallography and Diffraction,* 2nd edition, Oxford University Press, New York, 2001.

Hauth, W. E., "Crystal Chemistry in Ceramics," *American Ceramic Society Bulletin* 30, no. 1 (1951): 5–7; no. 2 (1951): 47;

no. 3 (1951): 76–77; no. 4 (1951): 137–142; no. 5 (1951): 165–167; no. 6 (1951): 203–205. A good overview of silicate structures.

Kingery, W. D., H. K. Bowen, and D. R. Uhlmann, *Introduction to Ceramics,* 2nd edition, Wiley, New York, 1976. Chapters 1–4.

Massa, W., *Crystal Structure Determination,* Springer, New York, 2004.

Richerson, D. W., *The Magic of Ceramics,* 2nd edition, American Ceramic Society, Westerville, OH, 2012.

Richerson, D. W., *Modern Ceramic Engineering,* 3rd edition, CRC Press, Boca Raton, FL, 2006.

Sands, D. E., *Introduction to Crystallography,* Dover, Mineola, NY, 1975.

QUESTIONS AND PROBLEMS

The Face-Centered Cubic Crystal Structure

4.1 If the atomic radius of aluminum is 0.143 nm, calculate the volume of its unit cell in cubic meters.

The Body-Centered Cubic Crystal Structure

4.2 Show for the body-centered cubic crystal structure that the unit cell edge length a and the atomic radius R are related through $a = 4R/\sqrt{3}$.

4.3 Show that the atomic packing factor for BCC is 0.68.

The Hexagonal Close-Packed Crystal Structure

4.4 For the HCP crystal structure, show that the ideal c/a ratio is 1.633.

4.5 Show that the atomic packing factor for HCP is 0.74.

Density Computations–Metals

4.6 Iron has a BCC crystal structure, an atomic radius of 0.124 nm, and an atomic weight of

55.85 g/mol. Compute and compare its theoretical density with the experimental value found inside the front cover.

4.7 Calculate the radius of an iridium atom, given that Ir has an FCC crystal structure, a density of 22.4 g/cm^3, and an atomic weight of 192.2 g/mol.

4.8 Calculate the radius of a vanadium atom, given that V has a BCC crystal structure, a density of 5.96 g/cm^3, and an atomic weight of 50.9 g/mol.

4.9 A hypothetical metal has the simple cubic crystal structure shown in Figure 4.2. If its atomic weight is 70.6 g/mol and the atomic radius is 0.128 nm, compute its density.

4.10 Zirconium has an HCP crystal structure and a density of 6.51 g/cm^3.

(a) What is the volume of its unit cell in cubic meters?

(b) If the c/a ratio is 1.593, compute the values of c and a.

4.11 Using atomic weight, crystal structure, and atomic radius data tabulated inside the front cover, compute the theoretical densities of lead, chromium, copper, and cobalt, and then compare these values with the measured densities listed in this same table. The c/a ratio for cobalt is 1.623.

4.12 Rhodium has an atomic radius of 0.1345 nm and a density of 12.41 g/cm^3. Determine whether it has an FCC or BCC crystal structure.

4.13 The atomic weight, density, and atomic radius for three hypothetical alloys are listed in the following table. For each, determine whether its crystal structure is FCC, BCC, or simple cubic and then justify your determination.

Alloy	Atomic Weight (g/mol)	Density (g/cm^3)	Atomic Radius (nm)
A	77.4	8.22	0.125
B	107.6	13.42	0.133
C	127.3	9.23	0.142

4.14 The unit cell for tin has tetragonal symmetry, with a and b lattice parameters of 0.583 and 0.318 nm, respectively. If its density, atomic weight, and atomic radius are 7.27 g/cm^3, 118.27 g/mol, and 0.152 nm, respectively, compute the atomic packing factor.

4.15 Iodine has an orthorhombic unit cell for which the a, b, and c lattice parameters are 0.479, 0.725, and 0.981 nm, respectively.

(a) If the atomic packing factor and atomic radius are 0.547 and 0.177 nm, respectively, determine the number of atoms in each unit cell.

(b) The atomic weight of iodine is 126.91 g/mol; compute its theoretical density.

4.16 Titanium has an HCP unit cell for which the ratio of the lattice parameters c/a is 1.58. If the radius of the Ti atom is 0.1442 nm, (a) determine the unit cell volume, and (b) calculate the density of Ti and compare it with the literature value.

4.17 Zinc has an HCP crystal structure, a c/a ratio of 1.856, and a density of 7.17 g/cm^3. Compute the atomic radius for Zn.

4.18 Rhenium has an HCP crystal structure, an atomic radius of 0.134 nm, and a c/a ratio of 1.615. Compute the volume of the unit cell for Re.

Ionic Arrangement Geometries

4.19 For a ceramic compound, what are the two characteristics of the component ions that determine the crystal structure?

4.20 Show that the minimum cation-to-anion radius ratio for a coordination number of 4 is 0.225.

4.21 Show that the minimum cation-to-anion radius ratio for a coordination number of 6 is 0.414. [*Hint*: use the NaCl crystal structure (Figure 4.5), and assume that anions and cations are just touching along cube edges and across face diagonals.]

4.22 Demonstrate that the minimum cation-to-anion radius ratio for a coordination number of 8 is 0.732.

4.23 On the basis of ionic charge and ionic radii given in Table 4.4, predict crystal structures for the following materials:

(a) CsI

(b) NiO

(c) KI

(d) NiS

Justify your selections.

AX-Type Crystal Structures

4.24 Which of the cations in Table 4.4 would you predict to form iodides having the cesium chloride crystal structure? Justify your choices.

4.25 Compute the atomic packing factor for the cesium chloride crystal structure in which $r_C/r_A = 0.735$.

4.26 The zinc blende crystal structure is one that may be generated from close-packed planes of anions.

(a) Will the stacking sequence for this structure be FCC or HCP? Why?

(b) Will cations fill tetrahedral or octahedral positions? Why?

(c) What fraction of the positions will be occupied?

4.27 Iron sulfide (FeS) may form a crystal structure that consists of an HCP arrangement of S^{2-} ions.

(a) Which type of interstitial site will the Fe^{2+} ions occupy?

(b) What fraction of these available interstitial sites will be occupied by Fe^{21} ions?

A$_m$X$_p$-Type Crystal Structures

4.28 The corundum crystal structure, found for Al_2O_3, consists of an HCP arrangement of O^{2-} ions; the Al^{3+} ions occupy octahedral positions.

(a) What fraction of the available octahedral positions are filled with Al^{3+} ions?

(b) Sketch two close-packed O^{2-} planes stacked in an *AB* sequence, and note octahedral positions that will be filled with the Al^{3+} ions.

4.29 Using the Molecule Definition Utility found in both "Metallic Crystal Structures and Crystallography" and "Ceramic Crystal Structures" modules of *VMSE*, available in WileyPlus, generate (and print out) a three-dimensional unit cell for titanium dioxide, TiO_2, given the following: (1) The unit cell is tetragonal with $a = 0.459$ nm and $c = 0.296$ nm, (2) oxygen atoms are located at the following point coordinates:

$$0.356 \ 0.356 \ 0 \qquad 0.856 \ 0.144 \ \tfrac{1}{2}$$
$$0.664 \ 0.664 \ 0 \qquad 0.144 \ 0.856 \ \tfrac{1}{2}$$

and (3) Ti atoms are located at the following point coordinates:

$$0\ 0\ 0 \qquad\qquad 1\ 0\ 1$$
$$1\ 0\ 0 \qquad\qquad 0\ 1\ 1$$
$$0\ 1\ 0 \qquad\qquad 1\ 1\ 1$$
$$0\ 0\ 1 \qquad\qquad \tfrac{1}{2}\ \tfrac{1}{2}\ \tfrac{1}{2}$$
$$1\ 1\ 0$$

$A_mB_nX_p$-Type Crystal Structures

4.30 Magnesium silicate, Mg_2SiO_4, forms in the olivine crystal structure that consists of an HCP arrangement of O^{2-} ions.

(a) Which type of interstitial site will the Mg^{2+} ions occupy? Why?

(b) Which type of interstitial site will the Si^{4+} ions occupy? Why?

(c) What fraction of the total tetrahedral sites will be occupied?

(d) What fraction of the total octahedral sites will be occupied?

Density Computations–Ceramics

4.31 Calculate the theoretical density of FeO, given that it has the rock salt crystal structure.

4.32 Magnesium oxide has the rock salt crystal structure and a density of 3.58 g/cm^3.

(a) Determine the unit cell edge length.

(b) How does this result compare with the edge length as determined from the radii in Table 4.4, assuming that the Mg^{2+} and O^{2-} ions just touch each other along the edges?

4.33 Compute the theoretical density of diamond given that the $C-C$ distance and bond angle are 0.154 nm and $109.5°$, respectively. How does this value compare with the measured density?

4.34 Compute the theoretical density of ZnS given that the $Zn-S$ distance and bond angle are 0.240 nm and $109.5°$, respectively. How does this value compare with the measured density?

4.35 Cadmium sulfide (CdS) has a cubic unit cell, and from x-ray diffraction data it is known that the cell edge length is 0.585 nm. If the measured density is 4.82 g/cm^3, how many Cd^{2+} and S^{2-} ions are there per unit cell?

4.36 (a) Using the ionic radii in Table 4.4, compute the theoretical density of CsCl. (*Hint*: use a modification of the result of Problem 4.2.)

(b) The measured density is 3.99 g/cm^3. How do you explain the slight discrepancy between your calculated value and the measured one?

4.37 From the data in Table 4.4, compute the theoretical density of CaF_2, which has the fluorite structure.

4.38 A hypothetical AX type of ceramic material is known to have a density of 2.31 g/cm^3 and a unit cell of cubic symmetry with a cell edge length of 0.45 nm. The atomic weights of the A and X elements are 86.6 and 40.3 g/mol, respectively. On the basis of this information, which of the following crystal structures is (are) possible for this material: rock salt, cesium chloride, or zinc blende? Justify your choice(s).

4.39 The unit cell for $MgFe_2O_4$ ($MgO-Fe_2O_3$) has cubic symmetry with a unit cell edge length of 0.850 nm. If the density of this material is 4.52 g/cm^3, compute its atomic packing factor. For this computation, you will need to use the ionic radii listed in Table 4.4.

4.40 The unit cell for Cr_2O_3 has hexagonal symmetry with lattice parameters $a = 0.5$ nm and $c = 1.34$ nm. If the density of this material is 5.22 g/cm^3, calculate its atomic packing factor. For this computation assume ionic radii of 0.062 nm and 0.140 nm, respectively, for Cr^{3+} and O^{2-}.

4.41 Compute the atomic packing factor for the diamond cubic crystal structure (Figure 4.17). Assume that bonding atoms touch one another, that the angle between adjacent bonds is $109.5°$, and that each atom internal to the unit cell is positioned $a/4$ of the distance away from the two nearest cell faces (a is the unit cell edge length).

4.42 Compute the atomic packing factor for cesium chloride using the ionic radii in Table 4.4 and assuming that the ions touch along the cube diagonals.

Silicate Ceramics

4.43 In terms of bonding, explain why silicate materials have relatively low densities.

4.44 Determine the angle between covalent bonds in an SiO_4^{4-} tetrahedron.

Polymer Crystallinity

4.45 Explain briefly why the tendency of a polymer to crystallize decreases with increasing molecular weight.

4.46 For each of the following pairs of polymers, do the following: (1) state whether it is possible to determine whether one polymer is more likely to crystallize than the other; (2) if it is possible, note which is the more likely and then cite reason(s) for your choice; and (3) if it is not possible to decide, then state why.

(a) Linear and syndiotactic poly(vinyl chloride); linear and isotactic polystyrene

(b) Network phenol-formaldehyde; linear and heavily crosslinked *cis*-isoprene

(c) Linear polyethylene; lightly branched isotactic polypropylene

(d) Alternating poly(styrene-ethylene) copolymer; random poly(vinyl chloride-tetrafluoroethylene) copolymer

4.47 The density of totally crystalline polypropylene at room temperature is 0.946 g/cm³. Also, at room temperature the unit cell for this material is monoclinic with the following lattice parameters:

$$a = 0.666 \text{ nm} \qquad \alpha = 90°$$
$$b = 2.078 \text{ nm} \qquad \beta = 99.62°$$
$$c = 0.650 \text{ nm} \qquad \gamma = 90°$$

If the volume of a monoclinic unit cell, V_{mono}, is a function of these lattice parameters as

$$V_{mono} = abc \sin \beta$$

determine the number of repeat units per unit cell.

4.48 The density and associated percent crystallinity for two polytetrafluoroethylene materials are as follows:

ρ (g/cm³)	crystallinity (%)
2.144	51.3
2.215	74.2

(a) Compute the densities of totally crystalline and totally amorphous polytetrafluoroethylene.

(b) Determine the percent crystallinity of a specimen having a density of 2.26 g/cm³.

4.49 The density and associated percent crystallinity for two nylon 6,6 materials are as follows:

ρ (g/cm³)	crystallinity (%)
1.188	67.3
1.152	43.7

(a) Compute the densities of totally crystalline and totally amorphous nylon 6,6.

(b) Determine the density of a specimen having 55.4% crystallinity.

Atomic Arrangements

4.50 Sketch the atomic packing of (a) the (100) plane for the BCC crystal structure, and (b) the (201) plane for the FCC crystal structure (similar to Figures 4.20b and 4.21b).

4.51 Consider the reduced-sphere unit cell shown in Problem 3.2, having an origin of the coordinate system positioned at the atom labeled O. For the following sets of planes, determine which are equivalent:

(a) $(00\bar{1})$, (010), and, $(\bar{1}00)$

(b) $(1\bar{1}0)$, $(10\bar{1})$, $(0\bar{1}1)$, and $(\bar{1}\bar{1}0)$

(c) $(\bar{1}\bar{1}\bar{1})$, $(\bar{1}1\bar{1})$, $(\bar{1}\bar{1}1)$, and $(1\bar{1}1)$

4.52 The accompanying figure shows the atomic packing schemes for several different crystallographic directions for a hypothetical metal. For each direction, the circles represent only the atoms contained within a unit cell; the circles are reduced from their actual size.

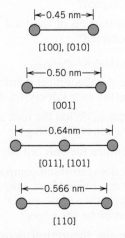

(a) To what crystal system does the unit cell belong?

(b) What would this crystal structure be called?

4.53 The accompanying figure shows three different crystallographic planes for a unit cell of a hypothetical metal. The circles represent atoms.

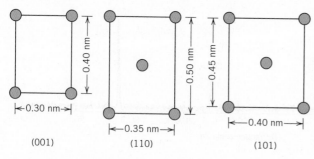

(001)   (110)   (101)

(a) To what crystal system does the unit cell belong?

(b) What would this crystal structure be called?

(c) If the density of this metal is 8.90 g/cm³, determine its atomic weight.

4.54 For each of the following crystal structures, represent the indicated plane in the manner of Figures 4.20 and 4.21, showing both anions and cations:

(a) (100) plane for the rock salt crystal structure

(b) (110) plane for the cesium chloride crystal structure

(c) (111) plane for the zinc blende crystal structure

(d) (110) plane for the perovskite crystal structure

Linear and Planar Densities

4.55 (a) Derive linear density expressions for FCC [100] and [111] directions in terms of the atomic radius R.

(b) Compute and compare linear density values for these same two directions for silver.

4.56 (a) Derive linear density expressions for BCC [110] and [111] directions in terms of the atomic radius R.

(b) Compute and compare linear density values for these same two direction for tungsten.

4.57 (a) Derive planar density expressions for FCC (100) and (111) planes in terms of the atomic radius R.

(b) Compute and compare planar density values for these same two planes for nickel.

4.58 (a) Derive planar density expressions for BCC (100) and (110) planes in terms of the atomic radius R.

(b) Compute and compare planar density values for these same two planes for vanadium.

4.59 (a) Derive the planar density expression for the HCP (0001) plane in terms of the atomic radius R.

(b) Compute the planar density value for this same plane for magnesium.

The Diffraction Phenomenon

X–Ray Diffraction and Bragg's Law

Diffraction Techniques

4.60 Using the data for molybdenum in Table 4.1, compute the interplanar spacing for the (111) set of planes.

4.61 Determine the expected diffraction angle for the first-order reflection from the (113) set of planes for FCC platinum when monochromatic radiation of wavelength 0.154 nm is used.

4.62 Using the data for aluminum in Table 4.1, compute the interplanar spacings for the (110) and (221) sets of planes.

4.63 The metal iridium has an FCC crystal structure. If the angle of diffraction for the (220) set of planes occurs at 69.20° (first-order reflection) when monochromatic x-radiation having a wavelength of 0.154 nm is used, compute **(a)** the interplanar spacing for this set of planes and **(b)** the atomic radius for an iridium atom.

4.64 The metal rubidium has a BCC crystal structure. If the angle of diffraction for the (321) set of planes occurs at 27.00° (first-order reflection) when monochromatic x-radiation having a wavelength of 0.071 nm is used, compute **(a)** the interplanar spacing for this set of planes and **(b)** the atomic radius for the rubidium atom.

4.65 For which set of crystallographic planes will a first-order diffraction peak occur at a diffraction angle of 46.21° for BCC iron when monochromatic radiation having a wavelength of 0.071 nm is used?

4.66 Figure 4.32 shows an x-ray diffraction pattern for α-iron taken using a diffractometer and monochromatic x-radiation having a wavelength of 0.154 nm; each diffraction peak on the pattern has been indexed. Compute the interplanar spacing for each set of planes indexed; also determine the lattice parameter of Fe for each of the peaks.

4.67 The diffraction peaks shown in Figure 4.32 are indexed according to the reflection rules for BCC (i.e., the sum $h + k + l$ must be even). Cite the $h, k,$ and l indices for the first four diffraction peaks for FCC crystals consistent with $h, k,$ and l all being either odd or even.

Figure 4.32
Diffraction pattern for polycrystalline α-iron.

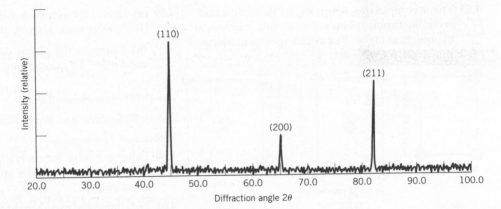

4.68 Figure 4.33 shows the first four peaks of the x-ray diffraction pattern for copper, which has an FCC crystal structure; monochromatic x-radiation having a wavelength of 0.154 nm was used.

(a) Index (i.e., give *h*, *k*, and *l* indices for) each of these peaks.

(b) Determine the interplanar spacing for each of the peaks.

(c) For each peak, determine the atomic radius for Cu and compare these with the value presented in Table 4.1.

Spreadsheet Problems

4.1SS For an x-ray diffraction pattern (having all peaks plane-indexed) of a metal that has a unit cell of cubic symmetry, generate a spreadsheet that allows the user to input the x-ray wavelength, and then determine, for each plane, **(a)** d_{hkl} and **(b)** the lattice parameter, *a*.

4.2SS For a specific polymer, given at least two density values and their corresponding percents crystallinity, develop a spreadsheet that allows the user to determine the following: **(a)** the density of the totally crystalline polymer, **(b)** the density of the totally amorphous polymer, **(c)** the percent crystallinity of a specified density, and **(d)** the density for a specified percent crystallinity.

Figure 4.33
Diffraction pattern for polycrystalline copper.

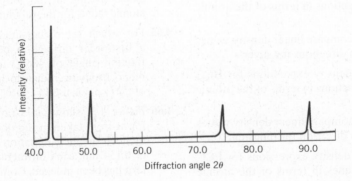

DESIGN PROBLEM

Density Computations–Ceramics

4.D1 Gallium arsenide (GaAs) and gallium phosphide (GaP) both have the zinc blende crystal structure and are soluble in one another at all concentrations. Determine the concentration in weight percent of GaP that must be added to GaAs to yield a unit cell edge length of 0.5580 nm. The densities of GaAs and GaP are 4.130 and 5.668 g/cm³, respectively.

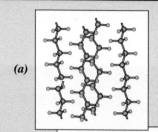

(a)

(*a*) Schematic representation of the arrangement of molecular chains for a crystalline region of polyethylene. Black and gray balls represent, respectively, carbon and hydrogen atoms.

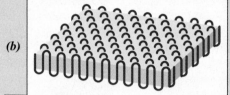

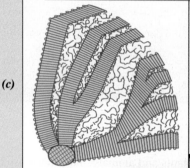

(b)

(*b*) Schematic diagram of a polymer chain-folded crystallite—a plate-shaped crystalline region in which the molecular chains (red lines/curves) fold back and forth on themselves; these folds occur at the crystallite faces.

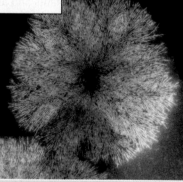

(c)

(*c*) Structure of a spherulite found in some semicrystalline polymers (schematic). Chain-folded crystallites radiate outward from a common center. Separating and connecting these crystallites are regions of amorphous material, wherein the molecular chains (red curves) assume misaligned and disordered configurations.

(*d*) Transmission electron micrograph showing the spherulite structure. Chain-folded lamellar crystallites (white lines) approximately 10 nm thick extend in radial directions from the center. 12,000×.

(*e*) A polyethylene produce bag containing some fruit.

(d)

[Photograph of Figure (*d*) supplied by P. J. Phillips. First published in R. Bartnikas and R. M. Eichhorn, *Engineering Dielectrics*, Vol. IIA, *Electrical Properties of Solid Insulating Materials: Molecular Structure and Electrical Behavior*, 1983. Copyright ASTM, 1916 Race Street, Philadelphia, PA 19103. Reprinted with permission.]

Glow Images

(e)

WHY STUDY *Polymer Structures?*

A relatively large number of chemical and structural characteristics affect the properties and behaviors of polymeric materials. Some of these influences are as follows:
1. Degree of crystallinity of semicrystalline polymers—on density, stiffness, strength, and ductility (Sections 4.13 and 15.8).

2. Degree of crosslinking—on the stiffness of rubber-like materials (Section 15.9).
3. Polymer chemistry—on melting and glass-transition temperatures (Section 15.14).

Learning Objectives

After studying this chapter, you should be able to do the following:

1. Describe a typical polymer molecule in terms of its chain structure and, in addition, how the molecule may be generated from repeat units.
2. Draw repeat units for polyethylene, poly(vinyl chloride), polytetrafluoroethylene, polypropylene, and polystyrene.
3. Calculate number-average and weight-average molecular weights and degree of polymerization for a specified polymer.
4. Name and briefly describe:
   (a) the four general types of polymer molecular structures,
   (b) the three types of stereoisomers,
   (c) the two kinds of geometrical isomers, and
   (d) the four types of copolymers.
5. Cite the differences in behavior and molecular structure for thermoplastic and thermosetting polymers.
6. Briefly describe/diagram the spherulitic structure for a semicrystalline polymer.

5.1 INTRODUCTION

Naturally occurring polymers—those derived from plants and animals—have been used for many centuries; these materials include wood, rubber, cotton, wool, leather, and silk. Other natural polymers, such as proteins, enzymes, starches, and cellulose, are important in biological and physiological processes in plants and animals. Modern scientific research tools have made possible the determination of the molecular structures of this group of materials and the development of numerous polymers that are synthesized from small organic molecules. Many of our useful plastics, rubbers, and fiber materials are synthetic polymers. In fact, since the conclusion of World War II, the field of materials has been virtually revolutionized by the advent of synthetic polymers. The synthetics can be produced inexpensively, and their properties may be managed to the degree that many are superior to their natural counterparts. In some applications, metal and wood parts have been replaced by plastics, which have satisfactory properties and can be produced at a lower cost.

As with metals and ceramics, the properties of polymers are intricately related to the structural elements of the material. This chapter explores molecular and crystal structures of polymers; Chapter 15 discusses the relationships between structure and some of the physical and chemical properties, along with typical applications and forming methods.

5.2 HYDROCARBON MOLECULES

Because most polymers are organic in origin, we briefly review some of the basic concepts relating to the structure of their molecules. First, many organic materials are *hydrocarbons*—that is, they are composed of hydrogen and carbon. Furthermore, the intramolecular bonds are covalent. Each carbon atom has four electrons that may participate in covalent bonding, whereas every hydrogen atom has only one bonding electron. A single covalent bond exists when each of the two bonding atoms contributes one electron, as represented schematically in Figure 2.12 for a molecule of hydrogen (H_2). Double and triple bonds between two carbon atoms involve the sharing of two and three pairs of electrons, respectively.[1] For example, in ethylene, which has the chemical formula C_2H_4, the two carbon atoms are doubly bonded together, and each is also singly bonded to two hydrogen atoms, as represented by the structural formula

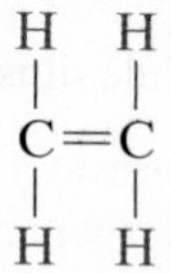

$$
\begin{array}{c c}
H & H \\
| & | \\
C & = C \\
| & | \\
H & H
\end{array}
$$

where — and = denote single and double covalent bonds, respectively. An example of a triple bond is found in acetylene, C_2H_2:

$$H-C\equiv C-H$$

unsaturated

saturated

Molecules that have double and triple covalent bonds are termed **unsaturated** — that is, each carbon atom is not bonded to the maximum (four) other atoms. Therefore, it is possible for another atom or group of atoms to become attached to the original molecule. Furthermore, for a **saturated** hydrocarbon, all bonds are single ones, and no new atoms may be joined without the removal of others that are already bonded.

Some of the simple hydrocarbons belong to the paraffin family; the chainlike paraffin molecules include methane (CH_4), ethane (C_2H_6), propane (C_3H_8), and butane (C_4H_{10}). Compositions and molecular structures for paraffin molecules are contained in Table 5.1. The covalent bonds in each molecule are strong, but only weak hydrogen and van der Waals bonds exist between molecules, and thus these hydrocarbons have relatively low melting and boiling points. However, boiling temperatures rise with increasing molecular weight (Table 5.1).

Table 5.1

Compositions and Molecular Structures for Some Paraffin Compounds: C_nH_{2n+2}

Name	Composition	Structure	Boiling Point (°C)
Methane	CH_4	$\begin{array}{c} H \\ \| \\ H-C-H \\ \| \\ H \end{array}$	−164
Ethane	C_2H_6	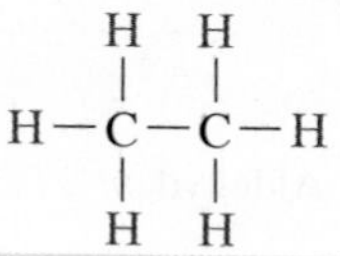 $\begin{array}{cc} H & H \\ \| & \| \\ H-C-&C-H \\ \| & \| \\ H & H \end{array}$	−88.6
Propane	C_3H_8	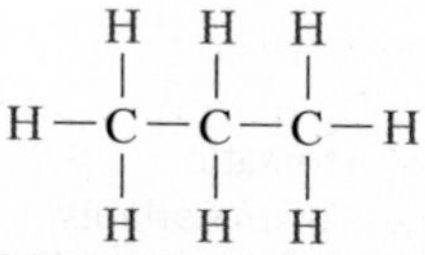 $\begin{array}{ccc} H & H & H \\ \| & \| & \| \\ H-C-&C-&C-H \\ \| & \| & \| \\ H & H & H \end{array}$	−42.1
Butane	C_4H_{10}		−0.5
Pentane	C_5H_{12}		36.1
Hexane	C_6H_{14}		69.0

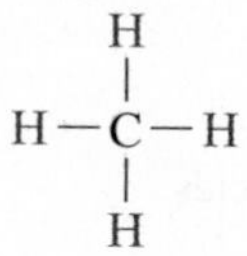

---

[1] In the hybrid bonding scheme for carbon (Section 2.6), a carbon atom forms sp^3 hybrid orbitals when all its bonds are single ones; a carbon atom with a double bond has sp^2 hybrid orbitals; and a carbon atom with a triple bond has sp hybridization.

isomerism

Hydrocarbon compounds with the same composition may have different atomic arrangements, a phenomenon termed **isomerism**. For example, there are two isomers for butane; normal butane has the structure

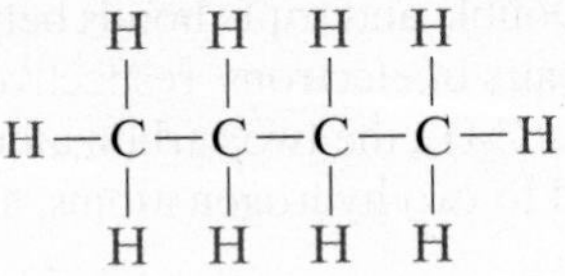

$$H-\underset{\underset{H}{|}}{\overset{\overset{H}{|}}{C}}-\underset{\underset{H}{|}}{\overset{\overset{H}{|}}{C}}-\underset{\underset{H}{|}}{\overset{\overset{H}{|}}{C}}-\underset{\underset{H}{|}}{\overset{\overset{H}{|}}{C}}-H$$

whereas a molecule of isobutane is represented as follows:

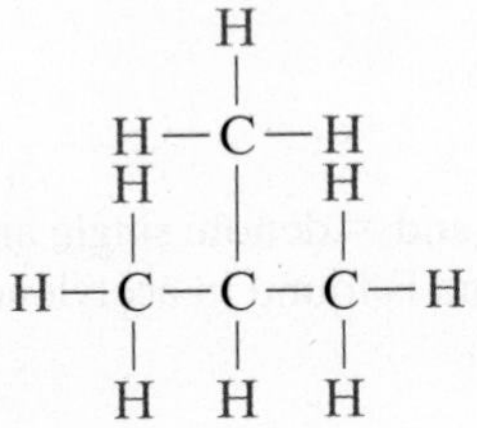

Some of the physical properties of hydrocarbons will depend on the isomeric state; for example, the boiling temperatures for normal butane and isobutane are −0.5°C and −12.3°C, respectively.

Table 5.2

Some Common Hydrocarbon Groups

Family	Characteristic Unit		Representative Compound
Alcohols	R—OH	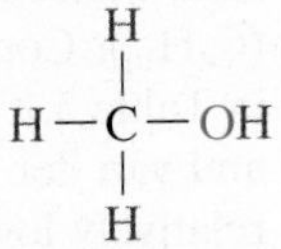	Methyl alcohol
Ethers	R—O—R′	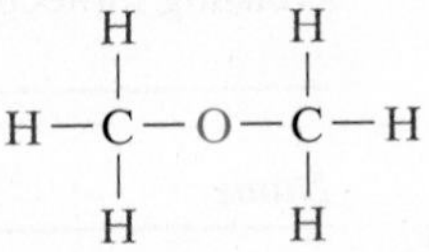	Dimethyl ether
Acids	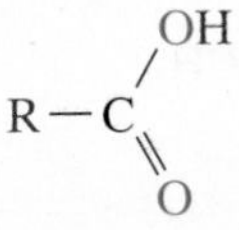 R—C(OH)=O	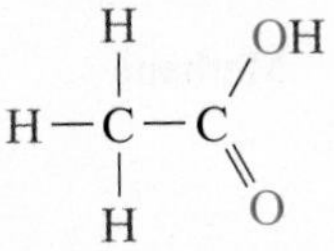	Acetic acid
Aldehydes	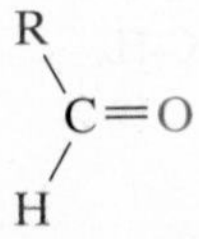 R,H C=O	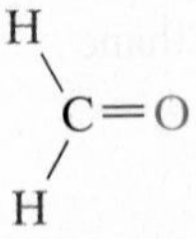	Formaldehyde
Aromatic hydrocarbons[a]	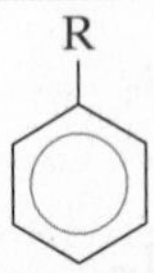	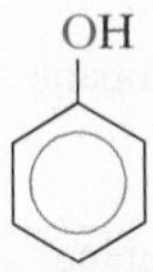	Phenol

[a]The simplified structure ⬡ denotes a phenyl group,

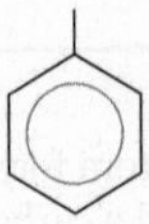

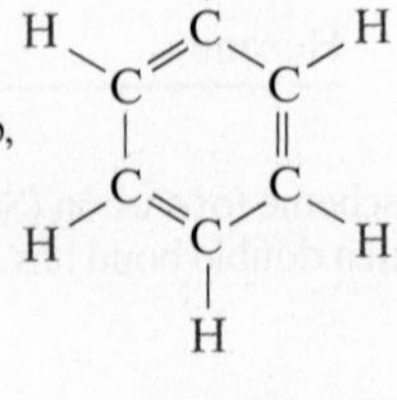

There are numerous other organic groups, many of which are involved in polymer structures. Several of the more common groups are presented in Table 5.2, where R and R′ represent organic groups such as CH_3, C_2H_5, and C_6H_5 (methyl, ethyl, and phenyl).

Concept Check 5.1 Differentiate between polymorphism (see Chapter 4) and isomerism.

[*The answer may be found at* www.wiley.com/college/callister *(Student Companion Site).*]

5.3 POLYMER MOLECULES

macromolecule

The molecules in polymers are gigantic in comparison to the hydrocarbon molecules already discussed; because of their size they are often referred to as **macromolecules**. Within each molecule, the atoms are bound together by covalent interatomic bonds. For carbon-chain polymers, the backbone of each chain is a string of carbon atoms. Many times each carbon atom singly bonds to two adjacent carbon atoms on either side, represented schematically in two dimensions as follows:

$$-\overset{|}{\underset{|}{C}}-\overset{|}{\underset{|}{C}}-\overset{|}{\underset{|}{C}}-\overset{|}{\underset{|}{C}}-\overset{|}{\underset{|}{C}}-\overset{|}{\underset{|}{C}}-\overset{|}{\underset{|}{C}}-$$

Each of the two remaining valence electrons for every carbon atom may be involved in side bonding with atoms or radicals that are positioned adjacent to the chain. Of course, both chain and side double bonds are also possible.

repeat unit
monomer

These long molecules are composed of structural entities called **repeat units**, which are successively repeated along the chain.[2] The term **monomer** refers to the small molecule from which a polymer is synthesized. Hence, *monomer* and *repeat unit* mean different things, but sometimes the term *monomer* or *monomer unit* is used instead of the more proper term *repeat unit*.

5.4 THE CHEMISTRY OF POLYMER MOLECULES

Consider again the hydrocarbon ethylene (C_2H_4), which is a gas at ambient temperature and pressure and has the following molecular structure:

$$\overset{\displaystyle H \quad\; H}{\underset{\displaystyle H \quad\; H}{\overset{|}{\underset{|}{C}}=\overset{|}{\underset{|}{C}}}}$$

If the ethylene gas is reacted under appropriate conditions, it will transform to polyethylene (PE), which is a solid polymeric material. This process begins when an active center is formed by the reaction between an initiator or catalyst species (R·) and the ethylene monomer, as follows:

$$R\cdot + \overset{\displaystyle H \quad\; H}{\underset{\displaystyle H \quad\; H}{\overset{|}{\underset{|}{C}}=\overset{|}{\underset{|}{C}}}} \longrightarrow R-\overset{\displaystyle H \quad\; H}{\underset{\displaystyle H \quad\; H}{\overset{|}{\underset{|}{C}}-\overset{|}{\underset{|}{C}}\cdot}} \qquad (5.1)$$

polymer

---

[2]A repeat unit is also sometimes called a *mer*. *Mer* originates from the Greek word *meros*, which means "part"; the term **polymer** was coined to mean "many mers."

The polymer chain then forms by the sequential addition of monomer units to this actively growing chain molecule. The active site, or unpaired electron (denoted by ·), is transferred to each successive end monomer as it is linked to the chain. This may be represented schematically as follows:

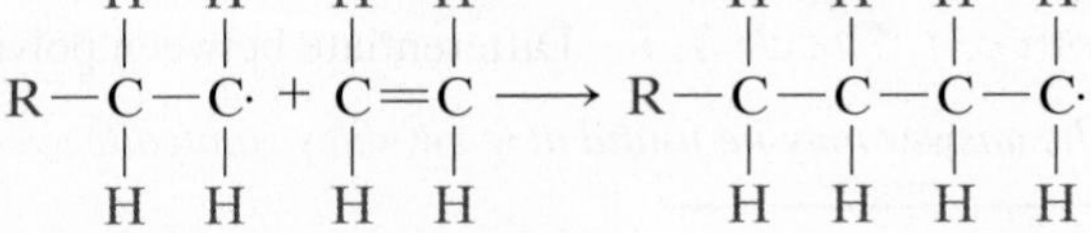

$$
\begin{array}{c}
\text{H} \quad \text{H} \quad\quad \text{H} \quad \text{H} \\
| \quad\; | \quad\quad\; | \quad\; | \\
\text{R} - \text{C} - \text{C} \cdot\; + \text{C} = \text{C} \longrightarrow \text{R} - \text{C} - \text{C} - \text{C} - \text{C} \cdot \\
| \quad\; | \quad\quad\; | \quad\; | \\
\text{H} \quad \text{H} \quad\quad \text{H} \quad \text{H}
\end{array}
\tag{5.2}
$$

WileyPLUS: VMSE
Repeat Unit Structures

The final result, after the addition of many ethylene monomer units, is the polyethylene molecule.[3] A portion of one such molecule and the polyethylene repeat unit are shown in Figure 5.1a. This polyethylene chain structure can also be represented as

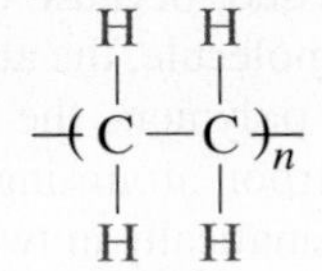

$$
-\!\!\left(\!\!\begin{array}{c} \text{H} \;\; \text{H} \\ | \;\;\; | \\ \text{C} - \text{C} \\ | \;\;\; | \\ \text{H} \;\; \text{H} \end{array}\!\!\right)_{\!n}
$$

or alternatively as

$$
-\!\!\left(\text{CH}_2 - \text{CH}_2\right)_{\!n}
$$

Here, the repeat units are enclosed in parentheses, and the subscript n indicates the number of times it repeats.[4]

The representation in Figure 5.1a is not strictly correct, in that the angle between the singly bonded carbon atoms is not 180° as shown, but rather is close to 109°. A more accurate three-dimensional model is one in which the carbon atoms form a zigzag pattern (Figure 5.1b), the C—C bond length being 0.154 nm. In this discussion, depiction of polymer molecules is frequently simplified using the linear chain model shown in Figure 5.1a.

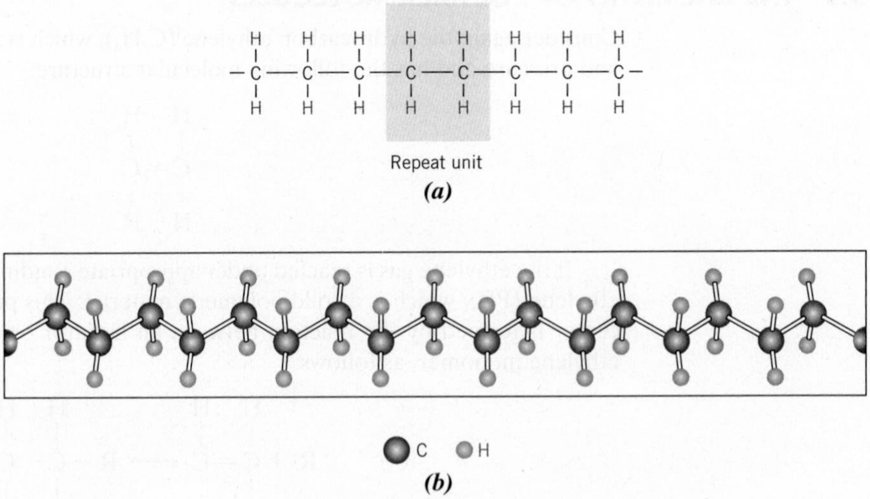

Figure 5.1 For polyethylene, (a) a schematic representation of repeat unit and chain structures, and (b) a perspective of the molecule, indicating the zigzag backbone structure.

Repeat unit

(a)

○ C ○ H

(b)

[3]A more detailed discussion of polymerization reactions, including both addition and condensation mechanisms, is given in Section 17.12.

[4]Chain ends/end groups (i.e., the Rs in Equation 5.2) are not normally represented in chain structures.

WileyPLUS: VMSE
Repeat Unit Structures

Of course, polymer structures having other chemistries are possible. For example, the tetrafluoroethylene monomer, $CF_2 = CF_2$, can polymerize to form *polytetrafluoroethylene* (PTFE) as follows:

$$n \begin{bmatrix} F & F \\ | & | \\ C = C \\ | & | \\ F & F \end{bmatrix} \longrightarrow \left(\begin{matrix} F & F \\ | & | \\ C - C \\ | & | \\ F & F \end{matrix} \right)_n \tag{5.3}$$

Polytetrafluoroethylene (having the trade name Teflon) belongs to a family of polymers called the *fluorocarbons*.

WileyPLUS: VMSE
Repeat Unit Structures

The vinyl chloride monomer ($CH_2 = CHCl$) is a slight variant of that for ethylene, in which one of the four H atoms is replaced with a Cl atom. Its polymerization is represented as

$$n \begin{bmatrix} H & H \\ | & | \\ C = C \\ | & | \\ H & Cl \end{bmatrix} \longrightarrow \left(\begin{matrix} H & H \\ | & | \\ C - C \\ | & | \\ H & Cl \end{matrix} \right)_n \tag{5.4}$$

and leads to *poly(vinyl chloride)* (PVC), another common polymer.

Some polymers may be represented using the following generalized form:

WileyPLUS: VMSE
Repeat Unit Structures

where the R depicts either an atom [i.e., H or Cl, for polyethylene or poly(vinyl chloride), respectively] or an organic group such as CH_3, C_2H_5, and C_6H_5 (methyl, ethyl, and phenyl). For example, when R represents a CH_3 group, the polymer is *polypropylene* (PP). Poly(vinyl chloride) and polypropylene chain structures are also represented in Figure 5.2. Table 5.3 lists repeat units for some of the more common polymers; as may be noted, some of them—for example, nylon, polyester, and polycarbonate—are relatively complex. Repeat units for a large number of relatively common polymers are given in Appendix D.

When all of the repeating units along a chain are of the same type, the resulting polymer is called a **homopolymer**. Chains may be composed of two or more different repeat units, in what are termed **copolymers** (see Section 5.10).

homopolymer

copolymer

bifunctional

functionality

trifunctional

The monomers discussed thus far have an active bond that may react to form two covalent bonds with other monomers forming a two-dimensional chainlike molecular structure, as indicated earlier for ethylene. Such a monomer is termed **bifunctional**. In general, the **functionality** is the number of bonds that a given monomer can form. For example, monomers such as phenol–formaldehyde (Table 5.3) are **trifunctional**; they have three active bonds, from which a three-dimensional molecular network structure results.

✓
Concept Check 5.2 On the basis of the structures presented in the previous section, sketch the repeat unit structure for poly(vinyl fluoride).

[*The answer may be found at* www.wiley.com/college/callister (*Student Companion Site*).]

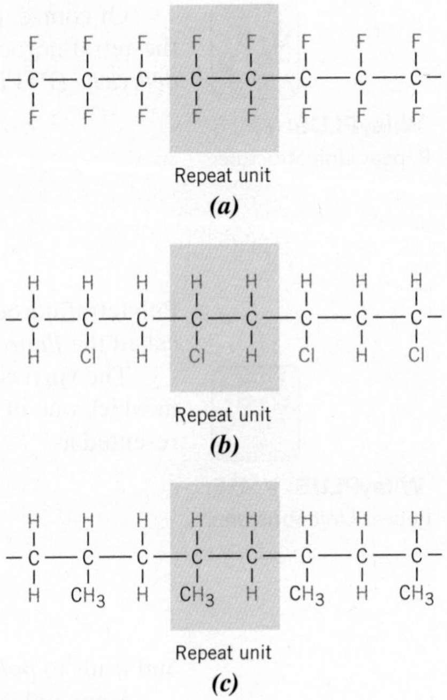

Figure 5.2 Repeat unit and chain structures for (*a*) polytetrafluoroethylene, (*b*) poly(vinyl chloride), and (*c*) polypropylene.

Repeat unit

(a)

Repeat unit

(b)

Repeat unit

(c)

Table 5.3 Repeat Units for Ten of the More Common Polymeric Materials

Polymer	Repeat Unit
Polyethylene (PE)	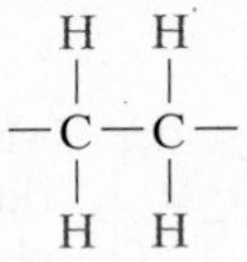
Poly(vinyl chloride) (PVC)	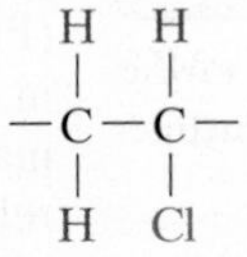
Polytetrafluoroethylene (PTFE)	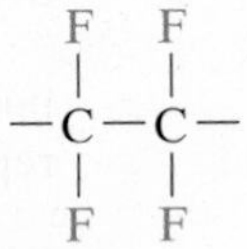
Polypropylene (PP)	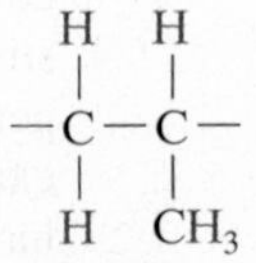
Polystyrene (PS)	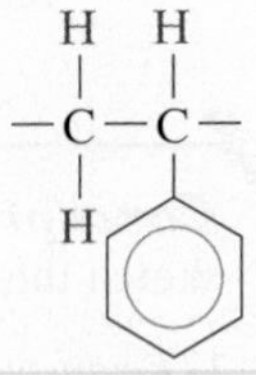

WileyPLUS: VMSE
Repeat Unit Structures

WileyPLUS: VMSE

WileyPLUS: VMSE

WileyPLUS: VMSE

WileyPLUS: VMSE

(continued)

Table 5.3 (Continued)

Polymer	Repeat Unit
Poly(methyl methacrylate) (PMMA)	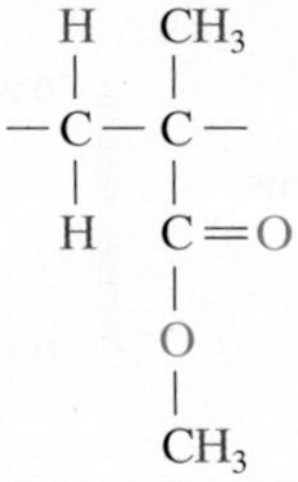 (structure)
Phenol-formaldehyde (Bakelite)	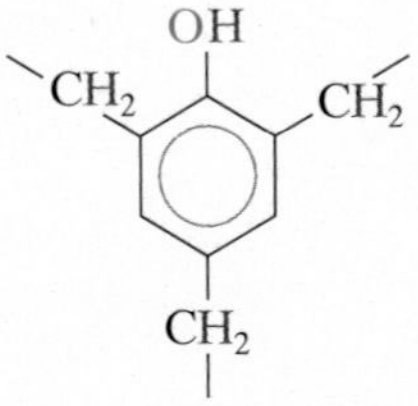 (structure)
Poly(hexamethylene adipamide) (nylon 6,6)	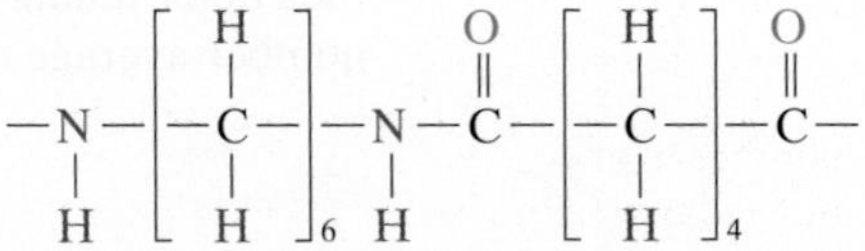 (structure)
Poly(ethylene terephthalate) (PET, a polyester)	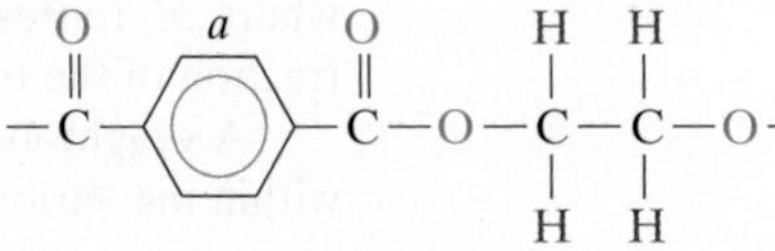 (structure)
Polycarbonate (PC)	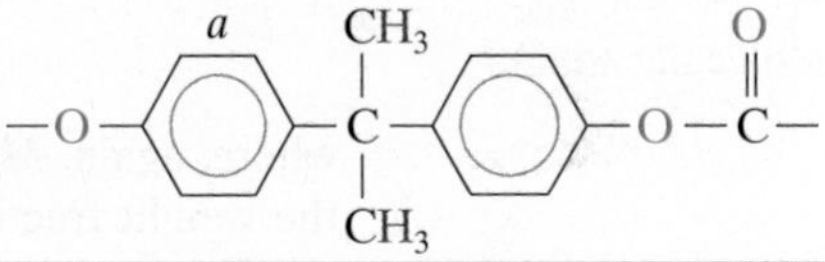 (structure)

[a] The ⬡ symbol in the backbone chain denotes an aromatic ring as (structure)

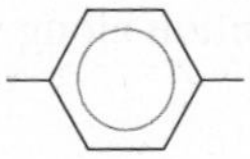

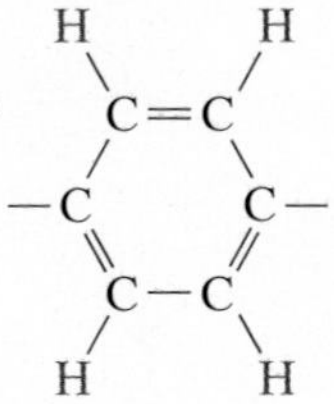

5.5 MOLECULAR WEIGHT

Extremely large molecular weights[5] are observed in polymers with very long chains. During the polymerization process, not all polymer chains will grow to the same length; this results in a distribution of chain lengths or molecular weights. Ordinarily, an average molecular weight is specified, which may be determined by the measurement of various physical properties such as viscosity and osmotic pressure.

There are several ways of defining average molecular weight. The number-average molecular weight $\overline{M}_n$ is obtained by dividing the chains into a series of size ranges and

---

[5]*Molecular mass, molar mass,* and *relative molecular mass* are sometimes used and are really more appropriate terms than *molecular weight* in the context of the present discussion—in actual fact, we are dealing with masses and not weights. However, molecular weight is most commonly found in the polymer literature and thus is used throughout this book.

Figure 5.3
Hypothetical polymer molecule size distributions on the basis of (*a*) number and (*b*) weight fractions of molecules.

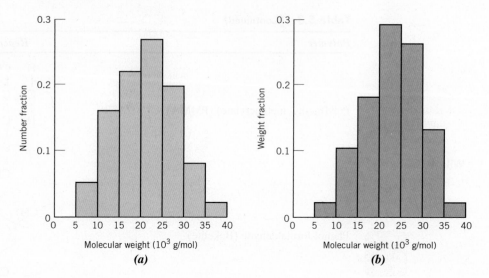

(a)   *(b)*

then determining the number fraction of chains within each size range (Figure 5.3*a*). The number-average molecular weight is expressed as

Number-average molecular weight

$$\overline{M}_n = \Sigma x_i M_i \tag{5.5a}$$

where M_i represents the mean (middle) molecular weight of size range i, and x_i is the fraction of the total number of chains within the corresponding size range.

A weight-average molecular weight $\overline{M}_w$ is based on the weight fraction of molecules within the various size ranges (Figure 5.3*b*). It is calculated according to

Weight-average molecular weight

$$\overline{M}_w = \Sigma w_i M_i \tag{5.5b}$$

where, again, M_i is the mean molecular weight within a size range, whereas w_i denotes the weight fraction of molecules within the same size interval. Computations for both number-average and weight-average molecular weights are carried out in Example Problem 5.1. A typical molecular weight distribution along with these molecular weight averages is shown in Figure 5.4.

Figure 5.4   Distribution of molecular weights for a typical polymer.

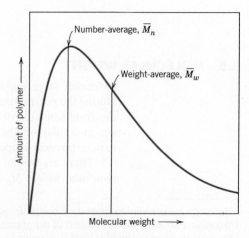

degree of polymerization

Degree of polymerization — dependence on number-average and repeat unit molecular weights

An alternate way of expressing average chain size of a polymer is as the **degree of polymerization**, DP, which represents the average number of repeat units in a chain. DP is related to the number-average molecular weight $\overline{M}_n$ by the equation

$$DP = \frac{\overline{M}_n}{m} \tag{5.6}$$

where m is the repeat unit molecular weight.

EXAMPLE PROBLEM 5.1

Computations of Average Molecular Weights and Degree of Polymerization

Assume that the molecular weight distributions shown in Figure 5.3 are for poly(vinyl chloride). For this material, compute **(a)** the number-average molecular weight, **(b)** the degree of polymerization, and **(c)** the weight-average molecular weight.

Solution

(a) The data necessary for this computation, as taken from Figure 5.3a, are presented in Table 5.4a. According to Equation 5.5a, summation of all the $x_i M_i$ products (from the right-hand column) yields the number-average molecular weight, which in this case is 21,150 g/mol.

(b) To determine the degree of polymerization (Equation 5.6), it is first necessary to compute the repeat unit molecular weight. For PVC, each repeat unit consists of two carbon atoms, three hydrogen atoms, and a single chlorine atom (Table 5.3). Furthermore, the atomic weights of C, H, and Cl are, respectively, 12.01, 1.01, and 35.45 g/mol. Thus, for PVC,

$$m = 2(12.01 \text{ g/mol}) + 3(1.01 \text{ g/mol}) + 35.45 \text{ g/mol}$$
$$= 62.50 \text{ g/mol}$$

and

$$DP = \frac{\overline{M}_n}{m} = \frac{21,150 \text{ g/mol}}{62.50 \text{ g/mol}} = 338$$

Table 5.4a Data Used for Number-Average Molecular Weight Computations in Example Problem 5.1

Molecular Weight Range (g/mol)	Mean M_i (g/mol)	x_i	$x_i M_i$
5,000–10,000	7,500	0.05	375
10,000–15,000	12,500	0.16	2000
15,000–20,000	17,500	0.22	3850
20,000–25,000	22,500	0.27	6075
25,000–30,000	27,500	0.20	5500
30,000–35,000	32,500	0.08	2600
35,000–40,000	37,500	0.02	750
			$\overline{M}_n = 21,150$

(c) Table 5.4b shows the data for the weight-average molecular weight, as taken from Figure 5.3b. The $w_i M_i$ products for the size intervals are tabulated in the right-hand column. The sum of these products (Equation 5.5b) yields a value of 23,200 g/mol for $\overline{M}_w$.

Table 5.4b Data Used for Weight-Average Molecular Weight Computations in Example Problem 5.1

Molecular Weight Range (g/mol)	Mean M_i (g/mol)	w_i	$w_i M_i$
5,000–10,000	7,500	0.02	150
10,000–15,000	12,500	0.10	1250
15,000–20,000	17,500	0.18	3150
20,000–25,000	22,500	0.29	6525
25,000–30,000	27,500	0.26	7150
30,000–35,000	32,500	0.13	4225
35,000–40,000	37,500	0.02	750
			$\overline{M}_w = 23{,}200$

Many polymer properties are affected by the length of the polymer chains. For example, the melting or softening temperature increases with increasing molecular weight (for $\overline{M}$ up to about 100,000 g/mol). At room temperature, polymers with very short chains (having molecular weights on the order of 100 g/mol) will generally exist as liquids. Those with molecular weights of approximately 1000 g/mol are waxy solids (such as paraffin wax) and soft resins. Solid polymers (sometimes termed *high polymers*), which are of prime interest here, commonly have molecular weights ranging between 10,000 and several million g/mol. Thus, the same polymer material can have quite different properties if it is produced with a different molecular weight. Other properties that depend on molecular weight include elastic modulus and strength (see Chapter 15).

5.6 MOLECULAR SHAPE

Previously, polymer molecules have been shown as linear chains, neglecting the zigzag arrangement of the backbone atoms (Figure 5.1b). Single chain bonds are capable of rotating and bending in three dimensions. Consider the chain atoms in Figure 5.5a; a third carbon atom may lie at any point on the cone of revolution and still subtend about a 109° angle with the bond between the other two atoms. A straight chain segment results when successive chain atoms are positioned as in Figure 5.5b. However, chain bending and twisting are possible when there is a rotation of the chain atoms into other positions, as illustrated in Figure 5.5c.[6] Thus, a single chain molecule composed of many chain atoms might assume a shape similar to that represented schematically in Figure 5.6, having a multitude of bends, twists, and kinks.[7] Also indicated in this figure

---

[6] For some polymers, rotation of carbon backbone atoms within the cone may be hindered by bulky side group elements on neighboring chain atoms.

[7] The term *conformation* is often used in reference to the physical outline of a molecule, or molecular shape, that can be altered only by rotation of chain atoms about single bonds.

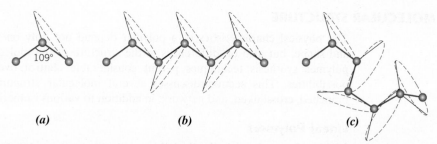

(a)   (b)   (c)

Figure 5.5 Schematic representations of how polymer chain shape is influenced by the positioning of backbone carbon atoms (gray circles). For (a), the rightmost atom may lie anywhere on the dashed circle and still subtend a 109° angle with the bond between the other two atoms. Straight and twisted chain segments are generated when the backbone atoms are situated as in (b) and (c), respectively.

is the end-to-end distance of the polymer chain r; this distance is much smaller than the total chain length.

Polymers consist of large numbers of molecular chains, each of which may bend, coil, and kink in the manner of Figure 5.6. This leads to extensive intertwining and entanglement of neighboring chain molecules, a situation similar to what is seen in a heavily tangled fishing line. These random coils and molecular entanglements are responsible for a number of important characteristics of polymers, to include the large elastic extensions displayed by the rubber materials.

Some of the mechanical and thermal characteristics of polymers are a function of the ability of chain segments to experience rotation in response to applied stresses or thermal vibrations. Rotational flexibility is dependent on repeat unit structure and chemistry. For example, the region of a chain segment that has a double bond (C=C) is rotationally rigid. Also, introduction of a bulky or large side group of atoms restricts rotational movement. For example, polystyrene molecules, which have a phenyl side group (Table 5.3), are more resistant to rotational motion than are polyethylene chains.

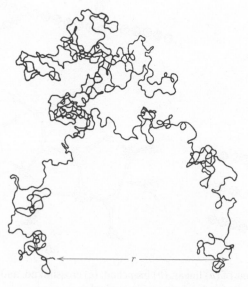

Figure 5.6 Schematic representation of a single polymer chain molecule that has numerous random kinks and coils produced by chain bond rotations.
(From *Physics of Rubber Elasticity*, 2nd edition, by Treloar (1958), Fig. 3.3, p. 47. By permission of Oxford University Press.)

5.7 MOLECULAR STRUCTURE

The physical characteristics of a polymer depend not only on its molecular weight and shape, but also on differences in the structure of the molecular chains. Modern polymer synthesis techniques permit considerable control over various structural possibilities. This section discusses several molecular structures including linear, branched, crosslinked, and network, in addition to various isomeric configurations.

Linear Polymers

linear polymer

Linear polymers are those in which the repeat units are joined together end to end in single chains. These long chains are flexible and may be thought of as a mass of "spaghetti," as represented schematically in Figure 5.7a, where each circle represents a repeat unit. For linear polymers, there may be extensive van der Waals and hydrogen bonding between the chains. Some of the common polymers that form with linear structures are polyethylene, poly(vinyl chloride), polystyrene, poly(methyl methacrylate), nylon, and the fluorocarbons.

Branched Polymers

branched polymer

Polymers may be synthesized in which side-branch chains are connected to the main ones, as indicated schematically in Figure 5.7b; these are fittingly called **branched polymers**. The branches, considered to be part of the main-chain molecule, may result from side reactions that occur during the synthesis of the polymer. The chain packing efficiency is reduced with the formation of side branches, which results in a lowering of the polymer density. Polymers that form linear structures may also be branched. For example, high-density polyethylene (HDPE) is primarily a linear polymer, whereas low-density polyethylene (LDPE) contains short-chain branches.

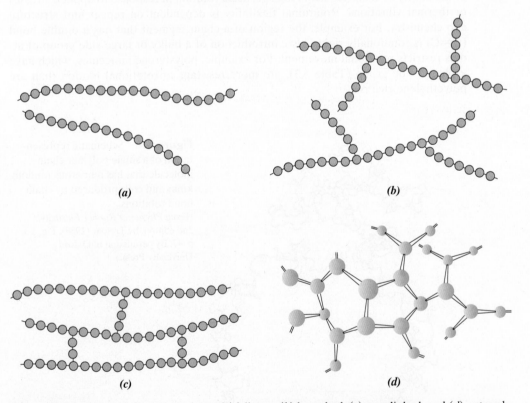

(a)   *(b)*

(c)   *(d)*

Figure 5.7 Schematic representations of (*a*) linear, (*b*) branched, (*c*) crosslinked, and (*d*) network (three-dimensional) molecular structures. Circles designate individual repeat units.

Crosslinked Polymers

crosslinked polymer

In **crosslinked polymers**, adjacent linear chains are joined one to another at various positions by covalent bonds, as represented in Figure 5.7c. The process of crosslinking is achieved either during synthesis or by a nonreversible chemical reaction. Often, this crosslinking is accomplished by additive atoms or molecules that are covalently bonded to the chains. Many of the rubber elastic materials are crosslinked; in rubbers, this is called vulcanization, a process described in Section 15.9.

Network Polymers

network polymer

Multifunctional monomers forming three or more active covalent bonds make three-dimensional networks (Figure 5.7d) and are termed **network polymers**. Actually, a polymer that is highly crosslinked may also be classified as a network polymer. These materials have distinctive mechanical and thermal properties; the epoxies, polyurethanes, and phenol-formaldehyde belong to this group.

Polymers are not usually of only one distinctive structural type. For example, a predominantly linear polymer may have limited branching and crosslinking.

5.8 MOLECULAR CONFIGURATIONS

For polymers having more than one side atom or group of atoms bonded to the main chain, the regularity and symmetry of the side group arrangement can significantly influence the properties. Consider the repeat unit

$$
\begin{array}{cc}
\text{H} & \text{H} \\
| & | \\
-\text{C}-\text{C}- \\
| & | \\
\text{H} & \text{R}
\end{array}
$$

in which R represents an atom or side group other than hydrogen (e.g., Cl, CH_3). One arrangement is possible when the R side groups of successive repeat units are bound to alternate carbon atoms as follows:

$$
\begin{array}{cccc}
\text{H} & \text{H} & \text{H} & \text{H} \\
| & | & | & | \\
-\text{C}-\text{C}-\text{C}-\text{C}- \\
| & | & | & | \\
\text{H} & \text{R} & \text{H} & \text{R}
\end{array}
$$

This is designated as a head-to-tail configuration.[8] Its complement, the head-to-head configuration, occurs when R groups are bound to adjacent chain atoms:

$$
\begin{array}{cccc}
\text{H} & \text{H} & \text{H} & \text{H} \\
| & | & | & | \\
-\text{C}-\text{C}-\text{C}-\text{C}- \\
| & | & | & | \\
\text{H} & \text{R} & \text{R} & \text{H}
\end{array}
$$

In most polymers, the head-to-tail configuration predominates; often a polar repulsion occurs between R groups for the head-to-head configuration.

Isomerism (Section 5.2) is also found in polymer molecules, wherein different atomic configurations are possible for the same composition. Two isomeric subclasses—stereoisomerism and geometrical isomerism—are topics of discussion in the succeeding sections.

---

[8]The term *configuration* is used in reference to arrangements of units along the axis of the chain, or atom positions that are not alterable except by the breaking and then re-forming of primary bonds.

Stereoisomerism

stereoisomerism

Stereoisomerism denotes the situation in which atoms are linked together in the same order (head-to-tail) but differ in their spatial arrangement. For one stereoisomer, all of the R groups are situated on the same side of the chain as follows:

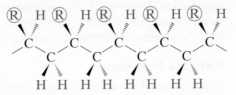

**isotactic
configuration**

This is called an **isotactic configuration**. This diagram shows the zigzag pattern of the carbon chain atoms. Furthermore, representation of the structural geometry in three dimensions is important, as indicated by the wedge-shaped bonds; solid wedges represent bonds that project out of the plane of the page, and dashed ones represent bonds that project into the page.[9]

**syndiotactic
configuration**

In a **syndiotactic configuration**, the R groups alternate sides of the chain:[10]

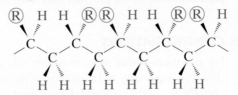

and for random positioning.

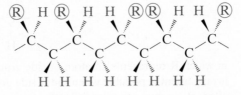

atactic configuration

the term **atactic configuration** is used.[11]

Conversion from one stereoisomer to another (e.g., isotactic to syndiotactic) is not possible by a simple rotation about single-chain bonds. These bonds must first be severed; then, after the appropriate rotation, they are re-formed.

---

[9]The isotactic configuration is sometimes represented using the following linear (i.e., nonzigzag) and two-dimensional schematic:

$$
\begin{array}{ccccccccc}
H & H & H & H & H & H & H & H & H \\
| & | & | & | & | & | & | & | & | \\
-C-&C-&C-&C-&C-&C-&C-&C-&C- \\
| & | & | & | & | & | & | & | & | \\
R & H & R & H & R & H & R & H & R
\end{array}
$$

[10]The linear and two-dimensional schematic for the syndiotactic configuration is represented as

$$
\begin{array}{ccccccccc}
H & H & R & H & H & H & R & H & H \\
| & | & | & | & | & | & | & | & | \\
-C-&C-&C-&C-&C-&C-&C-&C-&C- \\
| & | & | & | & | & | & | & | & | \\
R & H & H & H & R & H & H & H & R
\end{array}
$$

[11]For the atactic configuration, the linear and two-dimensional schematic is

$$
\begin{array}{ccccccccc}
H & H & H & H & R & H & H & H & R \\
| & | & | & | & | & | & | & | & | \\
-C-&C-&C-&C-&C-&C-&C-&C-&C- \\
| & | & | & | & | & | & | & | & | \\
R & H & R & H & H & H & R & H & H
\end{array}
$$

In reality, a specific polymer does not exhibit just one of these configurations; the predominant form depends on the method of synthesis.

Geometrical Isomerism

Other important chain configurations, or geometrical isomers, are possible within repeat units having a double bond between chain carbon atoms. Bonded to each of the carbon atoms participating in the double bond is a side group, which may be situated on one side of the chain or its opposite. Consider the isoprene repeat unit having the structure

WileyPLUS: VMSE
Stereo and Geometrical Isomers

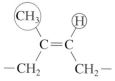

in which the CH_3 group and the H atom are positioned on the same side of the double bond. This is termed a **cis** structure, and the resulting polymer, *cis*-polyisoprene, is natural rubber. For the alternative isomer

cis (structure)

WileyPLUS: VMSE
Stereo and Geometrical Isomers

trans (structure)

the **trans** structure, the CH_3 and H reside on opposite sides of the double bond.[12] *Trans*-polyisoprene, sometimes called gutta percha, has properties that are distinctly different from those of natural rubber as a result of this configurational alteration. Conversion of *trans* to *cis*, or vice versa, is not possible by a simple chain bond rotation because the chain double bond is extremely rigid.

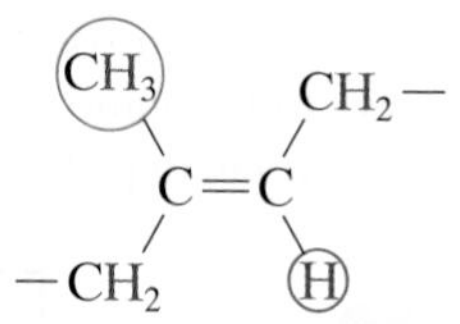

To summarize the preceding sections: polymer molecules may be characterized in terms of their size, shape, and structure. Molecular size is specified in terms of molecular weight (or degree of polymerization). Molecular shape relates to the degree of chain twisting, coiling, and bending. Molecular structure depends on the manner in which structural units are joined together. Linear, branched, crosslinked, and network structures are all possible, in addition to several isomeric configurations (isotactic, syndiotactic, atactic, cis, and trans). These molecular characteristics are presented in the taxonomic chart shown in Figure 5.8. Note that some of the structural elements are not mutually exclusive, and it may be necessary to specify molecular structure in terms of more than one. For example, a linear polymer may also be isotactic.

[12]For *cis*-polyisoprene the linear chain representation is as follows:

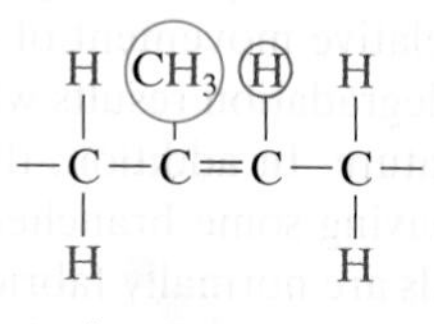

whereas the linear schematic for the trans structure is

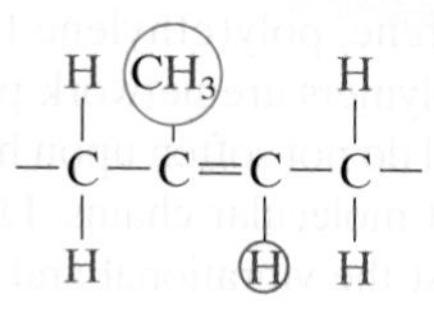

random copolymer

alternating copolymer

block copolymer

graft copolymer

of these repeat unit types, different sequencing arrangements along the polymer chains are possible. For one, as depicted in Figure 5.9a, the two different units are randomly dispersed along the chain in what is termed a **random copolymer**. For an **alternating copolymer**, as the name suggests, the two repeat units alternate chain positions, as illustrated in Figure 5.9b. A **block copolymer** is one in which identical repeat units are clustered in blocks along the chain (Figure 5.9c). Finally, homopolymer side branches of one type may be grafted to homopolymer main chains that are composed of a different repeat unit; such a material is termed a **graft copolymer** (Figure 5.9d).

When calculating the degree of polymerization for a copolymer, the value m in Equation 5.6 is replaced with the average value $\overline{m}$ that is determined from

Average repeat unit
molecular weight for
a copolymer

$$\overline{m} = \Sigma f_j m_j \tag{5.7}$$

In this expression, f_j and m_j are, respectively, the mole fraction and molecular weight of repeat unit j in the polymer chain.

Synthetic rubbers, discussed in Section 15.16, are often copolymers; chemical repeat units that are employed in some of these rubbers are shown in Table 5.5. Styrene–butadiene rubber (SBR) is a common random copolymer from which automobile tires are made. Nitrile rubber (NBR) is another random copolymer composed of acrylonitrile and butadiene. It is also highly elastic and, in addition, resistant to swelling in organic solvents; gasoline hoses are made of NBR. Impact-modified polystyrene is a block copolymer that consists of alternating blocks of styrene and butadiene. The rubbery isoprene blocks act to slow cracks propagating through the material.

Table 5.5 Chemical Repeat Units That Are Employed in Copolymer Rubbers

Repeat Unit Name	Repeat Unit Structure	Repeat Unit Name	Repeat Unit Structure
Acrylonitrile (VMSE Repeat Units for Rubbers)	$\begin{array}{cc} H & H \\ | & | \\ -C-C- \\ | & | \\ H & C\equiv N \end{array}$	Isoprene (VMSE)	$\begin{array}{cccc} H & CH_3 & H & H \\ | & | & | & | \\ -C-C=C-C- \\ | & & & | \\ H & & & H \end{array}$
Styrene (VMSE)	$\begin{array}{cc} H & H \\ | & | \\ -C-C- \\ | \\ H \\ \bigcirc \end{array}$	Isobutylene (VMSE)	$\begin{array}{cc} H & CH_3 \\ | & | \\ -C-C- \\ | & | \\ H & CH_3 \end{array}$
Butadiene (VMSE)	$\begin{array}{cccc} H & H & H & H \\ | & | & | & | \\ -C-C=C-C- \\ | & & & | \\ H & & & H \end{array}$	Dimethylsiloxane (VMSE)	$\begin{array}{c} CH_3 \\ | \\ -Si-O- \\ | \\ CH_3 \end{array}$
Chloroprene (VMSE)	$\begin{array}{cccc} H & Cl & H & H \\ | & | & | & | \\ -C-C=C-C- \\ | & & & | \\ H & & & H \end{array}$		

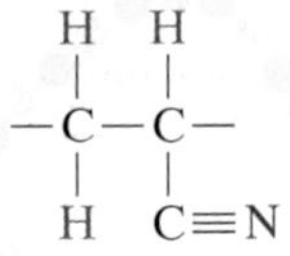

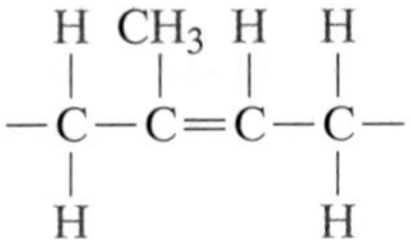

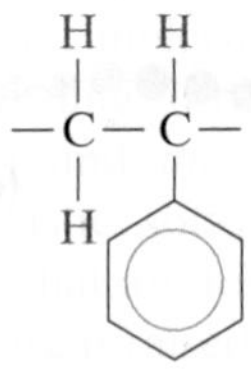

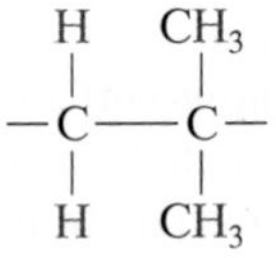

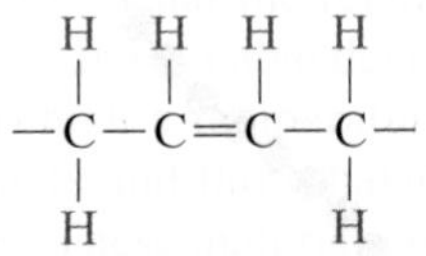

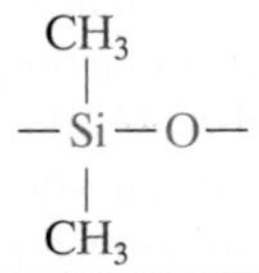

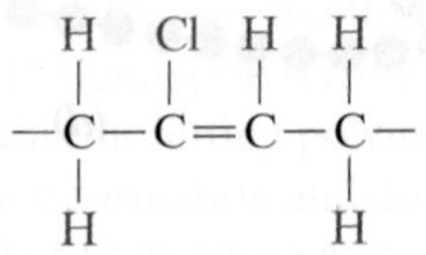

Figure 5.10 Electron micrograph of a polyethylene single crystal. 20,000×. (From A. Keller, R. H. Doremus, B. W. Roberts, and D. Turnbull (Editors), *Growth and Perfection of Crystals.* General Electric Company and John Wiley & Sons, Inc., 1958, p. 498. Reprinted with permission of John Wiley & Sons, Inc.)

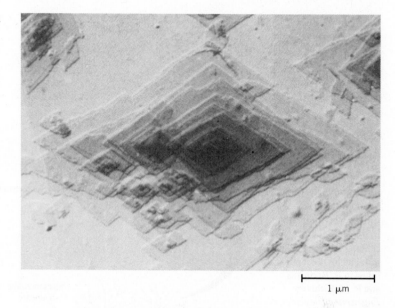

1 μm

5.11 POLYMER CRYSTALS

crystallite

It has been proposed that a semicrystalline polymer consists of small crystalline regions (**crystallites**), each having a precise alignment, which are interspersed with amorphous regions composed of randomly oriented molecules. The structure of the crystalline regions may be deduced by examination of polymer single crystals, which may be grown from dilute solutions. These crystals are regularly shaped, thin platelets (or *lamellae*) approximately 10 to 20 nm thick, and on the order of 10 μm long. Frequently, these platelets form a multilay-ered structure, like that shown in the electron micrograph of a single crystal of polyethylene in Figure 5.10. The molecular chains within each platelet fold back and forth on themselves, with folds occurring at the faces; this structure, aptly termed the **chain-folded model**, is illustrated schematically in Figure 5.11. Each platelet consists of a number of molecules; however, the average chain length is much greater than the thickness of the platelet.

chain-folded model

spherulite

Many bulk polymers that are crystallized from a melt are semicrystalline and form a **spherulite** structure. As implied by the name, each spherulite may grow to be roughly spherical in shape; one of them, as found in natural rubber, is shown in the transmission electron micrograph in the margin photograph on this page [and the chapter-opening photograph (*d*) for this chapter]. The spherulite consists of an aggregate of ribbon-like chain-folded crystallites (lamellae) approximately 10 nm thick that radiate outward from a single nucleation site in the center. In this electron micrograph, these lamellae appear as thin white lines. The detailed structure of a spherulite is illustrated schematically in

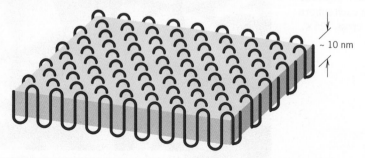

~ 10 nm

Figure 5.11 The chain-folded structure for a plate-shaped polymer crystallite.

Transmission electron micrograph showing the spherulite structure in a natural rubber specimen.
(Photograph supplied by P. J. Phillips. First published in R. Bartnikas and R. M. Eichhorn, *Engineering Dielectrics,* Vol. IIA, *Electrical Properties of Solid Insulating Materials: Molecular Structure and Electrical Behavior,* 1983. Copyright ASTM, 1916 Race Street, Philadelphia, PA. Reprinted with permission.)

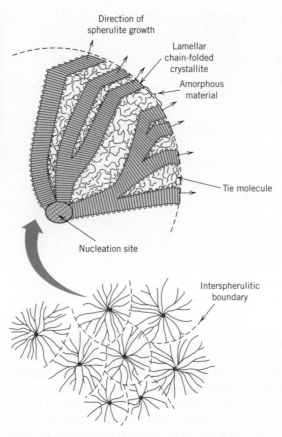

Direction of spherulite growth

Lamellar chain-folded crystallite

Amorphous material

Tie molecule

Nucleation site

Interspherulitic boundary

Figure 5.12 Schematic representation of the detailed structure of a spherulite.

Figure 5.12. Shown here are the individual chain-folded lamellar crystals that are separated by amorphous material. Tie-chain molecules that act as connecting links between adjacent lamellae pass through these amorphous regions.

As the crystallization of a spherulitic structure nears completion, the extremities of adjacent spherulites begin to impinge on one another, forming more or less planar boundaries; prior to this time, they maintain their spherical shape. These boundaries are evident in Figure 5.13, which is a photomicrograph of polyethylene using cross-polarized

Figure 5.13 A transmission photomicrograph (using cross-polarized light) showing the spherulite structure of polyethylene. Linear boundaries form between adjacent spherulites, and within each spherulite appears a Maltese cross. 525×.

100 μm

Courtesy F. P. Price, General Electric Company

light. A characteristic Maltese cross pattern appears within each spherulite. The bands or rings in the spherulite image result from twisting of the lamellar crystals as they extend like ribbons from the center.

Spherulites are considered to be the polymer analogue of grains in polycrystalline metals and ceramics. However, as discussed earlier, each spherulite is really composed of many different lamellar crystals and, in addition, some amorphous material. Polyethylene, polypropylene, poly(vinyl chloride), polytetrafluoroethylene, and nylon form a spherulitic structure when they crystallize from a melt.

SUMMARY

Polymer Molecules
- Most polymeric materials are composed of very large molecular chains with side groups of various atoms (O, Cl, etc.) or organic groups such as methyl, ethyl, or phenyl groups.
- These macromolecules are composed of repeat units, smaller structural entities, which are repeated along the chain.

The Chemistry of Polymer Molecules
- Repeat units for some of the chemically simple polymers [polyethylene, polytetrafluoroethylene, poly(vinyl chloride), polypropylene, etc.] are presented in Table 5.3.
- A *homopolymer* is one for which all of the repeat units are the same type. The chains for copolymers are composed of two or more kinds of repeat units.
- Repeat units are classified according to the number of active bonds (i.e., functionality):
    For bifunctional monomers, a two-dimensional chainlike structure results from a monomer that has two active bonds.
    Trifunctional monomers have three active bonds, from which three-dimensional network structures form.

Molecular Weight
- Molecular weights for high polymers may be in excess of a million. Because all molecules are not of the same size, there is a distribution of molecular weights.
- Molecular weight is often expressed in terms of number and weight averages; values for these parameters may be determined using Equations 5.5a and 5.5b, respectively.
- Chain length may also be specified by degree of polymerization—the number of repeat units per average molecule (Equation 5.6).

Molecular Shape
- Molecular entanglements occur when the chains assume twisted, coiled, and kinked shapes or contours as a consequence of chain bond rotations.
- Rotational flexibility is diminished when double chain bonds are present and also when bulky side groups are part of the repeat unit.

Molecular Structure
- Four different polymer molecular chain structures are possible: linear (Figure 5.7a), branched (Figure 5.7b), crosslinked (Figure 5.7c), and network (Figure 5.7d).

Molecular Configurations
- For repeat units that have more than one side atom or groups of atoms bonded to the main chain:
    Head-to-head and head-to-tail configurations are possible.
    Differences in spatial arrangements of these side atoms or groups of atoms lead to isotactic, syndiotactic, and atactic stereoisomers.

- When a repeat unit contains a double chain bond, both cis and trans geometrical isomers are possible.

Thermoplastic and Thermosetting Polymers

- With regard to behavior at elevated temperatures, polymers are classified as either thermoplastic or thermosetting.

  Thermoplastic polymers have linear and branched structures; they soften when heated and harden when cooled.

  In contrast, *thermosetting* polymers, once they have hardened, will not soften upon heating; their structures are crosslinked and network.

Copolymers

- The copolymers include random (Figure 5.9*a*), alternating (Figure 5.9*b*), block (Figure 5.9*c*), and graft (Figure 5.9*d*) types.
- Repeat units that are employed in copolymer rubber materials are presented in Table 5.5.

Polymer Crystals

- Crystalline regions (or crystallites) are plate-shape and have a chain-folded structure (Figure 5.11)—chains within the platelet are aligned and fold back and forth on themselves, with folds occurring at the faces.
- Many semicrystalline polymers form spherulites; each spherulite consists of a collection of ribbon-like chain-folded lamellar crystallites that radiate outward from its center.

Important Terms and Concepts

alternating copolymer	graft copolymer	random copolymer
atactic configuration	homopolymer	repeat unit
bifunctional	isomerism	saturated
block copolymer	isotactic configuration	spherulite
branched polymer	linear polymer	stereoisomerism
chain-folded model	macromolecule	syndiotactic configuration
cis (structure)	molecular chemistry	thermoplastic polymer
copolymer	molecular structure	thermosetting polymer
crosslinked polymer	molecular weight	trans (structure)
crystallite	monomer	trifunctional
degree of polymerization	network polymer	unsaturated
functionality	polymer	

REFERENCES

Brazel, C. S., and S. L. Rosen, *Fundamental Principles of Polymeric Materials*, 3rd edition, Wiley, Hoboken, NJ, 2012.

Carraher, C. E., Jr., *Seymour/Carraher's Polymer Chemistry*, 8th edition, CRC Press, Boca Raton, FL, 2010.

Cowie, J. M. G., and V. Arrighi, *Polymers: Chemistry and Physics of Modern Materials*, 3rd edition, CRC Press, Boca Raton, FL, 2007.

Engineered Materials Handbook, Vol. 2, *Engineering Plastics*, ASM International, Materials Park, OH, 1988.

McCrum, N. G., C. P. Buckley, and C. B. Bucknall, *Principles of Polymer Engineering*, 2nd edition, Oxford University Press, Oxford, 1997. Chapters 0–6.

Painter, P. C., and M. M. Coleman, *Fundamentals of Polymer Science: An Introductory Text*, 2nd edition, CRC Press, Boca Raton, FL, 1997.

Rodriguez, F., C. Cohen, C. K. Ober, and L. Archer, *Principles of Polymer Systems*, 5th edition, Taylor & Francis, New York, 2003.

Sperling, L. H., *Introduction to Physical Polymer Science*, 4th edition, Wiley, Hoboken, NJ, 2006.

Young, R. J., and P. Lovell, *Introduction to Polymers*, 3rd edition, CRC Press, Boca Raton, FL, 2011.

QUESTIONS AND PROBLEMS

Hydrocarbon Molecules
Polymer Molecules
The Chemistry of Polymer Molecules

5.1 On the basis of the structures presented in this chapter, sketch repeat unit structures for the following polymers: **(a)** polychlorotrifluoroethylene, and **(b)** poly(vinyl alcohol).

Molecular Weight

5.2 Compute repeat unit molecular weights for the following: **(a)** poly(vinyl chloride), **(b)** poly(ethylene terephthalate), **(c)** polycarbonate, and **(d)** polydimethylsiloxane.

5.3 The number-average molecular weight of a polypropylene is 1,000,000 g/mol. Compute the degree of polymerization.

5.4 (a) Compute the repeat unit molecular weight of polystyrene.

(b) Compute the number-average molecular weight for a polystyrene for which the degree of polymerization is 26,000.

5.5 The following table lists molecular weight data for a polypropylene material. Compute **(a)** the number-average molecular weight, **(b)** the weight-average molecular weight, and **(c)** the degree of polymerization.

Molecular Weight Range (g/mol)	x_i	w_i
8,000–16,000	0.05	0.02
16,000–24,000	0.16	0.10
24,000–32,000	0.24	0.20
32,000–40,000	0.28	0.30
40,000–48,000	0.20	0.27
48,000–56,000	0.07	0.11

5.6 Molecular weight data for some polymer are tabulated here. Compute **(a)** the number-average molecular weight and **(b)** the weight-average molecular weight. **(c)** If it is known that this material's degree of polymerization is 750, which one of the polymers listed in Table 5.3 is this polymer? Why?

Molecular Weight Range (g/mol)	x_i	w_i
15,000–30,000	0.04	0.01
30,000–45,000	0.07	0.04
45,000–60,000	0.16	0.11
60,000–75,000	0.26	0.24
75,000–90,000	0.24	0.27
90,000–105,000	0.12	0.16
105,000–120,000	0.08	0.12
120,000–135,000	0.03	0.05

5.7 Is it possible to have a poly(methyl methacrylate) homopolymer with the following molecular weight data and a degree of polymerization of 530? Why or why not?

Molecular Weight Range (g/mol)	w_i	x_i
8,000–20,000	0.02	0.05
20,000–32,000	0.08	0.15
32,000–44,000	0.17	0.21
44,000–56,000	0.29	0.28
56,000–68,000	0.23	0.18
68,000–80,000	0.16	0.10
80,000–92,000	0.05	0.03

5.8 High-density polyethylene may be chlorinated by inducing the random substitution of chlorine atoms for hydrogen.

(a) Determine the concentration of Cl (in wt%) that must be added if this substitution occurs for 8% of all the original hydrogen atoms.

(b) In what ways does this chlorinated polyethylene differ from poly(vinyl chloride)?

Molecular Shape

5.9 For a linear freely rotating polymer molecule, the total extended chain length L depends on the bond length between chain atoms d, the total number of bonds in the molecule N, and the angle between adjacent backbone chain atoms θ, as follows:

$$L = Nd \sin\left(\frac{\theta}{2}\right) \qquad (5.8)$$

Furthermore, the average end-to-end distance r for a randomly winding polymer molecule in Figure 5.6 is equal to

$$r = d\sqrt{N} \qquad (5.9)$$

A linear polytetrafluoroethylene has a number-average molecular weight of 500,000 g/mol; compute average values of L and r for this material.

5.10 Using the definitions for total chain mole-cule length L (Equation 5.8) and average chain end-to-end distance r (Equation 5.9), for a linear polyethylene determine the following:

(a) the number-average molecular weight for $L = 2600$ nm

(b) the number-average molecular weight for $r = 30$ nm

Molecular Configurations

5.11 Sketch portions of a linear polystyrene molecule that are **(a)** syndiotactic, **(b)** atactic, and **(c)** isotactic. Use two-dimensional schematics per footnotes 9, 10, and 11 of this chapter.

5.12 Sketch cis and trans structures for **(a)** butadiene, and **(b)** chloroprene. Use two-dimensional schematics per footnote 12 of this chapter.

Thermoplastic and Thermosetting Polymers

5.13 Make comparisons of thermoplastic and thermosetting polymers **(a)** on the basis of mechanical characteristics upon heating and **(b)** according to possible molecular structures.

5.14 **(a)** Is it possible to grind up and reuse phenol-formaldehyde? Why or why not?

(b) Is it possible to grind up and reuse polypropylene? Why or why not?

Copolymers

5.15 Sketch the repeat structure for each of the following alternating copolymers: **(a)** poly (butadiene-chloroprene), **(b)** poly(styrene-methyl methacrylate), and **(c)** poly(acrylonitrile-vinyl chloride).

5.16 The number-average molecular weight of a poly(styrene-butadiene) alternating copolymer is 1,350,000 g/mol; determine the average number of styrene and butadiene repeat units per molecule.

5.17 Calculate the number-average molecular weight of a random nitrile rubber [poly (acrylonitrile-butadiene) copolymer] in which the fraction of butadiene repeat units is 0.30; assume that this concentration corresponds to a degree of polymerization of 2500.

5.18 An alternating copolymer is known to have a number-average molecular weight of 250,000 g/mol and a degree of polymerization of 3500. If one of the repeat units is styrene, which of ethylene, propylene, tetrafluoroethylene, and vinyl chloride is the other repeat unit? Why?

5.19 **(a)** Determine the ratio of butadiene to styrene repeat units in a copolymer having a number-average molecular weight of 350,000 g/mol and degree of polymerization of 4500.

(b) Which type(s) of copolymer(s) will this copolymer be, considering the following possibilities: random, alternating, graft, and block? Why?

5.20 Crosslinked copolymers consisting of 60 wt% ethylene and 40 wt% propylene may have elastic properties similar to those for natural rubber. For a copolymer of this composition, determine the fraction of both repeat unit types.

5.21 A random poly(isobutylene-isoprene) copolymer has a number-average molecular weight of 200,000 g/mol and a degree of polymerization of 3000. Compute the fraction of isobutylene and isoprene repeat units in this copolymer.

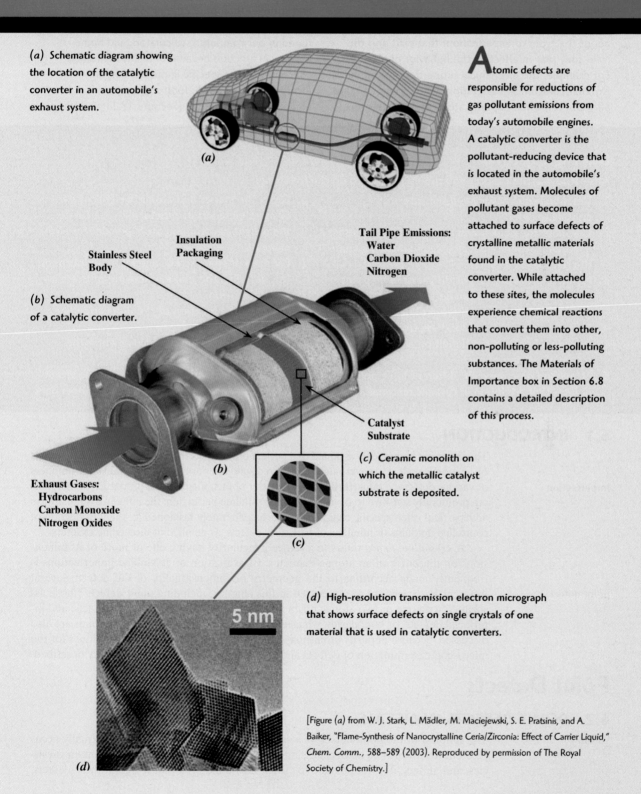

(a) Schematic diagram showing the location of the catalytic converter in an automobile's exhaust system.

(a)

Atomic defects are responsible for reductions of gas pollutant emissions from today's automobile engines. A catalytic converter is the pollutant-reducing device that is located in the automobile's exhaust system. Molecules of pollutant gases become attached to surface defects of crystalline metallic materials found in the catalytic converter. While attached to these sites, the molecules experience chemical reactions that convert them into other, non-polluting or less-polluting substances. The Materials of Importance box in Section 6.8 contains a detailed description of this process.

Insulation Packaging

Stainless Steel Body

(b) Schematic diagram of a catalytic converter.

Tail Pipe Emissions:
Water
Carbon Dioxide
Nitrogen

(b)

Catalyst Substrate

Exhaust Gases:
Hydrocarbons
Carbon Monoxide
Nitrogen Oxides

(c) Ceramic monolith on which the metallic catalyst substrate is deposited.

(c)

5 nm

(d) High-resolution transmission electron micrograph that shows surface defects on single crystals of one material that is used in catalytic converters.

(d)

[Figure (a) from W. J. Stark, L. Mädler, M. Maciejewski, S. E. Pratsinis, and A. Baiker, "Flame-Synthesis of Nanocrystalline Ceria/Zirconia: Effect of Carrier Liquid," *Chem. Comm.*, 588–589 (2003). Reproduced by permission of The Royal Society of Chemistry.]

The properties of some materials are profoundly influenced by the presence of imperfections. Consequently, it is important to have a knowledge about the types of imperfections that exist and the roles they play in affecting the behavior of materials. For example, the mechanical properties of pure metals experience significant alterations when the metals are alloyed (i.e., when impurity atoms are added)—for

example, brass (70% copper/30% zinc) is much harder and stronger than pure copper (Section 9.9).

Also, integrated-circuit microelectronic devices found in our computers, calculators, and home appliances function because of highly controlled concentrations of specific impurities that are incorporated into small, localized regions of semiconducting materials (Sections 19.11 and 19.15).

Learning Objectives

After studying this chapter you should be able to do the following:

1. Describe both vacancy and self-interstitial crystalline defects.
2. Calculate the equilibrium number of vacancies in a material at some specified temperature, given the relevant constants.
3. Name and describe eight different ionic point defects that are found in ceramic compounds (including Schottky and Frenkel defects).
4. Name the two types of solid solutions and provide a brief written definition and/or schematic sketch of each.
5. Name and describe eight different ionic point defects that are found in ceramic materials.
6. Given the masses and atomic weights of two or more elements in a metal alloy, calculate the weight percent and atom percent for each element.
7. For each of edge, screw, and mixed dislocations:
   (a) describe and make a drawing of the dislocation,
   (b) note the location of the dislocation line, and
   (c) indicate the direction along which the dislocation line extends.
8. Describe the atomic structure within the vicinity of (a) a grain boundary and (b) a twin boundary.

6.1 INTRODUCTION

imperfection

Thus far it has been tacitly assumed that perfect order exists throughout crystalline materials on an atomic scale. However, such an idealized solid does not exist; all contain large numbers of various defects or **imperfections**. As a matter of fact, many of the properties of materials are profoundly sensitive to deviations from crystalline perfection; the influence is not always adverse, and often specific characteristics are deliberately fashioned by the introduction of controlled amounts or numbers of particular defects, as detailed in succeeding chapters.

point defect

A *crystalline defect* refers to a lattice irregularity having one or more of its dimensions on the order of an atomic diameter. Classification of crystalline imperfections is frequently made according to the geometry or dimensionality of the defect. Several different imperfections are discussed in this chapter, including **point defects** (those associated with one or two atomic positions); linear (or one-dimensional) defects; and interfacial defects, or boundaries, which are two-dimensional. Impurities in solids are also discussed, because impurity atoms may exist as point defects. Finally, techniques for the microscopic examination of defects and the structure of materials are briefly described.

Point Defects

6.2 POINT DEFECTS IN METALS

vacancy

The simplest of the point defects is a **vacancy**, or vacant lattice site, one normally occupied but from which an atom is missing (Figure 6.1). All crystalline solids contain vacancies, and, in fact, it is not possible to create such a material that is free of these defects.

Figure 6.1 Two-dimensional representations of a vacancy and a self-interstitial.
(Adapted from W. G. Moffatt, G. W. Pearsall, and J. Wulff, *The Structure and Properties of Materials,* Vol. I, *Structure,* p. 77. Copyright © 1964 by John Wiley & Sons, New York, NY. Reprinted by permission of John Wiley & Sons, Inc.)

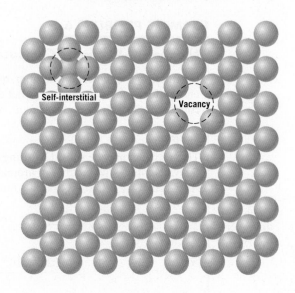

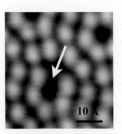

Scanning probe micrograph that shows a vacancy on a (111)-type surface plane for silicon. Approximately 7,000,000×.
(Micrograph courtesy of D. Huang, Stanford University.)

Temperature dependence of the equilibrium number of vacancies

Boltzmann's constant

self-interstitial

WileyPLUS

Tutorial Video:
Computation of the Equilibrium Number of Vacancies

The necessity of the existence of vacancies is explained using principles of thermodynamics; in essence, the presence of vacancies increases the entropy (i.e., the randomness) of the crystal.

The equilibrium number of vacancies N_v for a given quantity of material (usually per meter cubed) depends on and increases with temperature according to

$$N_v = N \exp\left(-\frac{Q_v}{kT}\right) \tag{6.1}$$

In this expression, N is the total number of atomic sites (most commonly per cubic meter), Q_v is the energy required for the formation of a vacancy (J/mol or eV/atom), T is the absolute temperature in kelvins,[1] and k is the gas or **Boltzmann's constant**. The value of k is 1.38×10^{-23} J/atom·K, or 8.62×10^{-5} eV/atom·K, depending on the units of Q_v.[2] Thus, the number of vacancies increases exponentially with temperature—that is, as T in Equation 6.1 increases, so also does the term $\exp(-Q_v/kT)$. For most metals, the fraction of vacancies N_v/N just below the melting temperature is on the order of 10^{-4}—that is, one lattice site out of 10,000 will be empty. As ensuing discussions indicate, a number of other material parameters have an exponential dependence on temperature similar to that in Equation 6.1.

A **self-interstitial** is an atom from the crystal that is crowded into an *interstitial site*—a small void space that under ordinary circumstances is not occupied. This kind of defect is also represented in Figure 6.1. In metals, a self-interstitial introduces relatively large distortions in the surrounding lattice because the atom is substantially larger than the interstitial position in which it is situated. Consequently, the formation of this defect is not highly probable, and it exists in very small concentrations that are significantly lower than for vacancies.

---

[1]Absolute temperature in kelvins (K) is equal to °C + 273.

[2]Boltzmann's constant per mole of atoms becomes the gas constant R; in such a case, $R = 8.31$ J/mol·K.

EXAMPLE PROBLEM 6.1

Number-of-Vacancies Computation at a Specified Temperature

Calculate the equilibrium number of vacancies per cubic meter for copper at 1000°C. The energy for vacancy formation is 0.9 eV/atom; the atomic weight and density (at 1000°C) for copper are 63.5 g/mol and 8.4 g/cm^3, respectively.

Solution

This problem may be solved by using Equation 6.1; it is first necessary, however, to determine the value of N—the number of atomic sites per cubic meter for copper, from its atomic weight A_{Cu}, its density ρ, and Avogadro's number N_A, according to

Number of atoms per unit volume for a metal

$$N = \frac{N_A \rho}{A_{Cu}} \tag{6.2}$$

$$= \frac{(6.022 \times 10^{23} \text{ atoms/mol})(8.4 \text{ g/cm}^3)(10^6 \text{ cm}^3/\text{m}^3)}{63.5 \text{ g/mol}}$$

$$= 8.0 \times 10^{28} \text{ atoms/m}^3$$

WileyPLUS

Tutorial Video

Thus, the number of vacancies at 1000°C (1273 K) is equal to

$$N_v = N \exp\left(-\frac{Q_v}{kT}\right)$$

$$= (8.0 \times 10^{28} \text{ atoms/m}^3) \exp\left[-\frac{(0.9 \text{ eV})}{(8.62 \times 10^{-5} \text{ eV/K})(1273 \text{ K})}\right]$$

$$= 2.2 \times 10^{25} \text{ vacancies/m}^3$$

6.3 POINT DEFECTS IN CERAMICS

Atomic Point Defects

Atomic defects involving host atoms may exist in ceramic compounds. As with metals, both vacancies and interstitials are possible; however, because ceramic materials contain ions of at least two kinds, defects for each ion type may occur. For example, in NaCl, Na interstitials and vacancies and Cl interstitials and vacancies may exist. It is highly improbable that there would be appreciable concentrations of anion interstitials. The anion is relatively large, and to fit into a small interstitial position, substantial strains on the surrounding ions must be introduced. Anion and cation vacancies and a cation interstitial are represented in Figure 6.2.

defect structure

electroneutrality

Frenkel defect

The expression **defect structure** is often used to designate the types and concentrations of atomic defects in ceramics. Because the atoms exist as charged ions, when defect structures are considered, conditions of electroneutrality must be maintained. **Electroneutrality** is the state that exists when there are equal numbers of positive and negative charges from the ions. As a consequence, defects in ceramics do not occur alone. One such type of defect involves a cation–vacancy and a cation–interstitial pair. This is called a **Frenkel defect** (Figure 6.3). It might be thought of as being formed by a cation leaving its normal position and moving into an interstitial site. There is no change in charge because the cation maintains the same positive charge as an interstitial.

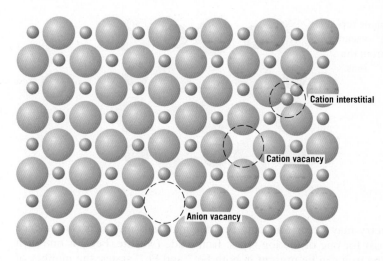

Figure 6.2 Schematic representations of cation and anion vacancies and a cation interstitial.
(From W. G. Moffatt, G. W. Pearsall, and J. Wulff, *The Structure and Properties of Materials,* Vol. I, *Structure,* p. 78. Copyright © 1964 by John Wiley & Sons, New York. Reprinted by permission of John Wiley & Sons, Inc.)

Schottky defect

Another type of defect found in AX materials is a cation vacancy–anion vacancy pair known as a **Schottky defect**, also schematically diagrammed in Figure 6.3. This defect might be thought of as being created by removing one cation and one anion from the interior of the crystal and then placing them both at an external surface. Because the magnitude of the negative charge on the cation is equal to the magnitude of the positive charge on the anion, and because for every anion vacancy there exists a cation vacancy, the charge neutrality of the crystal is maintained.

stoichiometry

The ratio of cations to anions is not altered by the formation of either a Frenkel or a Schottky defect. If no other defects are present, the material is said to be stoichiometric. **Stoichiometry** may be defined as a state for ionic compounds wherein there is the exact ratio of cations to anions as predicted by the chemical formula. For example, NaCl is stoichiometric if the ratio of Na^+ ions to Cl^- ions is exactly 1:1. A ceramic compound is *nonstoichiometric* if there is any deviation from this exact ratio.

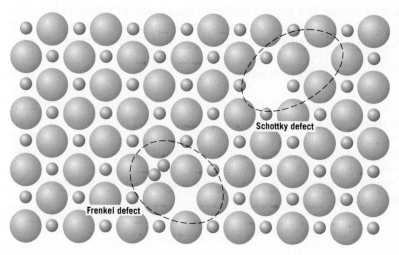

Figure 6.3 Schematic diagram showing Frenkel and Schottky defects in ionic solids.
(From W. G. Moffatt, G. W. Pearsall, and J. Wulff, *The Structure and Properties of Materials,* Vol. I, *Structure,* p. 78. Copyright © 1964 by John Wiley & Sons, New York. Reprinted by permission of John Wiley & Sons, Inc.)

Figure 6.4 Schematic representation of an Fe^{2+} vacancy in FeO that results from the formation of two Fe^{3+} ions.

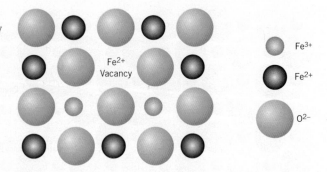

Fe³⁺

Fe²⁺

O²⁻

Nonstoichiometry may occur for some ceramic materials in which two valence (or ionic) states exist for one of the ion types. Iron oxide (wüstite, FeO) is one such material because the iron can be present in both Fe^{2+} and Fe^{3+} states; the number of each of these ion types depends on temperature and the ambient oxygen pressure. The formation of an Fe^{3+} ion disrupts the electroneutrality of the crystal by introducing an excess +1 charge, which must be offset by some type of defect. This may be accomplished by the formation of one Fe^{2+} vacancy (or the removal of two positive charges) for every two Fe^{3+} ions that are formed (Figure 6.4). The crystal is no longer stoichiometric because there is one more O ion than Fe ion; however, the crystal remains electrically neutral. This phenomenon is fairly common in iron oxide, and, in fact, its chemical formula is often written as $Fe_{1-x}O$ (where x is some small and variable fraction substantially less than unity) to indicate a condition of nonstoichiometry with a deficiency of Fe.

Concept Check 6.1   Can Schottky defects exist in K_2O? If so, briefly describe this type of defect. If they cannot exist, then explain why.

[*The answer may be found at* www.wiley.com/college/callister *(Student Companion Site).*]

The equilibrium numbers of both Frenkel and Schottky defects increase with and depend on temperature in a manner similar to the number of vacancies in metals (Equation 6.1). For Frenkel defects, the number of cation–vacancy/cation–interstitial defect pairs (N_{fr}) depends on temperature according to the following expression:

$$N_{fr} = N \exp\left(-\frac{Q_{fr}}{2kT}\right) \qquad (6.3)$$

Here, Q_{fr} is the energy required for the formation of each Frenkel defect, and N is the total number of lattice sites. (As in previous discussions, k and T represent Boltzmann's constant and the absolute temperature, respectively.) The factor 2 is present in the denominator of the exponential because two defects (a missing cation and an interstitial cation) are associated with each Frenkel defect.

Similarly, for Schottky defects, in an AX-type compound, the equilibrium number (N_s) is a function of temperature as

$$N_s = N \exp\left(-\frac{Q_s}{2kT}\right) \qquad (6.4)$$

where Q_s represents the Schottky defect energy of formation.

EXAMPLE PROBLEM 6.2

Computation of the Number of Schottky Defects in KCl

Calculate the number of Schottky defects per cubic meter in potassium chloride at 500°C. The energy required to form each Schottky defect is 2.6 eV, whereas the density for KCl (at 500°C) is 1.955 g/cm³.

Solution

To solve this problem it is necessary to use Equation 6.4. However, we must first compute the value of N (the number of lattice sites per cubic meter); this is possible using a modified form of Equation 6.2:

$$N = \frac{N_A \rho}{A_K + A_{Cl}} \tag{6.5}$$

where N_A is Avogadro's number (6.022×10^{23} atoms/mol), ρ is the density, and A_K and A_{Cl} are the atomic weights for potassium and chlorine (i.e., 39.10 and 35.45 g/mol), respectively. Therefore,

$$N = \frac{(6.022 \times 10^{23} \text{ atoms/mol})(1.955 \text{ g/cm}^3)(10^6 \text{ cm}^3/\text{m}^3)}{39.10 \text{ g/mol} + 35.45 \text{ g/mol}}$$

$$= 1.58 \times 10^{28} \text{ lattice sites/m}^3$$

Now, incorporating this value into Equation 6.4 leads to the following value for N_s:

$$N_s = N \exp\left(-\frac{Q_s}{2kT}\right)$$

$$= (1.58 \times 10^{28} \text{ lattice sites/m}^3) \exp\left[-\frac{2.6 \text{ eV}}{(2)(8.62 \times 10^{-5} \text{ eV/K})(500 + 273 \text{ K})}\right]$$

$$= 5.31 \times 10^{19} \text{ defects/m}^3$$

6.4 IMPURITIES IN SOLIDS

Impurities in Metals

alloy

A pure metal consisting of only one type of atom just isn't possible; impurity or foreign atoms are always present, and some exist as crystalline point defects. In fact, even with relatively sophisticated techniques, it is difficult to refine metals to a purity in excess of 99.9999%. At this level, on the order of 10^{22} to 10^{23} impurity atoms are present in 1 m³ of material. Most familiar metals are not highly pure; rather, they are **alloys**, in which impurity atoms have been added intentionally to impart specific characteristics to the material. Ordinarily, alloying is used in metals to improve mechanical strength and corrosion resistance. For example, sterling silver is a 92.5% silver/7.5% copper alloy. In normal ambient environments, pure silver is highly corrosion resistant, but also very soft. Alloying with copper significantly enhances the mechanical strength without depreciating the corrosion resistance appreciably.

solid solution

The addition of impurity atoms to a metal results in the formation of a **solid solution** and/or a new *second phase,* depending on the kinds of impurity, their concentrations, and the temperature of the alloy. The present discussion is concerned with the notion of a solid solution; treatment of the formation of a new phase is deferred to Chapter 11.

solute, solvent

Several terms relating to impurities and solid solutions deserve mention. With regard to alloys, **solute** and **solvent** are terms that are commonly employed. *Solvent* is the element or compound that is present in the greatest amount; on occasion, solvent atoms

Figure 6.5 Two-dimensional schematic representations of substitutional and interstitial impurity atoms. (Adapted from W. G. Moffatt, G. W. Pearsall, and J. Wulff, *The Structure and Properties of Materials,* Vol. I, *Structure,* p. 77. Copyright © 1964 by John Wiley & Sons, New York, NY. Reprinted by permission of John Wiley & Sons, Inc.)

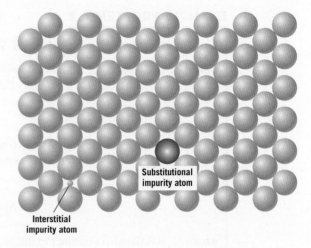

Substitutional impurity atom

Interstitial impurity atom

are also called *host atoms. Solute* is used to denote an element or compound present in a minor concentration.

Solid Solutions

A solid solution forms when, as the solute atoms are added to the host material, the crystal structure is maintained and no new structures are formed. Perhaps it is useful to draw an analogy with a liquid solution. If two liquids that are soluble in each other (such as water and alcohol) are combined, a liquid solution is produced as the molecules intermix, and its composition is homogeneous throughout. A solid solution is also compositionally homogeneous; the impurity atoms are randomly and uniformly dispersed within the solid.

Impurity point defects are found in solid solutions, of which there are two types: **substitutional** and **interstitial**. For the substitutional type, solute or impurity atoms replace or substitute for the host atoms (Figure 6.5). Several features of the solute and solvent atoms determine the degree to which the former dissolves in the latter. These are expressed as four *Hume–Rothery rules,* as follows:

substitutional solid solution

interstitial solid solution

1. *Atomic size factor.* Appreciable quantities of a solute may be accommodated in this type of solid solution only when the difference in atomic radii between the two atom types is less than about ±15%. Otherwise, the solute atoms create substantial lattice distortions and a new phase forms.

2. *Crystal structure.* For appreciable solid solubility, the crystal structures for metals of both atom types must be the same.

3. *Electronegativity factor.* The more electropositive one element and the more electronegative the other, the greater the likelihood that they will form an intermetallic compound instead of a substitutional solid solution.

4. *Valences.* Other factors being equal, a metal has more of a tendency to dissolve another metal of higher valency than to dissolve one of a lower valency.

WileyPLUS

Tutorial Video: Defects

What are the Differences between Interstitial and Substitutional Solid Solutions?

An example of a substitutional solid solution is found for copper and nickel. These two elements are completely soluble in one another at all proportions. With regard to the aforementioned rules that govern degree of solubility, the atomic radii for copper and nickel are 0.128 and 0.125 nm, respectively; both have the FCC crystal structure; and their electronegativities are 1.9 and 1.8 (Figure 2.9). Finally, the most common valences are +1 for copper (although it sometimes can be +2) and +2 for nickel.

For interstitial solid solutions, impurity atoms fill the voids or interstices among the host atoms (see Figure 6.5). For both FCC and BCC crystal structures, there are two

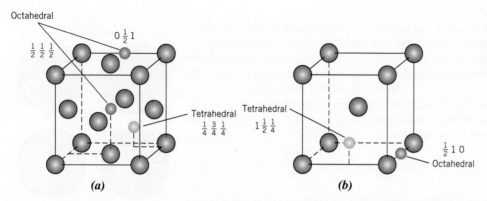

Figure 6.6
Locations of tetrahedral and octahedral interstitial sites within (a) FCC and (b) BCC unit cells.

types of interstitial sites—*tetrahedral* and *octahedral*—these are distinguished by the number of nearest neighbor host atoms—that is, the coordination number. Tetrahedral sites have a coordination number of 4; straight lines drawn from the centers of the surrounding host atoms form a four-sided tetrahedron. However, for octahedral sites the coordination number is 6; an octahedron is produced by joining these six sphere centers.[3] For FCC, there are two types of octahedral sites with representative point coordinates of $0\frac{1}{2}1$ and $\frac{1}{2}\frac{1}{2}\frac{1}{2}$. Representative coordinates for a single tetrahedral site type are $\frac{1}{4}\frac{3}{4}\frac{1}{4}$.[4] Locations of these sites within the FCC unit cell are noted in Figure 6.6a. One type of each of octahedral and tetrahedral interstitial sites is found for BCC. Representative coordinates are as follows: octahedral, $\frac{1}{2}10$ and tetrahedral, $1\frac{1}{2}\frac{1}{4}$. Figure 6.6b shows the positions of these sites within a BCC unit cell.[4]

Metallic materials have relatively high atomic packing factors, which means that these interstitial positions are relatively small. Consequently, the atomic diameter of an interstitial impurity must be substantially smaller than that of the host atoms. Normally, the maximum allowable concentration of interstitial impurity atoms is low (less than 10%). Even very small impurity atoms are ordinarily larger than the interstitial sites, and as a consequence, they introduce some lattice strains on the adjacent host atoms. Problem 6.10 calls for determination of the radii of impurity atoms r (in terms of R, the host atom radius) that just fit into tetrahedral and octahedral interstitial positions of both BCC and FCC without introducing any lattice strains.

Carbon forms an interstitial solid solution when added to iron; the maximum concentration of carbon is about 2%. The atomic radius of the carbon atom is much less than that of iron: 0.071 nm versus 0.124 nm.

EXAMPLE PROBLEM 6.3

Computation of Radius of BCC Interstitial Site

Compute the radius r of an impurity atom that just fits into a BCC octahedral site in terms of the atomic radius R of the host atom (without introducing lattice strains).

Solution

As Figure 6.6b notes, for BCC, the octahedral interstitial site is situated at the center of a unit cell edge. In order for an interstitial atom to be positioned in this site without introducing lattice

---

[3]The geometries of these site types may be observed in Figure 4.26.

[4]Other octahedral and tetrahedral interstices are located at positions within the unit cell that are equivalent to these representative ones.

strains, the atom just touches the two adjacent host atoms, which are corner atoms of the unit cell. The drawing shows atoms on the (100) face of a BCC unit cell; the large circles represent the host atoms—the small circle represents an interstitial atom that is positioned in an octahedral site on the cube edge.

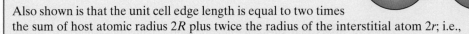

On this drawing is noted the unit cell edge length—the distance between the centers of the corner atoms—which, from Equation 4.4, is equal to

$$\text{Unit cell edge length} = \frac{4R}{\sqrt{3}}$$

Also shown is that the unit cell edge length is equal to two times the sum of host atomic radius $2R$ plus twice the radius of the interstitial atom $2r$; i.e.,

$$\text{Unit cell edge length} = 2R + 2r$$

Now, equating these two unit cell edge length expressions, we get

$$2R + 2r = \frac{4R}{\sqrt{3}}$$

and solving for r in terms of R

$$2r = \frac{4R}{\sqrt{3}} - 2R = \left(\frac{2}{\sqrt{3}} - 1\right)(2R)$$

or

$$r = \left(\frac{2}{\sqrt{3}} - 1\right)R = 0.155R$$

Concept Check 6.2   Is it possible for three or more elements to form a solid solution? Explain your answer.

[*The answer may be found at* www.wiley.com/college/callister *(Student Companion Site)*].

Concept Check 6.3   Explain why complete solid solubility may occur for substitutional solid solutions but not for interstitial solid solutions.

[*The answer may be found at* www.wiley.com/college/callister *(Student Companion Site)*].

Impurities in Ceramics

Impurity atoms can form solid solutions in ceramic materials much as they do in metals. Solid solutions of both substitutional and interstitial types are possible. For an interstitial, the ionic radius of the impurity must be relatively small in comparison to the anion. Because there are both anions and cations, a substitutional impurity substitutes for the host ion to which it is most similar in an electrical sense: If the impurity atom normally forms a cation in a ceramic material, it most probably will substitute for a host cation. For example, in sodium chloride, impurity Ca^{2+} and O^{2-} ions would most likely substitute for Na^+ and Cl^- ions, respectively. Schematic representations for cation and anion substitutional as well as interstitial impurities are shown in Figure 6.7. To achieve

Figure 6.7 Schematic representations of interstitial, anion-substitutional, and cation-substitutional impurity atoms in an ionic compound.
(Adapted from W. G. Moffatt, G. W. Pearsall, and J. Wulff, *The Structure and Properties of Materials,* Vol. I, *Structure,* p. 78. Copyright © 1964 by John Wiley & Sons, New York. Reprinted by permission of John Wiley & Sons, Inc.)

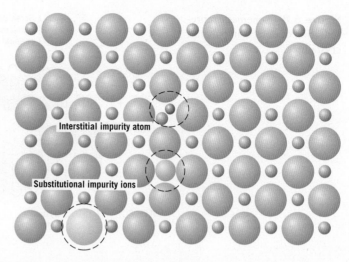

any appreciable solid solubility of substituting impurity atoms, the ionic size and charge must be very nearly the same as those of one of the host ions. For an impurity ion having a charge different from that of the host ion for which it substitutes, the crystal must compensate for this difference in charge so that electroneutrality is maintained with the solid. One way this is accomplished is by the formation of lattice defects—vacancies or interstitials of both ion types, as discussed previously.

EXAMPLE PROBLEM 6.4

Determination of Possible Point Defect Types in NaCl Due to the Presence of Ca^{2+} Ions

If electroneutrality is to be preserved, what point defects are possible in NaCl when a Ca^{2+} substitutes for an Na^+ ion? How many of these defects exist for every Ca^{2+} ion?

Solution

Replacement of an Na^+ by a Ca^{2+} ion introduces one extra positive charge. Electroneutrality is maintained when either a single positive charge is eliminated or another single negative charge is added. Removal of a positive charge is accomplished by the formation of one Na^+ vacancy. Alternatively, a Cl^- interstitial supplies an additional negative charge, negating the effect of each Ca^{2+} ion. However, as mentioned earlier, the formation of this defect is highly unlikely.

Concept Check 6.4 What point defects are possible for MgO as an impurity in Al_2O_3? How many Mg^{2+} ions must be added to form each of these defects?

[*The answer may be found at* www.wiley.com/college/callister *(Student Companion Site).*]

6.5 POINT DEFECTS IN POLYMERS

The point defect concept is different in polymers than in metals and ceramics as a consequence of the chainlike macromolecules and the nature of the crystalline state for polymers. Point defects similar to those found in metals have been observed in crystalline

Figure 6.8 Schematic representation of defects in polymer crystallites.

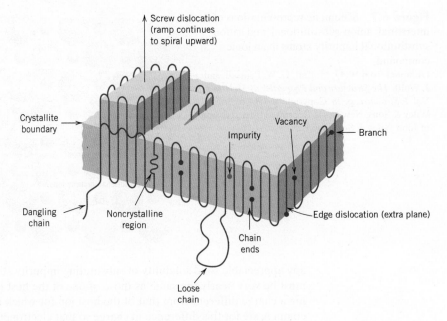

regions of polymeric materials; these include vacancies and interstitial atoms and ions. Chain ends are considered defects because they are chemically dissimilar to normal chain units. Vacancies are also associated with the chain ends (Figure 6.8). However, additional defects can result from branches in the polymer chain or chain segments that emerge from the crystal. A chain section can leave a polymer crystal and reenter it at another point, creating a loop, or can enter a second crystal to act as a tie molecule (see Figure 5.12). Screw dislocations also occur in polymer crystals (Figure 6.8). Impurity atoms/ions or groups of atoms/ions may be incorporated in the molecular structure as interstitials; they may also be associated with main chains or as short side branches.

Furthermore, the surfaces of chain-folded layers (Figure 5.12) are considered interfacial defects, as are also boundaries between two adjacent crystalline regions.

6.6 SPECIFICATION OF COMPOSITION

composition

weight percent

It is often necessary to express the **composition** (or *concentration*)[5] of an alloy in terms of its constituent elements. The two most common ways to specify composition are weight (or mass) percent and atom percent. The basis for **weight percent** (wt%) is the weight of a particular element relative to the total alloy weight. For an alloy that contains two hypothetical atoms denoted by 1 and 2, the concentration of 1 in wt%, C_1, is defined as

Computation of weight percent (for a two-element alloy)

$$C_1 = \frac{m_1}{m_1 + m_2} \times 100 \tag{6.6a}$$

---

[5]The terms *composition* and *concentration* will be assumed to have the same meaning in this book (i.e., the relative content of a specific element or constituent in an alloy) and will be used interchangeably.

where m_1 and m_2 represent the weight (or mass) of elements 1 and 2, respectively. The concentration of 2 is computed in an analogous manner.[6]

atom percent

The basis for **atom percent** (at%) calculations is the number of moles of an element in relation to the total moles of the elements in the alloy. The number of moles in some specified mass of a hypothetical element 1, n_{m1}, may be computed as follows:

$$n_{m1} = \frac{m_1'}{A_1} \tag{6.7}$$

Here, m_1' and A_1 denote the mass (in grams) and atomic weight, respectively, for element 1.

Concentration in terms of atom percent of element 1 in an alloy containing element 1 and element 2 atoms, C_1' is defined by[7]

Computation of atom percent (for a two-element alloy)

$$C_1' = \frac{n_{m1}}{n_{m1} + n_{m2}} \times 100 \tag{6.8a}$$

In like manner, the atom percent of element 2 is determined.[8]

Atom percent computations also can be carried out on the basis of the number of atoms instead of moles, because one mole of all substances contains the same number of atoms.

Composition Conversions

Sometimes it is necessary to convert from one composition scheme to another—for example, from weight percent to atom percent. We next present equations for making these conversions in terms of the two hypothetical elements 1 and 2. Using the convention of the previous section (i.e., weight percents denoted by C_1 and C_2, atom percents by C_1' and C_2', and atomic weights as A_1 and A_2), we express these conversion equations as follows:

$$C_1' = \frac{C_1 A_2}{C_1 A_2 + C_2 A_1} \times 100 \tag{6.9a}$$

Conversion of weight percent to atom percent (for a two-element alloy)

$$C_2' = \frac{C_2 A_1}{C_1 A_2 + C_2 A_1} \times 100 \tag{6.9b}$$

---

[6]When an alloy contains more than two (say n) elements, Equation (6.6a) takes the form

$$C_1 = \frac{m_1}{m_1 + m_2 + m_3 + \cdots + m_n} \times 100 \tag{6.6b}$$

[7]In order to avoid confusion in notations and symbols being used in this section, we should point out that the prime (as in C_1' and m_1') is used to designate both composition in atom percent and mass of material in grams.

[8]When an alloy contains more than two (say n) elements, Equation (6.8a) takes the form

$$C_1' = \frac{n_{m1}}{n_{m1} + n_{m2} + n_{m3} + \cdots + n_{mn}} \times 100 \tag{6.8b}$$

Conversion of
atom percent to
weight percent (for
a two-element alloy)

$$C_1 = \frac{C_1' A_1}{C_1' A_1 + C_2' A_2} \times 100 \tag{6.10a}$$

$$C_2 = \frac{C_2' A_2}{C_1' A_1 + C_2' A_2} \times 100 \tag{6.10b}$$

Because we are considering only two elements, computations involving the preceding equations are simplified when it is realized that

$$C_1 + C_2 = 100 \tag{6.11a}$$

$$C_1' + C_2' = 100 \tag{6.11b}$$

In addition, it sometimes becomes necessary to convert concentration from weight percent to mass of one component per unit volume of material (i.e., from units of wt% to kg/m^3); this latter composition scheme is often used in diffusion computations (Section 5.3). Concentrations in terms of this basis are denoted using a double prime (i.e., C_1'' and C_2''), and the relevant equations are as follows:

Conversion of weight
percent to mass per
unit volume (for a
two-element alloy)

$$C_1'' = \left(\frac{C_1}{\dfrac{C_1}{\rho_1} + \dfrac{C_2}{\rho_2}} \right) \times 10^3 \tag{6.12a}$$

$$C_2'' = \left(\frac{C_2}{\dfrac{C_1}{\rho_1} + \dfrac{C_2}{\rho_2}} \right) \times 10^3 \tag{6.12b}$$

WileyPLUS

Tutorial Video:
**Weight Percent
and Atom Percent
Calculations**

For density ρ in units of g/cm^3, these expressions yield C_1'' and C_2'' in kg/m^3.

Furthermore, on occasion we desire to determine the density and atomic weight of a binary alloy, given the composition in terms of either weight percent or atom percent. If we represent alloy density and atomic weight by ρ_{ave} and A_{ave}, respectively, then

Computation of
density (for a two-
element metal alloy)

$$\rho_{ave} = \frac{100}{\dfrac{C_1}{\rho_1} + \dfrac{C_2}{\rho_2}} \tag{6.13a}$$

$$\rho_{ave} = \frac{C_1' A_1 + C_2' A_2}{\dfrac{C_1' A_1}{\rho_1} + \dfrac{C_2' A_2}{\rho_2}} \tag{6.13b}$$

Computation of
atomic weight (for a
two-element metal
alloy)

$$A_{ave} = \frac{100}{\dfrac{C_1}{A_1} + \dfrac{C_2}{A_2}} \tag{6.14a}$$

$$A_{ave} = \frac{C_1' A_1 + C_2' A_2}{100} \tag{6.14b}$$

It should be noted that Equations 6.12 and 6.14 are not always exact. In their derivations, it is assumed that total alloy volume is exactly equal to the sum of the volumes of the individual elements. This normally is not the case for most alloys; however, it is a reasonably valid assumption and does not lead to significant errors for dilute solutions and over composition ranges where solid solutions exist.

EXAMPLE PROBLEM 6.5

Derivation of Composition-Conversion Equation

Derive Equation 6.9a.

Solution

To simplify this derivation, we assume that masses are expressed in units of grams and denoted with a prime (e.g., m_1'). Furthermore, the total alloy mass (in grams) M' is

$$M' = m_1' + m_2' \qquad (6.15)$$

Using the definition of C_1' (Equation 6.8a) and incorporating the expression for n_{m1}, Equation 6.7, and the analogous expression for n_{m2} yields

$$C_1' = \frac{n_{m1}}{n_{m1} + n_{m2}} \times 100$$

$$= \frac{\dfrac{m_1'}{A_1}}{\dfrac{m_1'}{A_1} + \dfrac{m_2'}{A_2}} \times 100 \qquad (6.16)$$

Rearrangement of the mass-in-grams equivalent of Equation 6.6a leads to

$$m_1' = \frac{C_1 M'}{100} \qquad (6.17)$$

Substitution of this expression and its m_2' equivalent into Equation 6.16 gives

$$C_1' = \frac{\dfrac{C_1 M'}{100 A_1}}{\dfrac{C_1 M'}{100 A_1} + \dfrac{C_2 M'}{100 A_2}} \times 100 \qquad (6.18)$$

Upon simplification, we have

$$C_1' = \frac{C_1 A_2}{C_1 A_2 + C_2 A_1} \times 100$$

which is identical to Equation 6.9a.

EXAMPLE PROBLEM 6.6

Composition Conversion—From Weight Percent to Atom Percent

Determine the composition, in atom percent, of an alloy that consists of 97 wt% aluminum and 3 wt% copper.

Solution

If we denote the respective weight percent compositions as $C_{Al} = 97$ and $C_{Cu} = 3$, substitution into Equations 6.9a and 6.9b yields

WileyPLUS

Tutorial Video:
How to
Convert from
Atom Percent
to Weight
Percent

and

$$C'_{Al} = \frac{C_{Al}A_{Cu}}{C_{Al}A_{Cu} + C_{Cu}A_{Al}} \times 100$$

$$= \frac{(97)(63.55 \text{ g/mol})}{(97)(63.55 \text{ g/mol}) + (3)(26.98 \text{ g/mol})} \times 100$$

$$= 98.7 \text{ at\%}$$

$$C'_{Cu} = \frac{C_{Cu}A_{Al}}{C_{Cu}A_{Al} + C_{Al}A_{Cu}} \times 100$$

$$= \frac{(3)(26.98 \text{ g/mol})}{(3)(26.98 \text{ g/mol}) + (97)(63.55 \text{ g/mol})} \times 100$$

$$= 1.30 \text{ at\%}$$

Miscellaneous Imperfections

6.7 DISLOCATIONS—LINEAR DEFECTS

edge dislocation

dislocation line

WileyPLUS: VMSE
Edge

screw dislocation

A *dislocation* is a linear or one-dimensional defect around which some of the atoms are misaligned. One type of dislocation is represented in Figure 6.9: an extra portion of a plane of atoms, or half-plane, the edge of which terminates within the crystal. This is termed an **edge dislocation**; it is a linear defect that centers on the line that is defined along the end of the extra half-plane of atoms. This is sometimes termed the **dislocation line**, which, for the edge dislocation in Figure 6.9, is perpendicular to the plane of the page. Within the region around the dislocation line there is some localized lattice distortion. The atoms above the dislocation line in Figure 6.9 are squeezed together, and those below are pulled apart; this is reflected in the slight curvature for the vertical planes of atoms as they bend around this extra half-plane. The magnitude of this distortion decreases with distance away from the dislocation line; at positions far removed, the crystal lattice is virtually perfect. Sometimes the edge dislocation in Figure 6.9 is represented by the symbol ⊥, which also indicates the position of the dislocation line. An edge dislocation may also be formed by an extra half-plane of atoms that is included in the bottom portion of the crystal; its designation is a ⊤.

Another type of dislocation, called a **screw dislocation**, may be thought of as being formed by a shear stress that is applied to produce the distortion shown in Figure 6.10a:

Figure 6.9 The atom positions around an edge dislocation; extra half-plane of atoms shown in perspective.
(Adapted from A. G. Guy, *Essentials of Materials Science*, McGraw-Hill Book Company, New York, NY, 1976, p. 153.)

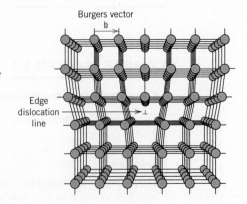

WileyPLUS: VMSE
Screw

the upper front region of the crystal is shifted one atomic distance to the right relative to the bottom portion. The atomic distortion associated with a screw dislocation is also linear and along a dislocation line, line *AB* in Figure 6.10*b*. The screw dislocation derives its name from the spiral or helical path or ramp that is traced around the dislocation line by the atomic planes of atoms. Sometimes the symbol $\complement$ is used to designate a screw dislocation.

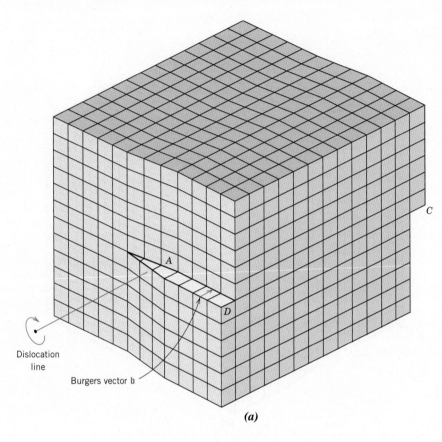

(a)

Figure 6.10 (*a*) A screw dislocation within a crystal. (*b*) The screw dislocation in (*a*) as viewed from above. The dislocation line extends along line *AB*. Atom positions above the slip plane are designated by open circles, those below by solid circles.
[Figure (*b*) from W. T. Read, Jr., *Dislocations in Crystals,* McGraw-Hill Book Company, New York, NY, 1953.]

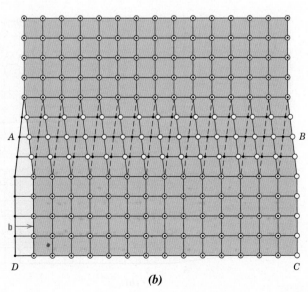

(b)

mixed dislocation

WileyPLUS: VMSE
Mixed

Most dislocations found in crystalline materials are probably neither pure edge nor pure screw but exhibit components of both types; these are termed **mixed dislocations**. All three dislocation types are represented schematically in Figure 6.11; the lattice distortion that is produced away from the two faces is mixed, having varying degrees of screw and edge character.

Figure 6.11 (*a*) Schematic representation of a dislocation that has edge, screw, and mixed character. (*b*) Top view, where open circles denote atom positions above the slip plane, and solid circles, atom positions below. At point *A*, the dislocation is pure screw, while at point *B*, it is pure edge. For regions in between where there is curvature in the dislocation line, the character is mixed edge and screw.
[Figure (*b*) from W. T. Read, Jr., *Dislocations in Crystals,* McGraw-Hill Book Company, New York, NY, 1953.]

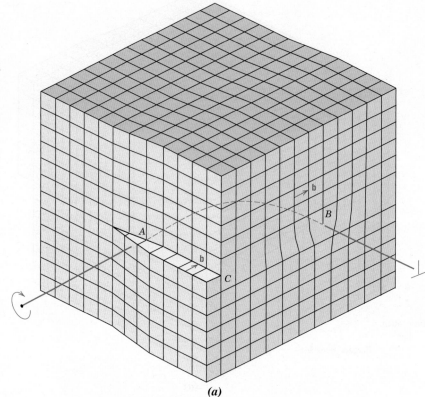

(a)

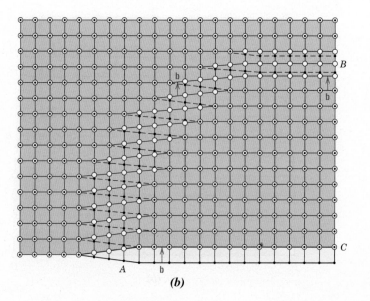

(b)

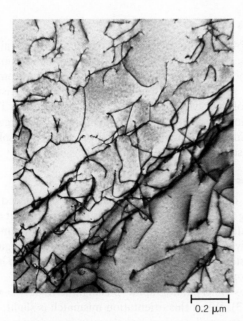

Figure 6.12 A transmission electron micrograph of a titanium alloy in which the dark lines are dislocations, 50,000×. (Courtesy of M. R. Plichta, Michigan Technological University.)

0.2 μm

Burgers vector

The magnitude and direction of the lattice distortion associated with a dislocation are expressed in terms of a **Burgers vector**, denoted by **b**. Burgers vectors are indicated in Figures 6.9 and 6.10 for edge and screw dislocations, respectively. Furthermore, the nature of a dislocation (i.e., edge, screw, or mixed) is defined by the relative orientations of dislocation line and Burgers vector. For an edge, they are perpendicular (Figure 6.9), whereas for a screw, they are parallel (Figure 6.10); they are neither perpendicular nor parallel for a mixed dislocation. Also, even though a dislocation changes direction and nature within a crystal (e.g., from edge to mixed to screw), the Burgers vector is the same at all points along its line. For example, all positions of the curved dislocation in Figure 6.11 have the Burgers vector shown. For metallic materials, the Burgers vector for a dislocation points in a close-packed crystallographic direction and is of magnitude equal to the interatomic spacing.

WileyPLUS

Tutorial Video:
Defects
Screw and Edge
Dislocations

As we note in Section 9.2, the permanent deformation of most crystalline materials is by the motion of dislocations. In addition, the Burgers vector is an element of the theory that has been developed to explain this type of deformation.

Dislocations can be observed in crystalline materials using electron-microscopic techniques. In Figure 6.12, a high-magnification transmission electron micrograph, the dark lines are the dislocations.

Virtually all crystalline materials contain some dislocations that were introduced during solidification, during plastic deformation, and as a consequence of thermal stresses that result from rapid cooling. Dislocations are involved in the plastic deformation of crystalline materials, both metals and ceramics, as discussed in Chapters 9 and 14. They have also been observed in polymeric materials; a screw dislocation is represented schematically in Figure 6.8.

6.8 INTERFACIAL DEFECTS

Interfacial defects are boundaries that have two dimensions and normally separate regions of the materials that have different crystal structures and/or crystallographic orientations. These imperfections include external surfaces, grain boundaries, phase boundaries, twin boundaries, and stacking faults.

External Surfaces

One of the most obvious boundaries is the external surface, along which the crystal structure terminates. Surface atoms are not bonded to the maximum number of nearest neighbors and are therefore in a higher energy state than the atoms at interior positions. The bonds of these surface atoms that are not satisfied give rise to a surface energy, expressed in units of energy per unit area (J/m^2 or erg/cm^2). To reduce this energy, materials tend to minimize, if at all possible, the total surface area. For example, liquids assume a shape having a minimum area—the droplets become spherical. Of course, this is not possible with solids, which are mechanically rigid.

Grain Boundaries

Another interfacial defect, the grain boundary, was introduced in Section 3.9 as the boundary separating two small grains or crystals having different crystallographic orientations in polycrystalline materials. A grain boundary is represented schematically from an atomic perspective in Figure 6.13. Within the boundary region, which is probably just several atom distances wide, there is some atomic mismatch in a transition from the crystalline orientation of one grain to that of an adjacent one.

Various degrees of crystallographic misalignment between adjacent grains are possible (Figure 6.13). When this orientation mismatch is slight, on the order of a few degrees, then the term *small- (or low-) angle grain boundary* is used. These boundaries can be described in terms of dislocation arrays. One simple small-angle grain boundary is formed when edge dislocations are aligned in the manner of Figure 6.14. This type is called a *tilt boundary;* the angle of misorientation, θ, is also indicated in the figure. When the angle of misorientation is parallel to the boundary, a *twist boundary* results, which can be described by an array of screw dislocations.

The atoms are bonded less regularly along a grain boundary (e.g., bond angles are longer), and consequently there is an interfacial or grain boundary energy similar to the surface energy just described. The magnitude of this energy is a function of the degree of misorientation, being larger for high-angle boundaries. Grain boundaries are more chemically reactive than the grains themselves as a consequence of this boundary energy. Furthermore, impurity atoms often preferentially segregate along these boundaries because of their higher

Figure 6.13 Schematic diagram showing small- and high-angle grain boundaries and the adjacent atom positions.

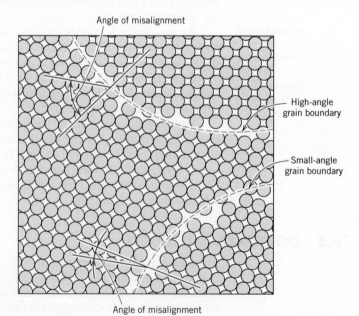

Angle of misalignment

High-angle grain boundary

Small-angle grain boundary

Angle of misalignment

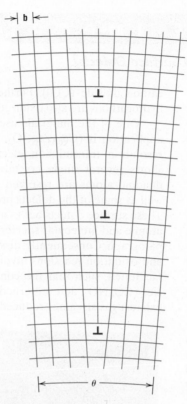

Figure 6.14 Demonstration of how a tilt boundary having an angle of misorientation θ results from an alignment of edge dislocations.

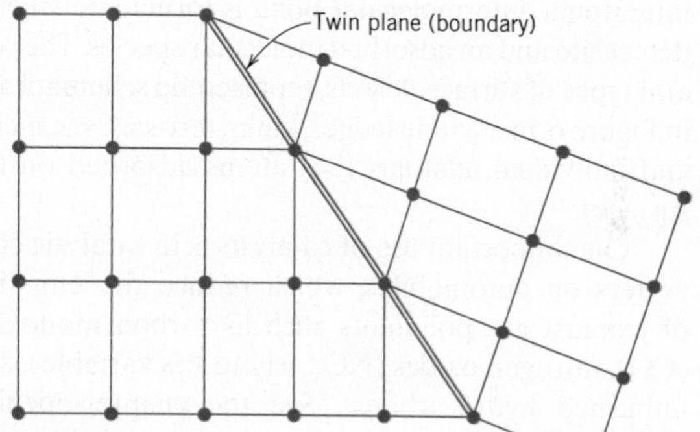

Figure 6.15 Schematic diagram showing a twin plane or boundary and the adjacent atom positions (colored circles).

energy state. The total interfacial energy is lower in large or coarse-grained materials than in fine-grained ones because there is less total boundary area in the former. Grains grow at elevated temperatures to reduce the total boundary energy, a phenomenon explained in Section 9.13.

In spite of this disordered arrangement of atoms and lack of regular bonding along grain boundaries, a polycrystalline material is still very strong; cohesive forces within and across the boundary are present. Furthermore, the density of a polycrystalline specimen is virtually identical to that of a single crystal of the same material.

Phase Boundaries

Phase boundaries exist in multiphase materials (Section 11.3), in which a different phase exists on each side of the boundary; furthermore, each of the constituent phases has its own distinctive physical and/or chemical characteristics. As we shall see in subsequent chapters, phase boundaries play an important role in determining the mechanical characteristics of some multiphase metal alloys.

Twin Boundaries

A *twin boundary* is a special type of grain boundary across which there is a specific mirror lattice symmetry; that is, atoms on one side of the boundary are located in mirror-image positions to those of the atoms on the other side (Figure 6.15). The region of

M A T E R I A L S   O F   I M P O R T A N C E

Catalysts (and Surface Defects)

A *catalyst* is a substance that speeds up the rate of a chemical reaction without participating in the reaction itself (i.e., it is not consumed). One type of catalyst exists as a solid; reactant molecules in a gas or liquid phase are adsorbed[9] onto the catalytic surface, at which point some type of interaction occurs that promotes an increase in their chemical reactivity rate.

Adsorption sites on a catalyst are normally surface defects associated with planes of atoms; an interatomic/intermolecular bond is formed between a defect site and an adsorbed molecular species. The several types of surface defects, represented schematically in Figure 6.16, include ledges, kinks, terraces, vacancies, and individual adatoms (i.e., atoms adsorbed on the surface).

One important use of catalysts is in catalytic converters on automobiles, which reduce the emission of exhaust gas pollutants such as carbon monoxide (CO), nitrogen oxides (NO_x, where x is variable), and unburned hydrocarbons. (See the chapter-opening diagrams and photograph for this chapter.) Air is introduced into the exhaust emissions from the automobile engine; this mixture of gases then passes over the catalyst, which on its surface adsorbs molecules of CO, NO_x, and O_2. The NO_x dissociates into N and O atoms, whereas the O_2 dissociates into its atomic species. Pairs of nitrogen atoms combine to form N_2 molecules, and carbon monoxide is oxidized to form

carbon dioxide (CO_2). Furthermore, any unburned hydrocarbons are also oxidized to CO_2 and H_2O.

One of the materials used as a catalyst in this application is $(Ce_{0.5}Zr_{0.5})O_2$. Figure 6.17 is a high-resolution transmission electron micrograph that shows several single crystals of this material. Individual atoms are resolved in this micrograph as well as some of the defects presented in Figure 6.16. These surface defects act as adsorption sites for the atomic and molecular species noted in the previous paragraph. Consequently, dissociation, combination, and oxidation reactions involving these species are facilitated, such that the content of pollutant species (CO, NO_x, and unburned hydrocarbons) in the exhaust gas stream is reduced significantly.

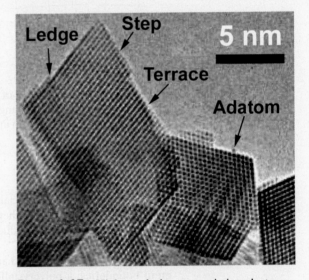

Figure 6.17 High-resolution transmission electron micrograph that shows single crystals of $(Ce_{0.5}Zr_{0.5})O_2$; this material is used in catalytic converters for automobiles. Surface defects represented schematically in Figure 6.16 are noted on the crystals.
[From W. J. Stark, L. Mädler, M. Maciejewski, S. E. Pratsinis, and A. Baiker, "Flame-Synthesis of Nanocrystalline Ceria/Zirconia: Effect of Carrier Liquid," *Chem. Comm.,* 588–589 (2003). Reproduced by permission of The Royal Society of Chemistry.]

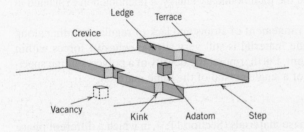

Figure 6.16 Schematic representations of surface defects that are potential adsorption sites for catalysis. Individual atom sites are represented as cubes.

---

[9]*Adsorption* is the adhesion of molecules of a gas or liquid to a solid surface. It should not be confused with *absorption*, which is the assimilation of molecules into a solid or liquid.

WileyPLUS

Tutorial Video:
Defects
Differences among
Point, Linear, and
Interfacial Defects

material between these boundaries is appropriately termed a *twin*. Twins result from atomic displacements that are produced from applied mechanical shear forces (mechanical twins) and also during annealing heat treatments following deformation (annealing twins). Twinning occurs on a definite crystallographic plane and in a specific direction, both of which depend on the crystal structure. Annealing twins are typically found in metals that have the FCC crystal structure, whereas mechanical twins are observed in BCC and HCP metals. The role of mechanical twins in the deformation process is discussed in Section 9.7. Annealing twins may be observed in the photomicrograph of the polycrystalline brass specimen shown in Figure 6.19c. The twins correspond to those regions having relatively straight and parallel sides and a different visual contrast than the untwinned regions of the grains within which they reside. An explanation for the variety of textural contrasts in this photomicrograph is provided in Section 6.12.

Miscellaneous Interfacial Defects

Other possible interfacial defects include stacking faults and ferromagnetic domain walls. Stacking faults are found in FCC metals when there is an interruption in the *ABCABCABC*... stacking sequence of close-packed planes (Section 4.17). For ferromagnetic and ferrimagnetic materials, the boundary that separates regions having different directions of magnetization is termed a *domain wall*, which is discussed in Section 21.7.

Associated with each of the defects discussed in this section is an interfacial energy, the magnitude of which depends on boundary type, and which varies from material to material. Normally, the interfacial energy is greatest for external surfaces and least for domain walls.

Concept Check 6.5   The surface energy of a single crystal depends on crystallographic orientation. Does this surface energy increase or decrease with an increase in planar density? Why?

[*The answer may be found at* www.wiley.com/college/callister *(Student Companion Site).*]

6.9 BULK OR VOLUME DEFECTS

Other defects exist in all solid materials that are much larger than those heretofore discussed. These include pores, cracks, foreign inclusions, and other phases. They are normally introduced during processing and fabrication steps. Some of these defects and their effects on the properties of materials are discussed in subsequent chapters.

6.10 ATOMIC VIBRATIONS

atomic vibration

Every atom in a solid material is vibrating very rapidly about its lattice position within the crystal. In a sense, these **atomic vibrations** may be thought of as imperfections or defects. At any instant of time, not all atoms vibrate at the same frequency and amplitude or with the same energy. At a given temperature, there exists a distribution of energies for the constituent atoms about an average energy. Over time, the vibrational energy of any specific atom also varies in a random manner. With rising temperature, this average energy increases, and, in fact, the temperature of a solid is really just a measure of the average vibrational activity of atoms and molecules. At room temperature, a typical vibrational frequency is on the order of 10^{13} vibrations per second, whereas the amplitude is a few thousandths of a nanometer.

Many properties and processes in solids are manifestations of this vibrational atomic motion. For example, melting occurs when the vibrations are vigorous enough to rupture large numbers of atomic bonds. A more detailed discussion of atomic vibrations and their influence on the properties of materials is presented in Chapter 20.

Microscopic Examination

6.11 BASIC CONCEPTS OF MICROSCOPY

On occasion it is necessary or desirable to examine the structural elements and defects that influence the properties of materials. Some structural elements are of *macroscopic* dimensions; that is, they are large enough to be observed with the unaided eye. For example, the shape and average size or diameter of the grains for a polycrystalline specimen are important structural characteristics. Macroscopic grains are often evident on aluminum streetlight posts and also on highway guardrails. Relatively large grains having different textures are clearly visible on the surface of the sectioned copper ingot shown in Figure 6.18. However, in most materials the constituent grains are of *microscopic* dimensions, having diameters that may be on the order of microns,[10] and their details must be investigated using some type of microscope. Grain size and shape are only two features of what is termed the **microstructure**; these and other microstructural characteristics are discussed in subsequent chapters.

microstructure

microscopy

Optical, electron, and scanning probe microscopes are commonly used in **microscopy**. These instruments aid in investigations of the microstructural features of all material types. Some of these techniques employ photographic equipment in conjunction with the microscope; the photograph on which the image is recorded is called a **photomicrograph**. In addition, many microstructural images are computer generated and/or enhanced.

photomicrograph

Microscopic examination is an extremely useful tool in the study and characterization of materials. Several important applications of microstructural examinations are as follows: to ensure that the associations between the properties and structure (and defects) are properly understood, to predict the properties of materials once these relationships have been established, to design alloys with new property combinations, to determine whether a material has been correctly heat-treated, and to ascertain the mode of mechanical fracture. Several techniques that are commonly used in such investigations are discussed next.

Figure 6.18 Cross-section of a cylindrical copper ingot. The small, needle-shape grains may be observed, which extend from the center radially outward.

1 in.

1 cm

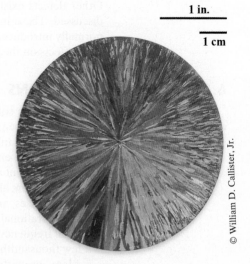

© William D. Callister, Jr.

---

[10]A micron (μm), sometimes called a *micrometer*, is 10^{-6} m.

6.12 MICROSCOPIC TECHNIQUES

Optical Microscopy

With *optical microscopy,* the light microscope is used to study the microstructure; optical and illumination systems are its basic elements. For materials that are opaque to visible light (all metals and many ceramics and polymers), only the surface is subject to observation, and the light microscope must be used in a reflecting mode. Contrasts in the image produced result from differences in reflectivity of the various regions of the microstructure. Investigations of this type are often termed *metallographic* because metals were first examined using this technique.

Normally, careful and meticulous surface preparations are necessary to reveal the important details of the microstructure. The specimen surface must first be ground and polished to a smooth and mirror-like finish. This is accomplished by using successively finer abrasive papers and powders. The microstructure is revealed by a surface treatment using an appropriate chemical reagent in a procedure termed *etching.* The chemical reactivity of the grains of some single-phase materials depends on crystallographic orientation. Consequently, in a polycrystalline specimen, etching characteristics vary from grain to grain. Figure 6.19*b* shows how normally incident light is reflected by three etched surface grains, each having a different orientation. Figure 6.19*a* depicts the surface structure as it might appear when viewed with the microscope; the luster or texture of each grain depends on its reflectance properties. A photomicrograph of a polycrystalline specimen exhibiting these characteristics is shown in Figure 6.19*c*.

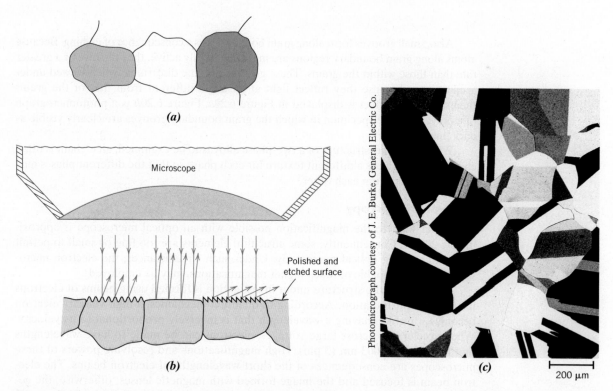

Photomicrograph courtesy of J. E. Burke, General Electric Co.

(a)

Microscope

Polished and etched surface

(b)

(c)

200 μm

Figure 6.19 (*a*) Polished and etched grains as they might appear when viewed with an optical microscope. (*b*) Section taken through these grains showing how the etching characteristics and resulting surface texture vary from grain to grain because of differences in crystallographic orientation. (*c*) Photomicrograph of a polycrystalline brass specimen, 60×.

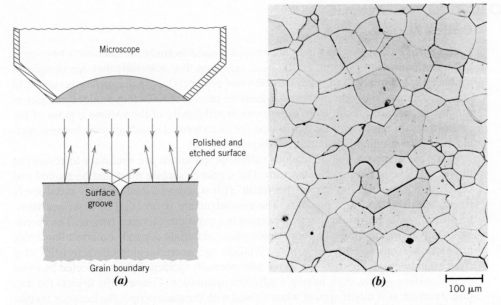

100 µm

Figure 6.20 (*a*) Section of a grain boundary and its surface groove produced by etching; the light reflection characteristics in the vicinity of the groove are also shown. (*b*) Photomicrograph of the surface of a polished and etched polycrystalline specimen of an iron–chromium alloy in which the grain boundaries appear dark, 100×.
[Photomicrograph courtesy of L. C. Smith and C. Brady, the National Bureau of Standards, Washington, DC (now the National Institute of Standards and Technology, Gaithersburg, MD.)]

Also, small grooves form along grain boundaries as a consequence of etching. Because atoms along grain boundary regions are more chemically active, they dissolve at a greater rate than those within the grains. These grooves become discernible when viewed under a microscope because they reflect light at an angle different from that of the grains themselves; this effect is displayed in Figure 6.20*a*. Figure 6.20*b* is a photomicrograph of a polycrystalline specimen in which the grain boundary grooves are clearly visible as dark lines.

When the microstructure of a two-phase alloy is to be examined, an etchant is often chosen that produces a different texture for each phase so that the different phases may be distinguished from each other.

Electron Microscopy

The upper limit to the magnification possible with an optical microscope is approximately 2000×. Consequently, some structural elements are too fine or small to permit observation using optical microscopy. Under such circumstances, the electron microscope, which is capable of much higher magnifications, may be employed.

An image of the structure under investigation is formed using beams of electrons instead of light radiation. According to quantum mechanics, a high-velocity electron becomes wavelike, having a wavelength that is inversely proportional to its velocity. When accelerated across large voltages, electrons can be made to have wavelengths on the order of 0.003 nm (3 pm). High magnifications and resolving powers of these microscopes are consequences of the short wavelengths of electron beams. The electron beam is focused and the image formed with magnetic lenses; otherwise, the geometry of the microscope components is essentially the same as with optical systems. Both transmission and reflection beam modes of operation are possible for electron microscopes.

Transmission Electron Microscopy

transmission electron microscope (TEM)

The image seen with a **transmission electron microscope (TEM)** is formed by an electron beam that passes through the specimen. Details of internal microstructural features are accessible to observation; contrasts in the image are produced by differences in beam scattering or diffraction produced between various elements of the microstructure or defect. Because solid materials are highly absorptive to electron beams, a specimen to be examined must be prepared in the form of a very thin foil; this ensures transmission through the specimen of an appreciable fraction of the incident beam. The transmitted beam is projected onto a fluorescent screen or a photographic film so that the image may be viewed. Magnifications approaching 1,000,000× are possible with transmission electron microscopy, which is frequently used to study dislocations.

Scanning Electron Microscopy

scanning electron microscope (SEM)

A more recent and extremely useful investigative tool is the **scanning electron microscope (SEM)**. The surface of a specimen to be examined is scanned with an electron beam, and the reflected (or *back-scattered*) beam of electrons is collected and then displayed at the same scanning rate on a cathode ray tube (CRT; similar to a CRT television screen). The image on the screen, which may be photographed, represents the surface features of the specimen. The surface may or may not be polished and etched, but it must be electrically conductive; a very thin metallic surface coating must be applied to nonconductive materials. Magnifications ranging from 10× to in excess of 50,000× are possible, as are also very great depths of field. Accessory equipment permits qualitative and semiquantitative analysis of the elemental composition of very localized surface areas.

Scanning Probe Microscopy

scanning probe microscope (SPM)

In the past two decades, the field of microscopy has experienced a revolution with the development of a new family of scanning probe microscopes. The **scanning probe microscope (SPM)**, of which there are several varieties, differs from optical and electron microscopes in that neither light nor electrons are used to form an image. Rather, the microscope generates a topographical map, on an atomic scale, that is a representation of surface features and characteristics of the specimen being examined. Some of the features that differentiate the SPM from other microscopic techniques are as follows:

- Examination on the nanometer scale is possible inasmuch as magnifications as high as $10^9\times$ are possible; much better resolutions are attainable than with other microscopic techniques.

- Three-dimensional magnified images are generated that provide topographical information about features of interest.

- Some SPMs may be operated in a variety of environments (e.g., vacuum, air, liquid); thus, a particular specimen may be examined in its most suitable environment.

Scanning probe microscopes employ a tiny probe with a very sharp tip that is brought into very close proximity (i.e., to within on the order of a nanometer) of the specimen surface. This probe is then raster-scanned across the plane of the surface. During scanning, the probe experiences deflections perpendicular to this plane in response to electronic or other interactions between the probe and specimen surface. The in-surface-plane and out-of-plane motions of the probe are controlled by piezoelectric (Section 19.25) ceramic components that have nanometer resolutions. Furthermore, these probe movements are monitored electronically and transferred to and stored in a computer, which then generates the three-dimensional surface image.

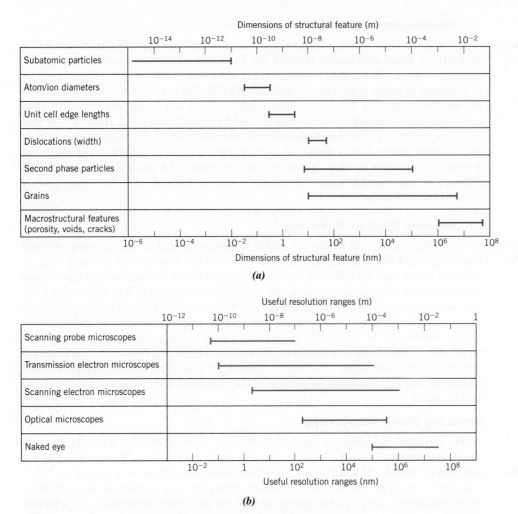

Figure 6.21 (*a*) Bar chart showing size ranges for several structural features found in materials. (*b*) Bar chart showing the useful resolution ranges for four microscopic techniques discussed in this chapter, in addition to the naked eye.
(Courtesy of Prof. Sidnei Paciornik, DCMM PUC-Rio, Rio de Janeiro, Brazil, and Prof. Carlos Pérez Bergmann, Federal University of Rio Grande do Sul, Porto Alegre, Brazil.)

These new SPMs, which allow examination of the surface of materials at the atomic and molecular level, have provided a wealth of information about a host of materials, from integrated circuit chips to biological molecules. Indeed, the advent of the SPMs has helped to usher in the era of *nanomaterials*—materials whose properties are designed by engineering atomic and molecular structures.

Figure 6.21*a* is a bar chart showing dimensional size ranges for several types of structures found in materials (note that the axes are scaled logarithmically). The useful dimensional resolution ranges for the several microscopic techniques discussed in this chapter (plus the naked eye) are presented in the bar chart of Figure 6.21*b*. For three of these techniques (SPM, TEM, and SEM), an upper resolution value is not imposed by the characteristics of the microscope and, therefore, is somewhat arbitrary and not well defined.

Furthermore, by comparing Figures 6.21*a* and 6.21*b*, it is possible to decide which microscopic technique(s) is (are) best suited for examination of each of the structure types.

6.13 GRAIN-SIZE DETERMINATION

grain size

The **grain size** is often determined when the properties of polycrystalline and single-phase materials are under consideration. In this regard, it is important to realize that for each material, the constituent grains have a variety of shapes and a distribution of sizes. Grain size may be specified in terms of average or mean grain diameter, and a number of techniques have been developed to measure this parameter.

Before the advent of the digital age, grain-size determinations were performed manually using photomicrographs. However, today, most techniques are automated and use digital images and image analyzers with the capacity to record, detect, and measure accurately features of the grain structure (i.e., grain intercept counts, grain boundary lengths, and grain areas).

We now briefly describe two common grain-size determination techniques: (1) *linear intercept*—counting numbers of grain boundary intersections by straight test lines; and (2) *comparison*—comparing grain structures with standardized charts, which are based upon grain areas (i.e., number of grains per unit area). Discussions of these techniques is from the manual perspective (using photomicrographs).

For the linear intercept method, lines are drawn randomly through several photomicrographs that show the grain structure (all taken at the same magnification). Grain boundaries intersected by all the line segments are counted. Let us represent the sum of the total number of intersections as P and the total length of all the lines by L_T. The mean intercept length $\bar{\ell}$ [in real space (at $1\times$—i.e., not magnified)], a measure of grain diameter, may be determined by the following expression:

$$\bar{\ell} = \frac{L_T}{PM} \tag{6.19}$$

where M is the magnification.

The comparison method of grain-size determination was devised by the American Society for Testing and Materials (ASTM).[11] The ASTM has prepared several standard comparison charts, all having different average grain sizes and referenced to photomicrographs taken at a magnification of $100\times$. To each chart is assigned a number ranging from 1 to 10, which is termed the *grain-size number*. A specimen must be prepared properly to reveal the grain structure, which is then photographed. Grain size is expressed as the grain-size number of the chart that most nearly matches the grains in the micrograph. Thus, a relatively simple and convenient visual determination of grain-size number is possible. Grain-size number is used extensively in the specification of steels.

Relationship between ASTM grain size number and number of grains per square inch (at 100×)

The rationale behind the assignment of the grain-size number to these various charts is as follows: Let G represent the grain-size number, and let n be the average number of grains per square inch at a magnification of $100\times$. These two parameters are related to each other through the expression[12]

$$n = 2^{G-1} \tag{6.20}$$

---

[11]ASTM Standard E112, "Standard Test Methods for Determining Average Grain Size."

[12]Please note that in this edition, the symbol n replaces N from previous editions; also, G in Equation 6.20 is used in place of the previous n. Equation 6.20 is the standard notation currently used in the literature.

For photomicrographs taken at magnifications other than 100×, use of the following modified form of Equation 6.20 is necessary:

$$n_M\left(\frac{M}{100}\right)^2 = 2^{G-1} \qquad (6.21)$$

In this expression, n_M is the number of grains per square inch at magnification M. In addition, the inclusion of the $\left(\frac{M}{100}\right)^2$ term makes use of the fact that, whereas magnification is a length parameter, area is expressed in terms of units of length squared. As a consequence, the number of grains per unit area increases with the square of the increase in magnification.

Relationships have been developed that relate mean intercept length to ASTM grain-size number; these are as follows:

$$G = -6.6457 \log \bar{\ell} - 3.298 \quad \text{(for } \bar{\ell} \text{ in mm)} \qquad (6.22)$$

At this point, it is worthwhile to discuss the representation of magnification (i.e., linear magnification) for a micrograph. Sometimes magnification is specified in the micrograph legend (e.g., "60×" for Figure 6.19b); this means the micrograph represents a 60 times enlargement of the specimen in real space. *Scale bars* are also used to express degree of magnification. A scale bar is a straight line (typically horizontal), either superimposed on or located near the micrograph image. Associated with the bar is a length, typically expressed in microns; this value represents the distance in magnified space corresponding to the scale line length. For example, in Figure 6.20b, a scale bar is located below the bottom right-hand corner of the micrograph; its "100 μm" notation indicates that 100 μm correlates with the scale bar length.

To compute magnification from a scale bar, the following procedure may be used:

1. Measure the length of the scale bar in millimeters using a ruler.

2. Convert this length into microns [i.e., multiply the value in step (1) by 1000 because there are 1000 microns in a millimeter].

3. Magnification M is equal to

$$M = \frac{\text{measured scale length (converted to microns)}}{\text{the number appearing by the scale bar (in microns)}} \qquad (6.23)$$

For example, for Figure 6.20b, the measured scale length is approximately 10 mm, which is equivalent to (10 mm)(1000 μm/mm) = 10,000 μm. Inasmuch as the scale bar length is 100 μm, the magnification is equal to

$$M = \frac{10,000 \ \mu m}{100 \ \mu m} = 100\times$$

This is the value given in the figure legend.

Concept Check 6.6   Does the grain-size number (G of Equation 6.20) increase or decrease with decreasing grain size? Why?

[*The answer may be found at* www.wiley.com/college/callister *(Student Companion Site).*]

EXAMPLE PROBLEM 6.7

Grain-Size Computations Using ASTM and Intercept Methods

The following is a schematic micrograph that represents the microstructure of some hypothetical metal.

    Determine the following:

(a) Mean intercept length
(b) ASTM grain-size number, G using Equation 6.22

Solution

(a) We first determine the magnification of the micrograph using Equation 6.23. The scale bar length is measured and found to be 16 mm, which is equal to 16,000 μm; and because the scale bar number is 100 μm, the magnification is

$$M = \frac{16{,}000\ \mu m}{100\ \mu m} = 160\times$$

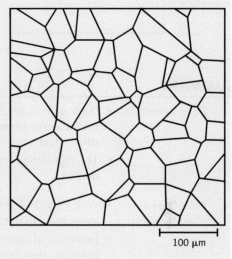

100 μm

The following sketch is the same micrograph on which have been drawn seven straight lines (in red), which have been numbered.

    The length of each line is 50 mm, and thus the total line length (L_T in Equation 6.19) is

$$(7\ \text{lines})(50\ \text{mm/line}) = 350\ \text{mm}$$

Tabulated next is the number of grain-boundary intersections for each line:

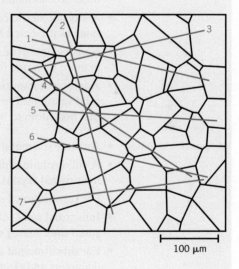

100 μm

Line Number	Number of Grain-Boundary Intersections
1	8
2	8
3	8
4	9
5	9
6	9
7	7
Total	58

Thus, inasmuch as $L_T = 350$ mm, $P = 58$ grain-boundary intersections, and the magnification $M = 160\times$, the mean intercept length $\bar{\ell}$ (in millimeters in real space), Equation 6.19, is equal to

$$\bar{\ell} = \frac{L_T}{PM}$$

$$= \frac{350\ \text{mm}}{(58\ \text{grain boundary intersections})(160\times)} = 0.0377\ \text{mm}$$

(b) The value of G is determined by substitution of this value for $\bar{\ell}$ into Equation 6.22; therefore,

$$G = -6.6457 \log \bar{\ell} - 3.298$$
$$= (-6.6457) \log(0.0377) - 3.298$$
$$= 6.16$$

SUMMARY

Point Defects in Metals
- Point defects are those associated with one or two atomic positions; these include vacancies (or vacant lattice sites) and self-interstitials (host atoms that occupy interstitial sites).
- The equilibrium number of vacancies depends on temperature according to Equation 6.1.

Point Defects in Ceramics
- With regard to atomic point defects in ceramics, interstitials and vacancies for each anion and cation type are possible (Figure 6.2).
- Inasmuch as electrical charges are associated with atomic point defects in ceramic materials, defects sometimes occur in pairs as (e.g., Frenkel and Schottky defects) in order to maintain charge neutrality.
- A stoichiometric ceramic is one in which the ratio of cations to anions is exactly the same as predicted by the chemical formula.
- Nonstoichiometric materials are possible in cases in which one of the ions may exist in more than one ionic state—for example, $Fe_{(1-x)}O$ for Fe^{2+} and Fe^{3+}.
- Addition of impurity atoms may result in the formation of substitutional or interstitial solid solutions. For substitutional solid solutions, an impurity atom will substitute for that host atom to which it is most similar in an electrical sense.

Impurities in Solids
- An alloy is a metallic substance that is composed of two or more elements.
- A solid solution may form when impurity atoms are added to a solid, in which case the original crystal structure is retained and no new phases are formed.
- For substitutional solid solutions, impurity atoms substitute for host atoms.
- Interstitial solid solutions form for relatively small impurity atoms that occupy interstitial sites among the host atoms.
- For substitutional solid solutions, appreciable solubility is possible only when atomic diameters and electronegativities for both atom types are similar, when both elements have the same crystal structure, and when the impurity atoms have a valence that is the same as or less than the host material.

Point Defects in Polymers
- Although the point defect concept in polymers is different than in metals and ceramics, vacancies, interstitial atoms, and impurity atoms/ions and groups of atoms/ions as interstitials have been found to exist in crystalline regions.
- Other defects include chains ends, dangling and loose chains, and dislocations. (Figure 6.8).

Specification of Composition
- Composition of an alloy may be specified in weight percent (on the basis of mass fraction; Equations 6.6a and 6.6b) or atom percent (on the basis of mole or atom fraction; Equations 6.8a and 6.8b).

- Expressions were provided that allow conversion of weight percent to atom percent (Equation 6.9a) and vice versa (Equation 6.10a).
- Computations of average density and average atomic weight for a two-phase alloy are possible using other equations cited in this chapter (Equations 6.13a, 6.13b, 6.14a, and 6.14b).

Dislocations—Linear Defects

- Dislocations are one-dimensional crystalline defects, of which there are two pure types: edge and screw.
  An edge may be thought of in terms of the lattice distortion along the end of an extra half-plane of atoms.
  A screw is as a helical planar ramp.
  For mixed dislocations, components of both pure edge and screw are found.
- The magnitude and direction of lattice distortion associated with a dislocation are specified by its Burgers vector.
- The relative orientations of Burgers vector and dislocation line are (1) perpendicular for edge, (2) parallel for screw, and (3) neither perpendicular nor parallel for mixed.

Interfacial Defects

- In the vicinity of a grain boundary (which is several atomic distances wide), there is some atomic mismatch between two adjacent grains that have different crystallographic orientations.
- For a high-angle grain boundary, the angle of misalignment between grains is relatively large; this angle is relatively small for small-angle grain boundaries.
- Across a twin boundary, atoms on one side reside in mirror-image positions to those of atoms on the other side.

Microscopic Techniques

- The microstructure of a material consists of defects and structural elements that are of microscopic dimensions. Microscopy is the observation of microstructure using some type of microscope.
- Both optical and electron microscopes are employed, usually in conjunction with photographic equipment.
- Transmissive and reflective modes are possible for each microscope type; preference is dictated by the nature of the specimen, as well as by the structural element or defect to be examined.
- In order to observe the grain structure of a polycrystalline material using an optical microscope, the specimen surface must be ground and polished to produce a very smooth and mirrorlike finish. Some type of chemical reagent (or etchant) must then be applied to either reveal the grain boundaries or produce a variety of light reflectance characteristics for the constituent grains.
- The two types of electron microscopes are transmission (TEM) and scanning (SEM).
  For TEM an image is formed from an electron beam that is scattered and/or diffracted while passing through the specimen.
  SEM employs an electron beam that raster-scans the specimen surface; an image is produced from backscattered or reflected electrons.
- A scanning probe microscope employs a small and sharp-tipped probe that raster-scans the specimen surface. Out-of-plane deflections of the probe result from interactions with surface atoms. A computer-generated and three-dimensional image of the surface results having nanometer resolution.

Grain-Size Determination

- With the intercept method used to measure grain size, a series of straight-line segments are drawn on the photomicrograph. The number of grain boundaries that are intersected by these lines are counted, and the *mean intercept length* (a measure of grain diameter) is computed using Equation 6.19.

- Comparison of a photomicrograph (taken at a magnification of 100×) with ASTM standard comparison charts may be used to specify grain size in terms of a grain-size number.
- The average number of grains per square inch at a magnification of 100× is related to grain-size number according to Equation 6.20; for magnifications other than 100×, Equation 6.21 is used.
- Grain-size number and mean intercept length are related per Equation 6.22.

Important Terms and Concepts

alloy	imperfection	screw dislocation
atom percent	interstitial solid solution	self-interstitial
atomic vibration	microscopy	solid solution
Boltzmann's constant	microstructure	solute
Burgers vector	mixed dislocation	solvent
composition	photomicrograph	stoichiometry
defect structure	point defect	substitutional solid solution
dislocation line	scanning electron microscope	transmission electron microscope
edge dislocation	(SEM)	(TEM)
electroneutrality	scanning probe microscope	vacancy
Frenkel defect	(SPM)	weight percent
grain size	Schottky defect	

REFERENCES

ASM Handbook, Vol. 9, *Metallography and Microstructures,* ASM International, Materials Park, OH, 2004.

Brandon, D., and W. D. Kaplan, *Microstructural Characterization of Materials,* 2nd edition, Wiley, Hoboken, NJ, 2008.

Chiang, Y. M., D. P. Birnie III, and W. D. Kingery, *Physical Ceramics: Principles for Ceramic Science and Engineering,* Wiley, New York, 1997.

Clarke, A. R., and C. N. Eberhardt, *Microscopy Techniques for Materials Science,* CRC Press, Boca Raton, FL, 2002.

Kingery, W. D., H. K. Bowen, and D. R. Uhlmann, *Introduction to Ceramics,* 2nd edition, Wiley, New York, 1976. Chapters 4 and 5.

Van Bueren, H.G., *Imperfections in Crystals,* North-Holland, Amsterdam, 1960.

Vander Voort, G. F., *Metallography, Principles and Practice,* ASM International, Materials Park, OH, 1999.

QUESTIONS AND PROBLEMS

Vacancies and Self-Interstitials

6.1 Calculate the fraction of atom sites that are vacant for lead at its melting temperature of 327°C (600 K). Assume an energy for vacancy formation of 0.52 eV/atom.

6.2 Calculate the number of vacancies per cubic meter in iron at 855°C. The energy for vacancy formation is 1.08 eV/atom. Furthermore, the density and atomic weight for Fe are 7.65 g/cm³ (at 850°C) and 55.85 g/mol, respectively.

6.3 Calculate the activation energy for vacancy formation in aluminum, given that the equilibrium number of vacancies at 500°C (773 K) is 7.55×10^{23} m^{-3}. The atomic weight and density

(at 500°C) for aluminum are, respectively, 26.98 g/mol and 2.62 g/cm³.

Imperfections in Ceramics

6.4 Would you expect Frenkel defects for anions to exist in ionic ceramics in relatively large concentrations? Why or why not?

6.5 Calculate the fraction of lattice sites that are Schottky defects for sodium chloride at its melting temperature (801°C). Assume an energy for defect formation of 2.3 eV.

6.6 Calculate the number of Frenkel defects per cubic meter in zinc oxide at 1000°C. The energy for defect formation is 2.51 eV, whereas the density for ZnO is 5.55 g/cm³ at (1000°C).

6.7 Using the following data that relate to the formation of Schottky defects in some oxide ceramic (having the chemical formula MO), determine the following:

(a) The energy for defect formation (in eV)

(b) The equilibrium number of Schottky defects per cubic meter at 1000°C

(c) The identity of the oxide (i.e., what is the metal M?)

T (°C)	ρ (g/cm³)	N_s (m⁻³)
750	5.50	9.21×10^{19}
1000	5.44	?
1250	5.37	5.0×10^{22}

6.8 In your own words, briefly define the term *stoichiometric*.

Impurities in Solids

6.9 Atomic radius, crystal structure, electronegativity, and the most common valence are tabulated in the following table for several elements; for those that are nonmetals, only atomic radii are indicated.

Element	Atomic Radius (nm)	Crystal Structure	Electro- negativity	Valence
Cu	0.1278	FCC	1.9	+2
C	0.071			
H	0.046			
O	0.060			
Ag	0.1445	FCC	1.9	+1
Al	0.1431	FCC	1.5	+3
Co	0.1253	HCP	1.8	+2
Cr	0.1249	BCC	1.6	+3
Fe	0.1241	BCC	1.8	+2
Ni	0.1246	FCC	1.8	+2
Pd	0.1376	FCC	2.2	+2
Pt	0.1387	FCC	2.2	+2
Zn	0.1332	HCP	1.6	+2

Which of these elements would you expect to form the following with copper:

(a) A substitutional solid solution having complete solubility

(b) A substitutional solid solution of incomplete solubility

(c) An interstitial solid solution

6.10 For both FCC and BCC crystal structures, there are two different types of interstitial sites. In each case, one site is larger than the other and is normally occupied by impurity atoms. For FCC, this larger one is located at the center of each edge of the unit cell; it is termed an *octahedral interstitial site*. On the other hand, with BCC the larger site type is found at $0\ \frac{1}{2}\ \frac{1}{4}$ positions—that is, lying on {100} faces and situated midway between two unit cell edges on this face and one-quarter of the distance between the other two unit cell edges; it is termed a *tetrahedral interstitial site*. For both FCC and BCC crystal structures, compute the radius r of an impurity atom that will just fit into one of these sites in terms of the atomic radius R of the host atom.

6.11 If cupric oxide (CuO) is exposed to reducing atmospheres at elevated temperatures, some of the Cu^{2+} ions will become Cu^+.

(a) Under these conditions, name one crystalline defect that you would expect to form in order to maintain charge neutrality.

(b) How many Cu^+ ions are required for the creation of each defect?

(c) How would you express the chemical formula for this nonstoichiometric material?

6.12 (a) Suppose that Li_2O is added as an impurity to CaO. If the Li^+ substitutes for Ca^{2+}, what kind of vacancies would you expect to form? How many of these vacancies are created for every Li^+ added?

(b) Suppose that $CaCl_2$ is added as an impurity to CaO. If the Cl substitutes for O^{2-} what kind of vacancies would you expect to form? How many of the vacancies are created for every Cl^- added?

6.13 What point defects are possible for Al_2O_3 as an impurity in MgO? How many Al^{3+} ions must be added to form each of these defects?

Specification of Composition

6.14 Derive the following equations:

(a) Equation 6.10a

(b) Equation 6.12a

(c) Equation 6.13a

(d) Equation 6.14a

6.15 What is the composition, in atom percent, of an alloy that consists of 30 wt% Zn and 70 wt% Cu?

6.16 What is the composition, in weight percent, of an alloy that consists of 6 at% Pb and 94 at% Sn?

6.17 Calculate the composition, in weight percent, of an alloy that contains 218.0 kg titanium, 15 kg aluminum, and 10 kg vanadium.

6.18 What is the composition, in atom percent, of an alloy that contains 100 g tin and 68 g lead?

6.19 What is the composition, in atom percent, of an alloy that contains 45.2 kg copper, 46.3 kg zinc, and 0.95 kg lead?

6.20 What is the composition, in atom percent, of an alloy that consists of 97 wt% Fe and 3 wt% Si?

6.21 Convert the atom percent composition in Problem 4.11 to weight percent.

6.22 Calculate the number of atoms per cubic meter in aluminum.

6.23 The concentration of carbon in an iron–carbon alloy is 0.15 wt%. What is the concentration in kilograms of carbon per cubic meter of alloy?

6.24 Determine the approximate density of a high-leaded brass that has a composition of 64.5 wt% Cu, 33.5 wt% Zn, and 2 wt% Pb.

6.25 Calculate the unit cell edge length for an 85 wt% Fe–15 wt% V alloy. All of the vanadium is in solid solution, and at room temperature the crystal structure for this alloy is BCC.

6.26 Some hypothetical alloy is composed of 12.5 wt% of metal A and 87.5 wt% of metal B. If the densities of metals A and B are 4.25 and 6.35 g/cm^3, respectively, whereas their respective atomic weights are 61.0 and 125.7 g/mol, determine whether the crystal structure for this alloy is simple cubic, face-centered cubic, or body-centered cubic. Assume a unit cell edge length of 0.395 nm.

6.27 For a solid solution consisting of two elements (designated as 1 and 2), sometimes it is desirable to determine the number of atoms per cubic centimeter of one element in a solid solution, N_1, given the concentration of that element specified in weight percent, C_1. This computation is possible using the following expression:

$$N_1 = \frac{N_A C_1}{\frac{C_1 A_1}{\rho_1} + \frac{A_1}{\rho_2}(100 - C_1)} \quad (6.24)$$

where

N_A = Avogadro's number
ρ_1 and ρ_2 = densities of the two elements
A_1 = the atomic weight of element 1

Derive Equation 6.24 using Equation 6.2 and expressions contained in Section 6.6.

6.28 Gold forms a substitutional solid solution with silver. Compute the number of gold atoms per cubic centimeter for a silver–gold alloy that contains 10 wt% Au and 90 wt% Ag. The densities of pure gold and silver are 19.32 and 10.49 g/cm3, respectively.

6.29 Germanium forms a substitutional solid solution with silicon. Compute the number of germanium atoms per cubic centimeter for a germanium–silicon alloy that contains 15 wt% Ge and 85 wt% Si. The densities of pure germanium and silicon are 5.32 and 2.33 g/cm^3, respectively.

6.30 Sometimes it is desirable to determine the weight percent of one element, C_1, that will produce a specified concentration in terms of the number of atoms per cubic centimeter, N_1, for an alloy composed of two types of atoms. This computation is possible using the following expression:

$$C_1 = \frac{100}{1 + \frac{N_A \rho_2}{N_1 A_1} - \frac{\rho_2}{\rho_1}} \quad (6.25)$$

where

N_A = Avogadro's number
ρ_1 and ρ_2 = densities of the two elements
A_1 and A_2 = the atomic weights of the two elements

Derive Equation 6.25 using Equation 6.2 and expressions contained in Section 6.6.

6.31 Molybdenum forms a substitutional solid solution with tungsten. Compute the weight percent of molybdenum that must be added to tungsten to yield an alloy that contains 1.0×10^{22} Mo atoms per cubic centimeter. The densities of pure Mo and W are 10.22 and 19.30 g/cm^3, respectively.

6.32 Niobium forms a substitutional solid solution with vanadium. Compute the weight percent of niobium that must be added to vanadium to yield an alloy that contains 1.55×10^{22} Nb atoms per cubic centimeter. The densities of pure Nb and V are 8.57 and 6.10 g/cm^3, respectively.

6.33 Silver and palladium both have the FCC crystal structure, and Pd forms a substitutional solid solution for all concentrations at room temperature. Compute the unit cell edge length for a 75 wt% Ag–25 wt% Pd alloy. The room-temperature density of Pd is 12.02 g/cm^3, and its atomic weight and atomic radius are 106.4 g/mol and 0.138 nm, respectively.

Dislocations—Linear Defects

6.34 Cite the relative Burgers vector–dislocation line orientations for edge, screw, and mixed dislocations.

Interfacial Defects

6.35 For an FCC single crystal, would you expect the surface energy for a (100) plane to be greater or less than that for a (111) plane? Why? (*Note:* You may want to consult the solution to Problem 4.57 at the end of Chapter 4.)

6.36 For a BCC single crystal, would you expect the surface energy for a (100) plane to be greater or less than that for a (110) plane? Why? (*Note:* You may want to consult the solution to Problem 4.58 at the end of Chapter 4.)

6.37 (a) For a given material, would you expect the surface energy to be greater than, the same as, or less than the grain boundary energy? Why?

(b) The grain boundary energy of a small-angle grain boundary is less than for a high-angle one. Why is this so?

6.38 (a) Briefly describe a twin and a twin boundary.

(b) Cite the difference between mechanical and annealing twins.

6.39 For each of the following stacking sequences found in FCC metals, cite the type of planar defect that exists:

(a) . . . *A B C A B C B A C B A* . . .

(b) . . . *A B C A B C B C A B C* . . .

Now, copy the stacking sequences and indicate the position(s) of planar defect(s) with a vertical dashed line.

Grain Size Determination

6.40 (a) Using the intercept method, determine the average grain size, in millimeters, of the specimen whose microstructure is shown in Figure 6.20*b*; use at least seven straight-line segments.

(b) Estimate the ASTM grain size number for this material.

6.41 (a) Employing the intercept technique, determine the average grain size for the steel specimen whose microstructure is shown in Figure 11.24*a*; use at least seven straight-line segments.

(b) Estimate the ASTM grain size number for this material.

6.42 For an ASTM grain size of 8, approximately how many grains would there be per square centimeter

(a) at a magnification of 100×, and

(b) without any magnification?

6.43 Determine the ASTM grain size number if 25 grains per square centimeter are measured at a magnification of 600×.

6.44 Determine the ASTM grain size number if 20 grains per square centimeter are measured at a magnification of 50×.

Spreadsheet Problems

6.1SS Generate a spreadsheet that allows the user to convert the concentration of one element of a two-element metal alloy from weight percent to atom percent.

6.2SS Generate a spreadsheet that allows the user to convert the concentration of one element of a two-element metal alloy from atom percent to weight percent.

6.3SS Generate a spreadsheet that allows the user to convert the concentration of one element of a two-element metal alloy from weight percent to number of atoms per cubic centimeter.

6.4SS Generate a spreadsheet that allows the user to convert the concentration of one element of a two-element metal alloy from number of atoms per cubic centimeter to weight percent.

DESIGN PROBLEMS

Specification of Composition

6.D1 Aluminum–lithium alloys have been developed by the aircraft industry to reduce the weight and improve the performance of its aircraft. A commercial aircraft skin material having a density of 2.50 g/cm³ is desired. Compute the concentration of Li (in wt%) that is required.

6.D2 Iron and vanadium both have the BCC crystal structure, and V forms a substitutional solid solution in Fe for concentrations up to approximately 20 wt% V at room temperature. Determine the concentration in weight percent of V that must be added to iron to yield a unit cell edge length of 0.290 nm.

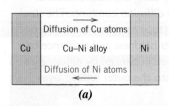
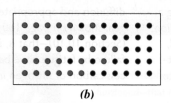
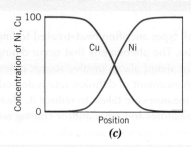

(a)　　　　　　　**(b)**　　　　　　　**(c)**

Figure 7.2 (*a*) A copper–nickel diffusion couple after a high-temperature heat treatment, showing the alloyed diffusion zone. (*b*) Schematic representations of Cu (red circles) and Ni (blue circles) atom locations within the couple. (*c*) Concentrations of copper and nickel as a function of position across the couple.

nickel at the two extremities of the couple, separated by an alloyed region. Concentrations of both metals vary with position as shown in Figure 7.2*c*. This result indicates that copper atoms have migrated or diffused into the nickel, and that nickel has diffused into copper. The process by which atoms of one metal diffuse into another is termed **interdiffusion**, or **impurity diffusion**.

interdiffusion

impurity diffusion

Interdiffusion may be discerned from a macroscopic perspective by changes in concentration that occur over time, as in the example for the Cu–Ni diffusion couple. There is a net drift or transport of atoms from high- to low-concentration regions. Diffusion also occurs for pure metals, but all atoms exchanging positions are of the same type; this is termed **self-diffusion**. Of course, self-diffusion is not normally subject to observation by noting compositional changes.

self-diffusion

7.2 DIFFUSION MECHANISMS

From an atomic perspective, diffusion is just the stepwise migration of atoms from lattice site to lattice site. In fact, the atoms in solid materials are in constant motion, rapidly changing positions. For an atom to make such a move, two conditions must be met: (1) there must be an empty adjacent site, and (2) the atom must have sufficient energy to break bonds with its neighbor atoms and then cause some lattice distortion during the displacement. This energy is vibrational in nature (Section 6.10). At a specific temperature, some small fraction of the total number of atoms is capable of diffusive motion, by virtue of the magnitudes of their vibrational energies. This fraction increases with rising temperature.

Several different models for this atomic motion have been proposed; of these possibilities, two dominate for metallic diffusion.

Vacancy Diffusion

One mechanism involves the interchange of an atom from a normal lattice position to an adjacent vacant lattice site or vacancy, as represented schematically in Figure 7.3*a*. This mechanism is aptly termed **vacancy diffusion**. Of course, this process necessitates the presence of vacancies, and the extent to which vacancy diffusion can occur is a function of the number of these defects that are present; significant concentrations of vacancies may exist in metals at elevated temperatures (Section 6.2). Because diffusing atoms and vacancies exchange positions, the diffusion of atoms in one direction corresponds to the motion of vacancies in the opposite direction. Both self-diffusion and interdiffusion occur by this mechanism; for the latter, the impurity atoms must substitute for host atoms.

vacancy diffusion

Interstitial Diffusion

The second type of diffusion involves atoms that migrate from an interstitial position to a neighboring one that is empty. This mechanism is found for interdiffusion of impurities

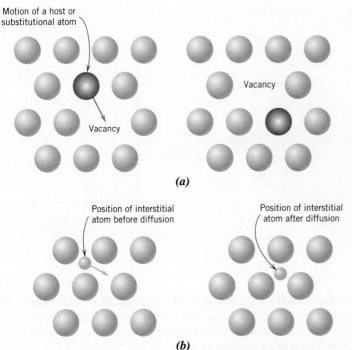

Motion of a host or
substitutional atom

Vacancy

Vacancy

(a)

Position of interstitial
atom before diffusion

Position of interstitial
atom after diffusion

(b)

Figure 7.3   Schematic representations of
(*a*) vacancy diffusion and (*b*) interstitial
diffusion.

such as hydrogen, carbon, nitrogen, and oxygen, which have atoms that are small enough to fit into the interstitial positions. Host or substitutional impurity atoms rarely form interstitials and do not normally diffuse via this mechanism. This phenomenon is appropriately termed **interstitial diffusion** (Figure 7.3*b*).

interstitial diffusion

In most metal alloys, interstitial diffusion occurs much more rapidly than diffusion by the vacancy mode, because the interstitial atoms are smaller and thus more mobile. Furthermore, there are more empty interstitial positions than vacancies; hence, the probability of interstitial atomic movement is greater than for vacancy diffusion.

7.3   STEADY-STATE DIFFUSION

Diffusion is a *time-dependent process*—that is, in a macroscopic sense, the quantity of an element that is transported within another is a function of time. Often it is necessary to know how fast diffusion occurs, or the *rate of mass transfer*. This rate is frequently expressed as a **diffusion flux** (*J*), defined as the mass (or, equivalently, the number of atoms) *M* diffusing through and perpendicular to a unit cross-sectional area of solid per unit of time. In mathematical form, this may be represented as

diffusion flux

Definition of
diffusion flux

$$J = \frac{M}{At}$$ (7.1)

where *A* denotes the area across which diffusion is occurring and *t* is the elapsed diffusion time. The units for *J* are kilograms or atoms per meter squared per second (kg/m^2·s or atoms/m^2·s).

The mathematics of steady-state diffusion in a single (*x*) direction is relatively simple, in that the flux is proportional to the concentration gradient, $\frac{dC}{dx}$ through the expression

Fick's first law—
diffusion flux for
steady-state diffusion
(in one direction)

$$J = -D\frac{dC}{dx}$$ (7.2)

Figure 7.4
(*a*) Steady-state
diffusion across a thin
plate. (*b*) A linear
concentration profile
for the diffusion
situation in (*a*).

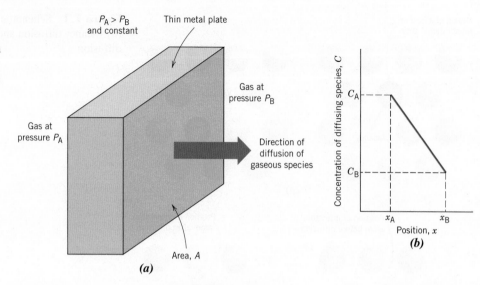

(a)

(b)

Fick's first law

diffusion coefficient

This equation is sometimes called **Fick's first law**. The constant of proportionality D is called the **diffusion coefficient**, which is expressed in square meters per second. The negative sign in this expression indicates that the direction of diffusion is down the concentration gradient, from a high to a low concentration.

Fick's first law may be applied to the diffusion of atoms of a gas through a thin metal plate for which the concentrations (or pressures) of the diffusing species on both surfaces of the plate are held constant, a situation represented schematically in Figure 7.4*a*. This diffusion process eventually reaches a state wherein the diffusion flux does not change with time—that is, the mass of diffusing species entering the plate on the high-pressure side is equal to the mass exiting from the low-pressure surface—such that there is no net accumulation of diffusing species in the plate. This is an example of what

steady-state diffusion

is termed **steady-state diffusion**.

When concentration C is plotted versus position (or distance) within the solid x, the

concentration profile

resulting curve is termed the **concentration profile**; furthermore, **concentration gradient**

concentration gradient

is the slope at a particular point on this curve. In the present treatment, the concentration profile is assumed to be linear, as depicted in Figure 7.4*b*, and

$$\text{concentration gradient} = \frac{dC}{dx} = \frac{\Delta C}{\Delta x} = \frac{C_A - C_B}{x_A - x_B} \tag{7.3}$$

For diffusion problems, it is sometimes convenient to express concentration in terms of mass of diffusing species per unit volume of solid (kg/m^3 or g/cm^3).[1]

driving force

Sometimes the term **driving force** is used in the context of what compels a reaction to occur. For diffusion reactions, several such forces are possible; but when diffusion is according to Equation 7.2, the concentration gradient is the driving force.[2]

WileyPLUS

Tutorial Video:
Steady-State and
Nonsteady-State
Diffusion

One practical example of steady-state diffusion is found in the purification of hydrogen gas. One side of a thin sheet of palladium metal is exposed to the impure gas composed of hydrogen and other gaseous species such as nitrogen, oxygen, and water vapor. The hydrogen selectively diffuses through the sheet to the opposite side, which is maintained at a constant and lower hydrogen pressure.

---

[1]Conversion of concentration from weight percent to mass per unit volume (kg/m^3) is possible using Equation 6.12.

[2]Another driving force is responsible for phase transformations. Phase transformations are topics of discussion in Chapters 11 and 12.

EXAMPLE PROBLEM 7.1

Diffusion Flux Computation

A plate of iron is exposed to a carburizing (carbon-rich) atmosphere on one side and a decarburizing (carbon-deficient) atmosphere on the other side at 700°C. If a condition of steady state is achieved, calculate the diffusion flux of carbon through the plate if the concentrations of carbon at positions of 5 and 10 mm (5×10^{-3} and 10^{-2} m) beneath the carburizing surface are 1.2 and 0.8 kg/m³, respectively. Assume a diffusion coefficient of 3×10^{-11} m²/s at this temperature.

Solution

Fick's first law, Equation 7.2, is used to determine the diffusion flux. Substitution of the values just given into this expression yields

$$J = -D \frac{C_A - C_B}{x_A - x_B} = -(3 \times 10^{-11} \text{ m}^2/\text{s}) \frac{(1.2 - 0.8) \text{ kg/m}^3}{(5 \times 10^{-3} - 10^{-2}) \text{ m}}$$

$$= 2.4 \times 10^{-9} \text{ kg/m}^2 \cdot \text{s}$$

7.4 NONSTEADY-STATE DIFFUSION

Most practical diffusion situations are nonsteady-state ones—that is, the diffusion flux and the concentration gradient at some particular point in a solid vary with time, with a net accumulation or depletion of the diffusing species resulting. This is illustrated in Figure 7.5, which shows concentration profiles at three different diffusion times. Under conditions of nonsteady state, use of Equation 7.2 is possible but not convenient; instead, the partial differential equation

$$\frac{\partial C}{\partial t} = \frac{\partial}{\partial x}\left(D \frac{\partial C}{\partial x} \right) \tag{7.4a}$$

Fick's second law

known as **Fick's second law**, is used. If the diffusion coefficient is independent of composition (which should be verified for each particular diffusion situation), Equation 7.4a simplifies to

Fick's second law—diffusion equation for nonsteady-state diffusion (in one direction)

$$\frac{\partial C}{\partial t} = D \frac{\partial^2 C}{\partial x^2} \tag{7.4b}$$

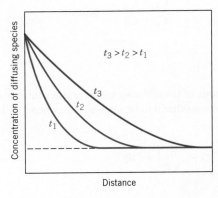

Figure 7.5 Concentration profiles for nonsteady-state diffusion taken at three different times, t_1, t_2, and t_3.

Concentration of diffusing species

$t_3 > t_2 > t_1$

t_3

t_2

t_1

Distance

Solutions to this expression (concentration in terms of both position and time) are possible when physically meaningful boundary conditions are specified. Comprehensive collections of these are given by Crank, and Carslaw and Jaeger (see References).

WileyPLUS

Tutorial Video:
Steady-State and
Nonsteady-State
Diffusion
What Are the
Differences between
Steady-State and
Nonsteady-State
Diffusion?

One practically important solution is for a semi-infinite solid[3] in which the surface concentration is held constant. Frequently, the source of the diffusing species is a gas phase, the partial pressure of which is maintained at a constant value. Furthermore, the following assumptions are made:

1. Before diffusion, any of the diffusing solute atoms in the solid are uniformly distributed with concentration of C_0.

2. The value of x at the surface is zero and increases with distance into the solid.

3. The time is taken to be zero the instant before the diffusion process begins.

These conditions are simply stated as follows:

Initial condition

$$\text{For } t = 0, C = C_0 \text{ at } 0 \leq x \leq \infty$$

Boundary conditions

$$\text{For } t > 0, C = C_s \text{ (the constant surface concentration) at } x = 0$$
$$\text{For } t > 0, C = C_0 \text{ at } x = \infty$$

Solution to Fick's
second law for the
condition of constant
surface concentration
(for a semi-infinite
solid)

Application of these conditions to Equation 7.4b yields the solution

$$\frac{C_x - C_0}{C_s - C_0} = 1 - \text{erf}\left(\frac{x}{2\sqrt{Dt}}\right) \tag{7.5}$$

where C_x represents the concentration at depth x after time t. The expression $\text{erf}(x/2\sqrt{Dt})$ is the Gaussian error function,[4] values of which are given in mathematical tables for various $x/2\sqrt{Dt}$ values; a partial listing is given in Table 7.1. The concentration parameters that appear in Equation 7.5 are noted in Figure 7.6, a concentration profile taken at a specific time. Equation 7.5 thus demonstrates the relationship between concentration, position, and time—namely, that C_x, being a function of the dimensionless parameter $x/\sqrt{Dt}$, may be determined at any time and position if the parameters C_0, C_s, and D are known.

Suppose that it is desired to achieve some specific concentration of solute, C_1, in an alloy; the left-hand side of Equation 7.5 now becomes

$$\frac{C_1 - C_0}{C_s - C_0} = \text{constant}$$

---

[3]A bar of solid is considered to be semi-infinite if none of the diffusing atoms reaches the bar end during the time over which diffusion takes place. A bar of length l is considered to be semi-infinite when $l > 10\sqrt{Dt}$.

[4]This Gaussian error function is defined by

$$\text{erf}(z) = \frac{2}{\sqrt{\pi}}\int_0^z e^{-y^2}dy$$

where $x/2\sqrt{Dt}$ has been replaced by the variable z.

Table 7.1

Tabulation of Error Function Values

z	$erf(z)$	z	$erf(z)$	z	$erf(z)$
0	0	0.55	0.5633	1.3	0.9340
0.025	0.0282	0.60	0.6039	1.4	0.9523
0.05	0.0564	0.65	0.6420	1.5	0.9661
0.10	0.1125	0.70	0.6778	1.6	0.9763
0.15	0.1680	0.75	0.7112	1.7	0.9838
0.20	0.2227	0.80	0.7421	1.8	0.9891
0.25	0.2763	0.85	0.7707	1.9	0.9928
0.30	0.3286	0.90	0.7970	2.0	0.9953
0.35	0.3794	0.95	0.8209	2.2	0.9981
0.40	0.4284	1.0	0.8427	2.4	0.9993
0.45	0.4755	1.1	0.8802	2.6	0.9998
0.50	0.5205	1.2	0.9103	2.8	0.9999

This being the case, the right-hand side of Equation 7.5 is also a constant, and subsequently

$$\frac{x}{2\sqrt{Dt}} = \text{constant} \tag{7.6a}$$

or

$$\frac{x^2}{Dt} = \text{constant} \tag{7.6b}$$

Some diffusion computations are facilitated on the basis of this relationship, as demonstrated in Example Problem 7.3.

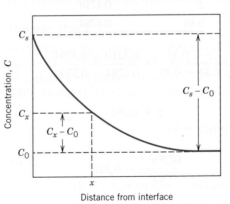

Figure 7.6 Concentration profile for nonsteady-state diffusion; concentration parameters relate to Equation 7.5.

EXAMPLE PROBLEM 7.2

Nonsteady-State Diffusion Time Computation I

For some applications, it is necessary to harden the surface of a steel (or iron–carbon alloy) above that of its interior. One way this may be accomplished is by increasing the surface concentration of carbon in a process termed **carburizing**; the steel piece is exposed, at an elevated temperature, to an atmosphere rich in a hydrocarbon gas, such as methane (CH_4).

Consider one such alloy that initially has a uniform carbon concentration of 0.25 wt% and is to be treated at 950°C. If the concentration of carbon at the surface is suddenly brought to and maintained at 1.20 wt%, how long will it take to achieve a carbon content of 0.80 wt% at a

carburizing

Table 7.2

A Tabulation of
Diffusion Data

WileyPLUS

Tutorial Video:
Diffusion Tables
How to Use
Diffusion Data
Found in a Table

Diffusing Species	Host Metal	$D_0(m^2/s)$	$Q_d(J/mol)$
Interstitial Diffusion			
C[b]	Fe (α or BCC)[a]	1.1×10^{-6}	87,400
C[c]	Fe (γ or FCC)[a]	2.3×10^{-5}	148,000
N[b]	Fe (α or BCC)[a]	5.0×10^{-7}	77,000
N[c]	Fe (γ or FCC)[a]	9.1×10^{-5}	168,000
Self-Diffusion			
Fe[c]	Fe (α or BCC)[a]	2.8×10^{-4}	251,000
Fe[c]	Fe (γ or FCC)[a]	5.0×10^{-5}	284,000
Cu[d]	Cu (FCC)	2.5×10^{-5}	200,000
Al[c]	Al (FCC)	2.3×10^{-4}	144,000
Mg[c]	Mg (HCP)	1.5×10^{-4}	136,000
Zn[c]	Zn (HCP)	1.5×10^{-5}	94,000
Mo[d]	Mo (BCC)	1.8×10^{-4}	461,000
Ni[d]	Ni (FCC)	1.9×10^{-4}	285,000
Interdiffusion (Vacancy)			
Zn[c]	Cu (FCC)	2.4×10^{-5}	189,000
Cu[c]	Zn (HCP)	2.1×10^{-4}	124,000
Cu[c]	Al (FCC)	6.5×10^{-5}	136,000
Mg[c]	Al (FCC)	1.2×10^{-4}	130,000
Cu[c]	Ni (FCC)	2.7×10^{-5}	256,000
Ni[d]	Cu (FCC)	1.9×10^{-4}	230,000

[a]There are two sets of diffusion coefficients for iron because iron experiences a phase transformation at 912°C; at temperatures less than 912°C, BCC α-iron exists; at temperatures higher than 912°C, FCC γ-iron is the stable phase.
[b]Y. Adda and J. Philibert, *Diffusion Dans Les Solides*, Universitaires de France, Paris, 1966.
[c]E. A. Brandes and G. B. Brook (Editors), *Smithells Metals Reference Book*, 7th edition, Butterworth-Heinemann, Oxford, 1992.
[d]J. Askill, *Tracer Diffusion Data for Metals, Alloys, and Simple Oxides*, IFI/Plenum, New York, 1970.

relatively small diffusion coefficient. Table 7.2 lists D_0 and Q_d values for several diffusion systems.

Taking natural logarithms of Equation 7.8 yields

$$\ln D = \ln D_0 - \frac{Q_d}{R}\left(\frac{1}{T}\right) \tag{7.9a}$$

or, in terms of logarithms to the base 10,

$$\log D = \log D_0 - \frac{Q_d}{2.3R}\left(\frac{1}{T}\right) \tag{7.9b}$$

Because D_0, Q_d, and R are all constants, Equation 7.9b takes on the form of an equation of a straight line:

$$y = b + mx$$

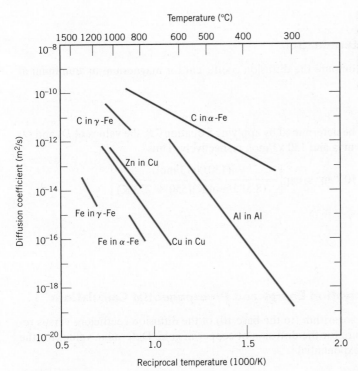

Figure 7.7 Plot of the logarithm of the diffusion coefficient versus the reciprocal of absolute temperature for several metals. [Data taken from E. A. Brandes and G. B. Brook (Editors), *Smithells Metals Reference Book*, 7th edition, Butterworth-Heinemann, Oxford, 1992.]

where y and x are analogous, respectively, to the variables $\log D$ and $1/T$. Thus, if $\log D$ is plotted versus the reciprocal of the absolute temperature, a straight line should result, having a slope and intercept of $-Q_d/2.3R$ and $\log D_0$, respectively. This is, in fact, the manner in which the values of Q_d and D_0 are determined experimentally. From such a plot for several alloy systems (Figure 7.7), it may be noted that linear relationships exist for all cases shown.

Concept Check 7.1 Rank the magnitudes of the diffusion coefficients from greatest to least for the following systems:

N in Fe at 700°C
Cr in Fe at 700°C
N in Fe at 900°C
Cr in Fe at 900°C

Now justify this ranking. (*Note*: Both Fe and Cr have the BCC crystal structure, and the atomic radii for Fe, Cr, and N are 0.124, 0.125, and 0.065 nm, respectively. You may also want to refer to Section 6.4.)

[*The answer may be found at* www.wiley.com/college/callister *(Student Companion Site).*]

Concept Check 7.2 Consider the self-diffusion of two hypothetical metals A and B. On a schematic graph of $\ln D$ versus $1/T$, plot (and label) lines for both metals, given that $D_0(A) > D_0(B)$ and $Q_d(A) > Q_d(B)$.

[*The answer may be found at* www.wiley.com/college/callister *(Student Companion Site).*]

EXAMPLE PROBLEM 7.4

Diffusion Coefficient Determination

Using the data in Table 7.2, compute the diffusion coefficient for magnesium in aluminum at 550°C.

Solution

This diffusion coefficient may be determined by applying Equation 7.8; the values of D_0 and Q_d from Table 7.2 are 1.2×10^{-4} m²/s and 130 kJ/mol, respectively. Thus,

$$D = (1.2 \times 10^{-4}\,\text{m}^2/\text{s}) \exp\left[-\frac{(130{,}000\,\text{J/mol})}{(8.31\,\text{J/mol·K})(550 + 273\,\text{K})}\right]$$

$$= 6.7 \times 10^{-13}\,\text{m}^2/\text{s}$$

EXAMPLE PROBLEM 7.5

Diffusion Coefficient Activation Energy and Preexponential Calculations

Figure 7.8 shows a plot of the logarithm (to the base 10) of the diffusion coefficient versus reciprocal of absolute temperature for the diffusion of copper in gold. Determine values for the activation energy and the preexponential.

Solution

From Equation 7.9b the slope of the line segment in Figure 7.8 is equal to $-Q_d/2.3R$, and the intercept at $1/T = 0$ gives the value of $\log D_0$. Thus, the activation energy may be determined as

$$Q_d = -2.3R(\text{slope}) = -2.3R\left[\frac{\Delta(\log D)}{\Delta\left(\dfrac{1}{T}\right)}\right]$$

$$= -2.3R\left[\frac{\log D_1 - \log D_2}{\dfrac{1}{T_1} - \dfrac{1}{T_2}}\right]$$

WileyPLUS: VMSE
D_0 and Q_d from Experimental Data

where D_1 and D_2 are the diffusion coefficient values at $1/T_1$ and $1/T_2$, respectively. Let us arbitrarily take $1/T_1 = 0.8 \times 10^{-3}$ $(\text{K})^{-1}$ and $1/T_2 = 1.1 \times 10^{-3}$ $(\text{K})^{-1}$. We may now read the corresponding $\log D_1$ and $\log D_2$ values from the line segment in Figure 7.8.

[Before this is done, however, a note of caution is offered: The vertical axis in Figure 7.8 is scaled logarithmically (to the base 10); however, the actual diffusion coefficient values are noted on this axis. For example, for $D = 10^{-14}$ m²/s, the logarithm of D is -14.0, *not* 10^{-14}. Furthermore, this logarithmic scaling affects the readings between decade values; for example, at a location midway between 10^{-14} and 10^{-15}, the value is not 5×10^{-15} but, rather, $10^{-14.5} = 3.2 \times 10^{-15}$.]

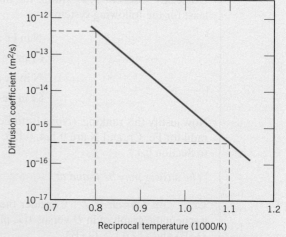

Figure 7.8 Plot of the logarithm of the diffusion coefficient versus the reciprocal of absolute temperature for the diffusion of copper in gold.

Thus, from Figure 7.8, at $1/T_1 = 0.8 \times 10^{-3}\,(\text{K})^{-1}$, $\log D_1 = -12.40$, whereas for $1/T_2 = 1.1 \times 10^{-3}\,(\text{K})^{-1}$, $\log D_2 = -15.45$, and the activation energy, as determined from the slope of the line segment in Figure 7.8, is

$$Q_d = -2.3R \left[\frac{\log D_1 - \log D_2}{\dfrac{1}{T_1} - \dfrac{1}{T_2}} \right]$$

$$= -2.3(8.31 \text{ J/mol·K}) \left[\frac{-12.40 - (-15.45)}{0.8 \times 10^{-3}(\text{K})^{-1} - 1.1 \times 10^{-3}(\text{K})^{-1}} \right]$$

$$= 194{,}000 \text{ J/mol} = 194 \text{ kJ/mol}$$

Now, rather than try to make a graphical extrapolation to determine D_0, we can obtain a more accurate value analytically using Equation 7.9b, and we obtain a specific value of D (or $\log D$) and its corresponding T (or $1/T$) from Figure 7.8. Because we know that $\log D = -15.45$ at $1/T = 1.1 \times 10^{-3}\,(\text{K})^{-1}$, then

$$\log D_0 = \log D + \frac{Q_d}{2.3R}\left(\frac{1}{T}\right)$$

$$= -15.45 + \frac{(194{,}000 \text{ J/mol})(1.1 \times 10^{-3}\,[\text{K}]^{-1})}{(2.3)(8.31 \text{ J/mol·K})}$$

$$= -4.28$$

Thus, $D_0 = 10^{-4.28} \text{ m}^2/\text{s} = 5.2 \times 10^{-5} \text{ m}^2/\text{s}$.

DESIGN EXAMPLE 7.1

Diffusion Temperature–Time Heat Treatment Specification

The wear resistance of a steel gear is to be improved by hardening its surface. This is to be accomplished by increasing the carbon content within an outer surface layer as a result of carbon diffusion into the steel; the carbon is to be supplied from an external carbon-rich gaseous atmosphere at an elevated and constant temperature. The initial carbon content of the steel is 0.20 wt%, whereas the surface concentration is to be maintained at 1.00 wt%. For this treatment to be effective, a carbon content of 0.60 wt% must be established at a position 0.75 mm below the surface. Specify an appropriate heat treatment in terms of temperature and time for temperatures between 900 and 1050°C. Use data in Table 7.2 for the diffusion of carbon in γ-iron.

Solution

Because this is a nonsteady-state diffusion situation, let us first employ Equation 7.5, using the following values for the concentration parameters:

$$C_0 = 0.20 \text{ wt\% C}$$

$$C_s = 1.00 \text{ wt\% C}$$

$$C_x = 0.60 \text{ wt\% C}$$

Therefore,

$$\frac{C_x - C_0}{C_s - C_0} = \frac{0.60 - 0.20}{1.00 - 0.20} = 1 - \text{erf}\left(\frac{x}{2\sqrt{Dt}}\right)$$

and thus,

$$0.5 = \text{erf}\left(\frac{x}{2\sqrt{Dt}}\right)$$

Using an interpolation technique as demonstrated in Example Problem 7.2 and the data presented in Table 7.1, we find

$$\frac{x}{2\sqrt{Dt}} = 0.4747 \tag{7.10}$$

The problem stipulates that $x = 0.75$ mm $= 7.5 \times 10^{-4}$ m. Therefore,

$$\frac{7.5 \times 10^{-4}\ \text{m}}{2\sqrt{Dt}} = 0.4747$$

This leads to

$$Dt = 6.24 \times 10^{-7}\ \text{m}^2$$

Furthermore, the diffusion coefficient depends on temperature according to Equation 7.8, and, from Table 7.2 for the diffusion of carbon in γ-iron, $D_0 = 2.3 \times 10^{-5}$ m²/s and $Q_d = 148,000$ J/mol. Hence,

$$Dt = D_0 \exp\!\left(-\frac{Q_d}{RT}\right)(t) = 6.24 \times 10^{-7}\ \text{m}^2$$

$$(2.3 \times 10^{-5}\,\text{m}^2/\text{s}) \exp\!\left[-\frac{148,000\ \text{J/mol}}{(8.31\ \text{J/mol·K})(T)}\right](t) = 6.24 \times 10^{-7}\ \text{m}^2$$

and, solving for the time t, we obtain

$$t\,(\text{in s}) = \frac{0.0271}{\exp\!\left(-\dfrac{17,810}{T}\right)}$$

Thus, the required diffusion time may be computed for some specified temperature (in K). The following table gives t values for four different temperatures that lie within the range stipulated in the problem.

Temperature	Time	
(°C)	s	h
900	106,400	29.6
950	57,200	15.9
1000	32,300	9.0
1050	19,000	5.3

7.6 DIFFUSION IN SEMICONDUCTING MATERIALS

One technology that applies solid-state diffusion is the fabrication of semiconductor integrated circuits (ICs) (Section 19.15). Each integrated circuit chip is a thin square wafer having dimensions on the order of 6 mm × 6 mm × 0.4 mm; furthermore, millions of interconnected electronic devices and circuits are embedded in one of the chip faces. Single-crystal silicon is the base material for most ICs. In order for these IC devices to function satisfactorily, very precise concentrations of an impurity (or impurities) must be incorporated into minute spatial regions in a very intricate and detailed pattern on the silicon chip; one way this is accomplished is by atomic diffusion.

Typically, two heat treatments are used in this process. In the first, or *predeposition step*, impurity atoms are diffused into the silicon, often from a gas phase, the partial pressure of which is maintained constant. Thus, the surface composition of the impurity also remains constant over time, such that impurity concentration within the silicon is a function of position and time according to Equation 7.5—that is,

$$\frac{C_x - C_0}{C_s - C_0} = 1 - \text{erf}\left(\frac{x}{2\sqrt{Dt}}\right)$$

Predeposition treatments are normally carried out within the temperature range of 900°C and 1000°C and for times typically less than 1 h.

The second treatment, sometimes called *drive-in diffusion,* is used to transport impurity atoms farther into the silicon in order to provide a more suitable concentration distribution without increasing the overall impurity content. This treatment is carried out at a higher temperature than the predeposition one (up to about 1200°C) and also in an oxidizing atmosphere so as to form an oxide layer on the surface. Diffusion rates through this SiO_2 layer are relatively slow, such that very few impurity atoms diffuse out of and escape from the silicon. Schematic concentration profiles taken at three different times for this diffusion situation are shown in Figure 7.9; these profiles may be compared and contrasted to those in Figure 7.5 for the case in which the surface concentration of diffusing species is held constant. In addition, Figure 7.10 compares (schematically) concentration profiles for predeposition and drive-in treatments.

If we assume that the impurity atoms introduced during the predeposition treatment are confined to a very thin layer at the surface of the silicon (which, of course, is only an approximation), then the solution to Fick's second law (Equation 7.4b) for drive-in diffusion takes the form

$$C(x,t) = \frac{Q_0}{\sqrt{\pi Dt}} \exp\left(-\frac{x^2}{4Dt}\right) \tag{7.11}$$

Here, Q_0 represents the total amount of impurities in the solid that were introduced during the predeposition treatment (in number of impurity atoms per unit area); all other parameters in this equation have the same meanings as previously. Furthermore, it can be shown that

$$Q_0 = 2C_s\sqrt{\frac{D_p t_p}{\pi}} \tag{7.12}$$

where C_s is the surface concentration for the predeposition step (Figure 7.10), which was held constant, D_p is the diffusion coefficient, and t_p is the predeposition treatment time.

Another important diffusion parameter is *junction depth,* x_j. It represents the depth (i.e., value of x) at which the diffusing impurity concentration is just equal to

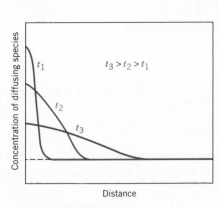

Figure 7.9 Schematic concentration profiles for drive-in diffusion of semiconductors at three different times, t_1, t_2, and t_3.

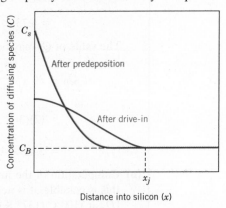

Figure 7.10 Schematic concentration profiles taken after (1) predeposition and (2) drive-in diffusion treatments for semiconductors. Also shown is the junction depth, x_j.

the background concentration of that impurity in the silicon (C_B) (Figure 7.10). For drive-in diffusion x_j may be computed using the following expression:

$$x_j = \left[(4D_d t_d) \ln \left(\frac{Q_0}{C_B \sqrt{\pi D_d t_d}} \right) \right]^{1/2} \tag{7.13}$$

Here, D_d and t_d represent, respectively, the diffusion coefficient and time for the drive-in treatment.

EXAMPLE PROBLEM 7.6

Diffusion of Boron into Silicon

Boron atoms are to be diffused into a silicon wafer using both predeposition and drive-in heat treatments; the background concentration of B in this silicon material is known to be 1×10^{20} atoms/m^3. The predeposition treatment is to be conducted at 900°C for 30 min; the surface concentration of B is to be maintained at a constant level of 3×10^{26} atoms/m^3. Drive-in diffusion will be carried out at 1100°C for a period of 2 h. For the diffusion coefficient of B in Si, values of Q_d and D_0 are 3.87 eV/atom and 2.4×10^{-3} m^2/s, respectively.

(a) Calculate the value of Q_0.

(b) Determine the value of x_j for the drive-in diffusion treatment.

(c) Also for the drive-in treatment, compute the concentration of B atoms at a position 1 μm below the surface of the silicon wafer.

Solution

(a) The value of Q_0 is calculated using Equation 7.12. However, before this is possible, it is first necessary to determine the value of D for the predeposition treatment [D_p at $T = T_p = $ 900°C (1173 K)] using Equation 7.8. (*Note:* For the gas constant R in Equation 7.8, we use Boltzmann's constant k, which has a value of 8.62×10^{-5} eV/atom·K). Thus,

$$D_p = D_0 \exp \left(-\frac{Q_d}{kT_p} \right)$$

$$= (2.4 \times 10^{-3} \text{ m}^2/\text{s}) \exp \left[-\frac{3.87 \text{ eV/atom}}{(8.62 \times 10^{-5} \text{ eV/atom·K})(1173 \text{ K})} \right]$$

$$= 5.73 \times 10^{-20} \text{ m}^2/\text{s}$$

The value of Q_0 may be determined as follows:

$$Q_0 = 2C_s \sqrt{\frac{D_p t_p}{\pi}}$$

$$= (2)(3 \times 10^{26} \text{ atoms/m}^3) \sqrt{\frac{(5.73 \times 10^{-20} \text{ m}^2/\text{s})(30 \text{ min})(60 \text{ s/min})}{\pi}}$$

$$= 3.44 \times 10^{18} \text{ atoms/m}^2$$

(b) Computation of the junction depth requires that we use Equation 7.13. However, before this is possible, it is necessary to calculate D at the temperature of the drive-in treatment [D_d at 1100°C (1373 K)]. Thus,

$$D_d = (2.4 \times 10^{-3} \text{ m}^2/\text{s}) \exp \left[-\frac{3.87 \text{ eV/atom}}{(8.62 \times 10^{-5} \text{ eV/atom·K})(1373 \text{ K})} \right]$$

$$= 1.51 \times 10^{-17} \text{ m}^2/\text{s}$$

Now, from Equation 7.13,

$$x_j = \left[(4D_d t_d)\ln\left(\frac{Q_0}{C_B \sqrt{\pi D_d t_d}} \right) \right]^{1/2}$$

$$= \left\{ (4)(1.51 \times 10^{-17}\,\text{m}^2/\text{s})(7200\,\text{s}) \times \right.$$

$$\left. \ln\left[\frac{3.44 \times 10^{18}\,\text{atoms/m}^2}{(1 \times 10^{20}\,\text{atoms/m}^3)\sqrt{(\pi)(1.51 \times 10^{-17}\,\text{m}^2/\text{s})(7200\,\text{s})}} \right] \right\}^{1/2}$$

$$= 2.19 \times 10^{-6}\,\text{m} = 2.19\,\mu\text{m}$$

(c) At $x = 1\,\mu$m for the drive-in treatment, we compute the concentration of B atoms using Equation 7.11 and values for Q_0 and D_d determined previously as follows:

$$C(x, t) = \frac{Q_0}{\sqrt{\pi D_d t}}\exp\left(-\frac{x^2}{4D_d t} \right)$$

$$= \frac{3.44 \times 10^{18}\,\text{atoms/m}^2}{\sqrt{(\pi)(1.51 \times 10^{-17}\,\text{m}^2/\text{s})(7200\,\text{s})}}\exp\left[-\frac{(1 \times 10^{-6}\,\text{m})^2}{(4)(1.51 \times 10^{-17}\,\text{m}^2/\text{s})(7200\,\text{s})} \right]$$

$$= 5.90 \times 10^{23}\,\text{atoms/m}^3$$

M A T E R I A L S   O F   I M P O R T A N C E

Aluminum for Integrated Circuit Interconnects

Subsequent to the predeposition and drive-in heat treatments just described, another important step in the IC fabrication process is the deposition of very thin and narrow conducting circuit paths to facilitate the passage of current from one device to another; these paths are called *interconnects,* and several are shown in Figure 7.11, a scanning electron micrograph of an IC chip. Of course, the material to be used for interconnects must have a high electrical conductivity—a metal, because, of all materials, metals have the highest conductivities. Table 7.3 gives values for silver, copper, gold, and aluminum, the most conductive metals. On the basis of these conductivities, and discounting material cost, Ag is the metal of choice, followed by Cu, Au, and Al.

Once these interconnects have been deposited, it is still necessary to subject the IC chip to other heat treatments, which may run as high as 500°C. If, during these treatments, there is significant diffusion of the interconnect metal into the silicon, the electrical functionality of the IC will be destroyed. Thus, because the extent of diffusion is dependent on the magnitude of the diffusion coefficient, it is necessary to select an interconnect metal that has a small value

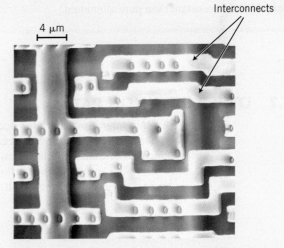

4 μm

Interconnects

Figure 7.11 Scanning electron micrograph of an integrated circuit chip, on which is noted aluminum interconnect regions. Approximately 2000×. (Photograph courtesy of National Semiconductor Corporation.)

of D in silicon. Figure 7.12 plots the logarithm of D versus $1/T$ for the diffusion, into silicon, of copper, gold, silver, and aluminum. Also, a dashed vertical line

(continued)

Table 7.3 Room-Temperature Electrical Conductivity Values for Silver, Copper, Gold, and Aluminum (the Four Most Conductive Metals)

Metal	Electrical Conductivity $[(ohm\text{-}m)^{-1}]$
Silver	6.8×10^7
Copper	6.0×10^7
Gold	4.3×10^7
Aluminum	3.8×10^7

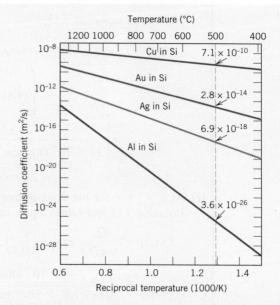

Figure 7.12 Logarithm of D-versus-$1/T$ (K) curves (lines) for the diffusion of copper, gold, silver, and aluminum in silicon. Also noted are D values at 500°C.

has been constructed at 500°C, from which values of D for the four metals are noted at this temperature. Here it may be seen that the diffusion coefficient for aluminum in silicon (3.6×10^{-26} m^2/s) is at least eight orders of magnitude (i.e., a factor of 10^8) lower than the values for the other three metals.

Aluminum is indeed used for interconnects in some integrated circuits; even though its electrical conductivity is slightly lower than the values for silver, copper, and gold, its extremely low diffusion coefficient makes it the material of choice for this application. An aluminum–copper–silicon alloy (94.5 wt% Al-4 wt% Cu-1.5 wt% Si) is sometimes also used for interconnects; it not only bonds easily to the surface of the chip, but is also more corrosion resistant than pure aluminum.

More recently, copper interconnects have also been used. However, it is first necessary to deposit a very thin layer of tantalum or tantalum nitride beneath the copper, which acts as a barrier to deter diffusion of copper into the silicon.

7.7 OTHER DIFFUSION PATHS

Atomic migration may also occur along dislocations, grain boundaries, and external surfaces. These are sometimes called *short-circuit diffusion paths* inasmuch as rates are much faster than for bulk diffusion. However, in most situations, short-circuit contributions to the overall diffusion flux are insignificant because the cross-sectional areas of these paths are extremely small.

7.8 DIFFUSION IN IONIC AND POLYMERIC MATERIALS

We now extrapolate some of the diffusion principles to ionic and polymeric materials.

Ionic Materials

For ionic compounds, the phenomenon of diffusion is more complicated than for metals inasmuch as it is necessary to consider the diffusive motion of two types of ions that have opposite charges. Diffusion in these materials usually occurs by a vacancy mechanism (Figure 7.3a). As noted in Section 6.4, in order to maintain charge neutrality in an ionic material, the following may be said about vacancies: (1) ion vacancies occur in pairs [as with Schottky defects (Figure 6.3)], (2) they form in nonstoichiometric compounds (Figure 6.4), and (3) they are created by substitutional impurity ions having different charge states than the host ions (Example Problem 6.4). In any event, associated with the diffusive motion of a single ion is a transference of electrical charge. In order to maintain localized charge neutrality in the vicinity of this moving ion, another species having an equal and opposite charge must accompany the ion's diffusive motion. Possible charged species include another vacancy, an impurity atom, or an electronic carrier [i.e., a free electron or hole (Section 19.6)]. It follows that the rate of diffusion of these electrically charged couples is limited by the diffusion rate of the slowest-moving species.

When an external electric field is applied across an ionic solid, the electrically charged ions migrate (i.e., diffuse) in response to forces that are brought to bear on them. As we discuss in Section 19.16, this ionic motion gives rise to an electric current. Furthermore, the electrical conductivity is a function of the diffusion coefficient (Equation 19.23). Consequently, much of the diffusion data for ionic solids come from electrical conductivity measurements.

Polymeric Materials

For polymeric materials, our interest is often in the diffusive motion of small foreign molecules (e.g., O_2, H_2O, CO_2, CH_4) between the molecular chains rather than in the diffusive motion of chain atoms within the polymer structure. A polymer's permeability and absorption characteristics relate to the degree to which foreign substances diffuse into the material. Penetration of these foreign substances can lead to swelling and/or chemical reactions with the polymer molecules and often a degradation of the material's mechanical and physical properties (Section 18.11).

Rates of diffusion are greater through amorphous regions than through crystalline regions; the structure of amorphous material is more "open." This diffusion mechanism may be considered analogous to interstitial diffusion in metals—that is, in polymers, diffusive movements occur through small voids between polymer chains from one open amorphous region to an adjacent open one.

Foreign molecule size also affects the diffusion rate: smaller molecules diffuse faster than larger ones. Furthermore, diffusion is more rapid for foreign molecules that are chemically inert than for those that interact with the polymer.

One step in diffusion through a polymer membrane is the dissolution of the molecular species in the membrane material. This dissolution is a time-dependent process, and, if slower than the diffusive motion, may limit the overall rate of diffusion. Consequently, the diffusion properties of polymers are often characterized in terms of a *permeability coefficient* (denoted by P_M), where for the case of steady-state diffusion through a polymer membrane, Fick's first law (Equation 7.2), is modified as

$$J = -P_M \frac{\Delta P}{\Delta x} \tag{7.14}$$

In this expression, J is the diffusion flux of gas through the membrane [(cm^3 STP)/(cm^2·s)], P_M is the permeability coefficient, Δx is the membrane thickness, and ΔP is the difference in pressure of the gas across the membrane. For small molecules in nonglassy polymers the permeability coefficient can be approximated as the product of the diffusion coefficient (D) and solubility of the diffusing species in the polymer (S)—that is,

$$P_M = DS \tag{7.15}$$

Table 7.4

Permeability Coefficient P_M at 25°C for Oxygen, Nitrogen, Carbon Dioxide, and Water Vapor in a Variety of Polymers

Polymer	Acronym	P_M [× 10^{-13} (cm^3 STP)(cm)/(cm^2·s·Pa)]			
		O_2	N_2	CO_2	H_2O
Polyethylene (low density)	LDPE	2.2	0.73	9.5	68
Polyethylene (high density)	HDPE	0.30	0.11	0.27	9.0
Polypropylene	PP	1.2	0.22	5.4	38
Poly(vinyl chloride)	PVC	0.034	0.0089	0.012	206
Polystyrene	PS	2.0	0.59	7.9	840
Poly(vinylidene chloride)	PVDC	0.0025	0.00044	0.015	7.0
Poly(ethylene terephthalate)	PET	0.044	0.011	0.23	—
Poly(ethyl methacrylate)	PEMA	0.89	0.17	3.8	2380

Source: Adapted from J. Brandrup, E. H. Immergut, E. A. Grulke, A. Abe, and D. R. Bloch (Editors), *Polymer Handbook*, 4th edition. Copyright © 1999 by John Wiley & Sons, New York. Reprinted by permission of John Wiley & Sons, Inc.

Table 7.4 presents the permeability coefficients of oxygen, nitrogen, carbon dioxide, and water vapor in several common polymers.[6]

For some applications, low permeability rates through polymeric materials are desirable, as with food and beverage packaging and automobile tires and inner tubes. Polymer membranes are often used as filters, to selectively separate one chemical species from another (or others) (i.e., the desalination of water). In such instances it is normally the case that the permeation rate of the substance to be filtered is significantly greater than for the other substance(s).

EXAMPLE PROBLEM 7.7

Computations of Diffusion Flux of Carbon Dioxide through a Plastic Beverage Container and Beverage Shelf Life

The clear plastic bottles used for carbonated beverages (sometimes also called *soda, pop,* or *soda pop*) are made from poly(ethylene terephthalate) (PET). The "fizz" in pop results from dissolved carbon dioxide (CO_2); because PET is permeable to CO_2, pop stored in PET bottles will eventually go "flat" (i.e., lose its fizz). A 0.6 L bottle of pop has a CO_2 pressure of about 400 kPa inside the bottle, and the CO_2 pressure outside the bottle is 0.4 kPa.

(a) Assuming conditions of steady state, calculate the diffusion flux of CO_2 through the wall of the bottle.

(b) If the bottle must lose 750 (cm^3 STP) of CO_2 before the pop tastes flat, what is the shelf life for a bottle of pop?

Note: Assume that each bottle has a surface area of 500 cm^2 and a wall thickness of 0.05 cm.

---

[6] The units for permeability coefficients in Table 7.4 are unusual, which are explained as follows: When the diffusing molecular species is in the gas phase, solubility is equal to

$$S = \frac{C}{P}$$

where C is the concentration of the diffusing species in the polymer [in units of (cm^3 STP)/cm^3 gas] and P is the partial pressure (in units of Pa). STP indicates that this is the volume of gas at standard temperature and pressure [273 K (0°C) and 101.3 kPa]. Thus, the units for S are (cm^3 STP)/Pa·cm^3. Because D is expressed in terms of cm^2/s, the units for the permeability coefficient are (cm^3 STP)(cm)/(cm^2·s·Pa).

Solution

(a) This is a permeability problem in which Equation 7.14 is employed. The permeability coefficient of CO_2 through PET (Table 7.4) is 0.23×10^{-13} (cm³ STP)(cm)/(cm²·s·Pa). Thus, the diffusion flux is

$$J = -P_M \frac{\Delta P}{\Delta x} = -P_M \frac{P_2 - P_1}{\Delta x}$$

$$= -0.23 \times 10^{-13} \frac{(\text{cm}^3 \text{ STP})(\text{cm})}{(\text{cm}^2)(\text{s})(\text{Pa})} \frac{(400 \text{ Pa} - 400{,}000 \text{ Pa})}{0.05 \text{ cm}}$$

$$= 1.8 \times 10^{-7} (\text{cm}^3 \text{ STP})/(\text{cm}^2 \cdot \text{s})$$

(b) The flow rate of CO_2 through the wall of the bottle $\dot{V}_{CO_2}$ is

$$\dot{V}_{CO_2} = JA$$

where A is the surface area of the bottle (i.e., 500 cm²); therefore,

$$\dot{V}_{CO_2} = \left[1.8 \times 10^{-7} (\text{cm}^3 \text{ STP})/(\text{cm}^2 \cdot \text{s})\right](500 \text{ cm}^2) = 9.0 \times 10^{-5} (\text{cm}^3 \text{ STP})/\text{s}$$

The time it will take for a volume (V) of 750 (cm³ STP) to escape is calculated as

$$\text{time} = \frac{V}{\dot{V}_{CO_2}} = \frac{750 \text{ (cm}^3 \text{ STP)}}{9.0 \times 10^{-5} \text{ (cm}^3 \text{ STP)/s}} = 8.3 \times 10^6 \text{ s}$$

$$= 97 \text{ days (or about 3 months)}$$

SUMMARY

Introduction

- Solid-state diffusion is a means of mass transport within solid materials by stepwise atomic motion.
- The term *interdiffusion* refers to the migration of impurity atoms; for host atoms, the term *self-diffusion* is used.

Diffusion Mechanisms

- Two mechanisms for diffusion are possible: vacancy and interstitial.
  Vacancy diffusion occurs via the exchange of an atom residing on a normal lattice site with an adjacent vacancy.
  For *interstitial diffusion*, an atom migrates from one interstitial position to an empty adjacent one.
- For a given host metal, interstitial atomic species generally diffuse more rapidly.

Fick's First Law

- *Diffusion flux* is defined in terms of mass of diffusing species, cross-sectional area, and time according to Equation 7.1.
- Diffusion flux is proportional to the negative of the concentration gradient according to Fick's first law, Equation 7.2.
- *Concentration profile* is represented as a plot of concentration versus distance into the solid material.
- *Concentration gradient* is the slope of the concentration profile curve at some specific point.

- The diffusion condition for which the flux is independent of time is known as *steady state*.
- The driving force for steady-state diffusion is the concentration gradient (dC/dx).

Fick's Second Law— Nonsteady-State Diffusion

- For nonsteady-state diffusion, there is a net accumulation or depletion of diffusing species, and the flux is dependent on time.
- The mathematics for nonsteady state in a single (x) direction (and when the diffusion coefficient is independent of concentration) may be described by Fick's second law, Equation 7.4b.
- For a constant surface composition boundary condition, the solution to Fick's second law (Equation 7.4b) is Equation 7.5, which involves the Gaussian error function (erf).

Factors That Influence Diffusion

- The magnitude of the diffusion coefficient is indicative of the rate of atomic motion and depends on both host and diffusing species as well as on temperature.
- The diffusion coefficient is a function of temperature according to Equation 7.8.

Diffusion in Semiconducting Materials

- The two heat treatments that are used to diffuse impurities into silicon during integrated circuit fabrication are predeposition and drive-in.
   During predeposition, impurity atoms are diffused into the silicon, often from a gas phase, the partial pressure of which is maintained constant.
   For the drive-in step, impurity atoms are transported deeper into the silicon so as to provide a more suitable concentration distribution without increasing the overall impurity content.
- Integrated circuit interconnects are normally made of aluminum—instead of metals such as copper, silver, and gold that have higher electrical conductivities—on the basis of diffusion considerations. During high-temperature heat treatments, interconnect metal atoms diffuse into the silicon; appreciable concentrations will compromise the chip's functionality.

Diffusion in Ionic Materials

- Diffusion in ionic materials normally occurs by a vacancy mechanism; localized charge neutrality is maintained by the coupled diffusive motion of a charged vacancy and some other charged entity.

Diffusion in Polymeric Materials

- With regard to diffusion in polymers, small molecules of foreign substances diffuse between molecular chains by an interstitial-type mechanism from one amorphous region to an adjacent one.
- Diffusion (or permeation) of gaseous species is often characterized in terms of the permeability coefficient, which is the product of the diffusion coefficient and solubility in the polymer (Equation 7.15).
- Permeation flow rates are expressed using a modified form of Fick's first law (Equation 7.14).

Important Terms and Concepts

activation energy	diffusion flux	nonsteady-state diffusion
carburizing	driving force	self-diffusion
concentration gradient	Fick's first law	steady-state diffusion
concentration profile	Fick's second law	vacancy diffusion
diffusion	interdiffusion (impurity diffusion)	
diffusion coefficient	interstitial diffusion	

REFERENCES

Carslaw, H. S., and J. C. Jaeger, *Conduction of Heat in Solids,* 2nd edition, Oxford University Press, Oxford, 1986.

Crank, J., *The Mathematics of Diffusion,* Oxford University Press, Oxford, 1980.

Gale, W. F., and T. C. Totemeier (Editors), *Smithells Metals Reference Book,* 8th edition, Elsevier Butterworth-Heinemann, Oxford, 2004.

Glicksman, M., *Diffusion in Solids,* Wiley-Interscience, New York, 2000.

Shewmon, P. G., *Diffusion in Solids,* 2nd edition, The Minerals, Metals and Materials Society, Warrendale, PA, 1989.

QUESTIONS AND PROBLEMS

Introduction

7.1 Briefly explain the difference between self-diffusion and interdiffusion.

7.2 Self-diffusion involves the motion of atoms that are all of the same type; therefore, it is not subject to observation by compositional changes, as with interdiffusion. Suggest one way in which self-diffusion may be monitored.

Diffusion Mechanisms

7.3 **(a)** Compare interstitial and vacancy atomic mechanisms for diffusion.

(b) Cite two reasons why interstitial diffusion is normally more rapid than vacancy diffusion.

Fick's First Law

7.4 Briefly explain the concept of steady state as it applies to diffusion.

7.5 **(a)** Briefly explain the concept of a driving force.

(b) What is the driving force for steady-state diffusion?

7.6 The purification of hydrogen gas by diffusion through a palladium sheet was discussed in Section 7.3. Compute the number of kilograms of hydrogen that pass per hour through a 5-mm-thick sheet of palladium having an area of 0.25 m^2 at 500°C. Assume a diffusion coefficient of 1.0 × 10^{-8} m^2/s, that the concentrations at the high- and low-pressure sides of the plate are 2.4 and 0.6 kg of hydrogen per cubic meter of palladium, and that steady-state conditions have been attained.

7.7 A sheet of steel 1.8 mm thick has nitrogen atmospheres on both sides at 1200°C and is permitted to achieve a steady-state diffusion condition. The diffusion coefficient for nitrogen in steel at this temperature is 6 × 10^{-11} m^2/s, and the diffusion flux is found to be 1.2 × 10^{-7} kg/m^2·s. Also, it is known that the concentration of nitrogen in the steel at the high-pressure surface

is 4 kg/m^3. How far into the sheet from this high-pressure side will the concentration be 2.0 kg/m^3? Assume a linear concentration profile.

7.8 A sheet of BCC iron 1 mm thick was exposed to a carburizing gas atmosphere on one side and a decarburizing atmosphere on the other side at 725°C. After reaching steady state, the iron was quickly cooled to room temperature. The carbon concentrations at the two surfaces of the sheet were determined to be 0.012 and 0.0075 wt%. Compute the diffusion coefficient if the diffusion flux is 1.5 × 10^{-8} kg/m^2·s. *Hint:* Use Equation 6.12 to convert the concentrations from weight percent to kilograms of carbon per cubic meter of iron.

7.9 When α-iron is subjected to an atmosphere of hydrogen gas, the concentration of hydrogen in the iron, C_H (in weight percent), is a function of hydrogen pressure, p_{H_2} (in MPa), and absolute temperature (T) according to

$$C_H = 1.34 \times 10^{-2}\sqrt{p_{H_2}} \exp\left(-\frac{27.2 \text{ kJ/mol}}{RT}\right) \quad (7.14)$$

Furthermore, the values of D_0 and Q_d for this diffusion system are 1.4 × 10^{-7} m^2/s and 13,400 J/mol, respectively. Consider a thin iron membrane 1 mm thick that is at 250°C. Compute the diffusion flux through this membrane if the hydrogen pressure on one side of the membrane is 0.15 MPa, and on the other side 7.5 MPa.

Fick's Second Law—Nonsteady-State Diffusion

7.10 Show that

$$C_x = \frac{B}{\sqrt{Dt}} \exp\left(-\frac{x^2}{4Dt}\right)$$

is also a solution to Equation 7.4b. The parameter B is a constant, being independent of both x and t.

7.11 Determine the carburizing time necessary to achieve a carbon concentration of 0.45 wt% at a position 2 mm into an iron–carbon alloy that initially contains 0.20 wt% C. The surface concentration is to be maintained at 1.30 wt% C, and the treatment is to be conducted at 1000°C. Use the diffusion data for γ-Fe in Table 7.2.

7.12 An FCC iron–carbon alloy initially containing 0.35 wt% C is exposed to an oxygen-rich and virtually carbon-free atmosphere at 1400 K (1127°C). Under these circumstances the carbon diffuses from the alloy and reacts at the surface with the oxygen in the atmosphere; that is, the carbon concentration at the surface position is maintained essentially at 0 wt% C. (This process of carbon depletion is termed *decarburization*.) At what position will the carbon concentration be 0.15 wt% after a 10-h treatment? The value of *D* at 1400 K is 6.9×10^{-11} m²/s.

7.13 Nitrogen from a gaseous phase is to be diffused into pure iron at 700°C. If the surface concentration is maintained at 0.1 wt% N, what will be the concentration 1 mm from the surface after 10 h? The diffusion coefficient for nitrogen in iron at 700°C is 2.5×10^{-11} m²/s.

7.14 Consider a diffusion couple composed of two semi-infinite solids of the same metal, and that each side of the diffusion couple has a different concentration of the same elemental impurity; furthermore, assume each impurity level is constant throughout its side of the diffusion couple. For this situation, the solution to Fick's second law (assuming that the diffusion coefficient for the impurity is independent of concentration) is as follows:

$$C_x = \left(\frac{C_1 + C_2}{2} \right) - \left(\frac{C_1 - C_2}{2} \right) \mathrm{erf}\left(\frac{x}{2\sqrt{Dt}} \right)$$
(7.15)

In this expression, when the $x = 0$ position is taken as the initial diffusion couple interface, then C_1 is the impurity concentration for $x < 0$; likewise, C_2 is the impurity content for $x > 0$.

A diffusion couple composed of two silver–gold alloys is formed; these alloys have compositions of 98 wt% Ag–2 wt% Au and 95 wt% Ag–5 wt% Au. Determine the time this diffusion couple must be heated at 750°C (1023 K) in order for the composition to be 2.5 wt% Au at the 50 μm position into the 2 wt% Au side of the diffusion couple. Preexponential and activation energy values for Au diffusion in Ag are 8.5×10^{-5} m²/s and 202,100 J/mol, respectively.

7.15 For a steel alloy it has been determined that a carburizing heat treatment of 10-h duration will raise the carbon concentration to 0.45 wt% at a point 2.5 mm from the surface. Estimate the time necessary to achieve the same concentration at a 5.0-mm position for an identical steel and at the same carburizing temperature.

Factors That Influence Diffusion

7.16 Cite the values of the diffusion coefficients for the interdiffusion of carbon in both α-iron (BCC) and γ-iron (FCC) at 900°C. Which is larger? Explain why this is the case.

7.17 Using the data in Table 7.2, compute the value of *D* for the diffusion of zinc in copper at 650°C.

7.18 At what temperature will the diffusion coefficient for the diffusion of copper in nickel have a value of 6.5×10^{-17} m²/s? Use the diffusion data in Table 7.2.

7.19 The preexponential and activation energy for the diffusion of iron in cobalt are 1.1×10^{-5} m²/s and 253,300 J/mol, respectively. At what temperature will the diffusion coefficient have a value of 2.1×10^{-14} m²/s?

7.20 The activation energy for the diffusion of carbon in chromium is 111,000 J/mol. Calculate the diffusion coefficient at 1100 K (827°C), given that *D* at 1400 K (1127°C) is 6.25×10^{-11} m²/s.

7.21 The diffusion coefficients for iron in nickel are given at two temperatures:

T (K)	*D* (m²/s)
1273	9.4×10^{-16}
1473	2.4×10^{-14}

(a) Determine the values of D_0 and the activation energy Q_d.

(b) What is the magnitude of *D* at 1100°C (1373 K)?

7.22 The diffusion coefficients for silver in copper are given at two temperatures:

T(°C)	*D*(m²/s)
650	5.5×10^{-16}
900	1.3×10^{-13}

(a) Determine the values of D_0 and Q_d.

(b) What is the magnitude of *D* at 875°C?

7.23 The following figure shows a plot of the logarithm (to the base 10) of the diffusion coefficient versus reciprocal of the absolute temperature, for the diffusion of iron in chromium. Determine values for the activation energy and preexponential.

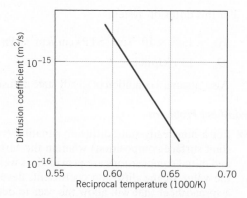

7.24 Carbon is allowed to diffuse through a steel plate 16 mm thick. The concentrations of carbon at the two faces are 0.65 and 0.30 kg C/m^3 Fe, which are maintained constant. If the preexponential and activation energy are 6.2×10^{-7} m^2/s and 80,000 J/mol, respectively, compute the temperature at which the diffusion flux is 1.43×10^{-9} kg/m$^2 \cdot$ s.

7.25 The steady-state diffusion flux through a metal plate is 5.4×10^{-10} kg/m$^2 \cdot$ s at a temperature of 727°C (1000 K) and when the concentration gradient is -350 kg/m^4. Calculate the diffusion flux at 1027°C (1300 K) for the same concentration gradient and assuming an activation energy for diffusion of 125,000 J/mol.

7.26 At approximately what temperature would a specimen of γ-iron have to be carburized for 2 h to produce the same diffusion result as at 900°C for 15 h?

7.27 **(a)** Calculate the diffusion coefficient for copper in aluminum at 500°C.

(b) What time will be required at 600°C to produce the same diffusion result (in terms of concentration at a specific point) as for 10 h at 500°C?

7.28 A copper–nickel diffusion couple similar to that shown in Figure 7.1a is fashioned. After a 700-h heat treatment at 1100°C (1373 K), the concentration of Cu is 2.5 wt% at the 3.0-mm position within the nickel. At what temperature must the diffusion couple be heated to produce this same concentration (i.e., 2.5 wt% Cu) at a 2.0-mm position after 700 h? The preexponential and activation energy for the diffusion of Cu in Ni are given in Table 7.2.

7.29 A diffusion couple similar to that shown in Figure 7.1a is prepared using two hypothetical metals A and B. After a 30-h heat treatment at 1000 K (and subsequently cooling to room temperature) the concentration of A in B is 3.2 wt% at the 15 mm position within metal B. If another heat treatment

is conducted on an identical diffusion couple, only at 800 K for 30 h, at what position will the composition be 3.2 wt% A? Assume that the preexponential and activation energy for the diffusion coefficient are 1.8×10^{-5} m^2/s and 152,000 J/mol, respectively.

7.30 The outer surface of a steel gear is to be hardened by increasing its carbon content. The carbon is to be supplied from an external carbon-rich atmosphere, which is maintained at an elevated temperature. A diffusion heat treatment at 850°C (1123 K) for 10 min increases the carbon concentration to 0.90 wt% at a position 1.5 mm below the surface. Estimate the diffusion time required at 650°C (923 K) to achieve this same concentration also at a 1.5 mm position. Assume that the surface carbon content is the same for both heat treatments, which is maintained constant. Use the diffusion data in Table 7.2 for C diffusion in α-Fe.

7.31 An FCC iron–carbon alloy initially containing 0.20 wt% C is carburized at an elevated temperature and in an atmosphere wherein the surface carbon concentration is maintained at 1.0 wt%. If after 49.5 h the concentration of carbon is 0.35 wt% at a position 3.5 mm below the surface, determine the temperature at which the treatment was carried out.

Diffusion in Semiconducting Materials

7.32 Phosphorus atoms are to be diffused into a silicon wafer using both predeposition and drive-in heat treatments; the background concentration of P in this silicon material is known to be 5×10^{19} atoms/m^3. The predeposition treatment is to be conducted at 950°C for 45 minutes; the surface concentration of P is to be maintained at a constant level of 1.5×10^{26} atoms/m^3. Drive-in diffusion will be carried out at 1200°C for a period of 2.5 h. For the diffusion of P in Si, values of Q_d and D_0 are 3.40 eV/atom and 1.1×10^{-4} m^2/s, respectively.

(a) Calculate the value of Q_0.

(b) Determine the value of x_j for the drive-in diffusion treatment.

(c) Also for the drive-in treatment, compute the position x at which the concentration of P atoms is 10^{24} m^{-3}.

7.33 Aluminum atoms are to be diffused into a silicon wafer using both predeposition and drive-in heat treatments; the background concentration of Al in this silicon material is known to be 3×10^{19} atoms/m^3. The drive-in diffusion treatment is to be carried out at 1050°C for a period of

4.0 h, which gives a junction depth x_j of 3.0 μm. Compute the predeposition diffusion time at 950°C if the surface concentration is maintained at a constant level of 2×10^{25} atoms/m³. For the diffusion of Al in Si, values of Q_d and D_0 are 3.41 eV/atom and 1.38×10^{-4} m²/s, respectively.

Diffusion in Ionic and Polymeric Materials

7.34 Consider the diffusion of water vapor through a polypropylene (PP) sheet 2 mm thick. The pressures of H_2O at the two faces are 1 kPa and 10 kPa, which are maintained constant. Assuming conditions of steady state, what is the diffusion flux [in (cm³ STP)/cm²·s] at 298 K?

7.35 Argon diffuses through a high-density polyethylene (HDPE) sheet 40 mm thick at a rate of 4.0×10^{-7} (cm³ STP)/cm²·s at 330 K. The pressures of argon at the two faces are 7000 kPa and 2000 kPa, which are maintained constant. Assuming conditions of steady state, what is the permeability coefficient at 330 K?

7.36 The permeability coefficient of a type of small gas molecule in a polymer is dependent on absolute temperature according to the following equation:

$$P_M = P_{M_0} \exp\left(-\frac{Q_p}{RT}\right)$$

where P_{M_0} and Qp are constants for a given gas-polymer pair. Consider the diffusion of hydrogen through a poly(dimethyl siloxane) (PDMSO) sheet 30 mm thick. The hydrogen pressures at the two faces are 10 kPa and 1 kPa, which are maintained constant. Compute the diffusion flux [in (cm3 STP)/cm2?s] at 360 K.

For this diffusion system

$$P_{M_0} = 1.45 \times 10^{-8}(\text{cm}^3 \text{ STP})(\text{cm})/\text{cm}^2 \cdot \text{s} \cdot \text{Pa}$$
$$Q_p = 13.7 \text{ kJ/mol}$$

Also, assume a condition of steady state diffusion.

Spreadsheet Problems

7.1SS For a nonsteady-state diffusion situation (constant surface composition) wherein the surface and initial compositions are provided, as well as the value of the diffusion coefficient, develop a spreadsheet that will allow the user to determine the diffusion time required to achieve a given composition at some specified distance from the surface of the solid.

7.2SS For a nonsteady-state diffusion situation (constant surface composition) wherein the surface and initial compositions are provided, as well as the value of the diffusion coefficient, develop a spreadsheet that will allow the user to determine the distance from the surface at which some specified composition is achieved for some specified diffusion time.

7.3SS For a nonsteady-state diffusion situation (constant surface composition) wherein the surface and initial compositions are provided, as well as the value of the diffusion coefficient, develop a spreadsheet that will allow the user to determine the composition at some specified distance from the surface for some specified diffusion time.

7.4SS Given a set of at least two diffusion coefficient values and their corresponding temperatures, develop a spreadsheet that will allow the user to calculate **(a)** the activation energy and **(b)** the preexponential.

DESIGN PROBLEMS

Fick's First Law
(Factors That Influence Diffusion)

7.D1 It is desired to enrich the partial pressure of hydrogen in a hydrogen–nitrogen gas mixture for which the partial pressures of both gases are 0.1013 MPa. It has been proposed to accomplish this by passing both gases through a thin sheet of some metal at an elevated temperature; inasmuch as hydrogen diffuses through the plate at a higher rate than does nitrogen, the partial pressure of hydrogen will be higher on the exit side of the sheet. The design calls for partial pressures of 0.0709 MPa and 0.02026 MPa, respectively,

for hydrogen and nitrogen. The concentrations of hydrogen and nitrogen (C_H and C_N, in mol/m³) in this metal are functions of gas partial pressures (p_{H_2} and p_{N_2}, in MPa) and absolute temperature and are given by the following expressions:

$$C_H = 2.5 \times 10^3 \sqrt{p_{H_2}} \exp\left(-\frac{27.8 \text{ kJ/mol}}{RT}\right)$$
$$\text{(7.16a)}$$

$$C_N = 2.75 \times 10^{-3} \sqrt{p_{N_2}} \exp\left(-\frac{37.6 \text{ kJ/mol}}{RT}\right)$$
$$\text{(7.16b)}$$

Furthermore, the diffusion coefficients for the diffusion of these gases in this metal are functions of the absolute temperature as follows:

$$D_H(m^2/s) = 1.4 \times 10^{-7} \exp\left(-\frac{13.4 \text{ kJ/mol}}{RT}\right)$$

$$\text{(7.17a)}$$

$$D_N(m^2/s) = 3.0 \times 10^{-7} \exp\left(-\frac{76.15 \text{ kJ/mol}}{RT}\right)$$

$$\text{(7.17b)}$$

Is it possible to purify hydrogen gas in this manner? If so, specify a temperature at which the process may be carried out, and also the thickness of metal sheet that would be required. If this procedure is not possible, then state the reason(s) why.

7.D2 A gas mixture is found to contain two diatomic A and B species for which the partial pressures of both are 0.05065 MPa. This mixture is to be enriched in the partial pressure of the A species by passing both gases through a thin sheet of some metal at an elevated temperature. The resulting enriched mixture is to have a partial pressure of 0.02026 MPa for gas A, and 0.01013 MPa for gas B. The concentrations of A and B (C_A and C_B, in mol/m^3) are functions of gas partial pressures (p_{A_2} and p_{B_2}, in MPa) and absolute temperature according to the following expressions:

$$C_A = 200\sqrt{p_{A_2}} \exp\left(-\frac{25.0 \text{ kJ/mol}}{RT}\right) \quad \text{(7.18a)}$$

$$C_B = 1.0 \times 10^3 \sqrt{p_{B_2}} \exp\left(-\frac{30.0 \text{ kJ/mol}}{RT}\right)$$

$$\text{(7.18b)}$$

Furthermore, the diffusion coefficients for the diffusion of these gases in the metal are functions of the absolute temperature as follows:

$$D_A(m^2/s) = 4.0 \times 10^{-7} \exp\left(-\frac{15.0 \text{ kJ/mol}}{RT}\right)$$

$$\text{(7.19a)}$$

$$D_B(m^2/s) = 2.5 \times 10^{-6} \exp\left(-\frac{24.0 \text{ kJ/mol}}{RT}\right)$$

$$\text{(7.19b)}$$

Is it possible to purify the A gas in this manner? If so, specify a temperature at which the process may be carried out, and also the thickness of metal sheet that would be required. If this procedure is not possible, then state the reason(s) why.

Fick's Second Law—Nonsteady-State Diffusion (Factors That Influence Diffusion)

7.D3 The wear resistance of a steel shaft is to be improved by hardening its surface. This is to be accomplished by increasing the nitrogen content within an outer surface layer as a result of nitrogen diffusion into the steel. The nitrogen is to be supplied from an external nitrogen-rich gas at an elevated and constant temperature. The initial nitrogen content of the steel is 0.002 wt%, whereas the surface concentration is to be maintained at 0.50 wt%. For this treatment to be effective, a nitrogen content of 0.10 wt% must be established at a position 0.45 mm below the surface. Specify appropriate heat treatments in terms of temperature and time for temperatures between 475°C and 625°C. The preexponential and activation energy for the diffusion of nitrogen in iron are 3×10^{-7} m^2/s and 76,150 J/mol, respectively, over this temperature range.

Diffusion in Semiconducting Materials

7.D4 One integrated circuit design calls for the diffusion of arsenic into silicon wafers; the background concentration of As in Si is 2.5×10^{20} atoms/m^3. The predeposition heat treatment is to be conducted at 1000°C for 45 minutes, with a constant surface concentration of 8×10^{26} As atoms/m^3. At a drive-in treatment temperature of 1100°C, determine the diffusion time required for a junction depth of 1.5 μm. For this system, values of Q_d and D_0 are 4.10 eV/atom and 2.29×10^{-3} m^2/s, respectively.

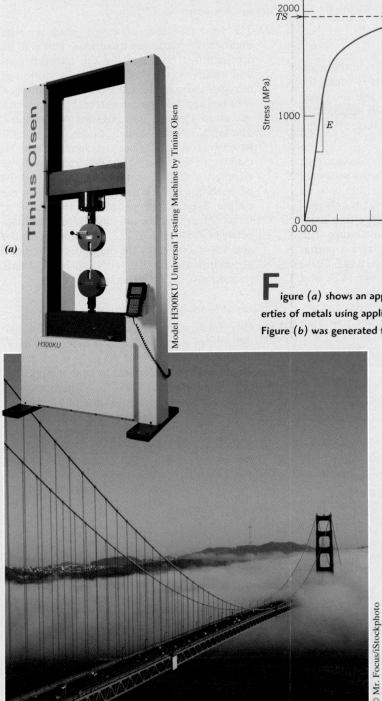

(a)

Model H300KU Universal Testing Machine by Tinius Olsen

(b)

(c)

© Mr. Focus/iStockphoto

Figure (*a*) shows an apparatus that measures the mechanical properties of metals using applied tensile forces (Sections 8.3 and 8.4). Figure (*b*) was generated from a tensile test performed by an apparatus such as this on a steel specimen. Data plotted are stress (vertical axis—a measure of applied force) versus strain (horizontal axis—related to the degree of specimen elongation). The mechanical properties of modulus of elasticity (stiffness, *E*), yield strength (σ_y), and tensile strength (*TS*) are determined as noted on these graphs.

Figure (*c*) shows a suspension bridge. The weight of the bridge deck and automobiles imposes tensile forces on the vertical suspender cables. These forces are transferred to the main suspension cable, which sags in a more-or-less parabolic shape. The metal alloy(s) from which these cables are constructed must meet certain stiffness and strength criteria. Stiffness and strength of the alloy(s) may be assessed from tests performed using a tensile-testing apparatus (and the resulting stress–strain plots) similar to those shown.

It is incumbent on engineers to understand how the various mechanical properties are measured and what these properties represent; they may be called upon to design structures/components using predetermined materials such that unacceptable levels of deformation and/or failure will not occur. We demonstrate this procedure with respect to the design of a tensile-testing apparatus in Design Example 8.1.

Learning Objectives

After studying this chapter, you should be able to do the following:

1. Define engineering stress and engineering strain.
2. State Hooke's law and note the conditions under which it is valid.
3. Define Poisson's ratio.
4. Given an engineering stress–strain diagram, determine (a) the modulus of elasticity, (b) the yield strength (0.002 strain offset), (c) the tensile strength and (d) estimate the percentage elongation.
5. For the tensile deformation of a ductile cylindrical specimen, describe changes in specimen profile to the point of fracture.
6. Compute ductility in terms of both percentage elongation and percentage reduction of area for a material that is loaded in tension to fracture.
7. Give brief definitions of and the units for modulus of resilience and toughness (static).
8. For a specimen being loaded in tension, given the applied load, the instantaneous cross-sectional dimensions, and original and instantaneous lengths, be able to compute true stress and true strain values.
9. Name the two most common hardness-testing techniques; note two differences between them.
10. (a) Name and briefly describe the two different microindentation hardness testing techniques, and (b) cite situations for which these techniques are generally used.
11. Compute the working stress for a ductile material.

8.1 INTRODUCTION

Many materials are subjected to forces or loads when in service; examples include the aluminum alloy from which an airplane wing is constructed and the steel in an automobile axle. In such situations it is necessary to know the characteristics of the material and to design the member from which it is made such that any resulting deformation will not be excessive and fracture will not occur. The mechanical behavior of a material reflects its response or deformation in relation to an applied load or force. Key mechanical design properties are stiffness, strength, hardness, ductility, and toughness.

The mechanical properties of materials are ascertained by performing carefully designed laboratory experiments that replicate as nearly as possible the service conditions. Factors to be considered include the nature of the applied load and its duration, as well as the environmental conditions. It is possible for the load to be tensile, compressive, or shear, and its magnitude may be constant with time, or it may fluctuate continuously. Application time may be only a fraction of a second, or it may extend over a period of many years. Service temperature may be an important factor.

Mechanical properties are of concern to a variety of parties (e.g., producers and consumers of materials, research organizations, government agencies) that have differing interests. Consequently, it is imperative that there be some consistency in the manner in which tests are conducted and in the interpretation of their results. This consistency is accomplished by using standardized testing techniques. Establishment and publication of these standards are often coordinated by professional societies. In the United States the most active organization is the American Society for Testing and Materials (ASTM). Its *Annual Book of ASTM Standards* (http://www.astm.org) comprises numerous volumes that are issued and updated yearly; a large number of these standards relate to mechanical testing techniques. Several of these are referenced by footnote in this and subsequent chapters.

The role of structural engineers is to determine stresses and stress distributions within members that are subjected to well-defined loads. This may be accomplished by experimental testing techniques and/or by theoretical and mathematical stress analyses. These topics are treated in traditional texts on stress analysis and strength of materials.

Materials and metallurgical engineers, however, are concerned with producing and fabricating materials to meet service requirements as predicted by these stress analyses. This necessarily involves an understanding of the relationships between the microstructure (i.e., internal features) of materials and their mechanical properties.

Materials are frequently chosen for structural applications because they have desirable combinations of mechanical characteristics. The present discussion is confined primarily to the mechanical behavior of metals; polymers and ceramics are treated separately because they are, to a large degree, mechanically different from metals. This chapter discusses the stress–strain behavior of metals and the related mechanical properties, and also examines other important mechanical characteristics. Discussions of the microscopic aspects of deformation mechanisms and methods to strengthen and regulate the mechanical behavior of metals are deferred to later chapters.

8.2 CONCEPTS OF STRESS AND STRAIN

If a load is static or changes relatively slowly with time and is applied uniformly over a cross section or surface of a member, the mechanical behavior may be ascertained by a simple stress–strain test; these are most commonly conducted for metals at room temperature. There are three principal ways in which a load may be applied: namely, tension, compression, and shear (Figures 8.1a, b, c). In engineering practice many loads are torsional rather than pure shear; this type of loading is illustrated in Figure 8.1d.

Tension Tests[1]

One of the most common mechanical stress–strain tests is performed in *tension*. As will be seen, the tension test can be used to ascertain several mechanical properties of materials that are important in design. A specimen is deformed, usually to fracture, with a gradually increasing tensile load that is applied uniaxially along the long axis of a specimen. A standard tensile specimen is shown in Figure 8.2. Normally, the cross section is circular, but rectangular specimens are also used. This "dogbone" specimen configuration was chosen so that, during testing, deformation is confined to the narrow center region (which has a uniform cross section along its length) and also to reduce the likelihood of fracture at the ends of the specimen. The standard diameter is approximately 12.8 mm, whereas the reduced section length should be at least four times this diameter; 60 mm is common. Gauge length is used in ductility computations, as discussed in Section 8.4; the standard value is 50 mm. The specimen is mounted by its ends into the holding grips of the testing apparatus (Figure 8.3). The tensile testing

---

[1]ASTM Standards E8 and E8M, "Standard Test Methods for Tension Testing of Metallic Materials."

Figure 8.1
(*a*) Schematic illustration of how a tensile load produces an elongation and positive linear strain. (*b*) Schematic illustration of how a compressive load produces contraction and a negative linear strain. (*c*) Schematic representation of shear strain γ, where $\gamma = \tan \theta$. (*d*) Schematic representation of torsional deformation (i.e., angle of twist ϕ) produced by an applied torque T.

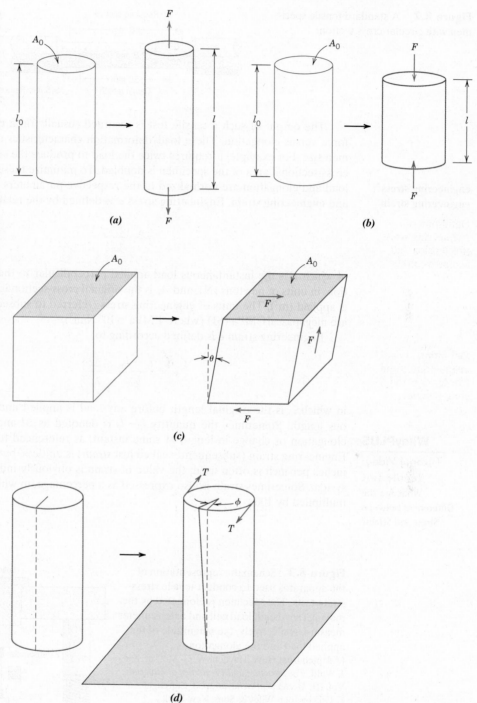

(*a*)

(*b*)

(*c*)

(*d*)

machine is designed to elongate the specimen at a constant rate, and to continuously and simultaneously measure the instantaneous applied load (with a load cell) and the resulting elongations (using an extensometer). A stress–strain test typically takes several minutes to perform and is destructive; that is, the test specimen is permanently deformed and usually fractured. [Photograph (*a*) opening this chapter is of a modern tensile-testing apparatus.]

Figure 8.2 A standard tensile specimen with circular cross section.

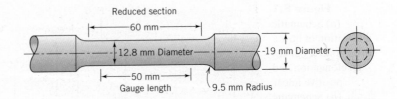

Reduced section

60 mm

12.8 mm Diameter

19 mm Diameter

50 mm

Gauge length

9.5 mm Radius

The output of such a tensile test is recorded (usually on a computer) as load or force versus elongation. These load–deformation characteristics depend on the specimen size. For example, it requires twice the load to produce the same elongation if the cross-sectional area of the specimen is doubled. To minimize these geometrical factors, load and elongation are normalized to the respective parameters of **engineering stress** and **engineering strain**. Engineering stress σ is defined by the relationship

engineering stress
engineering strain

Definition of engineering stress (for tension and compression)

$$\sigma = \frac{F}{A_0} \tag{8.1}$$

in which F is the instantaneous load applied perpendicular to the specimen cross section, in units of newtons (N), and A_0 is the original cross-sectional area before any load is applied (m^2). The units of engineering stress (referred to subsequently as just *stress*) are megapascals, MPa (SI) (where 1 MPa = 10^6 N/m^2).

Engineering strain ϵ is defined according to

Definition of engineering strain (for tension and compression)

$$\epsilon = \frac{l_i - l_0}{l_0} = \frac{\Delta l}{l_0} \tag{8.2}$$

in which l_0 is the original length before any load is applied and l_i is the instantaneous length. Sometimes the quantity $l_i - l_0$ is denoted as Δl and is the deformation elongation or change in length at some instant, as referenced to the original length. Engineering strain (subsequently called just *strain*) is unitless, but meters per meter or inches per inch is often used; the value of strain is obviously independent of the unit system. Sometimes strain is also expressed as a percentage, in which the strain value is multiplied by 100.

WileyPLUS

Tutorial Video:
Tensile Test
What Are the
Differences between
Stress and Strain?

Figure 8.3 Schematic representation of the apparatus used to conduct tensile stress–strain tests. The specimen is elongated by the moving crosshead; load cell and extensometer measure, respectively, the magnitude of the applied load and the elongation.
(Adapted from H. W. Hayden, W. G. Moffatt, and J. Wulff, *The Structure and Properties of Materials*, Vol. III, *Mechanical Behavior*, p. 2. Copyright © 1965 by John Wiley & Sons, New York.)

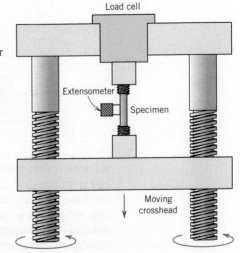

Load cell

Extensometer

Specimen

Moving crosshead

Compression Tests[2]

Compression stress–strain tests may be conducted if in-service forces are of this type. A compression test is conducted in a manner similar to the tensile test, except that the force is compressive and the specimen contracts along the direction of the stress. Equations 8.1 and 8.2 are utilized to compute compressive stress and strain, respectively. By convention, a compressive force is taken to be negative, which yields a negative stress. Furthermore, because l_0 is greater than l_i, compressive strains computed from Equation 8.2 are necessarily also negative. Tensile tests are more common because they are easier to perform; also, for most materials used in structural applications, very little additional information is obtained from compressive tests. Compressive tests are used when a material's behavior under large and permanent (i.e., plastic) strains is desired, as in manufacturing applications, or when the material is brittle in tension.

Shear and Torsional Tests[3]

For tests performed using a pure shear force as shown in Figure 8.1c, the shear stress τ is computed according to

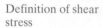
Definition of shear stress

$$\tau = \frac{F}{A_0} \tag{8.3}$$

where F is the load or force imposed parallel to the upper and lower faces, each of which has an area of A_0. The shear strain γ is defined as the tangent of the strain angle θ, as indicated in the figure. The units for shear stress and strain are the same as for their tensile counterparts.

   Torsion is a variation of pure shear in which a structural member is twisted in the manner of Figure 8.1d; torsional forces produce a rotational motion about the longitudinal axis of one end of the member relative to the other end. Examples of torsion are found for machine axles and drive shafts as well as for twist drills. Torsional tests are normally performed on cylindrical solid shafts or tubes. A shear stress τ is a function of the applied torque T, whereas shear strain γ is related to the angle of twist, ϕ in Figure 8.1d.

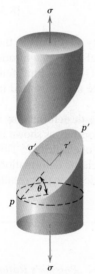

Figure 8.4
Schematic representation showing normal (σ') and shear (τ') stresses that act on a plane oriented at an angle θ relative to the plane taken perpendicular to the direction along which a pure tensile stress (σ) is applied.

Geometric Considerations of the Stress State

Stresses that are computed from the tensile, compressive, shear, and torsional force states represented in Figure 8.1 act either parallel or perpendicular to planar faces of the bodies represented in these illustrations. Note that the stress state is a function of the orientations of the planes upon which the stresses are taken to act. For example, consider the cylindrical tensile specimen of Figure 8.4 that is subjected to a tensile stress σ applied parallel to its axis. Furthermore, consider also the plane p-p' that is oriented at some arbitrary angle θ relative to the plane of the specimen end-face. Upon this plane p-p', the applied stress is no longer a pure tensile one. Rather, a more complex stress state is present that consists of a tensile (or normal) stress σ' that acts normal to the p-p' plane and, in addition, a shear stress τ' that acts parallel to this plane; both of these stresses are represented in the figure.

---

[2]ASTM Standard E9, "Standard Test Methods of Compression Testing of Metallic Materials at Room Temperature."

[3]ASTM Standard E143, "Standard Test Method for Shear Modulus at Room Temperature."

Using mechanics-of-materials principles,[4] it is possible to develop equations for σ' and τ' in terms of σ and θ, as follows:

$$\sigma' = \sigma \cos^2\theta = \sigma\left(\frac{1 + \cos 2\theta}{2}\right) \tag{8.4a}$$

$$\tau' = \sigma \sin\theta \cos\theta = \sigma\left(\frac{\sin 2\theta}{2}\right) \tag{8.4b}$$

These same mechanics principles allow the transformation of stress components from one coordinate system to another coordinate system with a different orientation. Such treatments are beyond the scope of the present discussion.

8.3 ELASTIC DEFORMATION

Stress–Strain Behavior

Hooke's law—
relationship between
engineering stress
and engineering
strain for elastic
deformation (tension
and compression)

modulus of elasticity

elastic deformation

The degree to which a structure deforms or strains depends on the magnitude of an imposed stress. For most metals that are stressed in tension and at relatively low levels, stress and strain are proportional to each other through the relationship

$$\sigma = E\epsilon \tag{8.5}$$

This is known as *Hooke's law*, and the constant of proportionality E (GPa)[5] is the **modulus of elasticity**, or *Young's modulus*. For most typical metals, the magnitude of this modulus ranges between 45 GPa, for magnesium, and 407 GPa, for tungsten. Modulus of elasticity values for several metals at room temperature are presented in Table 8.1.

Deformation in which stress and strain are proportional is called **elastic deformation**; a plot of stress (ordinate) versus strain (abscissa) results in a linear relationship, as shown in Figure 8.5. The slope of this linear segment corresponds to the modulus of elasticity E. This modulus may be thought of as stiffness, or a material's resistance to elastic deformation. The greater the modulus, the stiffer the material, or the smaller the elastic strain that results

WileyPLUS: VMSE
Metal Alloys

Table 8.1

Room-Temperature
Elastic and Shear
Moduli and Poisson's
Ratio for Various
Metal Alloys

Metal Alloy	Modulus of Elasticity GPa	Shear Modulus GPa	Poisson's Ratio
Aluminum	69	25	0.33
Brass	97	37	0.34
Copper	110	46	0.34
Magnesium	45	17	0.29
Nickel	207	76	0.31
Steel	207	83	0.30
Titanium	107	45	0.34
Tungsten	407	160	0.28

---

[4]See, for example, W. F. Riley, L. D. Sturges, and D. H. Morris, *Mechanics of Materials,* 6th edition, Wiley, Hoboken, NJ, 2006.

[5]The SI unit for the modulus of elasticity is *gigapascal* (GPa), where 1 GPa = 10^9 N/m^2 = 10^3 MPa.

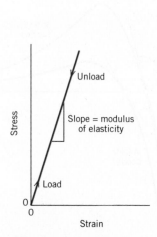

Figure 8.5 Schematic stress–strain diagram showing linear elastic deformation for loading and unloading cycles.

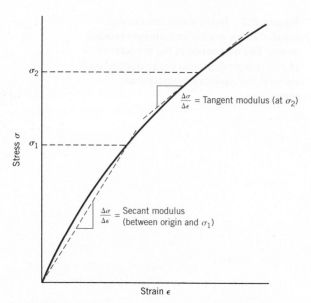

Figure 8.6 Schematic stress–strain diagram showing nonlinear elastic behavior and how secant and tangent moduli are determined.

WileyPLUS

Tutorial Video:
Tensile Test
Calculations
Calculating Elastic
Modulus Using a
Stress vs. Strain Curve

from the application of a given stress. The modulus is an important design parameter for computing elastic deflections.

Elastic deformation is *nonpermanent*, which means that when the applied load is released, the piece returns to its original shape. As shown in the stress–strain plot (Figure 8.5), application of the load corresponds to moving from the origin up and along the straight line. Upon release of the load, the line is traversed in the opposite direction, back to the origin.

There are some materials (i.e., gray cast iron, concrete, and many polymers) for which this elastic portion of the stress–strain curve is not linear (Figure 8.6); hence, it is not possible to determine a modulus of elasticity as described previously. For this nonlinear behavior, either the *tangent* or *secant modulus* is normally used. The tangent modulus is taken as the slope of the stress–strain curve at some specified level of stress, whereas the secant modulus represents the slope of a secant drawn from the origin to some given point of the σ-ϵ curve. The determination of these moduli is illustrated in Figure 8.6.

On an atomic scale, macroscopic elastic strain is manifested as small changes in the interatomic spacing and the stretching of interatomic bonds. As a consequence, the magnitude of the modulus of elasticity is a measure of the resistance to separation of adjacent atoms, that is, the interatomic bonding forces. Furthermore, this modulus is proportional to the slope of the interatomic force–separation curve (Figure 2.10*a*) at the equilibrium spacing:

$$E \propto \left(\frac{dF}{dr} \right)_{r_0} \tag{8.6}$$

Figure 8.7 shows the force–separation curves for materials having both strong and weak interatomic bonds; the slope at r_0 is indicated for each.

Values of the modulus of elasticity for ceramic materials are about the same as for metals; for polymers they are lower (Figure 1.5). These differences are a direct consequence of the different types of atomic bonding in the three materials types.

Figure 8.7 Force versus interatomic separation for weakly and strongly bonded atoms. The magnitude of the modulus of elasticity is proportional to the slope of each curve at the equilibrium interatomic separation r_0.

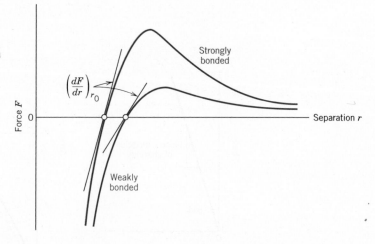

Furthermore, with increasing temperature, the modulus of elasticity decreases, as is shown for several metals in Figure 8.8.

As would be expected, the imposition of compressive, shear, or torsional stresses also evokes elastic behavior. The stress–strain characteristics at low stress levels are virtually the same for both tensile and compressive situations, to include the magnitude of the modulus of elasticity. Shear stress and strain are proportional to each other through the expression

Relationship between shear stress and shear strain for elastic deformation

$$\tau = G\gamma \tag{8.7}$$

where G is the *shear modulus*, the slope of the linear elastic region of the shear stress–strain curve. Table 8.1 also gives the shear moduli for a number of common metals.

Figure 8.8 Plot of modulus of elasticity versus temperature for tungsten, steel, and aluminum. (Adapted from K. M. Ralls, T. H. Courtney, and J. Wulff, *Introduction to Materials Science and Engineering.* Copyright © 1976 by John Wiley & Sons, New York. Reprinted by permission of John Wiley & Sons, Inc.)

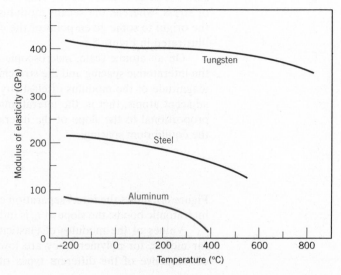

Anelasticity

To this point, it has been assumed that elastic deformation is time independent—that is, that an applied stress produces an instantaneous elastic strain that remains constant over the period of time the stress is maintained. It has also been assumed that upon release of the load, the strain is totally recovered—that is, that the strain immediately returns to zero. In most engineering materials, however, there will also exist a time-dependent elastic strain component—that is, elastic deformation will continue after the stress application, and upon load release, some finite time is required for complete recovery. This time-dependent elastic behavior is known as **anelasticity**, and it is due to time-dependent microscopic and atomistic processes that are attendant to the deformation. For metals, the anelastic component is normally small and is often neglected. However, for some polymeric materials, its magnitude is significant; in this case it is termed *viscoelastic behavior*, which is the discussion topic of Section 15.4.

anelasticity

EXAMPLE PROBLEM 8.1

Elongation (Elastic) Computation

A piece of copper originally 305 mm long is pulled in tension with a stress of 276 MPa. If the deformation is entirely elastic, what will be the resultant elongation?

Solution

Because the deformation is elastic, strain is dependent on stress according to Equation 8.5. Furthermore, the elongation Δl is related to the original length l_0 through Equation 8.2. Combining these two expressions and solving for Δl yields

$$\sigma = \epsilon E = \left(\frac{\Delta l}{l_0}\right) E$$

$$\Delta l = \frac{\sigma l_0}{E}$$

The values of σ and l_0 are given as 276 MPa and 305 mm, respectively, and the magnitude of E for copper from Table 8.1 is 110 GPa. Elongation is obtained by substitution into the preceding expression as

$$\Delta l = \frac{(276 \text{ MPa})(305 \text{ mm})}{110 \times 10^3 \text{ MPa}} = 0.77 \text{ mm}$$

Elastic Properties of Materials

When a tensile stress is imposed on a metal specimen, an elastic elongation and accompanying strain ϵ_z result in the direction of the applied stress (arbitrarily taken to be the z direction), as indicated in Figure 8.9. As a result of this elongation, there will be constrictions in the lateral (x and y) directions perpendicular to the applied stress; from these contractions, the compressive strains ϵ_x and ϵ_y may be determined. If the applied stress is uniaxial (only in the z direction) and the material is isotropic, then $\epsilon_x = \epsilon_y$. A parameter termed **Poisson's ratio** ν is defined as the ratio of the lateral and axial strains, or

Poisson's ratio

Definition of
Poisson's ratio in
terms of lateral
and axial strains

$$\nu = -\frac{\epsilon_x}{\epsilon_z} = -\frac{\epsilon_y}{\epsilon_z} \tag{8.8}$$

8.4 PLASTIC DEFORMATION

plastic deformation

For most metallic materials, elastic deformation persists only to strains of about 0.005. As the material is deformed beyond this point, the stress is no longer proportional to strain (Hooke's law, Equation 8.5, ceases to be valid), and permanent, nonrecoverable, or **plastic deformation** occurs. Figure 8.10*a* plots schematically the tensile stress–strain behavior into the plastic region for a typical metal. The transition from elastic to plastic is a gradual one for most metals; some curvature results at the onset of plastic deformation, which increases more rapidly with rising stress.

From an atomic perspective, plastic deformation corresponds to the breaking of bonds with original atom neighbors and then the re-forming of bonds with new neighbors as large numbers of atoms or molecules move relative to one another; upon removal of the stress, they do not return to their original positions. The mechanism of this deformation is different for crystalline and amorphous materials. For crystalline solids, deformation is accomplished by means of a process called *slip*, which involves the motion of dislocations as discussed in Section 9.2. Plastic deformation in noncrystalline solids (as well as liquids) occurs by a viscous flow mechanism, which is outlined in Section 14.8.

WileyPLUS: VMSE
Metal Alloys

Tensile Properties

Yielding and Yield Strength

yielding

Most structures are designed to ensure that only elastic deformation will result when a stress is applied. A structure or component that has plastically deformed—or experienced a permanent change in shape—may not be capable of functioning as intended. It is therefore desirable to know the stress level at which plastic deformation begins, or where the phenomenon of **yielding** occurs. For metals that experience this gradual

Figure 8.10 (*a*) Typical stress–strain behavior for a metal showing elastic and plastic deformations, the proportional limit *P*, and the yield strength σ_y, as determined using the 0.002 strain offset method. (*b*) Representative stress–strain behavior found for some steels demonstrating the yield point phenomenon.

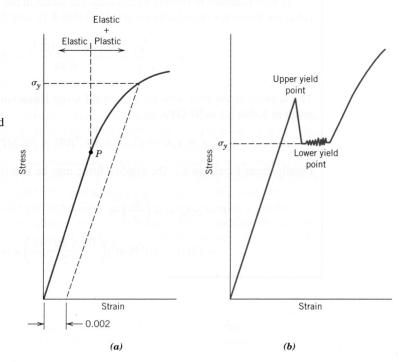

proportional limit

elastic–plastic transition, the point of yielding may be determined as the initial departure from linearity of the stress–strain curve; this is sometimes called the **proportional limit**, as indicated by point P in Figure 8.10a, and represents the onset of plastic deformation on a microscopic level. The position of this point P is difficult to measure precisely. As a consequence, a convention has been established by which a straight line is constructed parallel to the elastic portion of the stress–strain curve at some specified strain offset, usually 0.002. The stress corresponding to the inter-section of this line and the stress–strain curve as it bends over in the plastic region

yield strength

is defined as the **yield strength** σ_y.[7] This is demonstrated in Figure 8.10a. The units of yield strength are MPa.

For materials having a nonlinear elastic region (Figure 8.6), use of the strain offset method is not possible, and the usual practice is to define the yield strength as the stress required to produce some amount of strain (e.g., $\epsilon = 0.005$).

Some steels and other materials exhibit the tensile stress–strain behavior shown in Figure 8.10b. The elastic–plastic transition is very well defined and occurs abruptly in what is termed a *yield point phenomenon*. At the upper yield point, plastic deformation is initi-ated with an apparent decrease in engineering stress. Continued deformation fluctuates slightly about some constant stress value, termed the *lower yield point*; stress subse-quently rises with increasing strain. For metals that display this effect, the yield strength is taken as the average stress that is associated with the lower yield point because it is well defined and relatively insensitive to the testing procedure.[8] Thus, it is not necessary to employ the strain offset method for these materials.

The magnitude of the yield strength for a metal is a measure of its resistance to plastic deformation. Yield strengths may range from 35 MPa for a low-strength aluminum to greater than 1400 MPa for high-strength steels.

Concept Check 8.1 Cite the primary differences between elastic, anelastic, and plastic deformation behaviors.

[*The answer may be found at* www.wiley.com/college/callister *(Student Companion Site).*]

Tensile Strength

tensile strength

After yielding, the stress necessary to continue plastic deformation in metals increases to a maximum, point M in Figure 8.11, and then decreases to the eventual fracture, point F. The **tensile strength** TS (MPa) is the stress at the maximum on the engineer-ing stress–strain curve (Figure 8.11). This corresponds to the maximum stress that can be sustained by a structure in tension; if this stress is applied and maintained, fracture will result. All deformation to this point is uniform throughout the narrow region of the tensile specimen. However, at this maximum stress, a small constriction or neck be-gins to form at some point, and all subsequent deformation is confined at this neck, as indicated by the schematic specimen insets in Figure 8.11. This phenomenon is termed

---

[7]*Strength* is used in lieu of *stress* because strength is a property of the metal, whereas stress is related to the magni-tude of the applied load.

[8]Note that to observe the yield point phenomenon, a "stiff" tensile-testing apparatus must be used; by "stiff," it is meant that there is very little elastic deformation of the machine during loading.

Figure 8.11 Typical engineering stress–strain behavior to fracture, point *F*. The tensile strength *TS* is indicated at point *M*. The circular insets represent the geometry of the deformed specimen at various points along the curve.

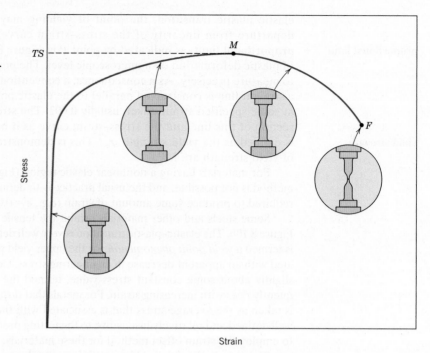

necking, and fracture ultimately occurs at the neck.[9] The fracture strength corresponds to the stress at fracture.

Tensile strengths vary from 50 MPa for an aluminum to as high as 3000 MPa for the high-strength steels. Typically, when the strength of a metal is cited for design purposes, the yield strength is used because by the time a stress corresponding to the tensile strength has been applied, often a structure has experienced so much plastic deformation that it is useless. Furthermore, fracture strengths are not normally specified for engineering design purposes.

EXAMPLE PROBLEM 8.3

Mechanical Property Determinations from Stress–Strain Plot

From the tensile stress–strain behavior for the brass specimen shown in Figure 8.12, determine the following:

(a) The modulus of elasticity
(b) The yield strength at a strain offset of 0.002
(c) The maximum load that can be sustained by a cylindrical specimen having an original diameter of 12.8 mm
(d) The change in length of a specimen originally 250 mm long that is subjected to a tensile stress of 345 MPa

Solution

(a) The modulus of elasticity is the slope of the elastic or initial linear portion of the stress–strain curve. The strain axis has been expanded in the inset of Figure 8.12 to facilitate

---

[9]The apparent decrease in engineering stress with continued deformation past the maximum point of Figure 8.11 is due to the necking phenomenon. As explained later in this section, the true stress (within the neck) actually increases.

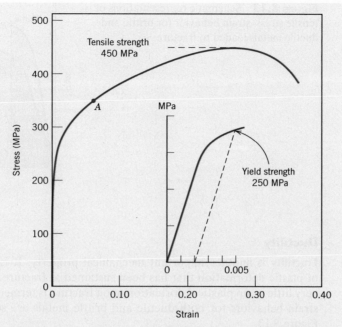

Figure 8.12 The stress–strain behavior for the brass specimen discussed in Example Problem 8.3.

this computation. The slope of this linear region is the rise over the run, or the change in stress divided by the corresponding change in strain; in mathematical terms,

$$E = \text{slope} = \frac{\Delta\sigma}{\Delta\epsilon} = \frac{\sigma_2 - \sigma_1}{\epsilon_2 - \epsilon_1} \tag{8.10}$$

Inasmuch as the line segment passes through the origin, it is convenient to take both σ_1 and ϵ_1 as zero. If σ_2 is arbitrarily taken as 150 MPa, then ϵ_2 will have a value of 0.0016. Therefore,

$$E = \frac{(150 - 0)\,\text{MPa}}{0.0016 - 0} = 93.8\,\text{GPa}$$

which is very close to the value of 97 GPa given for brass in Table 8.1.

(b) The 0.002 strain offset line is constructed as shown in the inset; its inter-section with the stress–strain curve is at approximately 250 MPa, which is the yield strength of the brass.

(c) The maximum load that can be sustained by the specimen is calculated by using Equation 8.1, in which σ is taken to be the tensile strength, from Figure 8.12, 450 MPa. Solving for F, the maximum load, yields

$$F = \sigma A_0 = \sigma \left(\frac{d_0}{2}\right)^2 \pi$$

$$= (450 \times 10^6\,\text{N/m}^2)\left(\frac{12.8 \times 10^{-3}\,\text{m}}{2}\right)^2 \pi = 57{,}900\,\text{N}$$

(d) To compute the change in length, Δl, in Equation 8.2, it is first necessary to determine the strain that is produced by a stress of 345 MPa. This is accomplished by locating the stress point on the stress–strain curve, point A, and reading the corresponding strain from the strain axis, which is approximately 0.06. Inasmuch as $l_0 = 250$ mm, we have

$$\Delta l = \epsilon l_0 = (0.06)(250\,\text{mm}) = 15\,\text{mm}$$

Figure 8.13 Schematic representations of tensile stress–strain behavior for brittle and ductile metals loaded to fracture.

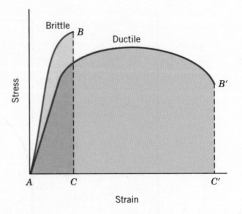

Ductility

ductility

Ductility is another important mechanical property. It is a measure of the degree of plastic deformation that has been sustained at fracture. A metal that experiences very little or no plastic deformation upon fracture is termed *brittle*. The tensile stress–strain behaviors for both ductile and brittle metals are schematically illustrated in Figure 8.13.

Ductility may be expressed quantitatively as either *percent elongation* or *percent reduction in area*. Percent elongation (%EL) is the percentage of plastic strain at fracture, or

Ductility, as percent elongation

$$\%EL = \left(\frac{l_f - l_0}{l_0} \right) \times 100 \tag{8.11}$$

where l_f is the fracture length[10] and l_0 is the original gauge length as given earlier. Inasmuch as a significant proportion of the plastic deformation at fracture is confined to the neck region, the magnitude of %EL will depend on specimen gauge length. The shorter l_0, the greater the fraction of total elongation from the neck and, consequently, the higher the value of %EL. Therefore, l_0 should be specified when percent elongation values are cited; it is commonly 50 mm.

Percent reduction in area (%RA) is defined as

Ductility, as percent reduction in area

$$\%RA = \left(\frac{A_0 - A_f}{A_0} \right) \times 100 \tag{8.12}$$

WileyPLUS

Tutorial Video:
Tensile Test
How do I determine ductility in percent elongation and percent reduction in area?

where A_0 is the original cross-sectional area and A_f is the cross-sectional area at the point of fracture.[10] Values of percent reduction in area are independent of both l_0 and A_0. Furthermore, for a given material, the magnitudes of %EL and %RA will, in general, be different. Most metals possess at least a moderate degree of ductility at room temperature; however, some become brittle as the temperature is lowered (Section 10.6).

Knowledge of the ductility of materials is important for at least two reasons. First, it indicates to a designer the degree to which a structure will deform plastically before

---

[10]Both l_f and A_f are measured subsequent to fracture and after the two broken ends have been repositioned back together.

Table 8.2

Typical Mechanical Properties of Several Metals and Alloys in an Annealed State

Metal Alloy	Yield Strength, MPa	Tensile Strength, MPa	Ductility, %EL (in 50 mm)
Aluminum	35	90	40
Copper	69	200	45
Brass (70Cu–30Zn)	75	300	68
Iron	130	262	45
Nickel	138	480	40
Steel (1020)	180	380	25
Titanium	450	520	25
Molybdenum	565	655	35

fracture. Second, it specifies the degree of allowable deformation during fabrication operations. We sometimes refer to relatively ductile materials as being "forgiving," in the sense that they may experience local deformation without fracture, should there be an error in the magnitude of the design stress calculation.

Brittle materials are *approximately* considered to be those having a fracture strain of less than about 5%.

Thus, several important mechanical properties of metals may be determined from tensile stress–strain tests. Table 8.2 presents some typical room-temperature values of yield strength, tensile strength, and ductility for several common metals. These properties are sensitive to any prior deformation, the presence of impurities, and/or any heat treatment to which the metal has been subjected. The modulus of elasticity is one mechanical parameter that is insensitive to these treatments. As with modulus of elasticity, the magnitudes of both yield and tensile strengths decline with increasing temperature; just the reverse holds for ductility—it usually increases with temperature. Figure 8.14 shows how the stress–strain behavior of iron varies with temperature.

Resilience

resilience

Resilience is the capacity of a material to absorb energy when it is deformed elastically and then, upon unloading, to have this energy recovered. The associated property is the *modulus of resilience, U_r*, which is the strain energy per unit volume required to stress a material from an unloaded state up to the point of yielding.

Computationally, the modulus of resilience for a specimen subjected to a uniaxial tension test is just the area under the engineering stress–strain curve taken to yielding (Figure 8.15), or

Definition of modulus of resilience

$$U_r = \int_0^{\epsilon_y} \sigma \, d\epsilon$$ (8.13a)

Assuming a linear elastic region, we have

Modulus of resilience for linear elastic behavior

$$U_r = \frac{1}{2} \sigma_y \epsilon_y$$ (8.13b)

in which ϵ_y is the strain at yielding.

true strain

Furthermore, it is occasionally more convenient to represent strain as **true strain** ϵ_T, defined by

Definition of true strain

$$\epsilon_T = \ln\frac{l_i}{l_0} \tag{8.16}$$

If no volume change occurs during deformation—that is, if

$$A_i l_i = A_0 l_0 \tag{8.17}$$

—then true and engineering stress and strain are related according to

Conversion of engineering stress to true stress

$$\sigma_T = \sigma(1 + \epsilon) \tag{8.18a}$$

Conversion of engineering strain to true strain

$$\epsilon_T = \ln(1 + \epsilon) \tag{8.18b}$$

Equations 8.18a and 8.18b are valid only to the onset of necking; beyond this point, true stress and strain should be computed from actual load, cross-sectional area, and gauge length measurements.

A schematic comparison of engineering and true stress–strain behaviors is made in Figure 8.16. It is worth noting that the true stress necessary to sustain increasing strain continues to rise past the tensile point M'.

Coincident with the formation of a neck is the introduction of a complex stress state within the neck region (i.e., the existence of other stress components in addition to the axial stress). As a consequence, the correct stress (*axial*) within the neck is slightly lower than the stress computed from the applied load and neck cross-sectional area. This leads to the "corrected" curve in Figure 8.16.

For some metals and alloys the region of the true stress–strain curve from the onset of plastic deformation to the point at which necking begins may be approximated by

True stress–true strain relationship in the plastic region of deformation (to the point of necking)

$$\sigma_T = K\epsilon_T^n \tag{8.19}$$

In this expression, K and n are constants; these values vary from alloy to alloy and also depend on the condition of the material (whether it has been plastically deformed, heat-treated, etc.). The parameter n is often termed the *strain-hardening exponent* and has a value less than unity. Values of n and K for several alloys are given in Table 8.4.

Figure 8.16 A comparison of typical tensile engineering stress–strain and true stress–strain behaviors. Necking begins at point M on the engineering curve, which corresponds to M' on the true curve. The "corrected" true stress–strain curve takes into account the complex stress state within the neck region.

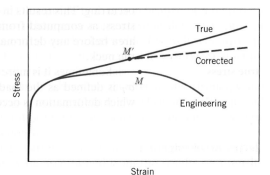

Table 8.4

The *n* and *K* Values (Equation 8.19) for Several Alloys

Material	n	K (MPa)
Low-carbon steel (annealed)	0.21	600
4340 steel alloy (tempered @ 315°C)	0.12	2650
304 stainless steel (annealed)	0.44	1400
Copper (annealed)	0.44	530
Naval brass (annealed)	0.21	585
2024 aluminum alloy (heat-treated—T3)	0.17	780
AZ-31B magnesium alloy (annealed)	0.16	450

EXAMPLE PROBLEM 8.4

Ductility and True-Stress-at-Fracture Computations

A cylindrical specimen of steel having an original diameter of 12.8 mm is tensile-tested to fracture and found to have an engineering fracture strength σ_f of 460 MPa. If its cross-sectional diameter at fracture is 10.7 mm, determine

(a) The ductility in terms of percentage reduction in area

(b) The true stress at fracture

Solution

(a) Ductility is computed using Equation 8.12, as

$$\% RA = \frac{\left(\dfrac{12.8 \text{ mm}}{2}\right)^2 \pi - \left(\dfrac{10.7 \text{ mm}}{2}\right)^2 \pi}{\left(\dfrac{12.8 \text{ mm}}{2}\right)^2 \pi} \times 100$$

$$= \frac{128.7 \text{ mm}^2 - 89.9 \text{ mm}^2}{128.7 \text{ mm}^2} \times 100 = 30\%$$

(b) True stress is defined by Equation 8.15, where, in this case, the area is taken as the fracture area A_f. However, the load at fracture must first be computed from the fracture strength as

$$F = \sigma_f A_0 = (460 \times 10^6 \text{ N/m}^2)(128.7 \text{ mm}^2)\left(\frac{1 \text{ m}^2}{10^6 \text{ mm}^2}\right) = 59,200 \text{ N}$$

Thus, the true stress is calculated as

$$\sigma_T = \frac{F}{A_f} = \frac{59,200 \text{ N}}{(89.9 \text{ mm}^2)\left(\dfrac{1 \text{ m}^2}{10^6 \text{ mm}^2}\right)}$$

$$= 6.6 \times 10^8 \text{ N/m}^2 = 660 \text{ MPa}$$

Figure 8.19 Relationships between hardness and tensile strength for steel, brass, and cast iron. [Adapted from *Metals Handbook: Properties and Selection: Irons and Steels,* Vol. 1, 9th edition, B. Bardes (Editor), 1978; and *Metals Handbook: Properties and Selection: Nonferrous Alloys and Pure Metals,* Vol. 2, 9th edition, H. Baker (Managing Editor), 1979. Reproduced by permission of ASM International, Materials Park, OH.]

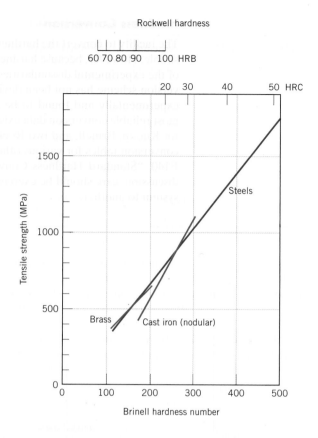

Correlation between Hardness and Tensile Strength

Both tensile strength and hardness are indicators of a metal's resistance to plastic deformation. Consequently, they are roughly proportional, as shown in Figure 8.19, for tensile strength as a function of the HB for cast iron, steel, and brass. The same proportionality relationship does not hold for all metals, as Figure 8.19 indicates. As a rule of thumb, for most steels, the HB and the tensile strength are related according to

For steel alloys, conversion of Brinell hardness to tensile strength

$$TS\,(\text{MPa}) = 3.45 \times \text{HB} \tag{8.20}$$

✓ *Concept Check 8.4*   Of those metals listed in Table 8.3, which is the hardest? Why?

[*The answer may be found at* www.wiley.com/college/callister (*Student Companion Site*).]

This concludes our discussion on the tensile properties of metals. By way of summary, Table 8.7 lists these properties, their symbols, and their characteristics (qualitatively).

Property	**Symbol**	**Measure of**
Modulus of elasticity	E	Stiffness—resistance to elastic deformation
Yield strength	σ_y	Resistance to plastic deformation
Tensile strength	TS	Maximum load-bearing capacity
Ductility	%EL, %RA	Degree of plastic deformation at fracture
Modulus of resilience	U_r	Energy absorption—elastic deformation
Toughness (static)	—	Energy absorption—plastic deformation
Hardness	e.g., HB, HRC	Resistance to localized surface deformation

Table 8.7
Summary of
Mechanical Properties
for Metals

Property Variability and Design/Safety Factors

8.6 VARIABILITY OF MATERIAL PROPERTIES

At this point, it is worthwhile to discuss an issue that sometimes proves troublesome to many engineering students—namely, that measured material properties are not exact quantities. That is, even if we have a most precise measuring apparatus and a highly controlled test procedure, there will always be some scatter or variability in the data that are collected from specimens of the same material. For example, consider a number of identical tensile samples that are prepared from a single bar of some metal alloy, and are subsequently stress–strain tested in the same apparatus. We would most likely observe that each resulting stress–strain plot is slightly different from the others. This would lead to a variety of modulus of elasticity, yield strength, and tensile strength values. A number of factors lead to uncertainties in measured data, including the test method, variations in specimen fabrication procedures, operator bias, and apparatus calibration. Furthermore, there might be inhomogeneities within the same lot of material and/or slight compositional and other differences from lot to lot. Of course, appropriate measures should be taken to minimize the possibility of measurement error and mitigate those factors that lead to data variability.

It should also be mentioned that scatter exists for other measured material properties, such as density, electrical conductivity, and coefficient of thermal expansion.

It is important for the design engineer to realize that scatter and variability of materials properties are inevitable and must be dealt with appropriately. On occasion, data must be subjected to statistical treatments and probabilities determined. For example, instead of asking, "What is the fracture strength of this alloy?" the engineer should become accustomed to asking, "What is the probability of failure of this alloy under these given circumstances?"

It is often desirable to specify a typical value and degree of dispersion (or scatter) for some measured property; this is commonly accomplished by taking the average and the standard deviation, respectively.

Computation of Average and Standard Deviation Values

An *average value* is obtained by dividing the sum of all measured values by the number of measurements taken. In mathematical terms, the average $\bar{x}$ of some parameter x is

Computation of
average value

$$\bar{x} = \frac{\sum_{i=1}^{n} x_i}{n}$$

(8.21)

where n is the number of observations or measurements and x_i is the value of a discrete measurement.

Furthermore, the standard deviation s is determined using the following expression:

Computation of standard deviation

$$s = \left[\frac{\sum\limits_{i=1}^{n}(x_i - \bar{x})^2}{n - 1} \right]^{1/2}$$

(8.22)

where x_i, $\bar{x}$, and n were defined earlier. A large value of the standard deviation corresponds to a high degree of scatter.

EXAMPLE PROBLEM 8.6

Average and Standard Deviation Computations

The following tensile strengths were measured for four specimens of the same steel alloy:

Sample Number	Tensile Strength (MPa)
1	520
2	512
3	515
4	522

(a) Compute the average tensile strength.
(b) Determine the standard deviation.

Solution

(a) The average tensile strength ($\overline{TS}$) is computed using Equation 8.21 with $n = 4$:

$$\overline{TS} = \frac{\sum\limits_{i=1}^{4}(TS)_i}{4}$$

$$= \frac{520 + 512 + 515 + 522}{4}$$

$$= 517 \text{ MPa}$$

(b) For the standard deviation, using Equation 8.22, we obtain

$$s = \left[\frac{\sum\limits_{i=1}^{4}\{(TS)_i - \overline{TS}\}^2}{4 - 1} \right]^{1/2}$$

$$= \left[\frac{(520 - 517)^2 + (512 - 517)^2 + (515 - 517)^2 + (522 - 517)^2}{4 - 1} \right]^{1/2}$$

$$= 4.6 \text{ MPa}$$

Figure 8.20 presents the tensile strength by specimen number for this example problem and also how the data may be represented in graphical form. The tensile strength data point (Figure 8.20b) corresponds to the average value $\overline{TS}$, and scatter is depicted by error bars (short horizontal lines) situated above and below the data point symbol and connected to this symbol by vertical lines. The upper error bar is positioned at a value of the average value plus the standard deviation ($\overline{TS} + s$), and the lower error bar corresponds to the average minus the standard deviation ($\overline{TS} - s$).

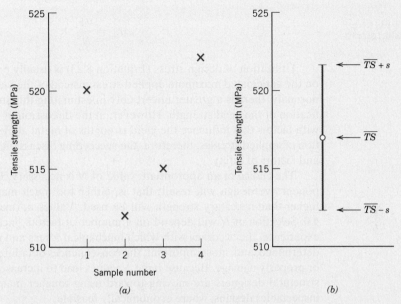

Figure 8.20 (*a*) Tensile strength data associated with Example Problem 8.6. (*b*) The manner in which these data could be plotted. The data point corresponds to the average value of the tensile strength ($\overline{TS}$); error bars that indicate the degree of scatter correspond to the average value plus and minus the standard deviation ($\overline{TS} \pm s$).

8.7 DESIGN/SAFETY FACTORS

There will always be uncertainties in characterizing the magnitude of applied loads and their associated stress levels for in-service applications; typically, load calculations are only approximate. Furthermore, as noted in Section 8.6, virtually all engineering materials exhibit a variability in their measured mechanical properties, have imperfections that were introduced during manufacture, and, in some instances, will have sustained damage during service. Consequently, design approaches must be employed to protect against unanticipated failure. During the 20th century, the protocol was to reduce the applied stress by a *design safety factor*. Although this is still an acceptable procedure for some structural applications, it does not provide adequate safety for critical applications such as those found in aircraft and bridge structural components. The current approach for these critical structural applications is to utilize materials that have adequate toughnesses and also offer redundancy in the structural design (i.e., excess or duplicate structures), provided there are regular inspections to detect the presence of flaws and, when necessary, safely remove or repair components. (These topics are discussed in Chapter 10, *Failure*—specifically Section 10.5.)

design stress   For less critical static situations and when tough materials are used, a **design stress**, σ_d, is taken as the calculated stress level σ_c (on the basis of the estimated maximum load) multiplied by a *design factor, N'*; that is,

$$\sigma_d = N'\sigma_c \tag{8.23}$$

where N' is greater than unity. Thus, the material to be used for the particular application is chosen so as to have a yield strength at least as high as this value of σ_d.

safe stress   Alternatively, a **safe stress** or *working stress, σ_w,* is used instead of design stress. This safe stress is based on the yield strength of the material and is defined as the yield strength divided by a *factor of safety, N*, or

Computation of safe
(or working) stress

$$\sigma_w = \frac{\sigma_y}{N} \qquad (8.24)$$

Utilization of design stress (Equation 8.23) is usually preferred because it is based on the anticipated maximum applied stress instead of the yield strength of the material; normally, there is a greater uncertainty in estimating this stress level than in the specification of the yield strength. However, in the discussion of this text, we are concerned with factors that influence the yield strengths of metal alloys and not in the determination of applied stresses; therefore, the succeeding discussion deals with working stresses and factors of safety.

The choice of an appropriate value of N is necessary. If N is too large, then component overdesign will result; that is, either too much material or an alloy having a higher-than-necessary strength will be used. Values normally range between 1.2 and 4.0. Selection of N will depend on a number of factors, including economics, previous experience, the accuracy with which mechanical forces and material properties may be determined, and, most important, the consequences of failure in terms of loss of life and/or property damage. Because large N values lead to increased material cost and weight, structural designers are moving toward using tougher materials with redundant (and inspectable) designs, where economically feasible.

DESIGN EXAMPLE 8.1

Specification of Support-Post Diameter

A tensile-testing apparatus is to be constructed that must withstand a maximum load of 220,000 N. The design calls for two cylindrical support posts, each of which is to support half of the maximum load. Furthermore, plain-carbon (1045) steel ground and polished shafting rounds are to be used; the minimum yield and tensile strengths of this alloy are 310 MPa and 565 MPa, respectively. Specify a suitable diameter for these support posts.

Solution

The first step in this design process is to decide on a factor of safety, N, which then allows determination of a working stress according to Equation 8.24. In addition, to ensure that the apparatus will be safe to operate, we also want to minimize any elastic deflection of the rods during testing; therefore, a relatively conservative factor of safety is to be used, say $N = 5$. Thus, the working stress σ_w is just

$$\sigma_w = \frac{\sigma_y}{N}$$

$$= \frac{310 \text{ MPa}}{5} = 62 \text{ MPa}$$

From the definition of stress, Equation 8.1,

$$A_0 = \left(\frac{d}{2}\right)^2 \pi = \frac{F}{\sigma_w}$$

where d is the rod diameter and F is the applied force; furthermore, each of the two rods must support half of the total force, or 110,000 N. Solving for d leads to

$$d = 2\sqrt{\frac{F}{\pi\sigma_w}}$$

$$= \sqrt{\frac{110{,}000 \text{ N}}{\pi(62 \times 10^6 \text{ N/m}^2)}}$$

$$= 4.75 \times 10^{-2} \text{ m} = 47.5 \text{ mm}$$

Therefore, the diameter of each of the two rods should be 47.5 mm.

DESIGN EXAMPLE 8.2

Materials Specification for a Pressurized Cylindrical Tube

(a) Consider a thin-walled cylindrical tube having a radius of 50 mm and wall thickness 2 mm that is to be used to transport pressurized gas. If inside and outside tube pressures are 2.027 and 0.057 MPa, respectively, which of the metals and alloys listed in Table 8.8 are suitable candidates? Assume a factor of safety of 4.0.

For a thin-walled cylinder, the circumferential (or "hoop") stress (σ) depends on pressure difference (Δp), cylinder radius (r_i), and tube wall thickness (t) as follows:

$$\sigma = \frac{r_i \Delta p}{t} \qquad (8.25)$$

These parameters are noted on the schematic sketch of a cylinder presented in Figure 8.21.

(b) Determine which of the alloys that satisfy the criterion of part (a) can be used to produce a tube with the lowest cost.

Solution

(a) In order for this tube to transport the gas in a satisfactory and safe manner, we want to minimize the likelihood of plastic deformation. To accomplish this, we replace the circumferential stress in Equation 8.25 with the yield strength of the tube material divided by the factor of safety, N—that is,

$$\frac{\sigma_y}{N} = \frac{r_i \Delta p}{t}$$

And solving this expression for σ_y leads to

$$\sigma_y = \frac{N r_i \Delta p}{t} \qquad (8.26)$$

Table 8.8 Yield Strengths, Densities, and Costs per Unit Mass for Metal Alloys That Are the Subjects of Design Example 8.2

Alloy	Yield Strength, σ_y (MPa)	Density, ρ (g/cm³)	Unit mass cost, $\bar{c}$ ($US/kg)
Steel	325	7.8	1.75
Aluminum	125	2.7	5.00
Copper	225	8.9	7.50
Brass	275	8.5	10.00
Magnesium	175	1.8	12.00
Titanium	700	4.5	85.00

8.9 Compute the elastic moduli for the following metal alloys, whose stress–strain behaviors may be observed in the "Tensile Tests" module of *Virtual Materials Science and Engineering (VMSE):* **(a)** titanium, **(b)** tempered steel, **(c)** aluminum, and **(d)** carbon steel. How do these values compare with those presented in Table 8.1 for the same metals?

8.10 Consider a cylindrical specimen of a steel alloy (Figure 8.22) 15.0 mm in diameter and 75 mm long that is pulled in tension. Determine its elongation when a load of 20,000 N is applied.

8.11 Figure 8.23 shows, for a gray cast iron, the tensile engineering stress–strain curve in the elastic region. Determine **(a)** the tangent modulus at 10.3 MPa and **(b)** the secant modulus taken to 6.9 MPa.

8.12 As noted in Section 3.10, for single crystals of some substances, the physical properties are anisotropic; that is, they are dependent on crystallographic direction. One such property is the modulus of elasticity. For cubic single crystals, the modulus of elasticity in a general [uvw] direction, E_{uvw}, is described by the relationship

$$\frac{1}{E_{uvw}} = \frac{1}{E_{(100)}} - 3\left(\frac{1}{E_{(100)}} - \frac{1}{E_{(111)}}\right)$$
$$(\alpha^2\beta^2 + \beta^2\gamma^2 + \gamma^2\alpha^2) \qquad (8.30)$$

where $E_{(100)}$ and $E_{(111)}$ are the moduli of elasticity in [100] and [111] directions, respectively; α, β, and γ are the cosines of the angles between [uvw] and the respective [100], [010], and [001]

directions. Verify that the $E_{(110)}$ values for aluminum, copper, and iron in Table 3.2 are correct.

8.13 In Section 2.6 it was noted that the net bonding energy E_N between two isolated positive and negative ions is a function of interionic distance r as follows:

$$E_N = -\frac{A}{r} + \frac{B}{r^n} \qquad (8.31)$$

where A, B, and n are constants for the particular ion pair. Equation 8.31 is also valid for the bonding energy between adjacent ions in solid materials. The modulus of elasticity E is proportional to the slope of the interionic force–separation curve at the equilibrium interionic separation; that is,

$$E \propto \left(\frac{dF}{dr}\right)_{r_0}$$

Derive an expression for the dependence of the modulus of elasticity on these A, B, and n parameters (for the two-ion system) using the following procedure:

1. Establish a relationship for the force F as a function of r, realizing that

$$F = \frac{dE_N}{dr}$$

2. Now take the derivative dF/dr.

3. Develop an expression for r_0, the equilibrium separation. Because r_0 corresponds to the value of r at the minimum of the E_N-versus-r curve (Figure 2.8b), take the derivative dE_N/dr, set it equal to zero, and solve for r, which corresponds to r_0.

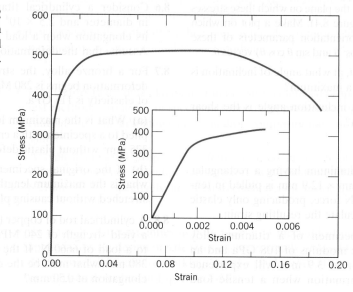

Figure 8.22 Tensile stress–strain behavior for a steel alloy.

Figure 8.23
Tensile stress–strain behavior for a gray cast iron.

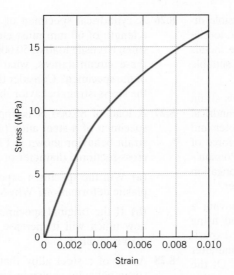

4. Finally, substitute this expression for r_0 into the relationship obtained by taking dF/dr.

8.14 Using the solution to Problem 8.13, rank the magnitudes of the moduli of elasticity for the following hypothetical X, Y, and Z materials from the greatest to the least. The appropriate A, B, and n parameters (Equation 8.31) for these three materials are shown in the following table; they yield E_N in units of electron volts and r in nanometers:

Material	A	B	n
X	2.5	2.0×10^{-5}	8
Y	2.3	8.0×10^{-6}	10.5
Z	3.0	1.5×10^{-5}	9

8.15 A cylindrical specimen of aluminum having a diameter of 20 mm and length of 210 mm is deformed elastically in tension with a force of 48,800 N. Using the data in Table 8.1, determine the following:

(a) The amount by which this specimen will elongate in the direction of the applied stress.

(b) The change in diameter of the specimen. Will the diameter increase or decrease?

8.16 A cylindrical bar of steel 15 mm in diameter is to be deformed elastically by application of a force along the bar axis. Using the data in Table 8.1, determine the force that will produce an elastic reduction of 4×10^{-3} mm in the diameter.

8.17 A cylindrical specimen of an alloy 8 mm in diameter is stressed elastically in tension. A force of 15,700 N produces a reduction in specimen diameter of 5×10^{-3} mm. Compute Poisson's ratio for this material if its modulus of elasticity is 140 GPa.

8.18 A cylindrical specimen of a hypothetical metal alloy is stressed in compression. If its original and final diameters are 20.000 and 20.025 mm, respectively, and its final length is 74.96 mm, compute its original length if the deformation is totally elastic. The elastic and shear moduli for this alloy are 105 GPa and 39.7 GPa, respectively.

8.19 Consider a cylindrical specimen of some hypothetical metal alloy that has a diameter of 8.0 mm. A tensile force of 1000 N produces an elastic reduction in diameter of 2.8×10^{-4} mm. Compute the modulus of elasticity for this alloy, given that Poisson's ratio is 0.30.

8.20 A brass alloy is known to have a yield strength of 275 MPa, a tensile strength of 380 MPa, and an elastic modulus of 103 GPa. A cylindrical specimen of this alloy 12.7 mm in diameter and 250 mm long is stressed in tension and found to elongate 7.6 mm. On the basis of the information given, is it possible to compute the magnitude of the load that is necessary to produce this change in length? If so, calculate the load. If not, explain why.

8.21 A cylindrical metal specimen 12.9 mm in diameter and 260 mm long is to be subjected to a tensile stress of 28 MPa; at this stress level the resulting deformation will be totally elastic.

(a) If the elongation must be less than 0.080 mm, which of the metals in Table 8.1 are suitable candidates? Why?

(b) If, in addition, the maximum permissible diameter decrease is 1.2×10^{-3} mm when the tensile stress of 28 MPa is applied, which of the metals that satisfy the criterion in part (a) are suitable candidates? Why?

8.22 Consider the brass alloy for which the stress–strain behavior is shown in Figure 8.12. A cylindrical specimen of this material 8 mm in diameter and 60 mm long is pulled in tension with a force of 5000 N. If it is known that this alloy has a Poisson's ratio of 0.30, compute **(a)** the specimen elongation and **(b)** the reduction in specimen diameter.

8.23 A cylindrical rod 100 mm long and having a diameter of 10.0 mm is to be deformed using a tensile load of 27,500 N. It must not experience either plastic deformation or a diameter reduction of more than 7.5×10^{-3} mm. Of the materials listed as follows, which are possible candidates? Justify your choice(s).

Material	Modulus of Elasticity (GPa)	Yield Strength (MPa)	Poisson's Ratio
Aluminum alloy	70	200	0.33
Brass alloy	101	300	0.34
Steel alloy	207	400	0.30
Titanium alloy	107	650	0.34

8.24 A cylindrical rod 380 mm long, having a diameter of 10.0 mm, is to be subjected to a tensile load. If the rod is to experience neither plastic deformation nor an elongation of more than 0.9 mm (0.035 in.) when the applied load is 24,500 N, which of the four metals or alloys listed in the following table are possible candidates? Justify your choice(s).

Material	Modulus of Elasticity (GPa)	Yield Strength (MPa)	Tensile Strength (MPa)
Aluminum alloy	70	255	420
Brass alloy	100	345	420
Copper	110	250	290
Steel alloy	207	450	550

Plastic Deformation

8.25 Figure 8.22 shows the tensile engineering stress–strain behavior for a steel alloy.

  (a) What is the modulus of elasticity?

  (b) What is the proportional limit?

  (c) What is the yield strength at a strain offset of 0.002?

  (d) What is the tensile strength?

8.26 A cylindrical specimen of a brass alloy having a length of 60 mm must elongate only 10.8 mm when a tensile load of 50,000 N is applied. Under these circumstances, what must be the radius of the specimen? Consider this brass alloy to have the stress–strain behavior shown in Figure 8.12.

8.27 A load of 85,000 N is applied to a cylindrical specimen of a steel alloy (displaying the stress–strain behavior shown in Figure 8.22) that has a cross-sectional diameter of 20 mm.

  (a) Will the specimen experience elastic and/or plastic deformation? Why?

  (b) If the original specimen length is 260 mm, how much will it increase in length when this load is applied?

8.28 A bar of a steel alloy that exhibits the stress–strain behavior shown in Figure 8.22 is subjected to a tensile load; the specimen is 310 mm long and has a square cross section 4.8 mm on a side.

  (a) Compute the magnitude of the load necessary to produce an elongation of 0.48 mm.

  (b) What will be the deformation after the load has been released?

8.29 A cylindrical specimen of aluminum having a diameter of 12.8 mm and a gauge length of 50.800 mm is pulled in tension. Use the load–elongation characteristics shown in the following table to complete parts (a) through (f).

Load	Length
N	mm
0	50.800
7,330	50.851
15,100	50.902
23,100	50.952
30,400	51.003
34,400	51.054
38,400	51.308
41,300	51.816
44,800	52.832
46,200	53.848
47,300	54.864
47,500	55.880
46,100	56.896
44,800	57.658
42,600	58.420
36,400	59.182
Fracture	

8.36 Calculate the moduli of resilience for the materials having the stress–strain behaviors shown in Figures 8.12 and 8.22.

8.37 Determine the modulus of resilience for each of the following alloys:

Material	Yield Strength MPa
Steel alloy	550
Brass alloy	350
Aluminum alloy	250
Titanium alloy	800

Use the modulus of elasticity values in Table 8.1.

8.38 A brass alloy to be used for a spring application must have a modulus of resilience of at least 0.80 MPa. What must be its minimum yield strength?

8.39 Show that Equations 8.18a and 8.18b are valid when there is no volume change during deformation.

8.40 Demonstrate that Equation 8.16, the expression defining true strain, may also be represented by

$$\epsilon_T = \ln\left(\frac{A_0}{A_i}\right)$$

when specimen volume remains constant during deformation. Which of these two expressions is more valid during necking? Why?

8.41 Using the data in Problem 8.29 and Equations 8.15, 8.16, and 8.18a, generate a true stress–true strain plot for aluminum. Equation 8.18a becomes invalid past the point at which necking begins; therefore, measured diameters are given in the following table for the last four data points, which should be used in true stress computations.

Load	Length	Diameter
N	mm	mm
46,100	56.896	11.71
42,400	57.658	10.95
42,600	58.420	10.62
36,400	59.182	9.40

8.42 A tensile test is performed on a metal specimen, and it is found that a true plastic strain of 0.20 is produced when a true stress of 575 MPa is applied; for the same metal, the value of K in

8.52 Estimate the Brinell and Rockwell hardnesses for the following:

(a) The naval brass for which the stress–strain behavior is shown in Figure 8.12.

(b) The steel alloy for which the stress–strain behavior is shown in Figure 8.22.

8.53 Using the data represented in Figure 8.19, specify equations relating tensile strength and Brinell hardness for brass and nodular cast iron, similar to Equation 8.20 for steels.

Variability of Material Properties

8.54 Cite five factors that lead to scatter in measured material properties.

8.55 The following table gives a number of Rockwell B hardness values that were measured on a single steel specimen. Compute average and standard deviation hardness values.

83.3	80.7	86.4
88.3	84.7	85.2
82.8	87.8	86.9
86.2	83.5	84.4
87.2	85.5	86.3

Design/Safety Factors

8.56 Upon what three criteria are factors of safety based?

8.57 Determine working stresses for the two alloys that have the stress–strain behaviors shown in Figures 7.12 and 7.22.

Spreadsheet Problem

8.1SS For a cylindrical metal specimen loaded in tension to fracture, given a set of load and corresponding length data, as well as the predeformation diameter and length, generate a spreadsheet that will allow the user to plot **(a)** engineering stress versus engineering strain, and **(b)** true stress versus true strain to the point of necking.

8.D2 (a) Gaseous hydrogen at a constant pressure of 1.015 MPa is to flow within the inside of a thin-walled cylindrical tube of nickel that has a radius of 0.2 m. The temperature of the tube is to be 300°C and the pressure of hydrogen outside of the tube will be maintained at 0.01013 MPa. Calculate the minimum wall thickness if the diffusion flux is to be no greater than 1×10^{-7} mol/m²·s. The concentration of hydrogen in the nickel, C_H (in moles hydrogen per m³ of Ni), is a function of hydrogen pressure, P_{H_2} (in MPa), and absolute temperature (T) according to

$$C_H = 30.8\sqrt{p_{H_2}} \, \exp\left(-\frac{12.3 \text{ kJ/mol}}{RT}\right) \quad (8.34)$$

Furthermore, the diffusion coefficient for the diffusion of H in Ni depends on temperature as

$$D_H(\text{m}^2/\text{s}) = 4.76 \times 10^{-7}\exp\left(-\frac{39.56 \text{ kJ/mol}}{RT}\right) \quad (8.35)$$

(b) For thin-walled cylindrical tubes that are pressurized, the circumferential stress is a function of the pressure difference across the wall (Δp), cylinder radius (r), and tube thickness (Δx) as

$$\sigma = \frac{r\Delta p}{4\Delta x} \quad (8.25a)$$

Compute the circumferential stress to which the walls of this pressurized cylinder are exposed.

(c) The room-temperature yield strength of Ni is 100 MPa and, furthermore, σ_y diminishes about 5 MPa for every 50°C rise in temperature. Would you expect the wall thickness computed in part (b) to be suitable for this Ni cylinder at 300°C? Why or why not?

(d) If this thickness is found to be suitable, compute the minimum thickness that could be used without any deformation of the tube walls. How much would the diffusion flux increase with this reduction in thickness? On the other hand, if the thickness determined in part (c) is found to be unsuitable, then specify a minimum thickness that you would use. In this case, how much of a diminishment in diffusion flux would result?

8.D3 Consider the steady-state diffusion of hydrogen through the walls of a cylindrical nickel tube as described in Problem 8.D2. One design calls for a diffusion flux of 5×10^{-8} mol/m²·s, a tube radius of 0.125 m, and inside and outside pressures of 2.026 MPa and 0.0203 MPa, respectively; the maximum allowable temperature is 450°C. Specify a suitable temperature and wall thickness to give this diffusion flux and yet ensure that the tube walls will not experience any permanent deformation.

Chapter 9

Dislocations and Strengthening Mechanisms

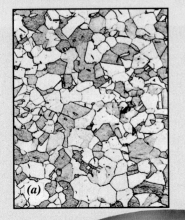

(a)

The photograph in Figure (b) is of a partially formed aluminum beverage can. The associated photomicrograph in Figure (a) represents the appearance of the aluminum's grain structure—that is, the grains are equiaxed (having approximately the same dimension in all directions).

Figure (c) shows a completely formed beverage can, fabrication of which is accomplished by a series of deep drawing operations during which the walls of the can are plastically deformed (i.e., are stretched). The grains of aluminum in these walls change shape—that is, they elongate in the direction of stretching. The resulting grain structure appears similar to that shown in the attendant photomicrograph, Figure (d). The magnification of Figures (a) and (d) is 150×.

(b)

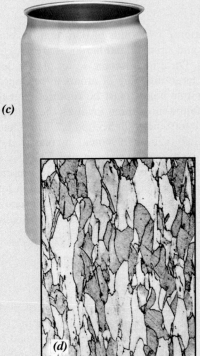

(c)

(d)

(The photomicrographs in figures (a) and (d) are taken from W. G. Moffatt, G. W. Pearsall, and J. Wulff, *The Structure and Properties of Materials*, Vol. I, *Structure*, p. 140. Copyright © 1964 by John Wiley & Sons, New York. Figures (b) and (c) © William D. Callister, Jr.)

With knowledge of the nature of dislocations and the role they play in the plastic deformation process, we are able to understand the underlying mechanisms of the techniques that are used to strengthen and harden metals and their alloys. Thus, it becomes possible to design and tailor the mechanical properties of materials—for example, the strength or toughness of a metal–matrix composite.

Learning Objectives

After studying this chapter, you should be able to do the following:

1. Describe edge and screw dislocation motion from an atomic perspective.
2. Describe how plastic deformation occurs by the motion of edge and screw dislocations in response to applied shear stresses.
3. Define slip system and cite one example.
4. Describe how the grain structure of a polycrystalline metal is altered when it is plastically deformed.
5. Explain how grain boundaries impede dislocation motion and why a metal having small grains is stronger than one having large grains.
6. Describe and explain solid-solution strengthening for substitutional impurity atoms in terms of lattice strain interactions with dislocations.
7. Describe and explain the phenomenon of strain hardening (or cold working) in terms of dislocations and strain field interactions.
8. Describe recrystallization in terms of both the alteration of microstructure and mechanical characteristics of the material.
9. Describe the phenomenon of grain growth from both macroscopic and atomic perspectives.

9.1 INTRODUCTION

Chapter 8 explained that materials may experience two kinds of deformation: elastic and plastic. Plastic deformation is permanent, and strength and hardness are measures of a material's resistance to this deformation. On a microscopic scale, plastic deformation corresponds to the net movement of large numbers of atoms in response to an applied stress. During this process, interatomic bonds must be ruptured and then re-formed. In crystalline solids, plastic deformation most often involves the motion of dislocations, linear crystalline defects that were introduced in Section 6.7. This chapter discusses the characteristics of dislocations and their involvement in plastic deformation. Twinning, another process by which some metals deform plastically, is also treated. In addition, and probably most important, several techniques are presented for strengthening single-phase metals, the mechanisms of which are described in terms of dislocations. Finally, the latter sections of this chapter are concerned with recovery and recrystallization—processes that occur in plastically deformed metals, normally at elevated temperatures—and, in addition, grain growth.

Dislocations and Plastic Deformation

Early materials studies led to the computation of the theoretical strengths of perfect crystals, which were many times greater than those actually measured. During the 1930s it was theorized that this discrepancy in mechanical strengths could be explained by a type of linear crystalline defect that has come to be known as a *dislocation*. Not until the 1950s, however, was the existence of such dislocation defects established by direct observation with the electron microscope. Since then, a theory of dislocations has evolved that explains many of the physical and mechanical phenomena in metals [as well as crystalline ceramics (Section 14.8)].

9.2 BASIC CONCEPTS

WileyPLUS: VMSE
Edge

Edge and screw are the two fundamental dislocation types. In an edge dislocation, localized lattice distortion exists along the end of an extra half-plane of atoms, which also defines the dislocation line (Figure 6.9). A screw dislocation may be thought of as resulting from shear distortion; its dislocation line passes through the center of a spiral, atomic plane ramp (Figure 6.10). Many dislocations in crystalline materials have both edge and screw components; these are mixed dislocations (Figure 6.11).

Plastic deformation corresponds to the motion of large numbers of dislocations. An edge dislocation moves in response to a shear stress applied in a direction perpendicular to its line; the mechanics of dislocation motion are represented in Figure 9.1. Let the initial extra half-plane of atoms be plane *A*. When the shear stress is applied as indicated (Figure 9.1*a*), plane *A* is forced to the right; this in turn pushes the top halves of planes *B*, *C*, *D*, and so on, in the same direction. If the applied shear stress is of sufficient magnitude, the interatomic bonds of plane *B* are severed along the shear plane, and the upper half of plane *B* becomes the extra half-plane as plane *A* links up with the bottom half of plane *B* (Figure 9.1*b*). This process is subsequently repeated for the other planes, such that the extra half-plane, by discrete steps, moves from left to right by successive and repeated breaking of bonds and shifting by interatomic distances of upper half-planes. Before and after the movement of a dislocation through some particular region of the crystal, the atomic arrangement is ordered and perfect; it is only during the passage of the extra half-plane that the lattice structure is disrupted. Ultimately, this extra half-plane may emerge from the right surface of the crystal, forming an edge that is one atomic distance wide; this is shown in Figure 9.1*c*.

slip

The process by which plastic deformation is produced by dislocation motion is termed **slip**; the crystallographic plane along which the dislocation line traverses is the *slip plane,* as indicated in Figure 9.1. Macroscopic plastic deformation simply corresponds to permanent deformation that results from the movement of dislocations, or slip, in response to an applied shear stress, as represented in Figure 9.2*a*.

Dislocation motion is analogous to the mode of locomotion employed by a caterpillar (Figure 9.3). The caterpillar forms a hump near its posterior end by pulling in its last pair of legs a unit leg distance. The hump is propelled forward by repeated lifting and shifting of leg pairs. When the hump reaches the anterior end, the entire caterpillar has

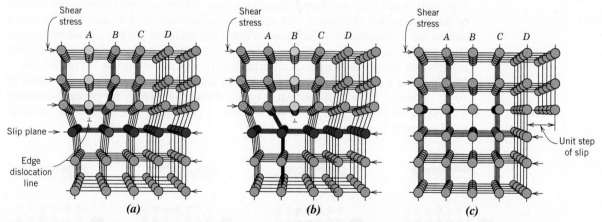

Figure 9.1 Atomic rearrangements that accompany the motion of an edge dislocation as it moves in response to an applied shear stress. (*a*) The extra half-plane of atoms is labeled *A*. (*b*) The dislocation moves one atomic distance to the right as *A* links up to the lower portion of plane *B*; in the process, the upper portion of *B* becomes the extra half-plane. (*c*) A step forms on the surface of the crystal as the extra half-plane exits.
(Adapted from A. G. Guy, *Essentials of Materials Science,* McGraw-Hill Book Company, New York, 1976, p. 153.)

Figure 9.2 The formation of a step on the surface of a crystal by the motion of (*a*) an edge dislocation and (*b*) a screw dislocation. Note that for an edge, the dislocation line moves in the direction of the applied shear stress τ; for a screw, the dislocation line motion is perpendicular to the stress direction. (Adapted from H. W. Hayden, W. G. Moffatt, and J. Wulff, *The Structure and Properties of Materials,* Vol. III, *Mechanical Behavior,* p. 70. Copyright © 1965 by John Wiley & Sons, New York.)

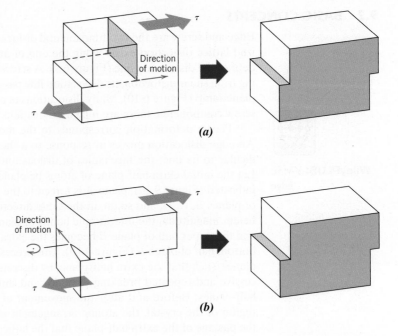

(*a*)

(*b*)

WileyPLUS: VMSE
Screw, Mixed

dislocation density

moved forward by the leg separation distance. The caterpillar hump and its motion correspond to the extra half-plane of atoms in the dislocation model of plastic deformation.

The motion of a screw dislocation in response to the applied shear stress is shown in Figure 9.2*b*; the direction of movement is perpendicular to the stress direction. For an edge, motion is parallel to the shear stress. However, the net plastic deformation for the motion of both dislocation types is the same (see Figure 9.2). The direction of motion of the mixed dislocation line is neither perpendicular nor parallel to the applied stress, but lies somewhere in between.

All metals and alloys contain some dislocations that were introduced during solidification, during plastic deformation, and as a consequence of thermal stresses that result from rapid cooling. The number of dislocations, or **dislocation density** in a material, is expressed as the total dislocation length per unit volume or, equivalently, the number of dislocations that intersect a unit area of a random section. The units of dislocation density are millimeters of dislocation per cubic millimeter or just per square millimeter. Dislocation densities as low as 10^3 mm^{-2} are typically found in carefully solidified metal crystals. For heavily deformed metals, the density may run as high as 10^9 to 10^{10} mm^{-2}. Heat-treating a deformed metal specimen can diminish the density to on the order of 10^5 to 10^6 mm^{-2}. By way of contrast, a typical dislocation density for ceramic materials is between 10^2 and 10^4 mm^{-2}; for silicon single crystals used in integrated circuits, the value normally lies between 0.1 and 1 mm^{-2}.

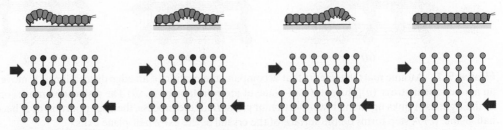

Figure 9.3 The analogy between caterpillar and dislocation motion.

9.3 CHARACTERISTICS OF DISLOCATIONS

lattice strain

Several characteristics of dislocations are important with regard to the mechanical properties of metals. These include strain fields that exist around dislocations, which are influential in determining the mobility of the dislocations, as well as their ability to multiply.

When metals are plastically deformed, some fraction of the deformation energy (approximately 5%) is retained internally; the remainder is dissipated as heat. The major portion of this stored energy is as strain energy associated with dislocations. Consider the edge dislocation represented in Figure 9.4. As already mentioned, some atomic lattice distortion exists around the dislocation line because of the presence of the extra half-plane of atoms. As a consequence, there are regions in which compressive, tensile, and shear **lattice strains** are imposed on the neighboring atoms. For example, atoms immediately above and adjacent to the dislocation line are squeezed together. As a result, these atoms may be thought of as experiencing a compressive strain relative to atoms positioned in the perfect crystal and far removed from the dislocation; this is illustrated in Figure 9.4. Directly below the half-plane, the effect is just the opposite; lattice atoms sustain an imposed tensile strain, which is as shown. Shear strains also exist in the vicinity of the edge dislocation. For a screw dislocation, lattice strains are pure shear only. These lattice distortions may be considered to be strain fields that radiate from the dislocation line. The strains extend into the surrounding atoms, and their magnitude decreases with radial distance from the dislocation.

The strain fields surrounding dislocations in close proximity to one another may interact such that forces are imposed on each dislocation by the combined interactions of all its neighboring dislocations. For example, consider two edge dislocations that have the same sign and the identical slip plane, as represented in Figure 9.5a. The compressive and tensile strain fields for both lie on the same side of the slip plane; the strain field interaction is such that there exists between these two isolated dislocations a mutual repulsive force that tends to move them apart. However, two dislocations of opposite sign and having the same slip plane are attracted to one another, as indicated in Figure 9.5b, and dislocation annihilation occurs when they meet. That is, the two extra half-planes of atoms align and become a complete plane. Dislocation interactions are possible among edge, screw, and/or mixed dislocations, and for a variety of orientations. These strain fields and associated forces are important in the strengthening mechanisms for metals.

During plastic deformation, the number of dislocations increases dramatically. The dislocation density in a metal that has been highly deformed may be as high as 10^{10} mm^{-2}. One important source of these new dislocations is existing dislocations, which multiply; furthermore, grain boundaries, as well as internal defects and surface irregularities such as scratches and nicks, which act as stress concentrations, may serve as dislocation formation sites during deformation.

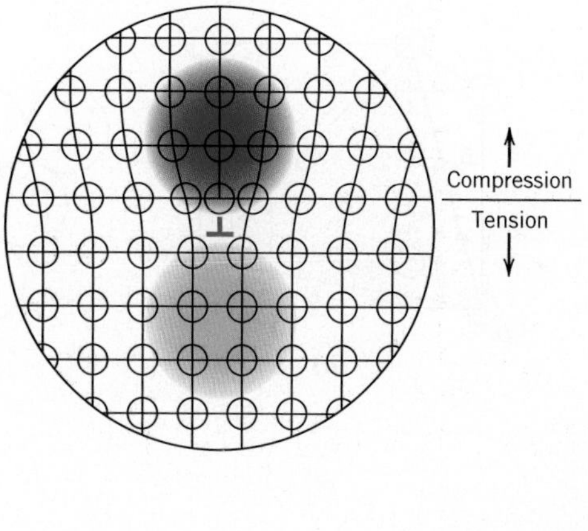

WileyPLUS

Tutorial Video:
Defects in Metals
Why Do Defects
Strengthen Metals?

Compression

Tension

Figure 9.4 Regions of compression (green) and tension (yellow) located around an edge dislocation.
(Adapted from W. G. Moffatt, G. W. Pearsall, and J. Wulff, *The Structure and Properties of Materials,* Vol. I, *Structure,* p. 85. Copyright © 1964 by John Wiley & Sons, New York.)

Figure 9.5 (*a*) Two edge dislocations of the same sign and lying on the same slip plane exert a repulsive force on each other; *C* and *T* denote compression and tensile regions, respectively. (*b*) Edge dislocations of opposite sign and lying on the same slip plane exert an attractive force on each other. Upon meeting, they annihilate each other and leave a region of perfect crystal.
(Adapted from H. W. Hayden, W. G. Moffatt, and J. Wulff, *The Structure and Properties of Materials,* Vol. III, *Mechanical Behavior,* p. 75. Copyright © 1965 by John Wiley & Sons, New York.)

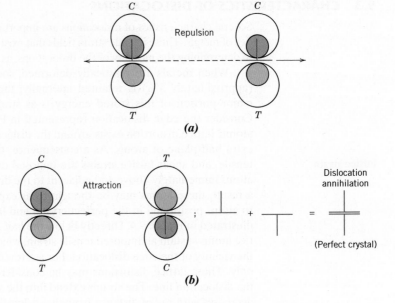

9.4 SLIP SYSTEMS

Dislocations do not move with the same degree of ease on all crystallographic planes of atoms and in all crystallographic directions. Typically, there is a preferred plane, and in that plane there are specific directions along which dislocation motion occurs. This plane is called the *slip plane;* it follows that the direction of movement is called the *slip direction.* This combination of the slip plane and the slip direction is termed the **slip system**. The slip system depends on the crystal structure of the metal and is such that the atomic distortion that accompanies the motion of a dislocation is a minimum. For a particular crystal structure, the slip plane is the plane that has the densest atomic packing—that is, has the greatest planar density. The slip direction corresponds to the direction in this plane that is most closely packed with atoms—that is, has the highest linear density. Planar and linear atomic densities were discussed in Section 4.16.

slip system

Consider, for example, the FCC crystal structure, a unit cell of which is shown in Figure 9.6*a*. There is a set of planes, the {111} family, all of which are closely packed. A (111)-type plane is indicated in the unit cell; in Figure 9.6*b*, this plane is positioned

Figure 9.6 (*a*) A {111}⟨110⟩ slip system shown within an FCC unit cell. (*b*) The (111) plane from (*a*) and three ⟨110⟩ slip directions (as indicated by arrows) within that plane constitute possible slip systems.

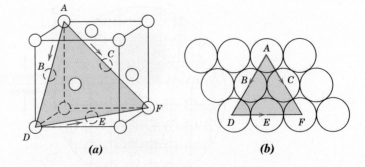

Table 9.1

Slip Systems for
Face-Centered Cubic,
Body-Centered
Cubic, and Hexagonal
Close-Packed Metals

Metals	*Slip Plane*	*Slip Direction*	*Number of Slip Systems*
Face-Centered Cubic			
Cu, Al, Ni, Ag, Au	$\{111\}$	$\langle 110 \rangle$	12
Body-Centered Cubic			
α-Fe, W, Mo	$\{110\}$	$\langle 111 \rangle$	12
α-Fe, W	$\{211\}$	$\langle 111 \rangle$	12
α-Fe, K	$\{321\}$	$\langle 111 \rangle$	24
Hexagonal Close-Packed			
Cd, Zn, Mg, Ti, Be	$\{0001\}$	$\langle 11\bar{2}0 \rangle$	3
Ti, Mg, Zr	$\{10\bar{1}0\}$	$\langle 11\bar{2}0 \rangle$	3
Ti, Mg	$\{10\bar{1}1\}$	$\langle 11\bar{2}0 \rangle$	6

within the plane of the page, in which atoms are now represented as touching nearest neighbors.

Slip occurs along $\langle 110 \rangle$-type directions within the $\{111\}$ planes, as indicated by arrows in Figure 9.6. Hence, $\{111\}\langle 110 \rangle$ represents the slip plane and direction combination, or the slip system for FCC. Figure 9.6b demonstrates that a given slip plane may contain more than a single slip direction. Thus, several slip systems may exist for a particular crystal structure; the number of independent slip systems represents the different possible combinations of slip planes and directions. For example, for face-centered cubic, there are 12 slip systems: four unique $\{111\}$ planes and, within each plane, three independent $\langle 110 \rangle$ directions.

The possible slip systems for BCC and HCP crystal structures are listed in Table 9.1. For each of these structures, slip is possible on more than one family of planes (e.g., $\{110\}$, $\{211\}$, and $\{321\}$ for BCC). For metals having these two crystal structures, some slip systems are often operable only at elevated temperatures.

Metals with FCC or BCC crystal structures have a relatively large number of slip systems (at least 12). These metals are quite ductile because extensive plastic deformation is normally possible along the various systems. Conversely, HCP metals, having few active slip systems, are normally quite brittle.

The Burgers vector, **b**, was introduced in Section 6.7, and shown for edge, screw, and mixed dislocations in Figures 6.9, 6.10, and 6.11, respectively. With regard to the process of slip, a Burgers vector's direction corresponds to a dislocation's slip direction, whereas its magnitude is equal to the unit slip distance (or interatomic separation in this direction). Of course, both the direction and the magnitude of **b** depends on crystal structure, and it is convenient to specify a Burgers vector in terms of unit cell edge length (a) and crystallographic direction indices. Burgers vectors for face-centered cubic, body-centered cubic, and hexagonal close-packed crystal structures are as follows:

$$\mathbf{b}(\text{FCC}) = \frac{a}{2}\langle 110 \rangle \tag{9.1a}$$

$$\mathbf{b}(\text{BCC}) = \frac{a}{2}\langle 111 \rangle \tag{9.1b}$$

$$\mathbf{b}(\text{HCP}) = \frac{a}{3}\langle 11\bar{2}0 \rangle \tag{9.1c}$$

✓

Concept Check 9.1   Which of the following is the slip system for the simple cubic crystal structure? Why?

$$\{100\}\langle 110\rangle$$
$$\{110\}\langle 110\rangle$$
$$\{100\}\langle 010\rangle$$
$$\{110\}\langle 111\rangle$$

(*Note:* A unit cell for the simple cubic crystal structure is shown in Figure 4.2.)

[*The answer may be found at* www.wiley.com/college/callister *(Student Companion Site).*]

9.5 SLIP IN SINGLE CRYSTALS

A further explanation of slip is simplified by treating the process in single crystals, then making the appropriate extension to polycrystalline materials. As mentioned previously, edge, screw, and mixed dislocations move in response to shear stresses applied along a slip plane and in a slip direction. As noted in Section 8.2, even though an applied stress may be pure tensile (or compressive), shear components exist at all but parallel or perpendicular alignments to the stress direction (Equation 8.4b). These are termed **resolved shear stresses**, and their magnitudes depend not only on the applied stress, but also on the orientation of both the slip plane and direction within that plane. Let ϕ represent the angle between the normal to the slip plane and the applied stress direction, and let λ be the angle between the slip and stress directions, as indicated in Figure 9.7; it can then be shown that for the resolved shear stress τ_R

resolved shear stress

Resolved shear stress—dependence on applied stress and orientation of stress direction relative to slip plane normal and slip direction

$$\tau_R = \sigma\cos\phi\,\cos\lambda \tag{9.2}$$

where σ is the applied stress. In general, $\phi + \lambda \neq 90°$ because it need not be the case that the tensile axis, the slip plane normal, and the slip direction all lie in the same plane.

A metal single crystal has a number of different slip systems that are capable of operating. The resolved shear stress normally differs for each one because the orientation of

Figure 9.7 Geometric relationships between the tensile axis, slip plane, and slip direction used in calculating the resolved shear stress for a single crystal.

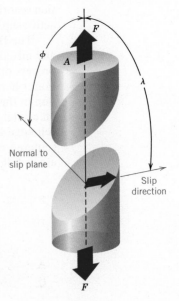

each relative to the stress axis (ϕ and λ angles) also differs. However, one slip system is generally oriented most favorably—that is, has the largest resolved shear stress, $\tau_R(\text{max})$:

$$\tau_R(\text{max}) = \sigma(\cos\phi \cos\lambda)_{\text{max}} \tag{9.3}$$

In response to an applied tensile or compressive stress, slip in a single crystal commences on the most favorably oriented slip system when the resolved shear stress reaches some critical value, termed the **critical resolved shear stress** τ_{crss}; it represents the minimum shear stress required to initiate slip and is a property of the material that determines when yielding occurs. The single crystal plastically deforms or yields when $\tau_R(\text{max}) = \tau_{\text{crss}}$, and the magnitude of the applied stress required to initiate yielding (i.e., the yield strength σ_y) is

critical resolved shear stress

Yield strength of a single crystal—dependence on the critical resolved shear stress and the orientation of the most favorably oriented slip system

$$\sigma_y = \frac{\tau_{\text{crss}}}{(\cos\phi \cos\lambda)_{\text{max}}} \tag{9.4}$$

The minimum stress necessary to introduce yielding occurs when a single crystal is oriented such that $\phi = \lambda = 45°$; under these conditions,

$$\sigma_y = 2\tau_{\text{crss}} \tag{9.5}$$

For a single-crystal specimen that is stressed in tension, deformation is as in Figure 9.8, where slip occurs along a number of equivalent and most favorably oriented planes and directions at various positions along the specimen length. This slip deformation forms as small steps on the surface of the single crystal that are parallel to one another and loop around the circumference of the specimen as indicated in Figure 9.8. Each step results from the movement of a large number of dislocations along the same slip plane. On the surface of a polished single-crystal specimen, these steps appear as lines, which are called *slip lines*. A zinc single crystal that has been plastically deformed to the degree that these slip markings are discernible is shown in Figure 9.9.

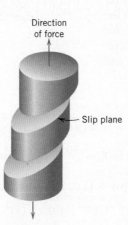

Direction
of force

Slip plane

Figure 9.8
Macroscopic slip in a single crystal.

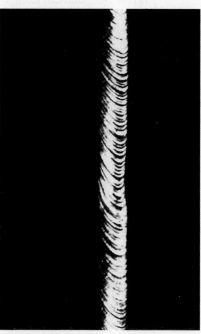

Figure 9.9 Slip in a zinc single crystal. (From C. F. Elam, *The Distortion of Metal Crystals,* Oxford University Press, London, 1935.)

With continued extension of a single crystal, both the number of slip lines and the slip step width increase. For FCC and BCC metals, slip may eventually begin along a second slip system, the system that is next most favorably oriented with the tensile axis. Furthermore, for HCP crystals having few slip systems, if the stress axis for the most favorable slip system is either perpendicular to the slip direction ($\lambda = 90°$) or parallel to the slip plane ($\phi = 90°$), the critical resolved shear stress is zero. For these extreme orientations, the crystal typically fractures rather than deforms plastically.

Concept Check 9.2 Explain the difference between resolved shear stress and critical resolved shear stress.

[*The answer may be found at* www.wiley.com/college/callister *(Student Companion Site).*]

EXAMPLE PROBLEM 9.1

Resolved Shear Stress and Stress-to-Initiate-Yielding Computations

Consider a single crystal of BCC iron oriented such that a tensile stress is applied along a [010] direction.

(a) Compute the resolved shear stress along a (110) plane and in a [$\bar{1}$11] direction when a tensile stress of 52 MPa is applied.

(b) If slip occurs on a (110) plane and in a [$\bar{1}$11] direction, and the critical resolved shear stress is 30 MPa, calculate the magnitude of the applied tensile stress necessary to initiate yielding.

Solution

(a) A BCC unit cell along with the slip direction and plane as well as the direction of the applied stress are shown in the accompanying diagram. In order to solve this problem, we must use Equation 9.2. However, it is first necessary to determine values for ϕ and λ, where, from this diagram, ϕ is the angle between the normal to the (110) slip plane (i.e., the [110] direction) and the [010] direction, and λ represents the angle between the [$\bar{1}$11] and [010] directions. In general, for cubic unit cells, the angle θ between directions 1 and 2, represented by [$u_1v_1w_1$] and [$u_2v_2w_2$], respectively, is given by

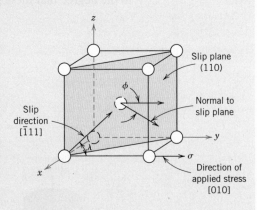

$$\theta = \cos^{-1}\left[\frac{u_1u_2 + v_1v_2 + w_1w_2}{\sqrt{(u_1^2 + v_1^2 + w_1^2)(u_2^2 + v_2^2 + w_2^2)}}\right] \quad (9.6)$$

For the determination of the value of ϕ, let [$u_1v_1w_1$] = [110] and [$u_2v_2w_2$] = [010], such that

$$\phi = \cos^{-1}\left\{\frac{(1)(0) + (1)(1) + (0)(0)}{\sqrt{[(1)^2 + (1)^2 + (0)^2][(0)^2 + (1)^2 + (0)^2]}}\right\}$$

$$= \cos^{-1}\left(\frac{1}{\sqrt{2}}\right) = 45°$$

However, for λ, we take $[u_1v_1w_1] = [\bar{1}11]$ and $[u_2v_2w_2] = [010]$, and

$$\lambda = \cos^{-1}\left[\frac{(-1)(0) + (1)(1) + (1)(0)}{\sqrt{[(-1)^2 + (1)^2 + (1)^2][(0)^2 + (1)^2 + (0)^2]}}\right]$$

$$= \cos^{-1}\left(\frac{1}{\sqrt{3}}\right) = 54.7°$$

Thus, according to Equation 9.2,

$$\tau_R = \sigma \cos\phi \cos\lambda = (52\ \text{MPa})(\cos 45°)(\cos 54.7°)$$

$$= (52\ \text{MPa})\left(\frac{1}{\sqrt{2}}\right)\left(\frac{1}{\sqrt{3}}\right)$$

$$= 21.3\ \text{MPa}$$

(b) The yield strength σ_y may be computed from Equation 9.4; ϕ and λ are the same as for part (a), and

$$\sigma_y = \frac{30\ \text{MPa}}{(\cos 45°)(\cos 54.7°)} = 73.4\ \text{MPa}$$

9.6 PLASTIC DEFORMATION OF POLYCRYSTALLINE MATERIALS

Deformation and slip in polycrystalline materials is somewhat more complex. Because of the random crystallographic orientations of the numerous grains, the direction of slip varies from one grain to another. For each, dislocation motion occurs along the slip system that has the most favorable orientation, as defined earlier. This is exemplified by a photomicrograph of a polycrystalline copper specimen that has been plastically deformed (Figure 9.10); before deformation, the surface was polished. Slip lines[1] are visible, and it appears that two slip systems operated for most of the grains, as evidenced by two sets of parallel yet intersecting sets of lines. Furthermore, variation in grain orientation is indicated by the difference in alignment of the slip lines for the several grains.

Gross plastic deformation of a polycrystalline specimen corresponds to the comparable distortion of the individual grains by means of slip. During deformation, mechanical integrity and coherency are maintained along the grain boundaries—that is, the grain boundaries usually do not come apart or open up. As a consequence, each individual grain is constrained, to some degree, in the shape it may assume by its neighboring grains. The manner in which grains distort as a result of gross plastic deformation is indicated in Figure 9.11. Before deformation the grains are *equiaxed,* or have approximately the same dimension in all directions. For this particular deformation, the grains become elongated along the direction in which the specimen was extended.

---

[1]These slip lines are microscopic ledges produced by dislocations (Figure 9.1c) that have exited from a grain and appear as lines when viewed with a microscope. They are analogous to the macroscopic steps found on the surfaces of deformed single crystals (Figures 9.8 and 9.9).

Figure 9.10 Slip lines on the surface of a polycrystalline specimen of copper that was polished and subsequently deformed. 173×. [Photomicrograph courtesy of C. Brady, National Bureau of Standards (now the National Institute of Standards and Technology, Gaithersburg, MD).]

100 µm

Polycrystalline metals are stronger than their single-crystal equivalents, which means that greater stresses are required to initiate slip and the attendant yielding. This is, to a large degree, also a result of geometric constraints that are imposed on the grains during deformation. Even though a single grain may be favorably oriented with the applied stress for slip, it cannot deform until the adjacent and less favorably oriented grains are capable of slip also; this requires a higher applied stress level.

Figure 9.11 Alteration of the grain structure of a polycrystalline metal as a result of plastic deformation. (*a*) Before deformation the grains are equiaxed. (*b*) The deformation has produced elongated grains. 170×. (From W. G. Moffatt, G. W. Pearsall, and J. Wulff, *The Structure and Properties of Materials,* Vol. I, *Structure,* p. 140. Copyright © 1964 by John Wiley & Sons, New York.)

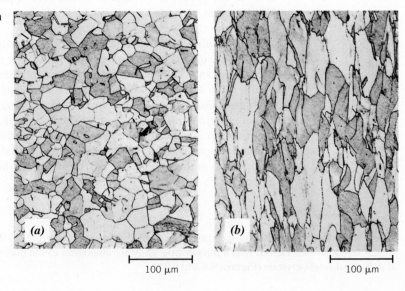

(a)

(b)

100 µm   100 µm

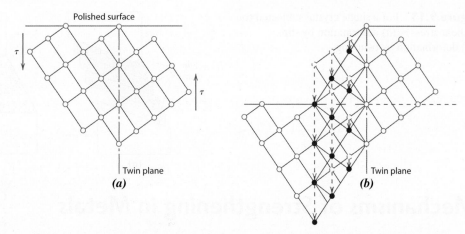

Figure 9.12 Schematic diagram showing how twinning results from an applied shear stress τ. In (b), open circles represent atoms that did not change position; dashed and solid circles represent original and final atom positions, respectively.

(From G. E. Dieter, *Mechanical Metallurgy,* 3rd edition. Copyright © 1986 by McGraw-Hill Book Company, New York. Reproduced with permission of McGraw-Hill Book Company.)

9.7 DEFORMATION BY TWINNING

In addition to slip, plastic deformation in some metallic materials can occur by the formation of mechanical twins, or *twinning*. The concept of a twin was introduced in Section 6.8—that is, a shear force can produce atomic displacements such that on one side of a plane (the twin boundary), atoms are located in mirror-image positions of atoms on the other side. The manner in which this is accomplished is demonstrated in Figure 9.12. Here, open circles represent atoms that did not move, and dashed and solid circles represent original and final positions, respectively, of atoms within the twinned region. As may be noted in this figure, the displacement magnitude within the twin region (indicated by arrows) is proportional to the distance from the twin plane. Furthermore, twinning occurs on a definite crystallographic plane and in a specific direction that depend on crystal structure. For example, for BCC metals, the twin plane and direction are (112) and [111], respectively.

Slip and twinning deformations are compared in Figure 9.13 for a single crystal that is subjected to a shear stress τ. Slip ledges are shown in Figure 9.13a; their formation was described in Section 9.5. For twinning, the shear deformation is homogeneous (Figure 9.13b). These two processes differ from each other in several respects. First, for slip, the crystallographic orientation above and below the slip plane is the same both before and after the deformation; for twinning, there is a reorientation across the twin plane. In addition, slip occurs in distinct atomic spacing multiples, whereas the atomic displacement for twinning is less than the interatomic separation.

Mechanical twinning occurs in metals that have BCC and HCP crystal structures, at low temperatures, and at high rates of loading (shock loading), conditions under which the slip process is restricted—that is, there are few operable slip systems. The amount of bulk plastic deformation from twinning is normally small relative to that resulting from slip. However, the real importance of twinning lies with the accompanying crystallographic reorientations; twinning may place new slip systems in orientations that are favorable relative to the stress axis such that the slip process can now take place.

Figure 9.13 For a single crystal subjected to a shear stress τ, (a) deformation by slip; (b) deformation by twinning.

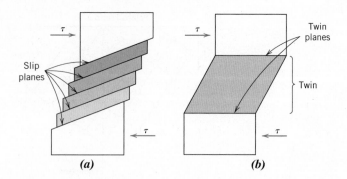

(a)   *(b)*

Mechanisms of Strengthening in Metals

Metallurgical and materials engineers are often called on to design alloys having high strengths yet some ductility and toughness; typically, ductility is sacrificed when an alloy is strengthened. Several hardening techniques are at the disposal of an engineer, and frequently alloy selection depends on the capacity of a material to be tailored with the mechanical characteristics required for a particular application.

Important to the understanding of strengthening mechanisms is the relation between dislocation motion and mechanical behavior of metals. Because macroscopic plastic deformation corresponds to the motion of large numbers of dislocations, *the ability of a metal to deform plastically depends on the ability of dislocations to move.* Because hardness and strength (both yield and tensile) are related to the ease with which plastic deformation can be made to occur, by reducing the mobility of dislocations, the mechanical strength may be enhanced—that is, greater mechanical forces are required to initiate plastic deformation. In contrast, the more unconstrained the dislocation motion, the greater is the facility with which a metal may deform, and the softer and weaker it becomes. Virtually all strengthening techniques rely on this simple principle: *Restricting or hindering dislocation motion renders a material harder and stronger.*

The present discussion is confined to strengthening mechanisms for single-phase metals by grain size reduction, solid-solution alloying, and strain hardening. Deformation and strengthening of multiphase alloys are more complicated, involving concepts beyond the scope of the present discussion; Chapter 12 and Section 17.7 treat techniques that are used to strengthen multiphase alloys.

WileyPLUS

Tutorial Video:
Defects in Metals
How Do Defects
Affect Metals?

9.8 STRENGTHENING BY GRAIN SIZE REDUCTION

The size of the grains, or average grain diameter, in a polycrystalline metal influences the mechanical properties. Adjacent grains normally have different crystallographic orientations and, of course, a common grain boundary, as indicated in Figure 9.14. During plastic deformation, slip or dislocation motion must take place across this common boundary—say, from grain A to grain B in Figure 9.14. The grain boundary acts as a barrier to dislocation motion for two reasons:

1. Because the two grains are of different orientations, a dislocation passing into grain B must change its direction of motion; this becomes more difficult as the crystallographic misorientation increases.

2. The atomic disorder within a grain boundary region results in a discontinuity of slip planes from one grain into the other.

It should be mentioned that, for high-angle grain boundaries, it may not be the case that dislocations traverse grain boundaries during deformation; rather, dislocations tend to

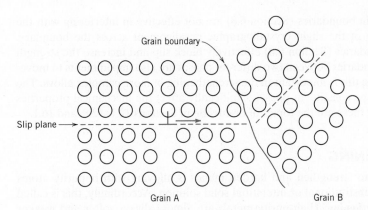

Grain boundary

Slip plane →

Grain A

Grain B

Figure 9.14 The motion of a dislocation as it encounters a grain boundary, illustrating how the boundary acts as a barrier to continued slip. Slip planes are discontinuous and change directions across the boundary. (From L. H. Van Vlack, *A Textbook of Materials Technology*, Addison-Wesley Publishing Co., 1973. Reproduced with the permission of the Estate of Lawrence H. Van Vlack.)

"pile up" (or back up) at grain boundaries. These pile-ups introduce stress concentrations ahead of their slip planes, which generate new dislocations in adjacent grains.

A fine-grained material (one that has small grains) is harder and stronger than one that is coarse grained because the former has a greater total grain boundary area to impede dislocation motion. For many materials, the yield strength σ_y varies with grain size according to

Hall–Petch equation—dependence of yield strength on grain size

$$\sigma_y = \sigma_0 + k_y d^{-1/2} \tag{9.7}$$

In this expression, termed the *Hall–Petch equation*, d is the average grain diameter, and σ_0 and k_y are constants for a particular material. Note that Equation 9.7 is not valid for both very large (i.e., coarse) grain and extremely fine grain polycrystalline materials. Figure 9.15 demonstrates the yield strength dependence on grain size for a brass alloy. Grain size may be regulated by the rate of solidification from the liquid phase, and also by plastic deformation followed by an appropriate heat treatment, as discussed in Section 9.13.

It should also be mentioned that grain size reduction improves not only the strength, but also the toughness of many alloys.

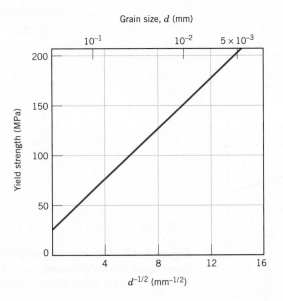

Grain size, d (mm)

Figure 9.15 The influence of grain size on the yield strength of a 70 Cu–30 Zn brass alloy. Note that the grain diameter increases from right to left and is not linear. (Adapted from H. Suzuki, "The Relation between the Structure and Mechanical Properties of Metals," Vol. II, *National Physical Laboratory, Symposium No. 15*, 1963, p. 524.)

Small-angle grain boundaries (Section 6.8) are not effective in interfering with the slip process because of the slight crystallographic misalignment across the boundary. However, twin boundaries (Section 6.8) effectively block slip and increase the strength of the material. Boundaries between two different phases are also impediments to movements of dislocations; this is important in the strengthening of more complex alloys. The sizes and shapes of the constituent phases significantly affect the mechanical properties of multiphase alloys; these are the topics of discussion in Sections 12.7, 12.8, and 16.1.

9.9 SOLID-SOLUTION STRENGTHENING

solid-solution strengthening

WileyPLUS: VMSE

Another technique to strengthen and harden metals is alloying with impurity atoms that go into either substitutional or interstitial solid solution. Accordingly, this is called **solid-solution strengthening**. High-purity metals are almost always softer and weaker than alloys composed of the same base metal. Increasing the concentration of the impurity results in an attendant increase in tensile and yield strengths, as indicated in Figures 9.16a and 9.16b, respectively, for nickel in copper; the dependence of ductility on nickel concentration is presented in Figure 9.16c.

Alloys are stronger than pure metals because impurity atoms that go into solid solution typically impose lattice strains on the surrounding host atoms. Lattice strain field

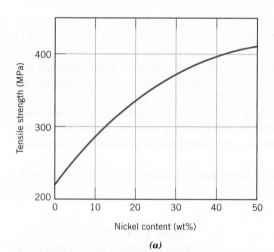

(a)

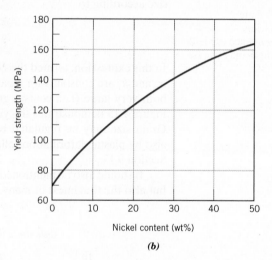

(b)

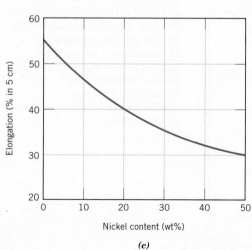

(c)

Figure 9.16 Variation with nickel content of (a) tensile strength, (b) yield strength, and (c) ductility (%EL) for copper–nickel alloys, showing strengthening.

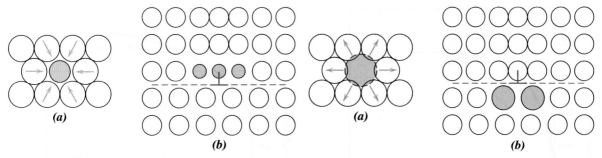

Figure 9.17 (*a*) Representation of tensile lattice strains imposed on host atoms by a smaller substitutional impurity atom. (*b*) Possible locations of smaller impurity atoms relative to an edge dislocation such that there is partial cancellation of impurity–dislocation lattice strains.

Figure 9.18 (*a*) Representation of compressive strains imposed on host atoms by a larger substitutional impurity atom. (*b*) Possible locations of larger impurity atoms relative to an edge dislocation such that there is partial cancellation of impurity–dislocation lattice strains.

interactions between dislocations and these impurity atoms result, and, consequently, dislocation movement is restricted. For example, an impurity atom that is smaller than a host atom for which it substitutes exerts tensile strains on the surrounding crystal lattice, as illustrated in Figure 9.17*a*. Conversely, a larger substitutional atom imposes compressive strains in its vicinity (Figure 9.18*a*). These solute atoms tend to diffuse to and segregate around dislocations in such a way as to reduce the overall strain energy—that is, to cancel some of the strain in the lattice surrounding a dislocation. To accomplish this, a smaller impurity atom is located where its tensile strain partially nullifies some of the dislocation's compressive strain. For the edge dislocation in Figure 9.18*b*, this would be adjacent to the dislocation line and above the slip plane. A larger impurity atom would be situated as in Figure 9.18*b*.

The resistance to slip is greater when impurity atoms are present because the overall lattice strain must increase if a dislocation is torn away from them. Furthermore, the same lattice strain interactions (Figures 9.17*b* and 9.18*b*) exist between impurity atoms and dislocations in motion during plastic deformation. Thus, a greater applied stress is necessary to first initiate and then continue plastic deformation for solid-solution alloys, as opposed to pure metals; this is evidenced by the enhancement of strength and hardness.

9.10 STRAIN HARDENING

strain hardening

Strain hardening is the phenomenon by which a ductile metal becomes harder and stronger as it is plastically deformed. Sometimes it is also called *work hardening,* or, because the temperature at which deformation takes place is "cold" relative to the absolute melting temperature of the metal, **cold working**. Most metals strain harden at room temperature.

cold working

It is sometimes convenient to express the degree of plastic deformation as *percent cold work* rather than as strain. Percent cold work (%CW) is defined as

Percent cold work— dependence on original and deformed cross-sectional areas

$$\%\text{CW} = \left(\frac{A_0 - A_d}{A_0}\right) \times 100 \qquad (9.8)$$

WileyPLUS

Tutorial Video: Defects in Metals What Is Cold Work?

where A_0 is the original area of the cross section that experiences deformation and A_d is the area after deformation.

Figures 9.19*a* and 9.19*b* demonstrate how steel, brass, and copper increase in yield and tensile strength with increasing cold work. The price for this enhancement of hardness

EXAMPLE PROBLEM 9.2

Tensile Strength and Ductility Determinations for Cold-Worked Copper

Compute the tensile strength and ductility (%EL) of a cylindrical copper rod if it is cold worked such that the diameter is reduced from 15.2 mm to 12.2 mm.

Solution

It is first necessary to determine the percent cold work resulting from the deformation. This is possible using Equation 9.8:

$$\%CW = \frac{\left(\dfrac{15.2 \text{ mm}}{2}\right)^2 \pi - \left(\dfrac{12.2 \text{ mm}}{2}\right)^2 \pi}{\left(\dfrac{15.2 \text{ mm}}{2}\right)^2 \pi} \times 100 = 35.6\%$$

The tensile strength is read directly from the curve for copper (Figure 9.19b) as 340 MPa. From Figure 9.19c, the ductility at 35.6%CW is about 7%EL.

In summary, we have discussed the three mechanisms that may be used to strengthen and harden single-phase metal alloys: strengthening by grain size reduction, solid-solution strengthening, and strain hardening. Of course, they may be used in conjunction with one another; for example, a solid-solution strengthened alloy may also be strain hardened.

It should also be noted that the strengthening effects due to grain size reduction and strain hardening can be eliminated or at least reduced by an elevated-temperature heat treatment (Sections 9.12 and 9.13). In contrast, solid-solution strengthening is unaffected by heat treatment.

As we shall see in Chapters 12 and 17, techniques other than those just discussed may be used to improve the mechanical properties of some metal alloys. These alloys are multiphase and property alterations result from phase transformations, which are induced by specifically designed heat treatments.

Recovery, Recrystallization, and Grain Growth

As outlined earlier in this chapter, plastically deforming a polycrystalline metal specimen at temperatures that are low relative to its absolute melting temperature produces microstructural and property changes that include (1) a change in grain shape (Section 9.6), (2) strain hardening (Section 9.10), and (3) an increase in dislocation density (Section 9.3). Some fraction of the energy expended in deformation is stored in the metal as strain energy, which is associated with tensile, compressive, and shear zones around the newly created dislocations (Section 9.3). Furthermore, other properties, such as electrical conductivity (Section 19.8) and corrosion resistance, may be modified as a consequence of plastic deformation.

WileyPLUS

Tutorial Video:
Annealing
What Is Annealing and
What Does It Do?

These properties and structures may revert back to the precold-worked states by appropriate heat treatment (sometimes termed an *annealing treatment*). Such restoration results from two different processes that occur at elevated temperatures: *recovery* and *recrystallization*, which may be followed by *grain growth*.

9.11 RECOVERY

recovery

During **recovery**, some of the stored internal strain energy is relieved by virtue of dislocation motion (in the absence of an externally applied stress), as a result of enhanced atomic diffusion at the elevated temperature. There is some reduction in the number of dislocations, and dislocation configurations (similar to that shown in Figure 6.11) are produced having low strain energies. In addition, physical properties such as electrical and thermal conductivities recover to their precold-worked states.

9.12 RECRYSTALLIZATION

recrystallization

WileyPLUS

Tutorial Video:
Annealing
What's the Difference
between Recovery and
Recrystallization?

**recrystallization
temperature**

Even after recovery is complete, the grains are still in a relatively high strain energy state. **Recrystallization** is the formation of a new set of strain-free and equiaxed grains (i.e., having approximately equal dimensions in all directions) that have low dislocation densities and are characteristic of the precold-worked condition. The driving force to produce this new grain structure is the difference in internal energy between the strained and unstrained material. The new grains form as very small nuclei and grow until they completely consume the parent material, processes that involve short-range diffusion. Several stages in the recrystallization process are represented in Figures 9.21a to 9.21d; in these photomicrographs, the small speckled grains are those that have recrystallized. Thus, recrystallization of cold-worked metals may be used to refine the grain structure.

Also, during recrystallization, the mechanical properties that were changed as a result of cold working are restored to their precold-worked values—that is, the metal becomes softer and weaker, yet more ductile. Some heat treatments are designed to allow recrystallization to occur with these modifications in the mechanical characteristics (Section 17.5).

The extent of recrystallization depends on both time and temperature. The degree (or fraction) of recrystallization increases with time, as may be noted in the photomicrographs shown in Figures 9.21a to 9.21d. The explicit time dependence of recrystallization is addressed in more detail near the end of Section 12.3.

The influence of temperature is demonstrated in Figure 9.22, which plots tensile strength and ductility (at room temperature) of a brass alloy as a function of the temperature and for a constant heat treatment time of 1 h. The grain structures found at the various stages of the process are also presented schematically.

The recrystallization behavior of a particular metal alloy is sometimes specified in terms of a **recrystallization temperature**, the temperature at which recrystallization just reaches completion in 1 h. Thus, the recrystallization temperature for the brass alloy of Figure 9.22 is about 450°C. Typically, it is between one-third and one-half of the absolute melting temperature of a metal or alloy and depends on several factors, including the amount of prior cold work and the purity of the alloy. Increasing the percent cold work enhances the rate of recrystallization, with the result that the recrystallization temperature is lowered, and approaches a constant or limiting value at high deformations; this effect is shown in Figure 9.23. Furthermore, it is this limiting or minimum recrystallization temperature that is normally specified in the literature. There exists some critical degree of cold work below which recrystallization cannot be made to occur, as shown in the figure; typically, this is between 2% and 20% cold work.

Recrystallization proceeds more rapidly in pure metals than in alloys. During recrystallization, grain-boundary motion occurs as the new grain nuclei form and then grow. It is believed that impurity atoms preferentially segregate at and interact with these recrystallized grain boundaries so as to diminish their (i.e., grain boundary) mobilities; this results in a decrease of the recrystallization rate and raises the recrystallization temperature, sometimes quite substantially. For pure metals, the recrystallization temperature is normally $0.4T_m$, where T_m is the absolute melting temperature; for some commercial alloys it may run as high as $0.7T_m$. Recrystallization and melting temperatures for a number of metals and alloys are listed in Table 9.2.

It should be noted that because recrystallization rate depends on several variables, as discussed previously, there is some arbitrariness to recrystallization temperatures cited in the literature. Furthermore, some degree of recrystallization may occur for an alloy that is heat treated at temperatures below its recrystallization temperature.

Figure 9.21
Photomicrographs showing several stages of the recrystallization and grain growth of brass. (*a*) Cold-worked (33%CW) grain structure. (*b*) Initial stage of recrystallization after heating for 3 s at 580°C; the very small grains are those that have recrystallized. (*c*) Partial replacement of cold-worked grains by recrystallized ones (4 s at 580°C). (*d*) Complete recrystallization (8 s at 580°C). (*e*) Grain growth after 15 min at 580°C. (*f*) Grain growth after 10 min at 700°C. All photomicrographs 70×. (Photomicrographs courtesy of J. E. Burke, General Electric Company.)

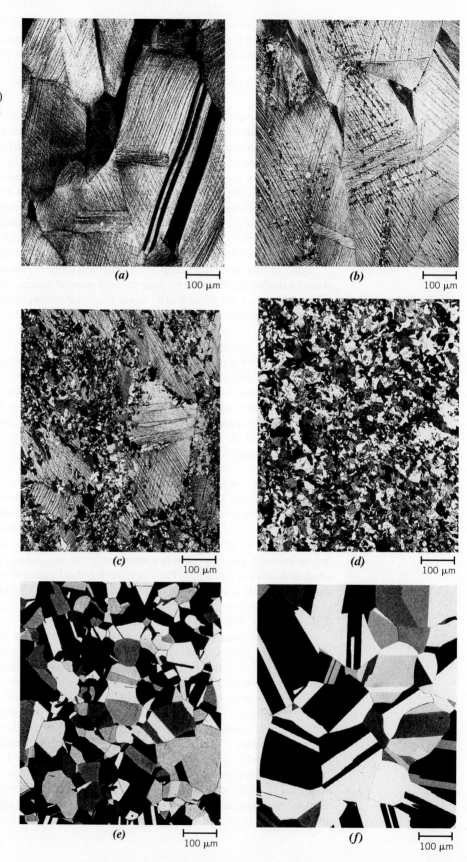

(*a*) 100 μm

(*b*) 100 μm

(*c*) 100 μm

(*d*) 100 μm

(*e*) 100 μm

(*f*) 100 μm

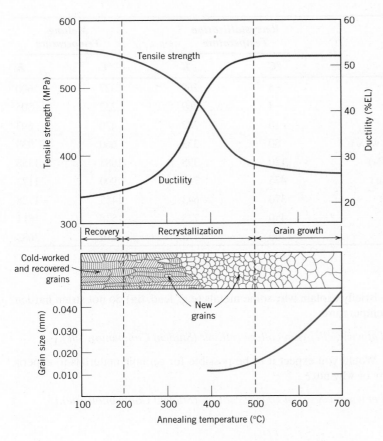

Figure 9.22 The influence of annealing temperature (for an annealing time of 1 h) on the tensile strength and ductility of a brass alloy. Grain size as a function of annealing temperature is indicated. Grain structures during recovery, recrystallization, and grain growth stages are shown schematically.
(Adapted from G. Sachs and K. R. Van Horn, *Practical Metallurgy, Applied Metallurgy and the Industrial Processing of Ferrous and Nonferrous Metals and Alloys,* 1940. Reproduced by permission of ASM International, Materials Park, OH.)

Plastic deformation operations are often carried out at temperatures above the recrystallization temperature in a process termed *hot working,* described in Section 17.2. The material remains relatively soft and ductile during deformation because it does not strain harden, and thus large deformations are possible.

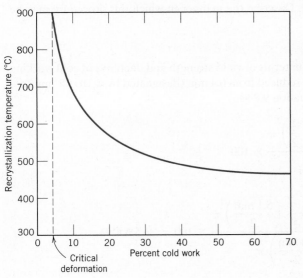

Figure 9.23 The variation of recrystallization temperature with percent cold work for iron. For deformations less than the critical (about 5%CW), recrystallization will not occur.

Table 9.2

Recrystallization and Melting Temperatures for Various Metals and Alloys

Metal	Recrystallization Temperature		Melting Temperature	
	°C	K	°C	K
Lead	−4	269	327	600
Tin	−4	269	232	505
Zinc	10	283	420	693
Aluminum (99.999 wt%)	80	353	660	933
Copper (99.999 wt%)	120	393	1085	1358
Brass (60 Cu–40 Zn)	475	748	900	1173
Nickel (99.99 wt%)	370	643	1455	1728
Iron	450	723	1538	1811
Tungsten	1200	1473	3410	3683

Concept Check 9.5   Briefly explain why some metals (e.g., lead, tin) do not strain harden when deformed at room temperature.

[*The answer may be found at* www.wiley.com/college/callister (*Student Companion Site*).]

Concept Check 9.6   Would you expect it to be possible for ceramic materials to experience recrystallization? Why or why not?

[*The answer may be found at* www.wiley.com/college/callister (*Student Companion Site*).]

DESIGN EXAMPLE 9.1

Description of Diameter Reduction Procedure

A cylindrical rod of noncold-worked brass having an initial diameter of 6.4 mm is to be cold worked by drawing such that the cross-sectional area is reduced. It is required to have a cold-worked yield strength of at least 345 MPa and a ductility in excess of 20%EL; in addition, a final diameter of 5.1 mm is necessary. Describe the manner in which this procedure may be carried out.

Solution

Let us first consider the consequences (in terms of yield strength and ductility) of cold working in which the brass specimen diameter is reduced from 6.4 mm (designated by d_0) to 5.1 mm (d_i). The %CW may be computed from Equation 9.8 as

$$\%CW = \frac{\left(\dfrac{d_0}{2}\right)^2 \pi - \left(\dfrac{d_i}{2}\right)^2 \pi}{\left(\dfrac{d_0}{2}\right)^2 \pi} \times 100$$

$$= \frac{\left(\dfrac{6.4 \text{ mm}}{2}\right)^2 \pi - \left(\dfrac{5.1 \text{ mm}}{2}\right)^2 \pi}{\left(\dfrac{6.4 \text{ mm}}{2}\right)^2 \pi} \times 100 = 36.5\%CW$$

From Figures 9.19a and 9.19c, a yield strength of 410 MPa and a ductility of 8%EL are attained from this deformation. According to the stipulated criteria, the yield strength is satisfactory; however, the ductility is too low.

Another processing alternative is a partial diameter reduction, followed by a recrystallization heat treatment in which the effects of the cold work are nullified. The required yield strength, ductility, and diameter are achieved through a second drawing step.

Again, reference to Figure 9.19a indicates that 20%CW is required to give a yield strength of 345 MPa. However, from Figure 9.19c, ductilities greater than 20%EL are possible only for deformations of 23%CW or less. Thus during the final drawing operation, deformation must be between 20%CW and 23%CW. Let's take the average of these extremes, 21.5%CW, and then calculate the final diameter for the first drawing d_0', which becomes the original diameter for the second drawing. Again, using Equation 9.8,

$$21.5\%\text{CW} = \frac{\left(\dfrac{d_0'}{2}\right)^2 \pi - \left(\dfrac{5.1\text{ mm}}{2}\right)^2 \pi}{\left(\dfrac{d_0'}{2}\right)^2 \pi} \times 100$$

Now, solving for d_0' from the preceding expression gives

$$d_0' = 5.8\text{ mm}$$

9.13 GRAIN GROWTH

grain growth

After recrystallization is complete, the strain-free grains will continue to grow if the metal specimen is left at the elevated temperature (Figures 9.21d to 9.21f); this phenomenon is called **grain growth**. Grain growth does not need to be preceded by recovery and recrystallization; it may occur in all polycrystalline materials, metals and ceramics alike.

An energy is associated with grain boundaries, as explained in Section 6.8. As grains increase in size, the total boundary area decreases, yielding an attendant reduction in the total energy; this is the driving force for grain growth.

Grain growth occurs by the migration of grain boundaries. Obviously, not all grains can enlarge, but large ones grow at the expense of small ones that shrink. Thus, the average grain size increases with time, and at any particular instant there exists a range of grain sizes. Boundary motion is just the short-range diffusion of atoms from one side of the boundary to the other. The directions of boundary movement and atomic motion are opposite to each other, as shown in Figure 9.24.

For many polycrystalline materials, the grain diameter d varies with time t according to the relationship

For grain growth, dependence of grain size on time

$$d^n - d_0^n = Kt \tag{9.9}$$

where d_0 is the initial grain diameter at $t = 0$, and K and n are time-independent constants; the value of n is generally equal to or greater than 2.

The dependence of grain size on time and temperature is demonstrated in Figure 9.25, a plot of the logarithm of grain size as a function of the logarithm of time for a brass alloy at several temperatures. At lower temperatures the curves are linear. Furthermore, grain growth proceeds more rapidly as temperature increases—that is, the curves are displaced upward to larger grain sizes. This is explained by the enhancement of diffusion rate with rising temperature.

The mechanical properties at room temperature of a fine-grained metal are usually superior (i.e., higher strength and toughness) to those of coarse-grained ones. If the

ferent $\langle 11\bar{2}0 \rangle$ slip directions within this plane. You may find Figure 3.8 helpful.

9.9 Equations 9.1a and 9.1b, expressions for Burgers vectors for FCC and BCC crystal structures, are of the form

$$\mathbf{b} = \frac{a}{2} \langle uvw \rangle$$

where a is the unit cell edge length. Also, because the magnitudes of these Burgers vectors may be determined from the following equation:

$$|\mathbf{b}| = \frac{a}{2}(u^2 + v^2 + w^2)^{1/2} \qquad (9.11)$$

determine values of $|\mathbf{b}|$ for copper and molybdenum. You may want to consult Table 4.1.

9.10 (a) In the manner of Equations 9.1a, 9.1b, and 9.1c, specify the Burgers vector for the simple cubic crystal structure. Its unit cell is shown in Figure 4.2. Also, simple cubic is the crystal structure for the edge dislocation of Figure 6.9, and for its motion as presented in Figure 9.1. You may also want to consult the answer to Concept Check 9.1.

(b) On the basis of Equation 9.11, formulate an expression for the magnitude of the Burgers vector, $|\mathbf{b}|$, for simple cubic.

Slip in Single Crystals

9.11 Sometimes $\cos \phi \cos \lambda$ in Equation 9.2 is termed the *Schmid factor*. Determine the magnitude of the Schmid factor for an FCC single crystal oriented with its [100] direction parallel to the loading axis.

9.12 Consider a metal single crystal oriented such that the normal to the slip plane and the slip direction are at angles of 43.1° and 47.9°, respectively, with the tensile axis. If the critical resolved shear stress is 22 MPa, will an applied stress of 50 MPa cause the single crystal to yield? If not, what stress will be necessary?

9.13 A single crystal of aluminum is oriented for a tensile test such that its slip plane normal makes an angle of 28.1° with the tensile axis. Three possible slip directions make angles of 62.4°, 72.0°, and 81.1° with the same tensile axis.

(a) Which of these three slip directions is most favored?

(b) If plastic deformation begins at a tensile stress of 1.8 MPa, determine the critical resolved shear stress for aluminum.

9.14 Consider a single crystal of silver oriented such that a tensile stress is applied along a [001] direction. If slip occurs on a (111) plane and in a $[\bar{1}01]$ direction, and is initiated at an applied tensile stress of 1.4 MPa, compute the critical resolved shear stress.

9.15 A single crystal of a metal that has the FCC crystal structure is oriented such that a tensile stress is applied parallel to the $[110]$ direction. If the critical resolved shear stress for this material is 2.2 MPa, calculate the magnitude(s) of applied stress(es) necessary to cause slip to occur on the (111) plane in each of the $[1\bar{1}0]$, $[10\bar{1}]$ and $[01\bar{1}]$ directions.

9.16 (a) A single crystal of a metal that has the BCC crystal structure is oriented such that a tensile stress is applied in the [010] direction. If the magnitude of this stress is 2.3 MPa, compute the resolved shear stress in the $[\bar{1}11]$ direction on each of the (110) and (101) planes.

(b) On the basis of these resolved shear stress values, which slip system(s) is (are) most favorably oriented?

9.17 Consider a single crystal of some hypothetical metal that has the FCC crystal structure and is oriented such that a tensile stress is applied along a $[\bar{1}02]$ direction. If slip occurs on a (111) plane and in a $[\bar{1}01]$ direction, compute the stress at which the crystal yields if its critical resolved shear stress is 3.2 MPa.

9.18 The critical resolved shear stress for iron is 27 MPa. Determine the maximum possible yield strength for a single crystal of Fe pulled in tension.

Deformation by Twinning

9.19 List four major differences between deformation by twinning and deformation by slip relative to mechanism, conditions of occurrence, and final result.

Strengthening by Grain Size Reduction

9.20 Briefly explain why small-angle grain boundaries are not as effective in interfering with the slip process as are high-angle grain boundaries.

9.21 Briefly explain why HCP metals are typically more brittle than FCC and BCC metals.

9.22 Describe in your own words the three strengthening mechanisms discussed in this chapter (i.e., grain size reduction, solid-solution strengthening, and strain hardening). Be sure to explain how dislocations are involved in each of the strengthening techniques.

9.23 **(a)** From the plot of yield strength versus (grain diameter)$^{-1/2}$ for a 70 Cu–30 Zn cartridge brass, Figure 9.15, determine values for the constants σ_0 and k_y in Equation 9.7.

(b) Now predict the yield strength of this alloy when the average grain diameter is 2.0×10^{-3} mm.

9.24 The lower yield point for an iron that has an average grain diameter of 6×10^{-2} mm is 135 MPa. At a grain diameter of 8×10^{-3} mm, the yield point increases to 260 MPa. At what grain diameter will the lower yield point be 205 MPa?

9.25 If it is assumed that the plot in Figure 9.15 is for noncold-worked brass, determine the grain size of the alloy in Figure 9.19; assume its composition is the same as the alloy in Figure 9.15.

Solid–Solution Strengthening

9.26 In the manner of Figures 9.17*b* and 9.18*b*, indicate the location in the vicinity of an edge dislocation at which an interstitial impurity atom would be expected to be situated. Now briefly explain in terms of lattice strains why it would be situated at this position.

Strain Hardening

9.27 **(a)** Show, for a tensile test, that

$$\%CW = \left(\frac{\epsilon}{\epsilon + 1} \right) \times 100$$

if there is no change in specimen volume during the deformation process (i.e., $A_0 l_0 = A_d l_d$).

(b) Using the result of part (a), compute the percent cold work experienced by naval brass (the stress–strain behavior of which is shown in Figure 8.12) when a stress of 400 MPa is applied.

9.28 Two previously undeformed cylindrical specimens of an alloy are to be strain hardened by reducing their cross-sectional areas (while maintaining their circular cross sections). For one specimen, the initial and deformed radii are 17 mm and 12 mm, respectively. The second specimen, with an initial radius of 13 mm, must have the same deformed hardness as the first specimen; compute the second specimen's radius after deformation.

9.29 Two previously undeformed specimens of the same metal are to be plastically deformed by reducing their cross-sectional areas. One has a circular cross section, and the other is rectangular; during deformation the circular cross section is to remain circular, and the rectangular is to remain as such. Their original and deformed dimensions are as follows:

	Circular (diameter, mm)	Rectangular (mm)
Original dimensions	15.2	125 × 175
Deformed dimensions	11.4	75 × 200

Which of these specimens will be the hardest after plastic deformation, and why?

9.30 A cylindrical specimen of cold-worked copper has a ductility (%EL) of 25%. If its cold-worked radius is 10 mm, what was its radius before deformation?

9.31 **(a)** What is the approximate ductility (%EL) of a brass that has a yield strength of 275 MPa?

(b) What is the approximate Brinell hardness of a 1040 steel having a yield strength of 690 MPa?

9.32 Experimentally, it has been observed for single crystals of a number of metals that the critical resolved shear stress τ_{crss} is a function of the dislocation density ρ_D as

$$\tau_{crss} = \tau_0 + A\sqrt{\rho_D}$$

where τ_0 and A are constants. For copper, the critical resolved shear stress is 2.10 MPa at a dislocation density of 10^5 mm^{-2}. If it is known that the value of A for copper is 6.35×10^{-3} MPa·mm, compute τ_{crss} at a dislocation density of 10^7 mm^{-2}.

Recovery
Recrystallization
Grain Growth

9.33 Briefly cite the differences between recovery and recrystallization processes.

9.34 Estimate the fraction of recrystallization from the photomicrograph in Figure 9.21*c*.

9.35 Explain the differences in grain structure for a metal that has been cold worked and one that has been cold worked and then recrystallized.

9.36 **(a)** What is the driving force for recrystallization?

(b) For grain growth?

9.37 **(a)** From Figure 9.25, compute the length of time required for the average grain diameter to increase from 0.03 to 0.12 mm at 600°C for this brass material.

(b) Repeat the calculation at 700°C.

9.38 The average grain diameter for a brass material was measured as a function of time at 650°C, which is shown in the following table at two different times:

Time (min)	Grain Diameter (mm)
30	3.9×10^{-2}
90	6.6×10^{-2}

(a) What was the original grain diameter?

(b) What grain diameter would you predict after 150 min at 650°C?

9.39 An undeformed specimen of some alloy has an average grain diameter of 0.045 mm. You are asked to reduce its average grain diameter to 0.010 mm. Is this possible? If so, explain the procedures you would use and name the processes involved. If it is not possible, explain why.

9.40 Grain growth is strongly dependent on temperature (i.e., rate of grain growth increases with increasing temperature), yet temperature is not explicitly given as a part of Equation 9.9.

(a) Into which of the parameters in this expression would you expect temperature to be included?

(b) On the basis of your intuition, cite an explicit expression for this temperature dependence.

9.41 An uncold-worked brass specimen of average grain size 0.009 mm has a yield strength of 160 MPa. Estimate the yield strength of this alloy after it has been heated to 600°C for 1000 s, if it is known that the value of k_y is 12.0 MPa·mm$^{1/2}$.

Spreadsheet Problem

9.1SS For crystals having cubic symmetry, generate a spreadsheet that will allow the user to determine the angle between two crystallographic directions, given their directional indices.

DESIGN PROBLEMS

Strain Hardening
Recrystallization

9.D1 Determine whether it is possible to cold work steel so as to give a minimum Brinell hardness of 225, and at the same time have a ductility of at least 12%EL. Justify your decision.

9.D2 Determine whether it is possible to cold work brass so as to give a minimum Brinell hardness of 120 and at the same time have a ductility of at least 20%EL. Justify your decision.

9.D3 A cylindrical specimen of cold-worked steel has a Brinell hardness of 250.

(a) Estimate its ductility in percent elongation.

(b) If the specimen remained cylindrical during deformation and its original radius was 5 mm, determine its radius after deformation.

9.D4 It is necessary to select a metal alloy for an application that requires a yield strength of at least 345 MPa while maintaining a minimum ductility (%EL) of 20%. If the metal may be cold worked, decide which of the following are candidates: copper, brass, and a 1040 steel. Why?

9.D5 A cylindrical rod of 1040 steel originally 15.2 mm in diameter is to be cold worked by drawing; the circular cross section will be maintained during deformation. A cold-worked tensile strength in excess of 840 MPa and a ductility of at least 12%EL are desired. Furthermore, the final diameter must be 10 mm. Explain how this may be accomplished.

9.D6 A cylindrical rod of copper originally 16.0 mm in diameter is to be cold worked by drawing; the circular cross section will be maintained during deformation. A cold-worked yield strength in excess of 250 MPa and a ductility of at least 12%EL are desired. Furthermore, the final diameter must be 11.3 mm. Explain how this may be accomplished.

9.D7 A cylindrical 1040 steel rod having a minimum tensile strength of 865 MPa, a ductility of at least 10%EL, and a final diameter of 6.3 mm is desired. Some 8 mm diameter 1040 steel stock, which has been cold worked 20%, is available. Describe the procedure you would follow to obtain this material. Assume that 1040 steel experiences cracking at 40%CW.

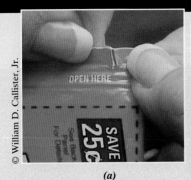

(a)

(b)

Have you ever experienced the aggravation of having to expend considerable effort to tear open a small plastic package that contains nuts, candy, or some other confection? You probably have also noticed that when a small incision (or cut) has been made into an edge, as appears in photograph (*a*), a minimal force is required to tear the package open. This phenomenon is related to one of the basic tenets of fracture mechanics: an applied tensile stress is amplified at the tip of a small incision or notch.

Photograph (*b*) is of an oil tanker that fractured in a brittle manner as a result of the propagation of a crack completely around its girth. This crack started as some type of small notch or sharp flaw. As the tanker was buffeted about while at sea, resulting stresses became amplified at the tip of this notch or flaw to the degree that a crack formed and rapidly elongated, which ultimately led to complete fracture of the tanker.

Photograph (*c*) is of a Boeing 737-200 commercial aircraft (Aloha Airlines flight 243) that experienced an explosive decompression and structural failure on April 28, 1988. An investigation of the accident concluded that the cause was metal fatigue aggravated by crevice corrosion (Section 18.7) inasmuch as the plane operated in a coastal (humid and salty) environment. Stress cycling of the fuselage resulted from compression and decompression of the cabin chamber during short hop flights. A properly executed maintenance program by the airline would have detected the fatigue damage and prevented this accident.

(c)

The design of a component or structure often calls upon the engineer to minimize the possibility of failure. Thus, it is important to understand the mechanics of the various failure modes—fracture, fatigue, and creep—and, in addition, be familiar with appropriate design principles that may be employed to prevent in-service failures. For example, in Sections M.14 through M.16 of the Mechanical Engineering Online Support Module, we discuss material selection and processing issues relating to the fatigue of an automobile valve spring.

Learning Objectives

After studying this chapter, you should be able to do the following:

1. Describe the mechanism of crack propagation for both ductile and brittle modes of fracture.
2. Explain why the strengths of brittle materials are much lower than predicted by theoretical calculations.
3. Define fracture toughness in terms of (a) a brief statement and (b) an equation; define all parameters in this equation.
4. Make a distinction between *fracture toughness* and *plane strain fracture toughness*.
5. Name and describe the two impact fracture testing techniques.
6. Define *fatigue* and specify the conditions under which it occurs.
7. From a fatigue plot for some material, determine (a) the fatigue lifetime (at a specified stress level) and (b) the fatigue strength (at a specified number of cycles).
8. Define *creep* and specify the conditions under which it occurs.
9. Given a creep plot for some material, determine (a) the steady-state creep rate and (b) the rupture lifetime.

10.1 INTRODUCTION

WileyPLUS

Tutorial Video:
Cyclical
Fatigue Failure
What Are Some
Real-World Examples
of Failure?

The failure of engineering materials is almost always an undesirable event for several reasons; these include putting human lives in jeopardy, causing economic losses, and interfering with the availability of products and services. Even though the causes of failure and the behavior of materials may be known, prevention of failures is difficult to guarantee. The usual causes are improper materials selection and processing and inadequate design of the component or its misuse. Also, damage can occur to structural parts during service, and regular inspection and repair or replacement are critical to safe design. It is the responsibility of the engineer to anticipate and plan for possible failure and, in the event that failure does occur, to assess its cause and then take appropriate preventive measures against future incidents.

The following topics are addressed in this chapter: simple fracture (both ductile and brittle modes), fundamentals of fracture mechanics, fracture toughness testing, the ductile-to-brittle transition, fatigue, and creep. These discussions include failure mechanisms, testing techniques, and methods by which failure may be prevented or controlled.

 Concept Check 10.1 Cite two situations in which the possibility of failure is part of the design of a component or product.

[*The answer may be found at www.wiley.com/college/callister (Student Companion Site).*]

Fracture

10.2 FUNDAMENTALS OF FRACTURE

Simple fracture is the separation of a body into two or more pieces in response to an imposed stress that is static (i.e., constant or slowly changing with time) and at temperatures that are low relative to the melting temperature of the material. Fracture can also occur from fatigue (when cyclic stresses are imposed) and creep (time-dependent deformation, normally at elevated temperatures); the topics of fatigue and creep are covered later in this chapter (Sections 10.7 through 10.15). Although applied stresses may be tensile, compressive, shear, or torsional (or combinations of these), the present discussion will be confined to fractures that result from uniaxial tensile loads. For metals, two fracture modes are possible: **ductile** and **brittle**. Classification is based on the ability of a material to experience plastic deformation. Ductile metals typically exhibit substantial plastic deformation with high energy absorption before fracture. However, there is normally little or no plastic deformation with low energy absorption accompanying a brittle fracture. The tensile stress–strain behaviors of both fracture types may be reviewed in Figure 8.13.

ductile fracture, brittle fracture

Ductile and *brittle* are relative terms; whether a particular fracture is one mode or the other depends on the situation. Ductility may be quantified in terms of percent elongation (Equation 8.11) and percent reduction in area (Equation 8.12). Furthermore, ductility is a function of temperature of the material, the strain rate, and the stress state. The disposition of normally ductile materials to fail in a brittle manner is discussed in Section 10.6.

Any fracture process involves two steps—crack formation and propagation—in response to an imposed stress. The mode of fracture is highly dependent on the mechanism of crack propagation. Ductile fracture is characterized by extensive plastic deformation in the vicinity of an advancing crack. Furthermore, the process proceeds relatively slowly as the crack length is extended. Such a crack is often said to be *stable*—that is, it resists any further extension unless there is an increase in the applied stress. In addition, there typically is evidence of appreciable gross deformation at the fracture surfaces (e.g., twisting and tearing). However, for brittle fracture, cracks may spread extremely rapidly, with very little accompanying plastic deformation. Such cracks may be said to be *unstable,* and crack propagation, once started, continues spontaneously without an increase in magnitude of the applied stress.

Ductile fracture is almost always preferred to brittle fracture for two reasons: First, brittle fracture occurs suddenly and catastrophically without any warning; this is a consequence of the spontaneous and rapid crack propagation. However, for ductile fracture, the presence of plastic deformation gives warning that failure is imminent, allowing preventive measures to be taken. Second, more strain energy is required to induce ductile fracture inasmuch as these materials are generally tougher. Under the action of an applied tensile stress, many metal alloys are ductile, whereas ceramics are typically brittle, and polymers may exhibit a range of behaviors.

10.3 DUCTILE FRACTURE

Ductile fracture surfaces have distinctive features on both macroscopic and microscopic levels. Figure 10.1 shows schematic representations for two characteristic macroscopic fracture profiles. The configuration shown in Figure 10.1*a* is found for extremely soft metals, such as pure gold and lead at room temperature, and other metals, polymers, and inorganic glasses at elevated temperatures. These highly ductile materials neck down to a point fracture, showing virtually 100% reduction in area.

The most common type of tensile fracture profile for ductile metals is that represented in Figure 10.1*b*, where fracture is preceded by only a moderate amount of necking. The fracture process normally occurs in several stages (Figure 10.2). First, after necking

(a)

(b)

Figure 10.5 (*a*) Photograph showing V-shaped "chevron" markings characteristic of brittle fracture. Arrows indicate origin of crack. Approximate actual size. (*b*) Photograph of a brittle fracture surface showing radial fan-shaped ridges. Arrow indicates origin of crack. Approximately 2×.
[(*a*) From R. W. Hertzberg, *Deformation and Fracture Mechanics of Engineering Materials,* 3rd edition. Copyright © 1989 by John Wiley & Sons, New York. Reprinted by permission of John Wiley & Sons, Inc. Photograph courtesy of Roger Slutter, Lehigh University. (*b*) From D. J. Wulpi, *Understanding How Components Fail,* 1985. Reproduced by permission of ASM International, Materials Park, OH.]

discerned with the naked eye. For very hard and fine-grained metals, there is no discernible fracture pattern. Brittle fracture in amorphous materials, such as ceramic glasses, yields a relatively shiny and smooth surface.

For most brittle crystalline materials, crack propagation corresponds to the successive and repeated breaking of atomic bonds along specific crystallographic planes (Figure 10.6*a*); such a process is termed *cleavage*. This type of fracture is said to be **transgranular** (or *transcrystalline*) because the fracture cracks pass through the grains. Macroscopically, the fracture surface may have a grainy or faceted texture (Figure 10.3*b*) as a result of changes in orientation of the cleavage planes from grain to grain. This cleavage feature is shown at a higher magnification in the scanning electron micrograph of Figure 10.6*b*.

transgranular fracture

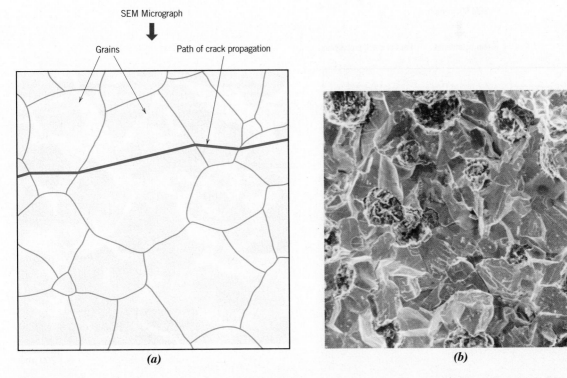

Figure 10.6 (*a*) Schematic cross-section profile showing crack propagation through the interior of grains for transgranular fracture. (*b*) Scanning electron fractograph of ductile cast iron showing a transgranular fracture surface. Magnification unknown.

[Figure (*b*) from V. J. Colangelo and F. A. Heiser, *Analysis of Metallurgical Failures,* 2nd edition. Copyright © 1987 by John Wiley & Sons, New York. Reprinted by permission of John Wiley & Sons, Inc.]

intergranular fracture

In some alloys, crack propagation is along grain boundaries (Figure 10.7*a*); this fracture is termed **intergranular**. Figure 10.7*b* is a scanning electron micrograph showing a typical intergranular fracture, in which the three-dimensional nature of the grains may be seen. This type of fracture normally results subsequent to the occurrence of processes that weaken or embrittle grain boundary regions.

10.5 PRINCIPLES OF FRACTURE MECHANICS[1]

fracture mechanics

Brittle fracture of normally ductile materials, such as that shown in the chapter-opening Figure *b* (of the oil barge), has demonstrated the need for a better understanding of the mechanisms of fracture. Extensive research endeavors over the past century have led to the evolution of the field of **fracture mechanics**. This subject allows quantification of the relationships among material properties, stress level, the presence of crack-producing flaws, and crack propagation mechanisms. Design engineers are now better equipped to anticipate, and thus prevent, structural failures. The present discussion centers on some of the fundamental principles of the mechanics of fracture.

[1]A more detailed discussion of the principles of fracture mechanics may be found in Section M.4 of the Mechanical Engineering (ME) Online Module.

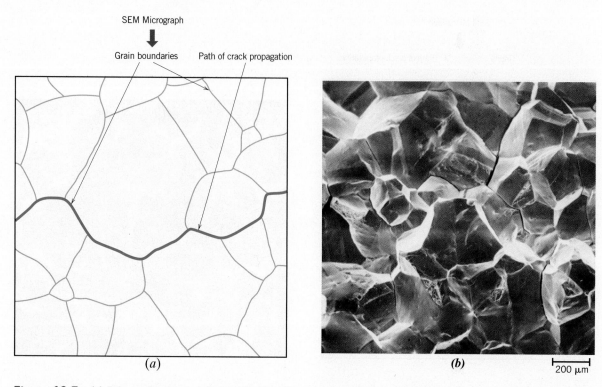

Figure 10.7 (*a*) Schematic cross-section profile showing crack propagation along grain boundaries for intergranular fracture. (*b*) Scanning electron fractograph showing an intergranular fracture surface. 50×.
[Figure (*b*) reproduced with permission from *ASM Handbook*, Vol. 12, *Fractography*, ASM International, Materials Park, OH, 1987.]

Stress Concentration

The measured fracture strengths for most materials are significantly lower than those predicted by theoretical calculations based on atomic bonding energies. This discrepancy is explained by the presence of microscopic flaws or cracks that always exist under normal conditions at the surface and within the interior of a body of material. These flaws are a detriment to the fracture strength because an applied stress may be amplified or concentrated at the tip, the magnitude of this amplification depending on crack orientation and geometry. This phenomenon is demonstrated in Figure 10.8—a stress profile across a cross section containing an internal crack. As indicated by this profile, the magnitude of this localized stress decreases with distance away from the crack tip. At positions far removed, the stress is just the nominal stress σ_0, or the applied load divided by the specimen cross-sectional area (perpendicular to this load). Because of their ability **stress raiser** to amplify an applied stress in their locale, these flaws are sometimes called **stress raisers**.

If it is assumed that a crack is similar to an elliptical hole through a plate and is oriented perpendicular to the applied stress, the maximum stress, σ_m, occurs at the crack tip and may be approximated by

For tensile loading, computation of maximum stress at a crack tip

$$\sigma_m = 2\sigma_0 \left(\frac{a}{\rho_t}\right)^{1/2} \tag{10.1}$$

where σ_0 is the magnitude of the nominal applied tensile stress, ρ_t is the radius of curvature of the crack tip (Figure 10.8*a*), and *a* represents the length of a surface crack, or half

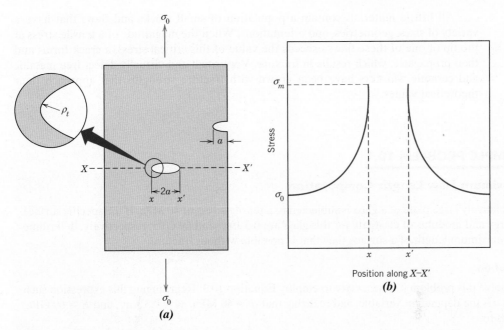

Figure 10.8 (a) The geometry of surface and internal cracks. (b) Schematic stress profile along the line X–X' in (a), demonstrating stress amplification at crack tip positions.

of the length of an internal crack. For a relatively long microcrack that has a small tip radius of curvature, the factor $(a/\rho_t)^{1/2}$ may be very large. This yields a value of σ_m that is many times the value of σ_0.

Sometimes the ratio σ_m/σ_0 is denoted the *stress concentration factor* K_t:

$$K_t = \frac{\sigma_m}{\sigma_0} = 2\left(\frac{a}{\rho_t}\right)^{1/2} \tag{10.2}$$

which is simply a measure of the degree to which an external stress is amplified at the tip of a crack.

Note that stress amplification is not restricted to these microscopic defects; it may occur at macroscopic internal discontinuities (e.g., voids or inclusions), sharp corners, scratches, and notches.

Furthermore, the effect of a stress raiser is more significant in brittle than in ductile materials. For a ductile metal, plastic deformation ensues when the maximum stress exceeds the yield strength. This leads to a more uniform distribution of stress in the vicinity of the stress raiser and to the development of a maximum stress concentration factor less than the theoretical value. Such yielding and stress redistribution do not occur to any appreciable extent around flaws and discontinuities in brittle materials; therefore, essentially the theoretical stress concentration results.

Using principles of fracture mechanics, it is possible to show that the critical stress σ_c required for crack propagation in a brittle material is described by the expression

Critical stress for crack propagation in a brittle material

$$\sigma_c = \left(\frac{2E\gamma_s}{\pi a}\right)^{1/2} \tag{10.3}$$

where E is modulus of elasticity, γ_s is the specific surface energy, and a is one-half the length of an internal crack.

All brittle materials contain a population of small cracks and flaws that have a variety of sizes, geometries, and orientations. When the magnitude of a tensile stress at the tip of one of these flaws exceeds the value of this critical stress, a crack forms and then propagates, which results in fracture. Very small and virtually defect-free metallic and ceramic whiskers have been grown with fracture strengths that approach their theoretical values.

EXAMPLE PROBLEM 10.1

Maximum Flaw Length Computation

A relatively large plate of a glass is subjected to a tensile stress of 40 MPa. If the specific surface energy and modulus of elasticity for this glass are 0.3 J/m² and 69 GPa, respectively, determine the maximum length of a surface flaw that is possible without fracture.

Solution

To solve this problem it is necessary to employ Equation 10.3. Rearranging this expression such that a is the dependent variable, and realizing that $\sigma = 40$ MPa, $\gamma_s = 0.3$ J/m², and $E = 69$ GPa, leads to

$$a = \frac{2E\gamma_s}{\pi\sigma^2}$$

$$= \frac{(2)(69 \times 10^9 \text{ N/m}^2)(0.3 \text{ N/m})}{\pi(40 \times 10^6 \text{ N/m}^2)^2}$$

$$= 8.2 \times 10^{-6} \text{ m} = 0.0082 \text{ mm} = 8.2 \text{ } \mu\text{m}$$

Fracture Toughness

Fracture toughness—dependence on critical stress for crack propagation and crack length

Using fracture mechanical principles, an expression has been developed that relates this critical stress for crack propagation (σ_c) and crack length (a) as

$$K_c = Y\sigma_c\sqrt{\pi a} \tag{10.4}$$

fracture toughness

In this expression K_c is the **fracture toughness**, a property that is a measure of a material's resistance to brittle fracture when a crack is present. K_c has the unusual units of MPa$\sqrt{\text{m}}$. Here, Y is a dimensionless parameter or function that depends on both crack and specimen sizes and geometries as well as on the manner of load application.

Relative to this Y parameter, for planar specimens containing cracks that are much shorter than the specimen width, Y has a value of approximately unity. For example, for a plate of infinite width having a through-thickness crack (Figure 10.9a), $Y = 1.0$, whereas for a plate of semi-infinite width containing an edge crack of length a (Figure 10.9b), $Y \cong 1.1$. Mathematical expressions for Y have been determined for a variety of crack-specimen geometries; these expressions are often relatively complex.

For relatively thin specimens, the value of K_c depends on specimen thickness. However, when specimen thickness is much greater than the crack dimensions, K_c becomes independent of thickness; under these conditions a condition of **plane strain** exists. By *plane strain*, we mean that when a load operates on a crack in the manner

plane strain

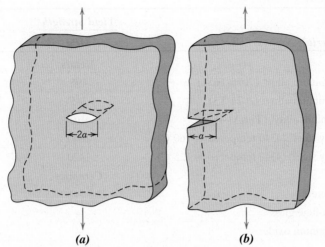

Figure 10.9 Schematic representations of (*a*) an interior crack in a plate of infinite width and (*b*) an edge crack in a plate of semi-infinite width.

(a) *(b)*

plane strain fracture toughness

Plane strain fracture toughness for mode I crack surface displacement

represented in Figure 10.9*a*, there is no strain component perpendicular to the front and back faces. The K_c value for this thick-specimen situation is known as the **plane strain fracture toughness**, K_{Ic}; it is also defined by

$$K_{Ic} = Y\sigma\sqrt{\pi a}$$

(10.5)

K_{Ic} is the fracture toughness cited for most situations. The *I* (i.e., Roman numeral "one") subscript for K_{Ic} denotes that the plane strain fracture toughness is for mode I crack displacement, as illustrated in Figure 10.10*a*.[2]

Brittle materials, for which appreciable plastic deformation is not possible in front of an advancing crack, have low K_{Ic} values and are vulnerable to catastrophic failure. However, K_{Ic} values are relatively large for ductile materials. Fracture mechanics is especially useful in predicting catastrophic failure in materials having intermediate ductilities. Plane strain fracture toughness values for a number of different materials are presented in Table 10.1 (and Figure 1.7); a more extensive list of K_{Ic} values is given in Table B.5, Appendix B.

The plane strain fracture toughness K_{Ic} is a fundamental material property that depends on many factors, the most influential of which are temperature, strain rate, and microstructure. The magnitude of K_{Ic} decreases with increasing strain rate and

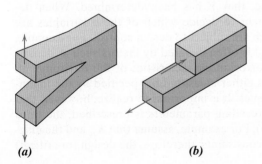 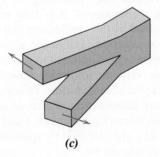

Figure 10.10 The three modes of crack surface displacement. (*a*) Mode I, opening or tensile mode; (*b*) mode II, sliding mode; and (*c*) mode III, tearing mode.

(a) *(b)* *(c)*

---

[2]Two other crack displacement modes, denoted II and III and illustrated in Figures 10.10*b* and 10.10*c*, are also possible; however, mode I is most commonly encountered.

Table 10.1

Room-Temperature
Yield Strength and
Plane Strain Fracture
Toughness Data for
Selected Engineering
Materials

Material	Yield Strength MPa	K_{Ic} MPa$\sqrt{m}$
Metals		
Aluminum alloy[a] (7075-T651)	495	24
Aluminum alloy[a] (2024-T3)	345	44
Titanium alloy[a] (Ti-6Al-4V)	910	55
Alloy steel[a] (4340 tempered @ 260°C)	1640	50.0
Alloy steel[a] (4340 tempered @ 425°C)	1420	87.4
Ceramics		
Concrete	—	0.2–1.4
Soda-lime glass	—	0.7–0.8
Aluminum oxide	—	2.7–5.0
Polymers		
Polystyrene (PS)	—	0.7–1.1
Poly(methyl methacrylate) (PMMA)	53.8–73.1	0.7–1.6
Polycarbonate (PC)	62.1	2.2

[a]**Source:** Reprinted with permission, *Advanced Materials and Processes,*
ASM International, © 1990.

decreasing temperature. Furthermore, an enhancement in yield strength wrought by
solid solution or dispersion additions or by strain hardening generally produces a corresponding decrease in K_{Ic}. Furthermore, K_{Ic} normally increases with reduction in grain
size as composition and other microstructural variables are maintained constant. Yield
strengths are included for some of the materials listed in Table 10.1.

Several different testing techniques are used to measure K_{Ic} (see Section 10.6).
Virtually any specimen size and shape consistent with mode I crack displacement may
be utilized, and accurate values will be realized, provided that the Y scale parameter in
Equation 10.5 has been determined properly.

Design Using Fracture Mechanics

According to Equations 10.4 and 10.5, three variables must be considered relative
to the possibility for fracture of some structural component—namely, the fracture
toughness (K_c) or plane strain fracture toughness (K_{Ic}), the imposed stress (σ), and
the flaw size (a)—assuming, of course, that Y has been determined. When designing a component, it is first important to decide which of these variables are
constrained by the application and which are subject to design control. For example,
material selection (and hence K_c or K_{Ic}) is often dictated by factors such as density
(for lightweight applications) or the corrosion characteristics of the environment.
Alternatively, the allowable flaw size is either measured or specified by the limitations of available flaw detection techniques. It is important to realize, however, that
once any combination of two of the preceding parameters is prescribed, the third
becomes fixed (Equations 10.4 and 10.5). For example, assume that K_{Ic} and the magnitude of a are specified by application constraints; therefore, the design (or critical)
stress σ_c is given by

Computation of
design stress

$$\sigma_c = \frac{K_{Ic}}{Y\sqrt{\pi a}} \tag{10.6}$$

Table 10.2

A List of Several Common Nondestructive Testing Techniques

Technique	Defect Location	Defect Size Sensitivity (mm)	Testing Location
Scanning electron microscopy (SEM)	Surface	>0.001	Laboratory
Dye penetrant	Surface	0.025–0.25	Laboratory/in-field
Ultrasonics	Subsurface	>0.050	Laboratory/in-field
Optical microscopy	Surface	0.1–0.5	Laboratory
Visual inspection	Surface	>0.1	Laboratory/in-field
Acoustic emission	Surface/subsurface	>0.1	Laboratory/in-field
Radiography (x-ray/ gamma ray)	Subsurface	>2% of specimen thickness	Laboratory/in-field

However, if stress level and plane strain fracture toughness are fixed by the design situation, then the maximum allowable flaw size a_c is given by

Computation of maximum allowable flaw length

$$a_c = \frac{1}{\pi}\left(\frac{K_{Ic}}{\sigma Y}\right)^2 \tag{10.7}$$

A number of nondestructive test (NDT) techniques have been developed that permit detection and measurement of both internal and surface flaws.[3] Such techniques are used to examine structural components that are in service for defects and flaws that could lead to premature failure; in addition, NDTs are used as a means of quality control for manufacturing processes. As the name implies, these techniques do not destroy the material/structure being examined. Furthermore, some testing methods must be conducted in a laboratory setting; others may be adapted for use in the field. Several commonly employed NDT techniques and their characteristics are listed in Table 10.2.[4]

One important example of the use of NDT is for the detection of cracks and leaks in the walls of oil pipelines in remote areas such as Alaska. Ultrasonic analysis is utilized in conjunction with a "robotic analyzer" that can travel relatively long distances within a pipeline.

DESIGN EXAMPLE 10.1

Material Specification for a Pressurized Spherical Tank

Consider a thin-walled spherical tank of radius r and thickness t (Figure 10.11) that may be used as a pressure vessel.

(a) One design of such a tank calls for yielding of the wall material prior to failure as a result of the formation of a crack of critical size and its subsequent rapid propagation. Thus, plastic distortion of the wall may be observed and the pressure within the tank released before the occurrence of catastrophic failure. Consequently, materials having large critical crack lengths

---

[3]Sometimes the terms *nondestructive evaluation* (NDE) and *nondestructive inspection* (NDI) are also used for these techniques.

[4]Section M.5 of the Mechanical Engineering (ME) Online Module discusses how NDT are used in the detection of flaws and cracks.

are desired. On the basis of this criterion, rank the metal alloys listed in Table B.5, Appendix B, as to critical crack size, from longest to shortest.

(b) An alternative design that is also often utilized with pressure vessels is termed *leak-before-break*. On the basis of principles of fracture mechanics, allowance is made for the growth of a crack through the thickness of the vessel wall prior to the occurrence of rapid crack propagation (Figure 10.11). Thus, the crack will completely penetrate the wall without catastrophic

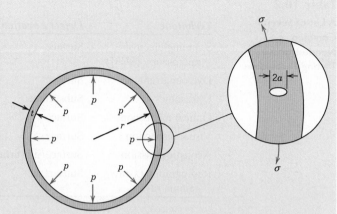

Figure 10.11 Schematic diagram showing the cross section of a spherical tank that is subjected to an internal pressure p and that has a radial crack of length $2a$ in its wall.

failure, allowing for its detection by the leaking of pressurized fluid. With this criterion the critical crack length a_c (i.e., one-half the total internal crack length) is taken to be equal to the pressure vessel thickness t. Allowance for $a_c = t$ instead of $a_c = t/2$ ensures that fluid leakage will occur prior to the buildup of dangerously high pressures. Using this criterion, rank the metal alloys in Table B.5, Appendix B, as to the maximum allowable pressure.

For this spherical pressure vessel, the circumferential wall stress σ is a function of the pressure p in the vessel and the radius r and wall thickness t according to

$$\sigma = \frac{pr}{2t} \tag{10.8}$$

For both parts (a) and (b), assume a condition of plane strain.

Solution

(a) For the first design criterion, it is desired that the circumferential wall stress be less than the yield strength of the material. Substitution of σ_y for σ in Equation 10.5 and incorporation of a factor of safety N leads to

$$K_{Ic} = Y\left(\frac{\sigma_y}{N}\right)\sqrt{\pi a_c} \tag{10.9}$$

where a_c is the critical crack length. Solving for a_c yields the following expression:

$$a_c = \frac{N^2}{Y^2 \pi}\left(\frac{K_{Ic}}{\sigma_y}\right)^2 \tag{10.10}$$

Therefore, the critical crack length is proportional to the square of K_{Ic}/σ_y, which is the basis for the ranking of the metal alloys in Table B.5. The ranking is provided in Table 10.3, where it may be seen that the medium carbon (1040) steel with the largest ratio has the longest critical crack length and, therefore, is the most desirable material on the basis of this criterion.

(b) As stated previously, the leak-before-break criterion is just met when one-half the internal crack length is equal to the thickness of the pressure vessel—that is, when $a = t$. Substitution of $a = t$ into Equation 10.5 gives

$$K_{Ic} = Y\sigma\sqrt{\pi t} \tag{10.11}$$

Table 10.3 Ranking of Several Metal Alloys Relative to Critical Crack Length (Yielding Criterion) for a Thin-Walled Spherical Pressure Vessel

Material	$\left(\dfrac{K_{Ic}}{\sigma_y}\right)^2 (mm)$
Medium carbon (1040) steel	43.1
AZ31B magnesium	19.6
2024 aluminum (T3)	16.3
Ti-5Al-2.5Sn titanium	6.6
4140 steel (tempered at 482°C)	5.3
4340 steel (tempered at 425°C)	3.8
Ti-6Al-4V titanium	3.7
17-7PH stainless steel	3.4
7075 aluminum (T651)	2.4
4140 steel (tempered at 370°C)	1.6
4340 steel (tempered at 260°C)	0.93

Table 10.4 Ranking of Several Metal Alloys Relative to Maximum Allowable Pressure (Leak-before-Break Criterion) for a Thin-Walled Spherical Pressure Vessel

Material	$\dfrac{K_{Ic}^2}{\sigma_y} (MPa \cdot m)$
Medium carbon (1040) steel	11.2
4140 steel (tempered at 482°C)	6.1
Ti-5Al-2.5Sn titanium	5.8
2024 aluminum (T3)	5.6
4340 steel (tempered at 425°C)	5.4
17-7PH stainless steel	4.4
AZ31B magnesium	3.9
Ti-6Al-4V titanium	3.3
4140 steel (tempered at 370°C)	2.4
4340 steel (tempered at 260°C)	1.5
7075 aluminum (T651)	1.2

and, from Equation 10.8,

$$t = \frac{pr}{2\sigma} \tag{10.12}$$

The stress is replaced by the yield strength because the tank should be designed to contain the pressure without yielding; furthermore, substitution of Equation 10.12 into Equation 10.11, after some rearrangement, yields the following expression:

$$p = \frac{2}{Y^2 \pi r}\left(\frac{K_{Ic}^2}{\sigma_y}\right) \tag{10.13}$$

Hence, for some given spherical vessel of radius r, the maximum allowable pressure consistent with this leak-before-break criterion is proportional to K_{Ic}^2/σ_y. The same several materials are ranked according to this ratio in Table 10.4; as may be noted, the medium carbon steel will contain the greatest pressures.

Of the 11 metal alloys listed in Table B.5, the medium carbon steel ranks first according to both yielding and leak-before-break criteria. For these reasons, many pressure vessels are constructed of medium carbon steels when temperature extremes and corrosion need not be considered.

10.6 FRACTURE TOUGHNESS TESTING

A number of different standardized tests have been devised to measure the fracture toughness values for structural materials.[5] In the United States, these standard test methods are developed by the ASTM. Procedures and specimen configurations for most

---

[5]See, for example, ASTM Standard E399, "Standard Test Method for Linear–Elastic Plane–Strain Fracture Toughness K_{Ic} of Metallic Materials." (This testing technique is described in Section M.6 of the Mechanical Engineering Online Module.) Two other fracture toughness testing techniques are ASTM Standard E561-05E1, "Standard Test Method for K–R Curve Determinations," and ASTM Standard E1290-08, "Standard Test Method for Crack-Tip Opening Displacement (CTOD) Fracture Toughness Measurement."

tests are relatively complicated, and we will not attempt to provide detailed explanations. In brief, for each test type, the specimen (of specified geometry and size) contains a preexisting defect, usually a sharp crack that has been introduced. The test apparatus loads the specimen at a specified rate, and also measures load and crack displacement values. Data are subjected to analyses to ensure that they meet established criteria before the fracture toughness values are deemed acceptable. Most tests are for metals, but some have also been developed for ceramics, polymers, and composites.

Impact Testing Techniques

Prior to the advent of fracture mechanics as a scientific discipline, impact testing techniques were established to ascertain the fracture characteristics of materials at high loading rates. It was realized that the results of laboratory tensile tests (at low loading rates) could not be extrapolated to predict fracture behavior. For example, under some circumstances, normally ductile metals fracture abruptly and with very little plastic deformation under high loading rates. Impact test conditions were chosen to represent those most severe relative to the potential for fracture—namely, (1) deformation at a relatively low temperature, (2) a high strain rate (i.e., rate of deformation), and (3) a triaxial stress state (which may be introduced by the presence of a notch).

Charpy, Izod tests

impact energy

Two standardized tests,[6] the **Charpy** and the **Izod**, are used to measure the **impact energy** (sometimes also termed *notch toughness*). The Charpy V-notch (CVN) technique is most commonly used in the United States. For both the Charpy and the Izod, the specimen is in the shape of a bar of square cross section, into which a V-notch is machined (Figure 10.12*a*). The apparatus for making V-notch impact tests is illustrated schematically in Figure 10.12*b*. The load is applied as an impact blow from a weighted pendulum hammer released from a cocked position at a fixed height *h*. The specimen is positioned at the base as shown. Upon release, a knife edge mounted on the pendulum strikes and fractures the specimen at the notch, which acts as a point of stress concentration for this high-velocity impact blow. The pendulum continues its swing, rising to a maximum height *h'*, which is lower than *h*. The energy absorption, computed from the difference between *h* and *h'*, is a measure of the impact energy. The primary difference between the Charpy and the Izod techniques lies in the manner of specimen support, as illustrated in Figure 10.12*b*. These are termed *impact tests* because of the manner of load application. Several variables, including specimen size and shape as well as notch configuration and depth, influence the test results.

Both plane strain fracture toughness and these impact tests have been used to determine the fracture properties of materials. The former are quantitative in nature, in that a specific property of the material is determined (i.e., K_{Ic}). The results of the impact tests, however, are more qualitative and are of little use for design purposes. Impact energies are of interest mainly in a relative sense and for making comparisons—absolute values are of little significance. Attempts have been made to correlate plane strain fracture toughnesses and CVN energies, with only limited success. Plane strain fracture toughness tests are not as simple to perform as impact tests; furthermore, equipment and specimens are more expensive.

Ductile-to-Brittle Transition

ductile-to-brittle transition

One of the primary functions of the Charpy and the Izod tests is to determine whether a material experiences a **ductile-to-brittle transition** with decreasing temperature and, if so, the range of temperatures over which it occurs. As may be noted in the chapter-opening photograph of the fractured oil tanker for this chapter (also the transport

---

[6]ASTM Standard E23, "Standard Test Methods for Notched Bar Impact Testing of Metallic Materials."

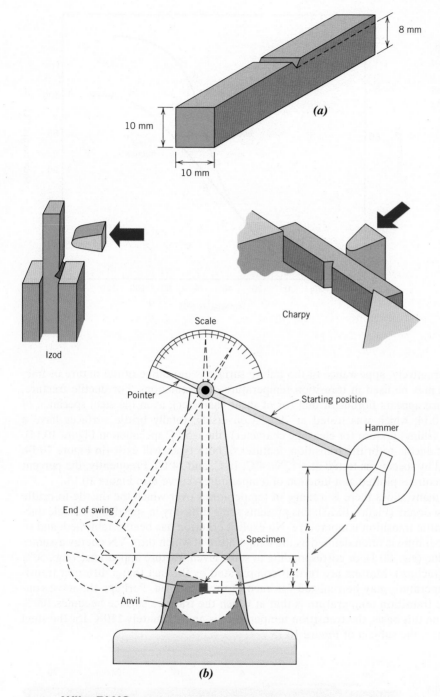

Figure 10.12 (*a*) Specimen used for Charpy and Izod impact tests. (*b*) A schematic drawing of an impact testing apparatus. The hammer is released from fixed height *h* and strikes the specimen; the energy expended in fracture is reflected in the difference between *h* and the swing height *h'*. Specimen placements for both the Charpy and the Izod tests are also shown.
[Figure (*b*) adapted from H. W. Hayden, W. G. Moffatt, and J. Wulff, *The Structure and Properties of Materials,* Vol. III, *Mechanical Behavior,* p. 13. Copyright © 1965 by John Wiley & Sons, New York.]

ship shown in Figure 1.3), widely used steels can exhibit this ductile-to-brittle transition with disastrous consequences. The ductile-to-brittle transition is related to the temperature dependence of the measured impact energy absorption. This transition is represented for a steel by curve *A* in Figure 10.13. At higher temperatures, the CVN energy is relatively large, corresponding to a ductile mode of fracture. As the temperature is lowered, the impact energy drops suddenly over a relatively narrow temperature range, below which the energy has a constant but small value—that is, the mode of fracture is brittle.

Figure 10.13 Temperature dependence of the Charpy V-notch impact energy (curve *A*) and percent shear fracture (curve *B*) for an A283 steel.
(Reprinted from *Welding Journal.* Used by permission of the American Welding Society.)

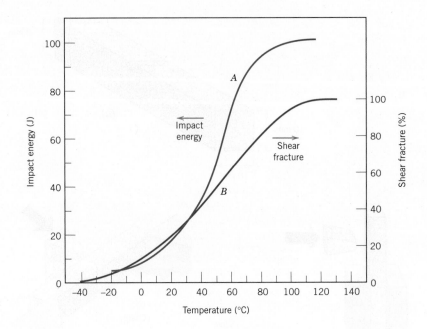

WileyPLUS

Tutorial Video:
Ductile-to-Brittle
Transition Failure
How Do I Interpret
the Ductile-to-Brittle
Transition Failure
Graphs and Equations?

Alternatively, appearance of the failure surface is indicative of the nature of fracture and may be used in transition temperature determinations. For ductile fracture, this surface appears fibrous or dull (or of shear character), as in the steel specimen of Figure 10.14, which was tested at 79°C. Conversely, totally brittle surfaces have a granular (shiny) texture (or cleavage character) (the −59°C specimen in Figure 10.14). Over the ductile-to-brittle transition, features of both types will exist (in Figure 10.14, displayed by specimens tested at −12°C, 4°C, 16°C, and 24°C). Frequently, the percent shear fracture is plotted as a function of temperature—curve *B* in Figure 10.13.

For many alloys there is a range of temperatures over which the ductile-to-brittle transition occurs (Figure 10.13); this presents some difficulty in specifying a single ductile-to-brittle transition temperature. No explicit criterion has been established, and so this temperature is often defined as the temperature at which the CVN energy assumes some value (e.g., 20 J), or corresponding to some given fracture appearance (e.g., 50% fibrous fracture). Matters are further complicated by the fact that a different transition temperature may be realized for each of these criteria. Perhaps the most conservative transition temperature is that at which the fracture surface becomes 100% fibrous; on this basis, the transition temperature is approximately 110°C for the steel alloy that is the subject of Figure 10.13.

Figure 10.14 Photograph of fracture surfaces of A36 steel Charpy V-notch specimens tested at indicated temperatures (in °C).
(From R. W. Hertzberg, *Deformation and Fracture Mechanics of Engineering Materials,* 3rd edition, Fig. 9.6, p. 329. Copyright © 1989 by John Wiley & Sons, Inc., New York. Reprinted by permission of John Wiley & Sons, Inc.)

Structures constructed from alloys that exhibit this ductile-to-brittle behavior should be used only at temperatures above the transition temperature to avoid brittle and catastrophic failure. Classic examples of this type of failure were discussed in the case study found in Chapter 1. During World War II, a number of welded transport ships away from combat suddenly split in half. The vessels were constructed of a steel alloy that possessed adequate toughness according to room-temperature tensile tests. The brittle fractures occurred at relatively low ambient temperatures, at about 4°C, in the vicinity of the transition temperature of the alloy. Each fracture crack originated at some point of stress concentration, probably a sharp corner or fabrication defect, and then propagated around the entire girth of the ship.

In addition to the ductile-to-brittle transition represented in Figure 10.13, two other general types of impact energy–versus–temperature behavior have been observed; these are represented schematically by the upper and lower curves of Figure 10.15. Here it may be noted that low-strength FCC metals (some aluminum and copper alloys) and most HCP metals do not experience a ductile-to-brittle transition (corresponding to the upper curve of Figure 10.15) and retain high impact energies (i.e., remain tough) with decreasing temperature. For high-strength materials (e.g., high-strength steels and titanium alloys), the impact energy is also relatively insensitive to temperature (the lower curve of Figure 10.15); however, these materials are also very brittle, as reflected by their low impact energies. The characteristic ductile-to-brittle transition is represented by the middle curve of Figure 10.15. As noted, this behavior is typically found in low-strength steels that have the BCC crystal structure.

For these low-strength steels, the transition temperature is sensitive to both alloy composition and microstructure. For example, decreasing the average grain size results in a lowering of the transition temperature. Hence, refining the grain size both strengthens (Section 9.8) and toughens steels. In contrast, increasing the carbon content, although it increases the strength of steels, also raises their CVN transition, as indicated in Figure 10.16.

Most ceramics and polymers also experience a ductile-to-brittle transition. For ceramic materials, the transition occurs only at elevated temperatures, ordinarily in excess of 1000°C. This behavior as related to polymers is discussed in Section 15.6.

WileyPLUS

Tutorial Video:
Impact Energy vs.
Temperature
and *S–N* Graph
Examples
How Do I Solve
Problems Using the
Impact Energy vs.
Temperature Graph?

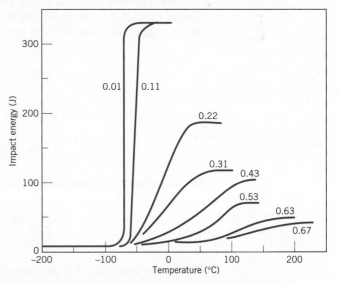

Figure 10.16 Influence of carbon content on the Charpy V-notch energy–versus–temperature behavior for steel.
(Reprinted with permission from ASM International, Materials Park, OH 44073-9989, USA; J. A. Reinbolt and W. J. Harris, Jr., "Effect of Alloying Elements on Notch Toughness of Pearlitic Steels," *Transactions of ASM*, Vol. 43, 1951.)

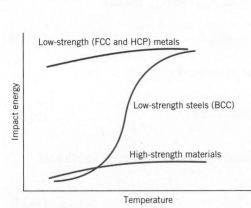

Figure 10.15 Schematic curves for the three general types of impact energy–versus–temperature behavior.

Fatigue

fatigue

Fatigue is a form of failure that occurs in structures subjected to dynamic and fluctuating stresses (e.g., bridges, aircraft, machine components). Under these circumstances, it is possible for failure to occur at a stress level considerably lower than the tensile or yield strength for a static load. The term *fatigue* is used because this type of failure normally occurs after a lengthy period of repeated stress or strain cycling. Fatigue is important inasmuch as it is the single largest cause of failure in metals, estimated to be involved in approximately 90% of all metallic failures; polymers and ceramics (except for glasses) are also susceptible to this type of failure. Furthermore, fatigue is catastrophic and insidious, occurring very suddenly and without warning.

Fatigue failure is brittle-like in nature even in normally ductile metals in that there is very little, if any, gross plastic deformation associated with failure. The process occurs by the initiation and propagation of cracks, and typically the fracture surface is perpendicular to the direction of an applied tensile stress.

10.7 CYCLIC STRESSES

WileyPLUS

Tutorial Video:
Cyclical Fatigue
Failure
What Is the Mechanism
of Cyclical Fatigue
Failure?

The applied stress may be axial (tension–compression), flexural (bending), or torsional (twisting) in nature. In general, three different fluctuating stress–time modes are possible. One is represented schematically by a regular and sinusoidal time dependence in Figure 10.17a, where the amplitude is symmetrical about a mean zero stress level, for example, alternating from a maximum tensile stress (σ_{max}) to a minimum compressive stress (σ_{min}) of equal magnitude; this is referred to as a *reversed stress cycle*. Another type, termed a *repeated stress cycle,* is illustrated in Figure 10.17b; the maxima and minima are asymmetrical relative to the zero stress level. Finally, the stress level may vary randomly in amplitude and frequency, as exemplified in Figure 10.17c.

Also indicated in Figure 10.17b are several parameters used to characterize the fluctuating stress cycle. The stress amplitude alternates about a *mean stress σ_m*, defined as the average of the maximum and minimum stresses in the cycle, or

Mean stress for cyclic loading—dependence on maximum and minimum stress levels

$$\sigma_m = \frac{\sigma_{max} + \sigma_{min}}{2} \tag{10.14}$$

The *range of stress σ_r* is the difference between σ_{max} and σ_{min}, namely,

Computation of range of stress for cyclic loading

$$\sigma_r = \sigma_{max} - \sigma_{min} \tag{10.15}$$

Stress amplitude σ_a is one-half of this range of stress, or

Computation of stress amplitude for cyclic loading

$$\sigma_a = \frac{\sigma_r}{2} = \frac{\sigma_{max} - \sigma_{min}}{2} \tag{10.16}$$

Finally, the *stress ratio R* is the ratio of minimum and maximum stress amplitudes:

Computation of stress ratio

$$R = \frac{\sigma_{min}}{\sigma_{max}} \tag{10.17}$$

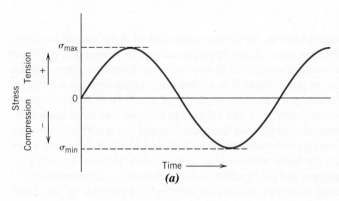

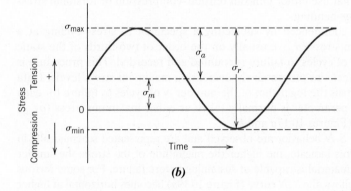

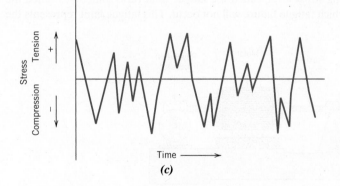

Figure 10.17 Variation of stress with time that accounts for fatigue failures. (*a*) Reversed stress cycle, in which the stress alternates from a maximum tensile stress (+) to a maximum compressive stress (−) of equal magnitude. (*b*) Repeated stress cycle, in which maximum and minimum stresses are asymmetrical relative to the zero-stress level; mean stress σ_m, range of stress σ_r, and stress amplitude σ_a are indicated. (*c*) Random stress cycle.

By convention, tensile stresses are positive and compressive stresses are negative. For example, for the reversed stress cycle, the value of R is −1.

Concept Check 10.2   Make a schematic sketch of a stress-versus-time plot for the situation when the stress ratio R has a value of +1.

Concept Check 10.3   Using Equations 10.16 and 10.17, demonstrate that increasing the value of the stress ratio R produces a decrease in stress amplitude σ_a.

[*The answers may be found at* www.wiley.com/college/callister (*Student Companion Site*).]

10.8 THE *S–N* CURVE

As with other mechanical characteristics, the fatigue properties of materials can be determined from laboratory simulation tests.[7] A test apparatus should be designed to duplicate as nearly as possible the service stress conditions (stress level, time frequency, stress pattern, etc.). The most common type of test conducted in a laboratory setting employs a rotating–bending beam: alternating tension and compression stresses of equal magnitude are imposed on the specimen as it is simultaneously bent and rotated. In this case, the stress cycle is reversed—that is, $R = -1$. Schematic diagrams of the apparatus and test specimen commonly used for this type of fatigue testing are shown in Figures 10.18a and 10.18b, respectively. From Figure 10.18a, during rotation, the lower surface of the specimen is subjected to a tensile (i.e., positive) stress, whereas the upper surface experiences compression (i.e., negative) stress.

Furthermore, anticipated in-service conditions may call for conducting simulated laboratory fatigue tests that use either uniaxial tension–compression or torsional stress cycling instead of rotating–bending.

A series of tests is commenced by subjecting a specimen to stress cycling at a relatively large maximum stress (σ_{max}), usually on the order of two-thirds of the static tensile strength; number of cycles to failure is counted and recorded. This procedure is repeated on other specimens at progressively decreasing maximum stress levels. Data are plotted as stress S versus the logarithm of the number N of cycles to failure for each of the specimens. The S parameter is normally taken as either maximum stress (σ_{max}) or stress amplitude (σ_a) (Figures 10.17a and b).

Two distinct types of *S–N* behavior are observed and are represented schematically in Figure 10.19. As these plots indicate, the higher the magnitude of the stress, the smaller the number of cycles the material is capable of sustaining before failure. For some ferrous (iron-base) and titanium alloys, the *S–N* curve (Figure 10.19a) becomes horizontal at higher

fatigue limit N values; there is a limiting stress level, called the **fatigue limit** (also sometimes called the *endurance limit*), below which fatigue failure will not occur. This fatigue limit represents the

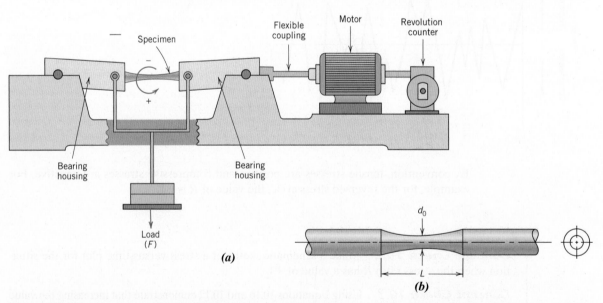

Figure 10.18 For rotating–bending fatigue tests, schematic diagrams of (*a*) a testing apparatus, and (*b*) a test specimen.

---

[7]See ASTM Standard E466, "Standard Practice for Conducting Force Controlled Constant Amplitude Axial Fatigue Tests of Metallic Materials," and ASTM Standard E468, "Standard Practice for Presentation of Constant Amplitude Fatigue Test Results for Metallic Materials."

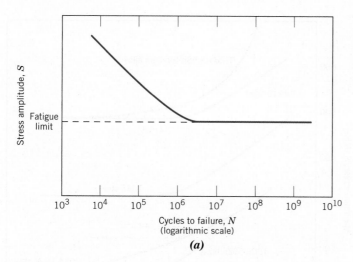

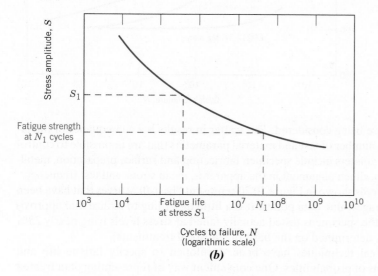

Figure 10.19 Stress amplitude (*S*) versus logarithm of the number of cycles to fatigue failure (*N*) for (*a*) a material that displays a fatigue limit and (*b*) a material that does not display a fatigue limit.

WileyPLUS

Tutorial Video:
Impact Energy vs.
Temperature
and *S–N* Graph
Examples
How Do I Interpret the
Cyclical Fatigue Failure
Graphs and Equations?

largest value of fluctuating stress that will *not* cause failure for essentially an infinite number of cycles. For many steels, fatigue limits range between 35% and 60% of the tensile strength.

Most nonferrous alloys (e.g., aluminum, copper) do not have a fatigue limit, in that the *S–N* curve continues its downward trend at increasingly greater *N* values (Figure 10.19*b*). Thus, fatigue ultimately occurs regardless of the magnitude of the stress. For these materials, the fatigue response is specified as **fatigue strength**, which is defined as the stress level at which failure will occur for some specified number of cycles (e.g., 10^7 cycles). The determination of fatigue strength is also demonstrated in Figure 10.19*b*.

Another important parameter that characterizes a material's fatigue behavior is **fatigue life** N_f. It is the number of cycles to cause failure at a specified stress level, as taken from the *S–N* plot (Figure 10.19*b*).

Fatigue *S–N* curves for several metal alloys are shown in Figure 10.20; data were generated using rotating-bending tests with reversed stress cycles (i.e., $R = -1$). Curves for the titanium, magnesium, and steel alloys as well as for cast iron display fatigue limits; curves for the brass and aluminum alloys do not have such limits.

Unfortunately, there always exists considerable scatter in fatigue data—that is, a variation in the measured *N* value for a number of specimens tested at the same stress level. This variation may lead to significant design uncertainties when fatigue life and/or fatigue

fatigue strength

fatigue life

Figure 10.20 Maximum stress (*S*) versus logarithm of the number of cycles to fatigue failure (*N*) for seven metal alloys. Curves were generated using rotating–bending and reversed-cycle tests. (Data taken from the following sources and reproduced with permission of ASM International, Materials Park, OH, 44073: *ASM Handbook, Vol. I, Properties and Selection: Irons, Steels, and High-Performance Alloys,* 1990; *ASM Handbook,* Vol. 2, *Properties and Selection; Nonferrous Alloys and Special-Purpose Materials,* 1990; G. M. Sinclair and W. J. Craig, "Influence of Grain Size on Work Hardening and Fatigue Characteristics of Alpha Brass," *Transactions of ASM,* Vol. 44, 1952.)

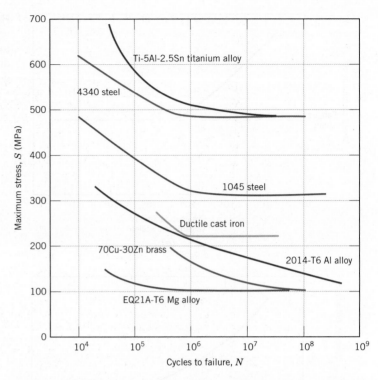

WileyPLUS

Tutorial Video:
Impact Energy vs.
Temperature
and *S–N* Graph
Examples
How Do I Solve
Problems Using the
S–N Graph?

limit (or strength) are being considered. The scatter in results is a consequence of the fatigue sensitivity to a number of test and material parameters that are impossible to control precisely. These parameters include specimen fabrication and surface preparation, metallurgical variables, specimen alignment in the apparatus, mean stress, and test frequency.

Fatigue *S–N* curves shown in Figure 10.20 represent "best-fit" curves that have been drawn through average-value data points. It is a little unsettling to realize that approximately one-half of the specimens tested actually failed at stress levels lying nearly 25% below the curve (as determined on the basis of statistical treatments).

Several statistical techniques have been developed to specify fatigue life and fatigue limit in terms of probabilities. One convenient way of representing data treated in this manner is with a series of constant probability curves, several of which are plotted in Figure 10.21. The *P* value associated with each curve represents the probability

Figure 10.21 Fatigue *S–N* probability of failure curves for a 7075-T6 aluminum alloy; *P* denotes the probability of failure. (From G. M. Sinclair and T. J. Dolan, *Trans. ASME,* 75, 1953, p. 867. Reprinted with permission of the American Society of Mechanical Engineers.)

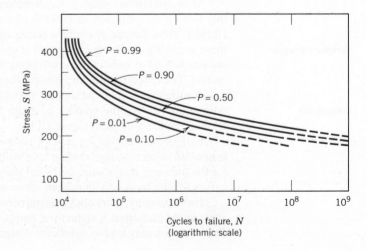

of failure. For example, at a stress of 200 MPa, we would expect 1% of the specimens to fail at about 10^6 cycles, 50% to fail at about 2×10^7 cycles, and so on. Remember that S–N curves represented in the literature are normally average values, unless noted otherwise.

The fatigue behaviors represented in Figures 10.19a and 10.19b may be classified into two domains. One is associated with relatively high loads that produce not only elastic strain but also some plastic strain during each cycle. Consequently, fatigue lives are relatively short; this domain is termed *low-cycle fatigue* and occurs at less than about 10^4 to 10^5 cycles. For lower stress levels wherein deformations are totally elastic, longer lives result. This is called *high-cycle fatigue* because relatively large numbers of cycles are required to produce fatigue failure. High-cycle fatigue is associated with fatigue lives greater than about 10^4 to 10^5 cycles.

EXAMPLE PROBLEM 10.2

Maximum Load Computation to Avert Fatigue for Rotating-Bending Tests

A cylindrical bar of 1045 steel having the S–N behavior shown in Figure 10.20 is subjected to rotating-bending tests with reversed-stress cycles (per Figure 10.18). If the bar diameter is 15.0 mm, determine the maximum cyclic load that may be applied to ensure that fatigue failure will not occur. Assume a factor of safety of 2.0 and that the distance between loadbearing points is 60.0 mm (0.0600 m).

Solution

From Figure 10.20, the 1045 steel has a fatigue limit (maximum stress) of magnitude 310 MPa. For a cylindrical bar of diameter d_0 (Figure 10.18b), maximum stress for rotating–bending tests may be determined using the following expression:

$$\sigma = \frac{16FL}{\pi d_0^3} \tag{10.18}$$

Here, L is equal to the distance between the two loadbearing points (Figure 10.18b), σ is the maximum stress (in our case the fatigue limit), and F is the maximum applied load. When σ is divided by the factor of safety (N), Equation 10.18 takes the form

$$\frac{\sigma}{N} = \frac{16FL}{\pi d_0^3} \tag{10.19}$$

and solving for F leads to

$$F = \frac{\sigma \pi d_0^3}{16NL} \tag{10.20}$$

Incorporating values for d_0, L, and N provided in the problem statement as well as the fatigue limit taken from Figure 10.20 (310 MPa, or 310×10^6 N/m²) yields the following:

$$F = \frac{(310 \times 10^6 \text{ N/m}^2)(\pi)(15 \times 10^{-3} \text{ m})^3}{(16)(2)(0.0600 \text{ m})}$$

$$= 1712 \text{ N}$$

Therefore, for cyclic reversed and rotating–bending, a maximum load of 1712 N may be applied without causing the 1045 steel bar to fail by fatigue.

EXAMPLE PROBLEM 10.3

Computation of Minimum Specimen Diameter to Yield a Specified Fatigue Lifetime for Tension-Compression Tests

A cylindrical 70Cu-30Zn brass bar (Figure 10.20) is subjected to axial tension–compression stress testing with reversed-cycling. If the load amplitude is 10,000 N, compute the minimum allowable bar diameter to ensure that fatigue failure will not occur at 10^7 cycles. Assume a factor of safety of 2.5, data in Figure 10.20 were taken for reversed axial tension–compression tests, and that S is stress amplitude.

Solution

From Figure 10.20, the fatigue strength for this alloy at 10^7 cycles is 115 MPa (115×10^6 N/m^2). Tensile and compressive stresses are defined in Equation 8.1 as

$$\sigma = \frac{F}{A_0} \tag{8.1}$$

Here, F is the applied load and A_0 is the cross-sectional area. For a cylindrical bar having a diameter of d_0,

$$A_0 = \pi \left(\frac{d_0}{2} \right)^2$$

Substitution of this expression for A_0 into Equation 8.1 leads to

$$\sigma = \frac{F}{A_0} = \frac{F}{\pi \left(\dfrac{d_0}{2} \right)^2} = \frac{4F}{\pi d_0^2} \tag{10.21}$$

We now solve for d_0, replacing stress with the fatigue strength divided by the factor of safety (i.e., σ/N). Thus,

$$d_0 = \sqrt{\frac{4F}{\pi \left(\dfrac{\sigma}{N} \right)}} \tag{10.22}$$

Incorporating values of F, N, and σ cited previously leads to

$$d_0 = \sqrt{\frac{(4)(10{,}000\text{ N})}{(\pi)\left(\dfrac{115 \times 10^6\text{ N/m}^2}{2.5} \right)}}$$

$$= 16.6 \times 10^{-3}\text{ m} = 16.6\text{ mm}$$

Hence, the brass bar diameter must be at least 16.6 mm to ensure that fatigue failure will not occur.

10.9 CRACK INITIATION AND PROPAGATION[8]

The process of fatigue failure is characterized by three distinct steps: (1) crack initiation, in which a small crack forms at some point of high stress concentration; (2) crack propagation, during which this crack advances incrementally with each stress cycle; and (3) final failure, which occurs very rapidly once the advancing crack has reached a critical size. Cracks associated with fatigue failure almost always initiate (or nucleate) on the

---

[8]More detailed and additional discussion on the propagation of fatigue cracks can be found in Sections M.10 and M.11 of the Mechanical Engineering (ME) Online Module.

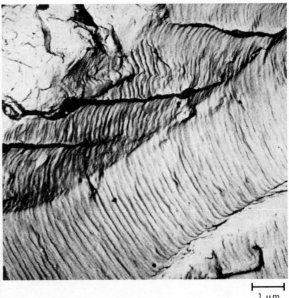

Figure 10.22 Fracture surface of a rotating steel shaft that experienced fatigue failure. Beachmark ridges are visible in the photograph.
(From D. J. Wulpi, *Understanding How Components Fail,* 1985. Reproduced by permission of ASM International, Materials Park, OH.)

Figure 10.23 Transmission electron fractograph showing fatigue striations in aluminum. 9000×.
(From V. J. Colangelo and F. A. Heiser, *Analysis of Metallurgical Failures,* 2nd edition. Copyright © 1987 by John Wiley & Sons, New York. Reprinted by permission of John Wiley & Sons, Inc.)

surface of a component at some point of stress concentration. Crack nucleation sites include surface scratches, sharp fillets, keyways, threads, dents, and the like. In addition, cyclic loading can produce microscopic surface discontinuities resulting from dislocation slip steps that may also act as stress raisers and therefore as crack initiation sites.

The region of a fracture surface that formed during the crack propagation step may be characterized by two types of markings termed *beachmarks* and *striations*. Both features indicate the position of the crack tip at some point in time and appear as concentric ridges that expand away from the crack initiation site(s), frequently in a circular or semi-circular pattern. Beachmarks (sometimes also called *clamshell marks*) are of macroscopic dimensions (Figure 10.22), and may be observed with the unaided eye. These markings are found for components that experienced interruptions during the crack propagation stage—for example, a machine that operated only during normal workshift hours. Each beachmark band represents a period of time over which crack growth occurred.

However, fatigue striations are microscopic in size and subject to observation with the electron microscope (either TEM or SEM). Figure 10.23 is an electron fractograph that shows this feature. Each striation is thought to represent the advance distance of a crack front during a single load cycle. Striation width depends on, and increases with, increasing stress range.

During the propagation of fatigue cracks and on a microscopic scale, there is very localized plastic deformation at crack tips, even though the maximum applied stress to which the object is exposed in each stress cycle lies below the yield strength of the metal. This applied stress is amplified at crack tips to the degree that local stress levels exceed the yield strength. The geometry of fatigue striations is a manifestation of this plastic deformation.[9]

It should be emphasized that although both beachmarks and striations are fatigue fracture surface features having similar appearances, they are nevertheless different in both origin and size. There may be thousands of striations within a single beachmark.

[9]The reader is referred to Section M.10 of the Mechanical Engineering (ME) Online Module, which explains and diagrams the proposed mechanism for the formation of fatigue striations.

Figure 10.24 Fatigue failure surface. A crack formed at the top edge. The smooth region also near the top corresponds to the area over which the crack propagated slowly. Rapid failure occurred over the area having a dull and fibrous texture (the largest area). Approximately 0.5×.
[From *Metals Handbook: Fractography and Atlas of Fractographs,* Vol. 9, 8th edition, H. E. Boyer (Editor), 1974. Reproduced by permission of ASM International, Materials Park, OH.]

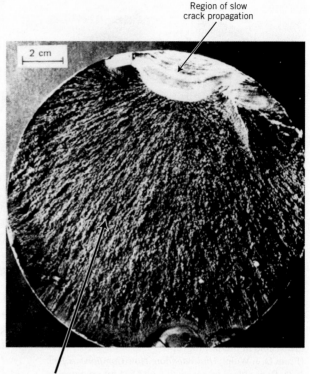

Region of slow
crack propagation

Region of rapid failure

Often the cause of failure may be deduced after examination of the failure surfaces. The presence of beachmarks and/or striations on a fracture surface confirms that the cause of failure was fatigue. Nevertheless, the absence of either or both does not exclude fatigue failure. Striations are not observed for all metals that experience fatigue. Furthermore, the likelihood of the appearance of striations may depend on stress state. Striation detectability decreases with the passage of time because of the formation of surface corrosion products and/or oxide films. Also, during stress cycling, striations may be destroyed by abrasive action as crack mating surfaces rub against one another.

One final comment regarding fatigue failure surfaces: Beachmarks and striations do not appear on the region over which the rapid failure occurs. Rather, the rapid failure may be either ductile or brittle; evidence of plastic deformation will be present for ductile failure and absent for brittle failure. This region of failure may be noted in Figure 10.24.

Concept Check 10.4 Surfaces for some steel specimens that have failed by fatigue have a bright crystalline or grainy appearance. Laymen may explain the failure by saying that the metal crystallized while in service. Offer a criticism for this explanation.

[*The answer may be found at* www.wiley.com/college/callister *(Student Companion Site).*]

10.10 FACTORS THAT AFFECT FATIGUE LIFE[10]

As mentioned in Section 10.8, the fatigue behavior of engineering materials is highly sensitive to a number of variables, including mean stress level, geometric design, surface effects, and metallurgical variables, as well as the environment. This section is devoted

---

[10]The case study on the automobile valve spring in Sections M.14 through M.16 of the Mechanical Engineering (ME) Online Module relates to the discussion of this section.

to a discussion of these factors and to measures that may be taken to improve the fatigue resistance of structural components.

Mean Stress

The dependence of fatigue life on stress amplitude is represented on the *S–N* plot. Such data are taken for a constant mean stress σ_m, often for the reversed cycle situation ($\sigma_m = 0$). Mean stress, however, also affects fatigue life; this influence may be represented by a series of *S–N* curves, each measured at a different σ_m, as depicted schematically in Figure 10.25. As may be noted, increasing the mean stress level leads to a decrease in fatigue life.

Surface Effects

For many common loading situations, the maximum stress within a component or structure occurs at its surface. Consequently, most cracks leading to fatigue failure originate at surface positions, specifically at stress amplification sites. Therefore, it has been observed that fatigue life is especially sensitive to the condition and configuration of the component surface. Numerous factors influence fatigue resistance, the proper management of which will lead to an improvement in fatigue life. These include design criteria as well as various surface treatments.

Design Factors

The design of a component can have a significant influence on its fatigue characteristics. Any notch or geometrical discontinuity can act as a stress raiser and fatigue crack initiation site; these design features include grooves, holes, keyways, threads, and so on. The sharper the discontinuity (i.e., the smaller the radius of curvature), the more severe the stress concentration. The probability of fatigue failure may be reduced by avoiding (when possible) these structural irregularities or by making design modifications by which sudden contour changes leading to sharp corners are eliminated—for example, calling for rounded fillets with large radii of curvature at the point where there is a change in diameter for a rotating shaft (Figure 10.26).

Surface Treatments

During machining operations, small scratches and grooves are invariably introduced into the workpiece surface by cutting-tool action. These surface markings can limit the fatigue life. It has been observed that improving the surface finish by polishing enhances fatigue life significantly.

One of the most effective methods of increasing fatigue performance is by imposing residual compressive stresses within a thin outer surface layer. Thus, a surface tensile

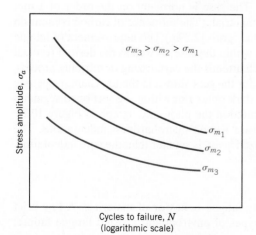

Figure 10.25 Demonstration of the influence of mean stress σ_m on *S–N* fatigue behavior.

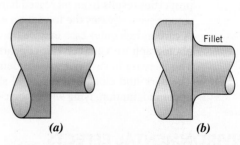

Figure 10.26 Demonstration of how design can reduce stress amplification. (*a*) Poor design: sharp corner. (*b*) Good design: fatigue lifetime is improved by incorporating a rounded fillet into a rotating shaft at the point where there is a change in diameter.

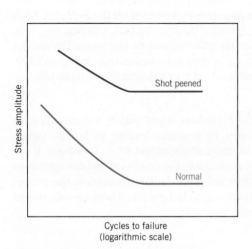

Figure 10.27 Schematic *S–N* fatigue curves for normal and shot-peened steel.

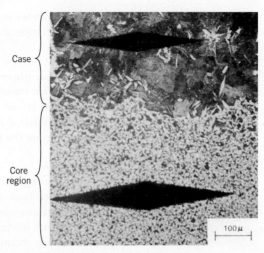

Figure 10.28 Photomicrograph showing both core (bottom) and carburized outer case (top) regions of a case-hardened steel. The case is harder, as attested by the smaller microhardness indentation. 100×. (From R. W. Hertzberg, *Deformation and Fracture Mechanics of Engineering Materials,* 3rd edition. Copyright © 1989 by John Wiley & Sons, New York. Reprinted by permission of John Wiley & Sons, Inc.)

stress of external origin is partially nullified and reduced in magnitude by the residual compressive stress. The net effect is that the likelihood of crack formation and therefore of fatigue failure is reduced.

Residual compressive stresses are commonly introduced into ductile metals mechanically by localized plastic deformation within the outer surface region. Commercially, this is often accomplished by a process termed *shot peening*. Small, hard particles (shot) having diameters within the range of 0.1 to 1.0 mm are projected at high velocities onto the surface to be treated. The resulting deformation induces compressive stresses to a depth of between one-quarter and one-half of the shot diameter. The influence of shot peening on the fatigue behavior of steel is demonstrated schematically in Figure 10.27.

case hardening   **Case hardening** is a technique by which both surface hardness and fatigue life are enhanced for steel alloys. This is accomplished by a carburizing or nitriding process by which a component is exposed to a carbonaceous or nitrogenous atmosphere at elevated temperature. A carbon- or nitrogen-rich outer surface layer (or *case*) is introduced by atomic diffusion from the gaseous phase. The case is normally on the order of 1 mm deep and is harder than the inner core of material. (The influence of carbon content on hardness for Fe–C alloys is demonstrated in Figure 12.29*a*.) The improvement of fatigue properties results from increased hardness within the case, as well as the desired residual compressive stresses the formation of which attends the carburizing or nitriding process. A carbon-rich outer case may be observed for the gear shown in the top chapter-opening photograph for Chapter 7; it appears as a dark outer rim within the sectioned segment. The increase in case hardness is demonstrated in the photomicrograph in Figure 10.28. The dark and elongated diamond shapes are Knoop microhardness indentations. The upper indentation, lying within the carburized layer, is smaller than the core indentation.

10.11 ENVIRONMENTAL EFFECTS

Environmental factors may also affect the fatigue behavior of materials. A few brief comments will be given relative to two types of environment-assisted fatigue failure: thermal fatigue and corrosion fatigue.

thermal fatigue

Thermal fatigue is normally induced at elevated temperatures by fluctuating thermal stresses; mechanical stresses from an external source need not be present. The origin of these thermal stresses is the restraint to the dimensional expansion and/or contraction that would normally occur in a structural member with variations in temperature. The magnitude of a thermal stress developed by a temperature change ΔT depends on the coefficient of thermal expansion α_l and the modulus of elasticity E according to

Thermal stress—dependence on coefficient of thermal expansion, modulus of elasticity, and temperature change

$$\sigma = \alpha_l E \, \Delta T \qquad (10.23)$$

(The topics of thermal expansion and thermal stresses are discussed in Sections 20.3 and 20.5.) Thermal stresses do not arise if this mechanical restraint is absent. Therefore, one obvious way to prevent this type of fatigue is to eliminate, or at least reduce, the restraint source, thus allowing unhindered dimensional changes with temperature variations, or to choose materials with appropriate physical properties.

corrosion fatigue

Failure that occurs by the simultaneous action of a cyclic stress and chemical attack is termed **corrosion fatigue**. Corrosive environments have a deleterious influence and produce shorter fatigue lives. Even normal ambient atmosphere affects the fatigue behavior of some materials. Small pits may form as a result of chemical reactions between the environment and the material, which may serve as points of stress concentration and therefore as crack nucleation sites. In addition, the crack propagation rate is enhanced as a result of the corrosive environment. The nature of the stress cycles influences the fatigue behavior; for example, lowering the load application frequency leads to longer periods during which the opened crack is in contact with the environment and to a reduction in the fatigue life.

Several approaches to corrosion fatigue prevention exist. On one hand, we can take measures to reduce the rate of corrosion by some of the techniques discussed in Chapter 18—for example, apply protective surface coatings, select a more corrosion-resistant material, and reduce the corrosiveness of the environment. On the other hand, it might be advisable to take actions to minimize the probability of normal fatigue failure, as outlined previously—for example, reduce the applied tensile stress level and impose residual compressive stresses on the surface of the member.

Creep

creep

Materials are often placed in service at elevated temperatures and exposed to static mechanical stresses (e.g., turbine rotors in jet engines and steam generators that experience centrifugal stresses; high-pressure steam lines). Deformation under such circumstances is termed **creep**. Defined as the time-dependent and permanent deformation of materials when subjected to a constant load or stress, creep is normally an undesirable phenomenon and is often the limiting factor in the lifetime of a part. It is observed in all materials types; for metals, it becomes important only for temperatures greater than about $0.4T_m$, where T_m is the absolute melting temperature. Amorphous polymers, which include plastics and rubbers, are especially sensitive to creep deformation, as discussed in Section 15.4.

10.12 GENERALIZED CREEP BEHAVIOR

A typical creep test[11] consists of subjecting a specimen to a constant load or stress while maintaining the temperature constant; deformation or strain is measured and plotted as a function of elapsed time. Most tests are the constant-load type, which yield information

---

[11]ASTM Standard E139, "Standard Test Methods for Conducting Creep, Creep-Rupture, and Stress-Rupture Tests of Metallic Materials."

of an engineering nature; constant-stress tests are employed to provide a better understanding of the mechanisms of creep.

Figure 10.29 is a schematic representation of the typical constant-load creep behavior of metals. Upon application of the load, there is an instantaneous deformation, as indicated in the figure, that is totally elastic. The resulting creep curve consists of three regions, each of which has its own distinctive strain–time feature. *Primary* or *transient creep* occurs first, typified by a continuously decreasing creep rate—that is, the slope of the curve decreases with time. This suggests that the material is experiencing an increase in creep resistance or strain hardening (Section 9.10)—deformation becomes more difficult as the material is strained. For *secondary creep,* sometimes termed *steady-state creep,* the rate is constant—that is, the plot becomes linear. This is often the stage of creep that is of the longest duration. The constancy of creep rate is explained on the basis of a balance between the competing processes of strain hardening and recovery, recovery (Section 9.11) being the process by which a material becomes softer and retains its ability to experience deformation. Finally, for *tertiary creep,* there is an acceleration of the rate and ultimate failure. This failure is frequently termed *rupture* and results from microstructural and/or metallurgical changes—for example, grain boundary separation, and the formation of internal cracks, cavities, and voids. Also, for tensile loads, a neck may form at some point within the deformation region. These all lead to a decrease in the effective cross-sectional area and an increase in strain rate.

For metallic materials, most creep tests are conducted in uniaxial tension using a specimen having the same geometry as for tensile tests (Figure 8.2). However, uniaxial compression tests are more appropriate for brittle materials; these provide a better measure of the intrinsic creep properties because there is no stress amplification and crack propagation, as with tensile loads. Compressive test specimens are usually right cylinders or parallelepipeds having length-to-diameter ratios ranging from about 2 to 4. For most materials, creep properties are virtually independent of loading direction.

Possibly the most important parameter from a creep test is the slope of the secondary portion of the creep curve ($\Delta\epsilon/\Delta t$ in Figure 10.29); this is often called the minimum or *steady-state creep rate* $\dot{\epsilon}_s$. It is the engineering design parameter that is considered for long-life applications, such as a nuclear power plant component that is scheduled to operate for several decades, and when failure or too much strain is not an option. However, for many relatively short-life creep situations (e.g., turbine blades in military aircraft and rocket motor nozzles), *time to rupture,* or the *rupture lifetime* t_r, is the dominant design consideration; it is also indicated in Figure 10.29. Of course, for its determination, creep tests must be conducted to the point of failure; these are termed *creep rupture* tests. Thus, knowledge of these creep characteristics of a material allows the design engineer to ascertain its suitability for a specific application.

Concept Check 10.5   Superimpose on the same strain-versus-time plot schematic creep curves for both constant tensile stress and constant tensile load, and explain the differences in behavior.

[*The answer may be found at* www.wiley.com/college/callister *(Student Companion Site).*]

10.13 STRESS AND TEMPERATURE EFFECTS

Both temperature and the level of the applied stress influence the creep characteristics (Figure 10.30). At a temperature substantially below $0.4T_m$, and after the initial deformation, the strain is virtually independent of time. With either increasing stress or temperature, the following will be noted: (1) the instantaneous strain at the time of stress application increases, (2) the steady-state creep rate increases, and (3) the rupture lifetime decreases.

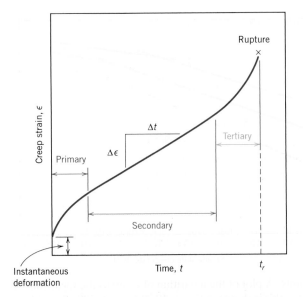

Figure 10.29 Typical creep curve of strain versus time at constant load and constant elevated temperature. The minimum creep rate $\Delta\epsilon/\Delta t$ is the slope of the linear segment in the secondary region. Rupture lifetime t_r is the total time to rupture.

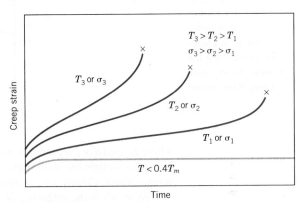

Figure 10.30 Influence of stress σ and temperature T on creep behavior.

The results of creep rupture tests are most commonly presented as the logarithm of stress versus the logarithm of rupture lifetime. Figure 10.31 is one such plot for an S-590 alloy in which a set of linear relationships can be seen to exist at each temperature. For some alloys and over relatively large stress ranges, nonlinearity in these curves is observed.

Empirical relationships have been developed in which the steady-state creep rate as a function of stress and temperature is expressed. Its dependence on stress can be written

Dependence of creep strain rate on stress

$$\dot{\epsilon}_s = K_1\sigma^n \qquad (10.24)$$

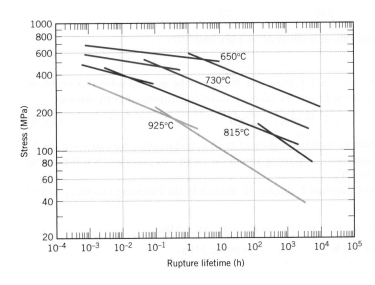

Figure 10.31 Stress (logarithmic scale) versus rupture lifetime (logarithmic scale) for an S-590 alloy at four temperatures. [The composition (in wt%) of S-590 is as follows: 20.0 Cr, 19.4 Ni, 19.3 Co, 4.0 W, 4.0 Nb, 3.8 Mo, 1.35 Mn, 0.43 C, and the balance Fe.]
(Reprinted with permission of ASM International.® All rights reserved. www.asminternational.org)

WileyPLUS

Tutorial Video:

Creep Examples
How Do I Solve Problems Using the Stress vs. Rupture Lifetime Graph?

Figure 10.33 Logarithm of stress versus the
Larson–Miller parameter for an S-590 alloy.
(From F. R. Larson and J. Miller, *Trans. ASME*, **74**, 765, 1952.
Reprinted by permission of ASME.)

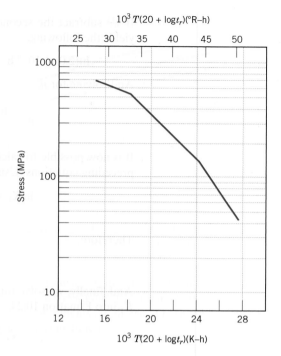

where C is a constant (usually on the order of 20), for T in Kelvin and the rupture lifetime t_r in hours. The rupture lifetime of a given material measured at some specific stress level varies with temperature such that this parameter remains constant. Alternatively, the data may be plotted as the logarithm of stress versus the Larson–Miller parameter, as shown in Figure 10.33. Use of this technique is demonstrated in the following design example.

DESIGN EXAMPLE 10.2

Rupture Lifetime Prediction

Using the Larson–Miller data for the S-590 alloy shown in Figure 10.33, predict the time to rupture for a component that is subjected to a stress of 140 MPa at 800°C (1073 K).

Solution

From Figure 10.33, at 140 MPa the value of the Larson–Miller parameter is 24.0×10^3 for T in K and t_r in h; therefore,

$$24.0 \times 10^3 = T(20 + \log t_r)$$
$$= 1073(20 + \log t_r)$$

and, solving for the time to rupture, we obtain

$$22.37 = 20 + \log t_r$$
$$t_r = 233 \text{ h (9.7 days)}$$

10.15 ALLOYS FOR HIGH-TEMPERATURE USE

Several factors affect the creep characteristics of metals. These include melting temperature, elastic modulus, and grain size. In general, the higher the melting temperature, the greater the elastic modulus; the larger the grain size, the better a material's resistance to

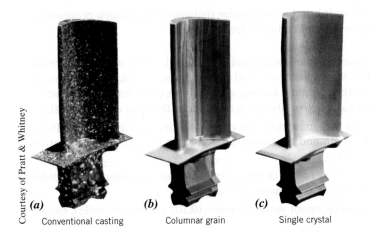

(a)   *(b)*   *(c)*

Conventional casting   Columnar grain   Single crystal

Figure 10.34   (*a*) Polycrystalline turbine blade that was produced by a conventional casting technique. High-temperature creep resistance is improved as a result of an oriented columnar grain structure (*b*) produced by a sophisticated directional solidification technique. Creep resistance is further enhanced when single-crystal blades (*c*) are used.

creep. Relative to grain size, smaller grains permit more grain boundary sliding, which results in higher creep rates. This effect may be contrasted to the influence of grain size on the mechanical behavior at low temperatures [i.e., increase in both strength (Section 9.8) and toughness (Section 10.6)].

Stainless steels (Section 13.2) and the superalloys (Section 13.9) are especially resilient to creep and are commonly employed in high-temperature service applications. The creep resistance of the superalloys is enhanced by solid-solution alloying and also by the formation of precipitate phases. In addition, advanced processing techniques have been utilized; one such technique is directional solidification, which produces either highly elongated grains or single-crystal components (Figure 10.34).

SUMMARY

Introduction
- The three usual causes of failure are
  - Improper materials selection and processing
  - Inadequate component design
  - Component misuse

Fundamentals of Fracture
- Fracture in response to tensile loading and at relatively low temperatures may occur by ductile and brittle modes.
- Ductile fracture is normally preferred because
  - Preventive measures may be taken inasmuch as evidence of plastic deformation indicates that fracture is imminent.
  - More energy is required to induce ductile fracture than for brittle fracture.
- Cracks in ductile materials are said to be *stable* (i.e., resist extension without an increase in applied stress).
- For brittle materials, cracks are *unstable*—that is, crack propagation, once started, continues spontaneously without an increase in stress level.

Ductile Fracture
- For ductile metals, two tensile fracture profiles are possible:
  - Necking down to a point fracture when ductility is high (Figure 10.1*a*)
  - Only moderate necking with a cup-and-cone fracture profile (Figure 10.1*b*) when the material is less ductile

Brittle Fracture
- For *brittle fracture*, the fracture surface is relatively flat and perpendicular to the direction of the applied tensile load (Figure 10.1*c*).

Data Extrapolation Methods
- Extrapolation of creep test data to lower-temperature/longer-time regimes is possible using a plot of logarithm of stress versus the Larson–Miller parameter for the particular alloy (Figure 10.33).

Alloys for High-Temperature Use
- Metal alloys that are especially resistant to creep have high elastic moduli and melting temperatures; these include the superalloys, the stainless steels, and the refractory metals. Various processing techniques are employed to improve the creep properties of these materials.

Important Terms and Concepts

brittle fracture
case hardening
Charpy test
corrosion fatigue
creep
ductile fracture
ductile-to-brittle transition
fatigue

fatigue life
fatigue limit
fatigue strength
fracture mechanics
fracture toughness
impact energy
intergranular fracture

Izod test
plane strain
plane strain fracture
  toughness
stress raiser
thermal fatigue
transgranular fracture

REFERENCES

ASM Handbook, Vol. 11, *Failure Analysis and Prevention,* ASM International, Materials Park, OH, 2002.

ASM Handbook, Vol. 12, *Fractography,* ASM International, Materials Park, OH, 1987.

ASM Handbook, Vol. 19, *Fatigue and Fracture,* ASM International, Materials Park, OH, 1996.

Boyer, H. E. (Editor), *Atlas of Creep and Stress–Rupture Curves,* ASM International, Materials Park, OH, 1988.

Boyer, H. E. (Editor), *Atlas of Fatigue Curves,* ASM International, Materials Park, OH, 1986.

Brooks, C. R., and A. Choudhury, *Failure Analysis of Engineering Materials,* McGraw-Hill, New York, 2002.

Colangelo, V. J., and F. A. Heiser, *Analysis of Metallurgical Failures,* 2nd edition, Wiley, New York, 1987.

Collins, J. A., *Failure of Materials in Mechanical Design,* 2nd edition, Wiley, New York, 1993.

Dennies, D. P., *How to Organize and Run a Failure Investigation,* ASM International, Materials Park, OH, 2005.

Dieter, G. E., *Mechanical Metallurgy,* 3rd edition, McGraw-Hill, New York, 1986.

Esaklul, K. A., *Handbook of Case Histories in Failure Analysis,* ASM International, Materials Park, OH, 1992 and 1993. In two volumes.

Hertzberg, R. W., R. P. Vinci, and J. L. Hertzberg, *Deformation and Fracture Mechanics of Engineering Materials,* 5th edition, Wiley, Hoboken, NJ, 2013.

Liu, A. F., *Mechanics and Mechanisms of Fracture: An Introduction,* ASM International, Materials Park, OH, 2005.

McEvily, A. J., *Metal Failures: Mechanisms, Analysis, Prevention,* Wiley, New York, 2002.

Stevens, R. I., A. Fatemi, R. R. Stevens, and H. O. Fuchs, *Metal Fatigue in Engineering,* 2nd edition, Wiley, New York, 2000.

Wulpi, D. J., *Understanding How Components Fail,* 2nd edition, ASM International, Materials Park, OH, 1999.

QUESTIONS AND PROBLEMS

Principles of Fracture Mechanics

10.1 What is the magnitude of the maximum stress that exists at the tip of an internal crack having a radius of curvature of 2.5×10^{-4} mm and a crack length of 2.6×10^{-2} mm when a tensile stress of 170 MPa is applied?

10.2 Estimate the theoretical fracture strength of a brittle material if it is known that fracture occurs by the propagation of an elliptically shaped surface crack of length 0.28 mm and having a tip radius of curvature of 1.2×10^{-3} mm when a stress of 1200 MPa is applied.

10.3 If the specific surface energy for soda-lime glass is 0.30 J/m², using data contained in Table 14.1, compute the critical stress required for the propagation of a surface crack of length 0.05 mm.

10.4 A polystyrene component must not fail when a tensile stress of 1.25 MPa is applied. Determine the maximum allowable surface crack length if the surface energy of polystyrene is 0.50 J/m². Assume a modulus of elasticity of 3.0 GPa.

10.5 A specimen of a 4340 steel alloy having a plane strain fracture toughness of 45 MPa$\sqrt{m}$ is exposed to a stress of 1000 MPa. Will this specimen experience fracture if it is known that the largest surface crack is 0.76 mm long? Why or why not? Assume that the parameter Y has a value of 1.0.

10.6 An aircraft component is fabricated from an aluminum alloy that has a plane strain fracture toughness of 35 MPa$\sqrt{m}$. It has been determined that fracture results at a stress of 250 MPa when the maximum (or critical) internal crack length is 2.0 mm. For this same component and alloy, will fracture occur at a stress level of 325 MPa when the maximum internal crack length is 1.1 mm? Why or why not?

10.7 Suppose that a wing component on an aircraft is fabricated from an aluminum alloy that has a plane strain fracture toughness of 40 MPa$\sqrt{m}$. It has been determined that fracture results at a stress of 365 MPa when the maximum internal crack length is 2.6 mm. For this same component and alloy, compute the stress level at which fracture will occur for a critical internal crack length of 4.0 mm.

10.8 A large plate is fabricated from a steel alloy that has a plane strain fracture toughness of 55 MPa$\sqrt{m}$. If, during service use, the plate is exposed to a tensile stress of 200 MPa, determine the minimum length of a surface crack that will lead to fracture. Assume a value of 1.0 for Y.

10.9 Calculate the maximum internal crack length allowable for a 7075-T651 aluminum alloy (Table 10.1) component that is loaded to a stress one-half of its yield strength. Assume that the value of Y is 1.35.

10.10 A structural component in the form of a wide plate is to be fabricated from a steel alloy that has a plane strain fracture toughness of 77.0 MPa$\sqrt{m}$ and a yield strength of 1400 MPa. The flaw size resolution limit of the flaw detection apparatus is 4.1 mm. If the design stress is one-half of the yield strength and the value of Y is 1.0, determine whether a critical flaw for this plate is subject to detection.

10.11 After consultation of other references, write a brief report on one or two nondestructive test techniques that are used to detect and measure internal and/or surface flaws in metal alloys.

Fracture Toughness Testing

10.12 Following is tabulated data that were gathered from a series of Charpy impact tests on a ductile cast iron.

Temperature (°C)	Impact Energy (J)
−25	124
−50	123
−75	115
−85	100
−100	73
−110	52
−125	26
−150	9
−175	6

(a) Plot the data as impact energy versus temperature.

(b) Determine a ductile-to-brittle transition temperature as that temperature corresponding to the average of the maximum and minimum impact energies.

(c) Determine a ductile-to-brittle transition temperature as that temperature at which the impact energy is 80 J.

10.13 Following is tabulated data that were gathered from a series of Charpy impact tests on a tempered 4140 steel alloy.

Temperature (°C)	Impact Energy (J)
100	89.3
75	88.6
50	87.6
25	85.4
0	82.9
−25	78.9
−50	73.1
−65	66.0
−75	59.3
−85	47.9
−100	34.3
−125	29.3
−150	27.1
−175	25.0

(a) Plot the data as impact energy versus temperature.

(b) Determine a ductile-to-brittle transition temperature as that temperature corresponding to the average of the maximum and minimum impact energies.

(c) Determine a ductile-to-brittle transition temperature as that temperature at which the impact energy is 70 J.

Cyclic Stresses
The S–N Curve

10.14 A fatigue test was conducted in which the mean stress was 50 MPa and the stress amplitude was 225 MPa.

(a) Compute the maximum and minimum stress levels.

(b) Compute the stress ratio.

(c) Compute the magnitude of the stress range.

10.15 A cylindrical 1045 steel bar (Figure 10.20) is subjected to repeated tension–compression stress cycling along its axis. If the load amplitude is 22,000 N, compute the minimum allowable bar diameter to ensure that fatigue failure will not occur. Assume a factor of safety of 2.0.

10.16 An 8.0-mm diameter cylindrical rod fabricated from a red brass alloy (Figure 10.20) is subjected to reversed tension–compression load cycling along its axis. If the maximum tensile and compressive loads are +7500 N and −7500 N, respectively, determine its fatigue life. Assume that the stress plotted in Figure 10.20 is stress amplitude.

10.17 A 12.5-mm diameter cylindrical rod fabricated from a 2014-T6 alloy (Figure 10.20) is subjected to a repeated tension– compression load cycling along its axis. Compute the maximum and minimum loads that will be applied to yield a fatigue life of 1.0×10^7 cycles. Assume that the stress plotted on the vertical axis is stress amplitude, and data were taken for a mean stress of 50 MPa.

10.18 The fatigue data for a brass alloy are given as follows:

Stress Amplitude (MPa)	Cycles to Failure
310	2×10^5
223	1×10^6
191	3×10^6
168	1×10^7
153	3×10^7
143	1×10^8
134	3×10^8
127	1×10^9

(a) Make an S–N plot (stress amplitude versus logarithm cycles to failure) using these data.

(b) Determine the fatigue strength at 5×10^5 cycles.

(c) Determine the fatigue life for 200 MPa.

10.19 Suppose that the fatigue data for the brass alloy in Problem 10.18 were taken from torsional tests, and that a shaft of this alloy is to be used for a coupling that is attached to an electric motor operating at 1500 rpm. Give the maximum torsional stress amplitude possible for each of the following lifetimes of the coupling: (a) 1 year, (b) 1 month, (c) 1 day, and (d) 2 hours.

10.20 The fatigue data for a ductile cast iron are given as follows:

Stress Amplitude [MPa]	Cycles to Failure
248	1×10^5
236	3×10^5
224	1×10^6
213	3×10^6
201	1×10^7
193	3×10^7
193	1×10^8
193	3×10^8

(a) Make an S–N plot (stress amplitude versus logarithm cycles to failure) using these data.

(b) What is the fatigue limit for this alloy?

(c) Determine fatigue lifetimes at stress amplitudes of 230 MPa and 175 MPa.

(d) Estimate fatigue strengths at 2×10^5 and 6×10^6 cycles.

10.21 Suppose that the fatigue data for the cast iron in Problem 10.20 were taken for bending-rotating tests, and that a rod of this alloy is to be used for an automobile axle that rotates at an average rotational velocity of 750 revolutions per minute. Give maximum lifetimes of continuous driving that are allowable for the following stress levels: (a) 250 MPa, (b) 215 MPa, (c) 200 MPa, and (d) 150 MPa.

10.22 Three identical fatigue specimens (denoted A, B, and C) are fabricated from a nonferrous alloy. Each is subjected to one of the maximum–minimum stress cycles listed in the following table; the frequency is the same for all three tests.

Specimen	σ_{max} (MPa)	σ_{min} (MPa)
A	1450	2350
B	1400	2300
C	+340	−340

(a) Rank the fatigue lifetimes of these three specimens from the longest to the shortest.

(b) Now justify this ranking using a schematic *S–N* plot.

10.23 Cite five factors that may lead to scatter in fatigue life data.

Crack Initiation and Propagation
Factors That Affect Fatigue Life

10.24 Briefly explain the difference between fatigue striations and beachmarks in terms of both **(a)** size and **(b)** origin.

10.25 List four measures that may be taken to increase the resistance to fatigue of a metal alloy.

Generalized Creep Behavior

10.26 Give the approximate temperature at which creep deformation becomes an important consideration for each of the following metals: nickel, copper, iron, tungsten, lead, and aluminum.

10.27 The following creep data were taken on an aluminum alloy at 400°C (673 K) and a constant stress of 25 MPa. Plot the data as strain versus time, then determine the steady-state or minimum creep rate. *Note:* The initial and instantaneous strain is not included.

Time (min)	Strain	Time (min)	Strain
0	0.000	16	0.135
2	0.025	18	0.153
4	0.043	20	0.172
6	0.065	22	0.193
8	0.078	24	0.218
10	0.092	26	0.255
12	0.109	28	0.307
14	0.120	30	0.368

Stress and Temperature Effects

10.28 A specimen 760 mm long of an S-590 alloy (Figure 10.32) is to be exposed to a tensile stress of 80 MPa at 815°C (1088 K). Determine its elongation after 5000 h. Assume that the total of both instantaneous and primary creep elongations is 1.5 mm.

10.29 For a cylindrical S-590 alloy specimen (Figure 10.32) originally 10 mm in diameter and 505 mm long, what tensile load is necessary to produce a total elongation of 145 mm after 2,000 h at 730°C (1003 K)? Assume that the sum of instantaneous and primary creep elongations is 8.6 mm.

10.30 If a component fabricated from an S-590 alloy (Figure 10.31) is to be exposed to a tensile stress of 300 MPa at 650°C (923 K), estimate its rupture lifetime.

10.31 A cylindrical component constructed from an S-590 alloy (Figure 10.31) has a diameter of 12.5 mm. Determine the maximum load that may be applied for it to survive 500 h at 925°C (1198 K).

10.32 From Equation 10.24, if the logarithm of $\dot{\epsilon}_s$ is plotted versus the logarithm of σ, then a straight line should result, the slope of which is the stress exponent n. Using Figure 10.32, determine the value of n for the S-590 alloy at 925°C, and for the initial (i.e., lower-temperature) straight line segments at each of 650°C, 730°C, and 815°C.

10.33 **(a)** Estimate the activation energy for creep (i.e., Q_c in Equation 10.25) for the S-590 alloy having the steady-state creep behavior shown in Figure 10.32. Use data taken at a stress level of 300 MPa and temperatures of 650°C and 730°C. Assume that the stress exponent n is independent of temperature. **(b)** Estimate $\dot{\epsilon}_s$ at 600°C (873 K) and 300 MPa.

10.34 Steady-state creep rate data are given in the following table for nickel at 1000°C (1273 K):

$\dot{\epsilon}_s$ (s^{-1})	σ [MPa]
10^{-4}	15
10^{-6}	4.5

If it is known that the activation energy for creep is 272,000 J/mol, compute the steady-state creep rate at a temperature of 850°C (1123 K) and a stress level of 25 MPa.

10.35 Steady-state creep data taken for a stainless steel at a stress level of 70 MPa are given as follows:

$\dot{\epsilon}_s$ (s^{-1})	T (K)
1.0×10^{-5}	977
2.5×10^{-3}	1089

If it is known that the value of the stress exponent n for this alloy is 7.0, compute the steady-state creep rate at 1250 K and a stress level of 50 MPa.

Alloys for High–Temperature Use

10.36 Cite three metallurgical/processing techniques that are employed to enhance the creep resistance of metal alloys.

10.1SS Given a set of fatigue stress amplitude and cycles-to-failure data, develop a spreadsheet that will allow the user to generate an *S*-versus-log *N* plot.

10.2SS Given a set of creep strain and time data, develop a spreadsheet that will allow the user to generate a strain-versus-time plot and then compute the steady-state creep rate.

DESIGN PROBLEMS

10.D1 Each student (or group of students) is to obtain an object/structure/component that has failed. It may come from your home, an automobile repair shop, a machine shop, and so on. Conduct an investigation to determine the cause and type of failure (i.e., simple fracture, fatigue, creep). In addition, propose measures that can be taken to prevent future incidents of this type of failure. Finally, submit a report that addresses these issues.

Principles of Fracture Mechanics

10.D2 (a) For the thin-walled spherical tank discussed in Design Example 10.1, on the basis of critical crack size criterion [as addressed in part (a)], rank the following polymers from longest to shortest critical crack length: nylon 6,6 (50% relative humidity), polycarbonate, poly(ethylene terephthalate), and poly(methyl methacrylate). Comment on the magnitude range of the computed values used in the ranking relative to those tabulated for metal alloys as provided in Table 10.3. For these computations, use data contained in Tables B.4 and B.5 in Appendix B.

(b) Now rank these same four polymers relative to maximum allowable pressure according to the leak-before-break criterion, as described in the (b) portion of Design Example 10.1. As previously, comment on these values in relation to those for the metal alloys that are tabulated in Table 10.4.

Data Extrapolation Methods

10.D3 An S-590 alloy component (Figure 10.33) must have a creep rupture lifetime of at least 100 days at 500°C (773 K). Compute the maximum allowable stress level.

10.D4 Consider an S-590 alloy component (Figure 10.33) that is subjected to a stress of 200 MPa. At what temperature will the rupture lifetime be 500 h?

10.D5 For an 18-8 Mo stainless steel (Figure 10.34), predict the time to rupture for a component that is subjected to a stress of 80 MPa at 700°C (973 K).

10.D6 Consider an 18-8 Mo stainless steel component (Figure 10.34) that is exposed to a temperature of 500°C (773 K). What is the maximum allowable stress level for a rupture lifetime of 5 years? 20 years?

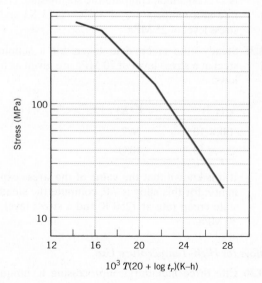

Figure 10.34 Logarithm stress versus the Larson–Miller parameter for an 18-8 Mo stainless steel. (From F. R. Larson and J. Miller, *Trans. ASME,* **74,** 765, 1952. Reprinted by permission of ASME.)

Chapter 11 Phase Diagrams

The accompanying graph is the phase diagram for pure H_2O. Parameters plotted are external pressure (vertical axis, scaled logarithmically) versus temperature. In a sense this diagram is a map in which regions for the three familiar phases—solid (ice), liquid (water), and vapor (steam)—are delineated. The three red curves represent phase boundaries that define the regions. A photograph located in each region shows an example of its phase—ice cubes, liquid water being poured into a glass, and steam spewing from a kettle. (Photographs left to right: © AlexStar/iStockphoto, © Canbalci/iStockphoto, © IJzendoorn/iStockphoto.)

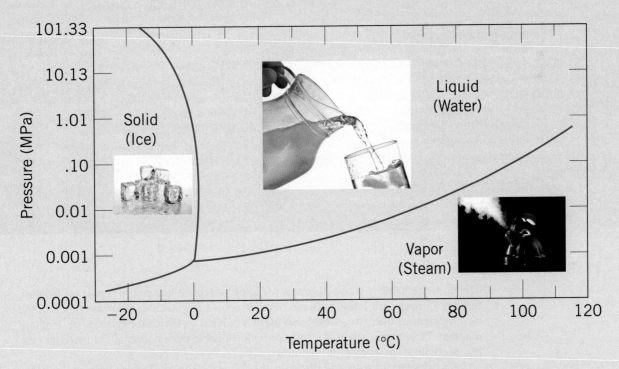

Three phases for the H_2O system are shown in this photograph: ice (the iceberg), water (the ocean or sea), and vapor (the clouds). These three phases are not in equilibrium with one another.

© Achim Baqué/Stockphoto/

One reason that a knowledge and understanding of phase diagrams is important to the engineer relates to the design and control of heat-treating procedures; some properties of materials are functions of their microstructures and, consequently, of their thermal histories. Even though most phase diagrams represent stable (or equilibrium) states and microstructures, they are nevertheless useful in understanding the development and preservation of nonequilibrium structures and their attendant properties; it is often the case that these properties are more desirable than those associated with the equilibrium state. This is aptly illustrated by the phenomenon of precipitation hardening (Section 17.7).

Learning Objectives

After studying this chapter, you should be able to do the following:

1. (a) Schematically sketch simple isomorphous and eutectic phase diagrams.
   (b) On these diagrams, label the various phase regions.
   (c) Label liquidus, solidus, and solvus lines.
2. Given a binary phase diagram, the composition of an alloy, and its temperature; and assuming that the alloy is at equilibrium, determine the following:
   (a) what phase(s) is (are) present,
   (b) the composition(s) of the phase(s), and
   (c) the mass fraction(s) of the phase(s).
3. For some given binary phase diagram, do the following:
   (a) locate the temperatures and compositions of all eutectic, eutectoid, peritectic, and congruent phase transformations; and
   (b) write reactions for all these transformations for either heating or cooling.
4. Given the composition of an iron–carbon alloy containing between 0.022 and 2.14 wt% C, be able to
   (a) specify whether the alloy is hypoeutectoid or hypereutectoid,
   (b) name the proeutectoid phase,
   (c) compute the mass fractions of proeutectoid phase and pearlite, and
   (d) make a schematic diagram of the microstructure at a temperature just below the eutectoid.

11.1 INTRODUCTION

The understanding of phase diagrams for alloy systems is extremely important because there is a strong correlation between microstructure and mechanical properties, and the development of microstructure of an alloy is related to the characteristics of its phase diagram. In addition, phase diagrams provide valuable information about melting, casting, crystallization, and other phenomena.

This chapter presents and discusses the following topics: (1) terminology associated with phase diagrams and phase transformations; (2) pressure–temperature phase diagrams for pure materials; (3) the interpretation of phase diagrams; (4) some of the common and relatively simple binary phase diagrams, including that for the iron–carbon system; and (5) the development of equilibrium microstructures upon cooling for several situations.

Definitions and Basic Concepts

It is necessary to establish a foundation of definitions and basic concepts relating to alloys, phases, and equilibrium before delving into the interpretation and utilization of phase diagrams. The term **component** is frequently used in this discussion; components are pure metals and/or compounds of which an alloy is composed. For example, in a copper–zinc brass, the components are Cu and Zn. *Solute* and *solvent,*

component

system

which are also common terms, were defined in Section 6.4. Another term used in this context is **system**, which has two meanings. *System* may refer to a specific body of material under consideration (e.g., a ladle of molten steel); or it may relate to the series of possible alloys consisting of the same components but without regard to alloy composition (e.g., the iron–carbon system).

The concept of a solid solution was introduced in Section 6.4. To review, a solid solution consists of atoms of at least two different types; the solute atoms occupy either substitutional or interstitial positions in the solvent lattice, and the crystal structure of the solvent is maintained.

11.2 SOLUBILITY LIMIT

solubility limit

For many alloy systems and at some specific temperature, there is a maximum concentration of solute atoms that may dissolve in the solvent to form a solid solution; this is called a **solubility limit**. The addition of solute in excess of this solubility limit results in the formation of another solid solution or compound that has a distinctly different composition. To illustrate this concept, consider the sugar–water ($C_{12}H_{22}O_{11}$–H_2O) system. Initially, as sugar is added to water, a sugar–water solution or syrup forms. As more sugar is introduced, the solution becomes more concentrated, until the solubility limit is reached or the solution becomes saturated with sugar. At this time, the solution is not capable of dissolving any more sugar, and further additions simply settle to the bottom of the container. Thus, the system now consists of two separate substances: a sugar–water syrup liquid solution and solid crystals of undissolved sugar.

WileyPLUS

Tutorial Video:
Phases and
Solubility Limits
What Is a
Solubility Limit?

This solubility limit of sugar in water depends on the temperature of the water and may be represented in graphical form on a plot of temperature along the ordinate and composition (in weight percent sugar) along the abscissa, as shown in Figure 11.1. Along the composition axis, increasing sugar concentration is from left to right, and percentage of water is read from right to left. Because only two components are involved (sugar and water), the sum of the concentrations at any composition will equal 100 wt%. The solubility limit is represented as the nearly vertical line in the figure. For compositions and temperatures to the left of the solubility line, only the syrup liquid solution exists; to the right of the line, syrup and solid sugar coexist. The solubility limit at some temperature is the composition that corresponds to the intersection of the given temperature coordinate and the solubility limit line. For example, at 20°C, the maximum solubility of sugar in water is 65 wt%. As Figure 11.1 indicates, the solubility limit increases slightly with rising temperature.

Figure 11.1 The solubility of sugar ($C_{12}H_{22}O_{11}$) in a sugar–water syrup.

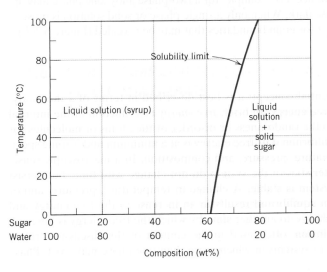

Figure 11.3 (*a*) The copper–nickel phase diagram. (*b*) A portion of the copper–nickel phase diagram for which compositions and phase amounts are determined at point *B*.
(Adapted from *Phase Diagrams of Binary Nickel Alloys,* P. Nash, Editor, 1991. Reprinted by permission of ASM International, Materials Park, OH.)

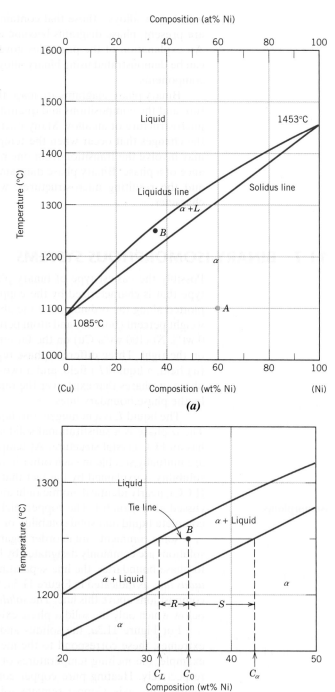

(*a*)

(*b*)

✓ *Concept Check 11.2*   The phase diagram for the cobalt–nickel system is an isomorphous one. On the basis of melting temperatures for these two metals, describe and/or draw a schematic sketch of the phase diagram for the Co–Ni system.

[*The answer may be found at* www.wiley.com/college/callister (*Student Companion Site*).]

11.8 INTERPRETATION OF PHASE DIAGRAMS

For a binary system of known composition and temperature at equilibrium, at least three kinds of information are available: (1) the phases that are present, (2) the compositions of these phases, and (3) the percentages or fractions of the phases. The procedures for making these determinations will be demonstrated using the copper–nickel system.

Phases Present

WileyPLUS: VMSE
Isomorphous
(Sb–Bi)

The establishment of what phases are present is relatively simple. One just locates the temperature–composition point on the diagram and notes the phase(s) with which the corresponding phase field is labeled. For example, an alloy of composition 60 wt% Ni–40 wt% Cu at 1100°C would be located at point A in Figure 11.3a; because this is within the α region, only the single α phase will be present. However, a 35 wt% Ni–65 wt% Cu alloy at 1250°C (point B) consists of both α and liquid phases at equilibrium.

Determination of Phase Compositions

WileyPLUS: VMSE
Isomorphous
(Sb–Bi)

tie line

The first step in the determination of phase compositions (in terms of the concentrations of the components) is to locate the temperature–composition point on the phase diagram. Different methods are used for single- and two-phase regions. If only one phase is present, the procedure is trivial: the composition of this phase is simply the same as the overall composition of the alloy. For example, consider the 60 wt% Ni–40 wt% Cu alloy at 1100°C (point A, Figure 11.3a). At this composition and temperature, only the α phase is present, having a composition of 60 wt% Ni–40 wt% Cu.

For an alloy having composition and temperature located in a two-phase region, the situation is more complicated. In all two-phase regions (and in two-phase regions only), one may imagine a series of horizontal lines, one at every temperature; each of these is known as a **tie line**, or sometimes as an *isotherm*. These tie lines extend across the two-phase region and terminate at the phase boundary lines on either side. To compute the equilibrium concentrations of the two phases, the following procedure is used:

1. A tie line is constructed across the two-phase region at the temperature of the alloy.
2. The intersections of the tie line and the phase boundaries on either side are noted.
3. Perpendiculars are dropped from these intersections to the horizontal composition axis, from which the composition of each of the respective phases is read.

For example, consider again the 35 wt% Ni–65 wt% Cu alloy at 1250°C, located at point B in Figure 11.3b and lying within the α + L region. Thus, the problem is to determine the composition (in wt% Ni and Cu) for both the α and liquid phases. The tie line is constructed across the α + L phase region, as shown in Figure 11.3b. The perpendicular from the intersection of the tie line with the liquidus boundary meets the composition axis at 31.5 wt% Ni–68.5 wt% Cu, which is the composition of the liquid phase, C_L. Likewise, for the solidus–tie line intersection, we find a composition for the α solid-solution phase, $C_α$, of 42.5 wt% Ni–57.5 wt% Cu.

Determination of Phase Amounts

WileyPLUS: VMSE
Isomorphous
(Sb–Bi)

The relative amounts (as fraction or as percentage) of the phases present at equilibrium may also be computed with the aid of phase diagrams. Again, the single- and two-phase situations must be treated separately. The solution is obvious in the single-phase region. Because only one phase is present, the alloy is composed entirely of that phase—that is, the phase fraction is 1.0, or, alternatively, the percentage is 100%. From the previous example for the 60 wt% Ni–40 wt% Cu alloy at 1100°C (point A in Figure 11.3a), only the α phase is present; hence, the alloy is completely, or 100%, α.

lever rule

If the composition and temperature position is located within a two-phase region, things are more complex. The tie line must be used in conjunction with a procedure that is often called the **lever rule** (or the *inverse lever rule*), which is applied as follows:

1. The tie line is constructed across the two-phase region at the temperature of the alloy.

2. The overall alloy composition is located on the tie line.

WileyPLUS

Tutorial Video:
Phase Diagram
Calculations and
Lever Rule
The Lever Rule

3. The fraction of one phase is computed by taking the length of tie line from the overall alloy composition to the phase boundary for the *other* phase and dividing by the total tie line length.

4. The fraction of the other phase is determined in the same manner.

5. If phase percentages are desired, each phase fraction is multiplied by 100. When the composition axis is scaled in weight percent, the phase fractions computed using the lever rule are *mass fractions*—the mass (or weight) of a specific phase divided by the total alloy mass (or weight). The mass of each phase is computed from the product of each phase fraction and the total alloy mass.

In the use of the lever rule, tie line segment lengths may be determined either by direct measurement from the phase diagram using a linear scale, preferably graduated in millimeters, or by subtracting compositions as taken from the composition axis.

Consider again the example shown in Figure 11.3b, in which at 1250°C both α and liquid phases are present for a 35 wt% Ni–65 wt% Cu alloy. The problem is to compute the fraction of each of the α and liquid phases. The tie line is constructed that was used for the determination of α and L phase compositions. Let the overall alloy composition be located along the tie line and denoted as C_0, and let the mass fractions be represented by W_L and W_α for the respective phases. From the lever rule, W_L may be computed according to

$$W_L = \frac{S}{R + S} \tag{11.1a}$$

or, by subtracting compositions,

Lever rule expression
for computation of
liquid mass fraction
(per Figure 11.3b)

$$W_L = \frac{C_\alpha - C_0}{C_\alpha - C_L} \tag{11.1b}$$

Composition need be specified in terms of only one of the constituents for a binary alloy; for the preceding computation, weight percent nickel is used (i.e., $C_0 = 35$ wt% Ni, $C_\alpha = 42.5$ wt% Ni, and $C_L = 31.5$ wt% Ni), and

$$W_L = \frac{42.5 - 35}{42.5 - 31.5} = 0.68$$

Similarly, for the α phase,

Lever rule expression
for computation of
α-phase mass fraction
(per Figure 11.3b)

$$W_\alpha = \frac{R}{R + S} \tag{11.2a}$$

$$= \frac{C_0 - C_L}{C_\alpha - C_L} \tag{11.2b}$$

$$= \frac{35 - 31.5}{42.5 - 31.5} = 0.32$$

Of course, identical answers are obtained if compositions are expressed in weight percent copper instead of nickel.

Thus, the lever rule may be employed to determine the relative amounts or fractions of phases in any two-phase region for a binary alloy if the temperature and composition are known and if equilibrium has been established. Its derivation is presented as an example problem.

It is easy to confuse the foregoing procedures for the determination of phase compositions and fractional phase amounts; thus, a brief summary is warranted. *Compositions* of phases are expressed in terms of weight percents of the components (e.g., wt% Cu, wt% Ni). For any alloy consisting of a single phase, the composition of that phase is the same as the total alloy composition. If two phases are present, the tie line must be employed, the extremes of which determine the compositions of the respective phases. With regard to *fractional phase amounts* (e.g., mass fraction of the α or liquid phase), when a single phase exists, the alloy is completely that phase. For a two-phase alloy, the lever rule is used, in which a ratio of tie line segment lengths is taken.

Concept Check 11.3   A copper–nickel alloy of composition 70 wt% Ni–30 wt% Cu is slowly heated from a temperature of 1300°C.

(a) At what temperature does the first liquid phase form?

(b) What is the composition of this liquid phase?

(c) At what temperature does complete melting of the alloy occur?

(d) What is the composition of the last solid remaining prior to complete melting?

Concept Check 11.4   Is it possible to have a copper–nickel alloy that, at equilibrium, consists of an α phase of composition 37 wt% Ni–63 wt% Cu and also a liquid phase of composition 20 wt% Ni–80 wt% Cu? If so, what will be the approximate temperature of the alloy? If this is not possible, explain why.

[*The answers may be found at* www.wiley.com/college/callister *(Student Companion Site).*]

EXAMPLE PROBLEM 11.1

Lever Rule Derivation

Derive the lever rule.

Solution

Consider the phase diagram for copper and nickel (Figure 11.3b) and alloy of composition C_0 at 1250°C, and let C_α, C_L, W_α, and W_L represent the same parameters as given earlier. This derivation is accomplished through two conservation-of-mass expressions. With the first, because only two phases are present, the sum of their mass fractions must be equal to unity; that is,

$$W_\alpha + W_L = 1 \tag{11.3}$$

For the second, the mass of one of the components (either Cu or Ni) that is present in both of the phases must be equal to the mass of that component in the total alloy, or

$$W_\alpha C_\alpha + W_L C_L = C_0 \tag{11.4}$$

Simultaneous solution of these two equations leads to the lever rule expressions for this particular situation,

$$W_L = \frac{C_\alpha - C_0}{C_\alpha - C_L} \qquad (11.1b)$$

$$W_\alpha = \frac{C_0 - C_L}{C_\alpha - C_L} \qquad (11.2b)$$

For multiphase alloys, it is often more convenient to specify relative phase amount in terms of volume fraction rather than mass fraction. Phase volume fractions are preferred because they (rather than mass fractions) may be determined from examination of the microstructure; furthermore, the properties of a multiphase alloy may be estimated on the basis of volume fractions.

For an alloy consisting of α and β phases, the volume fraction of the α phase, V_α, is defined as

α phase volume
fraction—dependence
on volumes of α and
β phases

$$V_\alpha = \frac{v_\alpha}{v_\alpha + v_\beta} \qquad (11.5)$$

where v_α and v_β denote the volumes of the respective phases in the alloy. An analogous expression exists for V_β, and, for an alloy consisting of just two phases, it is the case that $V_\alpha + V_\beta = 1$.

On occasion conversion from mass fraction to volume fraction (or vice versa) is desired. Equations that facilitate these conversions are as follows:

Conversion of mass
fractions of α and
β phases to volume
fractions

$$V_\alpha = \frac{\dfrac{W_\alpha}{\rho_\alpha}}{\dfrac{W_\alpha}{\rho_\alpha} + \dfrac{W_\beta}{\rho_\beta}} \qquad (11.6a)$$

$$V_\beta = \frac{\dfrac{W_\beta}{\rho_\beta}}{\dfrac{W_\alpha}{\rho_\alpha} + \dfrac{W_\beta}{\rho_\beta}} \qquad (11.6b)$$

and

Conversion of volume
fractions of α and
β phases to mass
fractions

$$W_\alpha = \frac{V_\alpha \rho_\alpha}{V_\alpha \rho_\alpha + V_\beta \rho_\beta} \qquad (11.7a)$$

$$W_\beta = \frac{V_\beta \rho_\beta}{V_\alpha \rho_\alpha + V_\beta \rho_\beta} \qquad (11.7b)$$

In these expressions, ρ_α and ρ_β are the densities of the respective phases; these may be determined approximately using Equations 6.13a and 6.13b.

When the densities of the phases in a two-phase alloy differ significantly, there will be quite a disparity between mass and volume fractions; conversely, if the phase densities are the same, mass and volume fractions are identical.

11.9 DEVELOPMENT OF MICROSTRUCTURE IN ISOMORPHOUS ALLOYS

WileyPLUS: VMSE
Isomorphous
(Sb–Bi)

At this point it is instructive to examine the development of microstructure that occurs for isomorphous alloys during solidification. We first treat the situation in which the cooling occurs very slowly, in that phase equilibrium is continuously maintained.

Let us consider the copper–nickel system (Figure 11.3a), specifically an alloy of composition 35 wt% Ni–65 wt% Cu as it is cooled from 1300°C. The region of the Cu–Ni phase diagram in the vicinity of this composition is shown in Figure 11.4. Cooling of an alloy of this composition corresponds to moving down the vertical

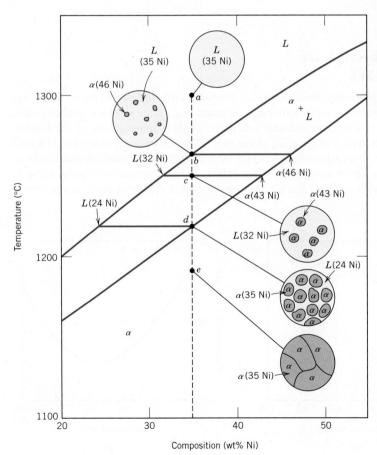

Figure 11.4 Schematic representation of the development of microstructure during the equilibrium solidification of a 35 wt% Ni–65 wt% Cu alloy.

dashed line. At 1300°C, point *a*, the alloy is completely liquid (of composition 35 wt% Ni–65 wt% Cu) and has the microstructure represented by the circle inset in the figure. As cooling begins, no microstructural or compositional changes will be realized until we reach the liquidus line (point *b*, ~1260°C). At this point, the first solid α begins to form, which has a composition dictated by the tie line drawn at this temperature [i.e., 46 wt% Ni–54 wt% Cu, noted as α(46 Ni)]; the composition of liquid is still approximately 35 wt% Ni–65 wt% Cu [*L*(35 Ni)], which is different from that of the solid α. With continued cooling, both compositions and relative amounts of each of the phases will change. The compositions of the liquid and α phases will follow the liquidus and solidus lines, respectively. Furthermore, the fraction of the α phase will increase with continued cooling. Note that the overall alloy composition (35 wt% Ni–65 wt% Cu) remains unchanged during cooling even though there is a redistribution of copper and nickel between the phases.

At 1250°C, point *c* in Figure 11.4, the compositions of the liquid and α phases are 32 wt% Ni–68 wt% Cu [*L*(32 Ni)] and 43 wt% Ni–57 wt% Cu [α(43 Ni)], respectively.

The solidification process is virtually complete at about 1220°C, point *d*; the composition of the solid α is approximately 35 wt% Ni–65 wt% Cu (the overall alloy composition), whereas that of the last remaining liquid is 24 wt% Ni–76 wt% Cu. Upon crossing the solidus line, this remaining liquid solidifies; the final product then is a polycrystalline α-phase solid solution that has a uniform 35 wt% Ni–65 wt% Cu composition (point *e*, Figure 11.4). Subsequent cooling produces no microstructural or compositional alterations.

11.10 MECHANICAL PROPERTIES OF ISOMORPHOUS ALLOYS

We now briefly explore how the mechanical properties of solid isomorphous alloys are affected by composition as other structural variables (e.g., grain size) are held constant. For all temperatures and compositions below the melting temperature of the lowest-melting component, only a single solid phase exists. Therefore, each component experiences solid-solution strengthening (Section 9.9) or an increase in strength and hardness by additions of the other component. This effect is demonstrated in Figure 11.5*a* as tensile strength versus composition for the copper–nickel system at room temperature;

Figure 11.5 For the copper–nickel system, (*a*) tensile strength versus composition and (*b*) ductility (%EL) versus composition at room temperature. A solid solution exists over all compositions for this system.

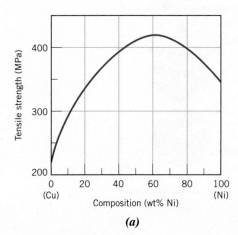

(*a*)

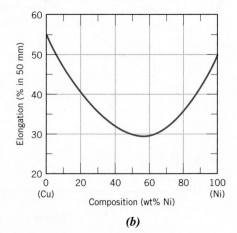

(*b*)

at some intermediate composition, the curve necessarily passes through a maximum. Plotted in Figure 11.5*b* is the ductility (%EL)–composition behavior, which is just the opposite of tensile strength—that is, ductility decreases with additions of the second component, and the curve exhibits a minimum.

11.11 BINARY EUTECTIC SYSTEMS

Another type of common and relatively simple phase diagram found for binary alloys is shown in Figure 11.6 for the copper–silver system; this is known as a *binary eutectic phase diagram*. A number of features of this phase diagram are important and worth noting. First, three single-phase regions are found on the diagram: α, β, and liquid. The α phase is a solid solution rich in copper; it has silver as the solute component and an FCC crystal structure. The β-phase solid solution also has an FCC structure, but copper is the solute. Pure copper and pure silver are also considered to be α and β phases, respectively.

Thus, the solubility in each of these solid phases is limited, in that at any temperature below line *BEG* only a limited concentration of silver dissolves in copper (for the α phase), and similarly for copper in silver (for the β phase). The solubility limit for the α phase corresponds to the boundary line, labeled *CBA*, between the α/(α + β) and α/(α + L) phase regions; it increases with temperature to a maximum (8.0 wt% Ag at 779°C) at point *B*, and decreases back to zero at the melting

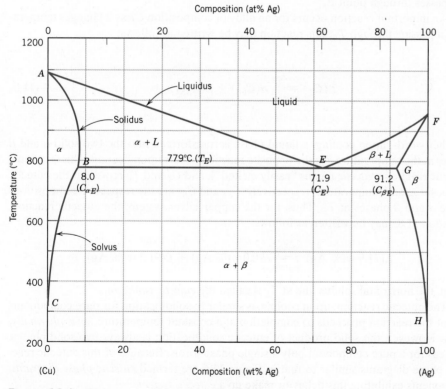

Figure 11.6 The copper–silver phase diagram.
[Adapted from *Binary Alloy Phase Diagrams,* 2nd edition, Vol. 1, T. B. Massalski (Editor-in-Chief), 1990. Reprinted by permission of ASM International, Materials Park, OH.]

solvus line
solidus line

temperature of pure copper, point A (1085°C). At temperatures below 779°C, the solid solubility limit line separating the α and $\alpha + \beta$ phase regions is termed a **solvus line**; the boundary AB between the α and $\alpha + L$ fields is the **solidus line**, as indicated in Figure 11.6. For the β phase, both solvus and solidus lines also exist, HG and GF, respectively, as shown. The maximum solubility of copper in the β phase, point G (8.8 wt% Cu), also occurs at 779°C. This horizontal line BEG, which is parallel to the composition axis and extends between these maximum solubility positions, may also be considered a solidus line; it represents the lowest temperature at which a liquid phase may exist for any copper–silver alloy that is at equilibrium.

There are also three two-phase regions found for the copper–silver system (Figure 11.6): $\alpha + L$, $\beta + L$, and $\alpha + \beta$. The α- and β-phase solid solutions coexist for all compositions and temperatures within the $\alpha + \beta$ phase field; the $\alpha +$ liquid and $\beta +$ liquid phases also coexist in their respective phase regions. Furthermore, compositions and relative amounts for the phases may be determined using tie lines and the lever rule as outlined previously.

liquidus line

As silver is added to copper, the temperature at which the alloys become totally liquid decreases along the **liquidus line**, line AE; thus, the melting temperature of copper is lowered by silver additions. The same may be said for silver: the introduction of copper reduces the temperature of complete melting along the other liquidus line, FE. These liquidus lines meet at the point E on the phase diagram, which point is designated by composition C_E and temperature T_E; for the copper–silver system, the values for these two parameters are 71.9 wt% Ag and 779°C, respectively. It should also be noted there is a horizontal isotherm at 779°C and represented by the line labeled BEG that also passes through point E.

An important reaction occurs for an alloy of composition C_E as it changes temperature in passing through T_E; this reaction may be written as follows:

The eutectic reaction
(per Figure 11.6)

$$L(C_E) \underset{\text{heating}}{\overset{\text{cooling}}{\rightleftharpoons}} \alpha(C_{\alpha E}) + \beta(C_{\beta E}) \tag{11.8}$$

eutectic reaction

In other words, upon cooling, a liquid phase is transformed into the two solid α and β phases at the temperature T_E; the opposite reaction occurs upon heating. This is called a **eutectic reaction** (*eutectic* means "easily melted"), and C_E and T_E represent the eutectic composition and temperature, respectively; $C_{\alpha E}$ and $C_{\beta E}$ are the respective compositions of the α and β phases at T_E. Thus, for the copper–silver system, the eutectic reaction, Equation 11.8, may be written as follows:

$$L(71.9 \text{ wt\% Ag}) \underset{\text{heating}}{\overset{\text{cooling}}{\rightleftharpoons}} \alpha(8.0 \text{ wt\% Ag}) + \beta(91.2 \text{ wt\% Ag})$$

Often, the horizontal solidus line at T_E is called the *eutectic isotherm*.

WileyPLUS

Tutorial Video:
Eutectic Reaction
Vocabulary and
Microstructures

Eutectic
Reaction Terms

The eutectic reaction, upon cooling, is similar to solidification for pure components in that the reaction proceeds to completion at a constant temperature, or *isothermally*, at T_E. However, the solid product of eutectic solidification is always two solid phases, whereas for a pure component only a single phase forms. Because of this eutectic reaction, phase diagrams similar to that in Figure 11.6 are termed *eutectic phase diagrams;* components exhibiting this behavior make up a *eutectic system.*

In the construction of binary phase diagrams, it is important to understand that one or at most two phases may be in equilibrium within a phase field. This holds true

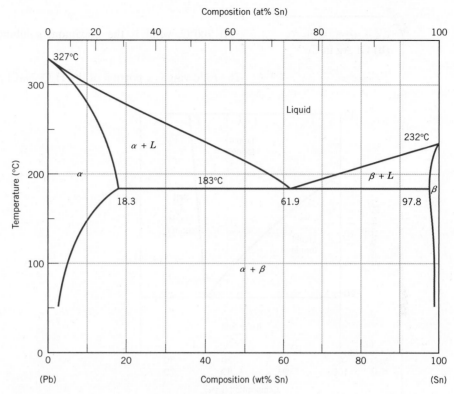

Figure 11.7 The lead–tin phase diagram.
[Adapted from *Binary Alloy Phase Diagrams,* 2nd edition, Vol. 3, T. B. Massalski (Editor-in-Chief), 1990. Reprinted by permission of ASM International, Materials Park, OH.]

for the phase diagrams in Figures 11.3*a* and 11.6. For a eutectic system, three phases (α, β, and *L*) may be in equilibrium, but only at points along the eutectic isotherm. Another general rule is that single-phase regions are always separated from each other by a two-phase region that consists of the two single phases that it separates. For example, the α + β field is situated between the α and β single-phase regions in Figure 11.6.

Another common eutectic system is that for lead and tin; the phase diagram (Figure 11.7) has a general shape similar to that for copper–silver. For the lead–tin system, the solid-solution phases are also designated by α and β; in this case, α represents a solid solution of tin in lead; for β, tin is the solvent and lead is the solute. The eutectic invariant point is located at 61.9 wt% Sn and 183°C. Of course, maximum solid solubility compositions as well as component melting temperatures are different for the copper–silver and lead–tin systems, as may be observed by comparing their phase diagrams.

On occasion, low-melting-temperature alloys are prepared having near-eutectic compositions. A familiar example is 60–40 solder, which contains 60 wt% Sn and 40 wt% Pb. Figure 11.7 indicates that an alloy of this composition is completely molten at about 185°C, which makes this material especially attractive as a low-temperature solder because it is easily melted.

Concept Check 11.5   At 700°C, what is the maximum solubility **(a)** of Cu in Ag? **(b)** Of Ag in Cu?

Concept Check 11.6   The following is a portion of the H_2O–NaCl phase diagram:

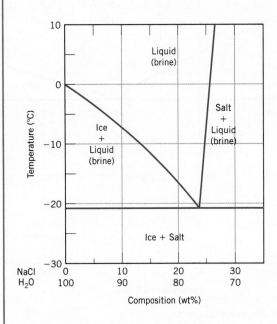

(a) Using this diagram, briefly explain how spreading salt on ice that is at a temperature below 0°C can cause the ice to melt.

(b) At what temperature is salt no longer useful in causing ice to melt?

[*The answers may be found at* www.wiley.com/college/callister *(Student Companion Site)*.]

EXAMPLE PROBLEM 11.2

Determination of Phases Present and Computation of Phase Compositions

For a 40 wt% Sn–60 wt% Pb alloy at 150°C, **(a)** what phase(s) is (are) present? **(b)** What is (are) the composition(s) of the phase(s)?

WileyPLUS

Tutorial Video:
Phase Diagram
Calculations and
Lever Rule
Calculations for
a Binary Eutectic
Phase Diagram

Solution

(a) Locate this temperature–composition point on the phase diagram (point *B* in Figure 11.8). Inasmuch as it is within the α + β region, both α and β phases will coexist.

(b) Because two phases are present, it becomes necessary to construct a tie line across the α + β phase field at 150°C, as indicated in Figure 11.8. The composition of the α phase corresponds to the tie line intersection with the α/(α + β) solvus phase boundary—about 11 wt% Sn–89 wt% Pb, denoted as C_α. This is similar for the β phase, which has a composition of approximately 98 wt% Sn–2 wt% Pb (C_β).

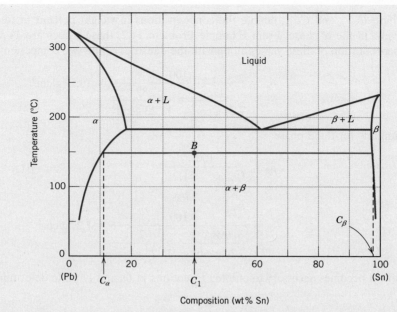

Figure 11.8 The lead–tin phase diagram. For a 40 wt% Sn–60 wt% Pb alloy at 150°C (point *B*), phase compositions and relative amounts are computed in Example Problems 11.2 and 11.3.

EXAMPLE PROBLEM 11.3

Relative Phase Amount Determinations—Mass and Volume Fractions

For the lead–tin alloy in Example Problem 11.2, calculate the relative amount of each phase present in terms of **(a)** mass fraction and **(b)** volume fraction. At 150°C, take the densities of Pb and Sn to be 11.23 and 7.24 g/cm³, respectively.

Solution

WileyPLUS

Tutorial Video:
Phase Diagram
Calculations
and Lever Rule
How Do I
Determine the
Volume Fraction
of Each Phase?

(a) Because the alloy consists of two phases, it is necessary to employ the lever rule. If C_1 denotes the overall alloy composition, mass fractions may be computed by subtracting compositions, in terms of weight percent tin, as follows:

$$W_\alpha = \frac{C_\beta - C_1}{C_\beta - C_\alpha} = \frac{98 - 40}{98 - 11} = 0.67$$

$$W_\beta = \frac{C_1 - C_\alpha}{C_\beta - C_\alpha} = \frac{40 - 11}{98 - 11} = 0.33$$

(b) To compute volume fractions it is first necessary to determine the density of each phase using Equation 6.13a. Thus

$$\rho_\alpha = \frac{100}{\dfrac{C_{Sn(\alpha)}}{\rho_{Sn}} + \dfrac{C_{Pb(\alpha)}}{\rho_{Pb}}}$$

where $C_{Sn(\alpha)}$ and $C_{Pb(\alpha)}$ denote the concentrations in weight percent of tin and lead, respectively, in the α phase. From Example Problem 11.2, these values are 11 wt% and 89 wt%. Incorporation of these values along with the densities of the two components leads to

$$\rho_\alpha = \frac{100}{\dfrac{11}{7.24 \text{ g/cm}^3} + \dfrac{89}{11.23 \text{ g/cm}^3}} = 10.59 \text{ g/cm}^3$$

Similarly for the β phase:

$$\rho_\beta = \frac{100}{\dfrac{C_{Sn(\beta)}}{\rho_{Sn}} + \dfrac{C_{Pb(\beta)}}{\rho_{Pb}}}$$

$$= \frac{100}{\dfrac{98}{7.24 \text{ g/cm}^3} + \dfrac{2}{11.23 \text{ g/cm}^3}} = 7.29 \text{ g/cm}^3$$

Now it becomes necessary to employ Equations 11.6a and 11.6b to determine V_α and V_β as

$$V_\alpha = \frac{\dfrac{W_\alpha}{\rho_\alpha}}{\dfrac{W_\alpha}{\rho_\alpha} + \dfrac{W_\beta}{\rho_\beta}}$$

$$= \frac{\dfrac{0.67}{10.59 \text{ g/cm}^3}}{\dfrac{0.67}{10.59 \text{ g/cm}^3} + \dfrac{0.33}{7.29 \text{ g/cm}^3}} = 0.58$$

$$V_\beta = \frac{\dfrac{W_\beta}{\rho_\beta}}{\dfrac{W_\alpha}{\rho_\alpha} + \dfrac{W_\beta}{\rho_\beta}}$$

$$= \frac{\dfrac{0.33}{7.29 \text{ g/cm}^3}}{\dfrac{0.67}{10.59 \text{ g/cm}^3} + \dfrac{0.33}{7.29 \text{ g/cm}^3}} = 0.42$$

11.12 DEVELOPMENT OF MICROSTRUCTURE IN EUTECTIC ALLOYS

Depending on composition, several different types of microstructures are possible for the slow cooling of alloys belonging to binary eutectic systems. These possibilities will be considered in terms of the lead–tin phase diagram, Figure 11.7.

The first case is for compositions ranging between a pure component and the maximum solid solubility for that component at room temperature (20°C). For the lead–tin system, this includes lead-rich alloys containing between 0 and about 2 wt% Sn (for the α-phase solid solution) and also between approximately 99 wt% Sn and pure tin (for the

MATERIALS OF IMPORTANCE

Lead-Free Solders

Solders are metal alloys that are used to bond or join two or more components (usually other metal alloys). They are used extensively in the electronics industry to physically hold assemblies together; they must allow expansion and contraction of the various components, transmit electrical signals, and dissipate any heat that is generated. The bonding action is accomplished by melting the solder material and allowing it to flow among and make contact with the components to be joined (which do not melt); finally, upon solidification, it forms a physical bond with all of these components.

In the past, the vast majority of solders have been lead–tin alloys. These materials are reliable and inexpensive and have relatively low melting temperatures. The most common lead–tin solder has a composition of 63 wt% Sn–37 wt% Pb. According to the lead–tin phase diagram, Figure 11.7, this composition is near the eutectic and has a melting temperature of about 183°C, the lowest temperature possible with the existence of a liquid phase (at equilibrium) for the lead–tin system. This alloy is often called a *eutectic lead–tin solder*.

Unfortunately, lead is a mildly toxic metal, and there is serious concern about the environmental impact of discarded lead-containing products that can leach into groundwater from landfills or pollute the air if incinerated. Consequently, in some countries legislation has been enacted that bans the use of lead-containing solders. This has forced the development of lead-free solders that, among other things, must have relatively low melting temperatures (or temperature ranges). Many of these are tin alloys that contain relatively low concentrations of copper, silver, bismuth, and/or antimony. Compositions as well as liquidus and solidus temperatures for several lead-free solders are listed in Table 11.1. Two lead-containing solders are also included in this table.

Melting temperatures (or temperature ranges) are important in the development and selection of these new solder alloys, information available from phase diagrams. For example, a portion of the tin-rich side of the silver–tin phase diagram is presented in Figure 11.9. Here, it may be noted that a eutectic exists at 96.5 wt% Sn and 221°C; these are indeed the composition and melting temperature, respectively, of the 96.5 Sn–3.5 Ag solder in Table 11.1.

Table 11.1 Compositions, Solidus Temperatures, and Liquidus Temperatures for Two Lead-Containing Solders and Five Lead-Free Solders

Composition (wt%)	Solidus Temperature (°C)	Liquidus Temperature (°C)
Solders Containing Lead		
63 Sn–37 Pb[a]	183	183
50 Sn–50 Pb	183	214
Lead-Free Solders		
99.3 Sn–0.7 Cu[a]	227	227
96.5 Sn–3.5 Ag[a]	221	221
95.5 Sn–3.8 Ag–0.7 Cu	217	220
91.8 Sn–3.4 Ag–4.8 Bi	211	213
97.0 Sn–2.0 Cu–0.85 Sb–0.2 Ag	219	235

[a]The compositions of these alloys are eutectic compositions; therefore, their solidus and liquidus temperatures are identical.

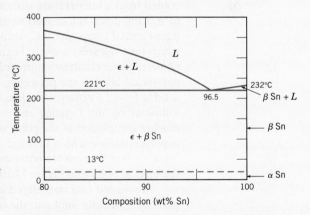

Figure 11.9 The tin-rich side of the silver–tin phase diagram.
[Adapted from *ASM Handbook,* Vol. 3, *Alloy Phase Diagrams,* H. Baker (Editor), ASM International, 1992. Reprinted by permission of ASM International, Materials Park, OH.]

Figure 11.10 Schematic representations of the equilibrium microstructures for a lead–tin alloy of composition C_1 as it is cooled from the liquid-phase region.

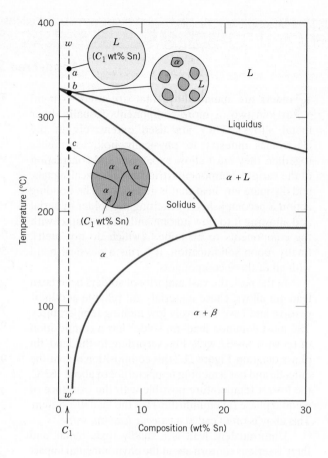

β phase). For example, consider an alloy of composition C_1 (Figure 11.10) as it is slowly cooled from a temperature within the liquid-phase region, say, 350°C; this corresponds to moving down the dashed vertical line ww' in the figure. The alloy remains totally liquid and of composition C_1 until we cross the liquidus line at approximately 330°C, at which time the solid α phase begins to form. While passing through this narrow $\alpha + L$ phase region, solidification proceeds in the same manner as was described for the copper–nickel alloy in the preceding section—that is, with continued cooling, more of the solid α forms. Furthermore, liquid- and solid-phase compositions are different, which follow along the liquidus and solidus phase boundaries, respectively. Solidification reaches completion at the point where ww' crosses the solidus line. The resulting alloy is polycrystalline with a uniform composition of C_1, and no subsequent changes occur upon cooling to room temperature. This microstructure is represented schematically by the inset at point c in Figure 11.10.

The second case considered is for compositions that range between the room temperature solubility limit and the maximum solid solubility at the eutectic temperature. For the lead–tin system (Figure 11.7), these compositions extend from about 2 to 18.3 wt% Sn (for lead-rich alloys) and from 97.8 to approximately 99 wt% Sn (for tin-rich alloys). Let us examine an alloy of composition C_2 as it is cooled along the vertical line xx' in Figure 11.11. Down to the intersection of xx' and the solvus line, changes that occur are similar to the previous case as we pass through the corresponding phase regions (as demonstrated by the insets at points d, e, and f). Just above the solvus intersection, point f, the microstructure consists of α grains of composition C_2. Upon crossing the solvus line, the α solid solubility is exceeded, which results in the formation of small β-phase particles; these are indicated in the microstructure inset at point g. With continued

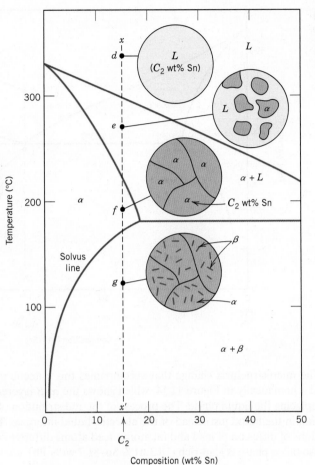

Figure 11.11 Schematic representations of the equilibrium microstructures for a lead–tin alloy of composition C_2 as it is cooled from the liquid-phase region.

WileyPLUS: VMSE
Eutectic (Pb-Sn)

cooling, these particles grow in size because the mass fraction of the β phase increases slightly with decreasing temperature.

The third case involves solidification of the eutectic composition, 61.9 wt% Sn (C_3 in Figure 11.12). Consider an alloy having this composition that is cooled from a temperature within the liquid-phase region (e.g., 250°C) down the vertical line yy' in Figure 11.12. As the temperature is lowered, no changes occur until we reach the eutectic temperature, 183°C. Upon crossing the eutectic isotherm, the liquid transforms into the two α and β phases. This transformation may be represented by the reaction

$$L(61.9 \text{ wt\% Sn}) \underset{\text{heating}}{\overset{\text{cooling}}{\rightleftarrows}} \alpha(18.3 \text{ wt\% Sn}) + \beta(97.8 \text{ wt\% Sn}) \tag{11.9}$$

in which the α- and β-phase compositions are dictated by the eutectic isotherm end points.

During this transformation, there must be a redistribution of the lead and tin components because the α and β phases have different compositions, neither of which is the same as that of the liquid (as indicated in Equation 11.9). This redistribution is accomplished by atomic diffusion. The microstructure of the solid that results from this transformation consists of alternating layers (sometimes called *lamellae*) of the α and β phases that form simultaneously during the transformation. This microstructure, represented schematically in Figure 11.12, point i, is called a **eutectic structure** and is characteristic of this reaction. A photomicrograph of this structure for the lead–tin eutectic is shown in Figure 11.13. Subsequent cooling of the alloy from just below the eutectic to room temperature results in only minor microstructural alterations.

eutectic structure

Figure 11.12 Schematic representations of the equilibrium microstructures for a lead–tin alloy of eutectic composition C_3 above and below the eutectic temperature.

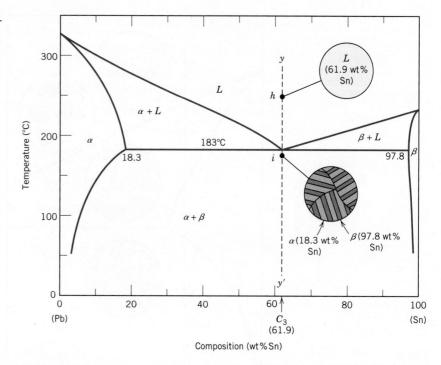

WileyPLUS

Tutorial Video:
Eutectic Reaction
Vocabulary and
Microstructures
How Do the Eutectic
Microstructures Form?

The microstructural change that accompanies this eutectic transformation is represented schematically in Figure 11.14, which shows the α–β layered eutectic growing into and replacing the liquid phase. The process of the redistribution of lead and tin occurs by diffusion in the liquid just ahead of the eutectic–liquid interface. The arrows indicate the directions of diffusion of lead and tin atoms; lead atoms diffuse toward the α-phase layers because this α phase is lead-rich (18.3 wt% Sn–81.7 wt% Pb); conversely, the direction of

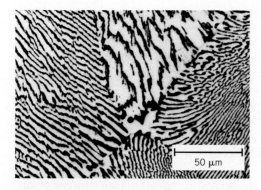

Figure 11.13 Photomicrograph showing the microstructure of a lead–tin alloy of eutectic composition. This microstructure consists of alternating layers of a lead-rich α-phase solid solution (dark layers), and a tin-rich β-phase solid solution (light layers). 375×.
(From *Metals Handbook,* 9th edition, Vol. 9, *Metallography and Microstructures,* 1985. Reproduced by permission of ASM International, Materials Park, OH.)

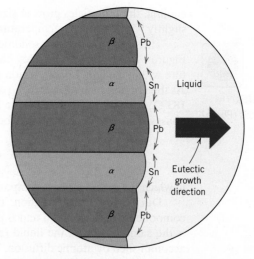

Figure 11.14 Schematic representation of the formation of the eutectic structure for the lead–tin system. Directions of diffusion of tin and lead atoms are indicated by blue and red arrows, respectively.

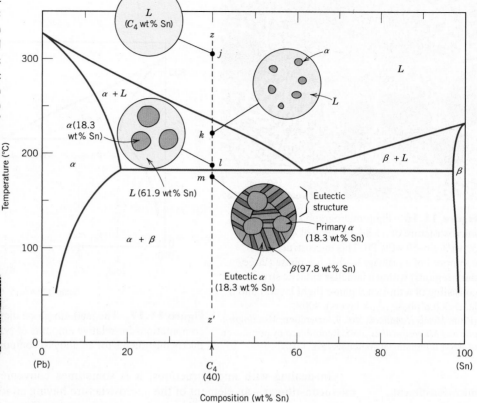

Photomicrograph
showing a reversible-
matrix interface (i.e.,
a black-on-white
to white-on-black
pattern reversal a
la Escher) for an
aluminum-copper
eutectic alloy.
Magnification un-
known.
(From *Metals
Handbook,* Vol. 9, 9th
edition, *Metallography
and Microstructures,*
1985. Reproduced by
permission of ASM
International, Materials
Park, OH.)

eutectic phase

primary phase

Figure 11.15 Schematic representations of the equilibrium microstructures for a lead–tin alloy of composition C_4 as it is cooled from the liquid-phase region.

diffusion of tin is in the direction of the β, tin-rich (97.8 wt% Sn–2.2 wt% Pb) layers. The eutectic structure forms in these alternating layers because, for this lamellar configuration, atomic diffusion of lead and tin need only occur over relatively short distances.

The fourth and final microstructural case for this system includes all compositions other than the eutectic that, when cooled, cross the eutectic isotherm. Consider, for example, the composition C_4 in Figure 11.15, which lies to the left of the eutectic; as the temperature is lowered, we move down the line zz', beginning at point j. The micro-structural development between points j and l is similar to that for the second case, such that just prior to crossing the eutectic isotherm (point l), the α and liquid phases are present with compositions of approximately 18.3 and 61.9 wt% Sn, respectively, as determined from the appropriate tie line. As the temperature is lowered to just below the eutectic, the liquid phase, which is of the eutectic composition, transforms into the eutectic structure (i.e., alternating α and β lamellae); insignificant changes occur with the α phase that formed during cooling through the $\alpha + L$ region. This microstructure is represented schematically by the inset at point m in Figure 11.15. Thus, the α phase is present both in the eutectic structure and also as the phase that formed while cooling through the $\alpha + L$ phase field. To distinguish one α from the other, that which resides in the eutectic structure is called **eutectic** α, whereas the other that formed prior to crossing the eutectic isotherm is termed **primary** α; both are labeled in Figure 11.15. The photomicrograph in Figure 11.16 is of a lead–tin alloy in which both primary α and eutectic structures are shown.

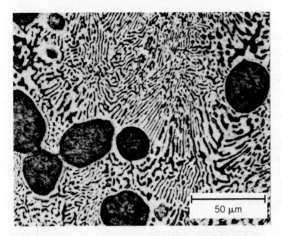

Figure 11.16 Photomicrograph showing the microstructure of a lead–tin alloy of composition 50 wt% Sn–50 wt% Pb. This microstructure is composed of a primary lead-rich α phase (large dark regions) within a lamellar eutectic structure consisting of a tin-rich β phase (light layers) and a lead-rich α phase (dark layers). 400×.
(From *Metals Handbook*, Vol. 9, 9th edition, *Metallography and Microstructures*, 1985. Reproduced by permission of ASM International, Materials Park, OH.)

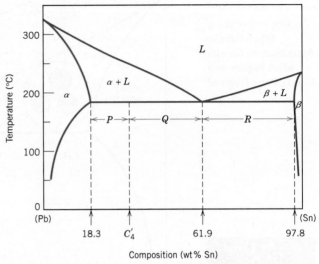

Figure 11.17 The lead–tin phase diagram used in computations for relative amounts of primary α and eutectic microconstituents for an alloy of composition C_4'.

In dealing with microstructures, it is sometimes convenient to use the term **microconstituent**—an element of the microstructure having an identifiable and characteristic structure. For example, in the point m inset in Figure 11.15, there are two microconstituents—primary α and the eutectic structure. Thus, the eutectic structure is a microconstituent even though it is a mixture of two phases because it has a distinct lamellar structure with a fixed ratio of the two phases.

It is possible to compute the relative amounts of both eutectic and primary α microconstituents. Because the eutectic microconstituent always forms from the liquid having the eutectic composition, this microconstituent may be assumed to have a composition of 61.9 wt% Sn. Hence, the lever rule is applied using a tie line between the α–$(\alpha + \beta)$ phase boundary (18.3 wt% Sn) and the eutectic composition. For example, consider the alloy of composition C_4' in Figure 11.17. The fraction of the eutectic microconstituent W_e is just the same as the fraction of liquid W_L from which it transforms, or

Lever rule expression for computation of eutectic microconstituent and liquid-phase mass fractions (composition C_4', Figure 11.17)

$$W_e = W_L = \frac{P}{P + Q}$$

$$= \frac{C_4' - 18.3}{61.9 - 18.3} = \frac{C_4' - 18.3}{43.6} \tag{11.10}$$

Furthermore, the fraction of primary α, $W_{\alpha'}$, is just the fraction of the α phase that existed prior to the eutectic transformation or, from Figure 11.17,

Lever rule expression for computation of primary α-phase mass fraction

$$W_{\alpha'} = \frac{Q}{P + Q}$$

$$= \frac{61.9 - C_4'}{61.9 - 18.3} = \frac{61.9 - C_4'}{43.6} \tag{11.11}$$

The fractions of *total α*, W_α (both eutectic and primary), and also of total *β*, W_β, are determined by use of the lever rule and a tie line that extends *entirely across the α + β phase field*. Again, for an alloy having composition C_4',

Lever rule expression for computation of total α-phase mass fraction

$$W_\alpha = \frac{Q + R}{P + Q + R}$$

$$= \frac{97.8 - C_4'}{97.8 - 18.3} = \frac{97.8 - C_4'}{79.5} \tag{11.12}$$

and

Lever rule expression for computation of total β-phase mass fraction

$$W_\beta = \frac{P}{P + Q + R}$$

$$= \frac{C_4' - 18.3}{97.8 - 18.3} = \frac{C_4' - 18.3}{79.5} \tag{11.13}$$

Analogous transformations and microstructures result for alloys having compositions to the right of the eutectic (i.e., between 61.9 and 97.8 wt% Sn). However, below the eutectic temperature, the microstructure will consist of the eutectic and primary *β* microconstituents because, upon cooling from the liquid, we pass through the *β* + liquid phase field.

When, for the fourth case represented in Figure 11.15, conditions of equilibrium are not maintained while passing through the α (or β) + liquid phase region, the following consequences will be realized for the microstructure upon crossing the eutectic isotherm: (1) grains of the primary microconstituent will be cored, that is, have a nonuniform distribution of solute across the grains; and (2) the fraction of the eutectic microconstituent formed will be greater than for the equilibrium situation.

11.13 EQUILIBRIUM DIAGRAMS HAVING INTERMEDIATE PHASES OR COMPOUNDS

terminal solid solution

intermediate solid solution

The isomorphous and eutectic phase diagrams discussed thus far are relatively simple, but those for many binary alloy systems are much more complex. The eutectic copper–silver and lead–tin phase diagrams (Figures 11.6 and 11.7) have only two solid phases, α and β; these are sometimes termed **terminal solid solutions** because they exist over composition ranges near the concentration extremes of the phase diagram. For other alloy systems, **intermediate solid solutions** (or *intermediate phases*) may be found at other than the two composition extremes. Such is the case for the copper–zinc system. Its phase diagram (Figure 11.18) may at first appear formidable because there are some invariant points and reactions similar to the eutectic that have not yet been discussed. In addition, there are six different solid solutions—two terminal (α and η) and four intermediate (β, γ, δ, and ε). (The β′ phase is termed an *ordered solid solution,* one in which the copper and zinc atoms are situated in a specific and ordered arrangement within each unit cell.) Some phase boundary lines near the bottom of Figure 11.18 are dashed to indicate that their positions have not been exactly determined. The reason for this is that at low temperatures, diffusion rates are very slow, and inordinately long times are required to attain equilibrium. Again, only single- and two-phase regions are found on the diagram, and the same rules outlined in Section 11.8 are used to compute phase compositions and relative

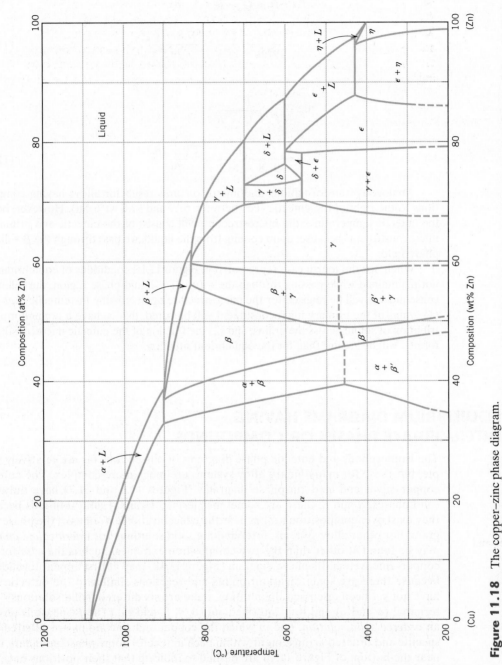

Figure 11.18 The copper–zinc phase diagram.
[Adapted from *Binary Alloy Phase Diagrams*, 2nd edition, Vol. 2, T. B. Massalski (Editor-in-Chief), 1990. Reprinted by permission of ASM International, Materials Park, OH.]

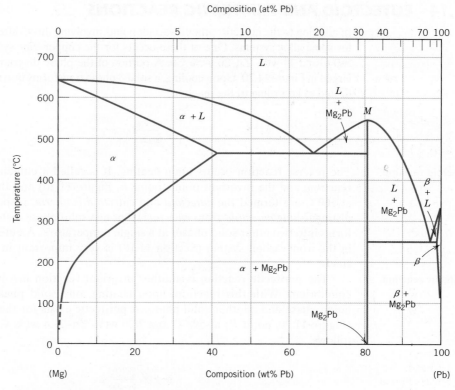

Figure 11.19 The magnesium–lead phase diagram.
[Adapted from *Phase Diagrams of Binary Magnesium Alloys*, A. A. Nayeb-Hashemi and J. B. Clark (Editors), 1988. Reprinted by permission of ASM International, Materials Park, OH.]

intermetallic compound

amounts. The commercial brasses are copper–rich copper–zinc alloys; for example, cartridge brass has a composition of 70 wt% Cu–30 wt% Zn and a microstructure consisting of a single α phase.

For some systems, discrete intermediate compounds rather than solid solutions may be found on the phase diagram, and these compounds have distinct chemical formulas; for metal–metal systems, they are called **intermetallic compounds**. For example, consider the magnesium–lead system (Figure 11.19). The compound Mg_2Pb has a composition of 19 wt% Mg–81 wt% Pb (33 at% Pb) and is represented as a vertical line on the diagram, rather than as a phase region of finite width; hence, Mg_2Pb can exist by itself only at this precise composition.

Several other characteristics are worth noting for this magnesium–lead system. First, the compound Mg_2Pb melts at approximately 550°C, as indicated by point M in Figure 11.19. Also, the solubility of lead in magnesium is rather extensive, as indicated by the relatively large composition span for the α-phase field. However, the solubility of magnesium in lead is extremely limited. This is evident from the very narrow β terminal solid-solution region on the right, or lead-rich, side of the diagram. Finally, this phase diagram may be thought of as two simple eutectic diagrams joined back to back, one for the $Mg–Mg_2Pb$ system and the other for $Mg_2Pb–Pb$; as such, the compound Mg_2Pb is really considered to be a component. This separation of complex phase diagrams into smaller-component units may simplify them and expedite their interpretation.

11.14 EUTECTOID AND PERITECTIC REACTIONS

In addition to the eutectic, other invariant points involving three different phases are found for some alloy systems. One of these occurs for the copper–zinc system (Figure 11.18) at 560°C and 74 wt% Zn–26 wt% Cu. A portion of the phase diagram in this vicinity is enlarged in Figure 11.20. Upon cooling, a solid δ phase transforms into two other solid phases (γ and ϵ) according to the reaction

The eutectoid reaction (per point *E*, Figure 11.20)

$$\delta \underset{\text{heating}}{\overset{\text{cooling}}{\rightleftharpoons}} \gamma + \epsilon \qquad (11.14)$$

eutectoid reaction

The reverse reaction occurs upon heating. It is called a **eutectoid** (or eutectic-like) **reaction**, and the invariant point (point *E*, Figure 11.20) and the horizontal tie line at 560°C are termed the *eutectoid* and *eutectoid isotherm,* respectively. The feature distinguishing *eutectoid* from *eutectic* is that one solid phase instead of a liquid transforms into two other solid phases at a single temperature. A eutectoid reaction found in the iron–carbon system (Section 11.18) is very important in the heat treating of steels.

peritectic reaction

The **peritectic reaction** is another invariant reaction involving three phases at equilibrium. With this reaction, upon heating, one solid phase transforms into a liquid phase and another solid phase. A peritectic exists for the copper–zinc system (Figure 11.20, point *P*) at 598°C and 78.6 wt% Zn–21.4 wt% Cu; this reaction is as follows:

The peritectic reaction (per point *P*, Figure 11.20)

$$\delta + L \underset{\text{heating}}{\overset{\text{cooling}}{\rightleftharpoons}} \epsilon \qquad (11.15)$$

The low-temperature solid phase may be an intermediate solid solution (e.g., ϵ in the preceding reaction), or it may be a terminal solid solution. One of the latter peritectics exists at about 97 wt% Zn and 435°C (see Figure 11.18), where the η phase, when heated, transforms into ϵ and liquid phases. Three other peritectics are found for the Cu–Zn system, the reactions of which involve β, δ, and γ intermediate solid solutions as the low-temperature phases that transform upon heating.

Figure 11.20 A region of the copper–zinc phase diagram that has been enlarged to show eutectoid and peritectic invariant points, labeled *E* (560°C, 74 wt% Zn) and *P* (598°C, 78.6 wt% Zn), respectively.
[Adapted from *Binary Alloy Phase Diagrams,* 2nd edition, Vol. 2, T. B. Massalski (Editor-in-Chief), 1990. Reprinted by permission of ASM International, Materials Park, OH.]

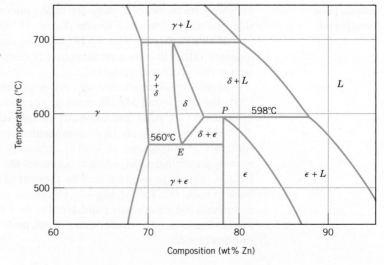

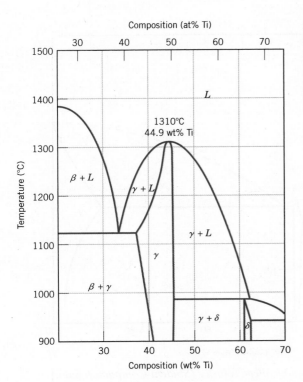

Composition (at% Ti)

Figure 11.21 A portion of the nickel–titanium phase diagram on which is shown a congruent melting point for the γ-phase solid solution at 1310°C and 44.9 wt% Ti. [Adapted from *Phase Diagrams of Binary Nickel Alloys*, P. Nash (Editor), 1991. Reprinted by permission of ASM International, Materials Park, OH.]

11.15 CONGRUENT PHASE TRANSFORMATIONS

congruent
transformation

Phase transformations may be classified according to whether there is any change in composition for the phases involved. Those for which there are no compositional alterations are said to be **congruent transformations**. Conversely, for *incongruent transformations*, at least one of the phases experiences a change in composition. Examples of congruent transformations include allotropic transformations (Section 4.14) and melting of pure materials. Eutectic and eutectoid reactions, as well as the melting of an alloy that belongs to an isomorphous system, all represent incongruent transformations.

Intermediate phases are sometimes classified on the basis of whether they melt congruently or incongruently. The intermetallic compound Mg_2Pb melts congruently at the point designated *M* on the magnesium–lead phase diagram, Figure 11.19. For the nickel–titanium system, Figure 11.21, there is a congruent melting point for the γ solid solution that corresponds to the point of tangency for the pairs of liquidus and solidus lines, at 1310°C and 44.9 wt% Ti. The peritectic reaction is an example of incongruent melting for an intermediate phase.

Concept Check 11.7 The following figure is the hafnium–vanadium phase diagram, for which only single-phase regions are labeled. Specify temperature–composition points at which all eutectics, eutectoids, peritectics, and congruent phase transformations occur. Also, for each, write the reaction upon cooling. [Phase diagram from *ASM Handbook,* Vol. 3, *Alloy Phase Diagrams,* H. Baker (Editor), 1992, p. 2.244. Reprinted by permission of ASM International, Materials Park, OH.]

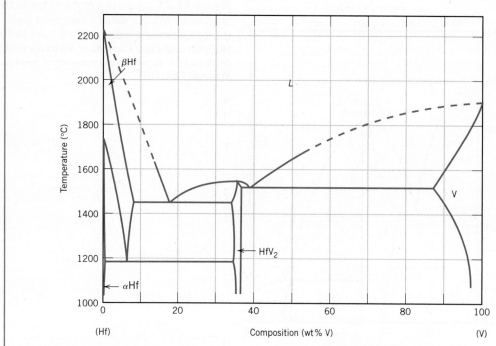

[*The answer may be found at* www.wiley.com/college/callister *(Student Companion Site).*]

11.16 CERAMIC AND TERNARY PHASE DIAGRAMS

It need not be assumed that phase diagrams exist only for metal–metal systems; in fact, phase diagrams that are very useful in the design and processing of ceramic systems have been experimentally determined for many of these materials. Ceramic phase diagrams are discussed in Sections 14.2–14.5.

Phase diagrams have also been determined for metallic (as well as ceramic) systems containing more than two components; however, their representation and interpretation may be exceedingly complex. For example, a ternary, or three-component, composition–temperature phase diagram in its entirety is depicted by a three-dimensional model. Portrayal of features of the diagram or model in two dimensions is possible, but somewhat difficult.

11.17 THE GIBBS PHASE RULE

Gibbs phase rule

The construction of phase diagrams—as well as some of the principles governing the conditions for phase equilibria—are dictated by laws of thermodynamics. One of these is the **Gibbs phase rule**, proposed by the nineteenth-century physicist J. Willard Gibbs. This rule represents a criterion for the number of phases that coexist within a system at equilibrium and is expressed by the simple equation

General form of the
Gibbs phase rule

$$P + F = C + N \tag{11.16}$$

where P is the number of phases present (the phase concept is discussed in Section 11.3). The parameter F is termed the *number of degrees of freedom* or the number of externally controlled variables (e.g., temperature, pressure, composition) that must be specified to define the state of the system completely. Expressed another way, F is the number of these variables that can be changed independently without altering the number of phases that coexist at equilibrium. The parameter C in Equation 11.16 represents the number of components in the system. Components are normally elements or stable compounds and, in the case of phase diagrams, are the materials at the two extremes of the horizontal compositional axis (e.g., H_2O and $C_{12}H_{22}O_{11}$, and Cu and Ni for the phase diagrams shown in Figures 11.1 and 11.3a, respectively). Finally, N in Equation 11.16 is the number of noncompositional variables (e.g., temperature and pressure).

Let us demonstrate the phase rule by applying it to binary temperature–composition phase diagrams, specifically the copper–silver system, Figure 11.6. Because pressure is constant (1 atm), the parameter N is 1—temperature is the only noncompositional variable. Equation 11.16 now takes the form

$$P + F = C + 1 \tag{11.17}$$

The number of components C is 2 (namely, Cu and Ag), and

$$P + F = 2 + 1 = 3$$

or

$$F = 3 - P$$

Consider the case of single-phase fields on the phase diagram (e.g., α, β, and liquid regions). Because only one phase is present, $P = 1$ and

$$F = 3 - P$$
$$= 3 - 1 = 2$$

This means that to completely describe the characteristics of any alloy that exists within one of these phase fields, we must specify two parameters—composition and temperature, which locate, respectively, the horizontal and vertical positions of the alloy on the phase diagram.

For the situation in which two phases coexist—for example, $\alpha + L$, $\beta + L$, and $\alpha + \beta$ phase regions (Figure 11.6)—the phase rule stipulates that we have but one degree of freedom because

$$F = 3 - P$$
$$= 3 - 2 = 1$$

Thus, it is necessary to specify either temperature or the composition of one of the phases to completely define the system. For example, suppose that we decide to specify temperature for the $\alpha + L$ phase region, say, T_1 in Figure 11.22. The compositions of the α and liquid phases (C_α and C_L) are thus dictated by the extremes of the tie line constructed at T_1 across the $\alpha + L$ field. Note that only the nature of the phases is important in this treatment and not the relative phase amounts. This is to say that the overall alloy composition could lie anywhere along this tie line constructed at temperature T_1 and still give C_α and C_L compositions for the respective α and liquid phases.

The second alternative is to stipulate the composition of one of the phases for this two-phase situation, which thereby fixes completely the state of the system. For example, if we specified C_α as the composition of the α phase that is in equilibrium with the liquid (Figure 11.22), then both the temperature of the alloy (T_1) and the composition of the liquid phase (C_L) are established, again by the tie line drawn across the $\alpha + L$ phase field so as to give this C_α composition.

Figure 11.22 Enlarged copper-rich section of the Cu–Ag phase diagram in which the Gibbs phase rule for the coexistence of two phases (α and L) is demonstrated. Once the composition of either phase (C_α or C_L) or the temperature (T_1) is specified, values for the two remaining parameters are established by construction of the appropriate tie line.

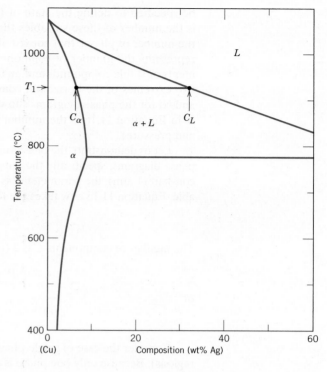

For binary systems, when three phases are present, there are no degrees of freedom because

$$F = 3 - P$$
$$= 3 - 3 = 0$$

This means that the compositions of all three phases—as well as the temperature—are fixed. This condition is met for a eutectic system by the eutectic isotherm; for the Cu–Ag system (Figure 11.6), it is the horizontal line that extends between points B and G. At this temperature, 779°C, the points at which each of the α, L, and β phase fields touch the isotherm line correspond to the respective phase compositions; namely, the composition of the α phase is fixed at 8.0 wt% Ag, that of the liquid at 71.9 wt% Ag, and that of the β phase at 91.2 wt% Ag. Thus, three-phase equilibrium is not represented by a phase field, but rather by the unique horizontal isotherm line. Furthermore, all three phases are in equilibrium for any alloy composition that lies along the length of the eutectic isotherm (e.g., for the Cu–Ag system at 779°C and compositions between 8.0 and 91.2 wt% Ag).

One use of the Gibbs phase rule is in analyzing for nonequilibrium conditions. For example, a microstructure for a binary alloy that developed over a range of temperatures and consists of three phases is a nonequilibrium one; under these circumstances, three phases exist only at a single temperature.

Concept Check 11.8   For a ternary system, three components are present; temperature is also a variable. What is the maximum number of phases that may be present for a ternary system, assuming that pressure is held constant?

[*The answer may be found at* www.wiley.com/college/callister (*Student Companion Site*).]

The Iron–Carbon System

Of all binary alloy systems, the
and carbon. Both steels and ca
logically advanced culture, are
to a study of the phase diagram
possible microstructures. The re
mechanical properties are explo

11.18 THE IRON–IRON CARBIDE (Fe
PHASE DIAGRAM

ferrite

austenite

A portion of the iron–carbon p
upon heating, experiences two
temperature, the stable form,
Ferrite experiences a polymor
912°C. This austenite persists

f iron, when alloyed with carbon alone, is not stable be-
e 11.23. The maximum solubility of carbon in austenite,
is solubility is approximately 100 times greater than the
use the FCC octahedral sites are larger than the BCC
of Problem 6.10), and, therefore, the strains imposed on
much lower. As the discussions that follow demonstrate,
g austenite are very important in the heat treating of
mentioned that austenite is nonmagnetic. Figure 11.24b
s austenite phase.[2]

same as α-ferrite, except for the range of temperatures
e the δ-ferrite is stable only at relatively high tempera-
nportance and is not discussed further.

hen the solubility limit of carbon in α-ferrite is ex-
ositions within the α + Fe₃C phase region). As indi-
coexists with the γ phase between 727°C and 1147°C.

hard and brittle; the strength of some steels is greatly

e is only metastable; that is, it remains as a compound
re. However, if heated to between 650°C and 700°C for
es or transforms into α-iron and carbon, in the form of
bsequent cooling to room temperature. Thus, the phase
true equilibrium one because cementite is not an equi-
ecause the decomposition rate of cementite is extremely
in steel is as Fe₃C instead of graphite, and the iron–iron
ll practical purposes, valid. As will be seen in Section
irons greatly accelerates this cementite decomposition

abeled in Figure 11.23. It may be noted that one eutectic
stem, at 4.30 wt% C and 1147°C; for this eutectic reaction,

$$L \underset{\text{heating}}{\overset{\text{cooling}}{\rightleftharpoons}} \gamma + Fe_3C \qquad (11.18)$$

enite and cementite phases. Subsequent cooling to room
al phase changes.

ctoid invariant point exists at a composition of 0.76 wt%
This eutectoid reaction may be represented by

$$\alpha(0.022 \text{ wt\% C}) + Fe_3C(6.7 \text{ wt\% C}) \qquad (11.19)$$

ase is transformed into α-iron and cementite. (Eutectoid
dressed in Section 11.14.) The eutectoid phase changes
e very important, being fundamental to the heat treat-
ubsequent discussions.

which iron is the prime component, but carbon as well
be present. In the classification scheme of ferrous alloys
are three types: iron, steel, and cast iron. Commercially
008 wt% C and, from the phase diagram, is composed

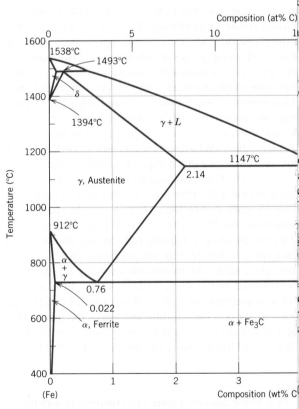

Figure 11.23 The iron–iron carbide phase diagram.
[Adapted from *Binary Alloy Phase Diagrams,* 2nd edition, Vol. 1
of ASM International, Materials Park, OH.]

structure (Section 6.8), may be observed in this photomi-
which explains their absence in the ferrite micrograph of

almost exclusively of the ferrite phase at room temperature. The iron–carbon alloys that contain between 0.008 and 2.14 wt% C are classified as steels. In most steels, the microstructure consists of both α and Fe_3C phases. Upon cooling to room temperature, an alloy within this composition range must pass through at least a portion of the γ-phase field; distinctive microstructures are subsequently produced, as discussed shortly. Although a steel alloy may contain as much as 2.14 wt% C, in practice, carbon concentrations rarely exceed 1.0 wt%. The properties and various classifications of steels are treated in Section 13.2. Cast irons are classified as ferrous alloys that contain between 2.14 and 6.70 wt% C. However, commercial cast irons normally contain less than 4.5 wt% C. These alloys are discussed further in Section 13.3.

11.19 DEVELOPMENT OF MICROSTRUCTURE IN IRON–CARBON ALLOYS

Several of the various microstructures that may be produced in steel alloys and their relationships to the iron–iron carbon phase diagram are now discussed, and it is shown that the microstructure that develops depends on both the carbon content and heat treatment. This discussion is confined to very slow cooling of steel alloys, in which equilibrium is continuously maintained. A more detailed exploration of the influence of heat treatment on microstructure, and ultimately on the mechanical properties of steels, is contained in Chapter 12.

Phase changes that occur upon passing from the γ region into the $\alpha + Fe_3C$ phase field (Figure 11.23) are relatively complex and similar to those described for the

Figure 11.25 Schematic representations of the microstructures for an iron–carbon alloy of eutectoid composition (0.76 wt% C) above and below the eutectoid temperature.

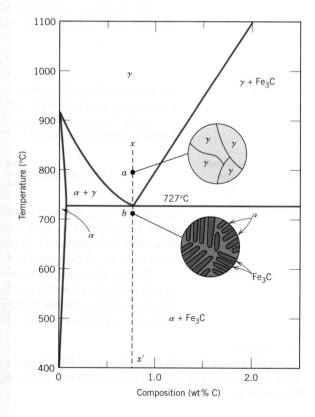

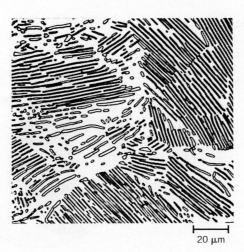

Figure 11.26 Photomicrograph of a eutectoid steel showing the pearlite microstructure consisting of alternating layers of α-ferrite (the light phase) and Fe_3C (thin layers most of which appear dark). 470×. (From *Metals Handbook,* Vol. 9, 9th edition, *Metallography and Microstructures,* 1985. Reproduced by permission of ASM International, Materials Park, OH.)

20 μm

eutectic systems in Section 11.12. Consider, for example, an alloy of eutectoid composition (0.76 wt% C) as it is cooled from a temperature within the γ-phase region, say, 800°C—that is, beginning at point *a* in Figure 11.25 and moving down the vertical line *xx'*. Initially, the alloy is composed entirely of the austenite phase having a composition of 0.76 wt% C and corresponding microstructure, also indicated in Figure 11.25. As the alloy is cooled, no changes occur until the eutectoid temperature (727°C) is reached. Upon crossing this temperature to point *b*, the austenite transforms according to Equation 11.19.

The microstructure for this eutectoid steel that is slowly cooled through the eutectoid temperature consists of alternating layers or lamellae of the two phases (α and Fe_3C) that form simultaneously during the transformation. In this case, the relative layer thickness is approximately 8 to 1. This microstructure, represented schematically in Figure 11.25, point *b*, is called **pearlite** because it has the appearance of mother-of-pearl when viewed under the microscope at low magnifications. Figure 11.26 is a photomicrograph of a eutectoid steel showing the pearlite. The pearlite exists as grains, often termed colonies; within each colony the layers are oriented in essentially the same direction, which varies from one colony to another. The thick light layers are the ferrite phase, and the cementite phase appears as thin lamellae, most of which appear dark. Many cementite layers are so thin that adjacent phase boundaries are so close together that they are indistinguishable at this magnification and, therefore, appear dark. Mechanically, pearlite has properties intermediate between those of the soft, ductile ferrite and the hard, brittle cementite.

The alternating α and Fe_3C layers in pearlite form for the same reason that the eutectic structure (Figures 11.12 and 11.13) forms—because the composition of the parent phase [in this case, austenite (0.76 wt% C)] is different from that of either of the product phases [ferrite (0.022 wt% C) and cementite (6.70 wt% C)], and the phase transformation requires that there be a redistribution of the carbon by diffusion. Figure 11.27 illustrates microstructural changes that accompany this eutectoid reaction; here, the directions of carbon diffusion are indicated by arrows. Carbon atoms diffuse away from the 0.022-wt% ferrite regions and to the 6.70-wt% cementite layers, as the pearlite extends from the grain boundary into the unreacted austenite grain. The layered pearlite forms because carbon atoms need diffuse only minimal distances with the formation of this structure.

Subsequent cooling of the pearlite from point *b* in Figure 11.25 produces relatively insignificant microstructural changes.

pearlite

Figure 11.27 Schematic representation of the formation of pearlite from austenite; direction of carbon diffusion indicated by arrows.

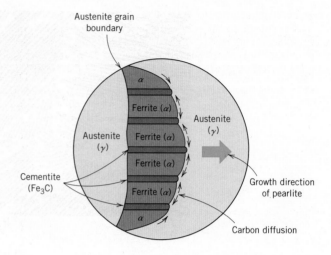

Hypoeutectoid Alloys

Microstructures for iron–iron carbide alloys having other than the eutectoid composition are now explored; these are analogous to the fourth case described in Section 11.12 and illustrated in Figure 11.15 for the eutectic system. Consider a composition C_0 to the left of the eutectoid, between 0.022 and 0.76 wt% C; this is termed a **hypoeutectoid** ("less than eutectoid") **alloy**. Cooling an alloy of this composition is represented by moving down the vertical line yy' in Figure 11.28. At about 875°C, point c, the microstructure consists

hypoeutectoid alloy

Figure 11.28 Schematic representations of the microstructures for an iron–carbon alloy of hypoeutectoid composition C_0 (containing less than 0.76 wt% C) as it is cooled from within the austenite phase region to below the eutectoid temperature.

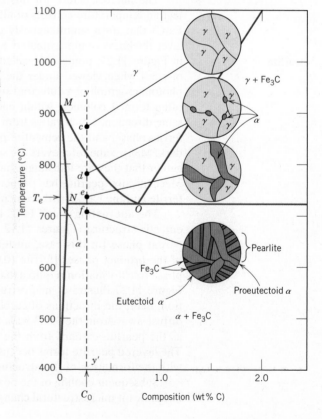

Scanning electron micrograph showing the microstructure of a steel that contains 0.44 wt% C. The large dark areas are proeutectoid ferrite. Regions having the alternating light and dark lamellar structure are pearlite; the dark and light layers in the pearlite correspond, respectively, to ferrite and cementite phases. 700×. (Micrograph courtesy of Republic Steel Corporation.)

proeutectoid ferrite

entirely of grains of the γ phase, as shown schematically in the figure. In cooling to point d, about 775°C, which is within the $\alpha + \gamma$ phase region, both these phases coexist as in the schematic microstructure. Most of the small α particles form along the original γ grain boundaries. The compositions of both α and γ phases may be determined using the appropriate tie line; these compositions correspond, respectively, to about 0.020 and 0.40 wt% C.

While cooling an alloy through the $\alpha + \gamma$ phase region, the composition of the ferrite phase changes with temperature along the $\alpha - (\alpha + \gamma)$ phase boundary, line MN, becoming slightly richer in carbon. However, the change in composition of the austenite is more dramatic, proceeding along the $(\alpha + \gamma) - \gamma$ boundary, line MO, as the temperature is reduced.

Cooling from point d to e, just above the eutectoid but still in the $\alpha + \gamma$ region, produces an increased fraction of the α phase and a microstructure similar to that also shown: the α particles will have grown larger. At this point, the compositions of the α and γ phases are determined by constructing a tie line at the temperature T_e; the α phase contains 0.022 wt% C, whereas the γ phase is of the eutectoid composition, 0.76 wt% C.

As the temperature is lowered just below the eutectoid, to point f, all of the γ phase that was present at temperature T_e (and having the eutectoid composition) transforms into pearlite, according to the reaction in Equation 11.19. There is virtually no change in the α phase that existed at point e in crossing the eutectoid temperature—it is normally present as a continuous matrix phase surrounding the isolated pearlite colonies. The microstructure at point f appears as the corresponding schematic inset of Figure 11.28. Thus the ferrite phase is present both in the pearlite and as the phase that formed while cooling through the $\alpha + \gamma$ phase region. The ferrite present in the pearlite is called *eutectoid ferrite*, whereas the other, which formed above T_e, is termed **proeutectoid** (meaning "pre- or before eutectoid") **ferrite**, as labeled in Figure 11.28. Figure 11.29 is a photomicrograph of a 0.38-wt% C steel; large, white regions correspond to the proeutectoid ferrite. For pearlite, the spacing between the α and Fe$_3$C layers varies from grain to grain; some

Figure 11.29 Photomicrograph of a 0.38 wt% C steel having a microstructure consisting of pearlite and proeutectoid ferrite. 635×. (Photomicrograph courtesy of Republic Steel Corporation.)

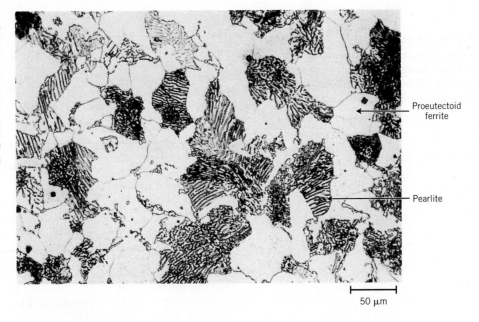

Proeutectoid ferrite

Pearlite

50 μm

of the pearlite appears dark because the many close-spaced layers are unresolved at the magnification of the photomicrograph. Note that two microconstituents are present in this micrograph—proeutectoid ferrite and pearlite—which appear in all hypoeutectoid iron–carbon alloys that are slowly cooled to a temperature below the eutectoid.

The relative amounts of the proeutectoid α and pearlite may be determined in a manner similar to that described in Section 11.12 for primary and eutectic microconstituents. We use the lever rule in conjunction with a tie line that extends from the $\alpha - (\alpha + Fe_3C)$ phase boundary (0.022-wt% C) to the eutectoid composition (0.76-wt% C) inasmuch as pearlite is the transformation product of austenite having this composition. For example, let us consider an alloy of composition C_0' in Figure 11.30. The fraction of pearlite, W_p, may be determined according to

Lever rule expression for computation of pearlite mass fraction (composition C_0', Figure 11.30)

$$W_p = \frac{T}{T + U}$$

$$= \frac{C_0' - 0.022}{0.76 - 0.022} = \frac{C_0' - 0.022}{0.74} \qquad (11.20)$$

The fraction of proeutectoid α, $W_{\alpha'}$, is computed as follows:

Lever rule expression for computation of proeutectoid ferrite mass fraction

$$W_{\alpha'} = \frac{U}{T + U}$$

$$= \frac{0.76 - C_0'}{0.76 - 0.022} = \frac{0.76 - C_0'}{0.74} \qquad (11.21)$$

Fractions of both total α (eutectoid and proeutectoid) and cementite are determined using the lever rule and a tie line that extends across the entirety of the $\alpha + Fe_3C$ phase region, from 0.022 to 6.70 wt% C.

Figure 11.30 A portion of the Fe–Fe$_3$C phase diagram used in computing the relative amounts of proeutectoid and pearlite microconstituents for hypoeutectoid C_0' and hypereutectoid C_1' compositions.

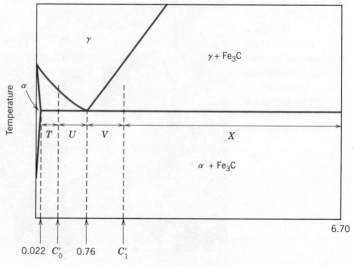

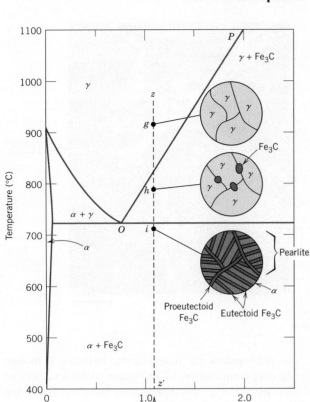

Figure 11.31 Schematic representations of the microstructures for an iron–carbon alloy of hypereutectoid composition C_1 (containing between 0.76 and 2.14 wt% C) as it is cooled from within the austenite-phase region to below the eutectoid temperature.

Hypereutectoid Alloys

hypereutectoid alloy

Analogous transformations and microstructures result for **hypereutectoid alloys**—those containing between 0.76 and 2.14 wt% C—that are cooled from temperatures within the γ-phase field. Consider an alloy of composition C_1 in Figure 11.31 that, upon cooling, moves down the line zz'. At point g, only the γ phase is present with a composition of C_1; the microstructure appears as shown, having only γ grains. Upon cooling into the $\gamma + Fe_3C$ phase field—say, to point h—the cementite phase begins to form along the initial γ grain boundaries, similar to the α phase in Figure 11.28,

proeutectoid cementite

point d. This cementite is called **proeutectoid cementite**—that which forms before the eutectoid reaction. The cementite composition remains constant (6.70 wt% C) as the temperature changes. However, the composition of the austenite phase moves along line PO toward the eutectoid. As the temperature is lowered through the eutectoid to point i, all remaining austenite of eutectoid composition is converted into pearlite; thus, the resulting microstructure consists of pearlite and proeutectoid cementite as microconstituents (Figure 11.31). In the photomicrograph of a 1.4-wt% C steel (Figure 11.32), note that the proeutectoid cementite appears light. Because it has much the same appearance as proeutectoid ferrite (Figure 11.29), there is some difficulty in distinguishing between hypoeutectoid and hypereutectoid steels on the basis of microstructure.

Relative amounts of both pearlite and proeutectoid Fe_3C microconstituents may be computed for hypereutectoid steel alloys in a manner analogous to that

Figure 11.32 Photomicrograph of a 1.4 wt% C steel having a microstructure consisting of a white proeutectoid cementite network surrounding the pearlite colonies. 1000×. (Copyright 1971 by United States Steel Corporation.)

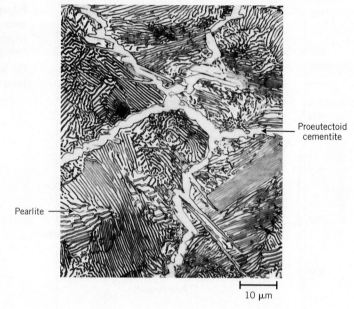

Proeutectoid cementite

Pearlite

10 μm

for hypoeutectoid materials; the appropriate tie line extends between 0.76 and 6.70 wt% C. Thus, for an alloy having composition C'_1 in Figure 10.30, fractions of pearlite W_p and proeutectoid cementite $W_{Fe_3C'}$ are determined from the following lever rule expressions:

$$W_p = \frac{X}{V + X} = \frac{6.70 - C'_1}{6.70 - 0.76} = \frac{6.70 - C'_1}{5.94} \tag{11.22}$$

and

$$W_{Fe_3C'} = \frac{V}{V + X} = \frac{C'_1 - 0.76}{6.70 - 0.76} = \frac{C'_1 - 0.76}{5.94} \tag{11.23}$$

 Concept Check 11.9   Briefly explain why a proeutectoid phase (ferrite or cementite) forms along austenite grain boundaries. *Hint*: Consult Section 6.8.

[*The answer may be found at* www.wiley.com/college/callister *(Student Companion Site).*]

EXAMPLE PROBLEM 11.4

Determination of Relative Amounts of Ferrite, Cementite, and Pearlite Microconstituents

For a 99.65 wt% Fe–0.35 wt% C alloy at a temperature just below the eutectoid, determine the following:

(a) The fractions of total ferrite and cementite phases

(b) The fractions of the proeutectoid ferrite and pearlite

(c) The fraction of eutectoid ferrite

Solution

(a) This part of the problem is solved by applying the lever rule expressions using a tie line that extends all the way across the $\alpha + Fe_3C$ phase field. Thus, C_0' is 0.35 wt% C, and

$$W_\alpha = \frac{6.70 - 0.35}{6.70 - 0.022} = 0.95$$

and

$$W_{Fe_3C} = \frac{0.35 - 0.022}{6.70 - 0.022} = 0.05$$

(b) The fractions of proeutectoid ferrite and pearlite are determined by using the lever rule and a tie line that extends only to the eutectoid composition (i.e., Equations 11.20 and 11.21). We have

$$W_p = \frac{0.35 - 0.022}{0.76 - 0.022} = 0.44$$

and

$$W_{\alpha'} = \frac{0.76 - 0.35}{0.76 - 0.022} = 0.56$$

(c) All ferrite is either as proeutectoid or eutectoid (in the pearlite). Therefore, the sum of these two ferrite fractions equals the fraction of total ferrite; that is,

$$W_{\alpha'} + W_{\alpha e} = W_\alpha$$

where $W_{\alpha e}$ denotes the fraction of the total alloy that is eutectoid ferrite. Values for W_α and $W_{\alpha'}$ were determined in parts (a) and (b) as 0.95 and 0.56, respectively. Therefore,

$$W_{\alpha e} = W_\alpha - W_{\alpha'} = 0.95 - 0.56 = 0.39$$

Nonequilibrium Cooling

In this discussion of the microstructural development of iron–carbon alloys, it has been assumed that, upon cooling, conditions of metastable equilibrium[3] have been continuously maintained; that is, sufficient time has been allowed at each new temperature for any necessary adjustment in phase compositions and relative amounts as predicted from the Fe–Fe$_3$C phase diagram. In most situations these cooling rates are impractically slow and unnecessary; in fact, on many occasions nonequilibrium conditions are desirable. Two nonequilibrium effects of practical importance are (1) the occurrence of phase changes or transformations at temperatures other than those predicted by phase boundary lines on the phase diagram, and (2) the existence at room temperature of nonequilibrium phases that do not appear on the phase diagram. Both are discussed in Chapter 12.

---

[3]The term *metastable equilibrium* is used in this discussion because Fe$_3$C is only a metastable compound.

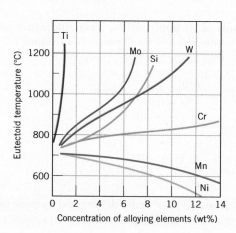

Figure 11.33 The dependence of eutectoid temperature on alloy concentration for several alloying elements in steel.
(From Edgar C. Bain, *Functions of the Alloying Elements in Steel,* 1939. Reproduced by permission of ASM International, Materials Park, OH.)

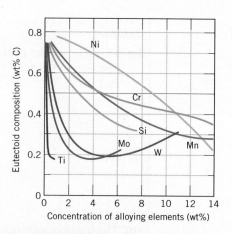

Figure 11.34 The dependence of eutectoid composition (wt% C) on alloy concentration for several alloying elements in steel.
(From Edgar C. Bain, *Functions of the Alloying Elements in Steel,* 1939. Reproduced by permission of ASM International, Materials Park, OH.)

11.20 THE INFLUENCE OF OTHER ALLOYING ELEMENTS

Additions of other alloying elements (Cr, Ni, Ti, etc.) bring about rather dramatic changes in the binary iron–iron carbide phase diagram, Figure 11.23. The extent of these alterations of the positions of phase boundaries and the shapes of the phase fields depends on the particular alloying element and its concentration. One of the important changes is the shift in position of the eutectoid with respect to temperature and carbon concentration. These effects are illustrated in Figures 11.33 and 11.34, which plot the eutectoid temperature and eutectoid composition (in wt% C), respectively, as a function of concentration for several other alloying elements. Thus, other alloy additions alter not only the temperature of the eutectoid reaction, but also the relative fractions of pearlite and the proeutectoid phase that form. Steels are normally alloyed for other reasons, however—usually either to improve their corrosion resistance or to render them amenable to heat treatment (see Section 17.6).

SUMMARY

Introduction
- Equilibrium phase diagrams are a convenient and concise way of representing the most stable relationships between phases in alloy systems.

Phases
- A *phase* is some portion of a body of material throughout which the physical and chemical characteristics are homogeneous.

Microstructure
- Three microstructural characteristics that are important for multiphase alloys are
  The number of phases present
  The relative proportions of the phases
  The manner in which the phases are arranged
- Three factors affect the microstructure of an alloy:
  What alloying elements are present
  The concentrations of these alloying elements
  The heat treatment of the alloy

Phase Equilibria
- A system at equilibrium is in its most stable state—that is, its phase characteristics do not change over time. Thermodynamically, the condition for phase equilibrium is that the free energy of a system is a minimum for some set combination of temperature, pressure, and composition.
- Metastable systems are nonequilibrium ones that persist indefinitely and experience imperceptible changes with time.

One-Component (or Unary) Phase Diagrams
- For one-component phase diagrams, the logarithm of the pressure is plotted versus the temperature; solid-, liquid-, and vapor-phase regions are found on this type of diagram.

Binary Phase Diagrams
- For binary systems, temperature and composition are variables, whereas external pressure is held constant. Areas, or phase regions, are defined on these temperature-versus-composition plots within which either one or two phases exist.

Binary Isomorphous Systems
- Isomorphous diagrams are those for which there is complete solubility in the solid phase; the copper–nickel system (Figure 11.3a) displays this behavior.

Interpretation of Phase Diagrams
- For an alloy of specified composition at a known temperature and that is at equilibrium, the following may be determined:

  What phase(s) is (are) present—from the location of the temperature–composition point on the phase diagram.

  Phase composition(s)—a horizontal tie line is used for the two-phase situation.

  Phase mass fraction(s)—the lever rule [which uses tie-line segment lengths (Equations 11.1 and 11.2)] is used in two-phase regions.

Binary Eutectic Systems
- In a eutectic reaction, as found in some alloy systems, a liquid phase transforms isothermally into two different solid phases upon cooling (i.e., $L \rightarrow \alpha + \beta$). Such a reaction is noted on the copper–silver and lead–tin phase diagrams (Figures 11.6 and 11.7, respectively).
- The solubility limit at some temperature corresponds to the maximum concentration of one component that will go into solution in a specific phase. For a binary eutectic system, solubility limits are to be found along solidus and solvus phase boundaries.

Development of Microstructure in Eutectic Alloys
- The solidification of an alloy (liquid) of eutectic composition yields a microstructure consisting of layers of the two solid phases that alternate.
- A primary (or pre-eutectic) phase and the layered eutectic structure are the solidification products for all compositions (other than the eutectic) that lie along the eutectic isotherm.
- Mass fractions of the primary phase and eutectic microconstituent may be computed using the lever rule and a tie line that extends to the eutectic composition (e.g., Equations 11.10 and 11.11).

Equilibrium Diagrams Having Intermediate Phases or Compounds
- Other equilibrium diagrams are more complex, in that they may have phases/solid solutions/compounds that do not lie at the concentration (i.e., horizontal) extremes on the diagram. These include intermediate solid solutions and intermetallic compounds.
- In addition to the eutectic, other reactions involving three phases may occur at invariant points on a phase diagram:

  For a eutectoid reaction, upon cooling, one solid phase transforms into two other solid phases (e.g., $\alpha \rightarrow \beta + \gamma$).

  For a peritectic reaction, upon cooling, a liquid and one solid phase transform into another solid phase (e.g., $L + \alpha \rightarrow \beta$).

- A transformation in which there is no change in composition for the phases involved is congruent.

The Gibbs Phase Rule
- The Gibbs phase rule is a simple equation (Equation 11.16 in its most general form) that relates the number of phases present in a system at equilibrium with the number of degrees of freedom, the number of components, and the number of noncompositional variables.

The Iron–Iron Carbide (Fe–Fe₃C) Phase Diagram
- Important phases found on the iron–iron carbide phase diagram (Figure 11.23) are α-ferrite (BCC), γ-austenite (FCC), and the intermetallic compound iron carbide [or cementite (Fe_3C)].
- On the basis of composition, ferrous alloys fall into three classifications:
  Irons (<0.008 wt% C)
  Steels (0.008 to 2.14 wt% C)
  Cast irons (>2.14 wt% C)

Development of Microstructure in Iron–Carbon Alloys
- The development of microstructure for many iron–carbon alloys and steels depends on a eutectoid reaction in which the austenite phase of composition 0.76 wt% C transforms isothermally (at 727°C) into α-ferrite (0.022 wt% C) and cementite (i.e., $\gamma \rightarrow \alpha + Fe_3C$).
- The microstructural product of an iron–carbon alloy of eutectoid composition is pearlite, a microconstituent consisting of alternating layers of ferrite and cementite.
- The microstructures of alloys having carbon contents less than the eutectoid (i.e., hypoeutectoid alloys) are composed of a proeutectoid ferrite phase in addition to pearlite.
- Pearlite and proeutectoid cementite constitute the microconstituents for hypereutectoid alloys—those with carbon contents in excess of the eutectoid composition.
- Mass fractions of a proeutectoid phase (ferrite or cementite) and pearlite may be computed using the lever rule and a tie line that extends to the eutectoid composition (0.76 wt% C) [e.g., Equations 11.20 and 11.21 (for hypoeutectoid alloys) and Equations 11.22 and 11.23 (for hypereutectoid alloys)].

Important Terms and Concepts

austenite
cementite
component
congruent transformation
equilibrium
eutectic phase
eutectic reaction
eutectic structure
eutectoid reaction
ferrite
free energy
Gibbs phase rule

hypereutectoid alloy
hypoeutectoid alloy
intermediate solid solution
intermetallic compound
isomorphous
lever rule
liquidus line
metastable
microconstituent
pearlite
peritectic reaction
phase

phase diagram
phase equilibrium
primary phase
proeutectoid cementite
proeutectoid ferrite
solidus line
solubility limit
solvus line
system
terminal solid solution
tie line

REFERENCES

ASM Handbook, Vol. 3, *Alloy Phase Diagrams,* ASM International, Materials Park, OH, 1992.

ASM Handbook, Vol. 9, *Metallography and Microstructures,* ASM International, Materials Park, OH, 2004.

Campbell, F. C., *Phase Diagrams: Understanding the Basics,* ASM International, Materials Park, OH, 2012.

Massalski, T. B., H. Okamoto, P. R. Subramanian, and L. Kacprzak (Editors), *Binary Phase Diagrams,* 2nd edition, ASM International, Materials Park, OH, 1990. Three volumes. Also on CD-ROM with updates.

Okamoto, H., *Desk Handbook: Phase Diagrams for Binary Alloys,* 2nd edition, ASM International, Materials Park, OH, 2010.

Villars, P., A. Prince, and H. Okamoto (Editors), *Handbook of Ternary Alloy Phase Diagrams,* ASM International, Materials Park, OH, 1995. Ten volumes. Also on CD-ROM.

QUESTIONS AND PROBLEMS

Solubility Limit

11.1 Consider the sugar–water phase diagram of Figure 11.1.

    (a) How much sugar will dissolve in 1500 g of water at 90°C (363 K)?

    (b) If the saturated liquid solution in part (a) is cooled to 20°C (273 K), some of the sugar will precipitate out as a solid. What will be the composition of the saturated liquid solution (in wt% sugar) at 20°C?

    (c) How much of the solid sugar will come out of solution upon cooling to 20°C?

11.2 At 500°C (773 K), what is the maximum solubility **(a)** of Cu in Ag? **(b)** Of Ag in Cu?

Microstructure

11.3 Cite three variables that determine the microstructure of an alloy.

Phase Equilibria

11.4 What thermodynamic condition must be met for a state of equilibrium to exist?

One-Component (or Unary) Phase Diagrams

11.5 Consider a specimen of ice that is at −10°C and 101.3 kPa pressure. Using Figure 11.2, the pressure–temperature phase diagram for H_2O, determine the pressure to which the specimen must be raised or lowered to cause it **(a)** to melt, and **(b)** to sublime.

11.6 At a pressure of 1 kPa, determine **(a)** the melting temperature for ice, and **(b)** the boiling temperature for water.

Binary Isomorphous Systems

11.7 Given here are the solidus and liquidus temperatures for the germanium–silicon system. Construct the phase diagram for this system and label each region.

Composition (wt% Si)	Solidus Temperature (°C)	Solidus Temperature (K)	Liquidus Temperature (°C)	Liquidus Temperature (K)
0	938	1211	938	1211
10	1005	1278	1147	1420
20	1065	1338	1226	1499
30	1123	1396	1278	1551
40	1178	1451	1315	1588
50	1232	1505	1346	1619
60	1282	1555	1367	1640
70	1326	1599	1385	1658
80	1359	1632	1397	1670
90	1390	1663	1408	1681
100	1414	1687	1414	1687

Assume that (1) α and β phases exist at the A and B extremities of the phase diagram, respectively; (2) the eutectic composition is 47 wt% B–53 wt% A; and (3) the composition of the β phase at the eutectic temperature is 92.6 wt% B–7.4 wt% A. Determine the composition of an alloy that will yield primary α and total α mass fractions of 0.356 and 0.693, respectively.

11.35 For an 85 wt% Pb–15 wt% Mg alloy, make schematic sketches of the microstructure that would be observed for conditions of very slow cooling at the following temperatures: 600°C (873 K), 500°C (773 K), 270°C (543 K), and 200°C (473 K). Label all phases and indicate their approximate compositions.

11.36 For a 68 wt% Zn–32 wt% Cu alloy, make schematic sketches of the microstructure that would be observed for conditions of very slow cooling at the following temperatures: 1000°C (1273 K), 760°C (1033 K), 600°C (1873 K), and 400°C (673 K). Label all phases and indicate their approximate compositions.

11.37 For a 30 wt% Zn–70 wt% Cu alloy, make schematic sketches of the microstructure that would be observed for conditions of very slow cooling at the following temperatures: 1100°C (1373 K), 950°C (1223 K), 900°C (1173 K), and 700°C (973 K). Label all phases and indicate their approximate compositions.

11.38 On the basis of the photomicrograph (i.e., the relative amounts of the microconstituents) for the lead–tin alloy shown in Figure 11.16 and the Pb–Sn phase diagram (Figure 11.7), estimate the composition of the alloy, and then compare this estimate with the composition given in the figure legend of Figure 11.16. Make the following assumptions: (1) The area

fraction of each phase and microconstituent in the photomicrograph is equal to its volume fraction; (2) the densities of the α and β phases as well as the eutectic structure are 11.2, 7.3, and 8.7 g/cm³, respectively; and (3) this photomicrograph represents the equilibrium microstructure at 180°C (453 K).

11.39 The room-temperature tensile strengths of pure lead and pure tin are 17.0 MPa and 14.7 MPa, respectively.

(a) Make a schematic graph of the room-temperature tensile strength versus composition for all compositions between pure lead and pure tin. (*Hint:* You may want to consult Sections 11.10 and 11.11, as well as Equation 11.24 in Problem 11.64.)

(b) On this same graph schematically plot tensile strength versus composition at 150°C.

(c) Explain the shapes of these two curves, as well as any differences between them.

Equilibrium Diagrams Having Intermediate Phases or Compounds

11.40 Two intermetallic compounds, AB and AB_2, exist for elements A and B. If the compositions for AB and AB_2 are 34.5 wt% A–65.5 wt% B and 20.5 wt% A–79.5 wt% B, respectively, and element A is potassium, identify element B.

Eutectoid and Peritectic Reactions
Congruent Phase Transformations

11.41 What is the principal difference between congruent and incongruent phase transformations?

11.42 Figure 11.35 is the aluminum–neodymium phase diagram, for which only single–phase regions are labeled. Specify temperature–composition

Figure 11.35 The aluminum–neodymium phase diagram. (Adapted from *ASM Handbook*, Vol. 3, *Alloy Phase Diagrams*, H. Baker, Editor, 1992. Reprinted by permission of ASM International, Materials Park, OH.)

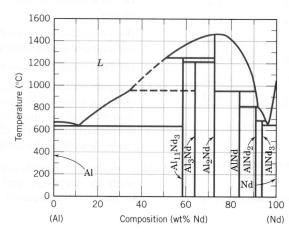

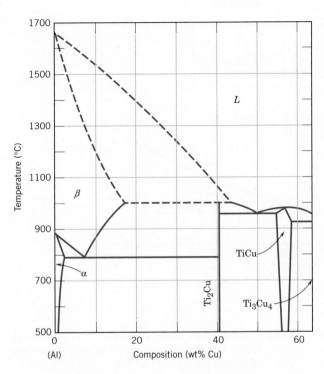

Figure 11.36 The titanium–copper phase diagram. (Adapted from *Phase Diagrams of Binary Titanium Alloys,* J. L. Murray, Editor, 1987. Reprinted by permission of ASM International, Materials Park, OH.)

points at which all eutectics, eutectoids, peritectics, and congruent phase transformations occur. Also, for each, write the reaction upon cooling.

11.43 Figure 11.36 is a portion of the titanium–copper phase diagram for which only single-phase regions are labeled. Specify all temperature–composition points at which eutectics, eutectoids, peritectics, and congruent phase transformations occur. Also, for each, write the reaction upon cooling.

11.44 Construct the hypothetical phase diagram for metals A and B between temperatures of 600°C and 1000°C given the following information:

- The melting temperature of metal A is 940°C.
- The solubility of B in A is negligible at all temperatures.
- The melting temperature of metal B is 830°C.
- The maximum solubility of A in B is 12 wt% A, which occurs at 700°C.
- At 600°C, the solubility of A in B is 8 wt% A.
- One eutectic occurs at 700°C and 75 wt% B–25 wt% A.
- A second eutectic occurs at 730°C and 60 wt% B–40 wt% A.

- A third eutectic occurs at 755°C and 40 wt% B–60 wt% A.
- One congruent melting point occurs at 780°C and 51 wt% B–49 wt% A.
- A second congruent melting point occurs at 780°C and 67 wt% B–33 wt% A.
- The intermetallic compound AB exists at 51 wt% B–49 wt% A.
- The intermetallic compound AB_2 exists at 67 wt% B–33 wt% A.

The Gibbs Phase Rule

11.45 Figure 11.37 shows the pressure–temperature phase diagram for H_2O. Apply the Gibbs phase rule at points *A*, *B*, and *C*; that is, specify the number of degrees of freedom at each of the points—that is, the number of externally controllable variables that need be specified to completely define the system.

The Iron–Iron Carbide (Fe–Fe₃C) Phase Diagram
Development of Microstructure in Iron–Carbon Alloys

11.46 Compute the mass fractions of α-ferrite and cementite in pearlite.

11.47 (a) What is the distinction between hypoeutectoid and hypereutectoid steels?

Figure 11.37
Logarithm pressure-versus-temperature phase diagram for H_2O.

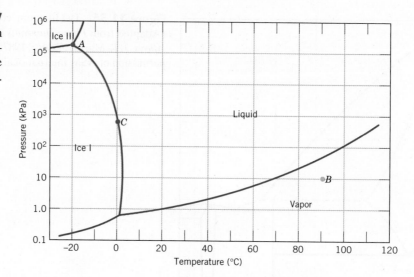

(b) In a hypoeutectoid steel, both eutectoid and proeutectoid ferrite exist. Explain the difference between them. What will be the carbon concentration in each?

11.48 What is the carbon concentration of an iron–carbon alloy for which the fraction of total ferrite is 0.94?

11.49 What is the proeutectoid phase for an iron–carbon alloy in which the mass fractions of total ferrite and total cementite are 0.92 and 0.08, respectively? Why?

11.50 Consider 1.0 kg of austenite containing 1.15 wt% C, cooled to below 725°C (998 K).

(a) What is the proeutectoid phase?

(b) How many kilograms each of total ferrite and cementite form?

(c) How many kilograms each of pearlite and the proeutectoid phase form?

(d) Schematically sketch and label the resulting microstructure.

11.51 Consider 2.5 kg of austenite containing 0.65 wt% C, cooled to below 727°C (1000 K).

(a) What is the proeutectoid phase?

(b) How many kilograms each of total ferrite and cementite form?

(c) How many kilograms each of pearlite and the proeutectoid phase form?

(d) Schematically sketch and label the resulting microstructure.

11.52 Compute the mass fractions of proeutectoid ferrite and pearlite that form in an iron–carbon alloy containing 0.30 wt% C.

11.53 The microstructure of an iron–carbon alloy consists of proeutectoid ferrite and pearlite; the mass fractions of these two microconstituents are 0.280 and 0.720, respectively. Determine the concentration of carbon in this alloy.

11.54 The mass fractions of total ferrite and total cementite in an iron–carbon alloy are 0.88 and 0.12, respectively. Is this a hypoeutectoid or hypereutectoid alloy? Why?

11.55 The microstructure of an iron–carbon alloy consists of proeutectoid ferrite and pearlite; the mass fractions of these microconstituents are 0.20 and 0.80, respectively. Determine the concentration of carbon in this alloy.

11.56 Consider 2.0 kg of a 99.6 wt% Fe–0.5 wt% C alloy that is cooled to a temperature just below the eutectoid.

(a) How many kilograms of proeutectoid ferrite form?

(b) How many kilograms of eutectoid ferrite form?

(c) How many kilograms of cementite form?

11.57 Compute the maximum mass fraction of proeutectoid cementite possible for a hypereutectoid iron–carbon alloy.

11.58 Is it possible to have an iron–carbon alloy for which the mass fractions of total ferrite and proeutectoid cementite are 0.850 and 0.045, respectively? Why or why not?

11.59 Is it possible to have an iron–carbon alloy for which the mass fractions of total cementite and pearlite are 0.040 and 0.420, respectively? Why or why not?

11.60 Compute the mass fraction of eutectoid ferrite in an iron–carbon alloy that contains 0.45 wt% C.

11.61 The mass fraction of *eutectoid* cementite in an iron–carbon alloy is 0.102. On the basis of this information, is it possible to determine the composition of the alloy? If so, what is its composition? If this is not possible, explain why.

11.62 The mass fraction of *eutectoid* ferrite in an iron–carbon alloy is 0.84. On the basis of this information, is it possible to determine the composition of the alloy? If so, what is its composition? If this is not possible, explain why.

11.63 For an iron–carbon alloy of composition 5 wt% C–95 wt% Fe, make schematic sketches of the microstructure that would be observed for conditions of very slow cooling at the following temperatures: 1180°C (1453 K), 1150°C (1423 K), and 700°C (973 K). Label the phases and indicate their compositions (approximate).

11.64 Often, the properties of multiphase alloys may be approximated by the relationship

$$E \text{ (alloy)} = E_\alpha V_\alpha + E_\beta V_\beta \qquad (11.24)$$

where E represents a specific property (modulus of elasticity, hardness, etc.), and V is the volume fraction. The subscripts α and β denote the existing phases or microconstituents. Employ this relationship to determine the approximate Brinell hardness of a 99.80 wt% Fe–0.20 wt% C alloy. Assume Brinell hardnesses of 80 and 280 for ferrite and pearlite, respectively, and that volume fractions may be approximated by mass fractions.

The Influence of Other Alloying Elements

11.65 A steel alloy contains 97.0 wt% Fe, 2.0 wt% Mo, and 1.0 wt% C.

(a) What is the eutectoid temperature of this alloy?

(b) What is the eutectoid composition?

(c) What is the proeutectoid phase?

Assume that there are no changes in the positions of other phase boundaries with the addition of Mo.

11.66 A steel alloy is known to contain 94.0 wt% Fe, 5.0 wt% Ni, and 0.1 wt% C.

(a) What is the approximate eutectoid temperature of this alloy?

(b) What is the proeutectoid phase when this alloy is cooled to a temperature just below the eutectoid?

(c) Compute the relative amounts of the proeutectoid phase and pearlite.

Assume that there are no alterations in the positions of other phase boundaries with the addition of Ni.

Chapter 12 Phase Transformations

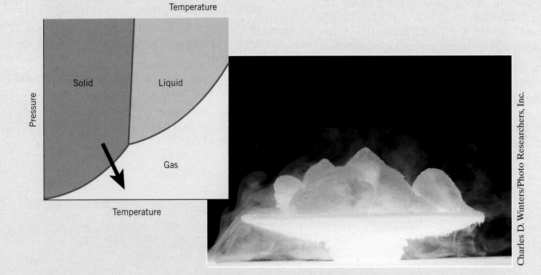

Two pressure–temperature phase diagrams are shown: for H_2O (top) and CO_2 (bottom). Phase transformations occur when phase boundaries (the red curves) on these plots are crossed as temperature and/or pressure is changed. For example, ice melts (transforms to liquid water) upon heating, which corresponds to crossing the solid-liquid phase boundary, as represented by the arrow on the H_2O phase diagram. Similarly, upon passing across the solid-gas phase boundary of the CO_2 phase diagram, dry ice (solid CO_2) sublimes (transforms into gaseous CO_2). Again, an arrow delineates this phase transformation.

Pressure

Solid
(Ice)

Liquid
(Water)

Vapor
(Steam)

Temperature

Pressure

Solid

Liquid

Gas

Temperature

The development of a set of desirable mechanical characteristics for a material often results from a phase transformation that is wrought by a heat treatment. The time and temperature dependencies of some phase transformations are conveniently represented on modified phase diagrams. It is important to know how to use these diagrams in order to design a heat treatment for some alloy that will yield the desired room-temperature mechanical properties. For example, the tensile strength of an iron–carbon alloy of eutectoid composition (0.76 wt% C) can be varied between approximately 700 MPa and 2000 MPa depending on the heat treatment employed.

Learning Objectives

After studying this chapter, you should be able to do the following:

1. Make a schematic fraction transformation-versus-logarithm of time plot for a typical solid–solid transformation; cite the equation that describes this behavior.
2. Briefly describe the microstructure for each of the following microconstituents that are found in steel alloys: fine pearlite, coarse pearlite, spheroidite, bainite, martensite, and tempered martensite.
3. Cite the general mechanical characteristics for each of the following microconstituents: fine pearlite, coarse pearlite, spheroidite, bainite, martensite, and tempered martensite; briefly explain these behaviors in terms of microstructure (or crystal structure).
4. Given the isothermal transformation (or continuous-cooling transformation) diagram for some iron–carbon alloy, design a heat treatment that will produce a specified microstructure.

12.1 INTRODUCTION

One reason metallic materials are so versatile is that their mechanical properties (strength, hardness, ductility, etc.) are subject to control and management over relatively large ranges. Three strengthening mechanisms were discussed in Chapter 9—namely grain size refinement, solid-solution strengthening, and strain hardening. Additional techniques are available in which the mechanical behavior of a metal alloy is influenced by its microstructure.

The development of microstructure in both single- and two-phase alloys typically involves some type of phase transformation—an alteration in the number and/or character of the phases. The first portion of this chapter is devoted to a brief discussion of some of the basic principles relating to transformations involving solid phases. Because most phase transformations do not occur instantaneously, consideration is given to the dependence of reaction progress on time, or the **transformation rate**. This is followed by a discussion of the development of two-phase microstructures for iron–carbon alloys. Modified phase diagrams are introduced that permit determination of the microstructure that results from a specific heat treatment. Finally, other microconstituents in addition to pearlite are presented and, for each, the mechanical properties are discussed.

transformation rate

Phase Transformations

12.2 BASIC CONCEPTS

phase transformation

A variety of **phase transformations** are important in the processing of materials, and usually they involve some alteration of the microstructure. For purposes of this discussion, these transformations are divided into three classifications. In one group are simple diffusion-dependent transformations in which there is no change in either the number or composition of the phases present. These include solidification of a pure metal, allotropic transformations, and recrystallization and grain growth (see Sections 9.12 and 9.13).

In another type of diffusion-dependent transformation, there is some alteration in phase compositions and often in the number of phases present; the final microstructure typically consists of two phases. The eutectoid reaction described by Equation 11.19 is of this type; it receives further attention in Section 12.5.

The third kind of transformation is diffusionless, in which a metastable phase is produced. As discussed in Section 12.5, a martensitic transformation, which may be induced in some steel alloys, falls into this category.

12.3 THE KINETICS OF PHASE TRANSFORMATIONS

With phase transformations, normally at least one new phase is formed that has different physical/chemical characteristics and/or a different structure than the parent phase. Furthermore, most phase transformations do not occur instantaneously. Rather, they begin by the formation of numerous small particles of the new phase(s), which increase in size until the transformation has reached completion. The progress of a phase **nucleation, growth** transformation may be broken down into two distinct stages: **nucleation** and **growth**. Nucleation involves the appearance of very small particles, or nuclei of the new phase (often consisting of only a few hundred atoms), which are capable of growing. During the growth stage, these nuclei increase in size, which results in the disappearance of some (or all) of the parent phase. The transformation reaches completion if the growth of these new-phase particles is allowed to proceed until the equilibrium fraction is attained. We now discuss the mechanics of these two processes and how they relate to solid-state transformations.

Nucleation

There are two types of nucleation: *homogeneous* and *heterogeneous*. The distinction between them is made according to the site at which nucleating events occur. For the homogeneous type, nuclei of the new phase form uniformly throughout the parent phase, whereas for the heterogeneous type, nuclei form preferentially at structural inhomogeneities, such as container surfaces, insoluble impurities, grain boundaries, dislocations, and so on. We begin by discussing homogeneous nucleation because its description and theory are simpler to treat. These principles are then extended to a discussion of the heterogeneous type.

Homogeneous Nucleation

A discussion of the theory of nucleation involves a thermodynamic parameter **free energy** called **free energy** (or *Gibbs free energy*), G. In brief, free energy is a function of other thermodynamic parameters, of which one is the internal energy of the system (i.e., the *enthalpy, H*) and another is a measurement of the randomness or disorder of the atoms or molecules (i.e., the *entropy, S*). It is not our purpose here to provide a detailed discussion of the principles of thermodynamics as they apply to materials systems. However, relative to phase transformations, an important thermodynamic parameter is the change in free energy ΔG; a transformation occurs spontaneously only when ΔG has a negative value.

For the sake of simplicity, let us first consider the solidification of a pure material, assuming that nuclei of the solid phase form in the interior of the liquid as atoms cluster together so as to form a packing arrangement similar to that found in the solid phase.

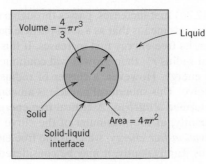

Figure 12.1 Schematic diagram showing the nucleation of a spherical solid particle in a liquid.

Furthermore, it will be assumed that each nucleus is spherical and has a radius r. This situation is represented schematically in Figure 12.1.

There are two contributions to the total free energy change that accompany a solidification transformation. The first is the free energy difference between the solid and liquid phases, or the volume free energy, ΔG_v. Its value is negative if the temperature is below the equilibrium solidification temperature, and the magnitude of its contribution is the product of ΔG_v and the volume of the spherical nucleus (i.e., $\frac{4}{3}\pi r^3$). The second energy contribution results from the formation of the solid–liquid phase boundary during the solidification transformation. Associated with this boundary is a surface free energy, γ, which is positive; furthermore, the magnitude of this contribution is the product of γ and the surface area of the nucleus (i.e., $4\pi r^2$). Finally, the total free energy change is equal to the sum of these two contributions:

Total free energy change for a solidification transformation

$$\Delta G = \tfrac{4}{3}\pi r^3\, \Delta G_v + 4\pi r^2 \gamma \qquad (12.1)$$

These volume, surface, and total free energy contributions are plotted schematically as a function of nucleus radius in Figures 12.2a and 12.2b. Figure 12.2a shows that for the curve corresponding to the first term on the right-hand side of Equation 12.1, the free energy (which is negative) decreases with the third power of r. Furthermore, for the curve resulting from the second term in Equation 12.1, energy values are positive and increase with the square of the radius. Consequently, the curve associated with the

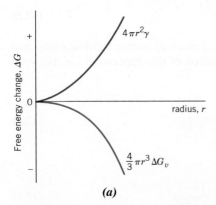

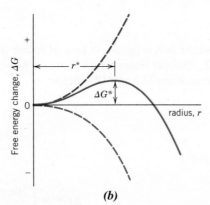

(a) (b)

Figure 12.2 (a) Schematic curves for volume free energy and surface free energy contributions to the total free energy change attending the formation of a spherical embryo/nucleus during solidification. (b) Schematic plot of free energy versus embryo/nucleus radius, on which is shown the critical free energy change (ΔG^*) and the critical nucleus radius (r^*).

sum of both terms (Figure 12.2b) first increases, passes through a maximum, and finally decreases. In a physical sense, this means that as a solid particle begins to form as atoms in the liquid cluster together, its free energy first increases. If this cluster reaches a size corresponding to the critical radius r^*, then growth will continue with the accompaniment of a decrease in free energy. However, a cluster of radius less than the critical value will shrink and redissolve. This subcritical particle is an *embryo,* and the particle of radius greater than r^* is termed a *nucleus.* A critical free energy, ΔG^*, occurs at the critical radius and, consequently, at the maximum of the curve in Figure 12.2b. This ΔG^* corresponds to an *activation free energy,* which is the free energy required for the formation of a stable nucleus. Equivalently, it may be considered an energy barrier to the nucleation process.

Because r^* and ΔG^* appear at the maximum on the free energy-versus-radius curve of Figure 12.2b, derivation of expressions for these two parameters is a simple matter. For r^*, we differentiate the ΔG equation (Equation 12.1) with respect to r, set the resulting expression equal to zero, and then solve for r (= r^*). That is,

$$\frac{d(\Delta G)}{dr} = \tfrac{4}{3}\pi\,\Delta G_v(3r^2) + 4\pi\gamma(2r) = 0 \tag{12.2}$$

which leads to the result

For homogeneous nucleation, critical radius of a stable solid particle nucleus

$$r^* = -\frac{2\gamma}{\Delta G_v} \tag{12.3}$$

Now, substitution of this expression for r^* into Equation 12.1 yields the following expression for ΔG^*:

For homogeneous nucleation, activation free energy required for the formation of a stable nucleus

$$\Delta G^* = \frac{16\pi\gamma^3}{3(\Delta G_v)^2} \tag{12.4}$$

This volume free energy change ΔG_v is the driving force for the solidification transformation, and its magnitude is a function of temperature. At the equilibrium solidification temperature T_m, the value of ΔG_v is zero, and with decreasing temperature its value becomes increasingly more negative.

It can be shown that ΔG_v is a function of temperature as

$$\Delta G_v = \frac{\Delta H_f(T_m - T)}{T_m} \tag{12.5}$$

where ΔH_f is the latent heat of fusion (i.e., the heat given up during solidification), and T_m and the temperature T are in Kelvin. Substitution of this expression for ΔG_v into Equations 12.3 and 12.4 yields

Dependence of critical radius on surface free energy, latent heat of fusion, melting temperature, and transformation temperature

$$r^* = \left(-\frac{2\gamma T_m}{\Delta H_f}\right)\left(\frac{1}{T_m - T}\right) \tag{12.6}$$

and

Activation free energy expression

$$\Delta G^* = \left(\frac{16\pi\gamma^3 T_m^2}{3\Delta H_f^2}\right)\frac{1}{(T_m - T)^2} \tag{12.7}$$

Thus, from these two equations, both the critical radius r^* and the activation free energy ΔG^* decrease as temperature T decreases. (The γ and ΔH_f parameters in these

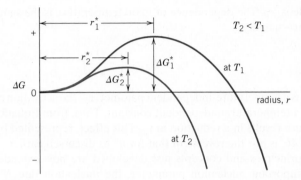

Figure 12.3 Schematic free energy-versus-embryo/nucleus-radius curves for two different temperatures. The critical free energy change (ΔG^*) and critical nucleus radius (r^*) are indicated for each temperature.

expressions are relatively insensitive to temperature changes.) Figure 12.3, a schematic ΔG-versus-r plot that shows curves for two different temperatures, illustrates these relationships. Physically, this means that with a lowering of temperature at temperatures below the equilibrium solidification temperature (T_m), nucleation occurs more readily. Furthermore, the number of stable nuclei n^* (having radii greater than r^*) is a function of temperature as

$$n^* = K_1 \exp\left(-\frac{\Delta G^*}{kT}\right) \tag{12.8}$$

where the constant K_1 is related to the total number of nuclei of the solid phase. For the exponential term of this expression, changes in temperature have a greater effect on the magnitude of the ΔG^* term in the numerator than the T term in the denominator. Consequently, as the temperature is lowered below T_m, the exponential term in Equation 12.8 also decreases, so that the magnitude of n^* increases. This temperature dependence (n^* versus T) is represented in the schematic plot of Figure 12.4a.

Another important temperature-dependent step is involved in and also influences nucleation: the clustering of atoms by short-range diffusion during the formation of nuclei. The influence of temperature on the rate of diffusion (i.e., magnitude of the diffusion coefficient, D) is given in Equation 7.8. Furthermore, this diffusion effect is related to the frequency at which atoms from the liquid attach themselves to the

Figure 12.4 For solidification, schematic plots of (a) number of stable nuclei versus temperature, (b) frequency of atomic attachment versus temperature, and (c) nucleation rate versus temperature (the dashed curves are reproduced from parts a and b).

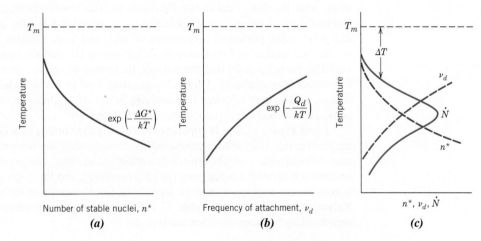

solid nucleus, v_d. The dependence of v_d on temperature is the same as for the diffusion coefficient—namely,

$$v_d = K_2 \exp\left(-\frac{Q_d}{kT}\right) \tag{12.9}$$

where Q_d is a temperature-independent parameter—the activation energy for diffusion—and K_2 is a temperature-independent constant. Thus, from Equation 12.9, a decrease of temperature results in a reduction in v_d. This effect, represented by the curve shown in Figure 12.4b, is just the reverse of that for n^* as discussed earlier.

The principles and concepts just developed are now extended to a discussion of another important nucleation parameter, the nucleation rate $\dot{N}$ (which has units of nuclei per unit volume per second). This rate is simply proportional to the product of n^* (Equation 12.8) and v_d (Equation 12.9)—that is,

Nucleation rate expression for homogeneous nucleation

$$\dot{N} = K_3 n^* v_d = K_1 K_2 K_3 \left[\exp\left(-\frac{\Delta G^*}{kT}\right) \exp\left(-\frac{Q_d}{kT}\right) \right] \tag{12.10}$$

Here, K_3 is the number of atoms on a nucleus surface. Figure 12.4c schematically plots nucleation rate as a function of temperature and, in addition, the curves of Figures 12.4a and 12.4b from which the $\dot{N}$ curve is derived. Figure 12.4c shows that, with a reduction of temperature from below T_m, the nucleation rate first increases, achieves a maximum, and subsequently diminishes.

The shape of this $\dot{N}$ curve is explained as follows: for the upper region of the curve (a sudden and dramatic increase in $\dot{N}$ with decreasing T), ΔG^* is greater than Q_d, which means that the $\exp(-\Delta G^*/kT)$ term of Equation 12.10 is much smaller than $\exp(-Q_d/kT)$. In other words, the nucleation rate is suppressed at high temperatures because of a small activation driving force. With continued reduction of temperature, there comes a point at which ΔG^* becomes smaller than the temperature-independent Q_d, with the result that $\exp(-Q_d/kT) < \exp(-\Delta G^*/kT)$, or that, at lower temperatures, a low atomic mobility suppresses the nucleation rate. This accounts for the shape of the lower curve segment (a precipitous reduction of $\dot{N}$ with a continued reduction of temperature). Furthermore, the $\dot{N}$ curve of Figure 12.4c necessarily passes through a maximum over the intermediate temperature range, where values for ΔG^* and Q_d are of approximately the same magnitude.

Several qualifying comments are in order regarding the preceding discussion. First, although we assumed a spherical shape for nuclei, this method may be applied to any shape with the same final result. Furthermore, this treatment may be used for types of transformations other than solidification (i.e., liquid–solid)—for example, solid–vapor and solid–solid. However, magnitudes of ΔG_v and γ, in addition to diffusion rates of the atomic species, will undoubtedly differ among the various transformation types. In addition, for solid–solid transformations, there may be volume changes attendant to the formation of new phases. These changes may lead to the introduction of microscopic strains, which must be taken into account in the ΔG expression of Equation 12.1 and, consequently, will affect the magnitudes of r^* and ΔG^*.

From Figure 12.4c it is apparent that during the cooling of a liquid, an appreciable nucleation rate (i.e., solidification) will begin only after the temperature has been low-ered to below the equilibrium solidification (or melting) temperature (T_m). This phe-nomenon is termed *supercooling* (or *undercooling*), and the degree of supercooling for homogeneous nucleation may be significant (on the order of several hundred degrees Kelvin) for some systems. Table 12.1 shows, for several materials, typical degrees of supercooling for homogeneous nucleation.

Table 12.1

Degree of
Supercooling (ΔT)
Values (Homogeneous
Nucleation) for
Several Metals

Metal	ΔT (°C)
Antimony	135
Germanium	227
Silver	227
Gold	230
Copper	236
Iron	295
Nickel	319
Cobalt	330
Palladium	332

Source: D. Turnbull and R. E. Cech, "Microscopic Observation of the Solidification of Small Metal Droplets," *J. Appl. Phys.,* **21**, 808 (1950).

EXAMPLE PROBLEM 12.1

Computation of Critical Nucleus Radius and Activation Free Energy

(a) For the solidification of pure gold, calculate the critical radius r^* and the activation free energy ΔG^* if nucleation is homogeneous. Values for the latent heat of fusion and surface free energy are -1.16×10^9 J/m^3 and 0.132 J/m^2, respectively. Use the supercooling value in Table 12.1.

(b) Now, calculate the number of atoms found in a nucleus of critical size. Assume a lattice parameter of 0.413 nm for solid gold at its melting temperature.

Solution

(a) In order to compute the critical radius, we employ Equation 12.6, using the melting temperature of 1064°C for gold, assuming a supercooling value of 230°C (Table 12.1), and realizing that ΔH_f is negative. Hence

$$r^* = \left(-\frac{2\gamma T_m}{\Delta H_f}\right)\left(\frac{1}{T_m - T}\right)$$

$$= \left[-\frac{(2)(0.132 \text{ J/m}^2)(1064 + 273 \text{ K})}{-1.16 \times 10^9 \text{ J/m}^3}\right]\left(\frac{1}{230 \text{ K}}\right)$$

$$= 1.32 \times 10^{-9} \text{ m} = 1.32 \text{ nm}$$

For computation of the activation free energy, Equation 12.7 is employed. Thus,

$$\Delta G^* = \left(\frac{16\pi\gamma^3 T_m^2}{3\Delta H_f^2}\right)\frac{1}{(T_m - T)^2}$$

$$= \left[\frac{(16)(\pi)(0.132 \text{ J/m}^2)^3(1064 + 273 \text{ K})^2}{(3)(-1.16 \times 10^9 \text{ J/m}^3)^2}\right]\left[\frac{1}{(230 \text{ K})^2}\right]$$

$$= 9.64 \times 10^{-19} \text{ J}$$

(b) In order to compute the number of atoms in a nucleus of critical size (assuming a spherical nucleus of radius r^*), it is first necessary to determine the number of unit cells, which we then multiply by the number of atoms per unit cell. The number of unit cells found in this critical nucleus is just the ratio of critical nucleus and unit cell volumes. Inasmuch as gold has the FCC crystal structure (and a cubic unit cell), its unit cell volume is just a^3, where a is the lattice parameter (i.e., unit cell edge length); its value is 0.413 nm, as cited in the problem statement. Therefore, the number of unit cells found in a radius of critical size is just

$$\# \text{ unit cells/particle} = \frac{\text{critical nucleus volume}}{\text{unit cell volume}} = \frac{\frac{4}{3}\pi r^{*3}}{a^3} \tag{12.11}$$

$$= \frac{\left(\frac{4}{3}\right)(\pi)(1.32 \text{ nm})^3}{(0.413 \text{ nm})^3} = 137 \text{ unit cells}$$

Because of the equivalence of four atoms per FCC unit cell (Section 4.2), the total number of atoms per critical nucleus is just

(137 unit cells/critical nucleus)(4 atoms/unit cell) = 548 atoms/critical nucleus

Heterogeneous Nucleation

Although levels of supercooling for homogeneous nucleation may be significant (on occasion several hundred degrees Celsius), in practical situations they are often on the order of only several degrees Celsius. The reason for this is that the activation energy (i.e., energy barrier) for nucleation (ΔG^* of Equation 12.4) is lowered when nuclei form on preexisting surfaces or interfaces, because the surface free energy (γ of Equation 12.4) is reduced. In other words, it is easier for nucleation to occur at surfaces and interfaces than at other sites. Again, this type of nucleation is termed *heterogeneous*.

In order to understand this phenomenon, let us consider the nucleation, on a flat surface, of a solid particle from a liquid phase. It is assumed that both the liquid and solid phases "wet" this flat surface—that is, both of these phases spread out and cover the surface; this configuration is depicted schematically in Figure 12.5. Also noted in the figure are three interfacial energies (represented as vectors) that exist at two-phase boundaries—γ_{SL}, γ_{SI}, and γ_{IL}—as well as the wetting angle θ (the angle between the γ_{SI} and γ_{SL} vectors). Taking a surface tension force balance in the plane of the flat surface leads to the following expression:

For heterogeneous nucleation of a solid particle, relationship among solid–surface, solid–liquid, and liquid–surface interfacial energies and the wetting angle

$$\gamma_{IL} = \gamma_{SI} + \gamma_{SL}\cos\theta \tag{12.12}$$

Now, using a somewhat involved procedure similar to the one presented for homogeneous nucleation (which we have chosen to omit), it is possible to derive equations for r^* and ΔG^*; these are as follows:

For heterogeneous nucleation, critical radius of a stable solid particle nucleus

$$r^* = -\frac{2\gamma_{SL}}{\Delta G_v} \tag{12.13}$$

For heterogeneous nucleation, activation free energy required for the formation of a stable nucleus

$$\Delta G^* = \left(\frac{16\pi\gamma_{SL}^3}{3\Delta G_v^2}\right)S(\theta) \tag{12.14}$$

Figure 12.5 Heterogeneous nucleation of a solid from a liquid. The solid–surface (γ_{SI}), solid–liquid (γ_{SL}), and liquid–surface (γ_{IL}), interfacial energies are represented by vectors. The wetting angle (θ) is also shown.

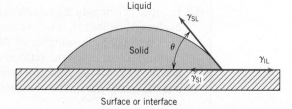

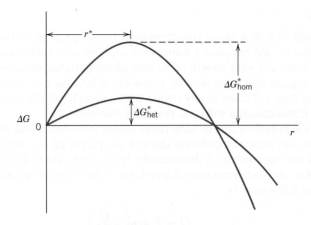

Figure 12.6 Schematic free-energy-versus-embryo/nucleus-radius plot on which are presented curves for both homogeneous and heterogeneous nucleation. Critical free energies and the critical radius are also shown.

The $S(\theta)$ term of this last equation is a function only of θ (i.e., the shape of the nucleus), which has a numerical value between zero and unity.[1]

From Equation 12.13, it is important to note that the critical radius r^* for heterogeneous nucleation is the same as for homogeneous nucleation, inasmuch as γ_{SL} is the same surface energy as γ in Equation 12.3. It is also evident that the activation energy barrier for heterogeneous nucleation (Equation 12.14) is smaller than the homogeneous barrier (Equation 12.4) by an amount corresponding to the value of this $S(\theta)$ function, or

$$\Delta G_{het}^* = \Delta G_{hom}^* S(\theta) \tag{12.15}$$

Figure 12.6, a schematic graph of ΔG versus nucleus radius, plots curves for both types of nucleation and indicates the difference in the magnitudes of ΔG_{het}^* and ΔG_{hom}^*, in addition to the constancy of r^*. This lower ΔG^* for heterogeneous nucleation means that a smaller energy must be overcome during the nucleation process (than for homogeneous nucleation), and, therefore, heterogeneous nucleation occurs more readily (Equation 12.10). In terms of the nucleation rate, the $\dot{N}$-versus-T curve (Figure 12.4c) is shifted to higher temperatures for heterogeneous. This effect is represented in Figure 12.7, which also shows that a much smaller degree of supercooling (ΔT) is required for heterogeneous nucleation.

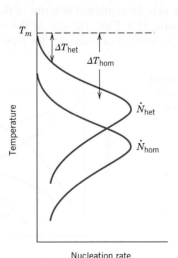

Figure 12.7 Nucleation rate versus temperature for both homogeneous and heterogeneous nucleation. Degree of supercooling (ΔT) for each is also shown.

---

[1]For example, for θ angles of 30° and 90°, values of $S(\theta)$ are approximately 0.01 and 0.5, respectively.

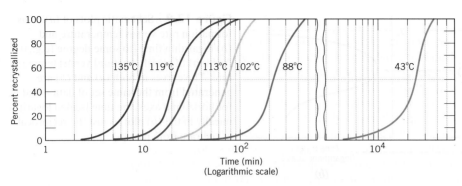

Figure 12.11 Percent recrystallization as a function of time and at constant temperature for pure copper. (Reprinted with permission from *Metallurgical Transactions,* Vol. 188, 1950, a publication of The Metallurgical Society of AIME, Warrendale, PA. Adapted from B. F. Decker and D. Harker, "Recrystallization in Rolled Copper," *Trans. AIME,* **188,** 1950, p. 888.)

Avrami equation—dependence of fraction of transformation on time

For solid-state transformations displaying the kinetic behavior in Figure 12.10, the fraction of transformation y is a function of time t as follows:

$$y = 1 - \exp(-kt^n) \tag{12.17}$$

where k and n are time-independent constants for the particular reaction. This expression is often referred to as the *Avrami equation.*

By convention, the rate of a transformation is taken as the reciprocal of time required for the transformation to proceed halfway to completion, $t_{0.5}$, or

Transformation rate—reciprocal of the halfway-to-completion transformation time

$$\text{rate} = \frac{1}{t_{0.5}} \tag{12.18}$$

Temperature has a profound influence on the kinetics and thus on the rate of a transformation. This is demonstrated in Figure 12.11, which shows y-versus-log$t$ S-shaped curves at several temperatures for the recrystallization of copper.

Section 12.5 gives a detailed discussion on the influence of both temperature and time on phase transformations.

EXAMPLE PROBLEM 12.2

Rate of Recrystallization Computation

It is known that the kinetics of recrystallization for some alloy obeys the Avrami equation and that the value of n is 3.1. If the fraction recrystallized is 0.30 after 20 min, determine the rate of recrystallization.

Solution

The rate of a reaction is defined by Equation 12.18 as

$$\text{rate} = \frac{1}{t_{0.5}}$$

Therefore, for this problem it is necessary to compute the value of $t_{0.5}$, the time it takes for the reaction to progress to 50% completion—or for the fraction of reaction y to equal 0.50. Furthermore, we may determine $t_{0.5}$ using the Avrami equation, Equation 12.17:

$$y = 1 - \exp(-kt^n)$$

The problem statement provides us with the value of y (0.30) at some time t (20 min), and also the value of n (3.1) from which data it is possible to compute the value of the constant k. In order to perform this calculation, some algebraic manipulation of Equation 12.17 is necessary. First, we rearrange this expression as follows:

$$\exp(-kt^n) = 1 - y$$

Taking natural logarithms of both sides leads to

$$-kt^n = \ln(1 - y) \qquad (12.17a)$$

Now, solving for k,

$$k = -\frac{\ln(1 - y)}{t^n}$$

Incorporating values cited above for y, n, and t yields the following value for k:

$$k = -\frac{\ln(1 - 0.30)}{(20\ \text{min})^{3.1}} = 3.30 \times 10^{-5}$$

At this point, we want to compute $t_{0.5}$—the value of t for $y = 0.5$—which means that it is necessary to establish a form of Equation 12.17 in which t is the dependent variable. This is accomplished using a rearranged form of Equation 12.17a as

$$t^n = -\frac{\ln(1 - y)}{k}$$

From which we solve for t

$$t = \left[-\frac{\ln(1 - y)}{k} \right]^{1/n}$$

And for $t = t_{0.5}$, this equation becomes

$$t_{0.5} = \left[-\frac{\ln(1 - 0.5)}{k} \right]^{1/n}$$

Now, substituting into this expression the value of k determined above, as well as the value of n cited in the problem statement (viz., 3.1), we calculate $t_{0.5}$ as follows:

$$t_{0.5} = \left[-\frac{\ln(1 - 0.5)}{3.30 \times 10^{-5}} \right]^{1/3.1} = 24.8\ \text{min}$$

And, finally, from Equation 12.18, the rate is equal to

$$\text{rate} = \frac{1}{t_{0.5}} = \frac{1}{24.8\ \text{min}} = 4.0 \times 10^{-2}\ (\text{min})^{-1}$$

12.4 METASTABLE VERSUS EQUILIBRIUM STATES

Phase transformations may be wrought in metal alloy systems by varying temperature, composition, and the external pressure; however, temperature changes by means of heat treatments are most conveniently utilized to induce phase transformations. This corresponds to crossing a phase boundary on the composition–temperature phase diagram as an alloy of given composition is heated or cooled.

During a phase transformation, an alloy proceeds toward an equilibrium state that is characterized by the phase diagram in terms of the product phases and their compositions and relative amounts. As Section 12.3 notes, most phase transformations require some finite time to go to completion, and the speed or rate is often important in the relationship between the heat treatment and the development of microstructure. One limitation of phase diagrams is their inability to indicate the time period required for the attainment of equilibrium.

The rate of approach to equilibrium for solid systems is so slow that true equilibrium structures are rarely achieved. When phase transformations are induced by temperature changes, equilibrium conditions are maintained only if heating or cooling is carried out at extremely slow and unpractical rates. For other-than-equilibrium cooling, transformations are shifted to lower temperatures than indicated by the phase diagram; for heating, the shift is to higher temperatures. These phenomena are termed **supercooling** and **superheating**, respectively. The degree of each depends on the rate of temperature change; the more rapid the cooling or heating, the greater the supercooling or superheating. For example, for normal cooling rates, the iron–carbon eutectoid reaction is typically displaced 10°C to 20°C below the equilibrium transformation temperature.[3]

For many technologically important alloys, the preferred state or microstructure is a metastable one, intermediate between the initial and equilibrium states; on occasion, a structure far removed from the equilibrium one is desired. It thus becomes imperative to investigate the influence of time on phase transformations. This kinetic information is, in many instances, of greater value than knowledge of the final equilibrium state.

supercooling

superheating

Microstructural and Property Changes in Iron–Carbon Alloys

Some of the basic kinetic principles of solid-state transformations are now extended and applied specifically to iron–carbon alloys in terms of the relationships among heat treatment, the development of microstructure, and mechanical properties. This system has been chosen because it is familiar and because a wide variety of microstructures and mechanical properties is possible for iron–carbon (or steel) alloys.

12.5 ISOTHERMAL TRANSFORMATION DIAGRAMS

Pearlite

Consider again the iron–iron carbide eutectoid reaction

Eutectoid reaction for the iron–iron carbide system

$$\gamma(0.76\,\text{wt\%}\,\text{C}) \underset{\text{heating}}{\overset{\text{cooling}}{\rightleftharpoons}} \alpha(0.022\,\text{wt\%}\,\text{C}) + \text{Fe}_3\text{C}\,(6.70\,\text{wt\%}\,\text{C}) \tag{12.19}$$

which is fundamental to the development of microstructure in steel alloys. Upon cooling, austenite, having an intermediate carbon concentration, transforms into a ferrite phase, which has a much lower carbon content, and also cementite, which has a much higher carbon concentration. Pearlite is one microstructural product of this transformation (Figure 11.26); the mechanism of pearlite formation was discussed previously (Section 11.19) and demonstrated in Figure 11.27.

---

[3]It is important to note that the treatments relating to the kinetics of phase transformations in Section 12.3 are constrained to the condition of constant temperature. By way of contrast, the discussion of this section pertains to phase transformations that occur with changing temperature. This same distinction exists between Sections 12.5 (Isothermal Transformation Diagrams) and 12.6 (Continuous-Cooling Transformation Diagrams).

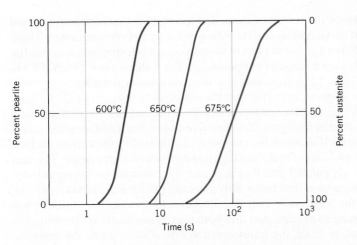

Figure 12.12 For an iron–carbon alloy of eutectoid composition (0.76 wt% C), isothermal fraction reacted versus the logarithm of time for the austenite-to-pearlite transformation.

Temperature plays an important role in the rate of the austenite-to-pearlite transformation. The temperature dependence for an iron–carbon alloy of eutectoid composition is indicated in Figure 12.12, which plots S-shaped curves of the percentage transformation versus the logarithm of time at three different temperatures. For each curve, data were collected after rapidly cooling a specimen composed of 100% austenite to the temperature indicated; that temperature was maintained constant throughout the course of the reaction.

A more convenient way of representing both the time and temperature dependence of this transformation is shown in the bottom portion of Figure 12.13. Here, the vertical and horizontal axes are, respectively, temperature and the logarithm of time. Two solid curves are plotted; one represents the time required at each temperature

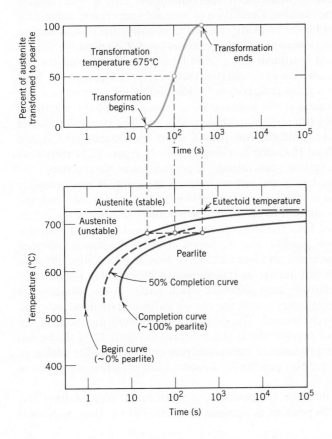

Figure 12.13 Demonstration of how an isothermal transformation diagram (bottom) is generated from percentage transformation-versus-logarithm of time measurements (top). [Adapted from H. Boyer (Editor), *Atlas of Isothermal Transformation and Cooling Transformation Diagrams*, 1977. Reproduced by permission of ASM International, Materials Park, OH.]

for the initiation or start of the transformation, and the other is for the transformation conclusion. The dashed curve corresponds to 50% of transformation completion. These curves were generated from a series of plots of the percentage transformation versus the logarithm of time taken over a range of temperatures. The S-shape curve (for 675°C) in the upper portion of Figure 12.13 illustrates how the data transfer is made.

In interpreting this diagram, note first that the eutectoid temperature (727°C) is indicated by a horizontal line; at temperatures above the eutectoid and for all times, only austenite exists, as indicated in the figure. The austenite-to-pearlite transformation occurs only if an alloy is supercooled to below the eutectoid; as indicated by the curves, the time necessary for the transformation to begin and then end depends on temperature. The start and finish curves are nearly parallel, and they approach the eutectoid line asymptotically. To the left of the transformation start curve, only austenite (which is unstable) is present, whereas to the right of the finish curve, only pearlite exists. In between, the austenite is in the process of transforming to pearlite, and thus both microconstituents are present.

According to Equation 12.18, the transformation rate at some particular temperature is inversely proportional to the time required for the reaction to proceed to 50% completion (to the dashed line in Figure 12.13). That is, the shorter this time, the higher is the rate. Thus, from Figure 12.13, at temperatures just below the eutectoid (corresponding to just a slight degree of undercooling) very long times (on the order of 10^5 s) are required for the 50% transformation, and therefore the reaction rate is very slow. The transformation rate increases with decreasing temperature such that at 540°C, only about 3 s is required for the reaction to go to 50% completion.

Several constraints are imposed on the use of diagrams like Figure 12.13. First, this particular plot is valid only for an iron–carbon alloy of eutectoid composition; for other compositions, the curves have different configurations. In addition, these plots are accurate only for transformations in which the temperature of the alloy is held constant throughout the duration of the reaction. Conditions of constant temperature are termed *isothermal;* thus, plots such as Figure 12.13 are referred to as **isothermal transformation diagrams** or sometimes as *time-temperature-transformation* (or *T-T-T*) plots.

An actual isothermal heat treatment curve (*ABCD*) is superimposed on the isothermal transformation diagram for a eutectoid iron–carbon alloy in Figure 12.14. Very rapid cooling of austenite to a given temperature is indicated by the near-vertical line *AB*, and the isothermal treatment at this temperature is represented by the horizontal segment *BCD*. Time increases from left to right along this line. The transformation of austenite to pearlite begins at the intersection, point *C* (after approximately 3.5 s), and has reached completion by about 15 s, corresponding to point *D*. Figure 12.14 also shows schematic microstructures at various times during the progression of the reaction.

The thickness ratio of the ferrite and cementite layers in pearlite is approximately 8 to 1. However, the absolute layer thickness depends on the temperature at which the isothermal transformation is allowed to occur. At temperatures just below the eutectoid, relatively thick layers of both the α-ferrite and Fe$_3$C phases are produced; this microstructure is called **coarse pearlite**, and the region at which it forms is indicated to the right of the completion curve on Figure 12.14. At these temperatures, diffusion rates are relatively high, such that during the transformation illustrated in Figure 11.27 carbon atoms can diffuse relatively long distances, which results in the formation of thick lamellae. With decreasing temperature, the carbon diffusion rate decreases, and the layers become progressively thinner. The thin-layered structure produced in the vicinity of 540°C is termed **fine pearlite**; this is also indicated in Figure 12.14. To be discussed in Section 12.7 is the dependence of mechanical properties on lamellar thickness. Photomicrographs of coarse and fine pearlite for a eutectoid composition are shown in Figure 12.15.

For iron–carbon alloys of other compositions, a proeutectoid phase (either ferrite or cementite) coexists with pearlite, as discussed in Section 11.19. Thus, additional

isothermal transformation diagram

coarse pearlite

fine pearlite

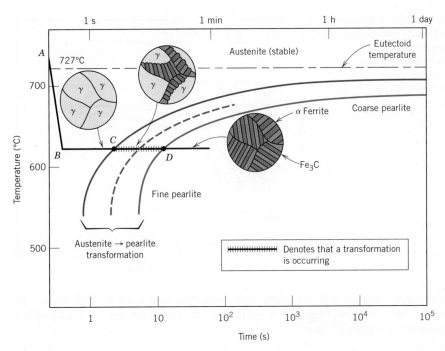

Figure 12.14 Isothermal transformation diagram for a eutectoid iron–carbon alloy, with superimposed isothermal heat treatment curve (*ABCD*). Microstructures before, during, and after the austenite-to-pearlite transformation are shown.
[Adapted from H. Boyer (Editor), *Atlas of Isothermal Transformation and Cooling Transformation Diagrams*, 1977. Reproduced by permission of ASM International, Materials Park, OH.]

Figure 12.15
Photomicrographs of (*a*) coarse pearlite and (*b*) fine pearlite. 3000×.
(From K. M. Ralls et al., *An Introduction to Materials Science and Engineering*, p. 361. Copyright © 1976 by John Wiley & Sons, New York. Reprinted by permission of John Wiley & Sons, Inc.)

WileyPLUS

Tutorial Video:
Iron–Carbon Alloy
Microstructures
What Is the
Appearance of the
Various Iron–Carbon
Alloys and How Can I
Draw Them?

Figure 12.16 Isothermal transformation diagram for a 1.13 wt% C iron–carbon alloy: A, austenite; C, proeutectoid cementite; P, pearlite.
[Adapted from H. Boyer (Editor), *Atlas of Isothermal Transformation and Cooling Transformation Diagrams,* 1977. Reproduced by permission of ASM International, Materials Park, OH.]

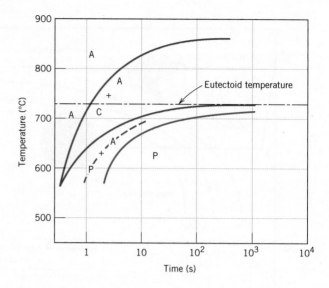

curves corresponding to a proeutectoid transformation also must be included on the isothermal transformation diagram. A portion of one such diagram for a 1.13 wt% C alloy is shown in Figure 12.16.

Bainite

bainite

In addition to pearlite, other microconstituents that are products of the austenitic transformation exist; one of these is called **bainite**. The microstructure of bainite consists of ferrite and cementite phases, and thus diffusional processes are involved in its formation. Bainite forms as needles or plates, depending on the temperature of the transformation; the microstructural details of bainite are so fine that their resolution is possible only using electron microscopy. Figure 12.17 is an electron micrograph that shows a grain of bainite (positioned diagonally from lower left to upper right). It is composed of a ferrite matrix and elongated particles of Fe_3C; the various phases in this micrograph have been labeled.

Figure 12.17 Transmission electron micrograph showing the structure of bainite. A grain of bainite passes from lower left to upper right corners; it consists of elongated and needle-shape particles of Fe_3C within a ferrite matrix. The phase surrounding the bainite is martensite. (From *Metals Handbook,* Vol. 8, 8th edition, *Metallography, Structures and Phase Diagrams,* 1973. Reproduced by permission of ASM International, Materials Park, OH.)

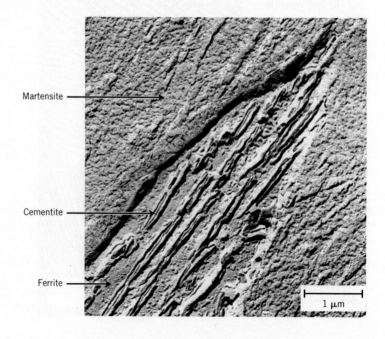

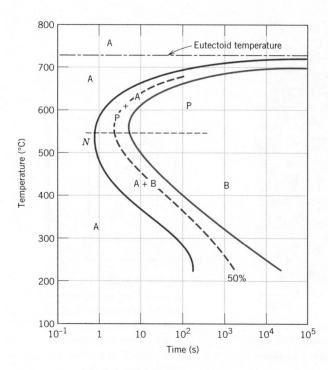

Figure 12.18 Isothermal transformation diagram for an iron–carbon alloy of eutectoid composition, including austenite-to-pearlite (A–P) and austenite-to-bainite (A–B) transformations.
[Adapted from H. Boyer (Editor), *Atlas of Isothermal Transformation and Cooling Transformation Diagrams,* 1977. Reproduced by permission of ASM International, Materials Park, OH.]

WileyPLUS

Tutorial Video:
Isothermal
Transformation
Diagrams
How Do I Read
a TTT Diagram?

In addition, the phase that surrounds the needle is martensite, the topic addressed by a subsequent section. Furthermore, no proeutectoid phase forms with bainite.

The time–temperature dependence of the bainite transformation may also be represented on the isothermal transformation diagram. It occurs at temperatures below those at which pearlite forms; begin-, end-, and half-reaction curves are just extensions of those for the pearlitic transformation, as shown in Figure 12.18, the isothermal transformation diagram for an iron–carbon alloy of eutectoid composition that has been extended to lower temperatures. All three curves are C-shaped and have a "nose" at point N, where the rate of transformation is a maximum. As may be noted, whereas pearlite forms above the nose (i.e., over the temperature range of about 540°C to 727°C), at temperatures between about 215°C and 540°C, bainite is the transformation product.

Note that the pearlitic and bainitic transformations are competitive with each other, and once some portion of an alloy has transformed into either pearlite or bainite, transformation to the other microconstituent is not possible without reheating to form austenite.

Spheroidite

spheroidite

If a steel alloy having either pearlitic or bainitic microstructures is heated to, and left at, a temperature below the eutectoid for a sufficiently long period of time—for example, at about 700°C for between 18 and 24 h—yet another microstructure will form called **spheroidite** (Figure 12.19). Instead of the alternating ferrite and cementite lamellae (pearlite) or the microstructure observed for bainite, the Fe_3C phase appears as sphere-like particles embedded in a continuous α–phase matrix. This transformation occurs by additional carbon diffusion with no change in the compositions or relative amounts of ferrite and cementite phases. The driving force for this transformation is the reduction in α–Fe_3C phase boundary area. The kinetics of spheroidite formation is not included on isothermal transformation diagrams.

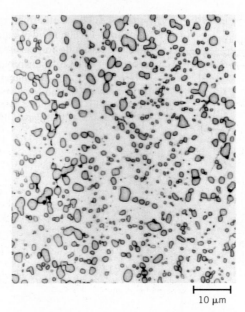

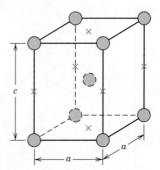

Figure 12.20 The body-centered tetragonal unit cell for martensitic steel showing iron atoms (circles) and sites that may be occupied by carbon atoms (×s). For this tetragonal unit cell, $c > a$.

Figure 12.19 Photomicrograph of a steel having a spheroidite microstructure. The small particles are cementite; the continuous phase is α-ferrite. 1000×.
(Copyright 1971 by United States Steel Corporation.)

Concept Check 12.1 Which is more stable, the pearlitic or the spheroiditic microstructure? Why?

[*The answer may be found at* www.wiley.com/college/callister *(Student Companion Site).*]

Martensite

martensite

Yet another microconstituent or phase called **martensite** is formed when austenitized iron–carbon alloys are rapidly cooled (or quenched) to a relatively low temperature (in the vicinity of the ambient). Martensite is a nonequilibrium single-phase structure that results from a diffusionless transformation of austenite. It may be thought of as a transformation product that is competitive with pearlite and bainite. The martensitic transformation occurs when the quenching rate is rapid enough to prevent carbon diffusion. Any diffusion whatsoever results in the formation of ferrite and cementite phases.

The martensitic transformation is not well understood. However, large numbers of atoms experience cooperative movements, in that there is only a slight displacement of each atom relative to its neighbors. This occurs in such a way that the FCC austenite experiences a polymorphic transformation to a body-centered tetragonal (BCT) martensite. A unit cell of this crystal structure (Figure 12.20) is simply a body-centered cube that has been elongated along one of its dimensions; this structure is distinctly different from that for BCC ferrite. All the carbon atoms remain as interstitial impurities in martensite; as such, they constitute a supersaturated solid solution that is capable of rapidly transforming to other structures if heated to temperatures at which diffusion rates become

Figure 12.21 Photomicrograph showing the martensitic microstructure. The needle-shape grains are the martensite phase, and the white regions are austenite that failed to transform during the rapid quench. 1220×. (Photomicrograph courtesy of United States Steel Corporation.)

10 μm

appreciable. Many steels, however, retain their martensitic structure almost indefinitely at room temperature.

The martensitic transformation is not, however, unique to iron–carbon alloys. It is found in other systems and is characterized, in part, by the diffusionless transformation.

Because the martensitic transformation does not involve diffusion, it occurs almost instantaneously; the martensite grains nucleate and grow at a very rapid rate—the velocity of sound within the austenite matrix. Thus the martensitic transformation rate, for all practical purposes, is time independent.

Martensite grains take on a platelike or needlelike appearance, as indicated in Figure 12.21. The white phase in the micrograph is austenite (retained austenite) that did not transform during the rapid quench. As already mentioned, martensite as well as other microconstituents (e.g., pearlite) can coexist.

Being a nonequilibrium phase, martensite does not appear on the iron–iron carbide phase diagram (Figure 11.23). The austenite-to-martensite transformation, however, is represented on the isothermal transformation diagram. Because the martensitic transformation is diffusionless and instantaneous, it is not depicted in this diagram as the pearlitic and bainitic reactions are. The beginning of this transformation is represented by a horizontal line designated M(start) (Figure 12.22). Two other horizontal and dashed lines, labeled M(50%) and M(90%), indicate percentages of the austenite-to-martensite transformation. The temperatures at which these lines are located vary with alloy composition, but they must be relatively low because carbon diffusion must be virtually nonexistent.[4] The horizontal and linear character of these lines indicates that the martensitic transformation is independent of time; it is a function only of the temperature to which the alloy is quenched or rapidly cooled. A transformation of this type is termed an **athermal transformation**.

athermal transformation

---

[4] The alloy that is the subject of Figure 12.21 is not an iron–carbon alloy of eutectoid composition; furthermore, its 100% martensite transformation temperature lies below room temperature. Because the photomicrograph was taken at room temperature, some austenite (i.e., the retained austenite) is present, having not transformed to martensite.

Figure 12.22 The complete isothermal transformation diagram for an iron–carbon alloy of eutectoid composition: A, austenite; B, bainite; M, martensite; P, pearlite.

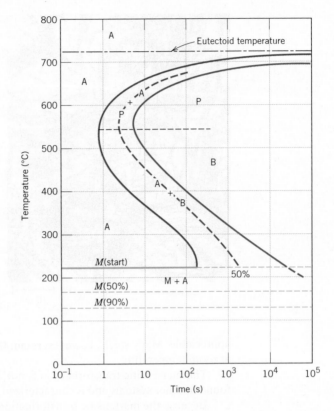

Consider an alloy of eutectoid composition that is very rapidly cooled from a temperature above 727°C to, say, 165°C. From the isothermal transformation diagram (Figure 12.22) it may be noted that 50% of the austenite will immediately transform into martensite; as long as this temperature is maintained, there will be no further transformation.

The presence of alloying elements other than carbon (e.g., Cr, Ni, Mo, and W) may cause significant changes in the positions and shapes of the curves in the isothermal transformation diagrams. These include (1) shifting to longer times the nose of the austenite-to-pearlite transformation (and also a proeutectoid phase nose, if such exists), and (2) the formation of a separate bainite nose. These alterations may be observed by comparing Figures 12.22 and 12.23, which are isothermal transformation diagrams for carbon and alloy steels, respectively.

plain carbon steel
alloy steel

Steels in which carbon is the prime alloying element are termed **plain carbon steels**, whereas **alloy steels** contain appreciable concentrations of other elements, including those cited in the preceding paragraph. Sections 13.2 and 13.3 discusses further the classification and properties of ferrous alloys.

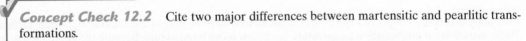

Concept Check 12.2 Cite two major differences between martensitic and pearlitic transformations.

[*The answer may be found at* www.wiley.com/college/callister (*Student Companion Site*).]

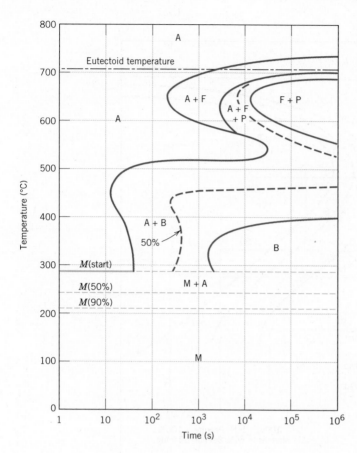

Figure 12.23 Isothermal transformation diagram for an alloy steel (type 4340): A, austenite; B, bainite; P, pearlite; M, martensite; F, proeutectoid ferrite. [Adapted from H. Boyer (Editor), *Atlas of Isothermal Transformation and Cooling Transformation Diagrams,* 1977. Reproduced by permission of ASM International, Materials Park, OH.]

EXAMPLE PROBLEM 12.3

Microstructural Determinations for Three Isothermal Heat Treatments

Using the isothermal transformation diagram for an iron–carbon alloy of eutectoid composition (Figure 12.22), specify the nature of the final microstructure (in terms of microconstituents present and approximate percentages) of a small specimen that has been subjected to the following time–temperature treatments. In each case, assume that the specimen begins at 760°C and that it has been held at this temperature long enough to have achieved a complete and homogeneous austenitic structure.

(a) Rapidly cool to 350°C, hold for 10^4 s, and quench to room temperature.
(b) Rapidly cool to 250°C, hold for 100 s, and quench to room temperature.
(c) Rapidly cool to 650°C, hold for 20 s, rapidly cool to 400°C, hold for 10^3 s, and quench to room temperature.

Solution

The time–temperature paths for all three treatments are shown in Figure 12.24. In each case, the initial cooling is rapid enough to prevent any transformation from occurring.

(a) At 350°C austenite isothermally transforms into bainite; this reaction begins after about 10 s and reaches completion at about 500 s elapsed time. Therefore, by 10^4 s, as stipulated in this problem, 100% of the specimen is bainite, and no further transformation is

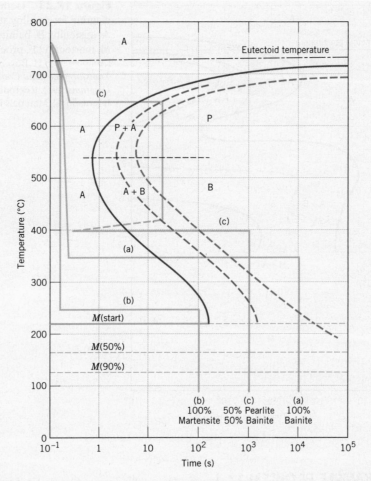

Figure 12.24 Isothermal transformation diagram for an iron–carbon alloy of eutectoid composition and the isothermal heat treatments (a), (b), and (c) in Example Problem 12.3.

possible, even though the final quenching line passes through the martensite region of the diagram.

(b) In this case, it takes about 150 s at 250°C for the bainite transformation to begin, so that at 100 s the specimen is still 100% austenite. As the specimen is cooled through the martensite region, beginning at about 215°C, progressively more of the austenite instantaneously transforms into martensite. This transformation is complete by the time room temperature is reached, such that the final microstructure is 100% martensite.

(c) For the isothermal line at 650°C, pearlite begins to form after about 7 s; by the time 20 s has elapsed, only approximately 50% of the specimen has transformed to pearlite. The rapid cool to 400°C is indicated by the vertical line; during this cooling, very little, if any, remaining austenite will transform to either pearlite or bainite, even though the cooling line passes through pearlite and bainite regions of the diagram. At 400°C, we begin timing at essentially zero time (as indicated in Figure 12.24); thus, by the time 10^3 s has elapsed, all of the remaining 50% austenite will have completely transformed to bainite. Upon quenching to room temperature, any further transformation is not possible inasmuch as no austenite remains, and so the final microstructure at room temperature consists of 50% pearlite and 50% bainite.

Concept Check 12.3   Make a copy of the isothermal transformation diagram for an iron–carbon alloy of eutectoid composition (Figure 12.22) and then sketch and label on this diagram a time–temperature path that will produce 100% fine pearlite.

[*The answer may be found at* www.wiley.com/college/callister (*Student Companion Site*).]

12.6   CONTINUOUS-COOLING TRANSFORMATION DIAGRAMS

Isothermal heat treatments are not the most practical to conduct because an alloy must be rapidly cooled to and maintained at an elevated temperature from a higher temperature above the eutectoid. Most heat treatments for steels involve the continuous cooling of a specimen to room temperature. An isothermal transformation diagram is valid only for conditions of constant temperature; this diagram must be modified for transformations that occur as the temperature is constantly changing. For continuous cooling, the time required for a reaction to begin and end is delayed. Thus the isothermal curves are shifted to longer times and lower temperatures, as indicated in Figure 12.25 for an iron–carbon alloy of eutectoid composition. A plot containing such modified beginning and ending reaction curves is termed a **continuous-cooling transformation (*CCT*) diagram**. Some control may be maintained over the rate of temperature change, depending on the cooling environment. Two cooling curves corresponding to moderately fast and

continuous-cooling transformation diagram

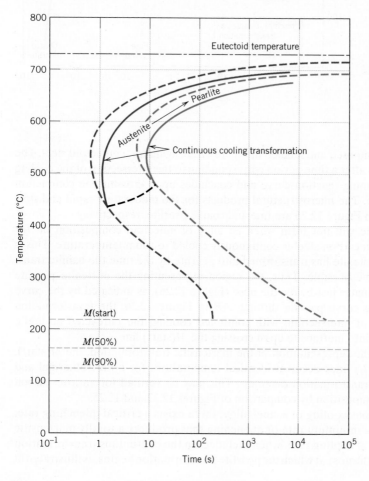

Figure 12.25 Superimposition of isothermal and continuous-cooling transformation diagrams for a eutectoid iron–carbon alloy.
[Adapted from H. Boyer (Editor), *Atlas of Isothermal Transformation and Cooling Transformation Diagrams*, 1977. Reproduced by permission of ASM International, Materials Park, OH.]

Figure 12.26 Moderately rapid and slow cooling curves superimposed on a continuous-cooling transformation diagram for a eutectoid iron–carbon alloy.

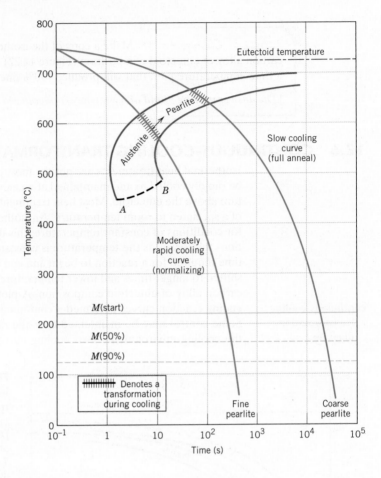

slow rates are superimposed and labeled in Figure 12.26, again for a eutectoid steel. The transformation starts after a time period corresponding to the intersection of the cooling curve with the beginning reaction curve and concludes upon crossing the completion transformation curve. The microstructural products for the moderately rapid and slow cooling rate curves in Figure 12.26 are fine and coarse pearlite, respectively.

Normally, bainite will not form when an alloy of eutectoid composition or, for that matter, any plain carbon steel is continuously cooled to room temperature. This is because all of the austenite has transformed into pearlite by the time the bainite transformation has become possible. Thus, the region representing the austenite–pearlite transformation terminates just below the nose (Figure 12.26), as indicated by the curve *AB*. For any cooling curve passing through *AB* in Figure 12.26, the transformation ceases at the point of intersection; with continued cooling, the unreacted austenite begins transforming into martensite upon crossing the *M*(start) line.

With regard to the representation of the martensitic transformation, the *M*(start), *M*(50%), and *M*(90%) lines occur at identical temperatures for both isothermal and continuous-cooling transformation diagrams. This may be verified for an iron–carbon alloy of eutectoid composition by comparison of Figures 12.22 and 12.25.

For the continuous cooling of a steel alloy, there exists a critical quenching rate, which represents the minimum rate of quenching that produces a totally martensitic structure. This critical cooling rate, when included on the continuous transformation diagram, just misses the nose at which the pearlite transformation begins, as illustrated in

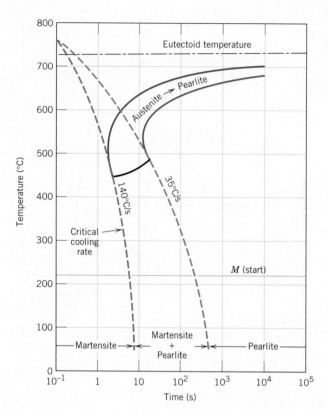

Figure 12.27 Continuous-cooling transformation diagram for a eutectoid iron–carbon alloy and superimposed cooling curves, demonstrating the dependence of the final microstructure on the transformations that occur during cooling.

Figure 12.27. As the figure also shows, only martensite exists for quenching rates greater than the critical one; in addition, there is a range of rates over which both pearlite and martensite are produced. Finally, a totally pearlitic structure develops for low cooling rates.

Carbon and other alloying elements also shift the pearlite (as well as the proeutectoid phase) and bainite noses to longer times, thus decreasing the critical cooling rate. In fact, one of the reasons for alloying steels is to facilitate the formation of martensite so that totally martensitic structures can develop in relatively thick cross sections. Figure 12.28 shows the continuous-cooling transformation diagram for the same alloy steel for which the isothermal transformation diagram is presented in Figure 12.23. The presence of the bainite nose accounts for the possibility of formation of bainite for a continuous-cooling heat treatment. Several cooling curves superimposed on Figure 12.28 indicate the critical cooling rate, and also how the transformation behavior and final microstructure are influenced by the rate of cooling.

Of interest, the critical cooling rate is decreased even by the presence of carbon. In fact, iron–carbon alloys containing less than about 0.25 wt% carbon are not normally heat-treated to form martensite because quenching rates too rapid to be practical are required. Other alloying elements that are particularly effective in rendering steels heat-treatable are chromium, nickel, molybdenum, manganese, silicon, and tungsten; however, these elements must be in solid solution with the austenite at the time of quenching.

In summary, isothermal and continuous-cooling transformation diagrams are, in a sense, phase diagrams in which the parameter of time is introduced. Each is experimentally determined for an alloy of specified composition, the variables being temperature and time. These diagrams allow prediction of the microstructure after some time period for constant-temperature and continuous-cooling heat treatments, respectively.

Figure 12.28 Continuous-cooling transformation diagram for an alloy steel (type 4340) and several super-imposed cooling curves demonstrating dependence of the final microstructure of this alloy on the transformations that occur during cooling. [Adapted from H. E. McGannon (Editor), *The Making, Shaping and Treating of Steel,* 9th edition, United States Steel Corporation, Pittsburgh, 1971, p. 1096.]

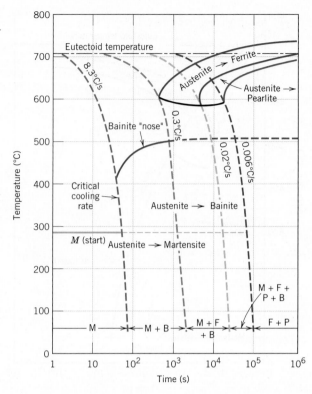

✓ *Concept Check 12.4* Briefly describe the simplest continuous cooling heat treatment procedure that would be used to convert a 4340 steel from (martensite + bainite) into (ferrite + pearlite).

[*The answer may be found at* www.wiley.com/college/callister (*Student Companion Site*).]

12.7 MECHANICAL BEHAVIOR OF IRON–CARBON ALLOYS

We now discuss the mechanical behavior of iron–carbon alloys having the microstructures discussed heretofore—namely, fine and coarse pearlite, spheroidite, bainite, and martensite. For all but martensite, two phases are present (ferrite and cementite), and so an opportunity is provided to explore several mechanical property–microstructure relationships that exist for these alloys.

Pearlite

Cementite is much harder but more brittle than ferrite. Thus, increasing the fraction of Fe_3C in a steel alloy while holding other microstructural elements constant will result in a harder and stronger material. This is demonstrated in Figure 12.29*a*, in which the tensile and yield strengths and the Brinell hardness number are plotted as a function of the weight percent carbon (or equivalently as the percentage of Fe_3C) for steels that are composed of fine pearlite. All three parameters increase with increasing carbon concentration. Inasmuch as cementite is more brittle, increasing its content results in a decrease in both ductility and toughness (or impact energy). These effects are shown in Figure 12.29*b* for the same fine pearlitic steels.

The layer thickness of each of the ferrite and cementite phases in the microstructure also influences the mechanical behavior of the material. Fine pearlite is harder and

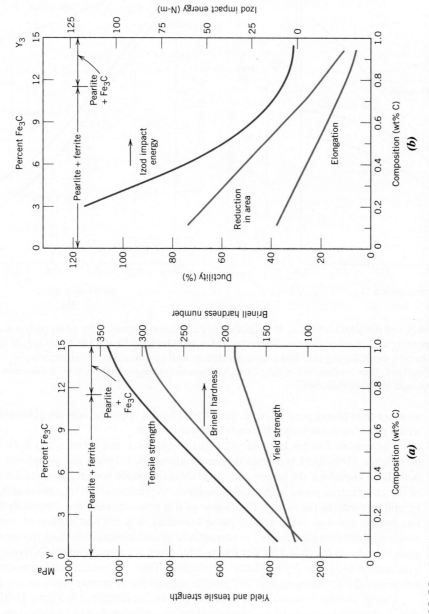

Figure 12.29 (a) Yield strength, tensile strength, and Brinell hardness versus carbon concentration for plain carbon steels having microstructures consisting of fine pearlite. (b) Ductility (%EL and %RA) and Izod impact energy versus carbon concentration for plain carbon steels having microstructures consisting of fine pearlite. [Data taken from *Metals Handbook: Heat Treating*, Vol. 4, 9th edition, V. Masseria (Managing Editor), 1981. Reproduced by permission of ASM International, Materials Park, OH.]

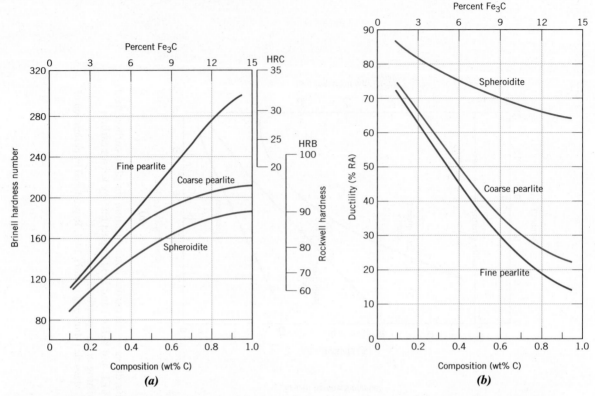

Figure 12.30 (*a*) Brinell and Rockwell hardness as a function of carbon concentration for plain carbon steels having fine and coarse pearlite as well as spheroidite microstructures. (*b*) Ductility (%RA) as a function of carbon concentration for plain carbon steels having fine and coarse pearlite as well as spheroidite microstructures.
[Data taken from *Metals Handbook: Heat Treating,* Vol. 4, 9th edition, V. Masseria (Managing Editor), 1981. Reproduced by permission of ASM International, Materials Park, OH.]

stronger than coarse pearlite, as demonstrated by the upper two curves of Figure 12.30*a*, which plots hardness versus the carbon concentration.

The reasons for this behavior relate to phenomena that occur at the α–Fe$_3$C phase boundaries. First, there is a large degree of adherence between the two phases across a boundary. Therefore, the strong and rigid cementite phase severely restricts deformation of the softer ferrite phase in the regions adjacent to the boundary; thus the cementite may be said to reinforce the ferrite. The degree of this reinforcement is substantially higher in fine pearlite because of the greater phase boundary area per unit volume of material. In addition, phase boundaries serve as barriers to dislocation motion in much the same way as grain boundaries (Section 9.8). For fine pearlite there are more boundaries through which a dislocation must pass during plastic deformation. Thus, the greater reinforcement and restriction of dislocation motion in fine pearlite account for its greater hardness and strength.

Coarse pearlite is more ductile than fine pearlite, as illustrated in Figure 12.30*b*, which plots percentage reduction in area versus carbon concentration for both microstructure types. This behavior results from the greater restriction to plastic deformation of the fine pearlite.

Spheroidite

Other elements of the microstructure relate to the shape and distribution of the phases. In this respect, the cementite phase has distinctly different shapes and arrangements in the pearlite and spheroidite microstructures (Figures 12.15 and 12.19). Alloys containing pearlitic microstructures have greater strength and hardness than do those with spheroidite. This is demonstrated in Figure 12.30*a*, which compares the hardness as a function of the weight percent carbon for spheroidite with both the pearlite structure types. This behavior is again explained in terms of reinforcement at, and impedance to, dislocation motion across the ferrite–cementite boundaries as discussed previously. There is less boundary area per unit volume in spheroidite, and consequently plastic deformation is not nearly as constrained, which gives rise to a relatively soft and weak material. In fact, of all steel alloys, those that are softest and weakest have a spheroidite microstructure.

As might be expected, spheroidized steels are extremely ductile, much more than either fine or coarse pearlite (Figure 12.30b). In addition, they are notably tough because any crack can encounter only a very small fraction of the brittle cementite particles as it propagates through the ductile ferrite matrix.

Bainite

Because bainitic steels have a finer structure (i.e., smaller α-ferrite and Fe_3C particles), they are generally stronger and harder than pearlitic steels; yet they exhibit a desirable combination of strength and ductility. Figures 12.31a and 12.31b show, respectively, the influence of transformation temperature on the strength/hardness and ductility for an iron–carbon alloy of eutectoid composition. Temperature ranges over which pearlite and bainite form (consistent with the isothermal transformation diagram for this alloy, Figure 12.18) are noted at the tops of Figures 12.31a and 12.31b.

Martensite

Of the various microstructures that may be produced for a given steel alloy, martensite is the hardest and strongest and, in addition, the most brittle; it has, in fact, negligible ductility. Its hardness is dependent on the carbon content, up to about 0.6 wt% as demonstrated in Figure 12.32, which plots the hardness of martensite and fine pearlite as a function of weight percent carbon. In contrast to pearlitic steels, the strength and hardness of martensite are not thought to be related to microstructure. Rather, these properties are attributed to the effectiveness of the interstitial carbon atoms in hindering dislocation motion (as a solid-solution effect, Section 9.9), and to the relatively few slip systems (along which dislocations move) for the BCT structure.

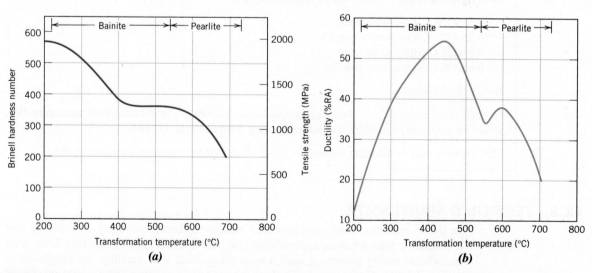

Figure 12.31 (a) Brinell hardness and tensile strength and (b) ductility (%RA) (at room temperature) as a function of isothermal transformation temperature for an iron–carbon alloy of eutectoid composition, taken over the temperature range at which bainitic and pearlitic microstructures form.
[Figure (a) Adapted from E. S. Davenport, "Isothermal Transformation in Steels," *Trans. ASM,* **27,** 1939, p. 847. Reprinted by permission of ASM International, Materials Park, OH.]

Figure 12.32 Hardness (at room temperature) as a function of carbon concentration for plain carbon martensitic, tempered martensitic (tempered at 371°C), and pearlitic steels. (Adapted from Edgar C. Bain, *Functions of the Alloying Elements in Steel,* 1939; and R. A. Grange, C. R. Hribal, and L. F. Porter, *Metall. Trans. A,* Vol. 8A. Reproduced by permission of ASM International, Materials Park, OH.)

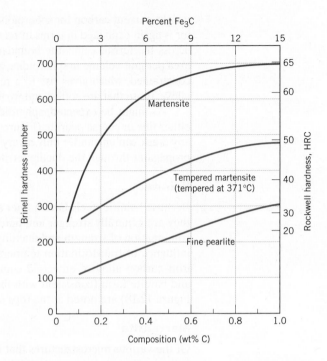

Austenite is slightly denser than martensite, and therefore, during the phase transformation upon quenching, there is a net volume increase. Consequently, relatively large pieces that are rapidly quenched may crack as a result of internal stresses; this becomes a problem especially when the carbon content is greater than about 0.5 wt%.

Concept Check 12.5   Rank the following iron–carbon alloys and associated microstructures from the highest to the lowest tensile strength:

   0.25 wt% C with spheroidite
   0.25 wt% C with coarse pearlite
   0.6 wt% C with fine pearlite
   0.6 wt% C with coarse pearlite

Justify this ranking.

Concept Check 12.6   For a eutectoid steel, describe an isothermal heat treatment that would be required to produce a specimen having a hardness of 93 HRB.

[*The answers may be found at* www.wiley.com/college/callister *(Student Companion Site).*]

12.8 TEMPERED MARTENSITE

In the as-quenched state, martensite, in addition to being very hard, is so brittle that it cannot be used for most applications; also, any internal stresses that may have been introduced during quenching have a weakening effect. The ductility and toughness of martensite may be enhanced and these internal stresses relieved by a heat treatment known as *tempering.*

Tempering is accomplished by heating a martensitic steel to a temperature below the eutectoid for a specified time period. Normally, tempering is carried out at temperatures

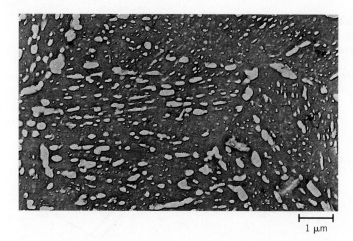

Figure 12.33 Electron micrograph of tempered martensite. Tempering was carried out at 594°C. The small particles are the cementite phase; the matrix phase is α-ferrite. 9300×. (Copyright 1971 by United States Steel Corporation.)

├─────┤
1 μm

between 250°C and 650°C; internal stresses, however, may be relieved at temperatures as low as 200°C. This tempering heat treatment allows, by diffusional processes, the formation of **tempered martensite**, according to the reaction

tempered martensite

Martensite to tempered martensite transformation reaction

$$\text{martensite (BCT, single phase)} \rightarrow \text{tempered martensite} (\alpha + \text{Fe}_3\text{C phases}) \quad (12.20)$$

where the single-phase BCT martensite, which is supersaturated with carbon, transforms into the tempered martensite, composed of the stable ferrite and cementite phases, as indicated on the iron–iron carbide phase diagram.

The microstructure of tempered martensite consists of extremely small and uniformly dispersed cementite particles embedded within a continuous ferrite matrix. This is similar to the microstructure of spheroidite except that the cementite particles are much, much smaller. An electron micrograph showing the microstructure of tempered martensite at a very high magnification is presented in Figure 12.33.

Tempered martensite may be nearly as hard and strong as martensite but with substantially enhanced ductility and toughness. For example, the hardness-versus-weight percent carbon plot of Figure 12.32 includes a curve for tempered martensite. The hardness and strength may be explained by the large ferrite–cementite phase boundary area per unit volume that exists for the very fine and numerous cementite particles. Again, the hard cementite phase reinforces the ferrite matrix along the boundaries, and these boundaries also act as barriers to dislocation motion during plastic deformation. The continuous ferrite phase is also very ductile and relatively tough, which accounts for the improvement of these two properties for tempered martensite.

The size of the cementite particles influences the mechanical behavior of tempered martensite; increasing the particle size decreases the ferrite–cementite phase boundary area and, consequently, results in a softer and weaker material yet one that is tougher and more ductile. Furthermore, the tempering heat treatment determines the size of the cementite particles. Heat treatment variables are temperature and time, and most treatments are constant-temperature processes. Because carbon diffusion is involved in the martensite-tempered martensite transformation, increasing the temperature accelerates diffusion, the rate of cementite particle growth, and, subsequently, the rate of softening. The dependence of tensile and yield strength and ductility on tempering temperature for an alloy steel is shown in Figure 12.34. Before tempering, the material was quenched in oil to produce the martensitic structure; the tempering time at each temperature was 1 h. This type of tempering data is ordinarily provided by the steel manufacturer.

The time dependence of hardness at several different temperatures is presented in Figure 12.35 for a water-quenched steel of eutectoid composition; the time scale is

Figure 12.34 Tensile and yield strengths and ductility (%RA) (at room temperature) versus tempering temperature for an oil-quenched alloy steel (type 4340). (Adapted from Edgar C. Bain, *Functions of the Alloying Elements in Steel,* 1939. Reproduced by permission of ASM International, Materials Park, OH.)

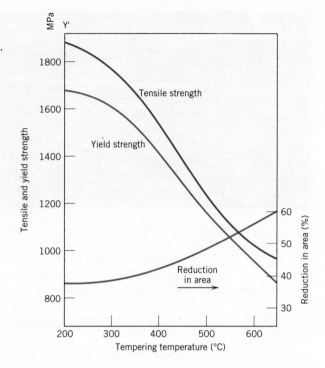

logarithmic. With increasing time the hardness decreases, which corresponds to the growth and coalescence of the cementite particles. At temperatures approaching the eutectoid (700°C) and after several hours, the microstructure will become spheroiditic (Figure 12.19), with large cementite spheroids embedded within the continuous ferrite phase. Correspondingly, overtempered martensite is relatively soft and ductile.

Figure 12.35
Hardness (at room temperature) versus tempering time for a water-quenched eutectoid plain carbon (1080) steel.
(Adapted from Edgar C. Bain, *Functions of the Alloying Elements in Steel,* American Society for Metals, 1939, p. 233.)

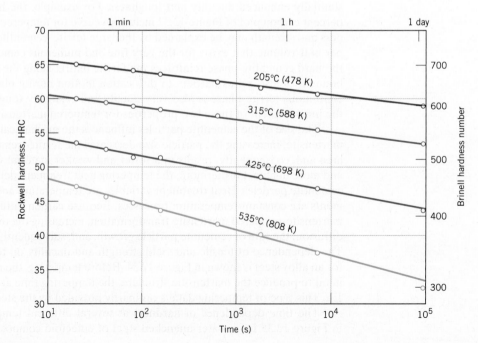

Concept Check 12.7 A steel alloy is quenched from a temperature within the austenite phase region into water at room temperature so as to form martensite; the alloy is subsequently tempered at an elevated temperature, which is held constant.

(a) Make a schematic plot showing how room-temperature ductility varies with the logarithm of tempering time at the elevated temperature. (Be sure to label your axes.)

(b) Superimpose and label on this same plot the room-temperature behavior resulting from tempering at a higher temperature and briefly explain the difference in behavior at these two temperatures.

[*The answer may be found at* www.wiley.com/college/callister *(Student Companion Site).*]

The tempering of some steels may result in a reduction of toughness as measured by impact tests (Section 10.6); this is termed *temper embrittlement*. The phenomenon occurs when the steel is tempered at a temperature above about 575°C followed by slow cooling to room temperature, or when tempering is carried out at between approximately 375°C and 575°C. Steel alloys that are susceptible to temper embrittlement have been found to contain appreciable concentrations of the alloying elements manganese, nickel, or chromium and, in addition, one or more of antimony, phosphorus, arsenic, and tin as impurities in relatively low concentrations. The presence of these alloying elements and impurities shifts the ductile-to-brittle transition to significantly higher temperatures; the ambient temperature thus lies below this transition in the brittle regime. It has been observed that crack propagation of these embrittled materials is *intergranular* (Figure 10.7)—that is, the fracture path is along the grain boundaries of the precursor austenite phase. Furthermore, alloy and impurity elements have been found to preferentially segregate in these regions.

Temper embrittlement may be avoided by (1) compositional control and/or (2) tempering above 575°C or below 375°C, followed by quenching to room temperature. Furthermore, the toughness of steels that have been embrittled may be improved significantly by heating to about 600°C and then rapidly cooling to below 300°C.

12.9 REVIEW OF PHASE TRANSFORMATIONS AND MECHANICAL PROPERTIES FOR IRON–CARBON ALLOYS

In this chapter, we discussed several different microstructures that may be produced in iron–carbon alloys depending on heat treatment. Figure 12.36 summarizes the

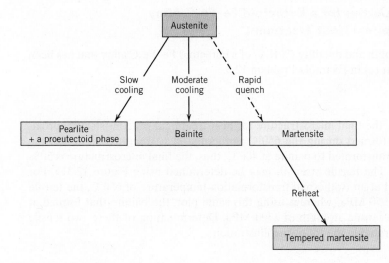

Figure 12.36 Possible transformations involving the decomposition of austenite. Solid arrows, transformations involving diffusion; dashed arrow, diffusionless transformation.

Table 12.2 Microstructures and Mechanical Properties for Iron–Carbon Alloys

Microconstituent	Phases Present	Arrangement of Phases	Mechanical Properties (Relative)
Spheroidite	α-Ferrite + Fe_3C	Relatively small Fe_3C spherelike particles in an α-ferrite matrix	Soft and ductile
Coarse pearlite	α-Ferrite + Fe_3C	Alternating layers of α-ferrite and Fe_3C that are relatively thick	Harder and stronger than spheroidite, but not as ductile as spheroidite
Fine pearlite	α-Ferrite + Fe_3C	Alternating layers of α-ferrite and Fe_3C that are relatively thin	Harder and stronger than coarse pearlite, but not as ductile as coarse pearlite
Bainite	α-Ferrite + Fe_3C	Very fine and elongated particles of Fe_3C in an α-ferrite matrix	Harder and stronger than fine pearlite; less hard than martensite; more ductile than martensite
Tempered martensite	α-Ferrite + Fe_3C	Very small Fe_3C spherelike particles in an α-ferrite matrix	Strong; not as hard as martensite, but much more ductile than martensite
Martensite	Body-centered, tetragonal, single phase	Needle-shaped grains	Very hard and very brittle

WileyPLUS

Tutorial Video:
Iron–Carbon Alloy
Microstructures
What Are the
Differences
between the Various
Iron–Carbon Alloy
Microstructures?

transformation paths that produce these various microstructures. Here, it is assumed that pearlite, bainite, and martensite result from continuous-cooling treatments; furthermore, the formation of bainite is possible only for alloy steels (not plain carbon ones), as outlined earlier.

Microstructural characteristics and mechanical properties of the several microconstituents for iron–carbon alloys are summarized in Table 12.2.

EXAMPLE PROBLEM 12.4

Determination of Properties for a Eutectoid Fe–Fe$_3$C Alloy Subjected to an Isothermal Heat Treatment

Determine the tensile strength and ductility (%RA) of a eutectoid Fe–Fe$_3$C alloy that has been subjected to heat treatment (c) in Example Problem 12.3.

Solution

According to Figure 12.24, the final microstructure for heat treatment (c) consists of approximately 50% pearlite that formed during the 650°C isothermal heat treatment, whereas the remaining 50% austenite transformed to bainite at 400°C; thus, the final microstructure is 50% pearlite and 50% bainite. The tensile strength may be determined using Figure 12.31a. For pearlite, which was formed at an isothermal transformation temperature of 650°C, the tensile strength is approximately 950 MPa, whereas using this same plot, the bainite that formed at 400°C has an approximate tensile strength of 1300 MPa. Determination of these two tensile strength values is demonstrated in the following illustration.

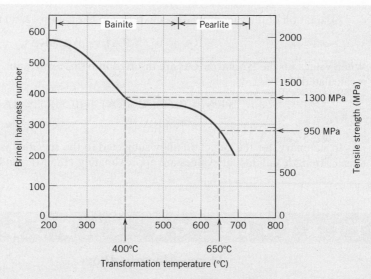

The tensile strength of this two-microconstituent alloy may be approximated using a "rule-of-mixtures" relationship—that is, the alloy tensile strength is equal to the fraction-weighted average of the two microconstituents, which may be expressed by the following equation:

$$\overline{TS} = W_p(TS)_p + W_b(TS)_b \tag{12.21}$$

Here,

$\overline{TS}$ = tensile strength of the alloy,

W_p and W_b = mass fractions of pearlite and bainite, respectively, and

$(TS)_p$ and $(TS)_b$ = tensile strengths of the respective microconstituents.

Thus, incorporating values for these four parameters into Equation 12.21 leads to the following alloy tensile strength:

$$\overline{TS} = (0.50)(950 \text{ MPa}) + (0.50)(1300 \text{ MPa})$$
$$= 1125 \text{ MPa}$$

This same technique is used for the computation of ductility. In this case, approximate ductility values for the two microconstituents, taken at 650°C (for pearlite) and 400°C (for bainite), are, respectively, 32%RA and 52%RA, as taken from the following adaptation of Figure 12.31b:

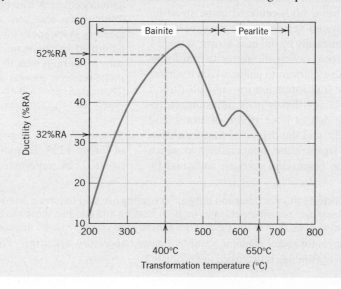

Adaptation of the rule-of-mixtures expression (Equation 12.21) for this case is as follows:

$$\%\overline{RA} = W_p(\%RA)_p + W_b(\%RA)_b$$

When values for the Ws and %RAs are inserted into this expression, the approximate ductility is calculated as

$$\%\overline{RA} = (0.50)(32\%RA) + (0.50)(52\%RA)$$
$$= 42\%RA$$

In summary, for the eutectoid alloy subjected to the specified isothermal heat treatment, tensile strength and ductility values are approximately 1125 MPa and 42%RA, respectively.

MATERIALS OF IMPORTANCE

Shape–Memory Alloys

A relatively new group of metals that exhibit an interesting (and practical) phenomenon are the *shape-memory alloys* (or *SMAs*). One of these materials, after being deformed, has the ability to return to its predeformed size and shape upon being subjected to an appropriate heat treatment—that is, the material "remembers" its previous size/shape. Deformation normally is carried out at a relatively low temperature, whereas shape memory occurs upon heating.[5] Materials that have been found to be capable of recovering significant amounts of deformation (i.e., strain) are nickel–titanium alloys (Nitinol,[6] is their trade name) and some copper-base alloys (Cu–Zn–Al and Cu–Al–Ni alloys).

A shape-memory alloy is polymorphic (Section 4.14)—that is, it may have two crystal structures (or phases), and the shape-memory effect involves phase transformations between them. One phase (termed an *austenite phase*) has a body-centered cubic structure that exists at elevated temperatures; its structure is represented schematically by the inset shown at stage 1 of Figure 12.37. Upon cooling, the austenite transforms spontaneously into a martensite phase, which is similar to the martensitic transformation for the iron–carbon system (Section 12.5)—that is, it is diffusionless, involves an orderly shift of large groups of atoms, and occurs very rapidly, and the degree of transformation is dependent on temperature; temperatures at which the transformation begins and ends are indicated by

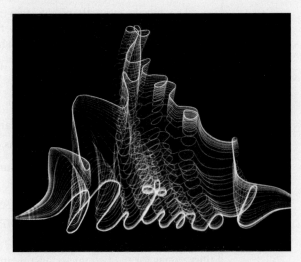

Time-lapse photograph that demonstrates the shape-memory effect. A wire of a shape-memory alloy (Nitinol) has been bent and treated such that its memory shape spells the word *Nitinol*. The wire is then deformed and, upon heating (by passage of an electric current), springs back to its predeformed shape; this shape recovery process is recorded on the photograph. [Photograph courtesy the Naval Surface Warfare Center (previously the Naval Ordnance Laboratory)].

M_s and M_f labels, respectively, on the left vertical axis of Figure 12.37. In addition, this martensite is heavily twinned,[7] as represented schematically by the stage 2

---

[5]Alloys that demonstrate this phenomenon only upon heating are said to have a *one-way* shape memory. Some of these materials experience size/shape changes on both heating and cooling; these are termed *two-way* shape memory alloys. In this discussion, we discuss the mechanism for only the one-way shape memory alloys.

[6]*Nitinol* is an acronym for *ni*ckel-*ti*tanium *N*aval *O*rdnance *L*aboratory, where this alloy was discovered.

[7]The phenomenon of twinning is described in Section 9.7.

inset of Figure 12.37. Under the influence of an applied stress, deformation of martensite (i.e., the passage from stage 2 to stage 3 in Figure 12.37) occurs by the migration of twin boundaries—some twinned regions grow while others shrink; this deformed martensitic structure is represented by the stage 3 inset. Furthermore, when the stress is removed, the deformed shape is retained at this temperature. Finally, upon subsequent heating to the initial temperature, the material reverts back to (i.e., "remembers") its original size and shape (stage 4). This stage 3–stage 4 process is accompanied by a phase transformation from the deformed martensite into the original high-temperature austenite phase. For these shape-memory alloys, the martensite-to-austenite transformation occurs over a temperature range, between the temperatures denoted by A_s (austenite start) and A_f (austenite finish) labels on the right vertical axis of Figure 12.37. This deformation–transformation cycle may be repeated for the shape-memory material.

The original shape (the one that is to be remembered) is created by heating to well above the A_f temperature (such that the transformation to austenite is complete) and then restraining the material to the desired memory shape for a sufficient time period. For example, for Nitinol alloys, a 1-h treatment at 500°C is necessary.

Although the deformation experienced by shape-memory alloys is semipermanent, it is not truly "plastic"

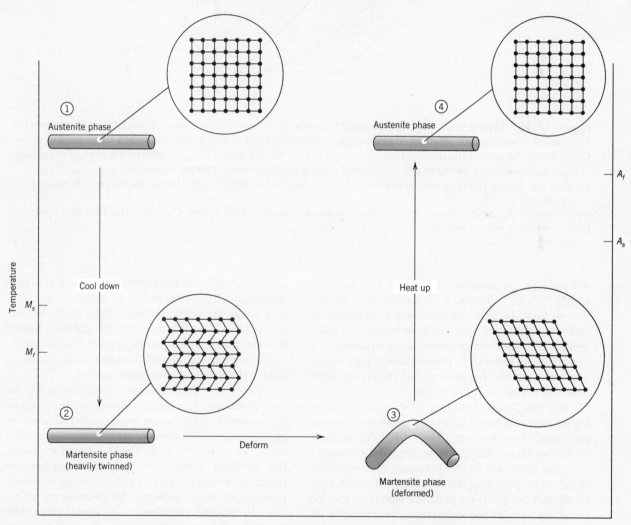

Figure 12.37 Diagram illustrating the shape-memory effect. The insets are schematic representations of the crystal structure at the four stages. M_s and M_f denote temperatures at which the martensitic transformation begins and ends, respectively. Likewise for the austenite transformation, A_s and A_f represent the respective beginning and end transformation temperatures.

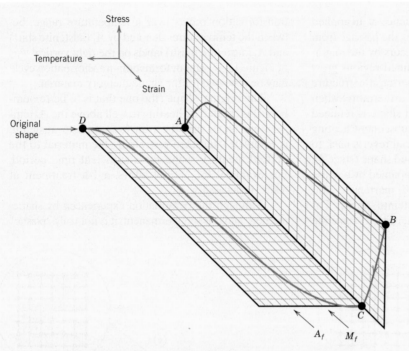

Figure 12.38 Typical stress-strain-temperature behavior of a shape-memory alloy, demonstrating its thermoelastic behavior. Specimen deformation, corresponding to the curve from A to B, is carried out at a temperature below that at which the martensitic transformation is complete (i.e., M_f of Figure 12.37). Release of the applied stress (also at M_f) is represented by the curve BC. Subsequent heating to above the completed austenite-transformation temperature (A_f, Figure 12.37) causes the deformed piece to resume its original shape (along the curve from point C to point D).
[From Helsen, J. A., and H. J. Breme (Editors), *Metals as Biomaterials,* John Wiley & Sons, Chichester, UK, 1998. Reprinted with permission of John Wiley & Sons Inc.]

deformation, as discussed in Section 8.4, nor is it strictly "elastic" (Section 8.3). Rather, it is termed *thermoelastic,* because deformation is nonpermanent when the deformed material is subsequently heat-treated. The stress-strain-temperature behavior of a thermoelastic material is presented in Figure 12.38. Maximum recoverable deformation strains for these materials are on the order of 8%.

For this Nitinol family of alloys, transformation temperatures can be made to vary over a wide temperature range (between about –200°C and 110°C) by altering the Ni–Ti ratio and also by adding other elements.

One important SMA application is in weldless, shrink-to-fit pipe couplers used for hydraulic lines on aircraft, for joints on undersea pipelines, and for plumbing on ships and submarines. Each coupler (in the form of a cylindrical sleeve) is fabricated so as to have an inside diameter slightly smaller than the outside diameter of the pipes to be joined. It is then stretched (circumferentially) at some temperature well below the ambient temperature. Next the coupler is fitted over the pipe junction and then heated to room temperature; heating causes the coupler to shrink back to its original diameter, thus creating a tight seal between the two pipe sections.

There is a host of other applications for alloys displaying this effect—for example, eyeglass frames, tooth-straightening braces, collapsible antennas, greenhouse window openers, antiscald control valves on showers, women's foundation-garments, fire sprinkler valves, and biomedical applications (such as blood-clot filters, self-extending coronary stents, and bone anchors). Shape-memory alloys also fall into the classification of "smart materials" (Section 1.5) because they sense and respond to environmental (i.e., temperature) changes.

SUMMARY

The Kinetics of Phase Transformations

- Nucleation and growth are the two steps involved in the production of a new phase.
- Two types of nucleation are possible: homogeneous and heterogeneous.
    For homogeneous nucleation, nuclei of the new phase form uniformly throughout the parent phase.
    For heterogeneous nucleation, nuclei form preferentially at the surfaces of structural inhomogeneities (e.g., container surfaces, insoluble impurities).
- For the homogeneous nucleation of a spherical solid particle in a liquid solution, expressions for the critical radius (r^*) and activation free energy (ΔG^*) are represented by Equations 12.3 and 12.4, respectively. These two parameters are indicated in the plot of Figure 12.2b.
- The activation free energy for heterogeneous nucleation (ΔG^*_{het}) is lower than that for homogeneous nucleation (ΔG^*_{hom}), as demonstrated on the schematic free energy–versus–nucleus radius curves of Figure 12.6.
- Heterogeneous nucleation occurs more easily than homogeneous nucleation, which is reflected in a smaller degree of supercooling (ΔT) for the former—that is, $\Delta T_{het} < \Delta T_{hom}$, Figure 12.7.
- The growth stage of phase particle formation begins once a nucleus has exceeded the critical radius (r^*).
- For typical solid transformations, a plot of fraction transformation versus logarithm of time yields an S-shaped curve, as depicted schematically in Figure 12.10.
- The time dependence of degree of transformation is represented by the Avrami equation, Equation 12.17.
- Transformation rate is taken as the reciprocal of time required for a transformation to proceed halfway to its completion, Equation 12.18.
- For transformations that are induced by temperature alterations, when the rate of temperature change is such that equilibrium conditions are not maintained, transformation temperature is raised (for heating) and lowered (for cooling). These phenomena are termed superheating and supercooling, respectively.

Isothermal Transformation Diagrams

Continuous-Cooling Transformation Diagrams

- Phase diagrams provide no information as to the time dependence of transformation progress. However, the element of time is incorporated into isothermal transformation diagrams. These diagrams do the following:
    Plot temperature versus the logarithm of time, with curves for beginning, as well as 50% and 100% transformation completion.
    Are generated from a series of plots of percentage transformation versus the logarithm of time taken over a range of temperatures (Figure 12.13).
    Are valid only for constant-temperature heat treatments.
    Permit determination of times at which a phase transformation begins and ends.
- Isothermal transformation diagrams may be modified for continuous-cooling heat treatments; isothermal transformation beginning and ending curves are shifted to longer times and lower temperatures (Figure 12.25). Intersections with these curves of continuous-cooling curves represent times at which the transformation starts and ceases.

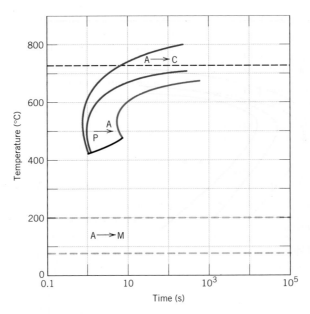

Figure 12.40 Continuous cooling transformation diagram for a 1.13 wt% C iron–carbon alloy.

12.27 Name the microstructural products of 4340 alloy steel specimens that are first completely transformed to austenite, then cooled to room temperature at the following rates:

(a) 10°C/s

(b) 1°C/s

(c) 0.1°C/s

(d) 0.01°C/s

12.28 Briefly describe the simplest continuous cooling heat treatment procedure that would be used in converting a 4340 steel from one microstructure to another.

(a) (Martensite + bainite) to (ferrite + pearlite)

(b) (Martensite + bainite) to spheroidite

(c) (Martensite + bainite) to (martensite + bainite + ferrite)

12.29 On the basis of diffusion considerations, explain why fine pearlite forms for the moderate cooling of austenite through the eutectoid temperature, whereas coarse pearlite is the product for relatively slow cooling rates.

Mechanical Behavior of Iron–Carbon Alloys
Tempered Martensite

12.30 Briefly explain why fine pearlite is harder and stronger than coarse pearlite, which in turn is harder and stronger than spheroidite.

12.31 Cite two reasons why martensite is so hard and brittle.

12.32 Rank the following iron–carbon alloys and associated microstructures from the highest to the lowest tensile strength:

(a) 0.25 wt% C with spheroidite

(b) 0.25 wt% C with coarse pearlite

(c) 0.60 wt% C with fine pearlite

(d) 0.60 wt% C with coarse pearlite

Justify this ranking.

12.33 Briefly explain why the hardness of tempered martensite diminishes with tempering time (at constant temperature) and with increasing temperature (at constant tempering time).

12.34 Briefly describe the simplest heat treatment procedure that would be used in converting a 0.76 wt% C steel from one microstructure to the other, as follows:

(a) Spheroidite to tempered martensite

(b) Tempered martensite to pearlite

(c) Bainite to martensite

(d) Martensite to pearlite

(e) Pearlite to tempered martensite

(f) Tempered martensite to pearlite

(g) Bainite to tempered martensite

(h) Tempered martensite to spheroidite

12.35 (a) Briefly describe the microstructural difference between spheroidite and tempered martensite.

(b) Explain why tempered martensite is much harder and stronger.

12.36 Estimate the Rockwell hardnesses for specimens of an iron–carbon alloy of eutectoid composition that have been subjected to the heat treatments described in parts (b), (d), (f), (g), and (h) of Problem 12.18.

12.37 Estimate the Brinell hardnesses for specimens of a 0.45 wt% C iron–carbon alloy that have been subjected to the heat treatments described in parts (a), (d), and (h) of Problem 12.20.

12.38 Determine the approximate tensile strengths for specimens of a eutectoid iron–carbon alloy that have experienced the heat treatments described in parts (a) and (c) of Problem 12.23.

12.39 For a eutectoid steel, describe isothermal heat treatments that would be required to

yield specimens having the following Rockwell hardnesses:

(a) 93 HRB

(b) 40 HRC

(c) 27 HRC

Spreadsheet Problem

12.1SS For some phase transformation, given at least two values of fraction transformation and their corresponding times, generate a spreadsheet that will allow the user to determine the following: **(a)** the values of n and k in the Avrami equation, **(b)** the time required for the transformation to proceed to some degree of fraction transformation, and **(c)** the fraction transformation after some specified time has elapsed.

DESIGN PROBLEMS

Continuous Cooling Transformation Diagrams
Mechanical Behavior of Iron–Carbon Alloys

12.D1 Is it possible to produce an iron–carbon alloy of eutectoid composition that has a minimum hardness of 100 HRB and a minimum ductility of 35%RA? If so, describe the continuous cooling heat treatment to which the alloy would be subjected to achieve these properties. If it is not possible, explain why.

12.D2 Is it possible to produce an iron–carbon alloy that has a minimum tensile strength of 690 MPa and a minimum ductility of 50%RA? If so, what will be its composition and microstructure (coarse and fine pearlites and spheroidite are alternatives)? If this is not possible, explain why.

12.D3 It is desired to produce an iron–carbon alloy that has a minimum hardness of 180 HB and a minimum ductility of 52%RA. Is such an alloy possible? If so, what will be its composition and microstructure (coarse and fine pearlites and spheroidite are alternatives)? If this is not possible, explain why.

Tempered Martensite

12.D4 **(a)** For a 1080 steel that has been water quenched, estimate the tempering time at 427°C (700 K) to achieve a hardness of 50 HRC.

(b) What will be the tempering time at 320°C (593 K) necessary to attain the same hardness?

12.D5 An alloy steel (4340) is to be used in an application requiring a minimum tensile strength of 1380 MPa and a minimum ductility of 43%RA. Oil quenching followed by tempering is to be used. Briefly describe the tempering heat treatment.

12.D6 Is it possible to produce an oil-quenched and tempered 4340 steel that has a minimum yield strength of 1500 MPa and a ductility of at least 42%RA? If this is possible, describe the tempering heat treatment. If it is not possible, explain why.

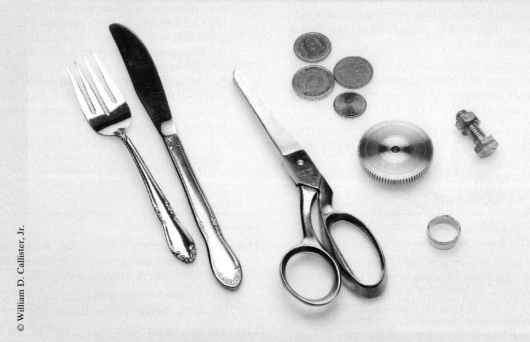

© William D. Callister, Jr.

The many available ferrous and nonferrous metal alloys provide the practicing engineer a wide range of physical properties; accordingly, materials in this class are used in a vast number and variety of applications. These include familiar objects such as those shown in this photograph: (from left to right) silverware (fork and knife), scissors, coins, a gear, a wedding ring, and a nut and bolt.

Engineers are often involved in materials selection decisions, which necessitates that they have some familiarity with the general characteristics of a wide variety of

metals and their alloys (as well as other material types). In addition, access to databases containing property values for a large number of materials may be required.

Learning Objectives

After studying this chapter, you should be able to do the following:

1. Name four different types of steels and cite compositional differences, distinctive properties, and typical uses for each.

2. Name the five cast iron types and describe the microstructure and note the general mechanical characteristics for each.

3. Name seven different types of nonferrous alloys and cite the distinctive physical and mechanical characteristics and list at least three typical applications for each.

13.1 INTRODUCTION

Often a materials problem is really one of selecting the material that has the right combination of characteristics for a specific application. Therefore, the people who are involved in the decision making should have some knowledge of the available options. The first portion of this chapter provides an abbreviated overview of some of the commercial alloys and their general properties and limitations.

Metal alloys, by virtue of composition, are often grouped into two classes—ferrous and nonferrous. Ferrous alloys, those in which iron is the principal constituent, include steels and cast irons. These alloys and their characteristics are the first topics of discussion of this section. The nonferrous ones—all alloys that are not iron based—are treated next.

Ferrous Alloys

ferrous alloy

Ferrous alloys—those in which iron is the prime constituent—are produced in larger quantities than any other metal type. They are especially important as engineering construction materials. Their widespread use is accounted for by three factors: (1) iron-containing compounds exist in abundant quantities within the Earth's crust; (2) metallic iron and steel alloys may be produced using relatively economical extraction, refining, alloying, and fabrication techniques; and (3) ferrous alloys are extremely versatile, in that they may be tailored to have a wide range of mechanical and physical properties. The principal disadvantage of many ferrous alloys is their susceptibility to corrosion. This section discusses compositions, microstructures, and properties of a number of different classes of steels and cast irons. A taxonomic classification scheme for the various ferrous alloys is presented in Figure 13.1.

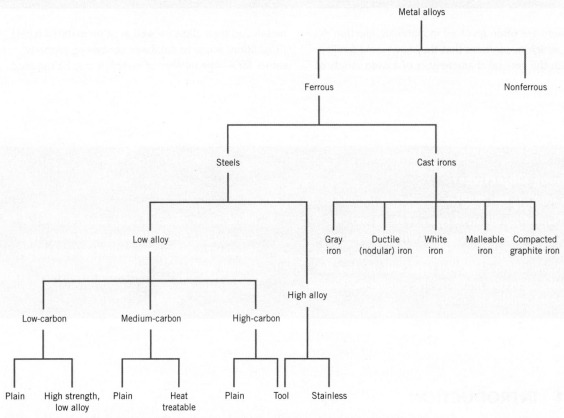

Figure 13.1 Classification scheme for the various ferrous alloys.

13.2 STEELS

Steels are iron–carbon alloys that may contain appreciable concentrations of other alloying elements; there are thousands of alloys that have different compositions and/or heat treatments. The mechanical properties are sensitive to the content of carbon, which is normally less than 1.0 wt%. Some of the more common steels are classified according to carbon concentration into low-, medium-, and high-carbon types. Subclasses also exist within each group according to the concentration of other alloying elements. **Plain carbon steels** contain only residual concentrations of impurities other than carbon and a little manganese. For **alloy steels**, more alloying elements are intentionally added in specific concentrations.

plain carbon steel

alloy steel

Low-Carbon Steels

Of the different steels, those produced in the greatest quantities fall within the low-carbon classification. These generally contain less than about 0.25 wt% C and are unresponsive to heat treatments intended to form martensite; strengthening is accomplished by cold work. Microstructures consist of ferrite and pearlite constituents. As a consequence, these alloys are relatively soft and weak but have outstanding ductility and toughness; in addition, they are machinable, weldable, and, of all steels, are the least expensive to produce. Typical applications include automobile body components, structural shapes (e.g., I-beams, channel and angle iron), and sheets that are used in pipelines, buildings, bridges, and tin cans. Tables 13.1a and 13.1b present the compositions and mechanical properties of several plain low-carbon steels. They typically have a yield strength of 275 MPa, tensile strengths between 415 and 550 MPa, and a ductility of 25%EL.

Table 13.1a

Compositions of Four Plain Low-Carbon Steels and Three High-Strength, Low-Alloy Steels

Designation[a]		Composition (wt%)[b]		
AISI/SAE or ASTM Number	UNS Number	C	Mn	Other
Plain Low-Carbon Steels				
1010	G10100	0.10	0.45	
1020	G10200	0.20	0.45	
A36	K02600	0.29	1.00	0.20 Cu (min)
A516 Grade 70	K02700	0.31	1.00	0.25 Si
High-Strength, Low-Alloy Steels				
A440	K12810	0.28	1.35	0.30 Si (max), 0.20 Cu (min)
A633 Grade E	K12002	0.22	1.35	0.30 Si, 0.08 V, 0.02 N, 0.03 Nb
A656 Grade 1	K11804	0.18	1.60	0.60 Si, 0.1 V, 0.20 Al, 0.015 N

[a]The codes used by the American Iron and Steel Institute (AISI), the Society of Automotive Engineers (SAE), and the American Society for Testing and Materials (ASTM), and in the Uniform Numbering System (UNS) are explained in the text.
[b]Also a maximum of 0.04 wt% P, 0.05 wt% S, and 0.30 wt% Si (unless indicated otherwise).
Source: Adapted from *Metals Handbook: Properties and Selection: Irons and Steels,* Vol. 1, 9th edition, B. Bardes (Editor), 1978. Reproduced by permission of ASM International, Materials Park, OH.

Table 13.1b

Mechanical Characteristics of Hot-Rolled Material and Typical Applications for Various Plain Low-Carbon and High-Strength, Low-Alloy Steels

AISI/SAE or ASTM Number	Tensile Strength (MPa)	Yield Strength (MPa)	Ductility (%EL in 50 mm)	Typical Applications
Plain Low-Carbon Steels				
1010	325	180	28	Automobile panels, nails, and wire
1020	380	205	25	Pipe; structural and sheet steel
A36	400	220	23	Structural (bridges and buildings)
A516 Grade 70	485	260	21	Low-temperature pressure vessels
High-Strength, Low-Alloy Steels				
A440	435	290	21	Structures that are bolted or riveted
A633 Grade E	520	380	23	Structures used at low ambient temperatures
A656 Grade 1	655	552	15	Truck frames and railway cars

high-strength, low-alloy steel

Another group of low-carbon alloys are the **high-strength, low-alloy (HSLA) steels.** They contain other alloying elements such as copper, vanadium, nickel, and molybdenum in combined concentrations as high as 10 wt%, and they possess higher strengths than the plain low-carbon steels. Most may be strengthened by heat treatment, giving tensile strengths in excess of 480 MPa; in addition, they are ductile, formable, and machinable. Several are listed in Tables 13.1a and 13.1b. In normal atmospheres, the HSLA steels are more resistant to corrosion than the plain carbon steels, which they have replaced in many applications where structural strength is critical (e.g., bridges, towers, support columns in high-rise buildings, pressure vessels).

Medium-Carbon Steels

The medium-carbon steels have carbon concentrations between about 0.25 and 0.60 wt%. These alloys may be heat-treated by austenitizing, quenching, and then tempering to improve their mechanical properties. They are most often utilized in the tempered condition, having microstructures of tempered martensite. The plain medium-carbon steels have low hardenabilities (Section 17.6) and can be successfully heat-treated only in very thin sections and with very rapid quenching rates. Additions of chromium, nickel, and molybdenum improve the capacity of these alloys to be heat-treated (Section 17.6), giving rise to a variety of strength–ductility combinations. These heat-treated alloys are stronger than the low-carbon steels, but at a sacrifice of ductility and toughness. Applications include railway wheels and tracks, gears, crankshafts, and other machine parts and high-strength structural components calling for a combination of high strength, wear resistance, and toughness.

The compositions of several of these alloyed medium-carbon steels are presented in Table 13.2a. Some comment is in order regarding the designation schemes that are also included. The Society of Automotive Engineers (SAE), the American Iron and Steel Institute (AISI), and the American Society for Testing and Materials (ASTM) are responsible for the classification and specification of steels as well as other alloys. The AISI/SAE designation for these steels is a four-digit number: the first two digits indicate the alloy content; the last two give the carbon concentration. For plain carbon steels, the first two digits are 1 and 0; alloy steels are designated by other initial two-digit combinations (e.g., 13, 41, 43). The third and fourth digits represent the weight percent carbon multiplied by 100. For example, a 1060 steel is a plain carbon steel containing 0.60 wt% C.

A unified numbering system (UNS) is used for uniformly indexing both ferrous and nonferrous alloys. Each UNS number consists of a single-letter prefix followed by

Table 13.2a

AISI/SAE and UNS Designation Systems and Composition Ranges for Plain Carbon Steel and Various Low-Alloy Steels

AISI/SAE Designation[a]	UNS Designation	Composition Ranges (wt% of Alloying Elements in Addition to C)[b]			
		Ni	Cr	Mo	Other
10xx, Plain carbon	G10xx0				
11xx, Free machining	G11xx0				0.08–0.33 S
12xx, Free machining	G12xx0				0.10–0.35 S, 0.04–0.12 P
13xx	G13xx0				1.60–1.90 Mn
40xx	G40xx0			0.20–0.30	
41xx	G41xx0		0.80–1.10	0.15–0.25	
43xx	G43xx0	1.65–2.00	0.40–0.90	0.20–0.30	
46xx	G46xx0	0.70–2.00		0.15–0.30	
48xx	G48xx0	3.25–3.75		0.20–0.30	
51xx	G51xx0		0.70–1.10		
61xx	G61xx0		0.50–1.10		0.10–0.15 V
86xx	G86xx0	0.40–0.70	0.40–0.60	0.15–0.25	
92xx	G92xx0				1.80–2.20 Si

[a]The carbon concentration, in weight percent times 100, is inserted in the place of "xx" for each specific steel.
[b]Except for 13xx alloys, manganese concentration is less than 1.00 wt%.
Except for 12xx alloys, phosphorus concentration is less than 0.35 wt%.
Except for 11xx and 12xx alloys, sulfur concentration is less than 0.04 wt%.
Except for 92xx alloys, silicon concentration varies between 0.15 and 0.35 wt%.

Table 13.2b Typical Applications and Mechanical Property Ranges for Oil-Quenched and Tempered Plain Carbon and Alloy Steels

AISI Number	UNS Number	Tensile Strength (MPa)	Yield Strength (MPa)	Ductility (%EL in 50 mm)	Typical Applications
		Plain Carbon Steels			
1040	G10400	605–780	430–585	33–19	Crankshafts, bolts
1080[a]	G10800	800–1310	480–980	24–13	Chisels, hammers
1095[a]	G10950	760–1280	510–830	26–10	Knives, hacksaw blades
		Alloy Steels			
4063	G40630	786–2380	710–1770	24–4	Springs, hand tools
4340	G43400	980–1960	895–1570	21–11	Bushings, aircraft tubing
6150	G61500	815–2170	745–1860	22–7	Shafts, pistons, gears

[a]Classified as high-carbon steels.

a five-digit number. The letter is indicative of the family of metals to which an alloy belongs. The UNS designation for these alloys begins with a G, followed by the AISI/SAE number; the fifth digit is a zero. Table 13.2b contains the mechanical characteristics and typical applications of several of these steels, which have been quenched and tempered.

High-Carbon Steels

The high-carbon steels, normally having carbon contents between 0.60 and 1.4 wt%, are the hardest, strongest, and yet least ductile of the carbon steels. They are almost always used in a hardened and tempered condition and, as such, are especially wear resistant and capable of holding a sharp cutting edge. The tool and die steels are high-carbon alloys, usually containing chromium, vanadium, tungsten, and molybdenum. These alloying elements combine with carbon to form very hard and wear-resistant carbide compounds (e.g., $Cr_{23}C_6$, V_4C_3, and WC). Some tool steel compositions and their applications are listed in Table 13.3. These steels are used as cutting tools and dies for

Table 13.3 Designations, Compositions, and Applications for Six Tool Steels

AISI Number	UNS Number	Composition (wt%)[a]						Typical Applications
		C	Cr	Ni	Mo	W	V	
M1	T11301	0.85	3.75	0.30 max	8.70	1.75	1.20	Drills, saws; lathe and planer tools
A2	T30102	1.00	5.15	0.30 max	1.15	—	0.35	Punches, embossing dies
D2	T30402	1.50	12	0.30 max	0.95	—	1.10 max	Cutlery, drawing dies
O1	T31501	0.95	0.50	0.30 max	—	0.50	0.30 max	Shear blades, cutting tools
S1	T41901	0.50	1.40	0.30 max	0.50 max	2.25	0.25	Pipe cutters, concrete drills
W1	T72301	1.10	0.15 max	0.20 max	0.10 max	0.15 max	0.10 max	Blacksmith tools, woodworking tools

[a]The balance of the composition is iron. Manganese concentrations range between 0.10 and 1.4 wt%, depending on alloy; silicon concentrations between 0.20 and 1.2 wt%, depending on the alloy.

Source: Adapted from *ASM Handbook*, Vol. 1, *Properties and Selection: Irons, Steels, and High-Performance Alloys*, 1990. Reprinted by permission of ASM International, Materials Park, OH.

Figure 13.2 The true equilibrium iron–carbon phase diagram with graphite instead of cementite as a stable phase. [Adapted from *Binary Alloy Phase Diagrams,* T. B. Massalski (Editor-in-Chief), 1990. Reprinted by permission of ASM International, Materials Park, OH.]

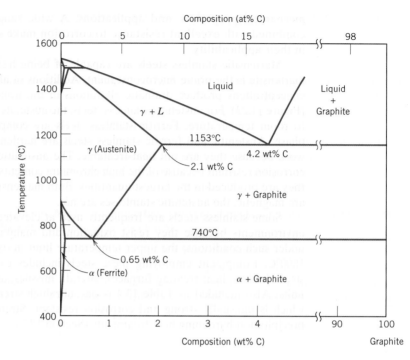

This tendency to form graphite is regulated by the composition and rate of cooling. Graphite formation is promoted by the presence of silicon in concentrations greater than about 1 wt%. Also, slower cooling rates during solidification favor graphitization (the formation of graphite). For most cast irons, the carbon exists as graphite, and both microstructure and mechanical behavior depend on composition and heat treatment. The most common cast iron types are gray, nodular, white, malleable, and compacted graphite.

Gray Iron

gray cast iron

The carbon and silicon contents of **gray cast irons** vary between 2.5 and 4.0 wt% and 1.0 and 3.0 wt%, respectively. For most of these cast irons, the graphite exists in the form of flakes (similar to corn flakes), which are normally surrounded by an α-ferrite or pearlite matrix; the microstructure of a typical gray iron is shown in Figure 13.3a. Because of these graphite flakes, a fractured surface takes on a gray appearance—hence its name.

Mechanically, gray iron is comparatively weak and brittle in tension as a consequence of its microstructure; the tips of the graphite flakes are sharp and pointed and may serve as points of stress concentration when an external tensile stress is applied. Strength and ductility are much higher under compressive loads. Typical mechanical properties and compositions of several common gray cast irons are listed in Table 13.5. Gray irons have some desirable characteristics and are used extensively. They are very effective in damping vibrational energy; this is represented in Figure 13.4, which compares the relative damping capacities of steel and gray iron. Base structures for machines and heavy equipment that are exposed to vibrations are frequently constructed of this material. In addition, gray irons exhibit a high resistance to wear. Furthermore, in the molten state they have a high fluidity at casting temperature, which permits casting pieces that have intricate shapes; also, casting shrinkage is low. Finally, and perhaps most important, gray cast irons are among the least expensive of all metallic materials.

Gray irons having microstructures different from that shown in Figure 13.3a may be generated by adjusting composition and/or using an appropriate treatment. For example, lowering the silicon content or increasing the cooling rate may prevent the complete

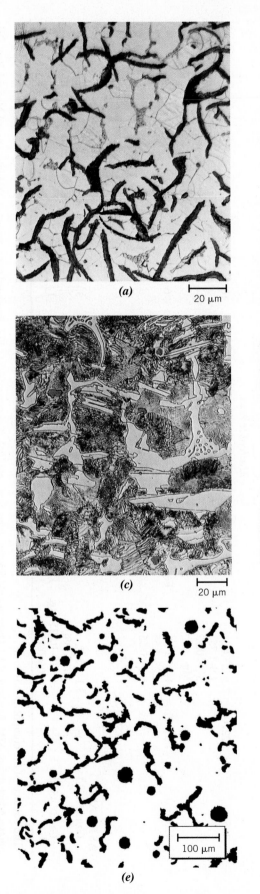

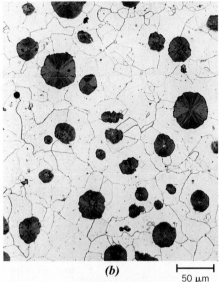

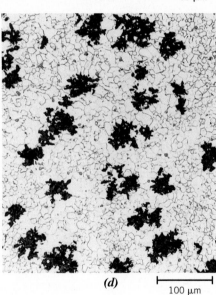

(a)

20 μm

(b)

50 μm

(c)

20 μm

(d)

100 μm

(e)

100 μm

Figure 13.3
Optical photo-micrographs of various cast irons. (*a*) Gray iron: the dark graphite flakes are embedded in an α-ferrite matrix. 500×. (*b*) Nodular (ductile) iron: the dark graphite nodules are surrounded by an α-ferrite matrix. 200×. (*c*) White iron: the light cementite regions are surrounded by pearlite, which has the ferrite–cementite layered structure. 400×. (*d*) Malleable iron: dark graphite rosettes (temper carbon) in an α-ferrite matrix. 150×. (*e*) Compacted graphite iron: dark graphite wormlike particles are embedded within an α-ferrite matrix. 100×.

[Figures (*a*) and (*b*) courtesy of C. H. Brady and L. C. Smith, National Bureau of Standards, Washington, DC (now the National Institute of Standards and Technology, Gaithersburg, MD). Figure (*c*) courtesy of Amcast Industrial Corporation. Figure (*d*) reprinted with permission of the Iron Castings Society, Des Plaines, IL. Figure (*e*) courtesy of SinterCast, Ltd.]

Table 13.5 Designations, Minimum Mechanical Properties, Approximate Compositions, and Typical Applications for Various Gray, Nodular, Malleable, and Compacted Graphite Cast Irons

Grade	UNS Number	Composition (wt%)[a]	Matrix Structure	Mechanical Properties			Typical Applications
				Tensile Strength (MPa)	Yield Strength (MPa)	Ductility (%EL in 50 mm)	
Gray Iron							
SAE G1800	F10004	3.40–3.7 C, 2.55 Si, 0.7 Mn	Ferrite + Pearlite	124	—	—	Miscellaneous soft iron castings in which strength is not a primary consideration
SAE G2500	F10005	3.2–3.5 C, 2.20 Si, 0.8 Mn	Ferrite + Pearlite	173	—	—	Small cylinder blocks, cylinder heads, pistons, clutch plates, transmission cases
SAE G4000	F10008	3.0–3.3 C, 2.0 Si, 0.8 Mn	Pearlite	276	—	—	Diesel engine castings, liners, cylinders, and pistons
Ductile (Nodular) Iron							
ASTM A536 60–40–18	F32800	3.5–3.8 C, 2.0–2.8 Si, 0.05 Mg, <0.20 Ni, <0.10 Mo	Ferrite	414	276	18	Pressure-containing parts such as valve and pump bodies
100–70–03	F34800		Pearlite	689	483	3	High-strength gears and machine components
120–90–02	F36200		Tempered martensite	827	621	2	Pinions, gears, rollers, slides
Malleable Iron							
32510	F22200	2.3–2.7 C, 1.0–1.75 Si, <0.55 Mn	Ferrite	345	224	10	General engineering service at normal and elevated temperatures
45006	F23131	2.4–2.7 C, 1.25–1.55 Si, <0.55 Mn	Ferrite + Pearlite	448	310	6	
Compacted Graphite Iron							
ASTM A842 Grade 250	—	3.1–4.0 C, 1.7–3.0 Si, 0.015–0.035 Mg, 0.06–0.13 Ti	Ferrite	250	175	3	Diesel engine blocks, exhaust manifolds, brake discs for high-speed trains
Grade 450	—		Pearlite	450	315	1	

[a]The balance of the composition is iron.

Source: Adapted from *ASM Handbook*, Vol. 1, *Properties and Selection: Irons, Steels, and High-Performance Alloys*, 1990. Reprinted by permission of ASM International, Materials Park, OH.

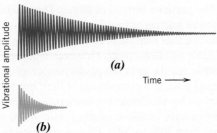

Figure 13.4 Comparison of the relative vibrational damping capacities of (*a*) steel and (*b*) gray cast iron.
(From *Metals Engineering Quarterly,* February 1961. Copyright © 1961. Reproduced by permission of ASM International, Materials Park, OH.)

dissociation of cementite to form graphite (Equation 13.1). Under these circumstances the microstructure consists of graphite flakes embedded in a pearlite matrix. Figure 13.5 compares schematically the several cast iron microstructures obtained by varying the composition and heat treatment.

Ductile (or Nodular) Iron

Adding a small amount of magnesium and/or cerium to the gray iron before casting produces a distinctly different microstructure and set of mechanical properties. Graphite still forms,

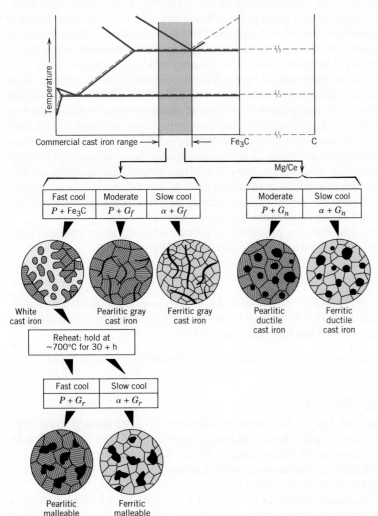

Figure 13.5 From the iron–carbon phase diagram, composition ranges for commercial cast irons. Also shown are schematic microstructures that result from a variety of heat treatments. G_f, flake graphite; G_r, graphite rosettes; G_n, graphite nodules; P, pearlite; α, ferrite.
(Adapted from W. G. Moffatt, G. W. Pearsall, and J. Wulff, *The Structure and Properties of Materials,* Vol. I, *Structure,* p. 195. Copyright © 1964 by John Wiley & Sons, New York.)

ductile (nodular)
iron

but as nodules or spherelike particles instead of flakes. The resulting alloy is called **ductile** or **nodular iron**, and a typical microstructure is shown in Figure 13.3b. The matrix phase surrounding these particles is either pearlite or ferrite, depending on heat treatment (Figure 13.5); it is normally pearlite for an as-cast piece. However, a heat treatment for several hours at about 700°C yields a ferrite matrix, as in this photomicrograph. Castings are stronger and much more ductile than gray iron, as a comparison of their mechanical properties in Table 13.5 shows. In fact, ductile iron has mechanical characteristics approaching those of steel. For example, ferritic ductile irons have tensile strengths between 380 and 480 MPa and ductilities (as percent elongation) from 10% to 20%. Typical applications for this material include valves, pump bodies, crankshafts, gears, and other automotive and machine components.

White Iron and Malleable Iron

white cast iron

For low-silicon cast irons (containing less than 1.0 wt% Si) and rapid cooling rates, most of the carbon exists as cementite instead of graphite, as indicated in Figure 13.5. A fracture surface of this alloy has a white appearance, and thus it is termed **white cast iron**. An optical photomicrograph showing the microstructure of white iron is presented in Figure 13.3c. Thick sections may have only a surface layer of white iron that was "chilled" during the casting process; gray iron forms at interior regions, which cool more slowly. As a consequence of large amounts of the cementite phase, white iron is extremely hard but also very brittle, to the point of being virtually unmachinable. Its use is limited to applications that necessitate a very hard and wear-resistant surface, without a high degree of ductility—for example, as rollers in rolling mills. Generally, white iron is

malleable iron

used as an intermediary in the production of yet another cast iron, **malleable iron**.

Heating white iron at temperatures between 800°C and 900°C for a prolonged time period and in a neutral atmosphere (to prevent oxidation) causes a decomposition of the cementite, forming graphite, which exists in the form of clusters or rosettes surrounded by a ferrite or pearlite matrix, depending on cooling rate, as indicated in Figure 13.5. A photomicrograph of a ferritic malleable iron is presented in Figure 13.3d. The microstructure is similar to that of nodular iron (Figure 13.3b), which accounts for relatively high strength and appreciable ductility or malleability. Some typical mechanical characteristics are also listed in Table 13.5. Representative applications include connecting rods, transmission gears, and differential cases for the automotive industry, and also flanges, pipe fittings, and valve parts for railroad, marine, and other heavy-duty services.

Gray and ductile cast irons are produced in approximately the same amounts; however, white and malleable cast irons are produced in smaller quantities.

✓

Concept Check 13.2   It is possible to produce cast irons that consist of a martensite matrix in which graphite is embedded in either flake, nodule, or rosette form. Briefly describe the treatment necessary to produce each of these three microstructures.

[*The answer may be found at* www.wiley.com/college/callister *(Student Companion Site).*]

Compacted Graphite Iron

compacted
graphite iron

A relatively recent addition to the family of cast irons is **compacted graphite iron** (abbreviated *CGI*). As with gray, ductile, and malleable irons, carbon exists as graphite, whose formation is promoted by the presence of silicon. Silicon content ranges between 1.7 and 3.0 wt%, whereas carbon concentration is normally between 3.1 and 4.0 wt%. Two CGI materials are included in Table 13.5.

Microstructurally, the graphite in CGI alloys has a wormlike (or vermicular) shape; a typical CGI microstructure is shown in the optical micrograph of Figure 13.3e. In a sense, this microstructure is intermediate between that of gray iron (Figure 13.3*a*) and ductile (nodular) iron (Figure 13.3*b*), and, in fact, some of the graphite (less than 20%) may be as nodules. However, sharp edges (characteristic of graphite flakes) should be avoided; the presence of this feature leads to a reduction in fracture and fatigue resistance of the material. Magnesium and/or cerium is also added, but concentrations are lower than for ductile iron. The chemistries of CGIs are more complex than for the other cast iron types; compositions of magnesium, cerium, and other additives must be controlled so as to produce a microstructure that consists of the wormlike graphite particles while at the same time limiting the degree of graphite nodularity, and preventing the formation of graphite flakes. Furthermore, depending on heat treatment, the matrix phase will be pearlite and/or ferrite.

As with the other types of cast irons, the mechanical properties of CGIs are related to microstructure: graphite particle shape, as well as the matrix phase/microconstituent. An increase in degree of nodularity of the graphite particles leads to enhancements of both strength and ductility. Furthermore, CGIs with ferritic matrices have lower strengths and higher ductilities than those with pearlitic matrices. Tensile and yield strengths for compacted graphite irons are comparable to values for ductile and malleable irons, yet are greater than those observed for the higher-strength gray irons (Table 13.5). In addition, ductilities for CGIs are intermediate between values for gray and ductile irons; moduli of elasticity range between 140 and 165 GPa.

Compared to the other cast iron types, desirable characteristics of CGIs include the following:

- Higher thermal conductivity
- Better resistance to thermal shock (i.e., fracture resulting from rapid temperature changes)
- Lower oxidation at elevated temperatures

Compacted graphite irons are now being used in a number of important applications, including diesel engine blocks, exhaust manifolds, gearbox housings, brake discs for high-speed trains, and flywheels.

Nonferrous Alloys

Steel and other ferrous alloys are consumed in exceedingly large quantities because they have such a wide range of mechanical properties, may be fabricated with relative ease, and are economical to produce. However, they have some distinct limitations chiefly (1) a relatively high density, (2) a comparatively low electrical conductivity, and (3) an inherent susceptibility to corrosion in some common environments. Thus, for many applications it is advantageous or even necessary to use other alloys that have more suitable property combinations. Alloy systems are classified either according to the base metal or according to some specific characteristic that a group of alloys share. This section discusses the following metal and alloy systems: copper, aluminum, magnesium, and titanium alloys; the refractory metals; the superalloys; the noble metals; and miscellaneous alloys, including those that have nickel, lead, tin, zirconium, and zinc as base metals.

nonferrous alloys Figure 13.6 represents a classification scheme for **nonferrous alloys** discussed in this section.

On occasion, a distinction is made between cast and wrought alloys. Alloys that are so brittle that forming or shaping by appreciable deformation is not possible typically are cast; these are classified as *cast alloys*. However, those that are amenable to

wrought alloy mechanical deformation are termed **wrought alloys**.

In addition, the heat-treatability of an alloy system is mentioned frequently. "Heat-treatable" designates an alloy whose mechanical strength is improved by precipitation hardening (Section 17.7) or a martensitic transformation (normally the former), both of which involve specific heat-treating procedures.

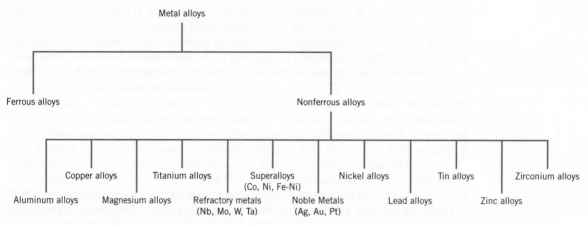

Figure 13.6 Classification scheme for the various nonferrous alloys.

13.4 COPPER AND ITS ALLOYS

Copper and copper-based alloys, possessing a desirable combination of physical properties, have been used in quite a variety of applications since antiquity. Unalloyed copper is so soft and ductile that it is difficult to machine; also, it has an almost unlimited capacity to be cold worked. Furthermore, it is highly resistant to corrosion in diverse environments including the ambient atmosphere, seawater, and some industrial chemicals. The mechanical and corrosion-resistance properties of copper may be improved by alloying. Most copper alloys cannot be hardened or strengthened by heat-treating procedures; consequently, cold working and/or solid-solution alloying must be used to improve these mechanical properties.

brass

The most common copper alloys are the **brasses**, for which zinc, as a substitutional impurity, is the predominant alloying element. As may be observed for the copper–zinc phase diagram (Figure 11.18), the α phase is stable for concentrations up to approximately 35 wt% Zn. This phase has an FCC crystal structure, and α-brasses are relatively soft, ductile, and easily cold worked. Brass alloys having a higher zinc content contain both α and β' phases at room temperature. The β' phase has an ordered BCC crystal structure and is harder and stronger than the α phase; consequently, $\alpha + \beta'$ alloys are generally hot worked.

Some of the common brasses are yellow, naval, and cartridge brass; muntz metal; and gilding metal. The compositions, properties, and typical uses of several of these alloys are listed in Table 13.6. Some of the common uses for brass alloys include costume jewelry, cartridge casings, automotive radiators, musical instruments, electronic packaging, and coins.

bronze

The **bronzes** are alloys of copper and several other elements, including tin, aluminum, silicon, and nickel. These alloys are somewhat stronger than the brasses, yet they still have a high degree of corrosion resistance. Table 13.6 lists several of the bronze alloys and their compositions, properties, and applications. Generally they are used when, in addition to corrosion resistance, good tensile properties are required.

The most common heat-treatable copper alloys are the beryllium coppers. They possess a remarkable combination of properties: tensile strengths as high as 1400 MPa, excellent electrical and corrosion properties, and wear resistance when properly lubricated; they may be cast, hot worked, or cold worked. High strengths are attained by precipitation-hardening heat treatments (Section 17.7). These alloys are costly because of the beryllium additions, which range between 1.0 and 2.5 wt%. Applications include jet aircraft landing gear bearings and bushings, springs, and surgical and dental instruments. One of these alloys (C17200) is included in Table 13.6.

Table 13.6 Compositions, Mechanical Properties, and Typical Applications for Eight Copper Alloys

Alloy Name	UNS Number	Composition (wt%)[a]	Condition	Mechanical Properties			Typical Applications
				Tensile Strength (MPa)	Yield Strength (MPa)	Ductility (%EL in 50 mm)	
Wrought Alloys							
Electrolytic tough pitch	C11000	0.04 O	Annealed	220	69	45	Electrical wire, rivets, screening, gaskets, pans, nails, roofing
Beryllium copper	C17200	1.9 Be, 0.20 Co	Precipitation hardened	1140–1310	690–860	4–10	Springs, bellows, firing pins, bushings, valves, diaphragms
Cartridge brass	C26000	30 Zn	Annealed	300	75	68	Automotive radiator cores, ammunition components, lamp fixtures, flashlight shells, kickplates
			Cold-worked (H04 hard)	525	435	8	
Phosphor bronze, 5% A	C51000	5 Sn, 0.2 P	Annealed	325	130	64	Bellows, clutch disks, diaphragms, fuse clips, springs, welding rods
			Cold-worked (H04 hard)	560	515	10	
Copper–nickel, 30%	C71500	30 Ni	Annealed	380	125	36	Condenser and heat-exchanger components, saltwater piping
			Cold-worked (H02 hard)	515	485	15	
Cast Alloys							
Leaded yellow brass	C85400	29 Zn, 3 Pb, 1 Sn	As cast	234	83	35	Furniture hardware, radiator fittings, light fixtures, battery clamps
Tin bronze	C90500	10 Sn, 2 Zn	As cast	310	152	25	Bearings, bushings, piston rings, steam fittings, gears
Aluminum bronze	C95400	4 Fe, 11 Al	As cast	586	241	18	Bearings, gears, worms, bushings, valve seats and guards, pickling hooks

[a]The balance of the composition is copper.

Source: Adapted from *ASM Handbook*, Vol. 2, *Properties and Selection: Nonferrous Alloys and Special-Purpose Materials*, 1990. Reprinted by permission of ASM International, Materials Park, OH.

> ✓ *Concept Check 13.3* What is the main difference between brass and bronze?
>
> [*The answer may be found at* www.wiley.com/college/callister *(Student Companion Site).*]

13.5 ALUMINUM AND ITS ALLOYS

Aluminum and its alloys are characterized by a relatively low density (2.7 g/cm³ as compared to 7.9 g/cm³ for steel), high electrical and thermal conductivities, and a resistance to corrosion in some common environments, including the ambient atmosphere. Many of these alloys are easily formed by virtue of high ductility; this is evidenced by the thin aluminum foil sheet into which the relatively pure material may be rolled. Because aluminum has an FCC crystal structure, its ductility is retained even at very low temperatures. The chief limitation of aluminum is its low melting temperature (660°C), which restricts the maximum temperature at which it can be used.

The mechanical strength of aluminum may be enhanced by cold work and by alloying; however, both processes tend to decrease resistance to corrosion. Principal alloying elements include copper, magnesium, silicon, manganese, and zinc. Non-heat-treatable alloys consist of a single phase, for which an increase in strength is achieved by solid-solution strengthening. Others are rendered heat-treatable (capable of being precipitation hardened) as a result of alloying. In several of these alloys, precipitation hardening is due to the precipitation of two elements other than aluminum to form an intermetallic compound such as $MgZn_2$.

temper designation

Generally, aluminum alloys are classified as either cast or wrought. Composition for both types is designated by a four-digit number that indicates the principal impurities and, in some cases, the purity level. For cast alloys, a decimal point is located between the last two digits. After these digits is a hyphen and the basic **temper designation**—a letter and possibly a one- to three-digit number, which indicates the mechanical and/or heat treatment to which the alloy has been subjected. For example, F, H, and O represent, respectively, the as-fabricated, strain-hardened, and annealed states. Table 13.7 presents the temper designation scheme for aluminum alloys. Furthermore, compositions, properties, and applications of several wrought and cast alloys are given in Table 13.8. Common applications of aluminum alloys include

Table 13.7 Temper Designation Scheme for Aluminum Alloys

Designation	Description
	Basic Tempers
F	As-fabricated—by casting or cold working
O	Annealed—lowest strength temper (wrought products only)
H	Strain-hardened (wrought products only)
W	Solution heat-treated—used only on products that precipitation harden naturally at room temperature over periods of months or years
T	Solution heat-treated—used on products that strength stabilize within a few weeks— followed by one or more digits
	Strain-Hardened Tempers[a]
H1	Strain hardened only
H2	Strain-hardened and then partially annealed
H3	Strain-hardened and then stabilized
	Heat-Treating Tempers[b]
T1	Cooled from an elevated-temperature shaping process and naturally aged
T2	Cooled from an elevated-temperature shaping process, cold-worked, and naturally aged
T3	Solution heat treated, cold worked, and naturally aged
T4	Solution heat treated and naturally aged
T5	Cooled from an elevated-temperature shaping process and artificially aged
T6	Solution heat treated and artificially aged
T7	Solution heat treated and overaged or stabilized
T8	Solution heat treated, cold worked, and artificially aged
T9	Solution heat treated, artificially aged, and cold worked
T10	Cooled from an elevated-temperature shaping process, cold worked, and artificially aged

[a]Two additional digits may be added to denote degree of strain hardening.
[b]Additional digits (the first of which cannot be zero) are used to denote variations of these 10 tempers.
Source: Adapted from *ASM Handbook*, Vol. 2, *Properties and Selection: Nonferrous Alloys and Special-Purpose Materials,* 1990. Reproduced with permission of ASM International, Materials Park, OH, 44073.

Table 13.8 Compositions, Mechanical Properties, and Typical Applications for Several Common Aluminum Alloys

Aluminum Association Number	UNS Number	Composition (wt%)[a]	Condition (Temper Designation)	Mechanical Properties			Typical Applications/ Characteristics
				Tensile Strength (MPa)	Yield Strength (MPa)	Ductility (%EL in 50 mm)	
Wrought, Nonheat-Treatable Alloys							
1100	A91100	0.12 Cu	Annealed (O)	90	35	35–45	Food/chemical handling and storage equipment, heat exchangers, light reflectors
3003	A93003	0.12 Cu, 1.2 Mn, 0.1 Zn	Annealed (O)	110	40	30–40	Cooking utensils, pressure vessels and piping
5052	A95052	2.5 Mg, 0.25 Cr	Strain hardened (H32)	230	195	12–18	Aircraft fuel and oil lines, fuel tanks, appliances, rivets, and wire
Wrought, Heat-Treatable Alloys							
2024	A92024	4.4 Cu, 1.5 Mg, 0.6 Mn	Heat-treated (T4)	470	325	20	Aircraft structures, rivets, truck wheels, screw machine products
6061	A96061	1.0 Mg, 0.6 Si, 0.30 Cu, 0.20 Cr	Heat-treated (T4)	240	145	22–25	Trucks, canoes, railroad cars, furniture, pipelines
7075	A97075	5.6 Zn, 2.5 Mg, 1.6 Cu, 0.23 Cr	Heat-treated (T6)	570	505	11	Aircraft structural parts and other highly stressed applications
Cast, Heat-Treatable Alloys							
295.0	A02950	4.5 Cu, 1.1 Si	Heat-treated (T4)	221	110	8.5	Flywheel and rear-axle housings, bus and aircraft wheels, crankcases
356.0	A03560	7.0 Si, 0.3 Mg	Heat-treated (T6)	228	164	3.5	Aircraft pump parts, automotive transmission cases, water-cooled cylinder blocks
Aluminum–Lithium Alloys							
2090	—	2.7 Cu, 0.25 Mg, 2.25 Li, 0.12 Zr	Heat-treated, cold-worked (T83)	455	455	5	Aircraft structures and cryogenic tankage structures
8090	—	1.3 Cu, 0.95 Mg, 2.0 Li, 0.1 Zr	Heat-treated, cold-worked (T651)	465	360	—	Aircraft structures that must be highly damage tolerant

[a]The balance of the composition is aluminum.

Source: Adapted from *ASM Handbook,* Vol. 2, *Properties and Selection: Nonferrous Alloys and Special-Purpose Materials,* 1990. Reprinted by permission of ASM International, Materials Park, OH.

aircraft structural parts, beverage cans, bus bodies, and automotive parts (engine blocks, pistons, and manifolds).

Recent attention has been given to alloys of aluminum and other low-density metals (e.g., Mg and Ti) as engineering materials for transportation, to effect reductions in fuel consumption. An important characteristic of these materials is **specific strength**, which is quantified by the tensile strength–specific gravity ratio. Even though an alloy of one of these metals may have a tensile strength that is inferior to that of a denser material (such as steel), on a weight basis it will be able to sustain a larger load.

A generation of new aluminum–lithium alloys have been developed recently for use by the aircraft and aerospace industries. These materials have relatively low densities (between about 2.5 and 2.6 g/cm^3), high specific moduli (elastic modulus–specific gravity ratios), and excellent fatigue and low-temperature toughness properties. Furthermore, some of them may be precipitation hardened. However, these materials are more costly to manufacture than the conventional aluminum alloys because special processing techniques are required as a result of lithium's chemical reactivity.

Concept Check 13.4   Explain why, under some circumstances, it is not advisable to weld a structure that is fabricated with a 3003 aluminum alloy. *Hint:* You may want to consult Section 9.12.

[*The answer may be found at* www.wiley.com/college/callister (*Student Companion Site*).]

13.6 MAGNESIUM AND ITS ALLOYS

Perhaps the most outstanding characteristic of magnesium is its density, 1.7 g/cm^3, which is the lowest of all the structural metals; therefore, its alloys are used where light weight is an important consideration (e.g., in aircraft components). Magnesium has an HCP crystal structure, is relatively soft, and has a low elastic modulus: 45 GPa. At room temperature, magnesium and its alloys are difficult to deform; in fact, only small degrees of cold work may be imposed without annealing. Consequently, most fabrication is by casting or hot working at temperatures between 200°C and 350°C. Magnesium, like aluminum, has a moderately low melting temperature (651°C). Chemically, magnesium alloys are relatively unstable and especially susceptible to corrosion in marine environments. However, corrosion or oxidation resistance is reasonably good in the normal atmosphere; it is believed that this behavior is due to impurities rather than being an inherent characteristic of Mg alloys. Fine magnesium powder ignites easily when heated in air; consequently, care should be exercised when handling it in this state.

These alloys are also classified as either cast or wrought, and some of them are heat-treatable. Aluminum, zinc, manganese, and some of the rare earths are the major alloying elements. A composition–temper designation scheme similar to that for aluminum alloys is also used. Table 13.9 lists several common magnesium alloys and their compositions, properties, and applications. These alloys are used in aircraft and missile applications, as well as in luggage. Furthermore, in recent years the demand for magnesium alloys has increased dramatically in a host of different industries. For many applications, magnesium alloys have replaced engineering plastics that have comparable densities because the magnesium materials are stiffer, more

Table 13.9 Compositions, Mechanical Properties, and Typical Applications for Six Common Magnesium Alloys

ASTM Number	UNS Number	Composition (wt%)[a]	Condition	Tensile Strength (MPa)	Yield Strength (MPa)	Ductility (%EL in 50 mm)	Typical Applications
				Mechanical Properties			
Wrought Alloys							
AZ31B	M11311	3.0 Al, 1.0 Zn, 0.2 Mn	As extruded	262	200	15	Structures and tubing, cathodic protection
HK31A	M13310	3.0 Th, 0.6 Zr	Strain hardened, partially annealed	255	200	9	High strength to 315°C (588 K)
ZK60A	M16600	5.5 Zn, 0.45 Zr	Artificially aged	350	285	11	Forgings of maximum strength for aircraft
Cast Alloys							
AZ91D	M11916	9.0 Al, 0.15 Mn, 0.7 Zn	As cast	230	150	3	Die-cast parts for automobiles, luggage, and electronic devices
AM60A	M10600	6.0 Al, 0.13 Mn	As cast	220	130	6	Automotive wheels
AS41A	M10410	4.3 Al, 1.0 Si, 0.35 Mn	As cast	210	140	6	Die castings requiring good creep resistance

[a]The balance of the composition is magnesium.

Source: Adapted from *ASM Handbook*, Vol. 2, *Properties and Selection: Nonferrous Alloys and Special-Purpose Materials*, 1990. Reprinted by permission of ASM International, Materials Park, OH.

recyclable, and less costly to produce. For example, magnesium is employed in a variety of handheld devices (e.g., chainsaws, powertools, hedge clippers), automobiles (e.g., steering wheels and columns, seat frames, transmission cases), and audio, video, computer, and communications equipment (e.g., laptop computers, camcorders, TV sets, cellular telephones).

Concept Check 13.5   On the basis of melting temperature, oxidation resistance, yield strength, and degree of brittleness, discuss whether it would be advisable to hot work or to cold work (a) aluminum alloys and (b) magnesium alloys. *Hint:* You may want to consult Sections 9.10 and 9.12.

[*The answer may be found at* www.wiley.com/college/callister *(Student Companion Site).*]

13.7 TITANIUM AND ITS ALLOYS

Titanium and its alloys are relatively new engineering materials that possess an extraordinary combination of properties. The pure metal has a relatively low density (4.5 g/cm³), a high melting point (1668°C), and an elastic modulus of 107 GPa. Titanium alloys are extremely strong; room-temperature tensile strengths as high as 1400 MPa are attainable, yielding remarkable specific strengths. Furthermore, the alloys are highly ductile and easily forged and machined.

Unalloyed (i.e., commercially pure) titanium has a hexagonal close-packed crystal structure, sometimes denoted as the α phase at room temperature. At 883°C, the HCP material transforms into a body-centered cubic (or β) phase. This transformation temperature is strongly influenced by the presence of alloying elements. For example, vanadium, niobium, and molybdenum decrease the α-to-β transformation temperature and promote the formation of the β phase (i.e., are β-phase stabilizers), which may exist at room temperature. In addition, for some compositions, both α and β phases coexist. On the basis of which phase(s) is (are) present after processing, titanium alloys fall into four classifications: α, β, $\alpha + \beta$, and near α.

The α-titanium alloys, often alloyed with aluminum and tin, are preferred for high-temperature applications because of their superior creep characteristics. Furthermore, strengthening by heat treatment is not possible because α is the stable phase; consequently, these materials are normally used in annealed or recrystallized states. Strength and toughness are satisfactory, whereas forgeability is inferior to that of the other Ti alloy types.

The β titanium alloys contain sufficient concentrations of β-stabilizing elements (V and Mo) such that, upon cooling at sufficiently rapid rates, the β (metastable) phase is retained at room temperature. These materials are highly forgeable and exhibit high fracture toughnesses.

The $\alpha + \beta$ materials are alloyed with stabilizing elements for both constituent phases. The strength of these alloys may be improved and controlled by heat treatment. A variety of microstructures is possible that consist of an α phase and a retained or transformed β phase. In general, these materials are quite formable.

Near-α alloys are also composed of both α and β phases, with only a small proportion of β—that is, they contain low concentrations of β stabilizers. Their properties and fabrication characteristics are similar to those of the α materials, except that a greater diversity of microstructures and properties are possible for near-α alloys.

The major limitation of titanium is its chemical reactivity with other materials at elevated temperatures. This property has necessitated the development of nonconventional refining, melting, and casting techniques; consequently, titanium alloys are quite expensive. In spite of this reactivity at high temperature, the corrosion resistance of titanium alloys at normal temperatures is unusually high; they are virtually immune to air, marine, and a variety of industrial environments. Table 13.10 presents several titanium alloys along with their typical properties and applications. They are commonly used in airplane structures, space vehicles, surgical implants, and in the petroleum and chemical industries.

13.8 THE REFRACTORY METALS

Metals that have extremely high melting temperatures are classified as refractory metals. Included in this group are niobium (Nb), molybdenum (Mo), tungsten (W), and tantalum (Ta). Melting temperatures range between 2468°C for niobium and 3410°C, the highest melting temperature of any metal, for tungsten. Interatomic bonding in these metals is extremely strong, which accounts for the melting temperatures, and, in addition, large elastic moduli and high strengths and hardnesses, at ambient as well as elevated temperatures. The applications of these metals are varied. For example, tantalum and molybdenum are alloyed with stainless steel to improve its corrosion resistance. Molybdenum alloys are used for extrusion dies and structural parts in space vehicles; incandescent light filaments, x-ray tubes, and welding electrodes employ

Table 13.10 Compositions, Mechanical Properties, and Typical Applications for Several Common Titanium Alloys

Alloy Type	Common Name (UNS Number)	Composition (wt%)	Condition	Tensile Strength (MPa)	Yield Strength (MPa)	Ductility (%EL in 50 mm)	Typical Applications
Commercially pure	Unalloyed (R50500)	99.5 Ti	Annealed	484	414	25	Jet engine shrouds, cases and airframe skins, corrosion-resistant equipment for marine and chemical processing industries
α	Ti-5Al-2.5Sn (R54520)	5 Al, 2.5 Sn, balance Ti	Annealed	826	784	16	Gas turbine engine casings and rings; chemical processing equipment requiring strength to temperatures of 480°C (753 K)
Near α	Ti-8Al-1Mo-1V (R54810)	8 Al, 1 Mo, 1 V, balance Ti	Annealed (duplex)	950	890	15	Forgings for jet engine components (compressor disks, plates, and hubs)
α – β	Ti-6Al-4V (R56400)	6 Al, 4 V, balance Ti	Annealed	947	877	14	High-strength prosthetic implants, chemical-processing equipment, airframe structural components
α – β	Ti-6Al-6V-2Sn (R56620)	6 Al, 2 Sn, 6 V, 0.75 Cu, balance Ti	Annealed	1050	985	14	Rocket engine case airframe applications and high-strength airframe structures
β	Ti-10V-2Fe-3Al	10 V, 2 Fe, 3 Al, balance Ti	Solution + aging	1223	1150	10	Best combination of high strength and toughness of any commercial titanium alloy; used for applications requiring uniformity of tensile properties at surface and center locations; high-strength airframe components

The Average Mechanical Properties span the Tensile Strength, Yield Strength, and Ductility columns.

Source: Adapted from *ASM Handbook*, Vol. 2, *Properties and Selection: Nonferrous Alloys and Special-Purpose Materials*, 1990. Reprinted by permission of ASM International, Materials Park, OH.

tungsten alloys. Tantalum is immune to chemical attack by virtually all environments at temperatures below 150°C and is frequently used in applications requiring such a corrosion-resistant material.

13.9 THE SUPERALLOYS

The superalloys have superlative combinations of properties. Most are used in aircraft turbine components, which must withstand exposure to severely oxidizing environments and high temperatures for reasonable time periods. Mechanical integrity under these conditions is critical; in this regard, density is an important consideration because centrifugal stresses are diminished in rotating members when the density is reduced. These materials are classified according to the predominant metal(s) in the alloy, of which there are three groups—iron–nickel, nickel, and cobalt. Other alloying elements include the refractory metals (Nb, Mo, W, Ta), chromium, and titanium. Furthermore, these alloys are also categorized as wrought or cast. Compositions of several of them are presented in Table 13.11.

In addition to turbine applications, superalloys are used in nuclear reactors and petrochemical equipment.

Table 13.11 Compositions for Several Superalloys

Alloy Name	Composition (wt%)									
	Ni	Fe	Co	Cr	Mo	W	Ti	Al	C	Other
Iron–Nickel (Wrought)										
A-286	26	55.2	—	15	1.25	—	2.0	0.2	0.04	0.005 B, 0.3 V
Incoloy 925	44	29	—	20.5	2.8	—	2.1	0.2	0.01	1.8 Cu
Nickel (Wrought)										
Inconel-718	52.5	18.5	—	19	3.0	—	0.9	0.5	0.08	5.1 Nb, 0.15 max Cu
Waspaloy	57.0	2.0 max	13.5	19.5	4.3	—		3.0	1.4	0.07	0.006 B, 0.09 Zr
Nickel (Cast)										
Rene 80	60	—	9.5	14	4	4	5	3	0.17	0.015 B, 0.03 Zr
Mar-M-247	59	0.5	10	8.25	0.7	10	1	5.5	0.15	0.015 B, 3 Ta, 0.05 Zr, 1.5 Hf
Cobalt (Wrought)										
Haynes 25 (L-605)	10	1	54	20	—	15	—	—	0.1	
Cobalt (Cast)										
X-40	10	1.5	57.5	22	—	7.5	—	—	0.50	0.5 Mn, 0.5 Si

Source: Reprinted with permission of ASM International.® All rights reserved. www.asminternational.org.

13.10 THE NOBLE METALS

The noble or precious metals are a group of eight elements that have some physical characteristics in common. They are expensive (precious) and are superior or notable (noble) in properties—characteristically soft, ductile, and oxidation resistant. The noble metals are silver, gold, platinum, palladium, rhodium, ruthenium, iridium, and osmium; the first three are most common and are used extensively in jewelry. Silver and gold may be strengthened by solid-solution alloying with copper; sterling silver is a silver–copper alloy containing approximately 7.5 wt% Cu. Alloys of both silver and gold are employed as dental restoration materials. Some integrated circuit electrical contacts are made of gold. Platinum is used for chemical laboratory equipment, as a catalyst (especially in the manufacture of gasoline), and in thermocouples to measure elevated temperatures.

13.11 MISCELLANEOUS NONFERROUS ALLOYS

The preceding discussion covers the vast majority of nonferrous alloys; however, a number of others are found in a variety of engineering applications, and a brief mention of these is worthwhile.

Nickel and its alloys are highly resistant to corrosion in many environments, especially those that are basic (alkaline). Nickel is often coated or plated on some metals that are susceptible to corrosion as a protective measure. Monel, a nickel-based alloy containing approximately 65 wt% Ni and 28 wt% Cu (the balance is iron), has very high strength and is extremely corrosion resistant; it is used in pumps, valves, and other components that are in contact with acid and petroleum solutions. As already mentioned, nickel is one of the principal alloying elements in stainless steels and one of the major constituents in the superalloys.

Lead, tin, and their alloys find some use as engineering materials. Both lead and tin are mechanically soft and weak, have low melting temperatures, are quite resistant to many corrosion environments, and have recrystallization temperatures below room temperature. Some common solders are lead–tin alloys, which have low melting temperatures. Applications for lead and its alloys include x-ray shields and storage batteries. The primary use of tin is as a very thin coating on the inside of plain carbon steel cans (tin cans) that are used for food containers; this coating inhibits chemical reactions between the steel and the food products.

Unalloyed zinc also is a relatively soft metal having a low melting temperature and a subambient recrystallization temperature. Chemically, it is reactive in a number of common environments and, therefore, susceptible to corrosion. Galvanized steel is just plain carbon steel that has been coated with a thin zinc layer; the zinc preferentially corrodes and protects the steel (Section 18.9). Typical applications of galvanized steel are familiar (sheet metal, fences, screen, screws, etc.). Common applications of zinc alloys include padlocks, plumbing fixtures, automotive parts (door handles and grilles), and office equipment.

Although zirconium is relatively abundant in the Earth's crust, not until quite recent times were commercial refining techniques developed. Zirconium and its alloys are ductile and have other mechanical characteristics that are comparable to those of titanium alloys and the austenitic stainless steels. However, the primary asset of these alloys is their resistance to corrosion in a host of corrosive media, including superheated water. Furthermore, zirconium is transparent to thermal neutrons, so that its alloys have been used as cladding for uranium fuel in water-cooled nuclear reactors. In terms of cost, these alloys are also often the materials of choice for heat exchangers, reactor vessels, and piping systems for the chemical-processing and nuclear industries. They are also used in incendiary ordnance and in sealing devices for vacuum tubes.

M A T E R I A L S   O F   I M P O R T A N C E

Metal Alloys Used for Euro Coins

On January 1, 2002, the euro became the single legal currency in twelve European countries; since that date, several other nations have also joined the European monetary union and have adopted the euro as their official currency. Euro coins are minted in eight different denominations: 1 and 2 euros, as well as 50, 20, 10, 5, 2, and 1 euro cent. Each coin has a common design on one face; the reverse face design is one of several chosen by the monetary union countries. Several of these coins are shown in Figure 13.7.

In deciding which metal alloys to use for these coins, a number of issues were considered, most of them centered on material properties.

- The ability to distinguish a coin of one denomination from that of another denomination is important. This may be accomplished by having coins of different sizes, colors, and shapes. With regard to color, alloys must be chosen that retain their distinctive colors, which means that they do not easily tarnish in the air and other commonly encountered environments.

- Security is an important issue—that is, producing coins that are difficult to counterfeit. Most vending machines use electrical conductivity to identify coins, to prevent false coins from being used. This means that each coin must have its own unique *electronic signature,* which depends on its alloy composition.

- The alloys chosen must be *coinable,* or easy to mint—that is, sufficiently soft and ductile to allow design reliefs to be stamped into the coin surfaces.

- The alloys must be wear resistant (i.e., hard and strong) for long-term use and so that the reliefs stamped into the coin surfaces are retained. Strain hardening (Section 9.10) occurs during the stamping operation, which enhances hardness.

- High degrees of corrosion resistance in common environments are required for the alloys selected, to ensure minimal material losses over the lifetimes of the coins.

- It is highly desirable to use alloys of a base metal (or metals) that retains (retain) its (their) intrinsic value(s).

- Alloy recyclability is another requirement for the alloy(s) used.

- The alloy(s) from which the coins are made should relate to human health considerations—that is, have antibacterial characteristics so that undesirable microorganisms will not grow on their surfaces.

Copper was selected as the base metal for all euro coins, because it and its alloys satisfy these criteria. Several different copper alloys and alloy combinations are used for the eight different coins, as follows:

- 2-euro coin: This coin is termed *bimetallic*—it consists of an outer ring and an inner disk. For the outer ring, a 75Cu–25Ni alloy is used, which has a silver color. The inner disk is composed of a three-layer structure—high-purity nickel that is clad on both sides with a nickel brass alloy (75Cu–20Zn–5Ni); this alloy has a gold color.

- 1-euro coin: This coin is also bimetallic, but the alloys used for its outer ring and inner disk are reversed from those for the 2-euro coin.

- 50-, 20-, and 10-euro-cent pieces: These coins are made of a "Nordic gold" alloy—89Cu–5Al–5Zn–1Sn.

- 5-, 2-, and 1-euro-cent pieces: Copper-plated steels are used for these coins.

Figure 13.7 Photograph showing 1-euro, 2-euro, 20-euro-cent, and 50-euro-cent coins. (Photograph courtesy of Outokumpu Copper.)

Appendix B tabulates a wide variety of properties (density, elastic modulus, yield and tensile strengths, electrical resistivity, coefficient of thermal expansion, etc.) for a large number of metals and alloys.

SUMMARY

Ferrous Alloys
- *Ferrous alloys* (steels and cast irons) are those in which iron is the prime constituent. Most steels contain less than 1.0 wt% C and, in addition, other alloying elements, which render them susceptible to heat treatment (and an enhancement of mechanical properties) and/or more corrosion resistant.
- Ferrous alloys are used extensively as engineering materials because
  Iron-bearing compounds are abundant.
  Economical extraction, refining, and fabrication techniques are available.
  They may be tailored to have a wide variety of mechanical and physical properties.
- Limitations of ferrous alloys include the following:
  Relatively high densities
  Comparatively low electrical conductivities
  Susceptibility to corrosion in common environments
- The most common types of steels are plain low-carbon, high-strength low-alloy, medium-carbon, tool, and stainless.
- Plain carbon steels contain (in addition to carbon) a little manganese and only residual concentrations of other impurities.
- Stainless steels are classified according to the main microstructural constituent. The three classes are ferritic, austenitic, and martensitic.
- Cast irons contain higher carbon contents than steels—normally between 3.0 and 4.5 wt% C—as well as other alloying elements, notably silicon. For these materials, most of the carbon exists in graphite form rather than combined with iron as cementite.
- Gray, ductile (or nodular), malleable, and compacted graphite irons are the four most widely used cast irons; the last three are reasonably ductile.

Nonferrous Alloys
- All other alloys fall within the nonferrous category, which is further subdivided according to base metal or some distinctive characteristic that is shared by a group of alloys.
- Nonferrous alloys may be further subclassified as either wrought or cast. Alloys that are amenable to forming by deformation are classified as wrought. Cast alloys are relatively brittle, and therefore fabrication by casting is most expedient.
- Seven classifications of nonferrous alloys were discussed—copper, aluminum, magnesium, titanium, the refractory metals, the superalloys, and the noble metals—as well as a miscellaneous category (nickel, lead, tin, zinc, and zirconium).

Important Terms and Concepts

alloy steel
brass
bronze
cast iron
compacted graphite iron
ductile (nodular) iron

ferrous alloy
gray cast iron
high-strength, low-alloy (HSLA) steel
malleable cast iron
nonferrous alloy

plain carbon steel
specific strength
stainless steel
temper designation
white cast iron
wrought alloy

REFERENCES

ASM Handbook, Vol. 1, *Properties and Selection: Irons, Steels, and High-Performance Alloys,* ASM International, Materials Park, OH, 1990.

ASM Handbook, Vol. 2, *Properties and Selection: Nonferrous Alloys and Special-Purpose Materials,* ASM International, Materials Park, OH, 1990.

Davis, J. R. (Editor), *Cast Irons,* ASM International, Materials Park, OH, 1996.

Dieter, G. E., *Mechanical Metallurgy,* 3rd edition, McGraw-Hill, New York, 1986. Chapters 15–21 provide an excellent discussion of various metal-forming techniques.

Frick, J. (Editor), *Woldman's Engineering Alloys,* 9th edition, ASM International, Materials Park, OH, 2000.

Henkel, D. P., and A. W. Pense, *Structures and Properties of Engineering Materials,* 5th edition, McGraw-Hill, New York, 2001.

Metals and Alloys in the Unified Numbering System, 12th edition, Society of Automotive Engineers and American Society for Testing and Materials, Warrendale, PA, 2012.

Worldwide Guide to Equivalent Irons and Steels, 5th edition, ASM International, Materials Park, OH, 2006.

Worldwide Guide to Equivalent Nonferrous Metals and Alloys, 4th edition, ASM International, Materials Park, OH, 2001.

QUESTIONS AND PROBLEMS

Ferrous Alloys

13.1 (a) List the four classifications of steels.

(b) For each, briefly describe the properties and typical applications.

13.2 (a) Cite three reasons why ferrous alloys are used so extensively.

(b) Cite three characteristics of ferrous alloys that limit their utilization.

13.3 What is the function of alloying elements in tool steels?

13.4 Compute the volume percent of graphite, V_{Gr}, in a 3.5 wt% C cast iron, assuming that all the carbon exists as the graphite phase. Assume densities of 8.0 and 2.5 g/cm^3 for ferrite and graphite, respectively.

13.5 On the basis of microstructure, briefly explain why gray iron is brittle and weak in tension.

13.6 Compare gray and malleable cast irons with respect to

(a) composition and heat treatment,

(b) microstructure, and

(c) mechanical characteristics.

13.7 Compare white and nodular cast irons with respect to

(a) composition and heat treatment,

(b) microstructure, and

(c) mechanical characteristics.

13.8 Is it possible to produce malleable cast iron in pieces having large cross-sectional dimensions? Why or why not?

Nonferrous Alloys

13.9 What is the principal difference between wrought and cast alloys?

13.10 Why must rivets of a 2017 aluminum alloy be refrigerated before they are used?

13.11 What is the chief difference between heat-treatable and non-heat-treatable alloys?

13.12 Give the distinctive features, limitations, and applications of the following alloy groups: titanium alloys, refractory metals, superalloys, and noble metals.

DESIGN PROBLEMS

Ferrous Alloys
Nonferrous Alloys

13.D1 The following is a list of metals and alloys:

Plain carbon steel	Magnesium
Brass	Zinc
Gray cast iron	Tool steel
Platinum	Aluminum
Stainless steel	Tungsten
Titanium alloy	

Select from this list the one metal or alloy that is best suited for each of the following applications, and cite at least one reason for your choice:

(a) The block of an internal combustion engine

(b) Condensing heat exchanger for steam

(c) Jet engine turbofan blades

(d) Drill bit

(e) Cryogenic (i.e., very low temperature) container

(f) As a pyrotechnic (i.e., in flares and fireworks)

(g) High-temperature furnace elements to be used in oxidizing atmospheres

13.D2 A group of new materials are the metallic glasses (or amorphous metals). Write an essay about these materials in which you address the following issues:

(a) compositions of some of the common metallic glasses,

(b) characteristics of these materials that make them technologically attractive,

(c) characteristics that limit their utilization,

(d) current and potential uses, and

(e) at least one technique that is used to produce metallic glasses.

13.D3 Of the following alloys, pick the one(s) that may be strengthened by heat treatment, cold work, or both: R50500 titanium, AZ31B magnesium, 6061 aluminum, C51000 phosphor bronze, lead, 6150 steel, 304 stainless steel, and C17200 beryllium copper.

13.D4 A structural member 100 mm long must be able to support a load of 50,000 N without experiencing any plastic deformation. Given the following data for brass, steel, aluminum, and titanium, rank them from least to greatest weight in accordance with these criteria.

Alloy	Yield Strength [MPa]	Density (g/cm³)
Brass	415	8.5
Steel	860	7.9
Aluminum	310	2.7
Titanium	550	4.5

13.D5 Discuss whether it would be advisable to hot work or cold work the following metals and alloys on the basis of melting temperature, oxidation resistance, yield strength, and degree of brittleness: tin, tungsten, aluminum alloys, magnesium alloys, and a 4140 steel.

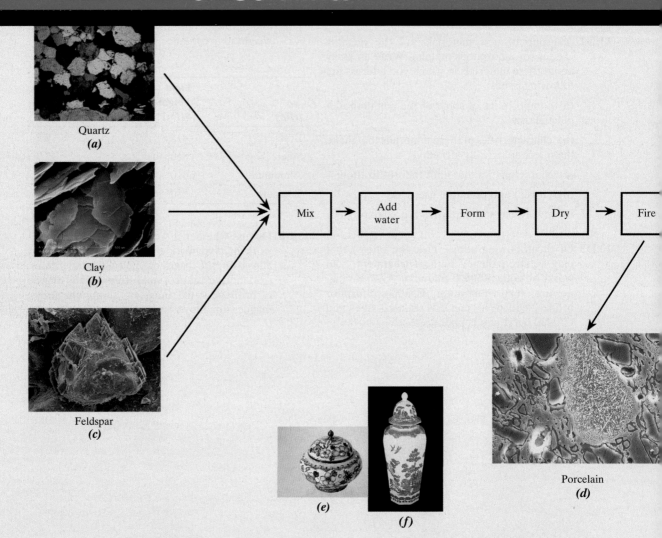

Quartz
(a)

Clay
(b)

Feldspar
(c)

Mix → Add water → Form → Dry → Fire

Porcelain
(d)

(e)

(f)

The above micrographs show particles of (*a*) quartz, (*b*) clay, and (*c*) feldspar—the primary constituents of porcelain. In order to produce a porcelain object, these three ingredients are mixed in the proper proportions, water is added, and an object having the desired shape is formed (either by slip casting or hydroplastic forming). Next, most of the water is removed during a drying operation, and the object is fired at an elevated temperature in order to improve its strength and impart other desirable properties. Decoration of this porcelain body is possible by applying a glaze to its surface.

(*d*) Scanning electron micrograph of a fired porcelain.

(*e*) and (*f*) Fired and glazed porcelain *objets d'art*.

[Figure (*a*) Courtesy Gregory C. Finn, Brock University; (*b*) courtesy of Hefa Cheng and Martin Reinhard, Stanford University; (*c*) courtesy of Martin Lee, University of Glasgow; (*d*) courtesy of H. G. Brinkies, Swinburne University of Technology, Hawthorn Campus, Hawthorn, Victoria, Australia; (*e*) © Maria Natalia Morales/iStockphoto and (*f*) © arturoli/iStockphoto.]

Some of the properties of ceramics may be explained by their structures. For example: (a) The optical transparency of inorganic glass materials is due, in part, to their non-crystallinity; (b) the hydroplasticity of clays (i.e., development of plasticity upon the addition of water) is related to interactions between water molecules and the clay structures (Sections 4.11 and 17.9 and Figure 4.15); and (c) the permanent magnetic and ferroelectric behaviors of some ceramic materials are explained by their crystal structures (Sections 21.5 and 19.24).

Learning Objectives

After studying this chapter, you should be able to do the following:

1. Briefly explain why there is normally significant scatter in the fracture strength for identical specimens of the same ceramic material.
2. Compute the flexural strength of ceramic rod specimens that have been bent to fracture in three-point loading.
3. On the basis of slip considerations, explain why crystalline ceramic materials are normally brittle.
4. Describe the process that is used to produce glass–ceramics.
5. Name the two types of clay products, and give two examples of each.
6. Cite three important requirements that normally must be met by refractory ceramics and abrasive ceramics.
7. Describe the mechanism by which cement hardens when water is added.
8. Name three forms of carbon discussed in this chapter and, for each, note at least two distinctive characteristics.

14.1 INTRODUCTION

Ceramic materials and their crystal structures were briefly discussed in Chapters 1 and 4. Up until the past 60 or so years, the most important materials in this class were termed the "traditional ceramics," those for which the primary raw material is clay; products considered to be traditional ceramics are china, porcelain, bricks, tiles, and, in addition, glasses and high-temperature ceramics. Of late, significant progress has been made in understanding the fundamental character of these materials and of the phenomena that occur in them that are responsible for their unique properties. Consequently, a new generation of these materials has evolved, and the term *ceramic* has taken on a much broader meaning. To one degree or another, these new materials have a rather dramatic effect on our lives; electronic, computer, communication, aerospace, and a host of other industries rely on their use.

This chapter discusses the phase diagrams and mechanical characteristics that are found for ceramic materials. Fabrication techniques for this class of materials are discussed in Chapter 17.

Ceramic Phase Diagrams

Phase diagrams have been experimentally determined for many ceramic systems. For binary or two-component phase diagrams, it is frequently the case that the two components are compounds that share a common element, often oxygen. These diagrams may have configurations similar to those for metal–metal systems, and they are interpreted in the same way. For a review of the interpretation of phase diagrams, the reader is referred to Section 11.8.

Figure 14.3 A portion of the zirconia–calcia phase diagram; ss denotes solid solution.
(Adapted from V. S. Stubican and S. P. Ray, "Phase Equilibria and Ordering in the System ZrO_2–CaO," *J. Am. Ceram. Soc.*, **60**[11–12] 535 (1977). Reprinted by permission of the American Ceramic Society.)

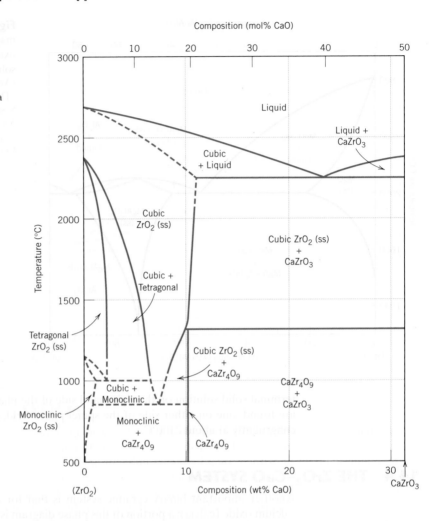

14.5 THE SiO_2–Al_2O_3 SYSTEM

Commercially, the silica–alumina system is an important one because the principal constituents of many ceramic refractories are these two materials. Figure 14.4 shows the SiO_2–Al_2O_3 phase diagram. The polymorphic form of silica that is stable at these temperatures is termed *cristobalite*, the unit cell for which is shown in Figure 4.11. Silica and alumina are not mutually soluble in one another, which is evidenced by the absence of terminal solid solutions at both extremes of the phase diagram. Also, it may be noted that the intermediate compound *mullite*, $3Al_2O_3$–$2SiO_2$, exists, which is represented as a narrow phase field in Figure 14.4; furthermore, mullite melts incongruently at 1890°C. A single eutectic exists at 1587°C and 7.7 wt% Al_2O_3. Section 14.13 discusses refractory ceramic materials, the prime constituents for which are silica and alumina.

✓ *Concept Check 14.1* **(a)** For the SiO_2–Al_2O_3 system, what is the maximum temperature that is possible without the formation of a liquid phase? **(b)** At what composition or over what range of compositions will this maximum temperature be achieved?

[*The answer may be found at* www.wiley.com/college/callister (*Student Companion Site*).]

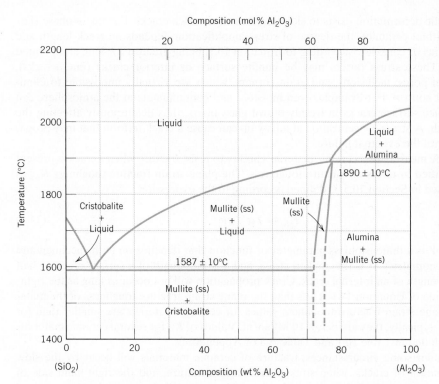

Composition (mol % Al_2O_3)

Figure 14.4
The silica–alumina
phase diagram; ss
denotes solid
solution.
(Adapted from F. J.
Klug, S. Prochazka,
and R. H. Doremus,
"Alumina–Silica Phase
Diagram in the Mullite
Region," *J. Am. Ceram.
Soc.,* **70**[10], 758 (1987).
Reprinted by permis-
sion of the American
Ceramic Society.)

Mechanical Properties

Prior to the Bronze Age, human tools and vessels were primarily made of stone (a ceramic). Between 3000 and 4000 years ago, metals came into widespread use because of
their toughness that is derived from their ductility. For most of that history, ceramic materials were somewhat limited in applicability because of their brittle nature. Their principal
drawback has been a disposition to catastrophic fracture in a brittle manner with very little
energy absorption. Although many new composites and other multiphase ceramics with
useful toughness are being developed (often mimicking naturally occurring composite
ceramics such as seashells), the bulk of ceramic materials currently in use are brittle.

14.6 BRITTLE FRACTURE OF CERAMICS

At room temperature, both crystalline and noncrystalline ceramics almost always fracture
before any plastic deformation can occur in response to an applied tensile load. The topics
of brittle fracture and fracture mechanics, as discussed previously in Sections 10.4 and 10.5,
also relate to the fracture of ceramic materials; they will be reviewed briefly in this context.

The brittle fracture process consists of the formation and propagation of cracks
through the cross section of material in a direction perpendicular to the applied load.
Crack growth in crystalline ceramics may be either transgranular (i.e., through the grains)
or intergranular (i.e., along grain boundaries); for transgranular fracture, cracks propagate along specific crystallographic (or cleavage) planes, planes of high atomic density.

The measured fracture strengths of most ceramic materials are substantially lower
than predicted by theory from interatomic bonding forces. This may be explained by
very small and omnipresent flaws in the material that serve as stress raisers—points at
which the magnitude of an applied tensile stress is amplified and no mechanism such

as plastic deformation exists to slow down or divert such cracks. For single-phase (i.e., monolithic) ceramics, the degree of stress amplification depends on crack length and tip radius of curvature according to Equation 10.1, being greatest for long and pointed flaws. These stress raisers may be minute surface or interior cracks (microcracks), internal pores, inclusions, and grain corners, which are virtually impossible to eliminate or control. For example, even moisture and contaminants in the atmosphere can introduce surface cracks in freshly drawn glass fibers, thus deleteriously affecting the strength. A stress concentration at a flaw tip can cause a crack to form that may propagate until the eventual failure.

The measure of a ceramic material's ability to resist fracture when a crack is present is specified in terms of fracture toughness. The plane strain fracture toughness K_{Ic}, as discussed in Section 10.5, is defined according to the expression

Plane strain fracture toughness for mode I crack surface displacement [Figure. 10.10*a*]

$$K_{Ic} = Y\sigma\sqrt{\pi a} \tag{14.1}$$

where Y is a dimensionless parameter or function that depends on both specimen and crack geometries, σ is the applied stress, and a is the length of a surface crack or half of the length of an internal crack. Crack propagation will not occur as long as the right-hand side of Equation 14.1 is less than the plane strain fracture toughness of the material. Plane strain fracture toughness values for ceramic materials are smaller than for metals; typically, they are below 10 MPa$\sqrt{m}$. Values of K_{Ic} for several ceramic materials are included in Table 10.1 and Table B.5 of Appendix B.

Under some circumstances, fracture of ceramic materials will occur by the slow propagation of cracks, when stresses are static in nature, and the right-hand side of Equation 14.1 is less than K_{Ic}. This phenomenon is called *static fatigue,* or *delayed fracture;* use of the term *fatigue* is somewhat misleading because fracture may occur in the absence of cyclic stresses (metal fatigue was discussed in Chapter 10). This type of fracture is especially sensitive to environmental conditions, specifically when moisture is present in the atmosphere. With regard to mechanism, a stress–corrosion process probably occurs at the crack tips. That is, the combination of an applied tensile stress and atmospheric moisture at crack tips causes ionic bonds to rupture; this leads to a sharpening and lengthening of the cracks until, ultimately, one crack grows to a size capable of rapid propagation according to Equation 10.3. Furthermore, the duration of stress application preceding fracture decreases with increasing stress. Consequently, when specifying the *static fatigue strength,* the time of stress application should also be stipulated. Silicate glasses are especially susceptible to this type of fracture; it has also been observed in other ceramic materials, including porcelain, Portland cement, high-alumina ceramics, barium titanate, and silicon nitride.

There is usually considerable variation and scatter in the fracture strength for many specimens of a specific brittle ceramic material. A distribution of fracture strengths for a silicon nitride material is shown in Figure 14.5. This phenomenon may be explained by the dependence of fracture strength on the probability of the existence of a flaw that is capable of initiating a crack. This probability varies from specimen to specimen of the same material and depends on fabrication technique and any subsequent treatment. Specimen size or volume also influences fracture strength; the larger the specimen, the greater this flaw existence probability, and the lower the fracture strength.

For compressive stresses, there is no stress amplification associated with any existent flaws. For this reason, brittle ceramics display much higher strengths in compression than in tension (on the order of a factor of 10), and they are generally used when load conditions are compressive. Also, the fracture strength of a brittle ceramic may be enhanced dramatically by imposing residual compressive stresses at its surface. One way this may be accomplished is by thermal tempering (see Section 17.8).

Statistical theories have been developed that in conjunction with experimental data are used to determine the risk of fracture for a given material; a discussion of these is

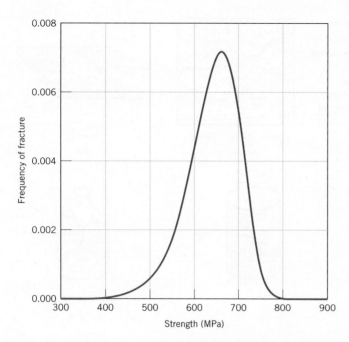

Figure 14.5 The frequency distribution of observed fracture strengths for a silicon nitride material.

beyond the scope of the present treatment. However, because of the dispersion in the measured fracture strengths of brittle ceramic materials, average values and factors of safety as discussed in Sections 8.6 and 8.7 typically are not used for design purposes.

Fractography of Ceramics

It is sometimes necessary to acquire information regarding the cause of a ceramic fracture so that measures may be taken to reduce the likelihood of future incidents. A failure analysis normally focuses on determination of the location, type, and source of the crack-initiating flaw. A fractographic study (Section 10.3) is normally a part of such an analysis, which involves examining the path of crack propagation, as well as microscopic features of the fracture surface. It is often possible to conduct an investigation of this type using simple and inexpensive equipment—for example, a magnifying glass and/or a low-power stereo binocular optical microscope in conjunction with a light source. When higher magnifications are required, the scanning electron microscope is used.

After nucleation and during propagation, a crack accelerates until a critical (or terminal) velocity is achieved; for glass, this critical value is approximately one-half of the speed of sound. Upon reaching this critical velocity, a crack may branch (or bifurcate), a process that may be successively repeated until a family of cracks is produced. Typical crack configurations for four common loading schemes are shown in Figure 14.6. The site of nucleation can often be traced back to the point where a set of cracks converges. Furthermore, the rate of crack acceleration increases with increasing stress level; correspondingly the degree of branching also increases with rising stress. For example, from experience we know that when a large rock strikes (and probably breaks) a window, more crack branching results [i.e., more and smaller cracks form (or more broken fragments are produced)] than for a small pebble impact.

During propagation, a crack interacts with the microstructure of the material, the stress, and with elastic waves that are generated; these interactions produce distinctive features on the fracture surface. Furthermore, these features provide important information on where the crack initiated and the source of the crack-producing defect. In addition, measurement of the approximate fracture-producing stress may be useful; stress magnitude is indicative of whether the ceramic piece was excessively weak or the in-service stress was greater than anticipated.

Figure 14.6 For brittle ceramic materials, schematic representations of crack origins and configurations that result from (*a*) impact (point contact) loading, (*b*) bending, (*c*) torsional loading, and (*d*) internal pressure. (From D. W. Richerson, *Modern Ceramic Engineering*, 2nd edition, Marcel Dekker, Inc., New York, 1992. Reprinted from *Modern Ceramic Engineering*, 2nd edition, p. 681, by courtesy of Marcel Dekker, Inc.)

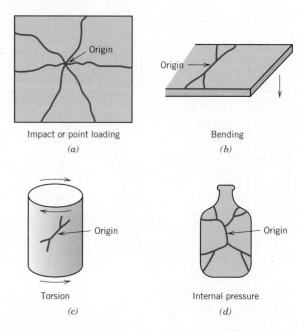

Impact or point loading
(*a*)

Bending
(*b*)

Torsion
(*c*)

Internal pressure
(*d*)

Several microscopic features normally found on the crack surfaces of failed ceramic pieces are shown in the schematic diagram of Figure 14.7 and the photomicrograph in Figure 14.8. The crack surface that formed during the initial acceleration stage of propagation is flat and smooth and is appropriately termed the *mirror* region (Figure 14.7). For glass fractures, this mirror region is extremely flat and highly reflective; for polycrystalline ceramics, the flat mirror surfaces are rougher and have a granular texture. The outer perimeter of the mirror region is roughly circular, with the crack origin at its center.

Upon reaching its critical velocity, the crack begins to branch—that is, the crack surface changes propagation direction. At this time there is a roughening of the crack interface on a microscopic scale and the formation of two more surface features—*mist* and *hackle;* these are also noted in Figures 14.7 and 14.8. The *mist* is a faint annular region just outside the mirror; it is often not discernible for polycrystalline ceramic pieces. Beyond the mist is the *hackle*, which has an even rougher texture. The hackle is composed of a set of striations or lines that radiate away from the crack source in the direction of crack propagation; they intersect near the crack initiation site and may be used to pinpoint its location.

Qualitative information regarding the magnitude of the fracture-producing stress is available from measurement of the mirror radius (r_m in Figure 14.7). This radius is a

Figure 14.7 Schematic diagram that shows typical features observed on the fracture surface of a brittle ceramic. (Adapted from J. J. Mecholsky, R. W. Rice, and S. W. Freiman, "Prediction of Fracture Energy and Flaw Size in Glasses from Measurements of Mirror Size," *J. Am. Ceram. Soc.,* **57**[10] 440 (1974). Reprinted with permission of The American Ceramic Society, *www.ceramics.org*. Copyright 1974. All rights reserved.)

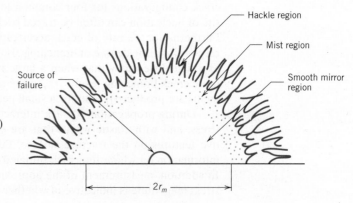

Hackle region

Mist region

Smooth mirror region

Source of failure

$2r_m$

Mist region

Hackle region

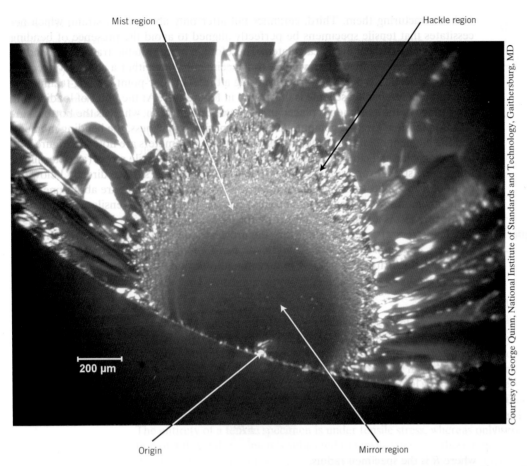

200 µm

Origin

Mirror region

Courtesy of George Quinn, National Institute of Standards and Technology, Gaithersburg, MD

Figure 14.8 Photomicrograph of the fracture surface of a 6-mm-diameter fused silica rod that was fractured in four-point bending. Features typical of this kind of fracture are noted—the origin as well as the mirror, mist, and hackle regions. 60×.

function of the acceleration rate of a newly formed crack—that is, the greater this acceleration rate, the sooner the crack reaches its critical velocity, and the smaller the mirror radius. Furthermore, the acceleration rate increases with stress level. Thus, as fracture stress level increases, the mirror radius decreases; experimentally it has been observed that

$$\sigma_f \propto \frac{1}{r_m^{0.5}} \tag{14.2}$$

Here, σ_f is the stress level at which fracture occurred.

Elastic (sonic) waves are also generated during a fracture event, and the locus of intersections of these waves with a propagating crack front gives rise to another type of surface feature known as a *Wallner line*. Wallner lines are arc shaped, and they provide information regarding stress distributions and directions of crack propagation.

14.7 STRESS–STRAIN BEHAVIOR

Flexural Strength

The stress–strain behavior of brittle ceramics is not usually ascertained by a tensile test as outlined in Section 8.2, for three reasons. First, it is difficult to prepare and test specimens having the required geometry. Second, it is difficult to grip brittle materials

Figure 14.16 Scanning electron micrograph showing the microstructure of a glass–ceramic material. The long, acicular, blade-shape particles yield a material with unusual strength and toughness. 40,000×.

Courtesy of L. R. Pinckney and G. J. Fine, Corning Incorporated

0.4 μm

14.12 CLAY PRODUCTS

One of the most widely used ceramic raw materials is clay. This inexpensive ingredient, found naturally in great abundance, often is used as mined without any upgrading of quality. Another reason for its popularity lies in the ease with which clay products may be formed; when mixed in the proper proportions, clay and water form a plastic mass that is very amenable to shaping. The formed piece is dried to remove some of the moisture, after which it is fired at an elevated temperature to improve its mechanical strength.

structural clay product

whiteware

firing

Most clay-based products fall within two broad classifications: the **structural clay products** and **whitewares**. Structural clay products include building bricks, tiles, and sewer pipes—applications in which structural integrity is important. Whiteware ceramics become white after high-temperature **firing**. Included in this group are porcelain, pottery, tableware, china, and plumbing fixtures (sanitary ware). In addition to clay, many of these products also contain nonplastic ingredients, which influence the changes that take place during the drying and firing processes and the characteristics of the finished piece (Section 17.9).

14.13 REFRACTORIES

refractory ceramic

Another important class of ceramics that are used in large tonnages is the **refractory ceramics**. The salient properties of these materials include the capacity to withstand high temperatures without melting or decomposing and the capacity to remain unreactive and inert when exposed to severe environments. In addition, the ability to provide thermal insulation is often an important consideration. Refractory materials are marketed in a variety of forms, but bricks are the most common. Typical applications include furnace linings for metal refining, glass manufacturing, metallurgical heat treatment, and power generation.

The performance of a refractory ceramic depends to a large degree on its composition. On this basis, there are several classifications—fireclay, silica, basic, and special refractories. Compositions for a number of commercial refractories are listed in Table 14.4. For many commercial materials, the raw ingredients consist of both large (or grog) particles and fine particles, which may have different compositions. Upon firing, the fine particles normally

Table 14.4 Compositions of Five Common Ceramic Refractory Materials

Refractory Type	Composition (wt%)							Apparent Porosity (%)
	Al_2O_3	SiO_2	MgO	Cr_2O_3	Fe_2O_3	CaO	TiO_2	
Fireclay	25–45	70–50	0–1		0–1	0–1	1–2	10–25
High-alumina fireclay	90–50	10–45	0–1		0–1	0–1	1–4	18–25
Silica	0.2	96.3	0.6			2.2		25
Periclase	1.0	3.0	90.0	0.3	3.0	2.5		22
Periclase–chrome ore	9.0	5.0	73.0	8.2	2.0	2.2		21

Source: From W. D. Kingery, H. K. Bowen, and D. R. Uhlmann, *Introduction to Ceramics,* 2nd edition. Copyright © 1976 by John Wiley & Sons, New York. Reprinted by permission of John Wiley & Sons, Inc.

are involved in the formation of a bonding phase, which is responsible for the increased strength of the brick; this phase may be predominantly either glassy or crystalline. The service temperature is normally below that at which the refractory piece was fired.

Porosity is one microstructural variable that must be controlled to produce a suitable refractory brick. Strength, load-bearing capacity, and resistance to attack by corrosive materials all increase with porosity reduction. At the same time, thermal insulation characteristics and resistance to thermal shock are decreased. The optimum porosity depends on the conditions of service.

Fireclay Refractories

The primary ingredients for the fireclay refractories are high-purity fireclays, alumina and silica mixtures usually containing between 25 and 45 wt% alumina. According to the SiO_2–Al_2O_3 phase diagram, Figure 14.4, over this composition range the highest temperature possible without the formation of a liquid phase is 1587°C. Below this temperature, the equilibrium phases present are mullite and silica (cristobalite). During refractory service use, the presence of a small amount of a liquid phase may be allowable without compromising mechanical integrity. Above 1587°C, the fraction of liquid phase present depends on refractory composition. Upgrading the alumina content increases the maximum service temperature, allowing for the formation of a small amount of liquid.

Fireclay bricks are used principally in furnace construction to confine hot atmospheres and to thermally insulate structural members from excessive temperatures. For fireclay brick, strength is not ordinarily an important consideration because support of structural loads is usually not required. Some control is normally maintained over the dimensional accuracy and stability of the finished product.

Silica Refractories

The prime ingredient for silica refractories, sometimes termed *acid refractories,* is silica. These materials, well known for their high-temperature load-bearing capacity, are commonly used in the arched roofs of steel- and glass-making furnaces; for these applications, temperatures as high as 1650°C may be realized. Under these conditions, some small portion of the brick actually exists as a liquid. The presence of even small concentrations of alumina has an adverse influence on the performance of these refractories, which may be explained by the silica–alumina phase diagram, Figure 14.4. Because the eutectic composition (7.7 wt% Al_2O_3) is very near the silica extremity of the phase diagram, even small additions of Al_2O_3 lower the liquidus temperature significantly, which means that substantial amounts of liquid may be present at temperatures in excess of 1600°C. Thus, the alumina content should be held to a minimum, normally to between 0.2 and 1.0 wt%.

These refractory materials are also resistant to slags that are rich in silica (called *acid slags*) and are often used as containment vessels for them. However, they are readily attacked by slags composed of a high proportion of CaO and/or MgO (basic slags), and contact with these oxide materials should be avoided.

Basic Refractories

The refractories that are rich in periclase, or magnesia (MgO), are termed *basic*; they may also contain calcium, chromium, and iron compounds. The presence of silica is deleterious to their high-temperature performance. Basic refractories are especially resistant to attack by slags containing high concentrations of MgO and CaO and find extensive use in some steel-making open hearth furnaces.

Special Refractories

Yet other ceramic materials are used for rather specialized refractory applications. Some of these are relatively high-purity oxide materials, many of which may be produced with very little porosity. Included in this group are alumina, silica, magnesia, beryllia (BeO), zirconia (ZrO_2), and mullite ($3Al_2O_3$-$2SiO_2$). Others include carbide compounds, in addition to carbon and graphite. Silicon carbide (SiC) has been used for electrical resistance heating elements, as a crucible material, and in internal furnace components. Carbon and graphite are very refractory, but find limited application because they are susceptible to oxidation at temperatures in excess of about 800°C. As would be expected, these specialized refractories are relatively expensive.

Concept Check 14.3 Upon consideration of the SiO_2–Al_2O_3 phase diagram (Figure 14.4) for the following pair of compositions, which would you judge to be the more desirable refractory? Justify your choice.

20 wt% Al_2O_3–80 wt% SiO_2
25 wt% Al_2O_3–75 wt% SiO_2

[*The answer may be found at www.wiley.com/college/callister (Student Companion Site).*]

14.14 ABRASIVES

abrasive ceramic

Abrasive ceramics are used to wear, grind, or cut away other material, which necessarily is softer. Therefore, the prime requisite for this group of materials is hardness or wear resistance; in addition, a high degree of toughness is essential to ensure that the abrasive particles do not easily fracture. Furthermore, high temperatures may be produced from abrasive frictional forces, so some refractoriness is also desirable.

Diamonds, both natural and synthetic, are used as abrasives; however, they are relatively expensive. The more common ceramic abrasives include silicon carbide, tungsten carbide (WC), aluminum oxide (or corundum), and silica sand.

Abrasives are used in several forms—bonded to grinding wheels, as coated abrasives, and as loose grains. In the first case, the abrasive particles are bonded to a wheel by means of a glassy ceramic or an organic resin. The surface structure should contain some porosity; a continual flow of air currents or liquid coolants within the pores that surround the refractory grains prevents excessive heating. Figure 14.17 shows the microstructure of a bonded abrasive, revealing abrasive grains, the bonding phase, and pores.

Coated abrasives are those in which an abrasive powder is coated on some type of paper or cloth material; sandpaper is probably the most familiar example. Wood, metals, ceramics, and plastics are all frequently ground and polished using this form of abrasive.

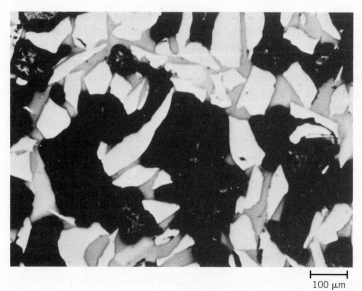

Figure 14.17 Photomicrograph of an aluminum oxide bonded ceramic abrasive. The light regions are the Al_2O_3 abrasive grains; the gray and dark areas are the bonding phase and porosity, respectively. 100×.
(From W. D. Kingery, H. K. Bowen, and D. R. Uhlmann, *Introduction to Ceramics,* 2nd edition, p. 568. Copyright © 1976 by John Wiley & Sons. Reprinted by permission of John Wiley & Sons, Inc.)

100 μm

Grinding, lapping, and polishing wheels often employ loose abrasive grains that are delivered in some type of oil- or water-based vehicle. Diamonds, corundum, silicon carbide, and rouge (an iron oxide) are used in loose form over a variety of grain size ranges.

14.15 CEMENTS

cement

Several familiar ceramic materials are classified as inorganic **cements**: cement, plaster of Paris, and lime, which, as a group, are produced in extremely large quantities. The characteristic feature of these materials is that when mixed with water, they form a paste that subsequently sets and hardens. This trait is especially useful in that solid and rigid structures having just about any shape may be formed expeditiously. Also, some of these materials act as a bonding phase that chemically binds particulate aggregates into a single cohesive structure. Under these circumstances, the role of the cement is similar to that of the glassy bonding phase that forms when clay products and some refractory bricks are fired. One important difference, however, is that the cementitious bond develops at room temperature.

Of this group of materials, Portland cement is consumed in the largest tonnages. It is produced by grinding and intimately mixing clay and lime-bearing minerals in the proper proportions and then heating the mixture to about 1400°C in a rotary kiln;

calcination

this process, sometimes called **calcination**, produces physical and chemical changes in the raw materials. The resulting "clinker" product is then ground into a very fine powder, to which is added a small amount of gypsum ($CaSO_4$–$2H_2O$) to retard the setting process. This product is Portland cement. The properties of Portland cement, including setting time and final strength, to a large degree depend on its composition.

Several different constituents are found in Portland cement, the principal ones being tricalcium silicate ($3CaO$–SiO_2) and dicalcium silicate ($2CaO$–SiO_2). The setting and hardening of this material result from relatively complicated hydration reactions that occur among the various cement constituents and the water that is added. For example, one hydration reaction involving dicalcium silicate is as follows:

$$2CaO\text{-}SiO_2 + xH_2O \rightarrow 2CaO\text{-}SiO_2\text{-}xH_2O \tag{14.7}$$

where x is variable that depends on how much water is available. These hydrated products are in the form of complex gels or crystalline substances that form the cementitious bond. Hydration reactions begin just as soon as water is added to the cement. These are first manifested as setting (the stiffening of the once-plastic paste), which takes place soon after mixing, usually within several hours. Hardening of the mass follows as a result of further hydration, a relatively slow process that may continue for as long as several years. It should be emphasized that the process by which cement hardens is not one of drying, but rather of hydration in which water actually participates in a chemical bonding reaction.

Portland cement is termed a *hydraulic cement* because its hardness develops by chemical reactions with water. It is used primarily in mortar and concrete to bind aggregates of inert particles (sand and/or gravel) into a cohesive mass; these are considered to be composite materials (see Section 16.2). Other cement materials, such as lime, are nonhydraulic—that is, compounds other than water (e.g., CO_2) are involved in the hardening reaction.

Concept Check 14.4   Explain why it is important to grind cement into a fine powder.

[*The answer may be found at* www.wiley.com/college/callister *(Student Companion Site).*]

14.16 CARBONS

Section 4.12 presented the crystal structures of two polymorphic forms of carbon—diamond and graphite. Furthermore, fibers are made of carbon materials that have other structures. In this section, we discuss these structures and, in addition, the important properties and applications for these three forms of carbon.

Diamond

The physical properties of diamond are extraordinary. Chemically, it is very inert and resistant to attack by a host of corrosive media. Of all known bulk materials, diamond is the hardest, as a result of its extremely strong interatomic sp^3 bonds. In addition, of all solids, it has the lowest sliding coefficient of friction. Its thermal conductivity is extremely high, its electrical properties are notable, and, optically, it is transparent in the visible and infrared regions of the electromagnetic spectrum—in fact, it has the widest spectral transmission range of all materials. The high index of refraction and optical brilliance of single crystals makes diamond a most highly valued gemstone. Several important properties of diamond, as well as other carbon materials, are listed in Table 14.5.

High-pressure high-temperature (HPHT) techniques to produce synthetic diamonds were developed beginning in the mid-1950s. These techniques have been refined to the degree that today a large proportion of industrial-quality diamonds are synthetic, as are some of those of gem quality.

Industrial-grade diamonds are used for a host of applications that exploit diamond's extreme hardness, wear resistance, and low coefficient of friction. These include diamond-tipped drill bits and saws, dies for wire drawing, and as abrasives used in cutting, grinding, and polishing equipment (Section 14.14).

Graphite

As a consequence of its structure (Figure 4.18), graphite is highly anisotropic—property values depend on crystallographic direction along which they are measured. For example, electrical resistivities parallel and perpendicular to the graphene plane are, respectively, on the order of 10^{-5} and 10^{-2} $\Omega \cdot m$. Delocalized electrons are highly mobile, and their motions in response to the presence of an electric field applied in a direction parallel

Table 14.5 Properties of Diamond, Graphite, and Carbon (for Fibers)

Property	Diamond	Graphite In-Plane	Graphite Out-of-Plane	Carbon (Fibers)
Density (g/cm^3)	3.51	2.26		1.78–2.15
Modulus of elasticity (GPa)	700–1200	350	36.5	230–725[a]
Strength (MPa)	1050	2500	—	1500–4500[a]
Thermal Conductivity (W/m·K)	2000–2500	1960	6.0	11–70[a]
Coefficient, Thermal Expansion (10^{-6} K^{-1})	0.11–1.2	−1	+29	−0.5--0.6[a] 7–10[b]
Electrical Resistivity (Ω·m)	10^{11}–10^{14}	1.4×10^{-5}	1×10^{-2}	9.5×10^{-6}–17×10^{-6}

[a]Longitudinal fiber direction.
[b]Transverse (radial) fiber direction.

to the plane are responsible for the relatively low resistivity (i.e., high conductivity) in that direction. Also, as a consequence of the weak interplanar van der Waals bonds, it is relatively easy for planes to slide past one another, which explains the excellent lubricative properties of graphite.

There is a significant disparity between the properties of graphite and diamond, as may be noted in Table 14.5. For example, mechanically, graphite is very soft and flaky, and has a significantly smaller modulus or elasticity. Its in-plane electrical conductivity is 10^{16} to 10^{19} times that of diamond, whereas thermal conductivities are approximately the same. Furthermore, whereas the coefficient of thermal expansion for diamond is relatively small and positive, graphite's in-plane value is small and negative, and the plane-perpendicular coefficient is positive and relatively large. Furthermore, graphite is optically opaque with a black–silver color. Other desirable properties of graphite include good chemical stability at elevated temperatures and in nonoxidizing atmospheres, high resistance to thermal shock, high adsorption of gases, and good machinability.

Applications for graphite are many, varied, and include lubricants, pencils, battery electrodes, friction materials (e.g., brake shoes), heating elements for electric furnaces, welding electrodes, metallurgical crucibles, high-temperature refractories and insulations, rocket nozzles, chemical reactor vessels, electrical contacts (e.g., brushes), and air purification devices.

Carbon Fibers

Small-diameter, high-strength, and high-modulus fibers composed of carbon are used as reinforcements in polymer-matrix composites (Section 16.8). Carbon in these fiber materials is in the form of graphene layers. However, depending on precursor (i.e., material from which the fibers are made) and heat treatment, different structural arrangements of these graphene layers exist. For what are termed *graphitic carbon* fibers, the graphene layers assume the ordered structure of graphite—planes are parallel to one another having relatively weak van der Waals interplanar bonds. Alternatively, a more disordered structure results when, during fabrication, graphene sheets become randomly folded, tilted, and crumpled to form what is termed *turbostratic carbon*.

microbeam. Compared to conventional air-bag systems, the MEMS units are smaller, lighter, and more reliable and are produced at a considerable cost reduction.

Potential MEMS applications include electronic displays, data storage units, energy conversion devices, chemical detectors (for hazardous chemical and biological agents and drug screening), and microsystems for DNA amplification and identification. There are undoubtedly many unforeseen uses of this MEMS technology that will have a profound impact on society; these will probably overshadow the effects that microelectronic integrated circuits have had during the past three decades.

Nanocarbons

nanocarbon

A class of recently discovered carbon materials, the **nanocarbons**, have novel and exceptional properties, are currently being used in some cutting-edge technologies, and will certainly play an important role in future high-tech applications. Three nanocarbons that belong to this class are fullerenes, carbon nanotubes, and graphene. The "nano" prefix denotes that the particle size is less than about 100 nanometers. In addition, the carbon atoms in each nanoparticle are bonded to one another through hybrid sp^2 orbitals.[5]

Fullerenes

One type of fullerene, discovered in 1985, consists of a hollow spherical cluster of 60 carbon atoms; a single molecule is denoted by C_{60}. Carbon atoms bond together so as to form both hexagonal (six-carbon atom) and pentagonal (five-carbon atom) geometrical configurations. One such molecule, shown in Figure 14.20, is found to consist of 20 hexagons and 12 pentagons, which are arrayed such that no two pentagons share a common side; the molecular surface thus exhibits the symmetry of a soccer ball. The material composed of C_{60} molecules is known as *buckminsterfullerene,* (or *buckyball* for short), named in honor of R. Buckminster Fuller, who invented the geodesic dome; each C_{60} is simply a molecular replica of such a dome. The term *fullerene* is used to denote the class of materials that are composed of this type of molecule.[6]

Figure 14.20 The structure of a C_{60} fullerene molecule (schematic).

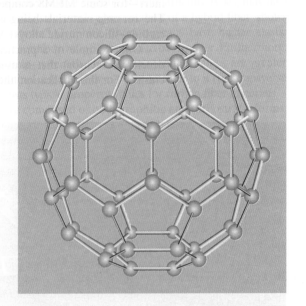

---

[5]As with graphite, delocalized electrons are associated with these sp^2 bonds; these bonds are confined to within the molecule.

[6]Fullerene molecules other than C_{60} exist (e.g., C_{50}, C_{70}, C_{76}, C_{84}) that also form hollow, spherelike clusters. Each of these is composed of twelve pentagons, whereas the number of hexagons is variable.

Table 14.6

Properties for Carbon Nanomaterials

	Material		
Property	C_{60} *(Fullerite)*	*Carbon Nanotubes (Single Walled)*	*Graphene (In-Plane)*
Density (g/cm³)	1.69	1.33–1.40	—
Modulus of elasticity (GPa)	—	1000	1000
Strength (MPa)	—	13,000–53,000	130,000
Thermal Conductivity (W/m·K)	0.4	~2000	3000–5000
Coefficient, Thermal Expansion (10^{-6} K^{-1})	—	—	~−6
Electrical Resistivity (Ω·m)	10^{14}	10^{-6}	10^{-8}

In the solid state, the C_{60} units form a crystalline structure and pack together in a face-centered cubic array. This material is called *fullerite,* and Table 14.6 lists some of its properties.

A number of fullerene compounds have been developed that have unusual chemical, physical, and biological characteristics, and are being used or have the potential to be used in a host of new applications. Some of these compounds involve atoms or groups of atoms that are encapsulated within the cage of carbon atoms (and are termed endo-hedral fullerenes). For other compounds, atoms, ions, or clusters of atoms are attached to the outside of the fullerene shell (exohedral fullerenes).

Uses and potential applications of fullerenes include antioxidants in personal care products, biopharmaceuticals, catalysts, organic solar cells, long-life batteries, high-temperature superconductors, and molecular magnets.

Carbon Nanotubes

Another molecular form of carbon has recently been discovered that has some unique and technologically promising properties. Its structure consists of a single sheet of graphite (i.e., graphene) that is rolled into a tube and represented schematically in Figure 14.21; the term *single-walled carbon nanotube* (abbreviated SWCNT) is used to denote this structure. Each nanotube is a single molecule composed of millions of atoms; the length of this molecule is much greater (on the order of thousands of times greater) than its diameter. Multiple-walled carbon nanotubes (MWCNTs) consisting of concentric cylinders also exist.

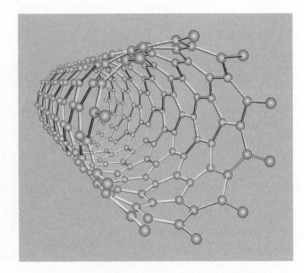

Figure 14.21 The structure of a single-walled carbon nanotube (schematic).

Nanotubes are extremely strong and stiff and relatively ductile. For single-walled nanotubes, measured tensile strengths range between 13 and 53 GPa (approximately an order of magnitude greater than for carbon fibers—viz. 2 to 6 GPa); this is one of the strongest known materials. Elastic modulus values are on the order of one terapascal [TPa (1 TPa = 10^3 GPa)], with fracture strains between about 5% and 20%. Furthermore, nanotubes have relatively low densities. Several properties of single-walled nanotubes are presented in Table 14.6.

On the basis of their exceedingly high strengths, carbon nanotubes have the potential to be used in structural applications. Most current applications, however, are limited to the use of bulk nanotubes—collections of unorganized tube segments. Thus, bulk nanotube materials will most likely never achieve strengths comparable to individual tubes. Bulk nanotubes are currently being used as reinforcements in polymer-matrix nanocomposites (Section 16.16) to improve not only mechanical strength, but also thermal and electrical properties.

Carbon nanotubes also have unique and structure-sensitive electrical characteristics. Depending on the orientation of the hexagonal units in the graphene plane (i.e., tube wall) with the tube axis, the nanotube may behave electrically as either a metal or a semiconductor. As a metal, they have the potential for use as wiring for small-scale circuits. In the semiconducting state they may be used for transistors and diodes. Furthermore, nanotubes are excellent electric field emitters. As such, they can be used for flat-screen displays (e.g., television screens and computer monitors).

Other potential applications are varied and numerous, and include the following:

- More efficient solar cells
- Better capacitors to replace batteries
- Heat removal applications
- Cancer treatments (target and destroy cancer cells)
- Biomaterial applications (e.g., artificial skin, monitor and evaluate engineered tissues)
- Body armor
- Municipal water-treatment plants (more efficient removal of pollutants and contaminants)

Graphene

Graphene, the newest member of the nanocarbons, is a single-atomic-layer of graphite, composed of hexagonally sp^2 bonded carbon atoms (Figure 14.22). These bonds are extremely strong, yet flexible, which allows the sheets to bend. The first

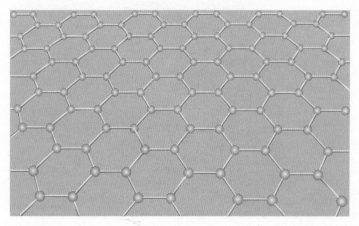

Figure 14.22 The structure of a graphene layer (schematic).

graphene material was produced by peeling apart a piece of graphite, layer by layer using plastic adhesive tape until only a single layer of carbon remained.[7] Although pristine graphene is still produced using this technique (which is very expensive), other processes have been developed that yield high-quality graphene at much lower costs.

Two characteristics of graphene make it an exceptional material. First is the perfect order found in its sheets—no atomic defects such as vacancies exist; also these sheets are extremely pure—only carbon atoms are present. The second characteristic relates to the nature of the unbonded electrons: at room temperature, they move much faster than conducting electrons in ordinary metals and semiconducting materials.[8]

In terms of its properties (some are listed in Table 14.6), graphene could be labeled the ultimate material. It is the strongest known material ($\sim$130 GPa), the best thermal conductor ($\sim$5000 W/m·K), and has the lowest electrical resistivity (10^{-8} Ω·m)—that is, is the best electrical conductor. Furthermore, it is transparent, chemically inert, and has a modulus of elasticity comparable to the other nanocarbons ($\sim$1 TPa).

Given this set of properties, the technological potential for graphene is enormous, and it is expected to revolutionize many industries to include electronics, energy, transportation, medicine/biotechnology, and aeronautics. However, before this revolution can begin to be realized, economical and reliable methods for the mass production of graphene must be devised.

The following is a short list of some of these potential applications for graphene: electronics—touch-screens, conductive ink for electronic printing, transparent conductors, transistors, heat sinks; energy—polymer solar cells, catalysts in fuel cells, battery electrodes, supercapacitors; medicine/biotechnology—artificial muscle, enzyme and DNA biosensors, photoimaging; aeronautics—chemical sensors (for explosives) and nanocomposites for aircraft structural components (Section 16.16).

SUMMARY

Ceramic Phase Diagrams

- The general characteristics of ceramic phase diagrams are similar to those for metallic systems.
- Diagrams for Al_2O_3–Cr_2O_3 (Figure 14.1), MgO–Al_2O_3 (Figure 14.2), ZrO_2–CaO (Figure 14.3), and SiO_2–Al_2O_3 (Figure 14.4) systems were discussed.
- These diagrams are especially useful in assessing the high-temperature performance of ceramic materials.

Brittle Fracture of Ceramics

- For ceramic materials, microcracks, the presence of which is very difficult to control, result in amplification of applied tensile stresses and account for relatively low fracture strengths (flexural strengths).
- Considerable variation in fracture strength for specimens of a specific material results inasmuch as the size of a crack-initiating flaw varies from specimen to specimen.
- This stress amplification does not occur with compressive loads; consequently, ceramics are stronger in compression.
- Fractographic analysis of the fracture surface of a ceramic material may reveal the location and source of the crack-producing flaw (Figure 14.8).

---

[7]This process is known as *micromechanical exfoliation*, or the *adhesive-tape method*.

[8]This phenomenon is called *ballistic conduction*.

Stress–Strain Behavior
- The stress–strain behaviors and fracture strengths of ceramic materials are determined using transverse bending tests.
- Flexural strengths as measured from three-point transverse bending tests may be determined for rectangular and circular cross-sections using, respectively, Equations 14.3a and 14.3b.

Mechanisms of Plastic Deformation
- Any plastic deformation of crystalline ceramics is a result of dislocation motion; the brittleness of these materials is explained, in part, by the limited number of operable slip systems.
- The mode of plastic deformation for noncrystalline materials is by viscous flow; a material's resistance to deformation is expressed as viscosity (in units of Pa·s). At room temperature, the viscosities of many noncrystalline ceramics are extremely high.

Influence of Porosity
- Many ceramic bodies contain residual porosity, which is deleterious to both their moduli of elasticity and fracture strengths.
  - Modulus of elasticity depends on and decreases with volume fraction porosity according to Equation 14.5.
  - The diminishment of flexural strength with volume fraction porosity is described by Equation 14.6.

Hardness
- Hardness of ceramic materials is difficult to measure because of their brittleness and susceptibility to cracking when indented.
- Microindentation Knoop and Vickers techniques are normally used, which employ pyramidal-shaped indenters.
- The hardest known materials are ceramics, which characteristic makes them especially attractive for use as abrasives (Section 14.14).

Glasses
- The familiar glass materials are noncrystalline silicates that contain other oxides. In addition to silica (SiO_2), the two other primary ingredients of a typical soda–lime glass are soda (Na_2O) and lime (CaO).
- The two prime assets of glass materials are optical transparency and ease of fabrication.

Glass–Ceramics
- Glass–ceramics are initially fabricated as glasses and then, by heat treatment, crystallized to form fine-grained polycrystalline materials.
- Two properties of glass–ceramics that make them superior to glass are improved mechanical strengths and lower coefficients of thermal expansion (which improves thermal shock resistance).

Clay Products
- Clay is the principal component of whitewares (e.g., pottery and tableware) and structural clay products (e.g., building bricks and tiles). Ingredients (in addition to clay) may be added, such as feldspar and quartz; these influence changes that occur during firing.

Refractories
- Materials that are employed at elevated temperatures and often in reactive environments are termed *refractory ceramics*.
- Requirements for this class of materials include high melting temperature, the ability to remain unreactive and inert when exposed to severe environments (often at elevated temperatures), and the ability to provide thermal insulation.
- On the basis of composition and application, the four main subdivisions are fire-clay (alumina–silica mixtures), silica (high silica contents), basic (rich in magnesia, MgO), and special.

Abrasives
- The abrasive ceramics are used to cut, grind, and polish other softer materials.
- This group of materials must be hard and tough and be able to withstand high temperatures that arise from frictional forces.
- Diamond, silicon carbide, tungsten carbide, corundum, and silica sand are the most common abrasive materials.

Cements
- Portland cement is produced by heating a mixture of clay and lime-bearing minerals in a rotary kiln. The resulting "clinker" is ground into very fine particles to which a small amount of gypsum is added.
- When mixed with water, inorganic cements form a paste that is capable of assuming just about any desired shape.
- Subsequent setting or hardening is a result of chemical reactions involving the cement particles and occurs at the ambient temperature. For hydraulic cements, of which Portland cement is the most common, the chemical reaction is one of hydration.

Carbons
- Two allotropic forms of carbon, diamond and graphite, have distinctively different sets of physical and chemical properties.
- Diamond is extremely hard, chemically inert, has a high thermal conductivity, a low electrical conductivity, and is transparent with a high index of refraction.
- Graphite is soft and flaky (i.e., has good lubricative properties), is optically opaque, and chemically stable at high temperatures and in nonoxidizing atmospheres. Some of its properties are highly isotropic, to include electrical conductivity.
- A form of carbon used as a fiber reinforcement was also discussed.
    Two structural arrangements of graphene layers may be found in carbon fibers— graphitic and turbostratic (Figure 14.18).
    High strengths and moduli of elasticity develop in the direction parallel to the fiber axis.

Advanced Ceramics
- Many modern technologies use and will continue to use advanced ceramics because of their unique mechanical, chemical, electrical, magnetic, and optical properties and property combinations.
- Microelectromechanical systems (MEMS)—these are smart systems that consist of miniaturized mechanical devices integrated with electrical elements on a substrate (normally silicon).
- Nanocarbons—carbon materials that have particle sizes less than about 100 nm. Three types of nanocarbons that can exist are as follows:
    Fullerenes (e.g., C_{60}, Figure 14.20)
    Carbon nanotubes (Figure 14.21)
    Graphene (Figure 14.22)
- Current and potential applications for the nanocarbons include the following:
    Fullerenes—high-temperature superconductors antioxidants (personal care products), organic solar cells
    Carbon nanotubes—electric field emitters, cancer treatments, photovoltaics (solar cells), better capacitors (to replace batteries)
    Graphene—transistors, supercapacitors, transparent electrical conductors, biosensors

Important Terms and Concepts

abrasive (ceramic)
calcination
cement
crystallization
    (glass–ceramics)

firing
flexural strength
glass–ceramic
microelectromechanical
    system (MEMS)

nanocarbon
refractory (ceramic)
structural clay product
viscosity
whiteware

REFERENCES

Barsoum, M. W., *Fundamentals of Ceramics*, Institute of Physics Publishing, Bristol, UK, 2003.

Bergeron, C. G., and S. H. Risbud, *Introduction to Phase Equilibria in Ceramics*, American Ceramic Society, Columbus, OH, 1984.

Carter, C. B., and M. G. Norton, Ceramic Materials Science and Engineering, Springer, New York, 2007.

Chiang, Y. M., D. P. Birnie, III, and W. D. Kingery, *Physical Ceramics: Principles for Ceramic Science and Engineering*, Wiley, New York, 1997.

Doremus, R. H., *Glass Science*, 2nd edition, Wiley, New York, 1994.

Engineered Materials Handbook, Vol. 4, *Ceramics and Glasses*, ASM International, Materials Park, OH, 1991.

Green, D. J., *An Introduction to the Mechanical Properties of Ceramics*, Cambridge University Press, Cambridge, 1998.

Hewlett, P. C., *Lea's Chemistry of Cement & Concrete*, 4th edition, Elsevier Butterworth-Heinemann, Oxford, 2004.

Hummel, F. A., *Introduction to Phase Equilibria in Ceramic Systems*, Marcel Dekker, New York, 1984.

Kingery, W. D., H. K. Bowen, and D. R. *Uhlmann, Introduction to Ceramics*, 2nd edition, Wiley, New York, 1976.

Phase Equilibria Diagrams (for Ceramists), American Ceramic Society, Westerville, OH. In fourteen volumes, published between 1964 and 2005. Also on CD-ROM.

Richerson, D. W., *The Magic of Ceramics*, 2nd edition, American Ceramic Society, Westerville, OH, 2012.

Richerson, D. W., *Modern Ceramic Engineering*, 3rd edition, CRC Press, Boca Raton, FL, 2006.

Schact, C. A. (Editor), *Refractories Handbook*, Marcel Dekker, New York, 2004.

Shelby, J. E., *Introduction to Glass Science and Technology*, 2nd edition, Royal Society of Chemistry, Cambridge, 2005.

Varshneya, A. K., *Fundamentals of Inorganic Glasses*, Elsevier, 1994.

Wachtman, J. B., W. R. Cannon, and M. J. Matthewson, *Mechanical Properties of Ceramics*, 2nd edition, Wiley, Hoboken, NJ, 2009.

QUESTIONS AND PROBLEMS

Ceramic Phase Diagrams

14.1 For the ZrO_2–CaO system (Figure 14.3), write all eutectic and eutectoid reactions for cooling.

14.2 From Figure 14.2, the phase diagram for the MgO–Al_2O_3 system, it may be noted that the spinel solid solution exists over a range of compositions, which means that it is nonstoichiometric at compositions other than 50 mol% MgO–50 mol% Al_2O_3.

(a) The maximum nonstoichiometry on the Al_2O_3-rich side of the spinel phase field exists at about 1998°C (2271 K), corresponding to approximately 82 mol% (92 wt%) Al_2O_3. Determine the type of vacancy defect that is produced and the percentage of vacancies that exist at this composition.

(b) The maximum nonstoichiometry on the MgO-rich side of the spinel phase field exists at about 1998°C (2271 K), corresponding to approximately 39 mol% (62 wt%) Al_2O_3. Determine the type of vacancy defect that is produced and the percentage of vacancies that exist at this composition.

14.3 When kaolinite clay $[Al_2(Si_2O_5)(OH)_4]$ is heated to a sufficiently high temperature, chemical water is driven off.

(a) Under these circumstances, what is the composition of the remaining product (in weight percent Al_2O_3)?

(b) What are the liquidus and solidus temperatures of this material?

Brittle Fracture of Ceramics

14.4 Briefly explain

(a) why there may be significant scatter in the fracture strength for some given ceramic material, and

(b) why fracture strength increases with decreasing specimen size.

14.5 The tensile strength of brittle materials may be determined using a variation of Equation 10.1. Compute the critical crack tip radius for an Al_2O_3 specimen that experiences tensile fracture at an applied stress of 275 MPa. Assume a critical surface crack length of 2×10^{-3} mm and a theoretical fracture strength of $E/10$, where E is the modulus of elasticity.

14.6 The fracture strength of glass may be increased by etching away a thin surface layer. It is believed that the etching may alter surface crack geometry (i.e., reduce crack length and increase the tip radius). Compute the ratio of the original and etched crack tip radii for an eightfold increase in fracture strength if two-thirds of the crack length is removed.

Stress–Strain Behavior

14.7 A three-point bending test is performed on a glass specimen having a rectangular cross section of height $d = 6$ mm and width $b = 12$ mm; the distance between support points is 45 mm.

(a) Compute the flexural strength if the load at fracture is 290 N.

(b) The point of maximum deflection Δy occurs at the center of the specimen and is described by

$$\Delta y = \frac{FL^3}{48EI} \qquad (14.8)$$

where E is the modulus of elasticity and I is the cross-sectional moment of inertia. Compute Δy at a load of 266 N.

14.8 A circular specimen of MgO is loaded using a three-point bending mode. Compute the minimum possible radius of the specimen without fracture, given that the applied load is 425 N, the flexural strength is 105 MPa, and the separation between load points is 60 mm.

14.9 A three-point bending test was performed on an aluminum oxide specimen having a circular cross section of radius 3.8 mm. The specimen fractured at a load of 950 N, when the distance between the support points was 60 mm. Another test is to be performed on a specimen of this same material, but one that has a square cross section of 12 mm length on each edge. At what load would you expect this specimen to fracture if the support point separation is 40 mm?

14.10 **(a)** A three-point transverse bending test is conducted on a cylindrical specimen of aluminum oxide having a reported flexural strength of 390 MPa. If the specimen radius is 2.8 mm and the support point separation distance is 40 mm, predict whether you would expect the specimen to fracture when a load of 620 N is applied. Justify your prediction.

(b) Would you be 100% certain of the prediction in part (a)? Why or why not?

Mechanisms of Plastic Deformation

14.11 Cite one reason why ceramic materials are, in general, harder yet more brittle than metals.

Miscellaneous Mechanical Considerations

14.12 The modulus of elasticity for beryllium oxide (BeO) having 5 vol% porosity is 310 GPa.

(a) Compute the modulus of elasticity for the nonporous material.

(b) Compute the modulus of elasticity for 10 vol% porosity.

14.13 The modulus of elasticity for boron carbide (B_4C) having 5 vol% porosity is 290 GPa.

(a) Compute the modulus of elasticity for the nonporous material.

(b) At what volume percent porosity will the modulus of elasticity be 235 GPa?

14.14 Using the data in Table 14.1, do the following:

(a) Determine the flexural strength for nonporous MgO, assuming a value of 3.75 for n in Equation 14.6.

(b) Compute the volume fraction porosity at which the flexural strength for MgO is 62 MPa.

14.15 The flexural strength and associated volume fraction porosity for two specimens of the same ceramic material are as follows:

σ_{fs} *(MPa)*	*P*
100	0.05
50	0.20

(a) Compute the flexural strength for a completely nonporous specimen of this material.

(b) Compute the flexural strength for a 0.10 volume fraction porosity.

Glasses
Glass–Ceramics

14.16 Cite the two desirable characteristics of glasses.

14.17 **(a)** What is crystallization?

(b) Cite two properties that may be improved by crystallization.

Refractories

14.18 For refractory ceramic materials, cite three characteristics that improve with and two characteristics that are adversely affected by increasing porosity.

14.19 Find the maximum temperature to which the following two magnesia–alumina refractory materials may be heated before a liquid phase will appear.

(a) A spinel-bonded alumina material of composition 94 wt% Al_2O_3–6 wt% MgO.

(b) A magnesia–alumina spinel of composition 60 wt% Al_2O_3–40 wt% MgO. Consult Figure 14.2.

14.20 Upon consideration of the SiO_2–Al_2O_3 phase diagram, Figure 14.4, for each pair of the following list of compositions, which would you judge to be the more desirable refractory? Justify your choices.

(a) 15 wt% Al_2O_3–85 wt% SiO_2 and 20 wt% Al_2O_3–80 wt% SiO_2

(b) 75 wt% Al_2O_3–25 wt% SiO_2 and 85 wt% Al_2O_3–15 wt% SiO_2

14.21 Compute the mass fractions of liquid in the following refractory materials at 1600°C (1873 K):

(a) 5 wt% Al_2O_3–95 wt% SiO_2

(b) 15 wt% Al_2O_3–85 wt% SiO_2

(c) 25 wt% Al_2O_3–75 wt% SiO_2

(d) 75 wt% Al_2O_3–25 wt% SiO_2

14.22 For the MgO–Al_2O_3 system, what is the maximum temperature that is possible without the formation of a liquid phase? At what composition or over what range of compositions will this maximum temperature be achieved?

Cements

14.23 Compare the manner in which the aggregate particles become bonded together in clay-based mixtures during firing and in cements during setting.

DESIGN PROBLEMS

Stress–Strain Behavior

14.D1 It is necessary to select a ceramic material to be stressed using a three-point loading scheme (Figure 14.9). The specimen must have a circular cross section and a radius of 2.8 mm, and must not experience fracture or a deflection of more than 6.2×10^{-2} mm at its center when a load of 275 N is applied. If the distance between support points is 50 mm, which of the materials in Table 14.1 are candidates? The magnitude of the centerpoint deflection may be computed using the equation supplied in Problem 14.7.

Glasses
Glass–Ceramics

14.D2 Some of our modern kitchen cookware is made of ceramic materials.

(a) List at least three important characteristics required of a material to be used for this application.

(b) Compare the relative properties and cost of three ceramic materials.

(c) On the basis of this comparison, select the material most suitable for the cookware.

Photograph (*a*) shows billiard balls made of phenol-formaldehyde (Bakelite). The Materials of Importance piece that follows Section 15.15 discusses the invention of phenol-formaldehyde and its replacement of ivory for billiard balls. Photograph (*b*) shows a woman playing billiards.

(*a*)

(*b*)

© William D. Callister, Jr.

© RapidEye/iStockphoto

There are several reasons why an engineer should know something about the characteristics, applications, and processing of polymeric materials. Polymers are used in a wide variety of applications, such as construction materials and microelectronics processing. Thus, most engineers will be required to work with polymers at some point in their careers. Understanding the mechanisms by which polymers elastically and plastically deform allows one to alter and control their moduli of elasticity and strengths (Sections 15.7 and 15.8).

Learning Objectives

After studying this chapter, you should be able to do the following:

1. Make schematic plots of the three characteristic stress–strain behaviors observed for polymeric materials.
2. Describe/sketch the various stages in the elastic and plastic deformations of a semicrystalline (spherulitic) polymer.
3. Discuss the influence of the following factors on polymer tensile modulus and/or strength: (a) molecular weight, (b) degree of crystallinity, (c) predeformation, and (d) heat treating of undeformed materials.
4. Describe the molecular mechanism by which elastomeric polymers deform elastically.
5. List four characteristics or structural components of a polymer that affect both its melting and glass transition temperatures.
6. Cite the seven different polymer application types and note the general characteristics of each type.

15.1 INTRODUCTION

This chapter discusses some of the characteristics important to polymeric materials and the various types and processing techniques.

Mechanical Behavior of Polymers

15.2 STRESS–STRAIN BEHAVIOR

The mechanical properties of polymers are specified with many of the same parameters that are used for metals—that is, modulus of elasticity and yield and tensile strengths. For many polymeric materials, the simple stress–strain test is used to characterize some of these mechanical parameters.[1] The mechanical characteristics of polymers, for the most part, are highly sensitive to the rate of deformation (strain rate), the temperature, and the chemical nature of the environment (the presence of water, oxygen, organic solvents, etc.). Some modifications of the testing techniques and specimen configurations used for metals (Chapter 8) are necessary with polymers, especially for highly elastic materials, such as rubbers.

WileyPLUS: VMSE
Polymers

Three typically different types of stress–strain behavior are found for polymeric materials, as represented in Figure 15.1. Curve *A* illustrates the stress–strain character for a brittle polymer, which fractures while deforming elastically. The behavior for a plastic material, curve *B*, is similar to that for many metallic materials; the initial deformation is elastic, which is followed by yielding and a region of plastic deformation. Finally, the deformation displayed by curve *C* is totally elastic; this rubber-like elasticity (large recoverable strains produced at low stress levels) is displayed by a class of polymers termed the **elastomers**.

elastomer

---

[1]ASTM Standard D638, "Standard Test Method for Tensile Properties of Plastics."

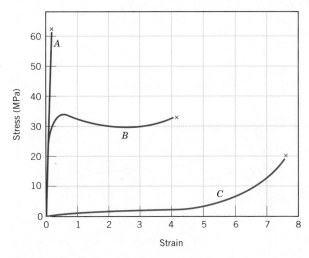

Figure 15.1 The stress–strain behavior for brittle (curve A), plastic (curve B), and highly elastic (elastomeric) (curve C) polymers.

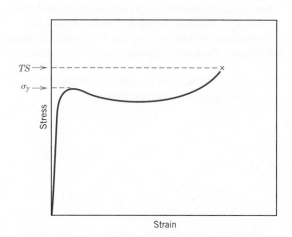

Figure 15.2 Schematic stress–strain curve for a plastic polymer showing how yield and tensile strengths are determined.

Modulus of elasticity (termed *tensile modulus* or sometimes just *modulus* for polymers) and ductility in percent elongation are determined for polymers in the same manner as for metals (Sections 8.3 and 8.4). For plastic polymers (curve B, Figure 15.1), the yield point is taken as a maximum on the curve, which occurs just beyond the termination of the linear-elastic region (Figure 15.2). The stress at this maximum is the yield strength (σ_y). Furthermore, tensile strength (*TS*) corresponds to the stress at which fracture occurs (Figure 15.2); *TS* may be greater than or less than σ_y. For these plastic polymers, strength is normally taken as tensile strength. Table 15.1 gives these mechanical properties for several polymeric materials; more comprehensive lists are provided in Tables B.2 to B.4, Appendix B.

Table 15.1 Room-Temperature Mechanical Characteristics of Some of the More Common Polymers

Material	Specific Gravity	Tensile Modulus (GPa)	Tensile Strength (MPa)	Yield Strength (MPa)	Elongation at Break (%)
Polyethylene (low density)	0.917–0.932	0.17–0.28	8.3–31.4	9.0–14.5	100–650
Polyethylene (high density)	0.952–0.965	1.06–1.09	22.1–31.0	26.2–33.1	10–1200
Poly(vinyl chloride)	1.30–1.58	2.4–4.1	40.7–51.7	40.7–44.8	40–80
Polytetrafluoroethylene	2.14–2.20	0.40–0.55	20.7–34.5	13.8–15.2	200–400
Polypropylene	0.90–0.91	1.14–1.55	31–41.4	31.0–37.2	100–600
Polystyrene	1.04–1.05	2.28–3.28	35.9–51.7	25.0–69.0	1.2–2.5
Poly(methyl methacrylate)	1.17–1.20	2.24–3.24	48.3–72.4	53.8–73.1	2.0–5.5
Phenol-formaldehyde	1.24–1.32	2.76–4.83	34.5–62.1	—	1.5–2.0
Nylon 6,6	1.13–1.15	1.58–3.80	75.9–94.5	44.8–82.8	15–300
Polyester (PET)	1.29–1.40	2.8–4.1	48.3–72.4	59.3	30–300
Polycarbonate	1.20	2.38	62.8–72.4	62.1	110–150

Source: *Modern Plastics Encyclopedia '96.* Copyright 1995, The McGraw-Hill Companies. Reprinted with permission.

Figure 15.3 The influence of temperature on the stress–strain characteristics of poly(methyl methacrylate). (From T. S. Carswell and H. K. Nason, "Effect of Environmental Conditions on the Mechanical Properties of Organic Plastics," *Symposium on Plastics,* American Society for Testing and Materials, Philadelphia, 1944. Copyright, ASTM, 1916 Race Street, Philadelphia, PA 19103. Reprinted with permission.)

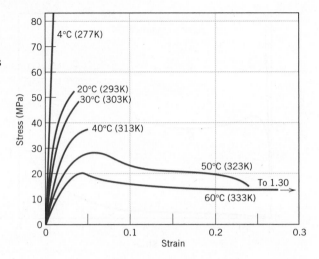

In many respects, polymers are mechanically dissimilar to metals (Figures 1.5, 1.6, and 1.7). For example, the modulus for highly elastic polymeric materials may be as low as 7 MPa, but it may run as high as 4 GPa for some very stiff polymers; modulus values for metals are much larger and range between 48 and 410 GPa. Maximum tensile strengths for polymers are about 100 MPa, whereas for some metal alloys they are 4100 MPa. Furthermore, whereas metals rarely elongate plastically to more than 100%, some highly elastic polymers may experience elongations to greater than 1000%.

In addition, the mechanical characteristics of polymers are much more sensitive to temperature changes near room temperature. Consider the stress–strain behavior for poly(methyl methacrylate) (Plexiglas) at several temperatures between 4°C and 60°C (Figure 15.3). Increasing the temperature produces (1) a decrease in elastic modulus, (2) a reduction in tensile strength, and (3) an enhancement of ductility—at 4°C the material is totally brittle, whereas there is considerable plastic deformation at both 50°C and 60°C.

The influence of strain rate on the mechanical behavior may also be important. In general, decreasing the rate of deformation has the same influence on the stress–strain characteristics as increasing the temperature; that is, the material becomes softer and more ductile.

15.3 MACROSCOPIC DEFORMATION

WileyPLUS: VMSE
Polymers

Some aspects of the macroscopic deformation of semicrystalline polymers deserve our attention. The tensile stress–strain curve for a semicrystalline material that was initially undeformed is shown in Figure 15.4; also included in the figure are schematic representations of the specimen profiles at various stages of deformation. Both upper and lower yield points are evident on the curve, which are followed by a near horizontal region. At the upper yield point, a small neck forms within the gauge section of the specimen. Within this neck, the chains become oriented (i.e., chain axes become aligned parallel to the elongation direction, a condition that is represented schematically in Figure 15.13*d*), which leads to localized strengthening. Consequently, there is a resistance to continued deformation at this point, and specimen elongation proceeds by the propagation of this neck region along the gauge length; the chain orientation phenomenon (Figure 15.13*d*) accompanies this neck extension. This tensile behavior may be contrasted to that found for ductile metals (Section 8.4), in which once a neck has formed, all subsequent deformation is confined to within the neck region.

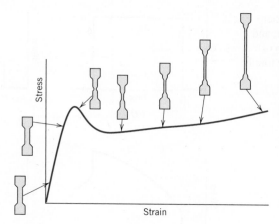

Figure 15.4 Schematic tensile stress–strain curve for a semicrystalline polymer. Specimen contours at several stages of deformation are included. (From Jerold M. Schultz, *Polymer Materials Science,* copyright © 1974, p. 488. Reprinted by permission of Prentice Hall, Inc., Englewood Cliffs, NJ.)

Concept Check 15.1   When citing the ductility as percent elongation for semicrystalline polymers, it is not necessary to specify the specimen gauge length, as is the case with metals. Why is this so?

[*The answer may be found at* www.wiley.com/college/callister *(Student Companion Site).*]

15.4 VISCOELASTIC DEFORMATION

An amorphous polymer may behave like a glass at low temperatures, a rubbery solid at intermediate temperatures [above the glass transition temperature (Section 15.12)], and a viscous liquid as the temperature is further raised. For relatively small deformations, the mechanical behavior at low temperatures may be elastic—that is, in conformity to Hooke's law, $\sigma = E\epsilon$. At the highest temperatures, viscous or liquid-like behavior prevails. For intermediate temperatures, the polymer is a rubbery solid that exhibits the combined mechanical characteristics of these two extremes; the condition is termed **viscoelasticity**.

viscoelasticity

Elastic deformation is instantaneous, which means that total deformation (or strain) occurs the instant the stress is applied or released (i.e., the strain is independent of time). In addition, upon release of the external stress, the deformation is totally recovered—the specimen assumes its original dimensions. This behavior is represented in Figure 15.5*b* as strain versus time for the instantaneous load–time curve, shown in Figure 15.5*a*.

By way of contrast, for totally viscous behavior, deformation or strain is not instantaneous; that is, in response to an applied stress, deformation is delayed or dependent on time. Also, this deformation is not reversible or completely recovered after the stress is released. This phenomenon is demonstrated in Figure 15.5*d*.

For the intermediate viscoelastic behavior, the imposition of a stress in the manner of Figure 15.5*a* results in an instantaneous elastic strain, which is followed by a viscous, time-dependent strain, a form of anelasticity (Section 8.3); this behavior is illustrated in Figure 15.5*c*.

A familiar example of these viscoelastic extremes is found in a silicone polymer that is sold as a novelty and known by some as Silly Putty. When rolled into a ball and dropped onto a horizontal surface, it bounces elastically—the rate of deformation during the bounce is very rapid. However, if pulled in tension with a gradually increasing applied stress, the material elongates or flows like a highly viscous liquid. For this and other viscoelastic materials, the rate of strain determines whether the deformation is elastic or viscous.

Figure 15.5 (a) Load versus time, where load is applied instantaneously at time t_a and released at t_r. For the load–time cycle in (a), the strain-versus-time responses are for totally elastic (b), viscoelastic (c), and viscous (d) behaviors.

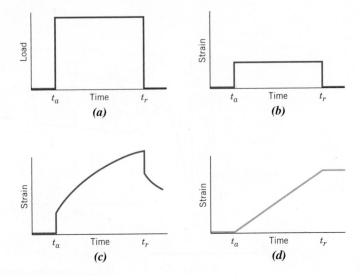

Viscoelastic Relaxation Modulus

The viscoelastic behavior of polymeric materials is dependent on both time and temperature; several experimental techniques may be used to measure and quantify this behavior. *Stress relaxation* measurements represent one possibility. With these tests, a specimen is initially strained rapidly in tension to a predetermined and relatively low strain level. The stress necessary to maintain this strain is measured as a function of time while temperature is held constant. Stress is found to decrease with time because of molecular relaxation processes that take place within the polymer. We may define a **relaxation modulus** $E_r(t)$, a time-dependent elastic modulus for viscoelastic polymers, as

relaxation modulus

Relaxation modulus—ratio of time-dependent stress and constant strain value

$$E_r(t) = \frac{\sigma(t)}{\epsilon_0} \tag{15.1}$$

where $\sigma(t)$ is the measured time-dependent stress and ϵ_0 is the strain level, which is maintained constant.

Furthermore, the magnitude of the relaxation modulus is a function of temperature; to more fully characterize the viscoelastic behavior of a polymer, isothermal stress relaxation measurements must be conducted over a range of temperatures. Figure 15.6 is a schematic log $E_r(t)$-versus-log time plot for a polymer that exhibits viscoelastic behavior. Curves generated at a variety of temperatures are included. Key features of this plot are that (1) the magnitude of $E_r(t)$ decreases with time (corresponding to the decay of stress, Equation 15.1), and (2) the curves are displaced to lower $E_r(t)$ levels with increasing temperature.

To represent the influence of temperature, data points are taken at a specific time from the log $E_r(t)$-versus-log time plot—for example, t_1 in Figure 15.6—and then cross-plotted as log $E_r(t_1)$ versus temperature. Figure 15.7 is such a plot for an amorphous (atactic) polystyrene; in this case, t_1 was arbitrarily taken 10 s after the load application. Several distinct regions may be noted on the curve shown in this figure. At the lowest temperatures, in the glassy region, the material is rigid and brittle, and the value of $E_r(10)$ is that of the elastic modulus, which initially is virtually independent of temperature. Over this temperature range, the strain–time characteristics are as represented in Figure 15.5b. On a molecular level, the long molecular chains are essentially frozen in position at these temperatures.

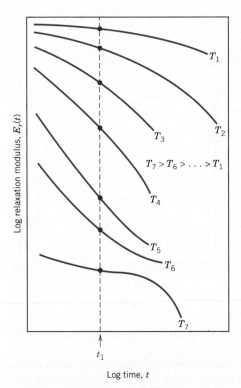

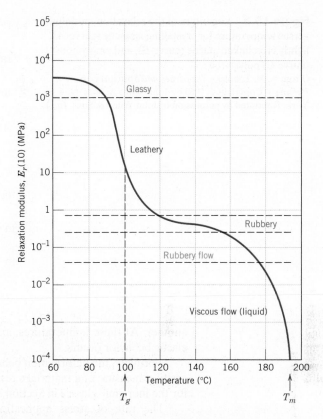

Figure 15.6 Schematic plot of logarithm of relaxation modulus versus logarithm of time for a viscoelastic polymer; isothermal curves are generated at temperatures T_1 through T_7. The temperature dependence of the relaxation modulus is represented as log $E_r(t_1)$ versus temperature.

Figure 15.7 Logarithm of the relaxation modulus versus temperature for amorphous polystyrene, showing the five different regions of viscoelastic behavior.
(From A. V. Tobolsky, *Properties and Structures of Polymers.* Copyright © 1960 by John Wiley & Sons, New York. Reprinted by permission of John Wiley & Sons, Inc.)

As the temperature is increased, $E_r(10)$ drops abruptly by about a factor of 10^3 within a 20°C temperature span; this is sometimes called the *leathery,* or *glass transition region,* and the glass transition temperature (T_g, Section 15.13) lies near the upper temperature extremity; for polystyrene (Figure 15.7), $T_g = 100°C$. Within this temperature region, a polymer specimen will be leathery; that is, deformation will be time dependent and not totally recoverable on release of an applied load, characteristics that are depicted in Figure 15.5c.

Within the rubbery plateau temperature region (Figure 15.7), the material deforms in a rubbery manner; here, both elastic and viscous components are present, and deformation is easy to produce because the relaxation modulus is relatively low.

The final two high-temperature regions are rubbery flow and viscous flow. Upon heating through these temperatures, the material experiences a gradual transition to a soft, rubbery state, and finally to a viscous liquid. In the rubbery flow region, the polymer is a very viscous liquid that exhibits both elastic and viscous flow components. Within the viscous flow region, the modulus decreases dramatically with increasing temperature; again, the strain–time behavior is as represented in Figure 15.5d. From a molecular standpoint, chain motion intensifies so greatly that for viscous flow, the chain segments experience vibration and rotational motion largely independent of one

Figure 15.8 Logarithm of the relaxation modulus versus temperature for crystalline isotactic (curve *A*), lightly crosslinked atactic (curve *B*), and amorphous (curve *C*) polystyrene.
(From A. V. Tobolsky, *Properties and Structures of Polymers.* Copyright © 1960 by John Wiley & Sons, New York. Reprinted by permission of John Wiley & Sons, Inc.)

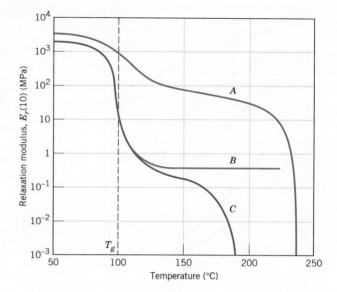

another. At these temperatures, any deformation is entirely viscous and essentially no elastic behavior occurs.

Normally, the deformation behavior of a viscous polymer is specified in terms of viscosity, a measure of a material's resistance to flow by shear forces. Viscosity is discussed for the inorganic glasses in Section 14.8.

The rate of stress application also influences the viscoelastic characteristics. Increasing the loading rate has the same influence as lowering the temperature.

The log $E_r(10)$-versus-temperature behavior for polystyrene materials having several molecular configurations is plotted in Figure 15.8. The curve for the amorphous material (curve *C*) is the same as in Figure 15.7. For a lightly crosslinked atactic polystyrene (curve *B*), the rubbery region forms a plateau that extends to the temperature at which the polymer decomposes; this material will not experience melting. For increased crosslinking, the magnitude of the plateau $E_r(10)$ value will also increase. Rubber or elastomeric materials display this type of behavior and are ordinarily used at temperatures within this plateau range.

Also shown in Figure 15.8 is the temperature dependence for an almost totally crystalline isotactic polystyrene (curve *A*). The decrease in $E_r(10)$ at T_g is much less pronounced than for the other polystyrene materials because only a small volume fraction of this material is amorphous and experiences the glass transition. Furthermore, the relaxation modulus is maintained at a relatively high value with increasing temperature until its melting temperature T_m is approached. From Figure 15.8, the melting temperature of this isotactic polystyrene is about 240°C.

Viscoelastic Creep

Many polymeric materials are susceptible to time-dependent deformation when the stress level is maintained constant; such deformation is termed *viscoelastic creep*. This type of deformation may be significant even at room temperature and under modest stresses that lie below the yield strength of the material. For example, automobile tires may develop flat spots on their contact surfaces when the automobile is parked for prolonged time periods. Creep tests on polymers are conducted in the same manner as for metals (Chapter 10); that is, a stress (normally tensile) is applied instantaneously and is maintained at a constant level while strain is measured as a function of time.

Furthermore, the tests are performed under isothermal conditions. Creep results are represented as a time-dependent *creep modulus* $E_c(t)$, defined by[2]

$$E_c(t) = \frac{\sigma_0}{\epsilon(t)} \qquad (15.2)$$

where σ_0 is the constant applied stress and $\epsilon(t)$ is the time-dependent strain. The creep modulus is also temperature sensitive and decreases with increasing temperature.

With regard to the influence of molecular structure on the creep characteristics, as a general rule the susceptibility to creep decreases [i.e., $E_c(t)$ increases] as the degree of crystallinity increases.

Concept Check 15.2   An amorphous polystyrene that is deformed at 120°C will exhibit which of the behaviors shown in Figure 15.5?

[*The answer may be found at* www.wiley.com/college/callister *(Student Companion Site).*]

15.5 FRACTURE OF POLYMERS

The fracture strengths of polymeric materials are low relative to those of metals and ceramics. As a general rule, the mode of fracture in thermosetting polymers (heavily crosslinked networks) is brittle. In simple terms, during the fracture process, cracks form at regions where there is a localized stress concentration (i.e., scratches, notches, and sharp flaws). As with metals (Section 10.5), the stress is amplified at the tips of these cracks, leading to crack propagation and fracture. Covalent bonds in the network or crosslinked structure are severed during fracture.

For thermoplastic polymers, both ductile and brittle modes are possible, and many of these materials are capable of experiencing a ductile-to-brittle transition. Factors that favor brittle fracture are a reduction in temperature, an increase in strain rate, the presence of a sharp notch, an increase in specimen thickness, and any modification of the polymer structure that raises the glass transition temperature (T_g) (see Section 15.14). Glassy thermoplastics are brittle below their glass transition temperatures. However, as the temperature is raised, they become ductile in the vicinity of their T_gs and experience plastic yielding prior to fracture. This behavior is demonstrated by the stress–strain characteristics of poly(methyl methacrylate) (PMMA) in Figure 15.3. At 4°C, PMMA is totally brittle, whereas at 60°C it becomes extremely ductile.

One phenomenon that frequently precedes fracture in some thermoplastic polymers is *crazing*. Associated with crazes are regions of very localized plastic deformation, which lead to the formation of small and interconnected microvoids (Figure 15.9a). Fibrillar bridges form between these microvoids wherein molecular chains become oriented as in Figure 15.13d. If the applied tensile load is sufficient, these bridges elongate and break, causing the microvoids to grow and coalesce. As the microvoids coalesce, cracks begin to form, as demonstrated in Figure 15.9b. A craze is different from a crack in that it can support a load across its face. Furthermore, this process of craze growth prior to cracking absorbs fracture energy and effectively increases the fracture toughness of the polymer. In glassy polymers, the cracks propagate with little craze formation, resulting in low fracture toughnesses. Crazes form at highly stressed regions associated

---

[2]*Creep compliance,* $J_c(t)$, the reciprocal of the creep modulus, is also sometimes used in this context.

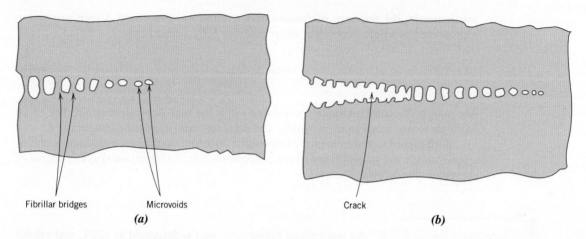

Fibrillar bridges   Microvoids   Crack
(a)   *(b)*

Figure 15.9 Schematic drawings of (*a*) a craze showing microvoids and fibrillar bridges and (*b*) a craze followed by a crack.
(From J. W. S. Hearle, *Polymers and Their Properties,* Vol. 1, *Fundamentals of Structure and Mechanics,* Ellis Horwood, Ltd., Chichester, West Sussex, England, 1982.)

with scratches, flaws, and molecular inhomogeneities; in addition, they propagate perpendicular to the applied tensile stress and typically are 5 μm or less thick. The photomicrograph in Figure 15.10 shows a craze.

Principles of fracture mechanics developed in Section 10.5 also apply to brittle and quasi-brittle polymers; the susceptibility of these materials to fracture when a crack is present may be expressed in terms of the plane strain fracture toughness. The magnitude of K_{Ic} depends on characteristics of the polymer (molecular weight, percent crystallinity, etc.), as well as on temperature, strain rate, and the external environment. Representative values of K_{Ic} for several polymers are given in Table 10.1 and Table B.5, Appendix B.

Figure 15.10 Photomicrograph of a craze in poly(phenylene oxide).
(From R. P. Kambour and R. E. Robertson, "The Mechanical Properties of Plastics," in *Polymer Science, A Materials Science Handbook,* A. D. Jenkins, Editor. Reprinted with permission of Elsevier Science Publishers.)

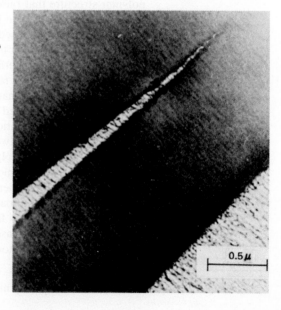

0.5 μ

15.6 MISCELLANEOUS MECHANICAL CHARACTERISTICS

Impact Strength

The degree of resistance of a polymeric material to impact loading may be of concern in some applications. Izod or Charpy tests are ordinarily used to assess impact strength (Section 10.6). As with metals, polymers may exhibit ductile or brittle fracture under impact loading conditions, depending on the temperature, specimen size, strain rate, and mode of loading, as discussed in the preceding section. Both semicrystalline and amorphous polymers are brittle at low temperatures, and both have relatively low impact strengths. However, they experience a ductile-to-brittle transition over a relatively narrow temperature range, similar to that shown for a steel in Figure 10.13. Of course, impact strength undergoes a gradual decrease at still higher temperatures as the polymer begins to soften. Ordinarily, the two impact characteristics most sought after are a high impact strength at the ambient temperature and a ductile-to-brittle transition temperature that lies below room temperature.

Fatigue

Polymers may experience fatigue failure under conditions of cyclic loading. As with metals, fatigue occurs at stress levels that are low relative to the yield strength. Fatigue testing in polymers has not been nearly as extensive as with metals; however, fatigue data are plotted in the same manner for both types of material, and the resulting curves have the same general shape. Fatigue curves for several common polymers are shown in Figure 15.11 as stress versus the number of cycles to failure (on a logarithmic scale). Some polymers have a *fatigue limit* (a stress level at which the stress at failure becomes independent of the number of cycles); others do not appear to have such a limit. As would be expected, fatigue strengths and fatigue limits for polymeric materials are much lower than for metals.

The fatigue behavior of polymers is much more sensitive to loading frequency than for metals. Cycling polymers at high frequencies and/or relatively large stresses can cause localized heating; consequently, failure may be due to a softening of the material rather than a result of typical fatigue processes.

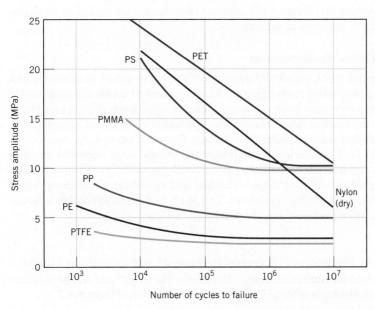

Figure 15.11 Fatigue curves (stress amplitude versus the number of cycles to failure) for poly(ethylene terephthalate) (PET), nylon, polystyrene (PS), poly (methyl methacrylate) (PMMA), polypropylene (PP), polyethylene (PE), and polytetrafluoroethylene (PTFE). The testing frequency was 30 Hz.
(From M. N. Riddell, "A Guide to Better Testing of Plastics," *Plast. Eng.*, Vol. 30, No. 4, p. 78, 1974.)

Tear Strength and Hardness

Other mechanical properties that are sometimes influential in the suitability of a polymer for some particular application include tear resistance and hardness. The ability to resist tearing is an important property of some plastics, especially those used for thin films in packaging. *Tear strength*, the mechanical parameter measured, is the energy required to tear apart a cut specimen that has a standard geometry. The magnitude of tensile and tear strengths are related.

As with metals, hardness represents a material's resistance to scratching, penetration, marring, and so on. Polymers are softer than metals and ceramics, and most hardness tests are conducted by penetration techniques similar to those described for metals in Section 8.5. Rockwell tests are frequently used for polymers.[3] Other indentation techniques employed are the Durometer and Barcol.[4]

Mechanisms of Deformation and for Strengthening of Polymers

An understanding of deformation mechanisms of polymers is important to be able to manage the mechanical characteristics of these materials. In this regard, deformation models for two different types of polymers—semicrystalline and elastomeric—deserve our attention. The stiffness and strength of semicrystalline materials are often important considerations; elastic and plastic deformation mechanisms are treated in the succeeding section, whereas methods used to stiffen and strengthen these materials are discussed in Section 15.8. However, elastomers are used on the basis of their unusual elastic properties; the deformation mechanism of elastomers is also treated.

15.7 DEFORMATION OF SEMICRYSTALLINE POLYMERS

Many semicrystalline polymers in bulk form have the spherulitic structure described in Section 5.11. By way of review, each spherulite consists of numerous chain-folded ribbons, or lamellae, that radiate outward from the center. Separating these lamellae are areas of amorphous material (Figure 5.12); adjacent lamellae are connected by tie chains that pass through these amorphous regions.

Mechanism of Elastic Deformation

As with other material types, elastic deformation of polymers occurs at relatively low stress levels on the stress–strain curve (Figure 15.1). The onset of elastic deformation for semicrystalline polymers results from chain molecules in amorphous regions elongating in the direction of the applied tensile stress. This process is represented schematically for two adjacent chain-folded lamellae and the interlamellar amorphous material as Stage 1 in Figure 15.12. Continued deformation in the second stage occurs by changes in both amorphous and lamellar crystalline regions. Amorphous chains continue to align and become elongated; in addition, there is bending and stretching of the strong chain covalent bonds within the lamellar crystallites. This leads to a slight, reversible increase in the lamellar crystallite thickness as indicated by Δt in Figure 15.12c.

---

[3]ASTM Standard D785, "Standard Testing Method for Rockwell Hardness of Plastics and Electrical Insulating Materials."

[4]ASTM Standard D2240, "Standard Test Method for Rubber Property—Durometer Hardness"; and ASTM Standard D2583, "Standard Test Method for Indentation Hardness of Rigid Plastics by Means of a Barcol Impressor."

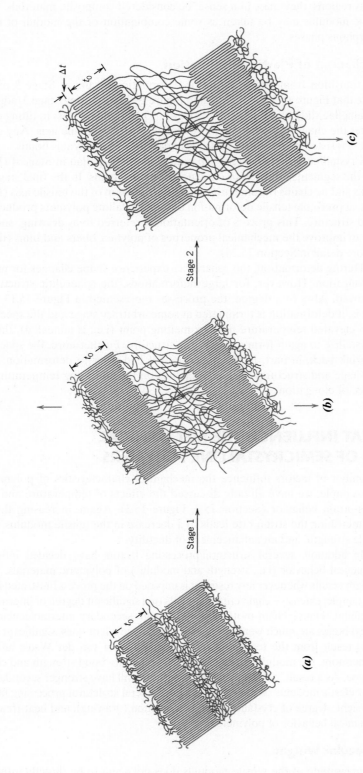

Figure 15.12 Stages in the elastic deformation of a semicrystalline polymer. (*a*) Two adjacent chain-folded lamellae and interlamellar amorphous material before deformation. (*b*) Elongation of amorphous tie chains during the first stage of deformation. (*c*) Increase in lamellar crystallite thickness (which is reversible) due to bending and stretching of chains in crystallite regions. (From SCHULTZ, POLYMER MATERIALS SCIENCE, 1st Edition, © 1974. Reprinted by permission of Pearson Education, Inc., Upper Saddle River, NJ.)

Inasmuch as semicrystalline polymers are composed of both crystalline and amorphous regions, they may, in a sense, be considered composite materials. Therefore, the elastic modulus may be taken as some combination of the moduli of crystalline and amorphous phases.

Mechanism of Plastic Deformation

The transition from elastic to plastic deformation occurs in Stage 3 of Figure 15.13. (Note that Figure 15.12c is identical to Figure 15.13a.) During Stage 3, adjacent chains in the lamellae slide past one another (Figure 15.13b); this results in tilting of the lamellae so that the chain folds become more aligned with the tensile axis. Any chain displacement is resisted by relatively weak secondary or van der Waals bonds.

Crystalline block segments separate from the lamellae in Stage 4 (Figure 15.13c), with the segments attached to one another by tie chains. In the final stage, Stage 5, the blocks and tie chains become oriented in the direction of the tensile axis (Figure 15.13d). Thus, appreciable tensile deformation of semicrystalline polymers produces a highly oriented structure. This process of orientation is referred to as **drawing**, and is commonly used to improve the mechanical properties of polymer fibers and films (this is discussed in more detail in Section 15.24).

During deformation, the spherulites experience shape changes for moderate levels of elongation. However, for large deformations, the spherulitic structure is virtually destroyed. Also, to a degree, the processes represented in Figure 15.13 are reversible. That is, if deformation is terminated at some arbitrary stage and the specimen is heated to an elevated temperature near its melting point (i.e., is annealed), the material will recrystallize to again form a spherulitic structure. Furthermore, the specimen will tend to shrink back, in part, to the dimensions it had prior to deformation. The extent of this shape and structural recovery depends on the annealing temperature and also the degree of elongation.

drawing

15.8 FACTORS THAT INFLUENCE THE MECHANICAL PROPERTIES OF SEMICRYSTALLINE POLYMERS

A number of factors influence the mechanical characteristics of polymeric materials. For example, we have already discussed the effects of temperature and strain rate on stress–strain behavior (Section 15.2, Figure 15.3). Again, increasing the temperature or diminishing the strain rate leads to a decrease in the tensile modulus, a reduction in tensile strength, and an enhancement of ductility.

In addition, several structural/processing factors have decided influences on the mechanical behavior (i.e., strength and modulus) of polymeric materials. An increase in strength results whenever any restraint is imposed on the process illustrated in Figure 15.13; for example, extensive chain entanglements or a significant degree of intermolecular bonding inhibit relative chain motions. Even though secondary intermolecular (e.g., van der Waals) bonds are much weaker than the primary covalent ones, significant intermolecular forces result from the formation of large numbers of van der Waals interchain bonds. Furthermore, the modulus rises as both the secondary bond strength and chain alignment increase. As a result, polymers with polar groups will have stronger secondary bonds and a larger elastic modulus. We now discuss how several structural/processing factors [molecular weight, degree of crystallinity, predeformation (drawing), and heat-treating] affect the mechanical behavior of polymers.

Molecular Weight

The magnitude of the tensile modulus does not seem to be directly influenced by molecular weight. On the other hand, for many polymers it has been observed that tensile

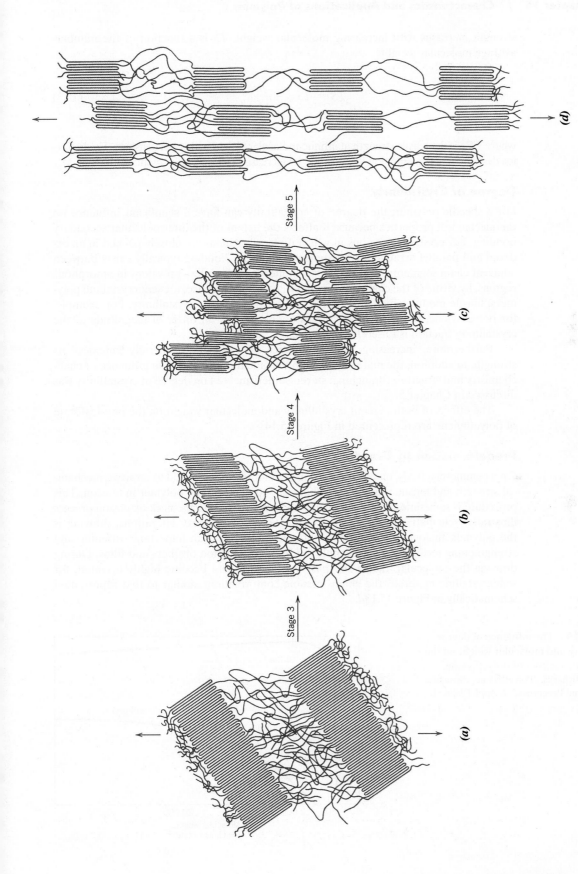

Figure 15.13 Stages in the plastic deformation of a semicrystalline polymer. (a) Two adjacent chain-folded lamellae and interlamellar amorphous material after elastic deformation (also shown as Figure 15.12c). (b) Tilting of lamellar chain folds. (c) Separation of crystalline block segments. (d) Orientation of block segments and tie chains with the tensile axis in the final plastic deformation stage. (From SCHULTZ, POLYMER MATERIALS SCIENCE, 1st Edition, © 1974. Reprinted by permission of Pearson Education, Inc., Upper Saddle River, NJ.)

strength increases with increasing molecular weight. *TS* is a function of the number-average molecular weight,

For some polymers, dependence of tensile strength on number-average molecular weight

$$TS = TS_\infty - \frac{A}{\overline{M}_n} \tag{15.3}$$

where TS_∞ is the tensile strength at infinite molecular weight and A is a constant. The behavior described by this equation is explained by increased chain entanglements with rising $\overline{M}_n$.

Degree of Crystallinity

For a specific polymer, the degree of crystallinity can have a significant influence on the mechanical properties because it affects the extent of the intermolecular secondary bonding. For crystalline regions in which molecular chains are closely packed in an ordered and parallel arrangement, extensive secondary bonding typically exists between adjacent chain segments. This secondary bonding is much less prevalent in amorphous regions, by virtue of the chain misalignment. As a consequence, for semicrystalline polymers, tensile modulus increases significantly with degree of crystallinity. For example, for polyethylene, the modulus increases approximately an order of magnitude as the crystallinity fraction is raised from 0.3 to 0.6.

Furthermore, increasing the crystallinity of a polymer generally enhances its strength; in addition, the material tends to become more brittle. The influence of chain chemistry and structure (branching, stereoisomerism, etc.) on degree of crystallinity was discussed in Chapter 5.

The effects of both percent crystallinity and molecular weight on the physical state of polyethylene are represented in Figure 15.14.

Predeformation by Drawing

On a commercial basis, one of the most important techniques used to improve mechanical strength and tensile modulus is to permanently deform the polymer in tension. This procedure is sometimes termed *drawing* and it corresponds to the neck extension process illustrated schematically in Figure 15.4. In terms of property alterations, drawing is the polymer analogue of strain hardening in metals. It is an important stiffening and strengthening technique that is employed in the production of fibers and films. During drawing the molecular chains slip past one another and become highly oriented; for semicrystalline materials the chains assume conformations similar to that represented schematically in Figure 15.13*d*.

Figure 15.14 The influence of degree of crystallinity and molecular weight on the physical characteristics of polyethylene. (From R. B. Richards, "Polyethylene—Structure, Crystallinity and Properties," *J. Appl. Chem.*, **1**, 370, 1951.)

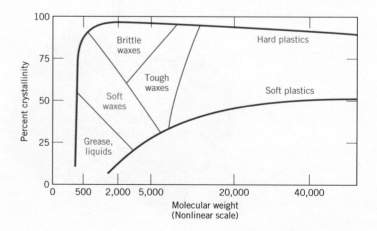

Degrees of strengthening and stiffening depend on the extent of deformation (or extension) of the material. Furthermore, the properties of drawn polymers are highly anisotropic. For materials drawn in uniaxial tension, tensile modulus and strength values are significantly greater in the direction of deformation than in other directions. Tensile modulus in the direction of drawing may be enhanced by up to approximately a factor of three relative to the undrawn material. At an angle of 45° from the tensile axis, the modulus is a minimum; at this orientation, the modulus has a value on the order of one-fifth that of the undrawn polymer.

Tensile strength parallel to the direction of orientation may be improved by a factor of at least two to five relative to that of the unoriented material. However, perpendicular to the alignment direction, tensile strength is reduced by on the order of one-third to one-half.

For an amorphous polymer that is drawn at an elevated temperature, the oriented molecular structure is retained only when the material is quickly cooled to the ambient temperature; this procedure gives rise to the strengthening and stiffening effects described in the previous paragraph. However, if, after stretching, the polymer is held at the temperature of drawing, molecular chains relax and assume random conformations characteristic of the predeformed state; as a consequence, drawing will have no effect on the mechanical characteristics of the material.

Heat-Treating

Heat-treating (or annealing) of semicrystalline polymers can lead to an increase in the percent crystallinity and crystallite size and perfection, as well as to modifications of the spherulite structure. For undrawn materials that are subjected to constant-time heat treatments, increasing the annealing temperature leads to the following: (1) an increase in tensile modulus, (2) an increase in yield strength, and (3) a reduction in ductility. Note that these annealing effects are opposite to those typically observed for metallic materials (Section 9.12)—weakening, softening, and enhanced ductility.

For some polymer fibers that have been drawn, the influence of annealing on the tensile modulus is contrary to that for undrawn materials—that is, the modulus decreases with increased annealing temperature because of a loss of chain orientation and strain-induced crystallinity.

Concept Check 15.3   For the following pair of polymers, do the following: (1) state whether it is possible to decide if one polymer has a higher tensile modulus than the other; (2) if this is possible, note which has the higher tensile modulus and then cite the reason(s) for your choice; and (3) if it is not possible to decide, state why not.

- Syndiotactic polystyrene having a number-average molecular weight of 400,000 g/mol
- Isotactic polystyrene having a number-average molecular weight of 650,000 g/mol

Concept Check 15.4   For the following pair of polymers, do the following: (1) state whether it is possible to decide if one polymer has a higher tensile strength than the other; (2) if this is possible, note which has the higher tensile strength and then cite the reason(s) for your choice; and (3) if it is not possible to decide, state why not.

- Syndiotactic polystyrene having a number-average molecular weight of 600,000 g/mol
- Isotactic polystyrene having a number-average molecular weight of 500,000 g/mol

[*The answers may be found at* www.wiley.com/college/callister *(Student Companion Site).*]

MATERIALS OF IMPORTANCE

Shrink-Wrap Polymer Films

An interesting application of heat treatment in polymers is the shrink-wrap used in packaging. Shrink-wrap is a polymer film, usually made of poly(vinyl chloride), polyethylene, or polyolefin (a multilayer sheet with alternating layers of polyethylene and polypropylene). It is initially plastically deformed (cold drawn) by about 20% to 300% to provide a prestretched (aligned) film. The film is wrapped around an object to be packaged and sealed at the edges. When heated to about 100°C to 150°C, this prestretched material shrinks to recover 80% to 90% of its initial deformation, which gives a tightly stretched, wrinkle-free, transparent polymer film. For example, CDs and many other consumer products are packaged in shrink-wrap.

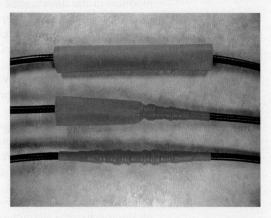

Top: An electrical connection positioned within a section of as-received polymer shrink-tubing. Center, Bottom: Application of heat to the tubing caused its diameter to shrink. In this constricted form, the polymer tubing stabilizes the connection and provides electrical insulation. (Photograph courtesy of Insulation Products Corporation.)

15.9 DEFORMATION OF ELASTOMERS

One of the fascinating properties of the elastomeric materials is their rubber-like elasticity—that is, they have the ability to be deformed to quite large deformations and then elastically spring back to their original form. This results from crosslinks in the polymer that provide a force to restore the chains to their undeformed conformations. Elastomeric behavior was probably first observed in natural rubber; however, the past several decades have brought about the synthesis of a large number of elastomers with a wide variety of properties. Typical stress–strain characteristics of elastomeric materials are displayed in Figure 15.1, curve *C*. Their moduli of elasticity are quite small, and, they vary with strain because the stress–strain curve is nonlinear.

In an unstressed state, an elastomer is amorphous and composed of crosslinked molecular chains that are highly twisted, kinked, and coiled. Elastic deformation upon application of a tensile load is simply the partial uncoiling, untwisting, and straightening and resultant elongation of the chains in the stress direction, a phenomenon represented in Figure 15.15. Upon release of the stress, the chains spring back to their prestressed conformations, and the macroscopic piece returns to its original shape.

Part of the driving force for elastic deformation is a thermodynamic parameter called *entropy,* which is a measure of the degree of disorder within a system; entropy increases with increasing disorder. As an elastomer is stretched and the chains straighten and become more aligned, the system becomes more ordered. From this state, the entropy increases if the chains return to their original kinked and coiled contours. Two intriguing phenomena result from this entropic effect. First, when stretched, an elastomer experiences a rise in temperature; second, the modulus of elasticity increases with increasing temperature, which is opposite to the behavior found in other materials (see Figure 8.8).

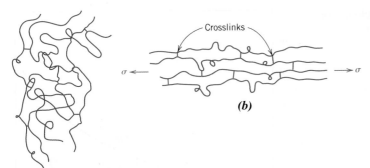

Figure 15.15 Schematic representation of crosslinked polymer chain molecules (*a*) in an unstressed state and (*b*) during elastic deformation in response to an applied tensile stress.
(Adapted from Z. D. Jastrzebski, *The Nature and Properties of Engineering Materials*, 3rd edition. Copyright © 1987 by John Wiley & Sons, New York. Reprinted by permission of John Wiley & Sons, Inc.)

Several criteria must be met for a polymer to be elastomeric: (1) It must not easily crystallize; elastomeric materials are amorphous, having molecular chains that are naturally coiled and kinked in the unstressed state. (2) Chain bond rotations must be relatively free for the coiled chains to readily respond to an applied force. (3) For elastomers to experience relatively large elastic deformations, the onset of plastic deformation must be delayed. Restricting the motions of chains past one another by crosslinking accomplishes this objective. The crosslinks act as anchor points between the chains and prevent chain slippage from occurring; the role of crosslinks in the deformation process is illustrated in Figure 15.15. Crosslinking in many elastomers is carried out in a process called vulcanization, to be discussed shortly. (4) Finally, the elastomer must be above its glass transition temperature (Section 15.13). The lowest temperature at which rubberlike behavior persists for many of the common elastomers is between −50°C and −90°C. Below its glass transition temperature, an elastomer becomes brittle, and its stress–strain behavior resembles curve *A* in Figure 15.1.

Vulcanization

vulcanization

The crosslinking process in elastomers is called **vulcanization**, which is achieved by a nonreversible chemical reaction, typically carried out at an elevated temperature. In most vulcanizing reactions, sulfur compounds are added to the heated elastomer; chains of sulfur atoms bond with adjacent polymer backbone chains and crosslink them, which is accomplished according to the following reaction:

$$
\begin{array}{c}
\text{H CH}_3\text{ H H} \\
\lvert\ \ \lvert\ \ \lvert\ \ \lvert \\
-\text{C}-\text{C}=\text{C}-\text{C}- \\
\lvert\qquad\quad\lvert \\
\text{H}\qquad\quad\text{H} \\[4pt]
\qquad\qquad\qquad + (m+n)\,\text{S} \longrightarrow \\[4pt]
\text{H}\qquad\quad\text{H} \\
\lvert\qquad\quad\lvert \\
-\text{C}-\text{C}=\text{C}-\text{C}- \\
\lvert\ \ \lvert\ \ \lvert\ \ \lvert \\
\text{H CH}_3\text{ H H}
\end{array}
\qquad
\begin{array}{c}
\text{H CH}_3\text{ H H} \\
\lvert\ \ \lvert\ \ \lvert\ \ \lvert \\
-\text{C}-\text{C}-\text{C}-\text{C}- \\
\lvert\ \ \lvert\ \ \lvert\ \ \lvert \\
\text{H}\ (\text{S})_m\ (\text{S})_n\ \text{H} \\[4pt]
\\[4pt]
\text{H}\ \lvert\ \ \lvert\ \text{H} \\
\lvert\ \ \lvert\ \ \lvert\ \ \lvert \\
-\text{C}-\text{C}-\text{C}-\text{C}- \\
\lvert\ \ \lvert\ \ \lvert\ \ \lvert \\
\text{H CH}_3\text{ H H}
\end{array}
\qquad (15.4)
$$

in which the two crosslinks shown consist of *m* and *n* sulfur atoms. Crosslink main-chain sites are carbon atoms that were doubly bonded before vulcanization but, after vulcanization, have become singly bonded.

Unvulcanized rubber, which contains very few crosslinks, is soft and tacky and has poor resistance to abrasion. Modulus of elasticity, tensile strength, and resistance to degradation by oxidation are all enhanced by vulcanization. The magnitude of the modulus of elasticity is directly proportional to the density of the crosslinks.

crystallization may be specified in the same manner as for the transformations discussed in Section 12.3, and according to Equation 12.18; that is, rate is equal to the reciprocal of time required for crystallization to proceed to 50% completion. This rate is dependent on crystallization temperature (Figure 15.17) and also on the molecular weight of the polymer; rate decreases with increasing molecular weight.

For polypropylene (as well as any polymer), the attainment of 100% crystallinity is not possible. Therefore, in Figure 15.17, the vertical axis is scaled as *normalized fraction crystallized*. A value of 1.0 for this parameter corresponds to the highest level of crystallization that is achieved during the tests, which, in reality, is less than complete crystallization.

15.11 MELTING

melting temperature

The melting of a polymer crystal corresponds to the transformation of a solid material, having an ordered structure of aligned molecular chains, into a viscous liquid in which the structure is highly random. This phenomenon occurs, upon heating, at the **melting temperature**, T_m. There are several features distinctive to the melting of polymers that are not normally observed with metals and ceramics; these are consequences of the polymer molecular structures and lamellar crystalline morphology. First, melting of polymers takes place over a range of temperatures; this phenomenon is discussed in more detail shortly. In addition, the melting behavior depends on the history of the specimen, in particular the temperature at which it crystallized. The thickness of chain-folded lamellae depends on crystallization temperature; the thicker the lamellae, the higher the melting temperature. Impurities in the polymer and imperfections in the crystals also decrease the melting temperature. Finally, the apparent melting behavior is a function of the rate of heating; increasing this rate results in an elevation of the melting temperature.

As Section 15.8 notes, polymeric materials are responsive to heat treatments that produce structural and property alterations. An increase in lamellar thickness may be induced by annealing just below the melting temperature. Annealing also raises the melting temperature by decreasing the vacancies and other imperfections in polymer crystals and increasing crystallite thickness.

15.12 THE GLASS TRANSITION

glass transition temperature

The glass transition occurs in amorphous (or glassy) and semicrystalline polymers and is due to a reduction in motion of large segments of molecular chains with decreasing temperature. Upon cooling, the glass transition corresponds to the gradual transformation from a liquid into a rubbery material and finally into a rigid solid. The temperature at which the polymer experiences the transition from rubbery into rigid states is termed the **glass transition temperature**, T_g. This sequence of events occurs in the reverse order when a rigid glass at a temperature below T_g is heated. In addition, abrupt changes in other physical properties accompany this glass transition: for example, stiffness (Figure 15.7), heat capacity, and coefficient of thermal expansion.

15.13 MELTING AND GLASS TRANSITION TEMPERATURES

Melting and glass transition temperatures are important parameters relative to in-service applications of polymers. They define, respectively, the upper and lower temperature limits for numerous applications, especially for semicrystalline polymers. The glass transition temperature may also define the upper use temperature for glassy amorphous materials. Furthermore, T_m and T_g also influence the fabrication and processing procedures for polymers and polymer–matrix composites. These issues are discussed in succeeding sections of this chapter.

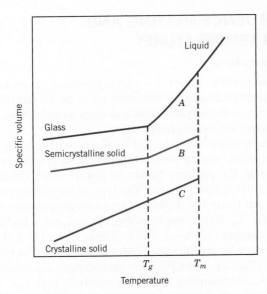

Figure 15.18 Specific volume versus temperature, upon cooling from the liquid melt, for totally amorphous (curve *A*), semicrystalline (curve *B*), and crystalline (curve *C*) polymers.

The temperatures at which melting and/or the glass transition occur for a polymer are determined in the same manner as for ceramic materials—from a plot of specific volume (the reciprocal of density) versus temperature. Figure 15.18 is such a plot, where curves *A* and *C*, for amorphous and crystalline polymers, respectively, have the same configurations as their ceramic counterparts (Figure 17.23).[5] For the crystalline material, there is a discontinuous change in specific volume at the melting temperature T_m. The curve for the totally amorphous material is continuous but experiences a slight decrease in slope at the glass transition temperature, T_g. The behavior is intermediate between these extremes for a semicrystalline polymer (curve *B*), in that both melting and glass transition phenomena are observed; T_m and T_g are properties of the respective crystalline and amorphous phases in this semicrystalline material. As discussed earlier, the behaviors represented in Figure 15.18 will depend on the rate of cooling or heating. Representative melting and glass transition temperatures of a number of polymers are given in Table 15.2 and Appendix E.

Table 15.2

Melting and Glass Transition Temperatures for Some of the More Common Polymeric Materials

Material	*Glass Transition Temperature* (°C)	*Melting Temperature* (°C)
Polyethylene (low density)	−110	115
Polytetrafluoroethylene	−97	327
Polyethylene (high density)	−90	137
Polypropylene	−18	175
Nylon 6,6	57	265
Poly(ethylene terephthalate) (PET)	69	265
Poly(vinyl chloride)	87	212
Polystyrene	100	240
Polycarbonate	150	265

[5]No engineering polymer is 100% crystalline; curve *C* is included in Figure 15.18 to illustrate the extreme behavior that would be displayed by a totally crystalline material.

15.14 FACTORS THAT INFLUENCE MELTING AND GLASS TRANSITION TEMPERATURES

Melting Temperature

During melting of a polymer there is a rearrangement of the molecules in the transformation from ordered to disordered molecular states. Molecular chemistry and structure influence the ability of the polymer chain molecules to make these rearrangements and, therefore, also affect the melting temperature.

Chain stiffness, which is controlled by the ease of rotation about the chemical bonds along the chain, has a pronounced effect. The presence of double bonds and aromatic groups in the polymer backbone lowers chain flexibility and causes an increase in T_m. Furthermore, the size and type of side groups influence chain rotational freedom and flexibility; bulky or large side groups tend to restrict molecular rotation and raise T_m. For example, polypropylene has a higher melting temperature than polyethylene (175°C vs. 115°C, Table 15.2); the CH_3 methyl side group for polypropylene is larger than the H atom found on polyethylene. The presence of polar groups (Cl, OH, and CN), even though not excessively large, leads to significant intermolecular bonding forces and relatively high T_ms. This may be verified by comparing the melting temperatures of polypropylene (175°C) and poly(vinyl chloride) (212°C).

The melting temperature of a polymer also depends on molecular weight. At relatively low molecular weights, increasing $\overline{M}$ (or chain length) raises T_m (Figure 15.19). Furthermore, the melting of a polymer takes place over a range of temperatures; thus, there is a range of T_ms rather than a single melting temperature. This is because every polymer is composed of molecules having a variety of molecular weights (Section 5.5) and because T_m depends on molecular weight. For most polymers, this melting temperature range is normally on the order of several degrees Celsius. The melting temperatures cited in Table 15.2 and Appendix E are near the high ends of these ranges.

Degree of branching also affects the melting temperature of a polymer. The introduction of side branches introduces defects into the crystalline material and lowers the melting temperature. High-density polyethylene, being a predominately linear polymer,

Figure 15.19 Dependence of polymer properties and melting and glass transition temperatures on molecular weight. (From F. W. Billmeyer, Jr., *Textbook of Polymer Science,* 3rd edition. Copyright © 1984 by John Wiley & Sons, New York. Reprinted by permission of John Wiley & Sons, Inc.)

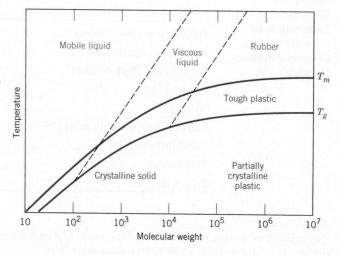

has a higher melting temperature (137°C, Table 15.2) than low-density polyethylene (115°C), which has some branching.

Glass Transition Temperature

Upon heating through the glass transition temperature, the amorphous solid polymer transforms from a rigid into a rubbery state. Correspondingly, the molecules that are virtually frozen in position below T_g begin to experience rotational and translational motions above T_g. Thus, the value of the glass transition temperature depends on molecular characteristics that affect chain stiffness; most of these factors and their influences are the same as for the melting temperature, as discussed earlier. Again, chain flexibility is decreased and T_g is increased by the presence of the following:

1. Bulky side groups; from Table 15.2, the respective T_g values for polypropylene and polystyrene are −18°C and 100°C.

2. Polar groups; for example, the T_g values for poly(vinyl chloride) and polypropylene are 87°C and −18°C, respectively.

3. Double bonds and aromatic groups in the backbone, which tend to stiffen the polymer chain.

Increasing the molecular weight also tends to raise the glass transition temperature, as noted in Figure 15.19. A small amount of branching tends to lower T_g; on the other hand, a high density of branches reduces chain mobility and elevates the glass transition temperature. Some amorphous polymers are crosslinked, which has been observed to elevate T_g; crosslinks restrict molecular motion. With a high density of crosslinks, molecular motion is virtually disallowed; long-range molecular motion is prevented, to the degree that these polymers do not experience a glass transition or its accompanying softening.

From the preceding discussion it is evident that essentially the same molecular characteristics raise and lower both melting and glass transition temperatures. Normally the value of T_g lies somewhere between $0.5T_m$ and $0.8T_m$ (in Kelvin). Consequently, for a homopolymer, it is not possible to independently vary both T_m and T_g. A greater degree of control over these two parameters is possible by the synthesis and use of copolymeric materials.

✔ *Concept Check 15.7* For each of the following two polymers, plot and label a schematic specific volume-versus-temperature curve (include both curves on the same graph):

• Spherulitic polypropylene of 25% crystallinity and having a weight-average molecular weight of 75,000 g/mol

• Spherulitic polystyrene of 25% crystallinity and having a weight-average molecular weight of 100,000 g/mol

[*The answer may be found at* www.wiley.com/college/callister (*Student Companion Site*).]

Concept Check 15.8 For the following two polymers, (1) state whether it is possible to determine whether one polymer has a higher melting temperature than the other; (2) if it is possible, note which has the higher melting temperature and then cite reason(s) for your choice; and (3) if it is not possible to decide, then state why not.

- Isotactic polystyrene that has a density of 1.12 g/cm^3 and a weight-average molecular weight of 150,000 g/mol

- Syndiotactic polystyrene that has a density of 1.10 g/cm^3 and a weight-average molecular weight of 125,000 g/mol

[*The answer may be found at* www.wiley.com/college/callister (*Student Companion Site*).]

Polymer Types

There are many different polymeric materials that are familiar and find a wide variety of applications; in fact, one way of classifying them is according to their end use. Within this scheme the various polymer types include plastics, elastomers (or rubbers), fibers, coatings, adhesives, foams, and films. Depending on its properties, a particular polymer may be used in two or more of these application categories. For example, a plastic, if crosslinked and used above its glass transition temperature, may make a satisfactory elastomer, or a fiber material may be used as a plastic if it is not drawn into filaments. This portion of the chapter includes a brief discussion of each of these types of polymer.

15.15 PLASTICS

plastic

Possibly the largest number of different polymeric materials come under the plastic classification. **Plastics** are materials that have some structural rigidity under load and are used in general-purpose applications. Polyethylene, polypropylene, poly(vinyl chloride), polystyrene, and the fluorocarbons, epoxies, phenolics, and polyesters may all be classified as plastics. They have a wide variety of combinations of properties. Some plastics are very rigid and brittle (Figure 15.1, curve *A*). Others are flexible, exhibiting both elastic and plastic deformations when stressed and sometimes experiencing considerable deformation before fracture (Figure 15.1, curve *B*).

Polymers falling within this classification may have any degree of crystallinity, and all molecular structures and configurations (linear, branched, isotactic, etc.) are possible. Plastic materials may be either thermoplastic or thermosetting; in fact, this is the manner in which they are usually subclassified. However, to be considered plastics, linear or branched polymers must be used below their glass transition temperatures (if amorphous) or below their melting temperatures (if semicrystalline), or they must be crosslinked enough to maintain their shape. The trade names, characteristics, and typical applications for a number of plastics are given in Table 15.3.

Several plastics exhibit especially outstanding properties. For applications in which optical transparency is critical, polystyrene and poly(methyl methacrylate) are especially well suited; however, it is imperative that the material be highly amorphous or, if semicrystalline, have very small crystallites. The fluorocarbons have a low coefficient of friction and are extremely resistant to attack by a host of chemicals, even at relatively high temperatures. They are used as coatings on nonstick cookware, in bearings and bushings, and for high-temperature electronic components.

Table 15.3 Trade Names, Characteristics, and Typical Applications for a Number of Plastic Materials

Material Type	Trade Names	Major Application Characteristics	Typical Applications
		Thermoplastics	
Acrylonitrile-butadiene-styrene (ABS)	Abson Cycolac Kralastic Lustran Lucon Novodur	Outstanding strength and toughness, resistant to heat distortion; good electrical properties; flammable and soluble in some organic solvents	Under-the-hood automotive applications, refrigerator linings, computer and television housings, toys, highway safety devices
Acrylics [poly(methyl methacrylate)]	Acrylite Diakon Lucite Paraloid Plexiglas	Outstanding light transmission and resistance to weathering; only fair mechanical properties	Lenses, transparent aircraft enclosures, drafting equipment, bathtub and shower enclosures
Fluorocarbons (PTFE or TFE)	Teflon Fluon Halar Hostaflon TF Neoflon	Chemically inert in almost all environments, excellent electrical properties; low coefficient of friction; may be used to 260°C; relatively weak and poor cold-flow properties	Anticorrosive seals, chemical pipes and valves, bearings, wire and cable insulation, antiadhesive coatings, high-temperature electronic parts
Polyamides (nylons)	Nylon Akulon Durethan Fostamid Nomex Ultramid Zytel	Good mechanical strength, abrasion resistance, and toughness; low coefficient of friction; absorbs water and some other liquids	Bearings, gears, cams, bushings, handles, and jacketing for wires and cables, fibers for carpet, hose, and belt reinforcement
Polycarbonates	Calibre Iupilon Lexan Makrolon Novarex	Dimensionally stable; low water absorption; transparent; very good impact resistance and ductility; chemical resistance not outstanding	Safety helmets, lenses, light globes, base for photographic film, automobile battery cases
Polyethylenes	Alathon Alkathene Fortiflex Hifax Petrothene Rigidex Zemid	Chemically resistant and electrically insulating; tough and relatively low coefficient of friction; low strength and poor resistance to weathering	Flexible bottles, toys, tumblers, battery parts, ice trays, film wrapping materials, automotive gas tanks
Polypropylenes	Hicor Meraklon Metocene Polypro Pro-fax Propak Propathene	Resistant to heat distortion; excellent electrical properties and fatigue strength; chemically inert; relatively inexpensive; poor resistance to ultraviolet light	Sterilizable bottles, packaging film, automotive kick panels, fibers, luggage
Polystyrenes	Avantra Dylene Innova Lutex Styron Vestyron	Excellent electrical properties and optical clarity; good thermal and dimensional stability; relatively inexpensive	Wall tile, battery cases, toys, indoor lighting panels, appliance housings, packaging

(continued)

VOCs react in the atmosphere to produce smog. Large users of coatings such as automobile manufacturers continue to reduce their VOC emissions to comply with environmental regulations.

Adhesives

adhesive

An **adhesive** is a substance used to bond together the surfaces of two solid materials (termed *adherends*). There are two types of bonding mechanisms: mechanical and chemical. In mechanical bonding there is actual penetration of the adhesive into surface pores and crevices. Chemical bonding involves intermolecular forces between the adhesive and adherend, which forces may be covalent and/or van der Waals; the degree of van der Waals bonding is enhanced when the adhesive material contains polar groups.

Although natural adhesives (animal glue, casein, starch, and rosin) are still used for many applications, a host of new adhesive materials based on synthetic polymers have been developed; these include polyurethanes, polysiloxanes (silicones), epoxies, polyimides, acrylics, and rubber materials. Adhesives may be used to join a large variety of materials—metals, ceramics, polymers, composites, skin, and so on—and the choice of which adhesive to use will depend on such factors as (1) the materials to be bonded and their porosities; (2) the required adhesive properties (i.e., whether the bond is to be temporary or permanent); (3) maximum/minimum exposure temperatures; and (4) processing conditions.

For all but the pressure-sensitive adhesives (discussed shortly), the adhesive material is applied as a low-viscosity liquid, so as to cover the adherend surfaces evenly and completely and allow for maximum bonding interactions. The actual bonding joint forms as the adhesive undergoes a liquid-to-solid transition (or cures), which may be accomplished through either a physical process (e.g., crystallization, solvent evaporation) or a chemical process [e.g., polymerization (Section 15.20), vulcanization]. Characteristics of a sound joint should include high shear, peel, and fracture strengths.

Adhesive bonding offers some advantages over other joining technologies (e.g., riveting, bolting, and welding), including lighter weight, the ability to join dissimilar materials and thin components, better fatigue resistance, and lower manufacturing costs. Furthermore, it is the technology of choice when exact positioning of components and processing speed are essential. The chief drawback of adhesive joints is service temperature limitation; polymers maintain their mechanical integrity only at relatively low temperatures, and strength decreases rapidly with increasing temperature. The maximum temperature possible for continuous use for some of the newly developed polymers is 300°C. Adhesive joints are found in a large number of applications, especially in the aerospace, automotive, and construction industries, in packaging, and in some household goods.

A special class of this group of materials is the pressure-sensitive adhesives (or self-adhesive materials), such as those found on self-stick tapes, labels, and postage stamps. These materials are designed to adhere to just about any surface by making contact and with the application of slight pressure. Unlike the adhesives described previously, bonding action does not result from a physical transformation or a chemical reaction. Rather, these materials contain polymer tackifying resins; during detachment of the two bonding surfaces, small fibrils form that are attached to the surfaces and tend to hold them together. Polymers used for pressure-sensitive adhesives include acrylics, styrenic block copolymers (Section 15.19), and natural rubber.

Films

Polymeric materials have found widespread use in the form of thin *films*. Films having thicknesses between 0.025 and 0.125 mm are fabricated and used extensively as bags for packaging food products and other merchandise, as textile products, and in a host of other uses. Important characteristics of the materials produced and used as films include low density, a high degree of flexibility, high tensile and tear strengths, resistance to attack by moisture and other chemicals, and low permeability to some gases, especially water vapor (Section 7.8). Some of the polymers that meet these criteria and are manufactured in film form are polyethylene, polypropylene, cellophane, and cellulose acetate.

Foams

foam

Foams are plastic materials that contain a relatively high volume percentage of small pores and trapped gas bubbles. Both thermoplastic and thermosetting materials are used as foams; these include polyurethane, rubber, polystyrene, and poly(vinyl chloride). Foams are commonly used as cushions in automobiles and furniture, as well as in packaging and thermal insulation. The foaming process is often carried out by incorporating into the batch of material a blowing agent that, upon heating, decomposes with the liberation of a gas. Gas bubbles are generated throughout the now-fluid mass, which remain in the solid upon cooling and give rise to a spongelike structure. The same effect is produced by dissolving an inert gas into a molten polymer under high pressure. When the pressure is rapidly reduced, the gas comes out of solution and forms bubbles and pores that remain in the solid as it cools.

15.19 ADVANCED POLYMERIC MATERIALS

A number of new polymers having unique and desirable combinations of properties have been developed over the past several years; many have found niches in new technologies and/or have satisfactorily replaced other materials. Some of these include ultra-high-molecular-weight polyethylene, liquid crystal polymers, and thermoplastic elastomers. Each of these will now be discussed.

Ultra-High-Molecular-Weight Polyethylene

ultra-high-molecular-
weight polyethylene

Ultra-high-molecular-weight polyethylene (*UHMWPE*) is a linear polyethylene that has an extremely high molecular weight. Its typical $\overline{M}_w$ is approximately 4×10^6 g/mol, which is an order of magnitude greater than that of high-density polyethylene. In fiber form, UHMWPE is highly aligned and has the trade name Spectra. Some of the extraordinary characteristics of this material are as follows:

1. An extremely high impact resistance
2. Outstanding resistance to wear and abrasion
3. A very low coefficient of friction
4. A self-lubricating and nonstick surface
5. Very good chemical resistance to normally encountered solvents
6. Excellent low-temperature properties
7. Outstanding sound damping and energy absorption characteristics
8. Electrically insulating and excellent dielectric properties

However, because this material has a relatively low melting temperature, its mechanical properties deteriorate rapidly with increasing temperature.

This unusual combination of properties leads to numerous and diverse applications for this material, including bulletproof vests, composite military helmets, fishing line, ski-bottom surfaces, golf-ball cores, bowling alley and ice-skating rink surfaces, biomedical prostheses, blood filters, marking-pen nibs, bulk material handling equipment (for coal, grain, cement, gravel, etc.), bushings, pump impellers, and valve gaskets.

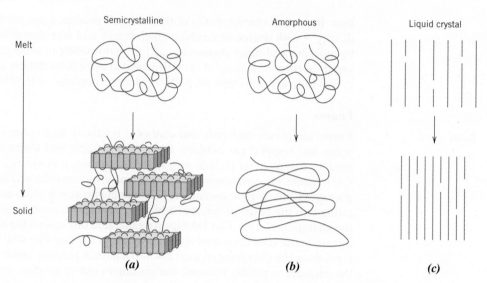

Figure 15.20 Schematic representations of the molecular structures in both melt and solid states for (*a*) semicrystalline, (*b*) amorphous, and (*c*) liquid crystal polymers.
(Adapted from G. W. Calundann and M. Jaffe, "Anisotropic Polymers, Their Synthesis and Properties," Chapter VII in *Proceedings of the Robert A. Welch Foundation Conferences on Polymer Research*, 26th Conference, Synthetic Polymers, Nov. 1982.)

Liquid Crystal Polymers

liquid crystal
polymer

Liquid crystal polymers (*LCPs*) are a group of chemically complex and structurally distinct materials that have unique properties and are used in diverse applications. Discussion of the chemistry of these materials is beyond the scope of this book. LCPs are composed of extended, rod-shaped, and rigid molecules. In terms of molecular arrangement, these materials do not fall within any of conventional liquid, amorphous, crystalline, or semicrystalline classifications but may be considered a new state of matter—the liquid crystalline state, being neither crystalline nor liquid. In the melt (or liquid) condition, whereas other polymer molecules are randomly oriented, LCP molecules can become aligned in highly ordered configurations. As solids, this molecular alignment remains, and, in addition, the molecules form in domain structures having characteristic intermolecular spacings. A schematic comparison of liquid crystals, amorphous polymers, and semicrystalline polymers in both melt and solid states is illustrated in Figure 15.20. There are three types of liquid crystals, based on orientation and positional ordering—smectic, nematic, and cholesteric; distinctions among these types are also beyond the scope of this discussion.

The principal use of liquid crystal polymers is in *liquid crystal displays* (*LCDs*) on digital watches, flat-panel computer monitors and televisions, and other digital displays. Here, cholesteric types of LCPs are used, which, at room temperature, are fluid liquids, transparent, and optically anisotropic. The displays are composed of two sheets of glass between which is sandwiched the liquid crystal material. The outer face of each glass sheet is coated with a transparent and electrically conductive film; in addition, the character-forming number/letter elements are etched into this film on the side that is to be viewed. A voltage applied through the conductive films (and thus between these two glass sheets) over one of these character-forming regions causes a disruption of the orientation of the LCP molecules in this region, a darkening of this LCP material, and, in turn, the formation of a visible character.

Some of the nematic type of liquid crystal polymers are rigid solids at room temperature and, on the basis of an outstanding combination of properties and processing characteristics, have found widespread use in a variety of commercial applications. For example, these materials exhibit the following behaviors:

1. Excellent thermal stability; they may be used to temperatures as high as 230°C.
2. Stiffness and strength; their tensile moduli range between 10 and 24 GPa, and their tensile strengths are from 125 to 255 MPa.
3. High impact strengths, which are retained upon cooling to relatively low temperatures.
4. Chemical inertness to a wide variety of acids, solvents, bleaches, and so on.
5. Inherent flame resistance and combustion products that are relatively nontoxic.

The thermal stability and chemical inertness of these materials are explained by extremely high intermolecular forces.

The following may be said about their processing and fabrication characteristics:

1. All conventional processing techniques available for thermoplastic materials may be used.
2. Extremely low shrinkage and warpage take place during molding.
3. There is exceptional dimensional repeatability from part to part.
4. Melt viscosity is low, which permits molding of thin sections and/or complex shapes.
5. Heats of fusion are low; this results in rapid melting and subsequent cooling, which shortens molding cycle times.
6. They have anisotropic finished-part properties; molecular orientation effects are produced from melt flow during molding.

These materials are used extensively by the electronics industry (in interconnect devices, relay and capacitor housings, brackets, etc.), by the medical equipment industry (in components that are sterilized repeatedly), and in photocopiers and fiber-optic components.

Thermoplastic Elastomers

thermoplastic elastomer

Thermoplastic elastomers (*TPEs* or *TEs*) are a type of polymeric material that, at ambient conditions, exhibits *elastomeric* (or rubbery) behavior yet is thermoplastic (Section 5.9). By way of contrast, most elastomers heretofore discussed are thermosets because they are crosslinked during vulcanization. Of the several varieties of TPEs, one of the best known and widely used is a block copolymer consisting of block segments of a hard and rigid thermoplastic (commonly styrene [S]) that alternate with block segments of a soft and flexible elastic material (often butadiene [B] or isoprene [I]). For a common TPE, hard, polymerized segments are located at chain ends, whereas each soft, central region consists of polymerized butadiene or isoprene units. These TPEs are frequently termed *styrenic block copolymers;* chain chemistries for the two (S-B-S and S-I-S) types are shown in Figure 15.21.

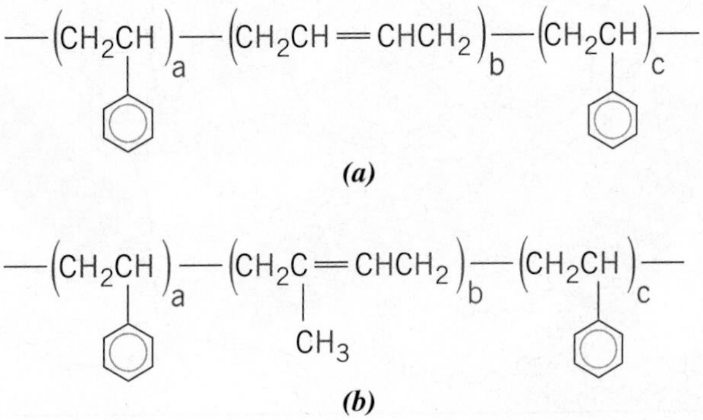

Figure 15.21
Representations of the chain chemistries for (*a*) styrene-butadiene-styrene (S-B-S), and (*b*) styrene-isoprene-styrene (S-I-S) thermoplastic elastomers.

At ambient temperatures, the soft, amorphous, central (butadiene or isoprene) segments impart rubbery, elastomeric behavior to the material. Furthermore, for temperatures below the T_m of the hard (styrene) component, hard chain-end segments from numerous adjacent chains aggregate together to form rigid crystalline domain regions. These domains are *physical crosslinks* that act as anchor points so as to restrict soft-chain segment motions; they function in much the same way as *chemical crosslinks* for the thermoset elastomers. A schematic illustration for the structure of this TPE type is presented in Figure 15.22.

The tensile modulus of this TPE material is subject to alteration; increasing the number of soft-component blocks per chain leads to a decrease in modulus and, therefore, a decrease of stiffness. Furthermore, the useful temperature range lies between T_g of the soft, flexible component and T_m of the hard, rigid one. For the styrenic block copolymers, this range is between about −70°C and 100°C.

In addition to the styrenic block copolymers, there are other types of TPEs, including thermoplastic olefins, copolyesters, thermoplastic polyurethanes, and elastomeric polyamides.

The chief advantage of the TPEs over the thermoset elastomers is that upon heating above T_m of the hard phase, they melt (i.e., the physical crosslinks disappear), and, therefore, they may be processed by conventional thermoplastic forming techniques [blow molding, injection molding, etc. (Section 15.22)]; thermoset polymers do not experience melting, and, consequently, forming is normally more difficult. Furthermore, because the melting–solidification process is reversible and repeatable for thermoplastic elastomers, TPE parts may be reformed into other shapes. In other words, they are recyclable; thermoset elastomers are, to a large degree, nonrecyclable. Scrap that is generated during forming procedures may also be recycled, which results in lower production costs than with thermosets. In addition, tighter controls may be maintained on part dimensions for TPEs, and TPEs have lower densities.

In quite a variety of applications, thermoplastic elastomers have replaced conventional thermoset elastomers. Typical uses for TPEs include automotive exterior trim (bumpers, fascia, etc.), automotive underhood components (electrical insulation and connectors, and gaskets), shoe soles and heels, sporting goods (e.g., bladders for footballs and soccer balls), medical barrier films and protective coatings, and components in sealants, caulking, and adhesives.

Figure 15.22
Schematic representation of the molecular structure for a thermoplastic elastomer. This structure consists of "soft" (i.e., butadiene or isoprene) repeat unit center-chain segments and "hard" (i.e., styrene) domains (chain ends), which act as physical crosslinks at room temperature.

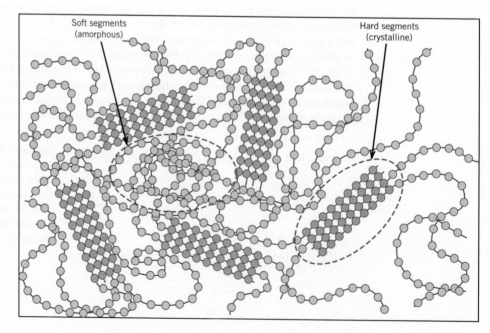

SUMMARY

Stress–Strain Behavior

- On the basis of stress–strain behavior, polymers fall within three general classifications (Figure 15.1): brittle (curve *A*), plastic (curve *B*), and highly elastic (curve *C*).
- Polymers are neither as strong nor as stiff as metals. However, their high flexibilities, low densities, and resistance to corrosion make them the materials of choice for many applications.
- The mechanical properties of polymers are sensitive to changes in temperature and strain rate. With either rising temperature or decreasing strain rate, modulus of elasticity diminishes, tensile strength decreases, and ductility increases.

Viscoelastic Deformation

- Viscoelastic mechanical behavior, intermediate between totally elastic and totally viscous, is displayed by a number of polymeric materials.
- This behavior is characterized by the relaxation modulus, a time-dependent modulus of elasticity.
- The magnitude of the relaxation modulus is very sensitive to temperature. Glassy, leathery, rubbery, and viscous flow regions may be identified on a plot of logarithm of relaxation modulus versus temperature (Figure 15.7).
- The logarithm of relaxation modulus versus temperature behavior depends on molecular configuration—degree of crystallinity, presence of crosslinking, and so on (Figure 15.8).

Fracture of Polymers

- Fracture strengths of polymeric materials are low relative to those of metals and ceramics.
- Both brittle and ductile fracture modes are possible.
- Some thermoplastic materials experience a ductile-to-brittle transition with a lowering of temperature, an increase in strain rate, and/or an alteration of specimen thickness or geometry.
- In some thermoplastics, the crack-formation process may be preceded by crazing; *crazes* are regions of localized deformation and microvoids (Figure 15.9).
- Crazing can lead to an increase in ductility and toughness of the material.

Deformation of Semicrystalline Polymers

- During the elastic deformation of a semicrystalline polymer having a spherulitic structure that is stressed in tension, the molecules in amorphous regions elongate in the stress direction (Figure 15.12).
- The tensile plastic deformation of spherulitic polymers occurs in several stages as both amorphous tie chains and chain-folded block segments (which separate from the ribbon-like lamellae) become oriented with the tensile axis (Figure 15.13).
- Also, during deformation the shapes of spherulites are altered (for moderate deformations); relatively large degrees of deformation lead to a complete destruction of the spherulites and formation of highly aligned structures.

Factors That Influence the Mechanical Properties of Semicrystalline Polymers

- The mechanical behavior of a polymer is influenced by both in-service and structural/processing factors.
- Increasing the temperature and/or diminishing the strain rate leads to reductions in tensile modulus and tensile strength and an enhancement of ductility.
- Other factors affect the mechanical properties:
  Molecular weight—Tensile modulus is relatively insensitive to molecular weight. However, tensile strength increases with increasing $\overline{M}_n$ (Equation 15.3).
  Degree of crystallinity—Both tensile modulus and strength increase with increasing percent crystallinity.

Predeformation by drawing—Stiffness and strength are enhanced by permanently deforming the polymer in tension.

Heat-treating—Heat-treating undrawn and semicrystalline polymers leads to increases in stiffness and strength and a decrease in ductility.

Deformation of Elastomers

- Large elastic extensions are possible for elastomeric materials that are amorphous and lightly crosslinked.
- Deformation corresponds to the unkinking and uncoiling of chains in response to an applied tensile stress.
- Crosslinking is often achieved during a vulcanization process; increased crosslinking enhances the modulus of elasticity and the tensile strength of the elastomer.
- Many elastomers are copolymers, whereas silicone elastomers are really inorganic materials.

Crystallization

- During the crystallization of a polymer, randomly oriented molecules in the liquid phase transform into chain-folded crystallites that have ordered and aligned molecular structures.

Melting

- The melting of crystalline regions of a polymer corresponds to the transformation of a solid material having an ordered structure of aligned molecular chains into a viscous liquid in which the structure is highly random.

The Glass Transition

- The glass transition occurs in amorphous regions of polymers.
- Upon cooling, this phenomenon corresponds to the gradual transformation from a liquid into a rubbery material, and finally into a rigid solid. With decreasing temperature there is a reduction in the motion of large segments of molecular chains.

Melting and Glass Transition Temperatures

- Melting and glass transition temperatures may be determined from plots of specific volume versus temperature (Figure 15.18).
- These parameters are important relative to the temperature range over which a particular polymer may be used and processed.

Factors That Influence Melting and Glass Transition Temperatures

- The magnitudes of T_m and T_g increase with increasing chain stiffness; stiffness is enhanced by the presence of chain double bonds and side groups that are either bulky or polar.
- At low molecular weights T_m and T_g increase with increasing $\overline{M}$.

Polymer Types

- One way of classifying polymeric materials is according to their end use. According to this scheme, the several types include plastics, fibers, coatings, adhesives, films, foams, and advanced materials.
- Plastic materials are perhaps the most widely used group of polymers and include the following: polyethylene, polypropylene, poly(vinyl chloride), polystyrene, and the fluorocarbons, epoxies, phenolics, and polyesters.
- Many polymeric materials may be spun into fibers, which are used primarily in textiles. Mechanical, thermal, and chemical characteristics of these materials are especially critical.
- Three advanced polymeric materials were discussed: ultra-high-molecular-weight polyethylene, liquid crystal polymers, and thermoplastic elastomers. These materials have unusual properties and are used in a host of high-technology applications.

Important Terms and Concepts

adhesive
drawing
elastomer
fiber
foam

glass transition temperature
liquid crystal polymer
melting temperature
plastic
relaxation modulus

thermoplastic elastomer
ultra-high-molecular-
 weight polyethylene
viscoelasticity
vulcanization

REFERENCES

Billmeyer, F. W., Jr., *Textbook of Polymer Science,* 3rd edition, Wiley-Interscience, New York, 1984.

Brazel, C. S., and S. L. Rosen, *Fundamental Principles of Polymeric Materials*, 3rd edition, Wiley, Hoboken, NJ, 2012.

Carraher, C. E., Jr., *Seymour/Carraher's Polymer Chemistry,* 8th edition, CRC Press, Boca Raton, FL, 2010.

Engineered Materials Handbook, Vol. 2, *Engineering Plastics,* ASM International, Materials Park, OH, 1988.

Harper, C. A. (Editor), *Handbook of Plastics, Elastomers and Composites,* 4th edition, McGraw-Hill, New York, 2002.

Lakes, R., *Viscoelastic Materials*, Cambridge University Press, New York, 2009.

Landel, R. F. (Editor), *Mechanical Properties of Polymers and Composites,* 2nd edition, Marcel Dekker, New York, 1994.

McCrum, N. G., C. P. Buckley, and C. B. Bucknall, *Principles of Polymer Engineering,* 2nd edition, Oxford University Press, Oxford, 1997. Chapters 7–8.

Muccio, E. A., *Plastic Part Technology,* ASM International, Materials Park, OH, 1991.

Ward, I. M., and J. Sweeney, *Mechanical Properties of Solid Polymers*, 3rd edition, Wiley, Chichester, UK, 2013.

QUESTIONS AND PROBLEMS

Stress–Strain Behavior

15.1 From the stress–strain data for poly(methyl methacrylate) shown in Figure 15.3, determine the modulus of elasticity and tensile strength at room temperature [20°C (293 K)], and compare these values with those given in Table 15.1.

15.2 Compute the elastic moduli for the following polymers, whose stress–strain behaviors may be observed in the "Tensile Tests" module of *Virtual Materials Science and Engineering* (*VMSE*): **(a)** high-density polyethylene, **(b)** nylon, and **(c)** phenol-formaldehyde (Bakelite). How do these values compare with those presented in Table 15.1 for the same polymers?

VMSE
Polymers

15.3 For the nylon polymer, whose stress-strain behavior may be observed in the "Tensile Tests" module of *Virtual Materials Science and Engineering* (*VMSE*), determine the following:

VMSE
Polymers

(a) the yield strength, and

(b) the approximate ductility, in percent elongation.

How do these values compare with those for the nylon material presented in Table 15.1?

15.4 For the phenol-formaldehyde (Bakelite) polymer, whose stress–strain behavior may be observed in the "Tensile Tests" module of *Virtual Material Science and Engineering (VMSE)*, determine the following:

VMSE
Polymers

(a) the tensile strength, and

(b) the approximate ductility, in percent elongation.

How do these values compare with those for the phenol-formaldehyde material presented in Table 15.1?

Viscoelastic Deformation

15.5 In your own words, briefly describe the phenomenon of viscoelasticity.

15.6 For some viscoelastic polymers that are subjected to stress relaxation tests, the stress decays with time according to

$$\sigma(t) = \sigma(0) \exp\left(-\frac{t}{\tau}\right) \qquad (15.5)$$

where $\sigma(t)$ and $\sigma(0)$ represent the time-dependent and initial (i.e., time = 0) stresses, respectively, and t and τ denote elapsed time and the relaxation time; τ is a time-independent constant characteristic of the material. A specimen of a viscoelastic polymer whose stress relaxation obeys Equation 15.5 was suddenly pulled in tension to a measured strain of 0.6; the stress necessary to maintain this constant strain was measured as a function of time. Determine $E_r(10)$ for this material if the initial stress level was 2.8 MPa, which dropped to 1.75 MPa after 60 s.

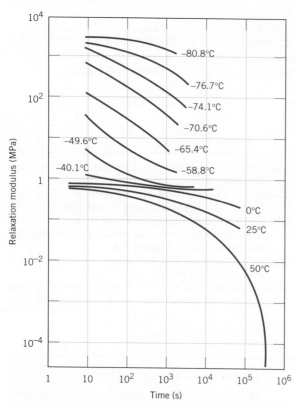

Figure 15.23 Logarithm of relaxation modulus versus logarithm of time for polyisobutylene between −80 and 50°C. (Adapted from E. Catsiff and A. V. Tobolsky, "Stress-Relaxation of Polyisobutylene in the Transition Region [1,2]," *J. Colloid Sci.,* **10,** 377 [1955]. Reprinted by permission of Academic Press, Inc.)

15.7 In Figure 15.23, the logarithm of $E_r(t)$ versus the logarithm of time is plotted for polyisobutylene at a variety of temperatures. Make a plot of $E_r(10)$ versus temperature and then estimate its T_g.

15.8 On the basis of the curves in Figure 15.5, sketch schematic strain–time plots for the following polystyrene materials at the specified temperatures:

(a) Amorphous at 125°C

(b) Crosslinked at 160°C

(c) Crystalline at 230°C

(d) Crosslinked at 50°C

15.9 (a) Contrast the manner in which stress relaxation and viscoelastic creep tests are conducted.

(b) For each of these tests, cite the experimental parameter of interest and how it is determined.

15.10 Make two schematic plots of the logarithm of relaxation modulus versus temperature for an amorphous polymer (curve *C* in Figure 15.8).

(a) On one of these plots demonstrate how the behavior changes with increasing molecular weight.

(b) On the other plot, indicate the change in behavior with increasing crosslinking.

Fracture of Polymers
Miscellaneous Mechanical Characteristics

15.11 For thermoplastic polymers, cite five factors that favor brittle fracture.

15.12 (a) Compare the fatigue limits for polystyrene (Figure 15.11) and the cast iron for which fatigue data are given in Problem 10.20.

(b) Compare the fatigue strengths at 10^6 cycles for poly(ethylene terephthalate) (PET, Figure 15.11) and red brass (Figure 10.20).

Deformation of Semicrystalline Polymers

15.13 In your own words, describe the mechanisms by which semicrystalline polymers **(a)** elastically deform and **(b)** plastically deform, and **(c)** by which elastomers elastically deform.

Factors That Influence the Mechanical Properties of Semicrystalline Polymers

15.14 Briefly explain how each of the following influences the tensile modulus of a semicrystalline polymer and why:

(a) Molecular weight

(b) Degree of crystallinity

(c) Deformation by drawing

(d) Annealing of an undeformed material

(e) Annealing of a drawn material

15.15 Briefly explain how each of the following influences the tensile or yield strength of a semicrystalline polymer and why:

(a) Molecular weight

(b) Degree of crystallinity

(c) Deformation by drawing

(d) Annealing of an undeformed material

15.16 Normal butane and isobutane have boiling temperatures of −0.5 and −12.3°C, respectively. Briefly explain this behavior on the basis of their molecular structures, as presented in Section 5.2.

15.17 The tensile strength and number-average molecular weight for two poly(methyl methacrylate) materials are as follows:

Tensile Strength (MPa)	Number-Average Molecular Weight (g/mol)
108	40,000
175	60,000

Estimate the tensile strength at a number-average molecular weight of 30,000 g/mol.

15.18 The tensile strength and number-average molecular weight for two polyethylene materials are as follows:

Tensile Strength (MPa)	Number-Average Molecular Weight (g/mol)
90	12,700
160	28,500

Estimate the number-average molecular weight that is required to give a tensile strength of 195 MPa.

15.19 For each of the following pairs of polymers, do the following: (1) state whether it is possible to decide whether one polymer has a higher tensile modulus than the other; (2) if this is possible, note which has the higher tensile modulus and then cite the reason(s) for your choice; and (3) if it is not possible to decide, then state why.

(a) Random acrylonitrile-butadiene copolymer with 10% of possible sites crosslinked; alternating acrylonitrile-butadiene copolymer with 5% of possible sites crosslinked

(b) Branched and syndiotactic polypropylene with a degree of polymerization of 5000; linear and isotactic polypropylene with a degree of polymerization of 3000

(c) Branched polyethylene with a number-average molecular weight of 250,000 g/mol; linear and isotactic poly(vinyl chloride) with a number-average molecular weight of 200,000 g/mol

15.20 For each of the following pairs of polymers, do the following: (1) state whether it is possible to decide whether one polymer has a higher tensile strength than the other; (2) if this is possible, note which has the higher tensile strength and then cite the reason(s) for your choice; and (3) if it is not possible to decide, then state why.

(a) Syndiotactic polystyrene having a number-average molecular weight of 600,000 g/mol; atactic polystyrene having a number-average molecular weight of 500,000 g/mol

(b) Random acrylonitrile-butadiene copolymer with 10% of possible sites crosslinked; block acrylonitrile-butadiene copolymer with 5% of possible sites crosslinked

(d) Network polyester; lightly branched polypropylene

15.21 Would you expect the tensile strength of polychlorotrifluoroethylene to be greater than, the same as, or less than that of a polytetrafluoroethylene specimen having the same molecular weight and degree of crystallinity? Why?

15.22 For each of the following pairs of polymers, plot and label schematic stress–strain curves on the same graph [i.e., make separate plots for parts (a), (b), and (c)].

(a) Isotactic and linear polypropylene having a weight-average molecular weight of 120,000 g/mol; atactic and linear polypropylene having a weight-average molecular weight of 100,000 g/mol

(b) Branched poly(vinyl chloride) having a degree of polymerization of 3000; heavily crosslinked poly(vinyl chloride) having a degree of polymerization of 3000

(c) Poly(styrene-butadiene) random copolymer having a number-average molecular weight of 100,000 g/mol and 10% of the available sites crosslinked and tested at 20°C; poly(styrene-butadiene) random copolymer having a number-average molecular weight of 120,000 g/mol and 15% of the available sites crosslinked and tested at −85°C *Hint:* Poly(styrene-butadiene) copolymers may exhibit elastomeric behavior.

15.23 List the two molecular characteristics that are essential for elastomers.

15.24 Which of the following would you expect to be elastomers and which thermosetting polymers at room temperature? Justify each choice.

(a) Epoxy having a network structure

(b) Lightly crosslinked poly(styrene-butadiene) random copolymer that has a glass transition temperature of −50°C

(c) Lightly branched and semicrystalline polytetrafluoroethylene that has a glass transition temperature of −100°C

(d) Heavily crosslinked poly(ethylene-propylene) random copolymer that has a glass transition temperature of 0°C

(e) Thermoplastic elastomer that has a glass transition temperature of 75°C

15.25 Ten kilograms of polybutadiene is vulcanized with 5.0 kg sulfur. What fraction of the possible crosslink sites is bonded to sulfur crosslinks, assuming that, on the average, 4.5 sulfur atoms participate in each crosslink?

15.26 Compute the weight percent sulfur that must be added to completely crosslink an alternating chloroprene-acrylonitrile copolymer, assuming that five sulfur atoms participate in each crosslink.

15.27 The vulcanization of polyisoprene is accomplished with sulfur atoms according to Equation 15.4. If 57 wt% sulfur is combined with polyisoprene, how many crosslinks will be associated with each isoprene repeat unit if it is assumed that, on the average, six sulfur atoms participate in each crosslink?

15.28 For the vulcanization of polyisoprene, compute the weight percent of sulfur that must be added to ensure that 10% of possible sites will be crosslinked; assume that, on the average, three sulfur atoms are associated with each crosslink.

15.29 Demonstrate, in a manner similar to Equation 15.4, how vulcanization may occur in a butadiene rubber.

Crystallization

15.30 Determine values for the constants n and k (Equation 12.17) for the crystallization of polypropylene (Figure 15.17) at 180°C.

Melting and Glass Transition Temperatures

15.31 Name the following polymer(s) that would be suitable for the fabrication of cups to contain hot coffee: polyethylene, polypropylene, poly(vinyl chloride), PET polyester, and polycarbonate. Why?

15.32 Of the polymers listed in Table 15.2, which polymer(s) would be best suited for use as ice cube trays? Why?

Factors That Influence Melting and
Glass Transition Temperatures

15.33 For each of the following pairs of polymers, plot and label schematic specific volume-versus-temperature curves on the same graph [i.e., make separate plots for parts (a), (b), and (c)].

(a) Spherulitic polypropylene, of 25% crystallinity, and having a weight-average molecular weight of 75,000 g/mol; spherulitic polystyrene, of 25% crystallinity, and having a weight-average molecular weight of 100,000 g/mol

(b) Graft poly(styrene-butadiene) copolymer with 10% of available sites crosslinked; random poly(styrene-butadiene) copolymer with 15% of available sites crosslinked

(c) Polyethylene having a density of 0.900 g/cm^3 and a degree of polymerization of 2500; polyethylene having a density of 0.920 g/cm^3 and a degree of polymerization of 3000

15.34 For each of the following pairs of polymers, (1) state whether it is possible to determine whether one polymer has a higher melting temperature than the other; (2) if it is possible, note which has the higher melting temperature and then cite reason(s) for your choice; and (3) if it is not possible to decide, then state why.

(a) Isotactic polystyrene that has a density of 1.15 g/cm^3 and a weight-average molecular weight of 150,000 g/mol; syndiotactic polystyrene that has a density of 1.15 g/cm^3 and a weight-average molecular weight of 125,000 g/mol

(b) Linear polyethylene that has a degree of polymerization of 6000; linear and isotactic polypropylene that has a degree of polymerization of 7000

(c) Branched and isotactic polystyrene that has a degree of polymerization of 5000; linear and isotactic polypropylene that has a degree of polymerization of 8000

15.35 Make a schematic plot showing how the modulus of elasticity of an amorphous polymer depends on the glass transition temperature. Assume that molecular weight is held constant.

Elastomers
Fibers
Miscellaneous Applications

15.36 Briefly explain the difference in molecular chemistry between silicone polymers and other polymeric materials.

15.37 List two important characteristics for polymers that are to be used in fiber applications.

15.38 Cite five important characteristics for polymers that are to be used in thin-film applications.

DESIGN QUESTIONS

15.D1 (a) List several advantages and disadvantages of using transparent polymeric materials for eyeglass lenses.

(b) Cite four properties (in addition to being transparent) that are important for this application.

(c) Note three polymers that may be candidates for eyeglass lenses, and then tabulate values of the properties noted in part (b) for these three materials.

15.D2 Write an essay on polymeric materials that are used in the packaging of food products and drinks. Include a list of the general requisite characteristics of materials that are used for these applications. Now cite a specific material that is used for each of three different container types and the rationale for each choice.

15.D3 Write an essay on the replacement of metallic automobile components by polymers and composite materials. Address the following issues: (1) Which automotive components (e.g., crankshaft) now use polymers and/or composites? (2) Specifically what materials (e.g., high-density polyethylene) are now being used? (3) What are the reasons for these replacements?

Courtesy of Black Diamond Equipment, Ltd.

Top sheet. Polyamide polymer that has a relatively low glass transition temperature and resists chipping.

Torsion box wrap. Fiber-reinforced composites that use glass, aramid, or carbon fibers. A variety of weaves and weights of reinforcement are possible that are utilized to "tune" the flexural characteristics of the ski.

Core. Foam, vertical laminates of wood, wood-foam laminates, honeycomb, and other materials. Commonly used woods include poplar, spruce, bamboo, balsa, and birch.

Vibration–absorbing material. Rubber is normally used.

Reinforcement layers. Fiber-reinforced composites that normally use glass fibers. A variety of weaves and weights of reinforcement are possible to provide longitudinal stiffness.

Base. Ultra-high-molecular-weight polyethylene is used because of its low coefficient of friction and abrasion resistance.

Edges. Carbon steel that has been treated to have a hardness of 48 HRC. Facilitates turning by "cutting" into the snow.

(a)

(a) One relatively complex composite structure is the modern ski. This illustration, a cross section of a high-performance snow ski, shows the various components. The function of each component is noted, as well as the material used in its construction.

(b) Photograph of a skier in fresh powder snow.

© Doug Berry/iStockphoto

(b)

With knowledge of the various types of composites, as well as an understanding of the dependence of their behaviors on the characteristics, relative amounts, geometry/distribution, and properties of the constituent phases, it is possible to design materials with property combinations that are better than those found in any monolithic metal alloys, ceramics, and polymeric materials. For example, in Design Example 16.1, we discuss how a tubular shaft is designed that meets specified stiffness requirements.

Learning Objectives

After studying this chapter, you should be able to do the following:

1. Name the four main divisions of composite materials and cite the distinguishing feature of each.
2. Cite the difference in strengthening mechanism for large-particle and dispersion-strengthened particle-reinforced composites.
3. Distinguish the three different types of fiber-reinforced composites on the basis of fiber length and orientation; comment on the distinctive mechanical characteristics for each type.
4. Calculate longitudinal modulus and longitudinal strength for an aligned and continuous fiber-reinforced composite.
5. Compute longitudinal strengths for discontinuous and aligned fibrous composite materials.
6. Note the three common fiber reinforcements used in polymer-matrix composites and, for each, cite both desirable characteristics and limitations.
7. Cite the desirable features of metal–matrix composites.
8. Note the primary reason for the creation of ceramic-matrix composites.
9. Name and briefly describe the two subclassifications of structural composites.

16.1 INTRODUCTION

The advent of the composites as a distinct classification of materials began during the mid-20th century with the manufacturing of deliberately designed and engineered multiphase composites such as fiberglass-reinforced polymers. Although multiphase materials, such as wood, bricks made from straw-reinforced clay, seashells, and even alloys such as steel had been known for millennia, recognition of this novel concept of combining dissimilar materials during manufacture led to the identification of composites as a new class that was separate from familiar metals, ceramics, and polymers. This concept of multiphase composites provides exciting opportunities for designing an exceedingly large variety of materials with property combinations that cannot be met by any of the monolithic conventional metal alloys, ceramics, and polymeric materials.[1]

Materials that have specific and unusual properties are needed for a host of high-technology applications such as those found in the aerospace, underwater, bioengineering, and transportation industries. For example, aircraft engineers are increasingly searching for structural materials that have low densities; are strong, stiff, and abrasion and impact resistant; and do not easily corrode. This is a rather formidable combination of characteristics. Among monolithic materials, strong materials are relatively dense; increasing the strength or stiffness generally results in a decrease in toughness.

Material property combinations and ranges have been, and are still being, extended by the development of composite materials. Generally speaking, a composite is considered to be any multiphase material that exhibits a significant proportion of the

---

[1]By *monolithic* we mean having a microstructure that is uniform and continuous and was formed from a single material; furthermore, more than one microconstituent may be present. In contrast, the microstructure of a composite is nonuniform, discontinuous, and multiphase, in the sense that it is a mixture of two or more distinct materials.

principle of combined action

properties of both constituent phases such that a better combination of properties is realized. According to this **principle of combined action**, better property combinations are fashioned by the judicious combination of two or more distinct materials. Property trade-offs are also made for many composites.

Composites of sorts have already been discussed; these include multiphase metal alloys, ceramics, and polymers. For example, pearlitic steels (Section 11.19) have a microstructure consisting of alternating layers of α-ferrite and cementite (Figure 11.26). The ferrite phase is soft and ductile, whereas cementite is hard and very brittle. The combined mechanical characteristics of the pearlite (reasonably high ductility and strength) are superior to those of either of the constituent phases. A number of composites also occur in nature. For example, wood consists of strong and flexible cellulose fibers surrounded and held together by a stiffer material called lignin. Also, bone is a composite of the strong yet soft protein collagen and the hard, brittle mineral apatite.

A composite, in the present context, is a multiphase material that is *artificially made,* as opposed to one that occurs or forms naturally. In addition, the constituent phases must be chemically dissimilar and separated by a distinct interface.

In designing composite materials, scientists and engineers have ingeniously combined various metals, ceramics, and polymers to produce a new generation of extraordinary materials. Most composites have been created to improve combinations of mechanical characteristics such as stiffness, toughness, and ambient and high-temperature strength.

matrix phase

dispersed phase

Many composite materials are composed of just two phases; one is termed the **matrix**, which is continuous and surrounds the other phase, often called the **dispersed phase**. The properties of composites are a function of the properties of the constituent phases, their relative amounts, and the geometry of the dispersed phase. *Dispersed phase geometry* in this context means the shape of the particles and the particle size, distribution, and orientation; these characteristics are represented in Figure 16.1.

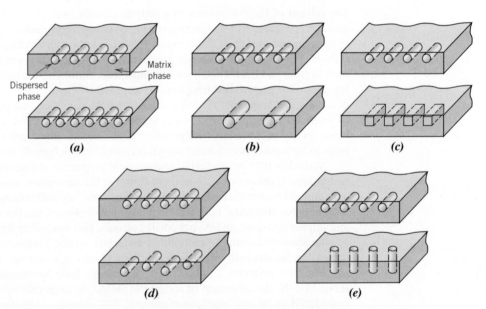

Figure 16.1 Schematic representations of the various geometrical and spatial characteristics of particles of the dispersed phase that may influence the properties of composites: (*a*) concentration, (*b*) size, (*c*) shape, (*d*) distribution, and (*e*) orientation.
(From Richard A. Flinn and Paul K. Trojan, *Engineering Materials and Their Applications,* 4th edition. Copyright © 1990 by John Wiley & Sons, Inc. Adapted by permission of John Wiley & Sons, Inc.)

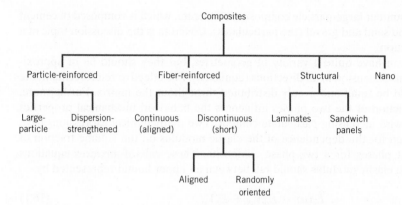

Figure 16.2 A classification scheme for the various composite types discussed in this chapter.

One simple scheme for the classification of composite materials is shown in Figure 16.2, which consists of four main divisions: particle-reinforced, fiber-reinforced, structural, and nanocomposites. The dispersed phase for particle-reinforced composites is *equiaxed* (i.e., particle dimensions are approximately the same in all directions); for fiber-reinforced composites, the dispersed phase has the geometry of a fiber (i.e., a large length-to-diameter ratio). Structural composites are multi-layered and designed to have low densities and high degrees of structural integrity. For nanocomposites dimensions of the dispersed phase particles are on the order of nanometers. The discussion of the remainder of this chapter is organized according to this classification scheme.

Particle-Reinforced Composites

large-particle composite

dispersion-strengthened composite

As noted in Figure 16.2, **large-particle** and **dispersion-strengthened composites** are the two subclassifications of particle-reinforced composites. The distinction between these is based on the reinforcement or strengthening mechanism. The term *large* is used to indicate that particle–matrix interactions cannot be treated on the atomic or molecular level; rather, continuum mechanics is used. For most of these composites, the particulate phase is harder and stiffer than the matrix. These reinforcing particles tend to restrain movement of the matrix phase in the vicinity of each particle. In essence, the matrix transfers some of the applied stress to the particles, which bear a fraction of the load. The degree of reinforcement or improvement of mechanical behavior depends on strong bonding at the matrix–particle interface.

For dispersion-strengthened composites, particles are normally much smaller, with diameters between 0.01 and 0.1 μm (10 and 100 nm). Particle–matrix interactions that lead to strengthening occur on the atomic or molecular level. The mechanism of strengthening is similar to that for precipitation hardening discussed in Section 17.7. Whereas the matrix bears the major portion of an applied load, the small dispersed particles hinder or impede the motion of dislocations. Thus, plastic deformation is restricted such that yield and tensile strengths, as well as hardness, improve.

16.2 LARGE-PARTICLE COMPOSITES

Some polymeric materials to which fillers have been added (Section 17.13) are really large-particle composites. Again, the fillers modify or improve the properties of the material and/or replace some of the polymer volume with a less expensive material—the filler.

Another familiar large-particle composite is concrete, which is composed of cement (the matrix) and sand and gravel (the particulates). Concrete is the discussion topic of a succeeding section.

Particles can have quite a variety of geometries, but they should be of approximately the same dimension in all directions (equiaxed). For effective reinforcement, the particles should be small and evenly distributed throughout the matrix. Furthermore, the volume fraction of the two phases influences the behavior; mechanical properties are enhanced with increasing particulate content. Two mathematical expressions have been formulated for the dependence of the elastic modulus on the volume fraction of the constituent phases for a two-phase composite. These **rule-of-mixtures** equations predict that the elastic modulus should fall between an upper bound represented by

rule of mixtures

For a two-phase composite, modulus of elasticity upper-bound expression

$$E_c(u) = E_m V_m + E_p V_p \tag{16.1}$$

and a lower bound, or limit,

For a two-phase composite, modulus of elasticity lower-bound expression

$$E_c(l) = \frac{E_m E_p}{V_m E_p + V_p E_m} \tag{16.2}$$

In these expressions, E and V denote the elastic modulus and volume fraction, respectively, and the subscripts c, m, and p represent composite, matrix, and particulate phases, respectively. Figure 16.3 plots upper- and lower-bound E_c-versus-V_p curves for a copper–tungsten composite, in which tungsten is the particulate phase; experimental data points fall between the two curves. Equations analogous to 16.1 and 16.2 for fiber-reinforced composites are derived in Section 16.5.

Large-particle composites are used with all three material types (metals, polymers, and ceramics). The **cermets** are examples of ceramic–metal composites. The most common cermet is cemented carbide, which is composed of extremely hard particles of a refractory carbide ceramic such as tungsten carbide (WC) or titanium carbide (TiC) embedded in a matrix of a metal such as cobalt or nickel. These composites are used extensively as cutting tools for hardened steels. The hard carbide particles provide the cutting surface but, being extremely brittle, are not capable of withstanding the cutting stresses. Toughness is enhanced by their inclusion in the ductile metal matrix, which isolates the carbide particles from one another and prevents particle-to-particle crack propagation. Both matrix and particulate phases are quite refractory to the high temper-

cermet

Figure 16.3 Modulus of elasticity versus volume percent tungsten for a composite of tungsten particles dispersed within a copper matrix. Upper and lower bounds are according to Equations 16.1 and 16.2, respectively; experimental data points are included. (Reprinted with permission from R. H. Krock, *ASTM Proceedings*, Vol. 63, 1963. Copyright ASTM International, 100 Barr Harbor Drive, West Conschohocken, PA 19428.)

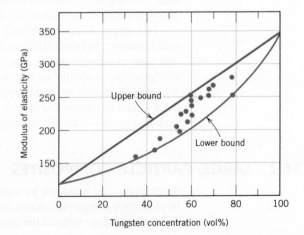

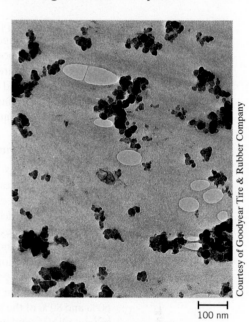

Courtesy of Carboloy Systems Department, General Electric Company

Courtesy of Goodyear Tire & Rubber Company

100 μm

Figure 16.4 Photomicrograph of a WC–Co cemented carbide. Light areas are the cobalt matrix; dark regions are the particles of tungsten carbide. 100×.

100 nm

Figure 16.5 Electron micrograph showing the spherical reinforcing carbon black particles in a synthetic rubber tire tread compound. The areas resembling water marks are tiny air pockets in the rubber. 80,000×.

atures generated by the cutting action on materials that are extremely hard. No single material could possibly provide the combination of properties possessed by a cermet. Relatively large volume fractions of the particulate phase may be used, often exceeding 90 vol%; thus the abrasive action of the composite is maximized. A photomicrograph of a WC–Co cemented carbide is shown in Figure 16.4.

Both elastomers and plastics are frequently reinforced with various particulate materials. Use of many modern rubbers would be severely restricted without reinforcing particulate materials such as *carbon black*. Carbon black consists of very small and essentially spherical particles of carbon, produced by the combustion of natural gas or oil in an atmosphere that has only a limited air supply. When added to vulcanized rubber, this extremely inexpensive material enhances tensile strength, toughness, and tear and abrasion resistance. Automobile tires contain on the order of 15 to 30 vol% carbon black. For the carbon black to provide significant reinforcement, the particle size must be extremely small, with diameters between 20 and 50 nm; also, the particles must be evenly distributed throughout the rubber and must form a strong adhesive bond with the rubber matrix. Particle reinforcement using other materials (e.g., silica) is much less effective because this special interaction between the rubber molecules and particle surfaces does not exist. Figure 16.5 is an electron micrograph of a carbon black–reinforced rubber.

Concrete

concrete

Concrete is a common large-particle composite in which both matrix and dispersed phases are ceramic materials. Because the terms *concrete* and *cement* are sometimes incorrectly interchanged, it is appropriate to make a distinction between them. In a broad sense, concrete implies a composite material consisting of an aggregate of particles that are bound together in a solid body by some type of binding medium, that is, a cement. The two most familiar concretes are those made with Portland and asphaltic cements, in which the aggregate is gravel and sand. Asphaltic concrete is widely used primarily as a paving material, whereas Portland cement concrete is employed extensively as a structural building material. Only the latter is treated in this discussion.

Portland Cement Concrete

The ingredients for this concrete are Portland cement, a fine aggregate (sand), a coarse aggregate (gravel), and water. The process by which Portland cement is produced and the mechanism of setting and hardening were discussed very briefly in Section 14.15. The aggregate particles act as a filler material to reduce the overall cost of the concrete product because they are cheap, whereas cement is relatively expensive. To achieve the optimum strength and workability of a concrete mixture, the ingredients must be added in the correct proportions. Dense packing of the aggregate and good interfacial contact are achieved by having particles of two different sizes; the fine particles of sand should fill the void spaces between the gravel particles. Typically, these aggregates constitute between 60% and 80% of the total volume. The amount of cement–water paste should be sufficient to coat all the sand and gravel particles; otherwise, the cementitious bond will be incomplete. Furthermore, all of the constituents should be thoroughly mixed. Complete bonding between cement and the aggregate particles is contingent on the addition of the correct quantity of water. Too little water leads to incomplete bonding, and too much results in excessive porosity; in either case, the final strength is less than the optimum.

The character of the aggregate particles is an important consideration. In particular, the size distribution of the aggregates influences the amount of cement–water paste required. Also, the surfaces should be clean and free from clay and silt, which prevent the formation of a sound bond at the particle surface.

Portland cement concrete is a major material of construction, primarily because it can be poured in place and hardens at room temperature, even when submerged in water. However, as a structural material, it has some limitations and disadvantages. Like most ceramics, Portland cement concrete is relatively weak and extremely brittle; its tensile strength is approximately one-fifteenth to one-tenth its compressive strength. Also, large concrete structures can experience considerable thermal expansion and contraction with temperature fluctuations. In addition, water penetrates into external pores, which can cause severe cracking in cold weather as a consequence of freeze–thaw cycles. Most of these inadequacies may be eliminated or at least reduced by reinforcement and/or the incorporation of additives.

Reinforced Concrete

The strength of Portland cement concrete may be increased by additional reinforcement. This is usually accomplished by means of steel rods, wires, bars (rebar), or mesh, which are embedded into the fresh and uncured concrete. Thus, the reinforcement renders the hardened structure capable of supporting greater tensile, compressive, and shear stresses. Even if cracks develop in the concrete, considerable reinforcement is maintained.

Steel serves as a suitable reinforcement material because its coefficient of thermal expansion is nearly the same as that of concrete. In addition, steel is not rapidly corroded in the cement environment, and a relatively strong adhesive bond is formed between it and the cured concrete. This adhesion may be enhanced by the incorporation of contours into the surface of the steel member, which permits a greater degree of mechanical interlocking.

Portland cement concrete may also be reinforced by mixing fibers of a high-modulus material such as glass, steel, nylon, or polyethylene into the fresh concrete. Care must be exercised in using this type of reinforcement because some fiber materials experience rapid deterioration when exposed to the cement environment.

Another reinforcement technique for strengthening concrete involves the introduction of residual compressive stresses into the structural member; the resulting material is called **prestressed concrete**. This method uses one characteristic of brittle ceramics— namely, that they are stronger in compression than in tension. Thus, to fracture a prestressed concrete member, the magnitude of the precompressive stress must be exceeded by an applied tensile stress.

prestressed concrete

In one such prestressing technique, high-strength steel wires are positioned inside the empty molds and stretched with a high tensile force, which is maintained constant. After the concrete has been placed and allowed to harden, the tension is released. As the wires contract, they put the structure in a state of compression because the stress is transmitted to the concrete via the concrete–wire bond that is formed.

Another technique, in which stresses are applied after the concrete hardens, is appropriately called *posttensioning*. Sheet metal or rubber tubes are situated inside and pass through the concrete forms, around which the concrete is cast. After the cement has hardened, steel wires are fed through the resulting holes, and tension is applied to the wires by means of jacks attached and abutted to the faces of the structure. Again, a compressive stress is imposed on the concrete piece, this time by the jacks. Finally, the empty spaces inside the tubing are filled with a grout to protect the wire from corrosion.

Concrete that is prestressed should be of high quality with low shrinkage and low creep rate. Prestressed concretes, usually prefabricated, are commonly used for highway and railway bridges.

16.3 DISPERSION-STRENGTHENED COMPOSITES

Metals and metal alloys may be strengthened and hardened by the uniform dispersion of several volume percent of fine particles of a very hard and inert material. The dispersed phase may be metallic or nonmetallic; oxide materials are often used. Again, the strengthening mechanism involves interactions between the particles and dislocations within the matrix, as with precipitation hardening. The dispersion strengthening effect is not as pronounced as with precipitation hardening; however, the strengthening is retained at elevated temperatures and for extended time periods because the dispersed particles are chosen to be unreactive with the matrix phase. For precipitation-hardened alloys, the increase in strength may disappear upon heat treatment as a consequence of precipitate growth or dissolution of the precipitate phase.

The high-temperature strength of nickel alloys may be enhanced significantly by the addition of about 3 vol% thoria (ThO_2) as finely dispersed particles; this material is known as *thoria-dispersed* (or TD) *nickel*. The same effect is produced in the aluminum– aluminum oxide system. A very thin and adherent alumina coating is caused to form on the surface of extremely small (0.1 to 0.2 μm thick) flakes of aluminum, which are dispersed within an aluminum metal matrix; this material is termed *sintered aluminum powder* (SAP).

Concept Check 16.1 Cite the general difference in strengthening mechanism between large-particle and dispersion-strengthened particle-reinforced composites.

[*The answer may be found at* www.wiley.com/college/callister (*Student Companion Site*).]

Fiber-Reinforced Composites

fiber-reinforced composite

specific strength

specific modulus

Technologically, the most important composites are those in which the dispersed phase is in the form of a fiber. Design goals of **fiber-reinforced composites** often include high strength and/or stiffness on a weight basis. These characteristics are expressed in terms of **specific strength** and **specific modulus** parameters, which correspond, respectively, to the ratios of tensile strength to specific gravity and modulus of elasticity to specific gravity. Fiber-reinforced composites with exceptionally high specific strengths and moduli have been produced that use low-density fiber and matrix materials.

As noted in Figure 16.2, fiber-reinforced composites are subclassified by fiber length. For short-fiber composites, the fibers are too short to produce a significant improvement in strength.

16.4 INFLUENCE OF FIBER LENGTH

The mechanical characteristics of a fiber-reinforced composite depend not only on the properties of the fiber, but also on the degree to which an applied load is transmitted to the fibers by the matrix phase. Important to the extent of this load transmittance is the magnitude of the interfacial bond between the fiber and matrix phases. Under an applied stress, this fiber–matrix bond ceases at the fiber ends, yielding a matrix deformation pattern as shown schematically in Figure 16.6; in other words, there is no load transmittance from the matrix at each fiber extremity.

Some critical fiber length is necessary for effective strengthening and stiffening of the composite material. This critical length l_c is dependent on the fiber diameter d and its ultimate (or tensile) strength σ_f^* and on the fiber–matrix bond strength (or the shear yield strength of the matrix, whichever is smaller) τ_c according to

Critical fiber length—dependence on fiber strength and diameter and fiber–matrix bond strength (or matrix shear yield strength)

$$l_c = \frac{\sigma_f^* d}{2\tau_c} \tag{16.3}$$

For a number of glass and carbon fiber–matrix combinations, this critical length is on the order of 1 mm, which ranges between 20 and 150 times the fiber diameter.

When a stress equal to σ_f^* is applied to a fiber having just this critical length, the stress–position profile shown in Figure 16.7a results—that is, the maximum fiber load is achieved only at the axial center of the fiber. As fiber length l increases, the fiber reinforcement becomes more effective; this is demonstrated in Figure 16.7b, a stress–axial position profile for $l > l_c$ when the applied stress is equal to the fiber strength. Figure 16.7c shows the stress–position profile for $l < l_c$.

Fibers for which $l \gg l_c$ (normally $l > 15l_c$) are termed *continuous; discontinuous* or *short fibers* have lengths shorter than this. For discontinuous fibers of lengths significantly less than l_c, the matrix deforms around the fiber such that there is virtually no

Figure 16.6 The deformation pattern in the matrix surrounding a fiber that is subjected to an applied tensile load.

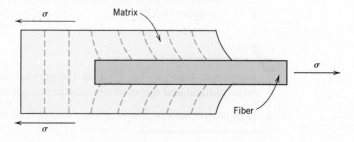

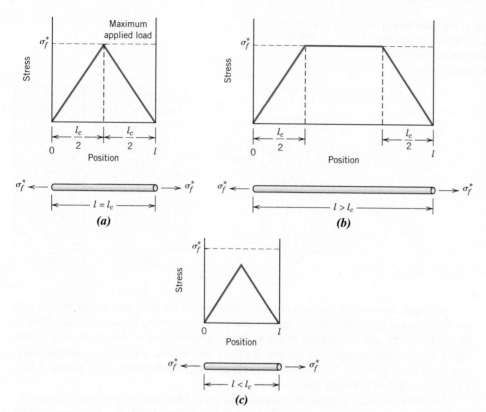

Figure 16.7
Stress–position profiles when the fiber length l (a) is equal to the critical length l_c, (b) is greater than the critical length, and (c) is less than the critical length for a fiber-reinforced composite that is subjected to a tensile stress equal to the fiber tensile strength σ_f^*.

stress transference and little reinforcement by the fiber. These are essentially the particulate composites as described earlier. To effect a significant improvement in strength of the composite, the fibers must be continuous.

16.5 INFLUENCE OF FIBER ORIENTATION AND CONCENTRATION

The arrangement or orientation of the fibers relative to one another, the fiber concentration, and the distribution all have a significant influence on the strength and other properties of fiber-reinforced composites. With respect to orientation, two extremes are possible: (1) a parallel alignment of the longitudinal axis of the fibers in a single direction, and (2) a totally random alignment. Continuous fibers are normally aligned (Figure 16.8a), whereas discontinuous fibers may be aligned (Figure 16.8b), randomly oriented (Figure 16.8c), or partially oriented. Better overall composite properties are realized when the fiber distribution is uniform.

Continuous and Aligned Fiber Composites

Tensile Stress–Strain Behavior—Longitudinal Loading

Mechanical responses of this type of composite depend on several factors, including the stress–strain behaviors of fiber and matrix phases, the phase volume fractions, and the direction in which the stress or load is applied. Furthermore, the properties of a composite having its fibers aligned are highly anisotropic, that is, they depend on the direction in which they are measured. Let us first consider the stress–strain behavior for the situation in which the stress is applied along the direction of alignment, the **longitudinal direction**, which is indicated in Figure 16.8a.

longitudinal direction

Figure 16.8 Schematic representations of (*a*) continuous and aligned, (*b*) discontinuous and aligned, and (*c*) discontinuous and randomly oriented fiber–reinforced composites.

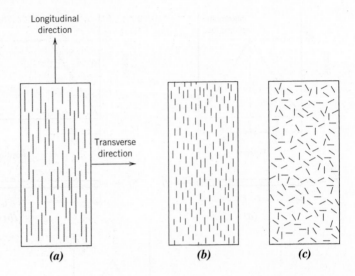

To begin, assume the stress-versus-strain behaviors for fiber and matrix phases that are represented schematically in Figure 16.9*a*; in this treatment we consider the fiber to be totally brittle and the matrix phase to be reasonably ductile. Also indicated in this figure are fracture strengths in tension for fiber and matrix, σ_f^*, and σ_m^*, respectively, and their corresponding fracture strains, ϵ_f^*, and ϵ_m^*; furthermore, it is assumed that $\epsilon_m^* > \epsilon_f^*$, which is normally the case.

A fiber-reinforced composite consisting of these fiber and matrix materials exhibits the uniaxial stress–strain response illustrated in Figure 16.9*b*; the fiber and matrix

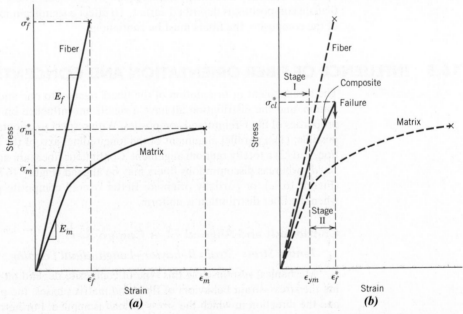

Figure 16.9 (*a*) Schematic stress–strain curves for brittle fiber and ductile matrix materials. Fracture stresses and strains for both materials are noted. (*b*) Schematic stress–strain curve for an aligned fiber–reinforced composite that is exposed to a uniaxial stress applied in the direction of alignment; curves for the fiber and matrix materials shown in part (*a*) are also superimposed.

behaviors from Figure 16.9a are included to provide perspective. In the initial Stage I region, both fibers and matrix deform elastically; normally this portion of the curve is linear. Typically, for a composite of this type, the matrix yields and deforms plastically (at ϵ_{ym}, Figure 16.9b) while the fibers continue to stretch elastically, inasmuch as the tensile strength of the fibers is significantly higher than the yield strength of the matrix. This process constitutes Stage II as noted in the figure; this stage is typically very nearly linear but of diminished slope relative to Stage I. In passing from Stage I to Stage II, the proportion of the applied load borne by the fibers increases.

The onset of composite failure begins as the fibers start to fracture, which corresponds to a strain of approximately ϵ_f^* as noted in Figure 16.9b. Composite failure is not catastrophic for a couple of reasons. First, not all fibers fracture at the same time, because there will always be considerable variations in the fracture strength of brittle fiber materials (Section 14.6). In addition, even after fiber failure, the matrix is still intact inasmuch as $\epsilon_f^* < \epsilon_m^*$ (Figure 16.9a). Thus, these fractured fibers, which are shorter than the original ones, are still embedded within the intact matrix and consequently are capable of sustaining a diminished load as the matrix continues to plastically deform.

Elastic Behavior—Longitudinal Loading

Let us now consider the elastic behavior of a continuous and oriented fibrous composite that is loaded in the direction of fiber alignment. First, it is assumed that the fiber–matrix interfacial bond is very good, such that deformation of both matrix and fibers is the same (an *isostrain* situation). Under these conditions, the total load sustained by the composite F_c is equal to the sum of the loads carried by the matrix phase F_m and the fiber phase F_f, or

$$F_c = F_m + F_f \tag{16.4}$$

From the definition of stress, Equation 6.1, $F = \sigma A$; thus expressions for F_c, F_m, and F_f in terms of their respective stresses (σ_c, σ_m, and σ_f) and cross-sectional areas (A_c, A_m, and A_f) are possible. Substitution of these into Equation 16.4 yields

$$\sigma_c A_c = \sigma_m A_m + \sigma_f A_f \tag{16.5}$$

Dividing through by the total cross-sectional area of the composite, A_c, we have

$$\sigma_c = \sigma_m \frac{A_m}{A_c} + \sigma_f \frac{A_f}{A_c} \tag{16.6}$$

where A_m/A_c and A_f/A_c are the area fractions of the matrix and fiber phases, respectively. If the composite, matrix, and fiber phase lengths are all equal, A_m/A_c is equivalent to the volume fraction of the matrix, V_m, and likewise for the fibers, $V_f = A_f/A_c$. Equation 16.6 becomes

$$\sigma_c = \sigma_m V_m + \sigma_f V_f \tag{16.7}$$

The previous assumption of an isostrain state means that

$$\epsilon_c = \epsilon_m = \epsilon_f \tag{16.8}$$

and when each term in Equation 16.7 is divided by its respective strain,

$$\frac{\sigma_c}{\epsilon_c} = \frac{\sigma_m}{\epsilon_m} V_m + \frac{\sigma_f}{\epsilon_f} V_f \tag{16.9}$$

Furthermore, if composite, matrix, and fiber deformations are all elastic, then $\sigma_c/\epsilon_c = E_c$, $\sigma_m/\epsilon_m = E_m$, and $\sigma_f/\epsilon_f = E_f$, the Es being the moduli of elasticity for the respective phases. Substitution into Equation 16.9 yields an expression for the modulus of elasticity of a continuous and aligned fibrous composite *in the direction of alignment* (or *longitudinal direction*), E_{cl}, as

For a continuous and aligned fiber–reinforced composite, modulus of elasticity in the longitudinal direction

or

$$E_{cl} = E_m V_m + E_f V_f \qquad (16.10a)$$

$$E_{cl} = E_m(1 - V_f) + E_f V_f \qquad (16.10b)$$

because the composite consists of only matrix and fiber phases; that is, $V_m + V_f = 1$.

Thus, E_{cl} is equal to the volume-fraction weighted average of the moduli of elasticity of the fiber and matrix phases. Other properties, including density, also have this dependence on volume fractions. Equation 16.10a is the fiber analogue of Equation 16.1, the upper bound for particle-reinforced composites.

It can also be shown, for longitudinal loading, that the ratio of the load carried by the fibers to that carried by the matrix is

Ratio of load carried by fibers and the matrix phase, for longitudinal loading

$$\frac{F_f}{F_m} = \frac{E_f V_f}{E_m V_m} \qquad (16.11)$$

The demonstration is left as a homework problem.

EXAMPLE PROBLEM 16.1

Property Determinations for a Glass Fiber–Reinforced Composite—Longitudinal Direction

A continuous and aligned glass fiber–reinforced composite consists of 40 vol% glass fibers having a modulus of elasticity of 69 GPa and 60 vol% polyester resin that, when hardened, displays a modulus of 3.4 GPa.

(a) Compute the modulus of elasticity of this composite in the longitudinal direction.

(b) If the cross-sectional area is 250 mm² and a stress of 50 MPa is applied in this longitudinal direction, compute the magnitude of the load carried by each of the fiber and matrix phases.

(c) Determine the strain that is sustained by each phase when the stress in part (b) is applied.

Solution

(a) The modulus of elasticity of the composite is calculated using Equation 16.10a:

$$E_{cl} = (3.4 \text{ GPa})(0.6) + (69 \text{ GPa})(0.4)$$
$$= 30 \text{ GPa}$$

(b) To solve this portion of the problem, first find the ratio of fiber load to matrix load, using Equation 16.11; thus,

$$\frac{F_f}{F_m} = \frac{(69 \text{ GPa})(0.4)}{(3.4 \text{ GPa})(0.6)} = 13.5$$

or $F_f = 13.5 \ F_m$.

In addition, the total force sustained by the composite F_c may be computed from the applied stress σ and total composite cross-sectional area A_c according to

$$F_c = A_c \sigma = (250 \text{ mm}^2)(50 \text{ MPa}) = 12{,}500 \text{ N}$$

However, this total load is just the sum of the loads carried by fiber and matrix phases; that is,

$$F_c = F_f + F_m = 12{,}500 \text{ N}$$

Substitution for F_f from the preceding equation yields

$$13.5 F_m + F_m = 12{,}500 \text{ N}$$

or

$$F_m = 860 \text{ N}$$

whereas

$$F_f = F_c - F_m = 12{,}500 \text{ N} - 860 \text{ N} = 11{,}640 \text{ N}$$

Thus, the fiber phase supports the vast majority of the applied load.

(c) The stress for both fiber and matrix phases must first be calculated. Then, by using the elastic modulus for each [from part (a)], the strain values may be determined.

For stress calculations, phase cross-sectional areas are necessary:

$$A_m = V_m A_c = (0.6)(250 \text{ mm}^2) = 150 \text{ mm}^2$$

and

$$A_f = V_f A_c = (0.4)(250 \text{ mm}^2) = 100 \text{ mm}^2$$

Thus,

$$\sigma_m = \frac{F_m}{A_m} = \frac{860 \text{ N}}{150 \text{ mm}^2} = 5.73 \text{ MPa}$$

$$\sigma_f = \frac{F_f}{A_f} = \frac{11{,}640 \text{ N}}{100 \text{ mm}^2} = 116.4 \text{ MPa}$$

Finally, strains are computed as

$$\epsilon_m = \frac{\sigma_m}{E_m} = \frac{5.73 \text{ MPa}}{3.4 \times 10^3 \text{ MPa}} = 1.69 \times 10^{-3}$$

$$\epsilon_f = \frac{\sigma_f}{E_f} = \frac{116.4 \text{ MPa}}{69 \times 10^3 \text{ MPa}} = 1.69 \times 10^{-3}$$

Therefore, strains for both matrix and fiber phases are identical, which they should be, according to Equation 16.8 in the previous development.

Elastic Behavior—Transverse Loading

transverse direction

A continuous and oriented fiber composite may be loaded in the **transverse direction**; that is, the load is applied at a 90° angle to the direction of fiber alignment as shown in Figure 16.8a. For this situation the stress σ to which the composite and both phases are exposed is the same, or

$$\sigma_c = \sigma_m = \sigma_f = \sigma \qquad (16.12)$$

This is termed an *isostress* state. The strain or deformation of the entire composite ϵ_c is

$$\epsilon_c = \epsilon_m V_m + \epsilon_f V_f \qquad (16.13)$$

but, because $\epsilon = \sigma/E$,

$$\frac{\sigma}{E_{ct}} = \frac{\sigma}{E_m}V_m + \frac{\sigma}{E_f}V_f \tag{16.14}$$

where E_{ct} is the modulus of elasticity in the transverse direction. Now, dividing through by σ yields

$$\frac{1}{E_{ct}} = \frac{V_m}{E_m} + \frac{V_f}{E_f} \tag{16.15}$$

For a continuous and aligned fiber–reinforced composite, modulus of elasticity in the transverse direction

which reduces to

$$E_{ct} = \frac{E_m E_f}{V_m E_f + V_f E_m} = \frac{E_m E_f}{(1 - V_f)E_f + V_f E_m} \tag{16.16}$$

Equation 16.16 is analogous to the lower-bound expression for particulate composites, Equation 16.2.

EXAMPLE PROBLEM 16.2

Elastic Modulus Determination for a Glass Fiber–Reinforced Composite—Transverse Direction

Compute the elastic modulus of the composite material described in Example Problem 16.1, but assume that the stress is applied perpendicular to the direction of fiber alignment.

Solution

According to Equation 16.16,

$$E_{ct} = \frac{(3.4\,\text{GPa})(69\,\text{GPa})}{(0.6)(69\,\text{GPa}) + (0.4)(3.4\,\text{GPa})}$$

$$= 5.5\,\text{GPa}$$

This value for E_{ct} is slightly greater than that of the matrix phase but, from Example Problem 16.1a, only approximately one-fifth of the modulus of elasticity along the fiber direction (E_{cl}), which indicates the degree of anisotropy of continuous and oriented fiber composites.

Longitudinal Tensile Strength

We now consider the strength characteristics of continuous and aligned fiber–reinforced composites that are loaded in the longitudinal direction. Under these circumstances, strength is normally taken as the maximum stress on the stress–strain curve, Figure 16.9b; often this point corresponds to fiber fracture and marks the on-set of composite failure. Table 16.1 lists typical longitudinal tensile strength values for three common fibrous composites. Failure of this type of composite material is a relatively complex process, and several different failure modes are possible. The mode that operates for a specific composite depends on fiber and matrix properties and the nature and strength of the fiber–matrix interfacial bond.

If we assume that $\epsilon_f^* < \epsilon_m^*$ (Figure 16.9a), which is the usual case, then fibers will fail before the matrix. Once the fibers have fractured, most of the load that was borne by the fibers will be transferred to the matrix. This being the case, it is possible to adapt

Table 16.1

Typical Longitudinal and Transverse Tensile Strengths for Three Unidirectional Fiber–Reinforced Composites.[a]

Material	Longitudinal Tensile Strength (MPa)	Transverse Tensile Strength (MPa)
Glass–polyester	700	20
Carbon (high modulus)–epoxy	1000	35
Kevlar–epoxy	1200	20

[a]The fiber content for each is approximately 50 vol%.
Source: D. Hull and T. W. Clyne, *An Introduction to Composite Materials,* 2nd edition, Cambridge University Press, New York, 1996, p. 179.

the expression for the stress on this type of composite, Equation 16.7, into the following expression for the longitudinal strength of the composite, σ_{cl}^*:

For a continuous and aligned fiber–reinforced composite, longitudinal strength in tension

$$\sigma_{cl}^* = \sigma_m'(1 - V_f) + \sigma_f^* V_f \qquad (16.17)$$

Here, σ_m' is the stress in the matrix at fiber failure (as illustrated in Figure 16.9a) and, as previously, σ_f^* is the fiber tensile strength.

Transverse Tensile Strength

The strengths of continuous and unidirectional fibrous composites are highly anisotropic, and such composites are normally designed to be loaded along the high-strength, longitudinal direction. However, during in-service applications, transverse tensile loads may also be present. Under these circumstances, premature failure may result inasmuch as transverse strength is usually extremely low—it sometimes lies below the tensile strength of the matrix. Thus, the reinforcing effect of the fibers is negative. Typical transverse tensile strengths for three unidirectional composites are listed in Table 16.1.

Whereas longitudinal strength is dominated by fiber strength, a variety of factors will have a significant influence on the transverse strength; these factors include properties of both the fiber and matrix, the fiber–matrix bond strength, and the presence of voids. Measures that have been used to improve the transverse strength of these composites usually involve modifying properties of the matrix.

✓ **Concept Check 16.2** The following table lists four hypothetical aligned fiber–reinforced composites (labeled A through D), along with their characteristics. On the basis of these data, rank the four composites from highest to lowest strength in the longitudinal direction, and then justify your ranking.

Composite	Fiber Type	Volume Fraction Fibers	Fiber Strength (MPa)	Average Fiber Length (mm)	Critical Length (mm)
A	Glass	0.20	3.5×10^3	8	0.70
B	Glass	0.35	3.5×10^3	12	0.75
C	Carbon	0.40	5.5×10^3	8	0.40
D	Carbon	0.30	5.5×10^3	8	0.50

[*The answer may be found at* www.wiley.com/college/callister *(Student Companion Site).*]

Discontinuous and Aligned–Fiber Composites

Even though reinforcement efficiency is lower for discontinuous than for continuous fibers, discontinuous and aligned-fiber composites (Figure 16.8b) are becoming increasingly more important in the commercial market. Chopped-glass fibers are used most extensively; however, carbon and aramid discontinuous fibers are also used. These short-fiber composites can be produced with moduli of elasticity and tensile strengths that approach 90% and 50%, respectively, of their continuous-fiber counterparts.

For a discontinuous and aligned-fiber composite having a uniform distribution of fibers and in which $l > l_c$, the longitudinal strength (σ_{cd}^*) is given by the relationship

For a discontinuous ($l > l_c$,) and aligned fiber–reinforced composite, longitudinal strength in tension

$$\sigma_{cd}^* = \sigma_f^* V_f \left(1 - \frac{l_c}{2l} \right) + \sigma_m'(1 - V_f) \tag{16.18}$$

where σ_f^* and σ_m' represent, respectively, the fracture strength of the fiber and the stress in the matrix when the composite fails (Figure 16.9a).

If the fiber length is less than critical ($l < l_c$), then the longitudinal strength ($\sigma_{cd'}^*$) is given by

For a discontinuous ($l < l_c$) and aligned fiber–reinforced composite, longitudinal strength in tension

$$\sigma_{cd'}^* = \frac{l\tau_c}{d} V_f + \sigma_m'(1 - V_f) \tag{16.19}$$

where d is the fiber diameter and τ_c is the smaller of either the fiber–matrix bond strength or the matrix shear yield strength.

Discontinuous and Randomly Oriented–Fiber Composites

Normally, when the fiber orientation is random, short and discontinuous fibers are used; reinforcement of this type is schematically demonstrated in Figure 16.8c. Under these circumstances, a *rule-of-mixtures* expression for the elastic modulus similar to Equation 16.10a may be used, as follows:

For a discontinuous and randomly oriented fiber–reinforced composite, modulus of elasticity

$$E_{cd} = K E_f V_f + E_m V_m \tag{16.20}$$

In this expression, K is a fiber efficiency parameter that depends on V_f and the E_f/E_m ratio. Its magnitude will be less than unity, usually in the range 0.1 to 0.6. Thus, for random-fiber reinforcement (as with oriented-fiber reinforcement), the modulus increases with increasing volume fraction of fiber. Table 16.2, which gives some of the mechanical properties of unreinforced and reinforced polycarbonates for discontinuous and randomly oriented glass fibers, provides an idea of the magnitude of the reinforcement that is possible.

By way of summary, then, we say that aligned fibrous composites are inherently anisotropic in that the maximum strength and reinforcement are achieved along the alignment (longitudinal) direction. In the transverse direction, fiber reinforcement is virtually nonexistent: fracture usually occurs at relatively low tensile stresses. For other stress orientations, composite strength lies between these extremes. The efficiency of fiber reinforcement for several situations is presented in Table 16.3; this efficiency is

Table 16.2

Properties of Unreinforced and Reinforced Polycarbonates with Randomly Oriented Glass Fibers

Property	Unreinforced	Value for Given Amount of Reinforcement (vol%)		
		20	30	40
Specific gravity	1.19–1.22	1.35	1.43	1.52
Tensile strength (MPa)	59–62	110	131	159
Modulus of elasticity (GPa)	2.24–2.345	5.93	8.62	11.6
Elongation (%)	90–115	4–6	3–5	3–5
Impact strength, notched Izod (N/cm)	21–28	3.5	3.5	4.4

Source: Adapted from Materials Engineering's *Materials Selector,* copyright © Penton/IPC.

taken to be unity for an oriented-fiber composite in the alignment direction and zero perpendicular to it.

When multidirectional stresses are imposed within a single plane, aligned layers that are fastened together on top of one another at different orientations are frequently used. These are termed *laminar composites,* which are discussed in Section 16.14.

Applications involving totally multidirectional applied stresses normally use discontinuous fibers, which are randomly oriented in the matrix material. Table 16.3 shows that the reinforcement efficiency is only one-fifth that of an aligned composite in the longitudinal direction; however, the mechanical characteristics are isotropic.

Consideration of orientation and fiber length for a particular composite depends on the level and nature of the applied stress as well as on the fabrication cost. Production rates for short-fiber composites (both aligned and randomly oriented) are rapid, and intricate shapes can be formed that are not possible with continuous fiber reinforcement. Furthermore, fabrication costs are considerably lower than for continuous and aligned fibers; fabrication techniques applied to short-fiber composite materials include compression, injection, and extrusion molding, which are described for unreinforced polymers in Section 17.14.

Concept Check 16.3 Cite one desirable characteristic and one less-desirable characteristic for (1) discontinuous- and oriented-fiber-reinforced composites and (2) discontinuous- and randomly oriented–fiber–reinforced composites.

[*The answer may be found at* www.wiley.com/college/callister *(Student Companion Site).*]

Table 16.3

Reinforcement Efficiency of Fiber–Reinforced Composites for Several Fiber Orientations and at Various Directions of Stress Application

Fiber Orientation	Stress Direction	Reinforcement Efficiency
All fibers parallel	Parallel to fibers	1
	Perpendicular to fibers	0
Fibers randomly and uniformly distributed within a specific plane	Any direction in the plane of the fibers	$\frac{3}{8}$
Fibers randomly and uniformly distributed within three dimensions in space	Any direction	$\frac{1}{5}$

Source: H. Krenchel, *Fibre Reinforcement,* Copenhagen: Akademisk Forlag, 1964 [33].

16.6 THE FIBER PHASE

An important characteristic of most materials, especially brittle ones, is that a small-diameter fiber is much stronger than the bulk material. As discussed in Section 14.6, the probability of the presence of a critical surface flaw that can lead to fracture decreases with decreasing specimen volume, and this feature is used to advantage in fiber-reinforced composites. Also, the materials used for reinforcing fibers have high tensile strengths.

whisker

On the basis of diameter and character, fibers are grouped into three different classifications: *whiskers, fibers,* and *wires.* **Whiskers** are very thin single crystals that have extremely large length-to-diameter ratios. As a consequence of their small size, they have a high degree of crystalline perfection and are virtually flaw-free, which accounts for their exceptionally high strengths; they are among the strongest known materials. In spite of these high strengths, whiskers are not used extensively as a reinforcement medium because they are extremely expensive. Moreover, it is difficult and often impractical to incorporate whiskers into a matrix. Whisker materials include graphite, silicon carbide, silicon nitride, and aluminum oxide; some mechanical characteristics of these materials are given in Table 16.4.

fiber

Materials that are classified as **fibers** are either polycrystalline or amorphous and have small diameters; fibrous materials are generally either polymers or ceramics (e.g., the polymer aramids, glass, carbon, boron, aluminum oxide, and silicon carbide). Table 16.4 also presents some data on a few materials that are used in fiber form.

Table 16.4 Characteristics of Several Fiber–Reinforcement Materials

Material	Specific Gravity	Tensile Strength (GPa)	Specific Strength (GPa)	Modulus of Elasticity (GPa)	Specific Modulus (GPa)
Whiskers					
Graphite	2.2	20	9.1	700	318
Silicon nitride	3.2	5–7	1.56–2.2	350–380	109–118
Aluminum oxide	4.0	10–20	2.5–5.0	700–1500	175–375
Silicon carbide	3.2	20	6.25	480	150
Fibers					
Aluminum oxide	3.95	1.38	0.35	379	96
Aramid (Kevlar 49)	1.44	3.6–4.1	2.5–2.85	131	91
Carbon[a]	1.78–2.15	1.5–4.8	0.70–2.70	228–724	106–407
E-glass	2.58	3.45	1.34	72.5	28.1
Boron	2.57	3.6	1.40	400	156
Silicon carbide	3.0	3.9	1.30	400	133
UHMWPE (Spectra 900)	0.97	2.6	2.68	117	121
Metallic Wires					
High-strength steel	7.9	2.39	0.30	210	26.6
Molybdenum	10.2	2.2	0.22	324	31.8
Tungsten	19.3	2.89	0.15	407	21.1

[a]As explained in Section 14.16, because these fibers are composed of both graphitic and turbostratic forms of carbon, the term *carbon* instead of *graphite* is used to denote these fibers.

Fine wires have relatively large diameters; typical materials include steel, molybdenum, and tungsten. Wires are used as a radial steel reinforcement in automobile tires, in filament-wound rocket casings, and in wire-wound high-pressure hoses.

16.7 THE MATRIX PHASE

The *matrix phase* of fibrous composites may be a metal, polymer, or ceramic. In general, metals and polymers are used as matrix materials because some ductility is desirable; for ceramic-matrix composites (Section 16.10), the reinforcing component is added to improve fracture toughness. The discussion of this section focuses on polymer and metal matrices.

For fiber-reinforced composites, the matrix phase serves several functions. First, it binds the fibers together and acts as the medium by which an externally applied stress is transmitted and distributed to the fibers; only a very small proportion of an applied load is sustained by the matrix phase. Furthermore, the matrix material should be ductile. In addition, the elastic modulus of the fiber should be much higher than that of the matrix. The second function of the matrix is to protect the individual fibers from surface damage as a result of mechanical abrasion or chemical reactions with the environment. Such interactions may introduce surface flaws capable of forming cracks, which may lead to failure at low tensile stress levels. Finally, the matrix separates the fibers and, by virtue of its relative softness and plasticity, prevents the propagation of brittle cracks from fiber to fiber, which could result in catastrophic failure; in other words, the matrix phase serves as a barrier to crack propagation. Even though some of the individual fibers fail, total composite fracture will not occur until large numbers of adjacent fibers fail and form a cluster of critical size.

It is essential that adhesive bonding forces between fiber and matrix be high to minimize fiber pullout. Bonding strength is an important consideration in the choice of the matrix–fiber combination. The ultimate strength of the composite depends to a large degree on the magnitude of this bond; adequate bonding is essential to maximize the stress transmittance from the weak matrix to the strong fibers.

16.8 POLYMER–MATRIX COMPOSITES

polymer–matrix
composite

Polymer–matrix composites (*PMCs*) consist of a polymer resin[2] as the matrix and fibers as the reinforcement medium. These materials are used in the greatest diversity of composite applications, as well as in the largest quantities, in light of their room-temperature properties, ease of fabrication, and cost. In this section the various classifications of PMCs are discussed according to reinforcement type (i.e., glass, carbon, and aramid), along with their applications and the various polymer resins that are employed.

Glass Fiber–Reinforced Polymer (GFRP) Composites

Fiberglass is simply a composite consisting of glass fibers, either continuous or discontinuous, contained within a polymer matrix; this type of composite is produced in the largest quantities. The composition of the glass that is most commonly drawn into fibers (sometimes

---

[2]The term *resin* is used in this context to denote a high-molecular-weight reinforcing plastic.

referred to as E-glass) is given in Table 14.3; fiber diameters normally range between 3 and 20 μm. Glass is popular as a fiber reinforcement material for several reasons:

1. It is easily drawn into high-strength fibers from the molten state.

2. It is readily available and may be fabricated into a glass-reinforced plastic economically using a wide variety of composite-manufacturing techniques.

3. As a fiber it is relatively strong, and when embedded in a plastic matrix, it produces a composite having a very high specific strength.

4. When coupled with the various plastics, it possesses a chemical inertness that renders the composite useful in a variety of corrosive environments.

The surface characteristics of glass fibers are extremely important because even minute surface flaws can deleteriously affect the tensile properties, as discussed in Section 14.6. Surface flaws are easily introduced by rubbing or abrading the surface with another hard material. Also, glass surfaces that have been exposed to the normal atmosphere for even short time periods generally have a weakened surface layer that interferes with bonding to the matrix. Newly drawn fibers are normally coated during drawing with a *size,* a thin layer of a substance that protects the fiber surface from damage and undesirable environmental interactions. This size is ordinarily removed before composite fabrication and replaced with a *coupling agent* or finish that produces a chemical bond between the fiber and matrix.

There are several limitations to this group of materials. In spite of having high strengths, they are not very stiff and do not display the rigidity that is necessary for some applications (e.g., as structural members for airplanes and bridges). Most fiberglass materials are limited to service temperatures below 200°C; at higher temperatures, most polymers begin to flow or to deteriorate. Service temperatures may be extended to approximately 300°C by using high-purity fused silica for the fibers and high-temperature polymers such as the polyimide resins.

Many fiberglass applications are familiar: automotive and marine bodies, plastic pipes, storage containers, and industrial floorings. The transportation industries are using increasing amounts of glass fiber–reinforced plastics in an effort to decrease vehicle weight and boost fuel efficiencies. A host of new applications is being used or currently investigated by the automotive industry.

Carbon Fiber–Reinforced Polymer (CFRP) Composites

Carbon is a high-performance fiber material that is the most commonly used reinforcement in advanced (i.e., nonfiberglass) polymer-matrix composites. The reasons for this are as follows:

1. Carbon fibers have high specific moduli and specific strengths.

2. They retain their high tensile modulus and high strength at elevated temperatures; high-temperature oxidation, however, may be a problem.

3. At room temperature, carbon fibers are not affected by moisture or a wide variety of solvents, acids, and bases.

4. These fibers exhibit a diversity of physical and mechanical characteristics, allowing composites incorporating these fibers to have specific engineered properties.

5. Fiber- and composite-manufacturing processes have been developed that are relatively inexpensive and cost effective.

A schematic representation of a typical carbon fiber is shown in Figure 14.18, where it may be noted that the fiber is composed of both graphitic (ordered) and turbostratic (disordered) structures.

Manufacturing techniques for producing carbon fibers are relatively complex and are not discussed. However, three different organic precursor materials are used: rayon, polyacrylonitrile (PAN), and pitch. Processing techniques vary from precursor to precursor, as do the resultant fiber characteristics.

One classification scheme for carbon fibers is by tensile modulus; on this basis, the four classes are standard, intermediate, high, and ultrahigh moduli. Fiber diameters normally range between 4 and 10 μm; both continuous and chopped forms are available. In addition, carbon fibers are normally coated with a protective epoxy size that also improves adhesion with the polymer matrix.

Carbon-reinforced polymer composites are currently being used extensively in sports and recreational equipment (fishing rods, golf clubs), filament-wound rocket motor cases, pressure vessels, and aircraft structural components—both military and commercial, both fixed-wing aircraft and helicopters (e.g., as wing, body, stabilizer, and rudder components).

Aramid Fiber–Reinforced Polymer Composites

Aramid fibers are high-strength, high-modulus materials that were introduced in the early 1970s. They are especially desirable for their outstanding strength-to-weight ratios, which are superior to those of metals. Chemically, this group of materials is known as poly(paraphenylene terephthalamide). There are a number of aramid materials; trade names for two of the most common are Kevlar and Nomex. For the former, there are several grades (Kevlar 29, 49, and 149) that have different mechanical behaviors. During synthesis, the rigid molecules are aligned in the direction of the fiber axis, as liquid crystal domains (Section 15.19); the repeat unit and the mode of chain alignment are represented in Figure 16.10. Mechanically, these fibers have longitudinal tensile strengths and tensile moduli (Table 16.4) that are higher than those of other polymeric fiber materials; however, they are relatively weak in compression. In addition, this material is known for its toughness, impact resistance, and resistance to creep and fatigue failure. Even though the aramids are thermoplastics, they are, nevertheless, resistant to combustion and stable to relatively high temperatures; the temperature range over which they retain their high mechanical properties is between −200°C and 200°C. Chemically, they are susceptible to degradation by strong acids and bases, but they are relatively inert in other solvents and chemicals.

The aramid fibers are most often used in composites having polymer matrices; common matrix materials are the epoxies and polyesters. Because the fibers are relatively flexible and somewhat ductile, they may be processed by most common textile operations. Typical applications of these aramid composites are in ballistic products (bulletproof

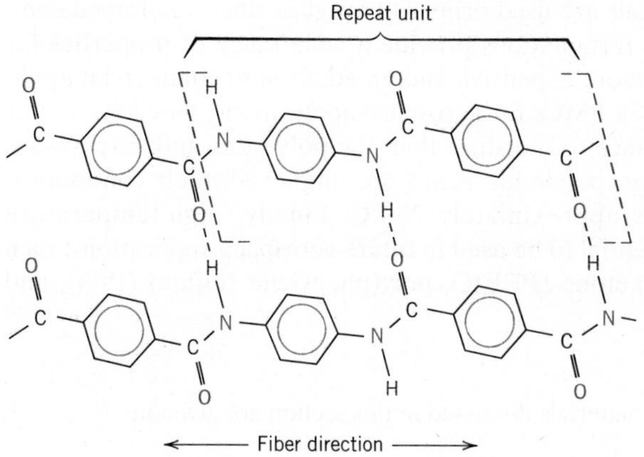

Repeat unit

Fiber direction

Figure 16.10 Schematic representation of repeat unit and chain structures for aramid (Kevlar) fibers. Chain alignment with the fiber direction and hydrogen bonds that form between adjacent chains are also shown. [From F. R. Jones (Editor), *Handbook of Polymer-Fibre Composites.* Copyright © 1994 by Addison-Wesley Longman. Reprinted with permission.]

Table 16.5

Properties of Continuous and Aligned Glass, Carbon, and Aramid Fiber–Reinforced Epoxy–Matrix Composites in Longitudinal and Transverse Directions[a]

Property	Glass (E-Glass)	Carbon (High Strength)	Aramid (Kevlar 49)
Specific gravity	2.1	1.6	1.4
Tensile modulus			
Longitudinal (GPa)	45	145	76
Transverse (GPa)	12	10	5.5
Tensile strength			
Longitudinal (MPa)	1020	1240	1380
Transverse (MPa)	40	41	30
Ultimate tensile strain			
Longitudinal	2.3	0.9	1.8
Transverse	0.4	0.4	0.5

[a]In all cases, the fiber volume fraction is 0.60.

Source: Adapted from R. F. Floral and S. T. Peters, "Composite Structures and Technologies," tutorial notes, 1989.

vests and armor), sporting goods, tires, ropes, missile cases, and pressure vessels and as a replacement for asbestos in automotive brake and clutch linings and gaskets.

The properties of continuous and aligned glass, carbon, and aramid fiber–reinforced epoxy composites are given in Table 16.5. A comparison of the mechanical characteristics of these three materials may be made in both longitudinal and transverse directions.

Other Fiber Reinforcement Materials

Glass, carbon, and the aramids are the most common fiber reinforcements incorporated into polymer matrices. Other fiber materials that are used to much lesser degrees are boron, silicon carbide, and aluminum oxide; tensile moduli, tensile strengths, specific strengths, and specific moduli of these materials in fiber form are given in Table 16.4. Boron fiber–reinforced polymer composites have been used in military aircraft components, helicopter rotor blades, and sporting goods. Silicon carbide and aluminum oxide fibers are used in tennis rackets, circuit boards, military armor, and rocket nose cones.

Polymer-Matrix Materials

The roles assumed by the polymer matrix are outlined in Section 16.7. In addition, the matrix often determines the maximum service temperature because it normally softens, melts, or degrades at a much lower temperature than the fiber reinforcement.

The most widely used and least expensive polymer resins are the polyesters and vinyl esters.[3] These matrix materials are used primarily for glass fiber–reinforced composites. A large number of resin formulations provide a wide range of properties for these polymers. The epoxies are more expensive and, in addition to commercial applications, are also used extensively in PMCs for aerospace applications; they have better mechanical properties and resistance to moisture than the polyesters and vinyl resins. For high-temperature applications, polyimide resins are employed; their continuous-use, upper-temperature limit is approximately 230°C. Finally, high-temperature thermoplastic resins offer the potential to be used in future aerospace applications; such materials include polyetheretherketone (PEEK), poly(phenylene sulfide) (PPS), and polyetherimide (PEI).

---

[3]The chemistry and typical properties of some of the matrix materials discussed in this section are given in Appendices B, D, and E.

DESIGN EXAMPLE 16.1

Design of a Tubular Composite Shaft

A tubular composite shaft is to be designed that has an outside diameter of 70 mm, an inside diameter of 50 mm, and a length of 1.0 m; such is represented schematically in Figure 16.11. The mechanical characteristic of prime importance is bending stiffness in terms of the longitudinal modulus of elasticity; strength and fatigue resistance are not significant parameters for this application when filament composites are used. Stiffness is to be specified as maximum allowable deflection in bending; when subjected to three-point bending as in Figure 14.9 (i.e., support points at both tube extremities and load application at the longitudinal midpoint), a load of 1000 N is to produce an elastic deflection of no more than 0.35 mm at the midpoint position.

Continuous fibers that are oriented parallel to the tube axis will be used; possible fiber materials are glass, and carbon in standard-, intermediate-, and high-modulus grades. The matrix material is to be an epoxy resin, and the maximum allowable fiber volume fraction is 0.60.

This design problem calls for us to do the following:

(a) Decide which of the four fiber materials, when embedded in the epoxy matrix, meet the stipulated criteria.
(b) Of these possibilities, select the one fiber material that will yield the lowest-cost composite material (assuming fabrication costs are the same for all fibers).

Elastic modulus, density, and cost data for the fiber and matrix materials are given in Table 16.6.

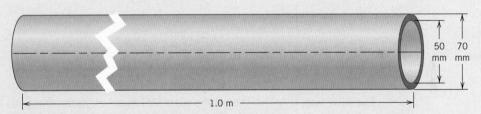

Figure 16.11 Schematic representation of a tubular composite shaft, the subject of Design Example 16.1.

Table 16.6 Elastic Modulus, Density, and Cost Data for Glass and Various Carbon Fibers and Epoxy Resin

Material	Elastic Modulus (GPa)	Density (g/cm³)	Cost ($US/kg)
Glass fibers	72.5	2.58	2.10
Carbon fibers (standard modulus)	230	1.80	60.00
Carbon fibers (intermediate modulus)	285	1.80	95.00
Carbon fibers (high modulus)	400	1.80	250.00
Epoxy resin	2.4	1.14	6.00

Solution

(a) It is first necessary to determine the required longitudinal modulus of elasticity for this composite material consistent with the stipulated criteria. This computation requires the use of the three-point deflection expression

$$\Delta y = \frac{FL^3}{48\,EI} \tag{16.21}$$

in which Δy is the midpoint deflection, F is the applied force, L is the support point separation distance, E is the modulus of elasticity, and I is the cross-sectional moment of inertia. For a tube having inside and outside diameters of d_i and d_o, respectively,

$$I = \frac{\pi}{64}(d_o^4 - d_i^4) \tag{16.22}$$

and

$$E = \frac{4FL^3}{3\pi\,\Delta y\,(d_o^4 - d_i^4)} \tag{16.23}$$

For this shaft design,

$$F = 1000\,\text{N}$$
$$L = 1.0\,\text{m}$$
$$\Delta y = 0.35\,\text{mm}$$
$$d_o = 70\,\text{mm}$$
$$d_i = 50\,\text{mm}$$

Thus, the required longitudinal modulus of elasticity for this shaft is

$$E = \frac{4(1000\,\text{N})(1.0\,\text{m})^3}{3\pi(0.35 \times 10^{-3}\,\text{m})\big[(70 \times 10^{-3}\,\text{m})^4 - (50 \times 10^{-3}\,\text{m})^4\big]}$$

$$= 69.3\,\text{GPa}$$

The next step is to determine the fiber and matrix volume fractions for each of the four candidate fiber materials. This is possible using the rule-of-mixtures expression, Equation 16.10b:

$$E_{cs} = E_m V_m + E_f V_f = E_m(1 - V_f) + E_f V_f$$

Table 16.7 lists the V_m and V_f values required for $E_{cs} = 69.3$ GPa; Equation 16.10b and the moduli data in Table 16.6 were used in these computations. Only the three carbon-fiber types are possible candidates because their V_f values are less than 0.6.

(b) At this point it becomes necessary to determine the volume of fibers and matrix for each of the three carbon types. The total tube volume V_c in centimeters is

$$V_c = \frac{\pi L}{4}(d_o^2 - d_i^2) \tag{16.24}$$

$$= \frac{\pi(100\,\text{cm})}{4}\big[(7.0\,\text{cm})^2 - (5.0\,\text{cm})^2\big]$$

$$= 1885\,\text{cm}^3$$

Thus, fiber and matrix volumes result from products of this value and the V_f and V_m values cited in Table 16.7. These volume values are presented in Table 16.8, which are then converted

Table 16.7 Fiber and Matrix Volume Fractions for Glass and Three Carbon-Fiber Types as Required to Give a Composite Modulus of 69.3 GPa

Fiber Type	V_f	V_m
Glass	0.954	0.046
Carbon (standard modulus)	0.293	0.707
Carbon (intermediate modulus)	0.237	0.763
Carbon (high modulus)	0.168	0.832

into masses using densities (Table 16.6), and finally into material costs, from the per unit mass cost (also given in Table 16.6).

As may be noted in Table 16.8, the material of choice (i.e., the least expensive) is the standard-modulus carbon-fiber composite; the relatively low cost per unit mass of this fiber material offsets its relatively low modulus of elasticity and required high volume fraction.

Table 16.8 Fiber and Matrix Volumes, Masses, and Costs and Total Material Cost for Three Carbon-Fiber Epoxy–Matrix Composites

Fiber Type	Fiber Volume (cm^3)	Fiber Mass (kg)	Fiber Cost ($US)	Matrix Volume (cm^3)	Matrix Mass (kg)	Matrix Cost ($US)	Total Cost ($US)
Carbon (standard modulus)	552	0.994	59.60	1333	1.520	9.10	68.70
Carbon (intermediate modulus)	447	0.805	76.50	1438	1.639	9.80	86.30
Carbon (high modulus)	317	0.571	142.80	1568	1.788	10.70	153.50

16.9 METAL-MATRIX COMPOSITES

metal-matrix composite

As the name implies, for **metal-matrix composites** (*MMC*s) the matrix is a ductile metal. These materials may be used at higher service temperatures than their base-metal counterparts; furthermore, the reinforcement may improve specific stiffness, specific strength, abrasion resistance, creep resistance, thermal conductivity, and dimensional stability. Some of the advantages of these materials over the polymer-matrix composites include higher operating temperatures, nonflammability, and greater resistance to degradation by organic fluids. Metal-matrix composites are much more expensive than PMCs, and, therefore, MMC use is somewhat restricted.

The superalloys, as well as alloys of aluminum, magnesium, titanium, and copper, are used as matrix materials. The reinforcement may be in the form of particulates, both continuous and discontinuous fibers, and whiskers; concentrations normally range between 10 and 60 vol%. Continuous-fiber materials include carbon, silicon carbide, boron, aluminum oxide, and the refractory metals. However, discontinuous reinforcements consist primarily of silicon carbide whiskers, chopped fibers of aluminum oxide and carbon, or particulates of silicon carbide and aluminum oxide. In a sense, the cermets (Section 16.2) fall within this MMC scheme. Table 16.9 presents the properties of several common metal-matrix, continuous and aligned fiber–reinforced composites.

Table 16.9 *Properties of Several Metal-Matrix Composites Reinforced with Continuous and Aligned Fibers*

Fiber	Matrix	Fiber Content (vol%)	Density (g/cm³)	Longitudinal Tensile Modulus (GPa)	Longitudinal Tensile Strength (MPa)
Carbon	6061 Al	41	2.44	320	620
Boron	6061 Al	48	—	207	1515
SiC	6061 Al	50	2.93	230	1480
Alumina	380.0 Al	24	—	120	340
Carbon	AZ31 Mg	38	1.83	300	510
Borsic	Ti	45	3.68	220	1270

Source: Adapted from J. W. Weeton, D. M. Peters, and K. L. Thomas, *Engineers' Guide to Composite Materials*, ASM International, Materials Park, OH, 1987.

Some matrix–reinforcement combinations are highly reactive at elevated temperatures. Consequently, composite degradation may be caused by high-temperature processing or by subjecting the MMC to elevated temperatures during service. This problem is commonly resolved either by applying a protective surface coating to the reinforcement or by modifying the matrix alloy composition.

Normally the processing of MMCs involves at least two steps: consolidation or synthesis (i.e., introduction of reinforcement into the matrix), followed by a shaping operation. A host of consolidation techniques are available, some of which are relatively sophisticated; discontinuous-fiber MMCs are amenable to shaping by standard metal-forming operations (e.g., forging, extrusion, rolling).

Automobile manufacturers have recently begun to use MMCs in their products. For example, some engine components have been introduced consisting of an aluminum-alloy matrix that is reinforced with aluminum oxide and carbon fibers; this MMC is light in weight and resists wear and thermal distortion. Metal-matrix composites are also employed in driveshafts (that have higher rotational speeds and reduced vibrational noise levels), extruded stabilizer bars, and forged suspension and transmission components.

The aerospace industry also employs MMCs in the form of advanced aluminum-alloy metal-matrix composites. These materials have low densities, and it is possible to control their properties (i.e., mechanical and thermal properties). Continuous graphite fibers are used as the reinforcement for an antenna boom on the Hubble Space Telescope; this boom stabilizes the antenna position during space maneuvers. In addition, Global Positioning System (GPS) satellites use silicon carbide–aluminum and graphite–aluminum MMCs for electronic packaging and thermal management systems. These MMCs have high thermal conductivities, and it is possible to match their coefficients of expansion with those of other electronic materials in the GPS components.

The high-temperature creep and rupture properties of some superalloys (Ni- and Co-based alloys) may be enhanced by fiber reinforcement using refractory metals such as tungsten. Excellent high-temperature oxidation resistance and impact strength are also maintained. Designs incorporating these composites permit higher operating temperatures and better efficiencies for turbine engines.

16.10 CERAMIC–MATRIX COMPOSITES

As discussed in Chapters 14, ceramic materials are inherently resilient to oxidation and deterioration at elevated temperatures; were it not for their disposition to brittle fracture, some of these materials would be ideal candidates for use in high-temperature and severe-stress applications, specifically for components in automobile and aircraft gas turbine engines. Fracture toughness values for ceramic materials are low and typically lie between 1 and 5 MPa$\sqrt{m}$; see Table 10.1 and Table B.5 in Appendix B. By way of contrast, K_{Ic} values for most metals are much higher (15 to greater than 150 MPa$\sqrt{m}$).

The fracture toughnesses of ceramics have been improved significantly by the development of a new generation of **ceramic-matrix composites** (*CMCs*)—particulates, fibers, or whiskers of one ceramic material that have been embedded into a matrix of another ceramic. Ceramic-matrix composite materials have extended fracture toughnesses to between about 6 and 20 MPa$\sqrt{m}$.

In essence, this improvement in the fracture properties results from interactions between advancing cracks and dispersed phase particles. Crack initiation normally occurs with the matrix phase, whereas crack propagation is impeded or hindered by the particles, fibers, or whiskers. Several techniques are used to retard crack propagation, which are discussed as follows.

One particularly interesting toughening technique uses a phase transformation to arrest the propagation of cracks and is aptly termed *transformation toughening*. Small particles of partially stabilized zirconia (Section 14.4) are dispersed within the matrix material, often Al_2O_3 or ZrO_2 itself. Typically, CaO, MgO, Y_2O_3, and CeO are used as stabilizers. Partial stabilization allows retention of the metastable tetragonal phase at ambient conditions rather than the stable monoclinic phase; these two phases are noted on the ZrO_2–$CaZrO_3$ phase diagram in Figure 14.3. The stress field in front of a propagating crack causes these metastably retained tetragonal particles to undergo transformation to the stable monoclinic phase. Accompanying this transformation is a slight particle volume increase, and the net result is that compressive stresses are established on the crack surfaces near the crack tip that tend to pinch the crack shut, thereby arresting its growth. This process is demonstrated schematically in Figure 16.12.

Other recently developed toughening techniques involve the use of ceramic whiskers, often SiC or Si_3N_4. These whiskers may inhibit crack propagation by (1) deflecting crack tips, (2) forming bridges across crack faces, (3) absorbing energy during pullout as the whiskers debond from the matrix, and/or (4) causing a redistribution of stresses in regions adjacent to the crack tips.

In general, increasing fiber content improves strength and fracture toughness; this is demonstrated in Table 16.10 for SiC whisker-reinforced alumina. Furthermore, there is a considerable reduction in the scatter of fracture strengths for whisker-reinforced ceramics relative to their unreinforced counterparts. In addition, these CMCs exhibit

ceramic-matrix composite

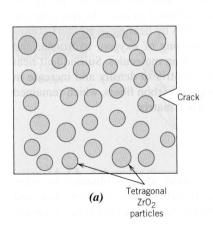

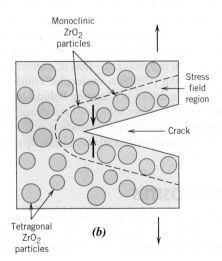

Figure 16.12
Schematic demonstration of transformation toughening. (*a*) A crack prior to inducement of the ZrO_2 particle phase transformation. (*b*) Crack arrestment due to the stress-induced phase transformation.

carbon, and aramid fibers, normally added in concentrations between 40 and 70 vol%. Commonly used matrix materials include polyesters, vinyl esters, and epoxy resins.

Pultrusion is a continuous process that is easily automated; production rates are relatively high, making it very cost effective. Furthermore, a wide variety of shapes are possible, and there is really no practical limit to the length of stock that may be manufactured.

Prepreg Production Processes

prepreg

Prepreg is the composite industry's term for continuous-fiber reinforcement preimpregnated with a polymer resin that is only partially cured. This material is delivered in tape form to the manufacturer, which then directly molds and fully cures the product without having to add any resin. It is probably the composite material form most widely used for structural applications.

The prepregging process, represented schematically for thermoset polymers in Figure 16.14, begins by collimating a series of spool-wound continuous-fiber tows. These tows are then sandwiched and pressed between sheets of release and carrier paper using heated rollers, a process termed *calendering*. The release paper sheet has been coated with a thin film of heated resin solution of relatively low viscosity so as to provide for its thorough impregnation of the fibers. A *doctor blade* spreads the resin into a film of uniform thickness and width. The final prepreg product—the thin tape consisting of continuous and aligned fibers embedded in a partially cured resin—is prepared for packaging by winding onto a cardboard core. As shown in Figure 16.14, the release paper sheet is removed as the impregnated tape is spooled. Typical tape thicknesses range between 0.08 and 0.25 mm and tape widths range between 25 and 1525 mm; resin content usually lies between about 35 and 45 vol%.

At room temperature, the thermoset matrix undergoes curing reactions; therefore, the prepreg is stored at 0°C or lower. Also, the time in use at room temperature (or *outtime*) must be minimized. If properly handled, thermoset prepregs have a lifetime of at least six months and usually longer.

Both thermoplastic and thermosetting resins are used; carbon, glass, and aramid fibers are the common reinforcements.

Actual fabrication begins with the *lay-up*—laying of the prepreg tape onto a tooled surface. Normally a number of plies are laid up (after removal from the carrier backing paper) to provide the desired thickness. The lay-up arrangement may be unidirectional, but more often the fiber orientation is alternated to produce a cross-ply or angle-ply laminate (Section 16.14). Final curing is accomplished by the simultaneous application of heat and pressure.

The lay-up procedure may be carried out entirely by hand (hand lay-up), in which the operator both cuts the lengths of tape and then positions them in the desired orientation on the tooled surface. Alternatively, tape patterns may be machine cut, then hand laid. Fabrication costs can be further reduced by automation of prepreg lay-up and other manufacturing procedures (e.g., filament winding, as discussed next), which virtually eliminates the need for hand labor. These automated methods are essential for many applications of composite materials to be cost effective.

Filament Winding

Filament winding is a process by which continuous reinforcing fibers are accurately positioned in a predetermined pattern to form a hollow (usually cylindrical) shape. The fibers, either as individual strands or as tows, are first fed through a resin bath and then are continuously wound onto a mandrel, usually using automated winding equipment (Figure 16.15). After the appropriate number of layers have been applied, curing is carried out either in an oven or at room temperature, after which the mandrel is removed. As an alternative, narrow and thin prepregs (i.e., tow pregs) 10 mm or less in width may be filament wound.

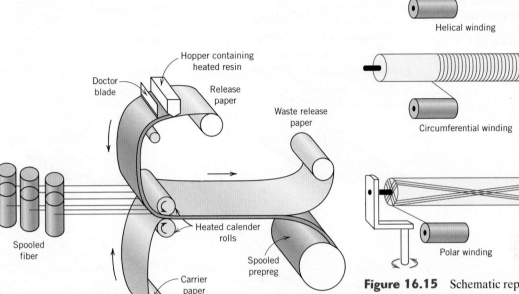

Figure 16.14 Schematic diagram illustrating the production of prepreg tape using a thermoset polymer.

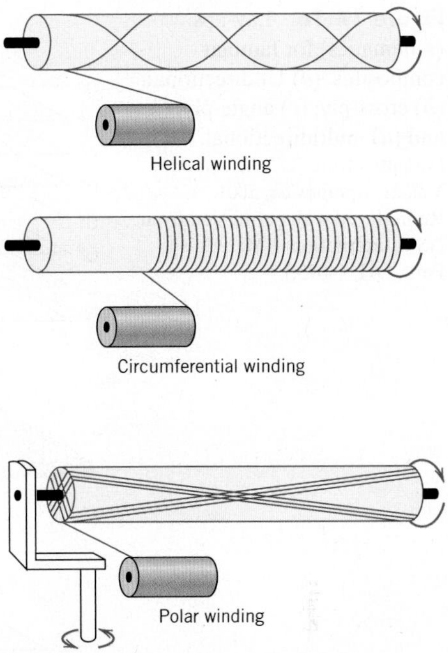

Figure 16.15 Schematic representations of helical, circumferential, and polar filament winding techniques.
[From N. L. Hancox, (Editor), *Fibre Composite Hybrid Materials,* The Macmillan Company, New York, 1981.]

Various winding patterns are possible (i.e., circumferential, helical, and polar) to give the desired mechanical characteristics. Filament-wound parts have very high strength-to-weight ratios. Also, a high degree of control over winding uniformity and orientation is afforded with this technique. Furthermore, when automated, the process is most economically attractive. Common filament-wound structures include rocket motor casings, storage tanks and pipes, and pressure vessels.

Manufacturing techniques are being used to produce a wide variety of structural shapes that are not necessarily limited to surfaces of revolution (e.g., I-beams). This technology is advancing very rapidly because it is very cost effective.

Structural Composites

structural composite A **structural composite** is a multi-layered and normally low-density composite used in applications requiring structural integrity, ordinarily high tensile, compressive, and torsional strengths and stiffnesses. The properties of these composites depend not only on the properties of the constituent materials, but also on the geometrical design of the structural elements. Laminar composites and sandwich panels are two of the most common structural composites.

16.14 LAMINAR COMPOSITES

laminar composite A **laminar composite** is composed of two-dimensional sheets or panels (*plies* or *laminae*) bonded to one another. Each ply has a preferred high-strength direction, such as is found in continuous and aligned fiber–reinforced polymers. A multi-layered structure such as

Figure 16.16 Lay-ups (schematics) for laminar composites. (*a*) Undirectional; (*b*) cross-ply; (*c*) angle-ply; and (*d*) multidirectional. (Adapted from *ASM Handbook,* Vol. 21, *Composites,* 2001. Reproduced with permission from ASM International, Materials Park, OH, 44073.)

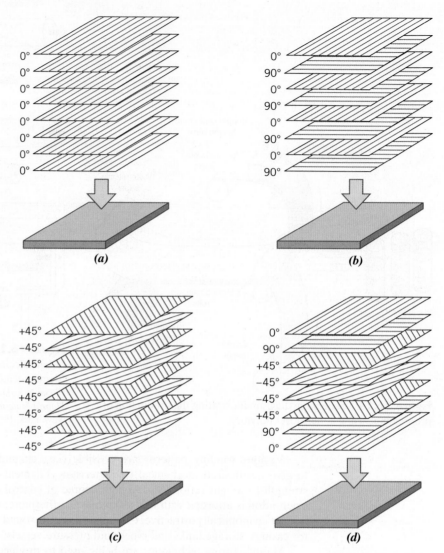

this is termed a *laminate*. Laminate properties depend on several factors to include how the high-strength direction varies from layer to layer. In this regard, there are four classes of laminar composites: *unidirectional, cross-ply, angle-ply,* and *multidirectional.* For unidirectional, the orientation of the high-strength direction for all laminae is the same (Figure 16.16*a*); cross-ply laminates have alternating high-strength layer orientations of 0° and 90° (Figure 16.16*b*); and for angle-ply, successive layers alternate between $+\theta$ and $-\theta$ high-strength orientations (e.g., ±45°) (Figure 16.16*c*). The multidirectional laminates have several high-strength orientations (Figure 16.16*d*). For virtually all laminates, layers are typically stacked such that fiber orientations are symmetric relative to the laminate midplane; this arrangement prevents any out-of-plane twisting or bending.

In-plane properties (e.g., modulus of elasticity and strength) of unidirectional laminates are highly anisotropic. Cross-, angle-, and multidirectional laminates are designed to increase the degree of in-plane isotropy; multidirectional can be fabricated to be most isotropic; degree of isotropy decreases with angle- and cross-ply materials.

Stress and strain relationships for laminates have been developed that are analogous to Equations 16.10 and 16.16 for continuous and aligned fiber–reinforced composites. However, these expressions use tensor algebra, which is beyond the scope of this discussion.

One of the most common laminate materials is unidirectional prepreg tape in an uncured matrix resin. A multi-layered structure having the desired configuration is produced during lay-up as a number of tapes are laid one upon another at a variety of predetermined high-strength orientations. Overall strength and degree of isotropy depends on fiber material, number of layers, as well as orientation sequence. Most laminate fiber materials are carbon, glass, and aramid. Subsequent to lay-up, the resin must be cured and layers bonded together; this is accomplished by heating the part while pressure is being applied. Techniques used for post-lay-up processing include autoclave molding, pressure-bag molding, and vacuum-bag molding.

Laminations may also be constructed using fabric material such as cotton, paper, or woven-glass fibers embedded in a plastic matrix. In-plane degree of isotropy is relatively high in this group of materials.

Applications that use laminate composites are primarily in aircraft, automotive, marine, and building/civil-infrastructure sectors. Specific applications include the following: aircraft—fuselage, vertical and horizontal stabilizers, landing-gear hatch, floors, fairings, and rotor blades for helicopters; automotive—automobile panels, sports car bodies, and drive shafts; marine—ship hulls, hatch covers, deckhouses, bulkheads, and propellers; building/civil-infrastructure—bridge components, long-span roof structures, beams, structural panels, roof panels, and tanks.

Laminates are also used extensively in sports and recreation equipment. For example, the modern ski (see the chapter-opening illustration) is a relatively complex laminated structure.

16.15 SANDWICH PANELS

sandwich panel

Sandwich panels, a class of structural composites, are designed to be lightweight beams or panels having relatively high stiffnesses and strengths. A sandwich panel consists of two outer sheets, faces, or skins that are separated by and adhesively bonded to a thicker core (Figure 16.17). The outer sheets are made of a relatively stiff and strong material, typically aluminum alloys, steel and stainless steel, fiber-reinforced plastics, and plywood; they carry bending loads that are applied to the panel. When a sandwich panel is bent, one face experiences compressive stresses, the other tensile stresses.

The core material is lightweight and normally has a low modulus of elasticity. Structurally, it serves several functions. First, it provides continuous support for the faces and holds them together. In addition, it must have sufficient shear strength to withstand transverse shear stresses and also be thick enough to provide high shear stiffness (to resist buckling of the panel). Tensile and compressive stresses on the core are much lower than on the faces. Panel stiffness depends primarily on the properties of the core material and core thickness; bending stiffness increases significantly with increasing core thickness. Furthermore, it is essential that faces be bonded strongly to the core.

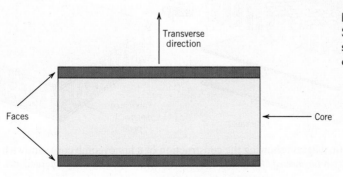

Figure 16.17
Schematic diagram showing the cross section of a sandwich panel.

The sandwich panel is a cost-effective composite because core materials are less expensive than materials used for the faces.

Core materials typically fall within three categories: rigid polymeric foams, wood, and honeycombs.

- Both thermoplastic and thermosetting polymers are used as rigid foam materials; these include (and are ranked from least to most expensive) polystyrene, phenol-formaldehyde (phenolic), polyurethane, poly(vinyl chloride), polypropylene, polyetherimide, and polymethacrylimide.

- Balsa wood is also commonly used as a core material for several reasons: (1) Its density is extremely low (0.10 to 0.25 g/cm^3), which, however, is higher than some other core materials; (2) it is relatively inexpensive; and (3) it has relatively high compression and shear strengths.

- Another popular core consists of a "honeycomb" structure—thin foils that have been formed into interlocking cells (having hexagonal as well as other configurations), with axes oriented perpendicular to the face planes; Figure 16.18 shows a cutaway view of a hexagonal honeycomb core sandwich panel. Mechanical properties of honeycombs are anisotropic: tensile and compressive strengths are greatest in a direction parallel to the cell axis; shear strength is highest in the plane of the panel. Strength and stiffness of honeycomb structures depend on cell size, cell wall thickness, and the material from which the honeycomb is made. Honeycomb structures also have excellent sound and vibration damping characteristics because of the high volume fraction of void space within each cell. Honeycombs are fabricated from thin sheets. Materials used for these core structures include metal alloys—aluminum, titanium, nickel-based, and stainless steels; and polymers—polypropylene, polyurethane, kraft paper (a tough brown paper used for heavy-duty shopping bags and cardboard), and aramid fibers.

Sandwich panels are used in a wide variety aircraft, construction, automotive, and marine applications, including the following: aircraft—leading and trailing edges, radomes, fairings, nacelles (cowlings and fan-duct sections around turbine engines), flaps, rudders, stabilizers, and rotor blades for helicopters; construction—architectural cladding for buildings, decorative facades and interior surfaces, insulated roof and wall systems, clean-room panels, and built-in cabinetry; automotive—headliners, luggage compartment floors, spare wheel covers, and cabin floors; marine—bulkheads, furniture, and wall, ceiling, and partition panels.

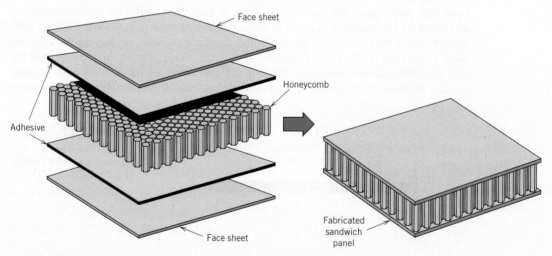

Figure 16.18 Schematic diagram showing the construction of a honeycomb core sandwich panel.
(Reprinted with permission from *Engineered Materials Handbook,* Vol. 1, *Composites,* ASM International, Materials Park, OH, 1987.)

CASE STUDY

Use of Composites in the Boeing 787 Dreamliner

A revolution in the use of composite materials for commercial aircraft has recently commenced with the advent of the Boeing 787 Dreamliner (Figure 16.19). This aircraft—a long-range, mid-size (210 to 290 passenger capacity), twin-engine jet airliner—is the first to use composite materials for the majority of its construction. Thus, it is lighter in weight than its predecessors, which leads to greater fuel efficiency (a reduction of approximately 20%), fewer emissions, and longer flying ranges. Furthermore, this composite construction makes for a more comfortable flying experience—cabin pressure and humidity levels are higher than for its predecessors and noise levels have been reduced. In addition, overhead bins are roomier and windows are larger.

Composite materials account for 50% (by weight) of the Dreamliner and aluminum alloys 20%. By way of contrast, the Boeing 777 consists of 11% composites and 70% aluminum alloys. These composite and aluminum contents as well as contents for other materials used in the construction of both 777 and 787 aircraft (i.e., titanium alloys, steel, and other) are listed in Table 16.11.

By far the most common composite structures are continuous carbon fiber–epoxy laminates, the majority of which are used in the fuselage (Figure 16.20). These laminates are composed of prepreg tapes that are laid one upon another in predetermined orientations using a continuous tape–laying machine. A single-piece section of fuselage (or *barrel*) is fashioned in this manner, which is subsequently cured under pressure in a huge autoclave. Six such barrel units are attached to one another to form the complete fuselage. For previous

Table 16.11 Material Types and Contents for Boeing 787 and 777 Aircraft

	Material Content (Weight Percent)				
Aircraft	*Composites*	*Al Alloys*	*Ti Alloys*	*Steel*	*Other*
787	50	20	15	10	5
777	11	70	7	11	1

commercial aircraft, the primary components of the fuselage structure were aluminum sheets fastened together using rivets. Advantages of this composite barrel structure over previous designs using aluminum alloys include the following:

- Reductions in assembly costs—approximately 1500 aluminum sheets that are fastened together with approximately 50,000 rivets are eliminated.

- Cost reductions for scheduled maintenance and inspections for corrosion and fatigue cracks.

- Reductions in aerodynamic drag—rivets protruding from surfaces increase wind resistance and decrease fuel efficiency.

The fuselage of the Dreamliner was the first attempt to mass produce extremely large composite structures composed of carbon fibers embedded in a thermosetting polymer (i.e., an epoxy). Thus, it became necessary for Boeing (and its subcontractors)

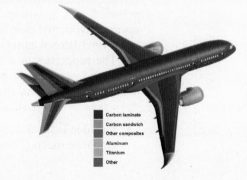

Figure 16.20 Locations of the various material types used in the Boeing 787 Dreamliner.
(Adapted from Ghabchi, Arash, "Thermal Spray at Boeing: Past, Present, and Future." *International Thermal Spray & Surface Engineering (iTSSe)*, Vol. 8, No, 1, February 2013, ASM International, Materials Park, OH.)

© Jens Wolf/dpa/Corbis

Figure 16.19 A Boeing 787 Dreamliner.

to develop and implement new and innovative manufacturing technologies.

As Figure 16.20 notes, carbon laminates are also used in wing and tail structures. The other composites indicated in this same illustration are glass fiber–reinforced epoxy and hybrid composites, which are composed of both glass and carbon fibers. These other composites are used primarily in tail and trailing wing structures.

Sandwich panels are used in *nacelles* (i.e., housing structures that surround the engines) as well as trailing tail components (Figure 16.20). Faces for most of these panels are carbon fiber–epoxy laminates, whereas cores are honeycomb structures typically made from aluminum alloy sheet. Noise reduction of some nacelle components is increased by embedding a nonmetallic (or "cap" material) within the honeycomb cells.

16.16 NANOCOMPOSITES

nanocomposite

The materials world is experiencing a revolution with the development of a new class of composite materials—**nanocomposites**. Nanocomposites are composed of nano-sized particles (or *nanoparticles*)[5] that are embedded in a matrix material. They can be designed to have mechanical, electrical, magnetic, optical, thermal, biological, and transport properties that are superior to conventional filler materials; furthermore, these properties can be tailored for use in specific applications. For these reasons, nanocomposites are becoming infused in a number of modern technologies.[6]

An interesting and novel phenomenon accompanies the decrease in size of a nanoparticle—its physical and chemical properties experience dramatic changes; furthermore, the degree of change depends on particle size (i.e., number of atoms). For example, the permanent magnetic behavior of some materials [e.g., iron, cobalt, and iron oxide (Fe_3O_4)] disappears for particles having diameters smaller than about 50 nm.[7]

Two factors account for these size-induced properties of nanoparticles: (1) the increase in ratio of particle surface area to volume; and (2) particle size. As Section 4.6 notes, surface atoms behave differently than atoms located in the interior of a material. Consequently, as the size of a particle decreases, the relative ratio of surface atoms to bulk atoms increases; this means that surface phenomena begin to dominate. Furthermore, for extremely small particles, quantum effects begin to appear.

Although nanocomposite matrix materials may be metals and ceramics, the most common matrices are polymers. For these *polymer nanocomposites,* a large number of thermoplastic, thermosetting, and elastomeric matrices are used, including epoxy resins, polyurethanes, polypropylene, polycarbonate, poly(ethylene terephthalate), silicone resins, poly(methyl methacrylate), polyamides (nylons), poly(vinylidene chloride), ethylene vinyl alcohol, butyl rubber, and natural rubber.

The properties of a nanocomposite depend not only on the properties of both matrix and nanoparticle, but also on nanoparticle shape and content as well as matrix–nanoparticle interfacial characteristics. Most of today's commercial nanocomposites use three general nanoparticle types: *nanocarbons, nanoclays,* and *particulate nanocrystals.*

- Included in the nanocarbon group are single- and multi-wall carbon nanotubes, graphene sheets (Section 14.17), and carbon nanofibers.

---

[5]To qualify as a nanoparticle, the largest particle dimension must on the order of at most 100 nm.

[6]Carbon-black reinforced rubber (Section 16.2) is an example of a nanocomposite; particle sizes typically range between 20 and 50 nm. Strength and toughness as well as tear and abrasion resistance are enhanced because of the presence of carbon-black particles.

[7]This phenomenon is termed *superparamagnetism;* superparamagnetic particles embedded in a matrix are used for magnetic storage, which is discussed in Section 21.11.

- The nanoclays are layered silicates (Section 4.11); the most common type is montmorillonite clay.

- Most particulate nanocrystals are inorganic oxides such as silica, alumina, zirconia, halfnia, and titania.

Nanoparticle loadings (i.e., contents) vary significantly and depend on application. For example, carbon nanotube concentrations on the order of 5 wt% can lead to significant increases in strength and stiffness. However, between 15 and 20 wt% of carbon nanotubes are required to produce electrical conductivities necessary for some applications (e.g., to protect a nanocomposite structure from experiencing electrostatic discharges).

One of the main challenges in the production of nanocomposite materials is processing. For most applications, the nanosize particles must be dispersed uniformly and homogeneously within the matrix. Novel dispersion and fabrication techniques have been and are continually being developed for producing nanocomposites with the desired properties.

These nanocomposite materials have carved out niches in a host of different technologies and industries, including the following:

- **Gas-barrier coatings**—The freshness and shelf lives of foods and beverages may be increased when they are packaged in nanocomposite thin film bags/containers. Normally, these films are composed of montmorillonite nanoclay particles that have been *exfoliated* (i.e., separated from one another) and during incorporation into the polymer matrix are aligned such that their lateral axes are parallel to the plane of the coating. Furthermore, the coatings may be transparent. The presence of nanoclay particles accounts for the ability of the film to effectively contain H_2O molecules in packaged foods (to preserve freshness) and CO_2 molecules in carbonated beverages (to retain "fizz"), and also keep O_2 molecules from the air outside (to protect packaged foods from oxidation). These platelet particles act as multilayer barriers to the diffusion of gas molecules—that is, they slow down the diffusion rate because the gas molecules must bypass the particles as they diffuse through the coating. Another asset of these coatings is their recyclability.

  Nanocomposite coatings are also used to increase air pressure retention for automobile tires and sports (e.g., tennis, soccer) balls. These coatings are composed of small and exfoliated vermiculite[8] platelets that are embedded in the tire/sports ball rubber. Furthermore, platelet particles are aligned in the same manner as for food/beverage coatings described previously such that diffusion of pressurized air molecules through the rubber walls is suppressed.

- **Energy storage**—Graphene nanocomposites are used in anodes for lithium-ion rechargeable batteries—batteries that store electrical energy in hybrid electric vehicles. Surface areas of nanocomposite electrodes that are in contact with the lithium electrolyte are greater than for conventional electrodes. Battery capacity is higher, life cycles are longer, and double the power is available at high charge/discharge rates when graphene nanocomposite anodes are used.

- **Flame-barrier coatings**—Thin coatings composed of multi-walled carbon nanotubes dispersed in silicone matrices exhibit outstanding flame barrier characteristics (i.e., protection from combustion and decomposition). In addition, they offer abrasion and scratch resistance; do not produce toxic gases; and are extremely adherent to most glass, metal, wood, plastic, and composite surfaces. Flame-barrier coatings are used in aerospace, aviation, electronic, and industrial applications, and are typically applied on wires and cables, foams, fuel tanks, and reinforced composites.

- **Dental restorations**—Some newly developed dental restoration (i.e., filling) materials are polymer nanocomposites. Nano-filler ceramic materials used include silica

---

[8]Vermiculite is another member of the layered silicates group discussed in Section 4.11.

nanoparticles (approximately 20 nm in diameter), and nanoclusters composed of loosely bound agglomerates composed of nano-size particles of both silica and zirconia. Most polymer matrix materials belong to the dimethacrylate family. These nanocomposite restoration materials have high fracture toughnesses, are wear resistant, have short curing times and curing shrinkages, and can be made to have the color and appearance of natural teeth.

- **Mechanical strength enhancements**—High-strength and lightweight polymer nanocomposites are produced by the addition of multi-walled carbon nanotubes into epoxy resins; nanotube contents that range between 20 and 30 wt% are normally required. These nanocomposites are used in wind turbine blades as well as some sports equipment (tennis rackets, baseball bats, golf clubs, skis, bicycle frames, and boat hulls and masts).

- **Electrostatic dissipation**—The motion of highly flammable fuels in automotive and aircraft polymer fuel lines can lead to the production of static charges. If not eliminated, these charges pose the risk of spark generation and the possibility of explosion. However, dissipation of such charge buildups can occur if the fuel lines are made to be electrically conductive. Adequate conductivities may be achieved by incorporating multi-walled carbon nanotubes into the polymer. Loading contents as high as 15 to 20 wt% are required, which normally do not compromise other polymer properties.

The number of commercial applications of nanocomposites is accelerating rapidly, and we can look forward to an explosion in the number and diversity of future nanocomposites. Production techniques will improve and, in addition to polymers, metallic- and ceramic-matrix nanoncomposite materials will undoubtedly be developed. Nanocomposite products will find their way into a number of commercial sectors [e.g., fuel cells, solar cells, drug delivery, biomedical, electronic, opto-electronic, and automotive (lubricants, body and under-hood structures, scratch-free paints)].

A can of Double Core tennis balls and an individual ball. Each ball retains its original pressure and bounces twice as long as a conventional one because the inner core has a nanocomposite barrier coating that consists of a matrix of butyl rubber, within which is embedded thin platelets of vermiculite. These particles inhibit the permeation of air molecules through the walls of the ball.
(Photograph courtesy of Wilson Sporting Goods Company.)

SUMMARY

Introduction
- Composites are artificially produced multiphase materials with desirable combinations of the best properties of the constituent phases.
- Usually, one phase (the matrix) is continuous and completely surrounds the other (the dispersed phase).
- In this discussion, composites were classified as particle-reinforced, fiber-reinforced, structural, and nanocomposites.

Large-Particle Composites

Dispersion-Strengthened Composites
- Large-particle and dispersion-strengthened composites fall within the particle-reinforced classification.
- For dispersion strengthening, improved strength is achieved by extremely small particles of the dispersed phase, which inhibit dislocation motion.
- The particle size is normally greater with large-particle composites, whose mechanical characteristics are enhanced by reinforcement action.
- For large-particle composites, upper and lower elastic modulus values depend on the moduli and volume fractions of matrix and particulate phases according to the rule-of-mixtures expressions Equations 16.1 and 16.2.
- Concrete, a type of large-particle composite, consists of an aggregate of particles bonded together with cement. In the case of Portland cement concrete, the aggregate consists of sand and gravel; the cementitious bond develops as a result of chemical reactions between the Portland cement and water.
- The mechanical strength of concrete may be improved by reinforcement methods (e.g., embedment into the fresh concrete of steel rods, wires).

Influence of Fiber Length
- Of the several composite types, the potential for reinforcement efficiency is greatest for those that are fiber reinforced.
- With fiber-reinforced composites, an applied load is transmitted to and distributed among the fibers via the matrix phase, which in most cases is at least moderately ductile.
- Significant reinforcement is possible only if the matrix–fiber bond is strong. Because reinforcement discontinues at the fiber extremities, reinforcement efficiency depends on fiber length.
- For each fiber–matrix combination, there exists some critical length (l_c), which depends on fiber diameter and strength and fiber–matrix bond strength according to Equation 16.3.
- The length of continuous fibers greatly exceeds this critical value (i.e., $l > 15l_c$), whereas shorter fibers are discontinuous.

Influence of Fiber Orientation and Concentration
- On the basis of fiber length and orientation, three different types of fiber-reinforced composites are possible:
  Continuous and aligned (Figure 16.8a)—mechanical properties are highly anisotropic. In the alignment direction, reinforcement and strength are a maximum; perpendicular to the alignment, they are a minimum.
  Discontinuous and aligned (Figure 16.8b)—significant strengths and stiffnesses are possible in the longitudinal direction.
  Discontinuous and randomly oriented (Figure 16.8c)—despite some limitations on reinforcement efficiency, properties are isotropic.
- For continuous and aligned composites, rule-of-mixtures expressions for the modulus in both longitudinal and transverse orientations were developed (Equations 16.10 and 16.16). In addition, an equation for longitudinal strength was also cited (Equation 16.17).

- For discontinuous and aligned composites, composite strength equations were presented for two different situations:
  When $l > l_c$, Equation 16.18 is valid.
  When $l < l_c$, it is appropriate to use Equation 16.19.
- The elastic modulus for discontinuous and randomly oriented fibrous composites may be determined using Equation 16.20.

The Fiber Phase
- On the basis of diameter and material type, fiber reinforcements are classified as follows:
  Whiskers—extremely strong single crystals that have very small diameters
  Fibers—normally polymers or ceramics that may be either amorphous or polycrystalline
  Wires—metals/alloys that have relatively large diameters

The Matrix Phase
- Although all three basic material types are used for matrices, the most common are polymers and metals.
- The matrix phase normally performs three functions:
  It binds the fibers together and transmits an externally applied load to the fibers.
  It protects the individual fibers from surface damage.
  It prevents the propagation of cracks from fiber to fiber.
- Fibrous reinforced composites are sometimes classified according to matrix type; within this scheme are three classifications: polymer-, metal-, and ceramic-matrix composites.

Polymer-Matrix Composites
- Polymer-matrix composites are the most common; they may be reinforced with glass, carbon, and aramid fibers.

Metal-Matrix Composites
- Service temperatures are higher for metal-matrix composites (MMCs) than for polymer-matrix composites. MMCs also use a variety of fiber and whisker types.

Ceramic-Matrix Composites
- With ceramic-matrix composites, the design goal is increased fracture toughness. This is achieved by interactions between advancing cracks and dispersed-phase particles.
- Transformation toughening is one such technique for improving K_{Ic}.

Carbon–Carbon Composites
- Carbon–carbon composites are composed of carbon fibers embedded in a pyrolyzed carbon matrix.
- These materials are expensive and used in applications requiring high strengths and stiffnesses (that are retained at elevated temperatures), resistance to creep, and good fracture toughnesses.

Hybrid Composites
- The hybrid composites contain at least two different fiber types. By using hybrids, it is possible to design composites having better all-around sets of properties.

Processing of Fiber-Reinforced Composites
- Several composite processing techniques have been developed that provide a uniform fiber distribution and a high degree of alignment.
- With pultrusion, components of continuous length and constant cross section are formed as resin-impregnated fiber tows are pulled through a die.
- Composites used for many structural applications are commonly prepared using a lay-up operation (either hand or automated), in which prepreg tape plies are laid down on a tooled surface and are subsequently fully cured by the simultaneous application of heat and pressure.
- Some hollow structures may be fabricated using automated filament winding procedures, by which resin-coated strands or tows or prepreg tape are continuously wound onto a mandrel, followed by a curing operation.

Structural Composites

- Two general kinds of structural composites were discussed: laminar composites and sandwich panels.
- Laminar composites are composed of a set of two-dimensional sheets that are bonded to one another; each sheet has a high-strength direction.

  In-plane laminate properties depend on layer-to-layer high-strength-direction sequencing—in this regard, there are four laminate types: unidirectional, cross-ply, angle-ply, and multidirectional. Multidirectional laminates are the most isotropic, whereas unidirectional laminates have the highest degree of anisotropy. One common laminate material is unidirectional prepreg tape, which can conveniently be laid-up in predetermined high-strength orientations.

- Sandwich panels consist of two strong and stiff sheet faces that are separated by a core material or structure. These structures combine relatively high strengths and stiffnesses with low densities.

  Common core types are rigid polymeric foams, low-density woods, and honeycomb structures.

  Honeycomb structures are composed of interlocking cells (often having hexagonal geometry) made of thin foils; cell axes are oriented perpendicular to the facing sheets.

- Most of the construction of the Boeing 787 Dreamliner uses low-density composite materials (i.e., honeycomb structures and continuous carbon fiber–epoxy resin laminates).

Nanocomposites

- Nanocomposites—nanomaterials embedded in a matrix (most often a polymer)—use the unusual properties of nanosized particles.
- Nanoparticle types include nanocarbons, nanoclays, and particulate nanocrystals.
- Uniform and homogeneous distribution of nanoparticles within the matrix is the major production challenge for nanocomposites.

Important Terms and Concepts

carbon–carbon composite
ceramic-matrix composite
cermet
concrete
dispersed phase
dispersion-strengthened composite
fiber
fiber-reinforced composite
hybrid composite

laminar composite
large-particle composite
longitudinal direction
matrix phase
metal-matrix composite
nanocomposite
polymer-matrix composite
prepreg
prestressed concrete
principle of combined action

reinforced concrete
rule of mixtures
sandwich panel
specific modulus
specific strength
structural composite
transverse direction
whisker

REFERENCES

Agarwal, B. D., L. J. Broutman, and K. Chandrashekhara, *Analysis and Performance of Fiber Composites,* 3rd edition, Wiley, Hoboken, NJ, 2006.

Ashbee, K. H., *Fundamental Principles of Fiber Reinforced Composites,* 2nd edition, CRC Press, Boca Raton, FL, 1993.

ASM Handbook, Vol. 21, *Composites,* ASM International, Materials Park, OH, 2001.

Barbero, E. J., *Introduction to Composite Materials Design,* 2nd edition, CRC Press, Boca Raton, FL, 2010.

Chawla, K. K., *Composite Materials Science and Engineering,* 3rd edition, Springer, New York, 2012.

Gerdeen, J. C., H. W. Lord, and R. A. L. Rorrer, *Engineering Design with Polymers and Composites,* 2nd edition, CRC Press, Boca Raton, FL, 2005.

Hull, D., and T. W. Clyne, *An Introduction to Composite Materials,* 2nd edition, Cambridge University Press, New York, 1996.

Mallick, P. K. (editor), *Composites Engineering Handbook,* Marcel Dekker, New York, 1997.

Mallick, P. K., *Fiber-Reinforced Composites: Materials, Manufacturing, and Design,* 3rd edition, CRC Press, Boca Raton, FL, 2008.

Strong, A. B., *Fundamentals of Composites: Materials, Methods, and Applications,* 2nd edition, Society of Manufacturing Engineers, Dearborn, MI, 2008.

QUESTIONS AND PROBLEMS

Large-Particle Composites

16.1 The mechanical properties of aluminum may be improved by incorporating fine particles of aluminum oxide (Al_2O_3). Given that the moduli of elasticity of these materials are, respectively, 69 GPa and 393 GPa, plot modulus of elasticity versus the volume percent of Al_2O_3 in Al from 0 to 100 vol%, using both upper- and lower-bound expressions.

16.2 Estimate the maximum and minimum thermal conductivity values for a cermet that contains 90 vol% titanium carbide (TiC) particles in a cobalt matrix. Assume thermal conductivities of 27 and 69 W/m·K for TiC and Co, respectively.

16.3 A large-particle composite consisting of tungsten particles within a copper matrix is to be prepared. If the volume fractions of tungsten and copper are 0.60 and 0.40, respectively, estimate the upper limit for the specific stiffness of this composite given the data that follow.

	Specific Gravity	*Modulus of Elasticity (GPa)*
Copper	8.9	110
Tungsten	19.3	407

16.4 (a) What is the distinction between cement and concrete?

(b) Cite three important limitations that restrict the use of concrete as a structural material.

(c) Briefly explain three techniques that are used to strengthen concrete by reinforcement.

Dispersion-Strengthened Composites

16.5 Cite one similarity and two differences between precipitation hardening and dispersion strengthening.

Influence of Fiber Length

16.6 For a glass fiber–epoxy matrix combination, the critical fiber length–fiber diameter ratio is 60. Using the data in Table 16.4, determine the fiber-matrix bond strength.

16.7 (a) For a fiber-reinforced composite, the efficiency of reinforcement η is dependent on fiber length l according to

$$\eta = \frac{l - 2x}{l}$$

where x represents the length of the fiber at each end that does not contribute to the load transfer. Make a plot of η versus l to $l = 50$ mm, assuming that $x = 0.80$ mm.

(b) What length is required for a 0.80 efficiency of reinforcement?

Influence of Fiber Orientation and Concentration

16.8 A continuous and aligned fiber–reinforced composite is to be produced consisting of 30 vol% silicon carbide fibers and 70 vol% of a polycarbonate matrix; mechanical characteristics of these two materials are as follows:

	Modulus of Elasticity [GPa]	*Tensile Strength [MPa]*
Silicon Carbide	400	3900
Polycarbonate	2.4	65

Also, the stress on the polycarbonate matrix when the aramid carbide fail is 45 MPa.

For this composite, compute

(a) the longitudinal tensile strength, and

(b) the longitudinal modulus of elasticity

16.9 Is it possible to produce a continuous and oriented aramid fiber–epoxy matrix composite having longitudinal and transverse moduli of elasticity of 57.1 GPa and 4.12 GPa, respectively? Why or why not? Assume that the modulus of elasticity of the epoxy is 2.4 GPa.

16.10 For a continuous and oriented fiber–reinforced composite, the moduli of elasticity in the longitudinal and transverse directions are 19.7 and 3.66 GPa, respectively. If the volume fraction of fibers is 0.25, determine the moduli of elasticity of fiber and matrix phases.

16.11 (a) Verify that Equation 16.11, the expression for the fiber load–matrix load ratio (F_f/F_m), is valid.

(b) What is the F_f/F_c ratio in terms of E_f, E_m, and V_f?

16.12 In an aligned and continuous glass fiber–reinforced nylon 6,6 composite, the fibers are to carry 96% of a load applied in the longitudinal direction.

(a) Using the data provided, determine the volume fraction of fibers that will be required.

(b) What will be the tensile strength of this composite? Assume that the matrix stress at fiber failure is 30 MPa.

	Modulus of Elasticity [GPa]	Tensile Strength [MPa]
Glass fiber	72.5	3400
Nylon 6,6	3.0	76

16.13 Assume that the composite described in Problem 16.8 has a cross-sectional area of 325 mm² and is subjected to a longitudinal load of 44,600 N.

(a) Calculate the fiber–matrix load ratio.

(b) Calculate the actual loads carried by both fiber and matrix phases.

(c) Compute the magnitude of the stress on each of the fiber and matrix phases.

(d) What strain is experienced by the composite?

16.14 A continuous and aligned fiber–reinforced composite having a cross-sectional area of 1140 mm² is subjected to an external tensile load. If the stresses sustained by the fiber and matrix phases are 156 MPa and 2.75 MPa, respectively; the force sustained by the fiber phase is 74,000 N; and the total longitudinal strain is 1.4×10^{-3}, determine:

(a) the force sustained by the matrix phase,

(b) the modulus of elasticity of the composite material in the longitudinal direction, and

(c) the moduli of elasticity for fiber and matrix phases.

16.15 Compute the longitudinal strength of an aligned carbon fiber–epoxy matrix composite having a 0.25 volume fraction of fibers, assuming the following: (1) an average fiber diameter of 20×10^{-3} mm, (2) an average fiber length of 5 mm, (3) a fiber fracture strength of 2.5 GPa, (4) a fiber-matrix bond strength of 90 MPa, (5) a matrix stress at fiber failure of 10.0 MPa, and (6) a matrix tensile strength of 80 MPa.

16.16 It is desired to produce an aligned carbon fiber–epoxy matrix composite having a longitudinal tensile strength of 750 MPa. Calculate the volume fraction of fibers necessary if (1) the average fiber diameter and length are 1.2×10^{-2} mm and 1 mm, respectively; (2) the fiber fracture strength is 5000 MPa; (3) the fiber-matrix bond strength is 25 MPa; and (4) the matrix stress at fiber failure is 10 MPa.

16.17 Compute the longitudinal tensile strength of an aligned glass fiber–epoxy matrix composite in which the average fiber diameter and length are 0.010 mm and 2.5 mm, respectively, and the volume fraction of fibers is 0.40. Assume that (1) the fiber-matrix bond strength is 75 MPa, (2) the fracture strength of the fibers is 3500 MPa, and (3) the matrix stress at fiber failure is 8.0 MPa.

16.18 **(a)** From the moduli of elasticity data in Table 16.2 for glass fiber–reinforced polycarbonate composites, determine the value of the fiber efficiency parameter for each of 20, 30, and 40 vol% fibers.

(b) Estimate the modulus of elasticity for 50 vol% glass fibers.

The Fiber Phase
The Matrix Phase

16.19 For a polymer-matrix fiber-reinforced composite:

(a) List three functions of the matrix phase.

(b) Compare the desired mechanical characteristics of matrix and fiber phases.

(c) Cite two reasons why there must be a strong bond between fiber and matrix at their interface.

16.20 **(a)** What is the distinction between matrix and dispersed phases in a composite material?

(b) Contrast the mechanical characteristics of matrix and dispersed phases for fiber-reinforced composites.

Polymer–Matrix Composites

16.21 **(a)** Calculate the specific longitudinal strengths of the glass fiber, carbon fiber, and aramid fiber–reinforced epoxy composites in Table 16.5 and compare them with those of the following alloys: tempered (315°C) 440A martensitic stainless steel, normalized 1020 plain-carbon steel, 2024-T3 aluminum alloy, cold-worked (HO2 temper) C36000 free-cutting brass, rolled AZ31B magnesium alloy, and annealed Ti-6Al-4V titanium alloy.

(b) Compare the specific moduli of the same three fiber-reinforced epoxy composites with the same metal alloys. Densities (i.e., specific gravities), tensile strengths, and moduli of elasticity for these metal alloys may be found in Tables B.1, B.4, and B.2, respectively, in Appendix B.

16.22 (a) List four reasons why glass fibers are most commonly used for reinforcement.

(b) Why is the surface perfection of glass fibers so important?

(c) What measures are taken to protect the surface of glass fibers?

16.23 Cite the distinction between carbon and graphite.

16.24 (a) Cite several reasons why fiberglass-reinforced composites are used extensively.

(b) Cite several limitations of this type of composite.

Hybrid Composites

16.25 (a) What is a hybrid composite?

(b) List two important advantages of hybrid composites over normal fiber composites.

16.26 (a) Write an expression for the modulus of elasticity for a hybrid composite in which all fibers of both types are oriented in the same direction.

(b) Using this expression, compute the longitudinal modulus of elasticity of a hybrid composite consisting of aramid and glass fibers in volume fractions of 0.30 and 0.40, respectively, within a polyester resin matrix [$E_m = 2.5$ GPa].

16.27 Derive a generalized expression analogous to Equation 16.16 for the transverse modulus of elasticity of an aligned hybrid composite consisting of two types of continuous fibers.

Processing of Fiber–Reinforced Composites

16.28 Briefly describe pultrusion, filament winding, and prepreg production fabrication processes; cite the advantages and disadvantages of each.

Laminar Composites
Sandwich Panels

16.29 Briefly describe laminar composites. What is the prime reason for fabricating these materials?

16.30 (a) Briefly describe sandwich panels.

(b) What is the prime reason for fabricating these structural composites?

(c) What are the functions of the faces and the core?

Spreadsheet Problems

16.1SS For an aligned polymer-matrix composite, develop a spreadsheet that will allow the user to compute the longitudinal tensile strength after inputting values for the following parameters: volume fraction of fibers, average fiber diameter, average fiber length, fiber fracture strength, fiber-matrix bond strength, matrix stress at composite failure, and matrix tensile strength.

16.2SS Generate a spreadsheet for the design of a tubular composite shaft (Design Example 16.1)—that is, which of available fiber materials provide the required stiffness, and, of these possibilities, which will cost the least. The fibers are continuous and are to be aligned parallel to the tube axis. The user is allowed to input values for the following parameters: inside and outside tube diameters, tube length, maximum deflection at the axial midpoint for some given applied load, maximum fiber volume fraction, elastic moduli of matrix and all fiber materials, densities of matrix and fiber materials, and cost per unit mass for the matrix and all fiber materials.

DESIGN PROBLEMS

16.D1 Composite materials are now being used extensively in sports equipment.

(a) List at least four different sports implements that are made of or contain composites.

(b) For one of these implements, write an essay in which you do the following: (1) Cite the materials that are used for matrix and dispersed phases, and, if possible, the proportions of each phase; (2) note the nature of the dispersed phase (i.e., continuous fibers); and (3) describe the process by which the implement is fabricated.

Influence of Fiber Orientation and Concentration

16.D2 It is desired to produce an aligned and continuous fiber–reinforced epoxy composite having a maximum of 50 vol% fibers. In addition, a minimum longitudinal modulus of elasticity of 55 GPa is required, as well as a minimum tensile strength of 1310 MPa. Of E-glass, carbon (PAN standard modulus), and aramid fiber materials, which are possible candidates and why? The epoxy has a modulus of elasticity of 3.1 GPa and a tensile strength of 75 MPa. In addition, assume the following stress levels on the epoxy matrix at fiber failure: E-glass—70 MPa; carbon (PAN standard modulus)—30 MPa; and aramid—50 MPa. Other fiber data are contained in Tables B.2 and B.4 in Appendix B. For aramid and carbon fibers, use average strengths computed from the minimum and maximum values provided in Table B.4.

16.D3 It is desired to produce a continuous and oriented carbon fiber–reinforced epoxy having a modulus of elasticity of at least 83 GPa in the direction of fiber alignment. The maximum permissible specific gravity is 1.40. Given the following data, is such a composite possible? Why or why not? Assume that composite specific gravity may be determined using a relationship similar to Equation 16.10a.

	Specific Gravity	Modulus of Elasticity [GPa]
Carbon fiber	1.80	260
Epoxy	1.25	2.4

16.D4 It is desired to fabricate a continuous and aligned glass fiber–reinforced polyester having a tensile strength of at least 1410 MPa in the longitudinal direction. The maximum possible specific gravity is 1.65. Using the following data, determine whether such a composite is possible. Justify your decision. Assume a value of 18 MPa for the stress on the matrix at fiber failure.

	Specific Gravity	Tensile Strength [MPa]
Glass fiber	2.50	3500
Polyester	1.35	50

16.D5 It is necessary to fabricate an aligned and discontinuous carbon fiber–epoxy matrix composite having a longitudinal tensile strength of 1900 MPa using 0.50 volume fraction of fibers. Compute the required fiber fracture strength assuming that the average fiber diameter and length are 10×10^{-3} mm and 3.5 mm, respectively. The fiber-matrix bond strength is 40 MPa, and the matrix stress at fiber failure is 12 MPa.

16.D6 A tubular shaft similar to that shown in Figure 16.11 is to be designed that has an outside diameter of 80 mm and a length of 0.75 m. The mechanical characteristic of prime importance is bending stiffness in terms of the longitudinal modulus of elasticity. Stiffness is to be specified as maximum allowable deflection in bending; when subjected to three-point bending as in Figure 14.11, a load of 1000 N is to produce an elastic deflection of no more than 0.40 mm at the midpoint position.

Continuous fibers that are oriented parallel to the tube axis will be used; possible fiber materials are glass, and carbon in standard-, intermediate-, and high-modulus grades. The matrix material is to be an epoxy resin, and fiber volume fraction is 0.35.

(a) Decide which of the four fiber materials are possible candidates for this application, and for each candidate determine the required inside diameter consistent with the preceding criteria.

(b) For each candidate, determine the required cost, and on this basis, specify the fiber that would be the least expensive to use.

Elastic modulus, density, and cost data for the fiber and matrix materials are contained in Table 16.6.

© William D. Callister, Jr.

(a)

(a) The aluminum beverage can in various stages of production. The can is formed from a single sheet of an aluminum alloy. Production operations include drawing, dome forming, trimming, cleaning, decorating, and neck and flange forming.

(b) A workman inspecting a roll of aluminum sheet.

Daniel R. Patmore/© AP/Wide World Photos.

(b)

On occasion, fabrication and processing procedures adversely affect some of the properties of materials. For example, in Section 17.6 we note that some steels may become embrittled during tempering heat treatments. Also, some stainless steels are made susceptible to intergranular corrosion (Section 16.7) when they are heated for long time periods within a specific temperature range. In addition, as discussed in Section 5.4, regions adjacent to weld junctions may experience decreases in strength and toughness as a result of undesirable microstructural alterations. It is important that engineers become familiar with possible consequences attendant to processing and fabricating procedures in order to prevent unanticipated material failures.

Learning Objectives

After studying this chapter, you should be able to do the following:

1. Name and describe four forming operations that are used to shape metal alloys.
2. Name and describe five casting techniques.
3. State the purposes of and describe procedures for the following heat treatments: process annealing, stress relief annealing, normalizing, full annealing, and spheroidizing.
4. Define *hardenability*.
5. Generate a hardness profile for a cylindrical steel specimen that has been austenitized and then quenched, given the hardenability curve for the specific alloy, as well as quenching rate–versus–bar diameter information.
6. Using a phase diagram, describe and explain the two heat treatments that are used to precipitation harden a metal alloy.
7. Make a schematic plot of room-temperature strength (or hardness) versus the logarithm of time for a precipitation heat treatment at constant temperature. Explain the shape of this curve in terms of the mechanism of precipitation hardening.
8. Name and briefly describe five forming methods that are used to fabricate glass pieces.
9. Briefly describe and explain the procedure by which glass pieces are thermally tempered.
10. Briefly describe processes that occur during the drying and firing of clay-based ceramic ware.
11. Briefly describe/diagram the sintering process of powder particle aggregates.
12. Briefly describe addition and condensation polymerization mechanisms.
13. Name the five types of polymer additives and, for each, indicate how it modifies polymer properties.
14. Name and briefly describe five fabrication techniques used for plastic polymers.

17.1 INTRODUCTION

Fabrication techniques are methods by which materials are formed or manufactured into components that may be incorporated into useful products. Sometimes it also may be necessary to subject the component to some type of processing treatment in order to achieve the required properties. In addition, on occasion, the suitability of a material for an application is dictated by economic considerations with respect to fabrication and processing operations. In this chapter we discuss various techniques that are used to fabricate and process metals, ceramics, and polymers (and also, for polymers, how they are synthesized).

Fabrication of Metals

Metal fabrication techniques are normally preceded by refining, alloying, and often heat-treating processes that produce alloys with the desired characteristics. The classifications of fabrication techniques include various metal-forming methods, casting,

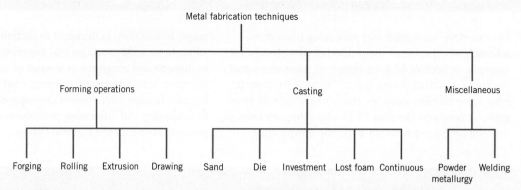

Figure 17.1 Classification scheme of metal fabrication techniques discussed in this chapter.

powder metallurgy, welding, and machining; often two or more must be used before a piece is finished. The methods chosen depend on several factors; the most important are the properties of the metal, the size and shape of the finished piece, and the cost. The metal fabrication techniques we discuss are classified according to the scheme illustrated in Figure 17.1.

17.2 FORMING OPERATIONS

Forming operations are those in which the shape of a metal piece is changed by plastic deformation; for example, forging, rolling, extrusion, and drawing are common forming techniques. The deformation must be induced by an external force or stress, the magnitude of which must exceed the yield strength of the material. Most metallic materials are especially amenable to these procedures, being at least moderately ductile and capable of some permanent deformation without cracking or fracturing.

hot working   When deformation is achieved at a temperature above that at which recrystallization occurs, the process is termed **hot working** (Section 9.12); otherwise, it is cold working. With most of the forming techniques, both hot- and cold-working procedures are possible. For hot-working operations, large deformations are possible, which may be successively repeated because the metal remains soft and ductile. Also, deformation energy requirements are less than for cold working. However, most metals experience some surface oxidation, which results in material loss and a poor final surface finish.

cold working   **Cold working** produces an increase in strength with the attendant decrease in ductility because the metal strain hardens; advantages over hot working include a higher-quality surface finish, better mechanical properties and a greater variety of them, and closer dimensional control of the finished piece. On occasion, the total deformation is accomplished in a series of steps in which the piece is successively cold worked a small amount and then process annealed (Section 17.5); however, this is an expensive and inconvenient procedure.

The forming operations to be discussed are illustrated schematically in Figure 17.2.

Forging

forging   **Forging** is mechanically working or deforming a single piece of a usually hot metal; this may be accomplished by the application of successive blows or by continuous squeezing. Forgings are classified as either closed or open die. For closed die, a force is

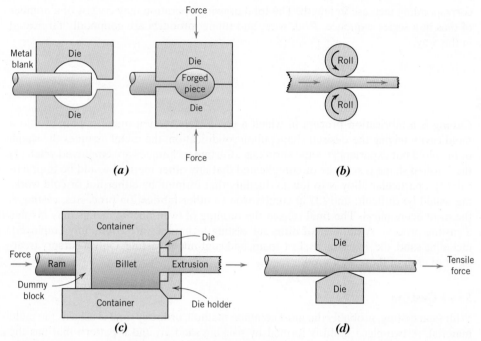

Figure 17.2 Metal deformation during (*a*) forging, (*b*) rolling, (*c*) extrusion, and (*d*) drawing.

brought to bear on two or more die halves having the finished shape such that the metal is deformed in the cavity between them (Figure 17.2*a*). For open die, two dies having simple geometric shapes (e.g., parallel flat, semicircular) are employed, normally on large workpieces. Forged articles have outstanding grain structures and the best combination of mechanical properties. Wrenches, automotive crankshafts, and piston connecting rods are typical articles formed using this technique.

Rolling

rolling

Rolling, the most widely used deformation process, consists of passing a piece of metal between two rolls; a reduction in thickness results from compressive stresses exerted by the rolls. Cold rolling may be used in the production of sheet, strip, and foil with a high-quality surface finish. Circular shapes, as well as I-beams and railroad rails, are fabricated using grooved rolls.

Extrusion

extrusion

For **extrusion**, a bar of metal is forced through a die orifice by a compressive force that is applied to a ram; the extruded piece that emerges has the desired shape and a reduced cross-sectional area. Extrusion products include rods and tubing that have rather complicated cross-sectional geometries; seamless tubing may also be extruded.

Drawing

drawing

Drawing is the pulling of a metal piece through a die having a tapered bore by means of a tensile force that is applied on the exit side. A reduction in cross section results, with a corresponding increase in length. The total drawing operation may consist of a number of dies in a series sequence. Rod, wire, and tubing products are commonly fabricated in this way.

17.3 CASTING

Casting is a fabrication process in which a completely molten metal is poured into a mold cavity having the desired shape; upon solidification, the metal assumes the shape of the mold but experiences some shrinkage. Casting techniques are employed when (1) the finished shape is so large or complicated that any other method would be impractical; (2) a particular alloy is so low in ductility that forming by either hot or cold working would be difficult; and (3) in comparison to other fabrication processes, casting is the most economical. The final step in the refining of even ductile metals may involve a casting process. A number of different casting techniques are commonly employed, including sand, die, investment, lost-foam, and continuous casting. Only a cursory treatment of each of these is offered.

Sand Casting

With sand casting, probably the most common method, ordinary sand is used as the mold material. A two-piece mold is formed by packing sand around a pattern that has the shape of the intended casting. A *gating system* is usually incorporated into the mold to expedite the flow of molten metal into the cavity and to minimize internal casting defects. Sand-cast parts include automotive cylinder blocks, fire hydrants, and large pipe fittings.

Die Casting

In die casting, the liquid metal is forced into a mold under pressure and at a relatively high velocity and allowed to solidify with the pressure maintained. A two-piece permanent steel mold or die is employed; when clamped together, the two pieces form the desired shape. When the metal has solidified completely, the die pieces are opened and the cast piece is ejected. Rapid casting rates are possible, making this an inexpensive method; furthermore, a single set of dies may be used for thousands of castings. However, this technique lends itself only to relatively small pieces and to alloys of zinc, aluminum, and magnesium, which have low melting temperatures.

Investment Casting

For investment (sometimes called *lost-wax*) casting, the pattern is made from a wax or plastic that has a low melting temperature. Around the pattern a fluid slurry is poured that sets up to form a solid mold or investment; plaster of Paris is usually used. The mold is then heated, such that the pattern melts and is burned out, leaving behind a mold cavity having the desired shape. This technique is employed when high dimensional accuracy, reproduction of fine detail, and an excellent finish are required—for example, in jewelry and dental crowns and inlays. In addition, blades for gas turbines and jet engine impellers are investment cast.

Lost-Foam Casting

A variation of investment casting is *lost-foam* (or *expendable pattern*) *casting*. Here, the expendable pattern is a foam that can be formed by compressing polystyrene beads into the desired shape and then bonding them together by heating. Alternatively, pattern shapes can be cut from sheets and assembled with glue. Sand is then packed around the pattern to form the mold. As the molten metal is poured into the mold, it replaces the pattern, which vaporizes. The compacted sand remains in place, and, upon solidification, the metal assumes the shape of the mold.

With lost-foam casting, complex geometries and tight tolerances are possible. Furthermore, in comparison to sand casting, lost-foam casting is a simpler, quicker, and less expensive process and there are fewer environmental wastes. Metal alloys that most commonly use this technique are cast irons and aluminum alloys; furthermore, applications include automobile engine blocks, cylinder heads, crankshafts, marine engine blocks, and electric motor frames.

Continuous Casting

At the conclusion of extraction processes, many molten metals are solidified by casting into large ingot molds. The ingots are normally subjected to a primary hot-rolling operation, the product of which is a flat sheet or slab; these are more convenient shapes as starting points for subsequent secondary metal-forming operations (forging, extrusion, drawing). These casting and rolling steps may be combined by a *continuous casting* (sometimes termed *strand casting*) process. Using this technique, the refined and molten metal is cast directly into a continuous strand that may have either a rectangular or circular cross section; solidification occurs in a water-cooled die having the desired cross-sectional geometry. The chemical composition and mechanical properties are more uniform throughout the cross sections for continuous castings than for ingot-cast products. Furthermore, continuous casting is highly automated and more efficient.

17.4 MISCELLANEOUS TECHNIQUES

Powder Metallurgy

powder metallurgy

Yet another fabrication technique involves the compaction of powdered metal followed by a heat treatment to produce a denser piece. The process is appropriately called **powder metallurgy**, frequently designated as P/M. Powder metallurgy makes it possible to produce a virtually nonporous piece having properties almost equivalent to those of the fully dense parent material. Diffusional processes during the heat treatment are central to the development of these properties. This method is especially suitable for metals having low ductilities, because only small plastic deformation of the powder particles need occur. Metals with high melting temperatures are difficult to melt and cast, and fabrication is expedited using P/M. Furthermore, parts that require very close dimensional tolerances (e.g., bushings and gears) may be economically produced using this technique.

Concept Check 17.1 **(a)** Cite two advantages of powder metallurgy over casting. **(b)** Cite two disadvantages.

[*The answer may be found at* www.wiley.com/college/callister *(Student Companion Site).*]

Welding

welding

In a sense, welding may be considered to be a fabrication technique. In **welding**, two or more metal parts are joined to form a single piece when one-part fabrication is expensive or inconvenient. Both similar and dissimilar metals may be welded. The joining bond is metallurgical (involving some diffusion) rather than just mechanical, as with riveting and bolting. A variety of welding methods exist, including arc and gas welding, as well as brazing and soldering.

Figure 17.4 The iron–iron carbide phase diagram in the vicinity of the eutectoid, indicating heat-treating temperature ranges for plain carbon steels. (Adapted from G. Krauss, *Steels: Heat Treatment and Processing Principles*, ASM International, 1990, page 108.)

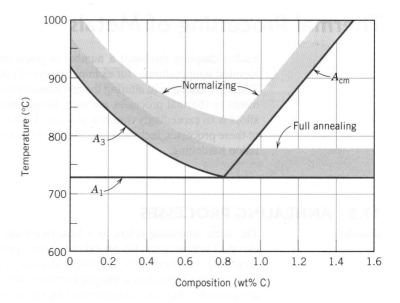

held there long enough to attain a uniform temperature, and finally cooled to room temperature in air. The annealing temperature is typically a relatively low one such that effects resulting from cold working and other heat treatments are not affected.

Annealing of Ferrous Alloys

Several different annealing procedures are employed to enhance the properties of steel alloys. However, before they are discussed, some comment relative to the labeling of phase boundaries is necessary. Figure 17.4 shows the portion of the iron–iron carbide phase diagram in the vicinity of the eutectoid. The horizontal line at the eutectoid temperature, conventionally labeled A_1, is termed the **lower critical temperature**, below which, under equilibrium conditions, all austenite has transformed into ferrite and cementite phases. The phase boundaries denoted as A_3 and A_{cm} represent the **upper critical temperature** lines for hypoeutectoid and hypereutectoid steels, respectively. For temperatures and compositions above these boundaries, only the austenite phase prevails. As explained in Section 11.20, other alloying elements shift the eutectoid and the positions of these phase boundary lines.

lower critical temperature

upper critical temperature

Normalizing

Steels that have been plastically deformed by, for example, a rolling operation, consist of grains of pearlite (and most likely a proeutectoid phase), which are irregularly shaped and relatively large and vary substantially in size. An annealing heat treatment called **normalizing** is used to refine the grains (i.e., to decrease the average grain size) and produce a more uniform and desirable size distribution; fine-grained pearlitic steels are tougher than coarse-grained ones. Normalizing is accomplished by heating at least 55°C above the upper critical temperature—that is, above A_3 for compositions less than the eutectoid (0.76 wt% C), and above A_{cm} for compositions greater than the eutectoid, as represented in Figure 17.4. After sufficient time has been allowed for the alloy to completely transform to austenite—a procedure termed **austenitizing**—the treatment is terminated by cooling in air. A normalizing cooling curve is superimposed on the continuous-cooling transformation diagram (Figure 12.26).

normalizing

austenitizing

Full Anneal

full annealing

A heat treatment known as **full annealing** is often used in low- and medium-carbon steels that will be machined or will experience extensive plastic deformation during a forming operation. In general, the alloy is treated by heating to a temperature of about 50°C above the A_3 line (to form austenite) for compositions less than the eutectoid, or, for compositions in excess of the eutectoid, 50°C above the A_1 line (to form austenite and Fe_3C phases), as noted in Figure 17.4. The alloy is then furnace cooled—that is, the heat-treating furnace is turned off, and both furnace and steel cool to room temperature at the same rate, which takes several hours. The microstructural product of this anneal is coarse pearlite (in addition to any proeutectoid phase) that is relatively soft and ductile. The full-anneal cooling procedure (also shown in Figure 12.26) is time consuming; however, a microstructure having small grains and a uniform grain structure results.

Spheroidizing

spheroidizing

Medium- and high-carbon steels having a microstructure containing even coarse pearlite may still be too hard to machine or plastically deform conveniently. These steels, and in fact any steel, may be heat-treated or annealed to develop the spheroidite structure as described in Section 12.5. Spheroidized steels have a maximum softness and ductility and are easily machined or deformed. The **spheroidizing** heat treatment, during which there is a coalescence of the Fe_3C to form the spheroid particles, can take place by several methods, as follows:

- Heating the alloy at a temperature just below the eutectoid (line A_1 in Figure 17.4, or at about 700°C) in the $\alpha + Fe_3C$ region of the phase diagram. If the precursor microstructure contains pearlite, spheroidizing times will typically range between 15 and 25 h.

- Heating to a temperature just above the eutectoid temperature and then either cooling very slowly in the furnace or holding at a temperature just below the eutectoid temperature.

- Heating and cooling alternately within about $\pm 50°C$ of the A_1 line of Figure 17.4.

To some degree, the rate at which spheroidite forms depends on prior microstructure. For example, it is slowest for pearlite, and the finer the pearlite, the more rapid the rate. Also, prior cold work increases the spheroidizing reaction rate.

Still other annealing treatments are possible. For example, glasses are annealed, as outlined in Section 17.8, to remove residual internal stresses that render the material excessively weak. In addition, microstructural alterations and the attendant modification of mechanical properties of cast irons, as discussed in Section 13.3, result from what are, in a sense, annealing treatments.

17.6 HEAT TREATMENT OF STEELS

Conventional heat treatment procedures for producing martensitic steels typically involve continuous and rapid cooling of an austenitized specimen in some type of quenching medium, such as water, oil, or air. The optimum properties of a steel that has been quenched and then tempered can be realized only if, during the quenching heat treatment, the specimen has been converted to a high content of martensite; the formation of any pearlite and/or bainite will result in other than the best combination of mechanical characteristics. During the quenching treatment, it is impossible to cool the specimen at a uniform rate throughout—the surface always cools more rapidly than interior regions. Therefore, the austenite transforms over a range of temperatures, yielding a possible variation of microstructure and properties with position within a specimen.

The successful heat treating of steels to produce a predominantly martensitic microstructure throughout the cross section depends mainly on three factors: (1) the composition of the alloy, (2) the type and character of the quenching medium, and (3) the size and shape of the specimen. The influence of each of these factors is now addressed.

Hardenability

hardenability

The influence of alloy composition on the ability of a steel alloy to transform to martensite for a particular quenching treatment is related to a parameter called **hardenability**. For every steel alloy, there is a specific relationship between the mechanical properties and the cooling rate. *Hardenability* is a term used to describe the ability of an alloy to be hardened by the formation of martensite as a result of a given heat treatment. Hardenability is not "hardness," which is the resistance to indentation; rather, hardenability is a qualitative measure of the rate at which hardness drops off with distance into the interior of a specimen as a result of diminished martensite content. A steel alloy that has a high hardenability is one that hardens, or forms martensite, not only at the surface, but also to a large degree throughout the entire interior.

The Jominy End–Quench Test

Jominy end-quench test

One standard procedure widely used to determine hardenability is the **Jominy end-quench test**.[1] With this procedure, except for alloy composition, all factors that may influence the depth to which a piece hardens (i.e., specimen size and shape and quenching treatment) are maintained constant. A cylindrical specimen 25 mm in diameter and 100 mm long is austenitized at a prescribed temperature for a prescribed time. After removal from the furnace, it is quickly mounted in a fixture as diagrammed in Figure 17.5a. The lower end is quenched by a jet of water of specified flow rate and temperature. Thus, the cooling rate is a maximum at the quenched end and diminishes with position from this point along the length of the specimen. After the piece has cooled to room temperature, shallow flats 0.4 mm deep are ground along the specimen length and Rockwell hardness measurements are made for the first 50 mm along each flat (Figure 17.5b); for the first 12.8 mm, hardness readings are taken at 1.6-mm intervals, and for the remaining 38.4 mm every 3.2 mm. A hardenability curve is produced when hardness is plotted as a function of position from the quenched end.

Hardenability Curves

A typical hardenability curve is represented in Figure 17.6. The quenched end is cooled most rapidly and exhibits the maximum hardness; 100% martensite is the product at this position for most steels. Cooling rate decreases with distance from the quenched end, and the hardness also decreases, as indicated in the figure. With diminishing cooling rate, more time is allowed for carbon diffusion and the formation of a greater proportion of the softer pearlite, which may be mixed with martensite

---

[1]ASTM Standard A255, "Standard Test Methods for Determining Hardenability of Steel."

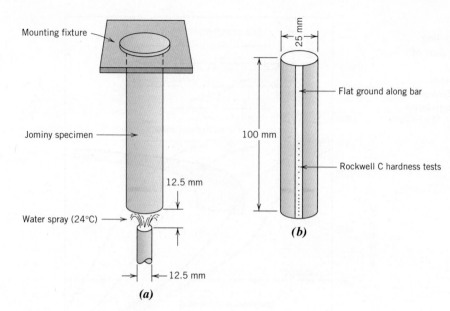

Figure 17.5 Schematic diagram of Jominy end-quench specimen (*a*) mounted during quenching and (*b*) after hardness testing from the quenched end along a ground flat.
(Adapted from A. G. Guy, *Essentials of Materials Science.* Copyright 1978 by McGraw-Hill Book Company, New York.)

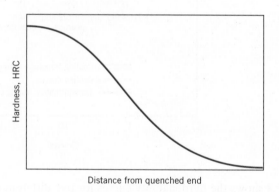

Figure 17.6 Typical hardenability plot of Rockwell C hardness as a function of distance from the quenched end.

and bainite. Thus, a steel that is highly hardenable retains large hardness values for relatively long distances; a steel with low hardenability does not. Also, each steel alloy has its own unique hardenability curve.

Sometimes, it is convenient to relate hardness to a cooling rate rather than to the location from the quenched end of a standard Jominy specimen. Cooling rate (taken at 700°C) is typically shown on the upper horizontal axis of a hardenability diagram; this scale is included with the hardenability plots presented here. This correlation between position and cooling rate is the same for plain carbon steels and many alloy steels because the rate of heat transfer is nearly independent of composition. On occasion, cooling rate or position from the quenched end is specified in terms of Jominy distance, one Jominy distance unit being 1.6 mm.

A correlation may be drawn between position along the Jominy specimen and continuous-cooling transformations. For example, Figure 17.7 is a continuous-cooling transformation diagram for a eutectoid iron–carbon alloy onto which are superimposed the cooling curves at four different Jominy positions together with the corresponding microstructures that result for each. The hardenability curve for this alloy is also included.

Figure 17.7 Correlation of hardenability and continuous-cooling information for an iron–carbon alloy of eutectoid composition. [Adapted from H. Boyer (Editor), *Atlas of Isothermal Transformation and Cooling Transformation Diagrams,* 1977. Reproduced by permission of ASM International, Materials Park, OH.]

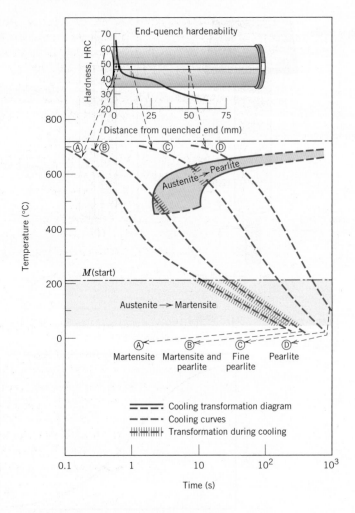

Figure 17.8 shows the hardenability curves for five different steel alloys all having 0.40 wt% C, but differing amounts of other alloying elements. One specimen is a plain carbon steel (1040); the other four (4140, 4340, 5140, and 8640) are alloy steels. The compositions of the four alloy steels are included within the figure. The significance of the alloy designation numbers (e.g., 1040) is explained in Section 13.2. Several details are worth noting from this figure. First, all five alloys have identical hardnesses at the quenched end (57 HRC); this hardness is a function of carbon content only, which is the same for all of these alloys.

Probably the most significant feature of these curves is shape, which relates to hardenability. The hardenability of the plain carbon 1040 steel is low because the hardness drops off precipitously (to about 30 HRC) after a relatively short Jominy distance (16.4 mm). By way of contrast, the decreases in hardness for the other four alloy steels are distinctly more gradual. For example, at a Jominy distance of 50 mm, the hardnesses of the 4340 and 8640 alloys are approximately 50 and 32 HRC, respectively; thus, of these two alloys, the 4340 is more hardenable. A water-quenched specimen of the 1040 plain carbon steel would harden only to a shallow depth below the surface, whereas for the other four alloy steels the high quenched hardness would persist to a much greater depth.

The hardness profiles in Figure 17.8 are indicative of the influence of cooling rate on the microstructure. At the quenched end, where the quenching rate is approximately

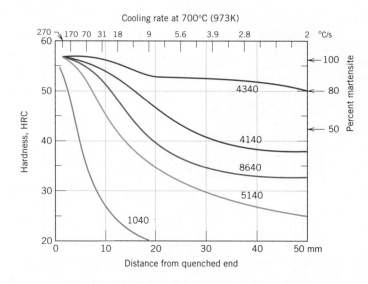

Figure 17.8 Hardenability curves for five different steel alloys, each containing 0.4 wt% C. Approximate alloy compositions (wt%) are as follows: 4340—1.85 Ni, 0.80 Cr, and 0.25 Mo; 4140—1.0 Cr and 0.20 Mo; 8640—0.55 Ni, 0.50 Cr, and 0.20 Mo; 5140—0.85 Cr; and 1040 is an unalloyed steel.
(Adapted from figure furnished courtesy Republic Steel Corporation.)

600°C/s, 100% martensite is present for all five alloys. For cooling rates less than about 70°C/s or Jominy distances greater than about 6.4 mm, the microstructure of the 1040 steel is predominantly pearlitic, with some proeutectoid ferrite. However, the microstructures of the four alloy steels consist primarily of a mixture of martensite and bainite; bainite content increases with decreasing cooling rate.

This disparity in hardenability behavior for the five alloys in Figure 17.8 is explained by the presence of nickel, chromium, and molybdenum in the alloy steels. These alloying elements delay the austenite-to-pearlite and/or bainite reactions, as explained previously; this permits more martensite to form for a particular cooling rate, yielding a greater hardness. The right-hand axis of Figure 17.8 shows the approximate percentage of martensite that is present at various hardnesses for these alloys.

The hardenability curves also depend on carbon content. This effect is demonstrated in Figure 17.9 for a series of alloy steels in which only the concentration of

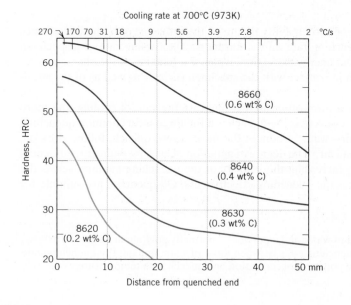

Figure 17.9 Hardenability curves for four 8600 series alloys of indicated carbon content.
(Adapted from figure furnished courtesy Republic Steel Corporation.)

Figure 17.10 The hardenability band for an 8640 steel indicating maximum and minimum limits.
(Adapted from figure furnished courtesy Republic Steel Corporation.)

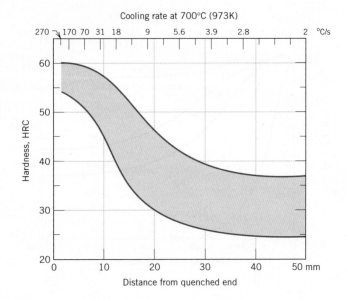

carbon is varied. The hardness at any Jominy position increases with the concentration of carbon.

Also, during the industrial production of steel, there is always a slight, unavoidable variation in composition and average grain size from one batch to another. This variation results in some scatter in measured hardenability data, which frequently are plotted as a band representing the maximum and minimum values that would be expected for the particular alloy. Such a hardenability band is plotted in Figure 17.10 for an 8640 steel. An H following the designation specification for an alloy (e.g., 8640H) indicates that the composition and characteristics of the alloy are such that its hardenability curve lies within a specified band.

Influence of Quenching Medium, Specimen Size, and Geometry

The preceding treatment of hardenability discussed the influence of both alloy composition and cooling or quenching rate on the hardness. The cooling rate of a specimen depends on the rate of heat energy extraction, which is a function of the characteristics of the quenching medium in contact with the specimen surface, as well as of the specimen size and geometry.

Severity of quench is a term often used to indicate the rate of cooling; the more rapid the quench, the more severe is the quench. Of the three most common quenching media—water, oil, and air—water produces the most severe quench, followed by oil, which is more effective than air.[2] The degree of agitation of each medium also influences the rate of heat removal. Increasing the velocity of the quenching medium across the specimen surface enhances the quenching effectiveness. Oil quenches are suitable for

---

[2]Aqueous polymer quenchants [solutions composed of water and a polymer—normally poly(alkylene glycol) or PAG] have recently been developed that provide quenching rates between those of water and oil. The quenching rate can be tailored to specific requirements by changing polymer concentration and quench bath temperature.

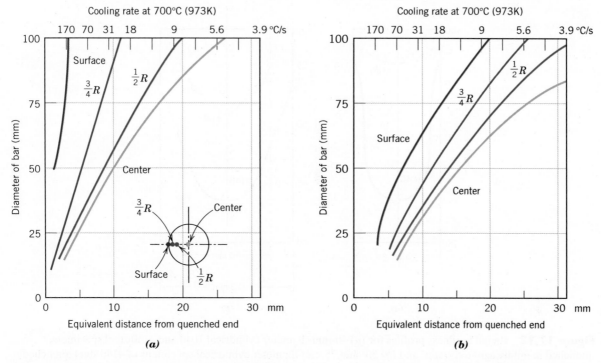

Figure 17.11 Cooling rate as a function of the diameter at the surface, the three-quarter radius ($\frac{3}{4}R$), the midradius ($\frac{1}{2}R$), and the center positions for cylindrical bars quenched in mildly agitated (a) water and (b) oil. Equivalent Jominy positions are included along the bottom axes.
[Adapted from *Metals Handbook: Properties and Selection: Irons and Steels*, Vol. 1, 9th edition, B. Bardes (Editor), 1978. Reproduced by permission of ASM International, Materials Park, OH.]

the heat treating of many alloy steels. In fact, for higher-carbon steels, a water quench is too severe because cracking and warping may be produced. Air cooling of austenitized plain carbon steels ordinarily produces an almost totally pearlitic structure.

During the quenching of a steel specimen, heat energy must be transported to the surface before it can be dissipated into the quenching medium. As a consequence, the cooling rate within and throughout the interior of a steel structure varies with position and depends on the geometry and size. Figures 17.11a and 17.11b show the quenching rate at 700°C as a function of diameter for cylindrical bars at four radial positions (surface, three-quarters radius, midradius, and center). Quenching is in mildly agitated water (Figure 17.11a) and oil (Figure 17.11b); cooling rate is also expressed as equivalent Jominy distance because these data are often used in conjunction with hardenability curves. Diagrams similar to those in Figure 17.11 have also been generated for geometries other than cylindrical (e.g., flat plates).

One utility of such diagrams is in the prediction of the hardness traverse along the cross section of a specimen. For example, Figure 17.12a compares the radial hardness distributions for cylindrical plain carbon (1040) and alloy (4140) steel specimens; both have a diameter of 50 mm and are water quenched. The difference in hardenability is evident from these two profiles. Specimen diameter also influences the hardness distribution, as demonstrated in Figure 17.12b, which plots the hardness profiles for oil-quenched 4140 cylinders 50 and 75 mm in diameter. Example Problem 17.1 illustrates how these hardness profiles are determined.

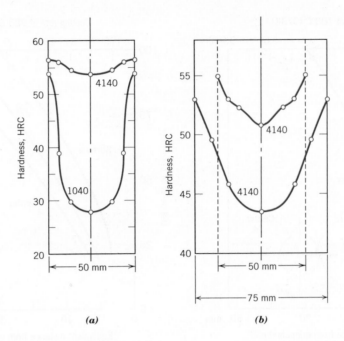

Figure 17.12 Radial hardness profiles for (*a*) 50-mm-diameter cylindrical 1040 and 4140 steel specimens quenched in mildly agitated water, and (*b*) 50- and 75-mm-diameter cylindrical specimens of 4140 steel quenched in mildly agitated oil.

As far as specimen shape is concerned, because the heat energy is dissipated to the quenching medium at the specimen surface, the rate of cooling for a particular quenching treatment depends on the ratio of surface area to the mass of the specimen. The larger this ratio, the more rapid is the cooling rate and, consequently, the deeper is the hardening effect. Irregular shapes with edges and corners have larger surface-to-mass ratios than regular and rounded shapes (e.g., spheres and cylinders) and are thus more amenable to hardening by quenching.

A multitude of steels are responsive to a martensitic heat treatment, and one of the most important criteria in the selection process is hardenability. Hardenability curves, when used in conjunction with plots such as those in Figure 17.11 for various quenching media, may be used to ascertain the suitability of a specific steel alloy for a particular application. Conversely, the appropriateness of a quenching procedure for an alloy may be determined. For parts that are to be involved in relatively high stress applications, a minimum of 80% martensite must be produced throughout the interior as a consequence of the quenching procedure. Only a 50% minimum is required for moderately stressed parts.

Concept Check 17.3 Name the three factors that influence the degree to which martensite is formed throughout the cross section of a steel specimen. For each, tell how the extent of martensite formation may be increased.

[*The answer may be found at* www.wiley.com/college/callister (*Student Companion Site*).]

EXAMPLE PROBLEM 17.1

Determination of Hardness Profile for Heat-Treated 1040 Steel

Determine the radial hardness profile for a cylindrical specimen of 1040 steel of diameter 50 mm that has been quenched in moderately agitated water.

Solution

First, evaluate the cooling rate (in terms of the Jominy end-quench distance) at center, surface, midradius, and three-quarter radius positions of the cylindrical specimen. This is accomplished using the cooling rate–versus–bar diameter plot for the appropriate quenching medium—in this case, Figure 17.11a. Then, convert the cooling rate at each of these radial positions into a hardness value from a hardenability plot for the particular alloy. Finally, determine the hardness profile by plotting the hardness as a function of radial position.

This procedure is demonstrated in Figure 17.13, for the center position. Note that for a water-quenched cylinder of 50 mm diameter, the cooling rate at the center is equivalent to that approximately 9.5 mm from the Jominy specimen quenched end (Figure 17.13a). This corresponds to a hardness of about 28 HRC, as noted from the hardenability plot for the 1040 steel alloy (Figure 17.13b). Finally, this data point is plotted on the hardness profile in Figure 17.13c.

Surface, midradius, and three-quarter radius hardnesses are determined in a similar manner. The complete profile has been included, and the data used are shown in the following table.

Radial Position	Equivalent Distance from Quenched End (mm)	Hardness (HRC)
Center	9.5	28
Midradius	8	30
Three-quarters radius	4.8	39
Surface	1.6	54

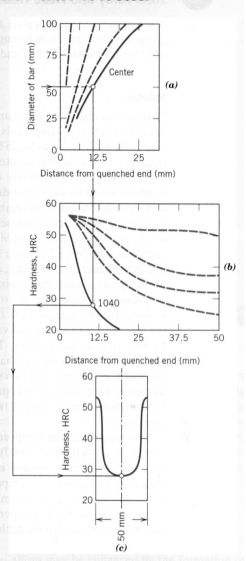

Figure 17.13 Use of hardenability data in the generation of hardness profiles. (*a*) The cooling rate is determined at the center of a water-quenched specimen of diameter 50 mm. (*b*) The cooling rate is converted into an HRC hardness for a 1040 steel. (*c*) The Rockwell hardness is plotted on the radial hardness profile.

Figure 17.16 Schematic temperature-versus-time plot showing both solution and precipitation heat treatments for precipitation hardening.

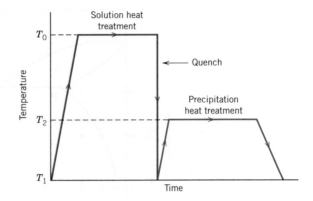

Precipitation Heat Treating

precipitation heat treatment

For the second or **precipitation heat treatment**, the supersaturated α solid solution is ordinarily heated to an intermediate temperature T_2 (Figure 17.15) within the $\alpha + \beta$ two-phase region, at which temperature diffusion rates become appreciable. The β precipitate phase begins to form as finely dispersed particles of composition C_β, the process of which is sometimes termed *aging*. After the appropriate aging time at T_2, the alloy is cooled to room temperature; normally, this cooling rate is not an important consideration. Both solution and precipitation heat treatments are represented on the temperature-versus-time plot in Figure 17.16. The character of these β particles, and subsequently the strength and hardness of the alloy, depend on both the precipitation temperature T_2 and the aging time at this temperature. For some alloys, aging occurs spontaneously at room temperature over extended time periods.

The dependence of the growth of the precipitate β particles on time and temperature under isothermal heat treatment conditions may be represented by C-shape curves similar to those in Figure 12.18 for the eutectoid transformation in steels. However, it is more useful and convenient to present the data as tensile strength, yield strength, or hardness at room temperature as a function of the logarithm of aging time, at constant temperature T_2. The behavior for a typical precipitation-hardenable alloy is represented schematically in Figure 17.17. With increasing time, the strength or hardness increases, reaches a maximum, and finally diminishes. This reduction in strength and hardness that occurs after long time periods is known as **overaging**. The influence of temperature is incorporated by the superposition, on a single plot, of curves at a variety of temperatures.

overaging

Figure 17.17 Schematic diagram showing strength and hardness as a function of the logarithm of aging time at constant temperature during the precipitation heat treatment.

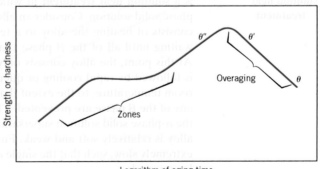

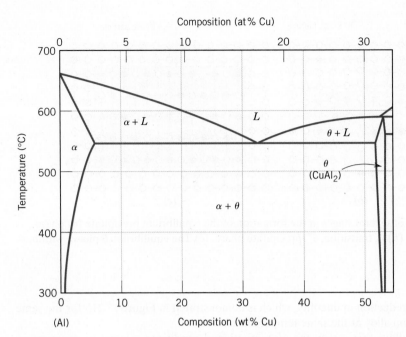

Figure 17.18 The aluminum-rich side of the aluminum–copper phase diagram.
(Adapted from J. L. Murray, *International Metals Review*, **30**, 5, 1985. Reprinted by permission of ASM International.)

Mechanism of Hardening

Precipitation hardening is commonly employed with high-strength aluminum alloys. Although a large number of these alloys have different proportions and combinations of alloying elements, the mechanism of hardening has perhaps been studied most extensively for the aluminum–copper alloys. Figure 17.18 presents the aluminum-rich portion of the aluminum–copper phase diagram. The α phase is a substitutional solid solution of copper in aluminum, whereas the intermetallic compound $CuAl_2$ is designated the θ phase. For an aluminum–copper alloy of, say, composition 96 wt% Al–4 wt% Cu, in the development of this equilibrium θ phase during the precipitation heat treatment, several transition phases are first formed in a specific sequence. The mechanical properties are influenced by the character of the particles of these transition phases. During the initial hardening stage (at short times, Figure 17.17), copper atoms cluster together in very small, thin discs that are only one or two atoms thick and approximately 25 atoms in diameter; these form at countless positions within the α phase. The clusters, sometimes called *zones,* are so small that they are really not regarded as distinct precipitate particles. However, with time and the subsequent diffusion of copper atoms, zones become particles as they increase in size. These precipitate particles then pass through two transition phases (denoted as θ'' and θ'), before the formation of the equilibrium θ phase (Figure 17.19c). Transition phase particles for a precipitation-hardened 7150 aluminum alloy are shown in the electron micrograph of Figure 11.20.

The strengthening and hardening effects shown in Figure 17.17 result from the innumerable particles of these transition and metastable phases. As shown in the figure, maximum strength coincides with the formation of the θ'' phase, which may be preserved upon cooling the alloy to room temperature. Overaging results from continued particle growth and the development of θ' and θ phases.

The strengthening process is accelerated as the temperature is increased. This is demonstrated in Figure 17.21a, a plot of yield strength versus the logarithm of time for a 2014 aluminum alloy at several different precipitation temperatures. Ideally, temperature and time for the precipitation heat treatment should be designed to produce a hardness or strength in the vicinity of the maximum. Associated with an increase in

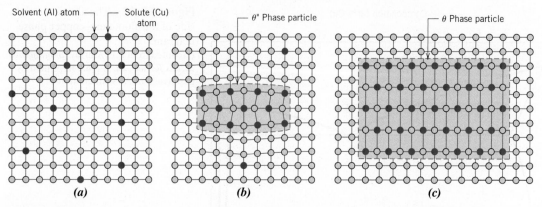

Solvent (Al) atom ⎯   ⎯ Solute (Cu) atom   ⎯ θ'' Phase particle   ⎯ θ Phase particle

(a)   *(b)*   *(c)*

Figure 17.19 Schematic depiction of several stages in the formation of the equilibrium precipitate (θ) phase. (*a*) A supersaturated α solid solution. (*b*) A transition, θ'', precipitate phase. (*c*) The equilibrium θ phase, within the α-matrix phase.

strength is a reduction in ductility, which is demonstrated in Figure 17.21*b* for the same 2014 aluminum alloy at the same temperatures.

Not all alloys that satisfy the aforementioned conditions relative to composition and phase diagram configuration are amenable to precipitation hardening. In addition, lattice strains must be established at the precipitate–matrix interface. For aluminum–copper alloys, there is a distortion of the crystal lattice structure around and within the vicinity of particles of these transition phases (Figure 17.19*b*). During plastic deformation, dislocation motions are effectively impeded as a result of these distortions, and, consequently, the alloy becomes harder and stronger. As the θ phase forms, the resultant overaging (softening and weakening) is explained by a reduction in the resistance to slip that is offered by these precipitate particles.

Alloys that experience appreciable precipitation hardening at room temperature and after relatively short time periods must be quenched to and stored under refrigerated conditions. Several aluminum alloys that are used for rivets exhibit this behavior.

Figure 17.20 A transmission electron micrograph showing the microstructure of a 7150–T651 aluminum alloy (6.2 wt% Zn, 2.3 wt% Cu, 2.3 wt% Mg, 0.12 wt% Zr, the balance Al) that has been precipitation hardened. The light matrix phase in the micrograph is an aluminum solid solution. The majority of the small plate-shaped dark precipitate particles are a transition η' phase, the remainder being the equilibrium η (MgZn$_2$) phase. Note that grain boundaries are "decorated" by some of these particles. 90,000×.
(Courtesy of G. H. Narayanan and A. G. Miller, Boeing Commercial Airplane Company.)

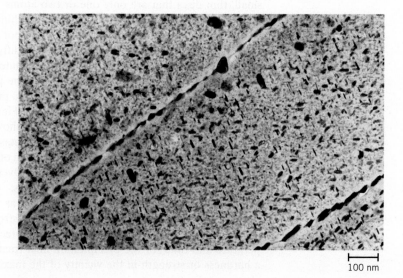

100 nm

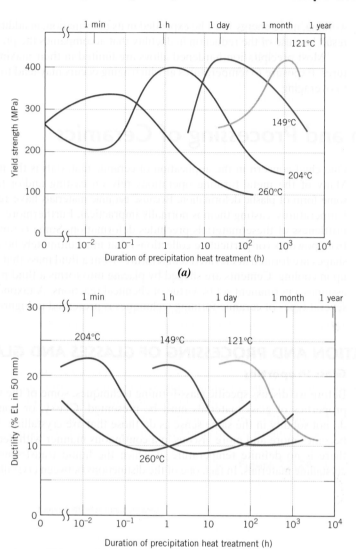

Figure 17.21   The precipitation hardening characteristics of a 2014 aluminum alloy (0.9 wt% Si, 4.4 wt% Cu, 0.8 wt% Mn, 0.5 wt% Mg) at four different aging temperatures: (*a*) yield strength, and (*b*) ductility (%EL).
[Adapted from *Metals Handbook: Properties and Selection: Nonferrous Alloys and Pure Metals,* Vol. 2, 9th edition, H. Baker (Managing Editor), 1979. Reproduced by permission of ASM International, Materials Park, OH.]

**natural aging,
    artificial aging**

They are driven while still soft, then allowed to age harden at the normal ambient temperature. This is termed **natural aging**; **artificial aging** is carried out at elevated temperatures.

Miscellaneous Considerations

The combined effects of strain hardening and precipitation hardening may be employed in high-strength alloys. The order of these hardening procedures is important in the production of alloys having the optimum combination of mechanical properties. Normally, the alloy is solution heat-treated and then quenched. This is followed by cold working and finally by the precipitation-hardening heat treatment. In the final treatment, little strength loss is sustained as a result of recrystallization. If the alloy is precipitation hardened before cold

for fused silica, high silica, borosilicate, and soda-lime glasses. On the viscosity scale, several specific points that are important in the fabrication and processing of glasses are labeled:

melting point

1. The **melting point** corresponds to the temperature at which the viscosity is 10 Pa·s (100 P); the glass is fluid enough to be considered a liquid.

working point

2. The **working point** represents the temperature at which the viscosity is 10^3 Pa·s (10^4 P); the glass is easily deformed at this viscosity.

softening point

3. The **softening point,** the temperature at which the viscosity is 4×10^6 Pa·s (4×10^7 P), is the maximum temperature at which a glass piece may be handled without causing significant dimensional alterations.

annealing point

4. The **annealing point** is the temperature at which the viscosity is 10^{12} Pa·s (10^{13} P); at this temperature, atomic diffusion is sufficiently rapid that any residual stresses may be removed within about 15 min.

strain point

5. The **strain point** corresponds to the temperature at which the viscosity becomes 3×10^{13} Pa·s (3×10^{14} P); for temperatures below the strain point, fracture will occur before the onset of plastic deformation. The glass transition temperature will be above the strain point.

Most glass-forming operations are carried out within the working range—between the working and softening temperatures.

The temperature at which each of these points occurs depends on glass composition. For example, from Figure 17.24, the softening points for soda-lime and 96% silica glasses are about 700°C and 1550°C, respectively. That is, forming operations may be carried out at significantly lower temperatures for the soda-lime glass. The formability of a glass is tailored to a large degree by its composition.

Glass Forming

Glass is produced by heating the raw materials to an elevated temperature above which melting occurs. Most commercial glasses are of the silica-soda-lime variety; the silica is usually supplied as common quartz sand, whereas Na_2O and CaO are added as soda ash (Na_2CO_3) and limestone ($CaCO_3$). For most applications, especially when optical transparency is important, it is essential that the glass product be homogeneous and pore free. Homogeneity is achieved by complete melting and mixing of the raw ingredients. Porosity results from small gas bubbles that are produced; these must be absorbed into the melt or otherwise eliminated, which requires proper adjustment of the viscosity of the molten material.

Five different forming methods are used to fabricate glass products: pressing, blowing, drawing, and sheet and fiber forming. Pressing is used in the fabrication of relatively thick-walled pieces such as plates and dishes. The glass piece is formed by pressure application in a graphite-coated cast iron mold having the desired shape; the mold is typically heated to ensure an even surface.

Although some glass blowing is done by hand, especially for art objects, the process has been completely automated for the production of glass jars, bottles, and light bulbs. The several steps involved in one such technique are illustrated in Figure 17.25. From a raw gob of glass, a *parison,* or temporary shape, is formed by mechanical pressing in a mold. This piece is inserted into a finishing or blow mold and forced to conform to the mold contours by the pressure created from a blast of air.

Drawing is used to form long glass pieces that have a constant cross section, such as sheet, rod, tubing, and fibers.

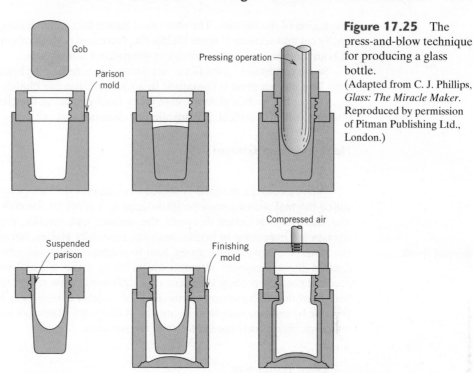

Figure 17.25 The press-and-blow technique for producing a glass bottle.
(Adapted from C. J. Phillips, *Glass: The Miracle Maker*. Reproduced by permission of Pitman Publishing Ltd., London.)

Until the late 1950s, sheet glass (or plate) was produced by casting (or drawing) the glass into a plate shape, grinding both faces to make them flat and parallel, and finally, polishing the faces to make the sheet transparent—a procedure that was relatively expensive. A more economical float process was patented in 1959 in England. With this technique (represented schematically in Figure 17.26), the molten glass passes (on rollers) from one furnace onto a bath of liquid tin located in a second furnace. Thus, as this continuous glass ribbon "floats" on the surface of the molten tin, gravitational and surface tension forces cause the faces to become perfectly flat and parallel and the resulting sheet to be of uniform thickness. Furthermore, sheet faces acquire a bright, "fire-polished" finish in

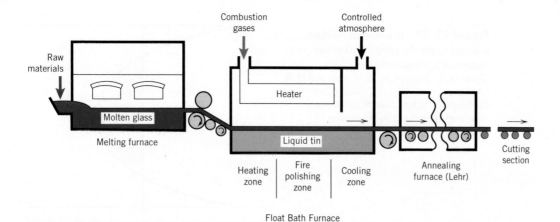

Figure 17.26 Schematic diagram showing the float process for making sheet glass.
(Courtesy of Pilkington Group Limited.)

one region of the furnace. The sheet next passes into an annealing furnace (lehr), and is finally cut into sections (Figure 17.26). The success of this operation requires rigid control of both temperature and chemistry of the gaseous atmosphere.

Some continuous glass fibers are formed in a rather sophisticated drawing operation. The molten glass is contained in a platinum heating chamber. Fibers are formed by drawing the molten glass through many small orifices at the chamber base. The glass viscosity, which is critical, is controlled by chamber and orifice temperatures.

Heat-Treating Glasses

Annealing

When a ceramic material is cooled from an elevated temperature, internal stresses, called thermal stresses, may be introduced as a result of the difference in cooling rate and thermal contraction between the surface and interior regions. These thermal stresses are important in brittle ceramics, especially glasses, because they may weaken the material or, in extreme cases, lead to fracture, which is termed **thermal shock** (see Section 20.5). Normally, attempts are made to avoid thermal stresses, which may be accomplished by cooling the piece at a sufficiently slow rate. Once such stresses have been introduced, however, elimination, or at least a reduction in their magnitude, is possible by an annealing heat treatment in which the glassware is heated to the annealing point, then slowly cooled to room temperature.

thermal shock

Glass Tempering

The strength of a glass piece may be enhanced by intentionally inducing compressive residual surface stresses. This can be accomplished by a heat treatment procedure called **thermal tempering**. With this technique, the glassware is heated to a temperature above the glass transition region yet below the softening point. It is then cooled to room temperature in a jet of air or, in some cases, an oil bath. The residual stresses arise from differences in cooling rates for surface and interior regions. Initially, the surface cools more rapidly and, once it has dropped to a temperature below the strain point, it becomes rigid. At this time, the interior, having cooled less rapidly, is at a higher temperature (above the strain point) and, therefore, is still plastic. With continued cooling, the interior attempts to contract to a greater degree than the now-rigid exterior will allow. Thus, the inside tends to draw in the outside, or to impose inward

thermal tempering

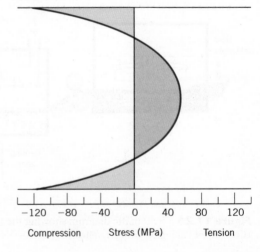

Figure 17.27 Room-temperature residual stress distribution over the cross section of a tempered glass plate. (From W. D. Kingery, H. K. Bowen, and D. R. Uhlmann, *Introduction to Ceramics,* 2nd edition. Copyright © 1976 by John Wiley & Sons, New York. Reprinted by permission of John Wiley & Sons, Inc.)

Compression   Stress (MPa)   Tension

−120   −80   −40   0   40   80   120

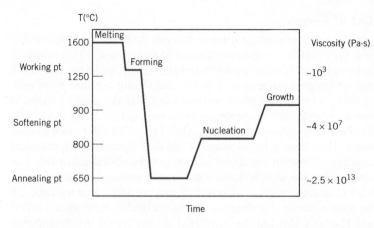

Figure 17.28 Typical time-versus-temperature processing cycle for a Li_2O–Al_2O_3–SiO_2 glass–ceramic. (Adapted from Y. M. Chiang, D. P. Birnie, III, and W. D. Kingery, *Physical Ceramics—Principles for Ceramic Science and Engineering.* Copyright © 1997 by John Wiley & Sons, New York. Reprinted by permission of John Wiley & Sons, Inc.)

radial stresses. As a consequence, after the glass piece has cooled to room temperature, it sustains compressive stresses on the surface and tensile stresses at interior regions. The room-temperature stress distribution over a cross section of a glass plate is represented schematically in Figure 17.27.

The failure of ceramic materials almost always results from a crack that is initiated at the surface by an applied tensile stress. To cause fracture of a tempered glass piece, the magnitude of an externally applied tensile stress must be great enough to first overcome the residual compressive surface stress and, in addition, to stress the surface in tension sufficient to initiate a crack, which may then propagate. For an untempered glass, a crack is introduced at a lower external stress level, and, consequently, the fracture strength is smaller.

Tempered glass is used for applications in which high strength is important; these include large doors and eyeglass lenses.

Concept Check 17.4 How does the thickness of a glassware affect the magnitude of the thermal stresses that may be introduced? Why?

[*The answer may be found at* www.wiley.com/college/callister *(Student Companion Site).*]

Fabrication and Heat-Treating of Glass–Ceramics

The first stage in the fabrication of a glass–ceramic ware is forming it into the desired shape as a glass. Forming techniques used are the same as for glass pieces, as described previously (e.g., pressing and drawing). Conversion of the glass into a glass–ceramic (i.e., crystallation, Section 14.11) is accomplished by appropriate heat treatments. One such set of heat treatments for a Li_2O–Al_2O_3–SiO_2 glass–ceramic is detailed in the time-versus-temperature plot of Figure 17.28. After melting and forming operations, nucleation and growth of the crystalline phase particles are carried out isothermally at two different temperatures.

17.9 FABRICATION AND PROCESSING OF CLAY PRODUCTS

As Section 14.12 noted, this class of materials includes the structural clay products and the whitewares. In addition to clay, many of these products also contain other ingredients. After being formed, pieces most often must be subjected to drying and firing operations; each of the ingredients influences the changes that take place during these processes and the characteristics of the finished piece.

The Characteristics of Clay

The clay minerals play two very important roles in ceramic bodies. First, when water is added, they become very plastic, a condition termed *hydroplasticity*. This property is very important in forming operations, as discussed shortly. In addition, clay fuses or melts over a range of temperatures; thus, a dense and strong ceramic piece may be produced during firing without complete melting such that the desired shape is maintained. This fusion temperature range depends on the composition of the clay.

Clays are aluminosilicates composed of alumina (Al_2O_3) and silica (SiO_2) that contain chemically bound water. They have a broad range of physical characteristics, chemical compositions, and structures; common impurities include compounds (usually oxides) of barium, calcium, sodium, potassium, iron, and also some organic matter. Crystal structures for the clay minerals are relatively complicated; however, one prevailing characteristic is a layered structure. The most common clay minerals that are of interest have what is called the kaolinite structure. Kaolinite clay $[Al_2(Si_2O_5)(OH)_4]$ has the crystal structure shown in Figure 4.15. When water is added, the water molecules fit between these layered sheets and form a thin film around the clay particles. The particles are thus free to move over one another, which accounts for the resulting plasticity of the water–clay mixture.

Compositions of Clay Products

In addition to clay, many of these products (in particular the whitewares) also contain some nonplastic ingredients; the nonclay minerals include flint, or finely ground quartz, and a flux such as feldspar.[3] The quartz is used primarily as a filler material, being inexpensive, relatively hard, and chemically unreactive. It experiences little change during high-temperature heat treatment because it has a melting temperature well above the normal firing temperature; when melted, however, quartz has the ability to form a glass.

When mixed with clay, a flux forms a glass that has a relatively low melting point. The feldspars are some of the more common fluxing agents; they are a group of aluminosilicate materials that contain K^+, Na^+, and Ca^{2+} ions.

As expected, the changes that take place during drying and firing processes, and also the characteristics of the finished piece, are influenced by the proportions of the three constituents: clay, quartz, and flux. A typical porcelain might contain approximately 50% clay, 25% quartz, and 25% feldspar.

Fabrication Techniques

The as-mined raw materials usually have to go through a milling or grinding operation in which particle size is reduced; this is followed by screening or sizing to yield a powdered product having a desired range of particle sizes. For multicomponent systems, powders must be thoroughly mixed with water and perhaps other ingredients to give flow characteristics that are compatible with the particular forming technique. The formed piece must have sufficient mechanical strength to remain intact during transporting, drying, and firing operations. Two common shaping techniques are used to form clay-based compositions: **hydroplastic forming** and **slip casting**.

hydroplastic forming

slip casting

Hydroplastic Forming

As mentioned previously, clay minerals, when mixed with water, become highly plastic and pliable and may be molded without cracking; however, they have extremely low yield strengths. The consistency (water–clay ratio) of the hydroplastic mass must give a yield strength sufficient to permit a formed ware to maintain its shape during handling and drying.

---

[3]*Flux*, in the context of clay products, is a substance that promotes the formation of a glassy phase during the firing heat treatment.

The most common hydroplastic forming technique is extrusion, in which a stiff plastic ceramic mass is forced through a die orifice having the desired cross-sectional geometry; it is similar to the extrusion of metals (Figure 17.2c). Brick, pipe, ceramic blocks, and tiles are all commonly fabricated using hydroplastic forming. Usually the plastic ceramic is forced through the die by means of a motor-driven auger, and often air is removed in a vacuum chamber to enhance the density. Hollow internal columns in the extruded piece (e.g., building brick) are formed by inserts situated within the die.

Slip Casting

Another forming process used for clay-based compositions is slip casting. A slip is a suspension of clay and/or other nonplastic materials in water. When poured into a porous mold (commonly made of plaster of Paris), water from the slip is absorbed into the mold, leaving behind a solid layer on the mold wall, the thickness of which depends on the time. This process may be continued until the entire mold cavity becomes solid (*solid casting*), as demonstrated in Figure 17.29a. Alternatively, it may be terminated when the solid shell wall reaches the desired thickness, by inverting the mold and pouring out the excess slip; this is termed *drain casting* (Figure 17.29b). As the cast piece dries and shrinks, it pulls away (or releases) from the mold wall; at this time, the mold may be disassembled and the cast piece removed.

The nature of the slip is extremely important; it must have a high specific gravity and yet be very fluid and pourable. These characteristics depend on the solid-to-water ratio and other agents that are added. A satisfactory casting rate is an essential requirement. In addition, the cast piece must be free of bubbles, and it must have low drying shrinkage and relatively high strength.

The properties of the mold influence the quality of the casting. Normally, plaster of Paris, which is economical, relatively easy to fabricate into intricate shapes, and

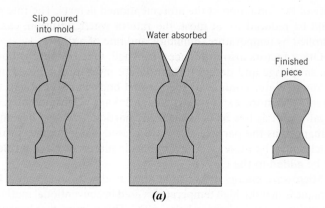

Figure 17.29 The steps in (*a*) solid and (*b*) drain slip casting using a plaster of Paris mold. (From W. D. Kingery, *Introduction to Ceramics.* Copyright © 1960 by John Wiley & Sons, New York. Reprinted by permission of John Wiley & Sons, Inc.)

(a)

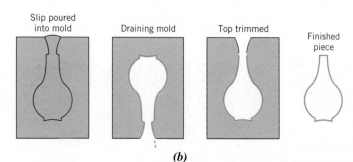

(b)

reusable, is used as the mold material. Most molds are multipiece items that must be assembled before casting. The mold porosity may be varied to control the casting rate. The rather complex ceramic shapes that may be produced by means of slip casting include sanitary lavatory ware, art objects, and specialized scientific laboratory ware such as ceramic tubes.

Drying and Firing

A ceramic piece that has been formed hydroplastically or by slip casting retains significant porosity and has insufficient strength for most practical applications. In addition, it may still contain some of the liquid (e.g., water) that was added to assist in the forming operation. This liquid is removed in a drying process; density and strength are enhanced as a result of a high-temperature heat treatment or firing procedure. A body that has been formed and dried but not fired is termed **green**. Drying and firing techniques are critical inasmuch as defects that ordinarily render the ware useless (e.g., warpage, distortion, cracks) may be introduced during the operation. These defects normally result from stresses that are set up from nonuniform shrinkage.

green ceramic body

Drying

As a clay-based ceramic body dries, it also experiences some shrinkage. In the early stages of drying, the clay particles are virtually surrounded by and separated from one another by a thin film of water. As drying progresses and water is removed, the interparticle separation decreases, which is manifested as shrinkage (Figure 17.30). During drying it is critical to control the rate of water removal. Drying at interior regions of a body is accomplished by the diffusion of water molecules to the surface, where evaporation occurs. If the rate of evaporation is greater than the rate of diffusion, the surface will dry (and as a consequence shrink) more rapidly than the interior, with a high probability of the formation of the aforementioned defects. The rate of surface evaporation should be reduced to, at most, the rate of water diffusion; evaporation rate may be controlled by temperature, humidity, and rate of airflow.

Other factors also influence shrinkage. One of these is body thickness; nonuniform shrinkage and defect formation are more pronounced in thick pieces than in thin ones. Water content of the formed body is also critical: the greater the water content, the more extensive is the shrinkage. Consequently, the water content is typically kept as low as possible. Clay particle size also has an influence; shrinkage is enhanced as the particle size is decreased. To minimize shrinkage, the size of the particles may be increased, or nonplastic materials having relatively large particles may be added to the clay.

Microwave energy may also be used to dry ceramic wares. One advantage of this technique is that the high temperatures used in conventional methods are avoided; drying temperatures may be kept to below 50°C. This is important because the drying temperature of some temperature-sensitive materials should be kept as low as possible.

Figure 17.30 Several stages in the removal of water from between clay particles during the drying process. (*a*) Wet body. (*b*) Partially dry body. (*c*) Completely dry body. (From W. D. Kingery, *Introduction to Ceramics.* Copyright © 1960 by John Wiley & Sons, New York. Reprinted by permission of John Wiley & Sons, Inc.)

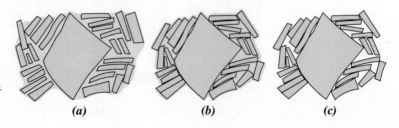

(*a*)   (*b*)   (*c*)

Concept Check 17.5 Thick ceramic wares are more likely to crack upon drying than thin wares. Why is this so?

[*The answer may be found at* www.wiley.com/college/callister (*Student Companion Site*).]

Firing

After drying, a body is usually fired at a temperature between 900°C and 1400°C; the firing temperature depends on the composition and desired properties of the finished piece. During the firing operation, the density is further increased (with an attendant decrease in porosity) and the mechanical strength is enhanced.

vitrification

When clay-based materials are heated to elevated temperatures, some rather complex and involved reactions occur. One of these is **vitrification**—the gradual formation of a liquid glass that flows into and fills some of the pore volume. The degree of vitrification depends on firing temperature and time, as well as on the composition of the body. The temperature at which the liquid phase forms is lowered by the addition of fluxing agents such as feldspar. This fused phase flows around the remaining unmelted particles and fills in the pores as a result of surface tension forces (or capillary action); shrinkage also accompanies this process. Upon cooling, this fused phase forms a glassy matrix that results in a dense, strong body. Thus, the final microstructure consists of the vitrified phase, any unreacted quartz particles, and some porosity. Figure 17.31 is a scanning electron micrograph of a fired porcelain in which these microstructural elements may be seen.

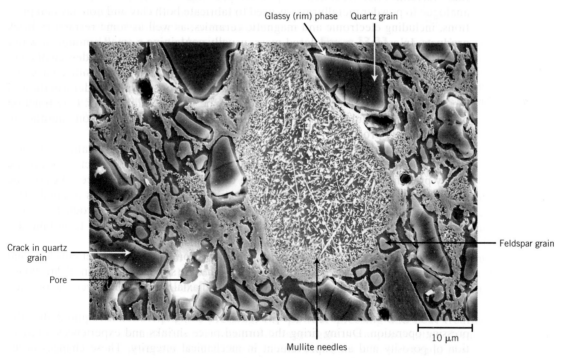

Figure 17.31 Scanning electron micrograph of a fired porcelain specimen (etched 15 s, 5°C, 10% HF) in which the following features may be seen: quartz grains (large dark particles), which are surrounded by dark glassy solution rims; partially dissolved feldspar regions (small unfeatured areas); mullite needles; and pores (dark holes with white border regions). Cracks within the quartz particles may be noted, which were formed during cooling as a result of the difference in shrinkage between the glassy matrix and the quartz. 1500×.
(Courtesy of H. G. Brinkies, Swinburne University of Technology, Hawthorn Campus, Hawthorn, Victoria, Australia.)

Figure 13.34 Scanning electron micrograph of an aluminum oxide powder compact that was sintered at 1700°C for 6 min. 5000×.
(From W. D. Kingery, H. K. Bowen, and D. R. Uhlmann, *Introduction to Ceramics,* 2nd edition, p. 483. Copyright © 1976 by John Wiley & Sons, New York. Reprinted by permission of John Wiley & Sons, Inc.)

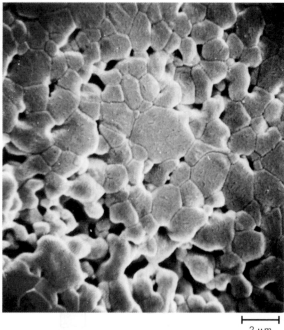

2 μm

17.11 TAPE CASTING

Tape casting is an important ceramic fabrication technique. As the name implies, in this technique, thin sheets of a flexible tape are produced by means of a casting process. These sheets are prepared from slips in many respects similar to those employed for slip casting (Section 17.9). This type of slip consists of a suspension of ceramic particles in an organic liquid that also contains binders and plasticizers, which are incorporated to impart strength and flexibility to the cast tape. De-airing in a vacuum may also be necessary to remove any entrapped air or solvent vapor bubbles, which may act as crack-initiation sites in the finished piece. The actual tape is formed by pouring the slip onto a flat surface (of stainless steel, glass, a polymeric film, or paper); a doctor blade spreads the slip into a thin tape of uniform thickness, as shown schematically in Figure 17.35. In the drying process, volatile slip components are removed by evaporation; this green product is a flexible tape that may be cut or into which holes may be punched prior to a firing operation. Tape thicknesses normally range between 0.1 and 2 mm. Tape casting is widely used in the production of ceramic substrates that are used for integrated circuits and for multilayered capacitors.

Figure 17.35 Schematic diagram showing the tape-casting process using a doctor blade.
(From D. W. Richerson, *Modern Ceramic Engineering,* 2nd edition, Marcel Dekker, Inc., NY, 1992. Reprinted from *Modern Ceramic Engineering,* 2nd edition, p. 472 by courtesy of Marcel Dekker, Inc.)

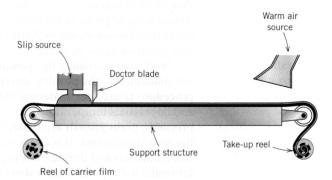

Cementation is also considered a ceramic fabrication process (Figure 17.22). The cement material, when mixed with water, forms a paste that, after being fashioned into a desired shape, subsequently hardens as a result of complex chemical reactions. Cements and the cementation process were discussed briefly in Section 14.15.

Synthesis and Processing of Polymers

The large macromolecules of the commercially useful polymers must be synthesized from substances having smaller molecules in a process termed polymerization. Furthermore, the properties of a polymer may be modified and enhanced by the inclusion of additive materials. Finally, a finished piece having a desired shape must be fashioned during a forming operation. This section treats polymerization processes and the various forms of additives, as well as specific forming procedures.

17.12 POLYMERIZATION

The synthesis of these large molecules (polymers) is termed *polymerization;* it is simply the process by which monomers are linked together to generate long chains composed of repeat units. Most generally, the raw materials for synthetic polymers are derived from coal, natural gas, and petroleum products. The reactions by which polymerization occur are grouped into two general classifications—addition and condensation—according to the reaction mechanism, as discussed next.

Addition Polymerization

addition polymerization

Addition polymerization (sometimes called *chain reaction polymerization*) is a process by which monomer units are attached one at a time in chainlike fashion to form a linear macromolecule. The composition of the resultant product molecule is an exact multiple of that of the original reactant monomer.

Three distinct stages—initiation, propagation, and termination—are involved in addition polymerization. During the initiation step, an active center capable of propagation is formed by a reaction between an initiator (or catalyst) species and the monomer unit. This process has already been demonstrated for polyethylene in Equation 5.1, which is repeated as follows:

$$\text{R}\cdot + \begin{array}{c} \text{H} \ \ \text{H} \\ | \ \ \ | \\ \text{C}=\text{C} \\ | \ \ \ | \\ \text{H} \ \ \text{H} \end{array} \longrightarrow \begin{array}{c} \text{H} \ \ \text{H} \\ | \ \ \ | \\ \text{R}-\text{C}-\text{C}\cdot \\ | \ \ \ | \\ \text{H} \ \ \text{H} \end{array} \tag{17.1}$$

R· represents the active initiator, and · is an unpaired electron.

Propagation involves the linear growth of the polymer chain by the sequential addition of monomer units to this active growing chain molecule. This may be represented, again for polyethylene, as follows:

$$\begin{array}{c} \text{H} \ \ \text{H} \\ | \ \ \ | \\ \text{R}-\text{C}-\text{C}\cdot + \\ | \ \ \ | \\ \text{H} \ \ \text{H} \end{array} \begin{array}{c} \text{H} \ \ \text{H} \\ | \ \ \ | \\ \text{C}=\text{C} \\ | \ \ \ | \\ \text{H} \ \ \text{H} \end{array} \longrightarrow \begin{array}{c} \text{H} \ \ \text{H} \ \ \text{H} \ \ \text{H} \\ | \ \ \ | \ \ \ | \ \ \ | \\ \text{R}-\text{C}-\text{C}-\text{C}-\text{C}\cdot \\ | \ \ \ | \ \ \ | \ \ \ | \\ \text{H} \ \ \text{H} \ \ \text{H} \ \ \text{H} \end{array} \tag{17.2}$$

Chain growth is relatively rapid; the period required to grow a molecule consisting of, say, 1000 repeat units is on the order of 10^{-2} to 10^{-3} s.

Propagation may end or terminate in different ways. First, the active ends of two propagating chains may link together to form one molecule according to the following reaction:[4]

$$
\text{R} \!\left(\text{C}-\text{C}\right)_{\!m}\!\text{C}-\text{C}\cdot + \cdot\text{C}-\text{C}\!\left(\text{C}-\text{C}\right)_{\!n}\!\text{R} \longrightarrow \text{R}\!\left(\text{C}-\text{C}\right)_{\!m}\!\text{C}-\text{C}-\text{C}-\text{C}\!\left(\text{C}-\text{C}\right)_{\!n}\!\text{R}
\tag{17.3}
$$

The other termination possibility involves two growing molecules that react to form two "dead chains" as[5]

$$
\text{R}\!\left(\text{C}-\text{C}\right)_{\!m}\!\text{C}-\text{C}\cdot + \cdot\text{C}-\text{C}\!\left(\text{C}-\text{C}\right)_{\!n}\!\text{R} \longrightarrow \text{R}\!\left(\text{C}-\text{C}\right)_{\!m}\!\text{C}-\text{C}-\text{H} + \text{C}=\text{C}\!\left(\text{C}-\text{C}\right)_{\!n}\!\text{R}
\tag{17.4}
$$

thus terminating the growth of each chain.

Molecular weight is governed by the relative rates of initiation, propagation, and termination. Typically, they are controlled to ensure the production of a polymer having the desired degree of polymerization.

Addition polymerization is used in the synthesis of polyethylene, polypropylene, poly(vinyl chloride), and polystyrene, as well as many of the copolymers.

Concept Check 17.7 State whether the molecular weight of a polymer that is synthesized by addition polymerization is relatively high, medium, or relatively low for the following situations:

(a) Rapid initiation, slow propagation, and rapid termination
(b) Slow initiation, rapid propagation, and slow termination
(c) Rapid initiation, rapid propagation, and slow termination
(d) Slow initiation, slow propagation, and rapid termination

[*The answer may be found at* www.wiley.com/college/callister *(Student Companion Site).*]

Condensation Polymerization

condensation polymerization

Condensation (or *step reaction*) **polymerization** is the formation of polymers by stepwise intermolecular chemical reactions that may involve more than one monomer species. There is usually a low-molecular-weight by-product such as water that is eliminated (or condensed). No reactant species has the chemical formula of the repeat unit, and the intermolecular reaction occurs every time a repeat unit is formed. For example, consider the formation of the polyester poly(ethylene terephthalate)

---

[4]This type of termination reaction is referred to as *combination*.
[5]This type of termination reaction is called *disproportionation*.

(PET) from the reaction between dimethyl terephthalate and ethylene glycol to form a linear PET molecule with methyl alcohol as a by-product; the intermolecular reaction is as follows:

Dimethyl terephthalate   Ethylene glycol

$$n \left(\text{DMT} \right) + n \left(\text{EG} \right) \longrightarrow$$

(17.5)

$$\text{Poly(ethylene terephthalate)} + 2n \left(\text{Methyl alcohol} \right)$$

This stepwise process is successively repeated, producing a linear molecule. Reaction times for condensation polymerization are generally longer than for addition polymerization.

For the previous condensation reaction, both ethylene glycol and dimethyl terephthalate are bifunctional. However, condensation reactions can include trifunctional or higher functional monomers capable of forming crosslinked and network polymers. The thermosetting polyesters and phenol-formaldehyde, the nylons, and the polycarbonates are produced by condensation polymerization. Some polymers, such as nylon, may be polymerized by either technique.

Concept Check 17.8   Nylon 6,6 may be formed by means of a condensation polymerization reaction in which hexamethylene diamine $[NH_2-(CH_2)_6-NH_2]$ and adipic acid react with one another with the formation of water as a by-product. Write out this reaction in the manner of Equation 17.5. *Note:* The structure for adipic acid is

$$HO-\overset{O}{\underset{}{C}}-\overset{H}{\underset{H}{C}}-\overset{H}{\underset{H}{C}}-\overset{H}{\underset{H}{C}}-\overset{H}{\underset{H}{C}}-\overset{O}{\underset{}{C}}-OH$$

[*The answer may be found at* www.wiley.com/college/callister *(Student Companion Site).*]

17.13 POLYMER ADDITIVES

Most of the properties of polymers discussed earlier in this chapter are intrinsic ones—that is, they are characteristic of or fundamental to the specific polymer. Some of these properties are related to and controlled by the molecular structure. Often, however, it is necessary to modify the mechanical, chemical, and physical properties to a much greater degree than is possible by the simple alteration of this fundamental molecular structure. Foreign substances called *additives* are intentionally introduced to enhance or modify many of these properties and thus render a polymer more serviceable. Typical additives include filler materials, plasticizers, stabilizers, colorants, and flame retardants.

Fillers

filler

Filler materials are most often added to polymers to improve tensile and compressive strengths, abrasion resistance, toughness, dimensional and thermal stability, and other properties. Materials used as particulate fillers include wood flour (finely powdered sawdust), silica flour and sand, glass, clay, talc, limestone, and even some synthetic polymers. Particle sizes range from 10 nm to macroscopic dimensions. Polymers that contain fillers may also be classified as composite materials, which are discussed in Chapter 16. Often the fillers are inexpensive materials that replace some volume of the more expensive polymer, reducing the cost of the final product.

Plasticizers

plasticizer

The flexibility, ductility, and toughness of polymers may be improved with the aid of additives called **plasticizers**. Their presence also produces reductions in hardness and stiffness. Plasticizers are generally liquids with low vapor pressures and low molecular weights. The small plasticizer molecules occupy positions between the large polymer chains, effectively increasing the interchain distance with a reduction in the secondary intermolecular bonding. Plasticizers are commonly used in polymers that are intrinsically brittle at room temperature, such as poly(vinyl chloride) and some of the acetate copolymers. The plasticizer lowers the glass transition temperature, so that at ambient conditions the polymers may be used in applications requiring some degree of pliability and ductility. These applications include thin sheets or films, tubing, raincoats, and curtains.

Concept Check 17.9

(a) Why must the vapor pressure of a plasticizer be relatively low?

(b) How will the crystallinity of a polymer be affected by the addition of a plasticizer? Why?

(c) How does the addition of a plasticizer influence the tensile strength of a polymer? Why?

[*The answer may be found at* www.wiley.com/college/callister (*Student Companion Site*).]

Stabilizers

stabilizer

Some polymeric materials, under normal environmental conditions, are subject to rapid deterioration, generally in terms of mechanical integrity. Additives that counteract deteriorative processes are called **stabilizers**.

One common form of deterioration results from exposure to light [in particular, ultraviolet (UV) radiation]. Ultraviolet radiation interacts with and causes a severance of some of the covalent bonds along the molecular chains, which may also result in some crosslinking. There are two primary approaches to UV stabilization. The first is to add a UV-absorbent material, often as a thin layer at the surface. This essentially acts as a sunscreen and blocks out the UV radiation before it can penetrate into and damage the polymer. The second approach is to add materials that react with the bonds broken by UV radiation before they can participate in other reactions that lead to additional polymer damage.

Another important type of deterioration is oxidation (Section 18.12). It is a consequence of the chemical interaction between oxygen [as either diatomic oxygen (O_2) or ozone (O_3)] and the polymer molecules. Stabilizers that protect against oxidation consume oxygen before it reaches the polymer and/or prevent the occurrence of oxidation reactions that would further damage the material.

Colorants

colorant

Colorants impart a specific color to a polymer; they may be added in the form of dyes or pigments. The molecules in a dye actually dissolve in the polymer. Pigments are filler materials that do not dissolve but remain as a separate phase; normally, they have a small particle size and a refractive index near that of the parent polymer. Others may impart opacity as well as color to the polymer.

Flame Retardants

flame retardant

The flammability of polymeric materials is a major concern, especially in the manufacture of textiles and children's toys. Most polymers are flammable in their pure form; exceptions include those containing significant contents of chlorine and/or fluorine, such as poly(vinyl chloride) and polytetrafluoroethylene. The flammability resistance of the remaining combustible polymers may be enhanced by additives called **flame retardants**. These retardants may function by interfering with the combustion process through the gas phase or by initiating a different combustion reaction that generates less heat, thereby reducing the temperature; this causes a slowing or cessation of burning.

17.14 FORMING TECHNIQUES FOR PLASTICS

Quite a variety of different techniques are employed in the forming of polymeric materials. The method used for a specific polymer depends on several factors: (1) whether the material is thermoplastic or thermosetting; (2) if thermoplastic, the temperature at which it softens; (3) the atmospheric stability of the material being formed; and (4) the geometry and size of the finished product. There are numerous similarities between some of these techniques and those used for fabricating metals and ceramics.

Fabrication of polymeric materials normally occurs at elevated temperatures and often by the application of pressure. Thermoplastics are formed above their glass transition temperatures, if amorphous, or above their melting temperatures, if semicrystalline. An applied pressure must be maintained as the piece is cooled so that the formed article retains its shape. One significant economic benefit of using thermoplastics is that they may be recycled; scrap thermoplastic pieces may be remelted and re-formed into new shapes.

Fabrication of thermosetting polymers is typically accomplished in two stages. First comes the preparation of a linear polymer (sometimes called a *prepolymer*) as a liquid having a low molecular weight. This material is converted into the final hard and stiff product during the second stage, which is normally carried out in a mold having the desired shape. This second stage, termed *curing*, may occur during heating and/or by the addition of catalysts and often under pressure. During curing, chemical and structural changes occur on a molecular level: a crosslinked or a network structure forms. After curing, thermoset polymers may be removed from a mold while still hot because they are now dimensionally stable. Thermosets are difficult to recycle, do not melt, are usable at higher temperatures than thermoplastics, and are often more chemically inert.

molding

Molding is the most common method for forming plastic polymers. The several molding techniques used include compression, transfer, blow, injection, and extrusion molding. For each, a finely pelletized or granulized plastic is forced, at an elevated temperature and by pressure, to flow into, fill, and assume the shape of a mold cavity.

Compression and Transfer Molding

For compression molding, the appropriate amounts of thoroughly mixed polymer and necessary additives are placed between male and female mold members, as illustrated in Figure 17.36. Both mold pieces are heated; however, only one is movable. The mold is closed, and heat and pressure are applied, causing the plastic to become viscous and flow to conform to the mold shape. Before molding, raw materials may be mixed and cold-pressed into a disc, which is called a *preform*. Preheating of the preform reduces molding

Figure 17.36 Schematic diagram of a compression molding apparatus. (From F. W. Billmeyer, Jr., *Textbook of Polymer Science,* 3rd edition. Copyright © 1984 by John Wiley & Sons, New York. Reprinted by permission of John Wiley & Sons, Inc.)

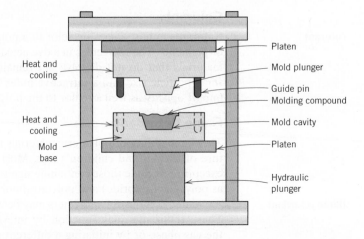

time and pressure, extends die lifetime, and produces a more uniform finished piece. This molding technique lends itself to the fabrication of both thermoplastic and thermosetting polymers; however, its use with thermoplastics is more time-consuming and expensive than the more commonly used extrusion or injection molding techniques discussed next.

In transfer molding—a variation of compression molding—the solid ingredients are first melted in a heated transfer chamber. As the molten material is injected into the mold chamber, the pressure is distributed more uniformly over all surfaces. This process is used with thermosetting polymers and for pieces having complex geometries.

Injection Molding

Injection molding—the polymer analogue of die casting for metals—is the most widely used technique for fabricating thermoplastic materials. A schematic cross section of the apparatus used is illustrated in Figure 17.37. The correct amount of pelletized material is fed from a feed hopper into a cylinder by the motion of a plunger or ram. This charge is pushed forward into a heating chamber, where it is forced around a spreader so as to make better contact with the heated wall. As a result, the thermoplastic material melts to form a viscous liquid. Next, the molten plastic is impelled, again by ram motion, through a nozzle into the enclosed mold cavity; pressure is maintained until the molding has solidified. Finally, the mold is opened, the piece is ejected, the mold is closed, and the entire cycle is repeated. Probably the most outstanding feature of this technique is the speed with which pieces may be produced. For thermoplastics, solidification of the injected charge is almost immediate; consequently, cycle times for this process are short (commonly within the range of 10 to 30 s). Thermosetting polymers may also be injection molded; curing takes place while the material is under pressure in a heated mold, which results in longer cycle times than for thermoplastics. This process is sometimes termed *reaction injection molding* (RIM) and is commonly used for materials such as polyurethane.

Figure 17.37 Schematic diagram of an injection molding apparatus. (Adapted from F. W. Billmeyer, Jr., *Textbook of Polymer Science,* 2nd edition. Copyright © 1971 by John Wiley & Sons, New York. Reprinted by permission of John Wiley & Sons, Inc.)

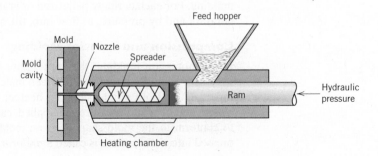

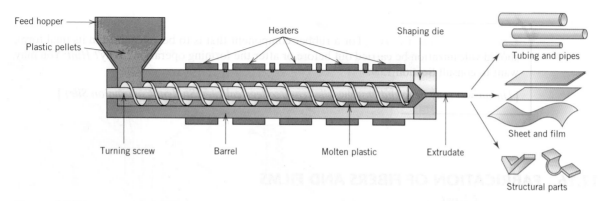

Figure 17.38 Schematic diagram of an extruder.
(Reprinted with permission from *Encyclopædia Britannica*, © 1997 by Encyclopædia Britannica, Inc.)

Extrusion

The extrusion process is the molding of a viscous thermoplastic under pressure through an open-ended die, similar to the extrusion of metals (Figure 17.2c). A mechanical screw or auger propels the pelletized material through a chamber, where it is successively compacted, melted, and formed into a continuous charge of viscous fluid (Figure 17.38). Extrusion takes place as this molten mass is forced through a die orifice. Solidification of the extruded length is expedited by blowers, a water spray, or a bath. The technique is especially adapted to producing continuous lengths having constant cross-sectional geometries—for example, rods, tubes, hose channels, sheets, and filaments.

Blow Molding

The blow-molding process for the fabrication of plastic containers is similar to that used for blowing glass bottles, as represented in Figure 17.25. First, a *parison*, or length of polymer tubing, is extruded. While still in a semimolten state, the parison is placed in a two-piece mold having the desired container configuration. The hollow piece is formed by blowing air or steam under pressure into the parison, forcing the tube walls to conform to the contours of the mold. The temperature and viscosity of the parison must be regulated carefully.

Casting

Like metals, polymeric materials may be cast, as when a molten plastic material is poured into a mold and allowed to solidify. Both thermoplastic and thermosetting plastics may be cast. For thermoplastics, solidification occurs upon cooling from the molten state; however, for thermosets, hardening is a consequence of the actual polymerization or curing process, which is usually carried out at an elevated temperature.

17.15 FABRICATION OF ELASTOMERS

Techniques used in the actual fabrication of rubber parts are essentially the same as those discussed for plastics as described previously—compression molding, extrusion, and so on. Furthermore, most rubber materials are vulcanized (Section 15.9), and some are reinforced with carbon black (Section 16.2).

- *Hardenability* is a parameter used to ascertain the influence of composition on the susceptibility to the formation of a predominantly martensitic structure for some specific heat treatment. Martensite content is determined using hardness measurements.
- Determination of hardenability is accomplished by the standard Jominy end-quench test (Figure 17.5), from which hardenability curves are generated.
- A *hardenability curve* plots hardness versus distance from the quenched end of a Jominy specimen. Hardness decreases with distance from the quenched end (Figure 17.6) because the quenching rate decreases with this distance, as does the martensite content. Each steel alloy has its own distinctive hardenability curve.
- The quenching medium also influences the extent to which martensite forms. Of the common quenching media, water is the most efficient, followed by aqueous polymers, oil, and air, in that order. Increasing the degree of medium agitation also enhances the quenching efficiency.
- Relationships among cooling rate and specimen size and geometry for a specific quenching medium frequently are expressed on empirical charts (Figures 17.11*a* and 17.11*b*). These plots may be used in conjunction with hardenability data to predict cross-sectional hardness profiles (Example Problem 17.1).

Precipitation Hardening
- Some alloys are amenable to *precipitation hardening*—that is, to strengthening by the formation of very small particles of a second, or precipitate, phase.
- Control of particle size and, subsequently, strength is accomplished by two heat treatments:
    In the first, or *solution,* heat treatment, all solute atoms are dissolved to form a single-phase solid solution; quenching to a relatively low temperature preserves this state.
    During the second, or *precipitation,* treatment (at constant temperature), precipitate particles form and grow; strength, hardness, and ductility depend on heat-treating time (and particle size).
- Strength and hardness increase with time to a maximum and then decrease during overaging (Figure 17.17). This process is accelerated with rising temperature (Figure 17.21*a*).
- The strengthening phenomenon is explained in terms of an increased resistance to dislocation motion by lattice strains that are established in the vicinity of these microscopically small precipitate particles.

Fabrication and Processing of Glasses and Glass–Ceramics
- Because glasses are formed at elevated temperatures, the temperature–viscosity behavior is an important consideration. Melting, working, softening, annealing, and strain points represent temperatures that correspond to specific viscosity values.
- Among the more common glass-forming techniques are pressing, blowing (Figure 17.25), drawing (Figure 17.26), and fiber forming.
- When glass pieces are cooled, internal thermal stresses may be generated because of differences in cooling rate (and degrees of thermal contraction) between interior and surfaces regions.
- After fabrication, glasses may be annealed and/or tempered to improve mechanical characteristics.

Fabrication and Processing of Clay Products
- Clay minerals assume two roles in the fabrication of ceramic bodies:
    When water is added to clay, it becomes pliable and amenable to forming.
    Clay minerals melt over a range of temperatures; thus, during firing, a dense and strong piece is produced without complete melting.
- For clay products, two common fabrication techniques are hydroplastic forming and slip casting.

For hydroplastic forming, a plastic and pliable mass is formed into a desired shape by forcing the mass through a die orifice.

With slip casting, a slip (suspension of clay and other minerals in water) is poured into a porous mold. As water is absorbed into the mold, a solid layer is deposited on the inside of the mold wall.

- After forming, a clay-based body must be first dried and then fired at an elevated temperature to reduce porosity and enhance strength.

Powder Pressing
- Some ceramic pieces are formed by powder compaction; uniaxial, isostatic, and hot pressing techniques are possible.
- Densification of pressed pieces takes place by a sintering mechanism (Figure 17.33) during a high-temperature firing procedure.

Tape Casting
- With tape casting, a thin sheet of ceramic of uniform thickness is formed from a slip that is spread onto a flat surface using a doctor blade (Figure 17.35). This tape is then subjected to drying and firing operations.

Polymerization
- Synthesis of high-molecular-weight polymers is attained by polymerization, of which there are two types: addition and condensation.

For addition polymerization, monomer units are attached one at a time in chain-like fashion to form a linear molecule.

Condensation polymerization involves stepwise intermolecular chemical reactions that may include more than a single molecular species.

Polymer Additives
- The properties of polymers may be further modified by using additives; these include fillers, plasticizers, stabilizers, colorants, and flame retardants.

Fillers are added to improve the strength, abrasion resistance, toughness, and or thermal/dimensional stability of polymers.

Flexibility, ductility, and toughness are enhanced by the addition of plasticizers.

Stabilizers counteract deteriorative processes due to exposure to light and gaseous species in the atmosphere.

Colorants are used to impart specific colors to polymers.

The flammability resistance of polymers is enhanced by the incorporation of flame retardants.

Forming Techniques for Plastics
- Fabrication of plastic polymers is usually accomplished by shaping the material in molten form at an elevated temperature, using at least one of several different molding techniques—compression (Figure 17.36), transfer, injection (Figure 17.37), and blow. Extrusion (Figure 17.38) and casting are also possible.

Fabrication of Fibers and Films
- Some fibers are spun from a viscous melt or solution, after which they are plastically elongated during a drawing operation, which improves the mechanical strength.
- Films are formed by extrusion and blowing (Figure 17.39) or by calendering.

Important Terms and Concepts

addition polymerization	colorant	forging
annealing	condensation polymerization	full annealing
annealing point (glass)	drawing	glass transition temperature
artificial aging	extrusion	green ceramic body
austenitizing	filler	hardenability
cold working	flame retardant	hot working

Important Terms and Concepts (Cont.)

hydroplastic forming	precipitation hardening	stabilizer
Jominy end-quench test	precipitation heat treatment	strain point (glass)
lower critical temperature	process annealing	stress relief
melting point (glass)	rolling	thermal shock
molding	sintering	thermal tempering
natural aging	slip casting	upper critical temperature
normalizing	softening point (glass)	vitrification
overaging	solution heat treatment	welding
plasticizer	spheroidizing	working point (glass)
powder metallurgy (P/M)	spinning	

REFERENCES

ASM Handbook, Vol. 4, *Heat Treating,* ASM International, Materials Park, OH, 1991.

ASM Handbook, Vol. 6, *Welding, Brazing and Soldering,* ASM International, Materials Park, OH, 1993.

ASM Handbook, Vol. 14A: *Metalworking: Bulk Forming,* ASM International, Materials Park, OH, 2005.

ASM Handbook, Vol. 14B: *Metalworking: Sheet Forming,* ASM International, Materials Park, OH, 2006.

ASM Handbook, Vol. 15, *Casting,* ASM International, Materials Park, OH, 2008.

Billmeyer, F. W., Jr., *Textbook of Polymer Science,* 3rd edition, Wiley-Interscience, New York, 1984.

Dieter, G. E., *Mechanical Metallurgy,* 3rd edition, McGraw-Hill, New York, 1986. Chapters 15–21 provide an excellent discussion of various metal-forming techniques.

Heat Treater's Guide: Standard Practices and Procedures for Irons and Steels, 2nd edition, ASM International, Materials Park, OH, 1995.

Kalpakjian, S., and S. R. Schmid, *Manufacturing Processes for Engineering Materials,* 5th edition, Pearson Education, Upper Saddle River, NJ, 2008.

Krauss, G., *Steels: Processing, Structure, and Performance,* ASM International, Materials Park, OH, 2005.

McCrum, N. G., C. P. Buckley, and C. B. Bucknall, *Principles of Polymer Engineering,* 2nd edition, Oxford University Press, Oxford, 1997.

Muccio, E. A., *Plastic Part Technology,* ASM International, Materials Park, OH, 1991.

Muccio, E. A., *Plastics Processing Technology,* ASM International, Materials Park, OH, 1994.

Powell, P. C., and A. J. Housz, *Engineering with Polymers,* 2nd edition, CRC Press, Boca Raton, FL, 1998.

Reed, J. S., *Principles of Ceramic Processing,* 2nd edition, Wiley, New York, 1995.

Richerson, D. W., *Modern Ceramic Engineering,* 3rd edition, CRC Press, Boca Raton, FL, 2006.

Strong, A. B., *Plastics: Materials and Processing,* 3rd edition, Pearson Education, Upper Saddle River, NJ, 2006.

QUESTIONS AND PROBLEMS

Forming Operations

17.1 Cite advantages and disadvantages of hot working and cold working.

17.2 **(a)** Cite advantages of forming metals by extrusion as opposed to rolling.

(b) Cite some disadvantages.

Casting

17.3 List four situations in which casting is the preferred fabrication technique.

17.4 Compare sand, die, investment, lost foam, and continuous casting techniques.

Miscellaneous Techniques

17.5 If it is assumed that, for steel alloys, the average cooling rate of the heat-affected zone in the vicinity

of a weld is 108C/s, compare the microstructures and associated properties that will result for 1080 (eutectoid) and 4340 alloys in their HAZs.

17.6 Describe one problem that might exist with a steel weld that was cooled very rapidly.

Annealing Processes

17.7 In your own words describe the following heat treatment procedures for steels and, for each, the intended final microstructure: full annealing, normalizing, quenching, and tempering.

17.8 Cite three sources of internal residual stresses in metal components. What are two possible adverse consequences of these stresses?

17.9 Give the approximate minimum temperature at which it is possible to austenitize each of the following iron–carbon alloys during a normal-izing heat treatment: **(a)** 0.25 wt% C, **(b)** 0.75 wt% C, and **(c)** 0.90 wt% C.

17.10 Give the approximate temperature at which it is desirable to heat each of the following iron–carbon alloys during a full anneal heat treatment: **(a)** 0.20 wt% C, **(b)** 0.50 wt% C, **(c)** 0.80 wt% C, and **(d)** 1.15 wt% C.

17.11 What is the purpose of a spheroidizing heat treatment? On what classes of alloys is it normally used?

Heat Treatment of Steels

17.12 Briefly explain the difference between hardness and hardenability.

17.13 What influence does the presence of alloying elements (other than carbon) have on the shape of a hardenability curve? Briefly explain this effect.

17.14 How would you expect a decrease in the austenite grain size to affect the hardenability of a steel alloy? Why?

17.15 Name two thermal properties of a liquid medium that will influence its quenching effectiveness.

17.16 Construct radial hardness profiles for the following:

(a) A 45-mm-diameter cylindrical specimen of an 8640 steel alloy that has been quenched in moderately agitated oil

(b) A 80-mm-diameter cylindrical specimen of a 5140 steel alloy that has been quenched in moderately agitated oil

(c) A 60-mm-diameter cylindrical specimen of an 8620 steel alloy that has been quenched in moderately agitated water

(d) A 75-mm-diameter cylindrical specimen of a 1040 steel alloy that has been quenched in moderately agitated water

17.17 Compare the effectiveness of quenching in moderately agitated water and oil by graphing, on a single plot, radial hardness profiles for 70-mm-diameter cylindrical specimens of an 8630 steel that have been quenched in both media.

Precipitation Hardening

17.18 Compare precipitation hardening (Section 17.7) and the hardening of steel by quenching and tempering (Sections 12.5, 12.6, and 12.8) with regard to the following:

(a) The total heat treatment procedure

(b) The microstructures that develop

(c) How the mechanical properties change during the several heat treatment stages

17.19 What is the principal difference between natural and artificial aging processes?

Fabrication and Processing of Glasses and Glass-Ceramics

17.20 Soda and lime are added to a glass batch in the form of soda ash (Na_2CO_3) and limestone ($CaCO_3$). During heating, these two ingredients decompose to give off carbon dioxide (CO_2), the resulting products being soda and lime. Compute the weight of soda ash and limestone that must be added to 45.3 kg of quartz (SiO_2) to yield a glass of composition 75 wt% SiO_2, 20 wt% Na_2O, and 5 wt% CaO.

17.21 What is the distinction between glass transition temperature and melting temperature?

17.22 Compare the temperatures at which soda–lime, borosilicate, 96% silica, and fused silica may be annealed.

17.23 Compare the softening points for 96% silica, borosilicate, and soda–lime glasses.

17.24 The viscosity η of a glass varies with temperature according to the relationship

$$\eta = A \exp\left(\frac{Q_{vis}}{RT}\right)$$

where Q_{vis} is the energy of activation for viscous flow, A is a temperature-independent constant, and R and T are, respectively, the gas constant and the absolute temperature. A plot of $\ln \eta$ versus $1/T$ should be nearly linear, and with a slope of Q_{vis}/R. Using the data in Figure 17.24, **(a)** make such a plot for the borosilicate glass, and **(b)** determine the activation energy between temperatures of 500 and 900°C.

17.25 For many viscous materials, the viscosity η may be defined in terms of the expression

$$\eta = \frac{\sigma}{d\epsilon/dt}$$

where σ and $d\epsilon/dt$ are, respectively, the tensile stress and the strain rate. A cylindrical specimen of a soda–lime glass of diameter 5 mm and length 100 mm is subjected to a tensile force of 1 N along its axis. If its deformation is to be less than 1 mm over a week's time, using Figure 17.24, determine the maximum temperature to which the specimen may be heated.

17.26 **(a)** Explain why residual thermal stresses are introduced into a glass piece when it is cooled.

(b) Are thermal stresses introduced upon heating? Why or why not?

17.27 Borosilicate glasses and fused silica are resistant to thermal shock. Why is this so?

17.28 In your own words, briefly describe what happens as a glass piece is thermally tempered.

17.29 Glass pieces may also be strengthened by chemical tempering. With this procedure, the glass surface is put in a state of compression by exchanging some of the cations near the surface with other cations having a larger diameter. Suggest one type of cation that, by replacing Na^+, will induce chemical tempering in a soda–lime glass.

Fabrication and Processing of Clay Products

17.30 Cite the two desirable characteristics of clay minerals relative to fabrication processes.

17.31 From a molecular perspective, briefly explain the mechanism by which clay minerals become hydroplastic when water is added.

17.32 (a) What are the three main components of a whiteware ceramic such as porcelain?

(b) What role does each component play in the forming and firing procedures?

17.33 (a) Why is it so important to control the rate of drying of a ceramic body that has been hydroplastically formed or slip cast?

(b) Cite three factors that influence the rate of drying, and explain how each affects the rate.

17.34 Cite one reason why drying shrinkage is greater for slip cast or hydroplastic products that have smaller clay particles.

17.35 (a) Name three factors that influence the degree to which vitrification occurs in clay-based ceramic wares.

(b) Explain how density, firing distortion, strength, corrosion resistance, and thermal conductivity are affected by the extent of vitrification.

Powder Pressing

17.36 Some ceramic materials are fabricated by hot isostatic pressing. Cite some of the limitations and difficulties associated with this technique.

Polymerization

17.37 Cite the primary differences between addition and condensation polymerization techniques.

17.38 (a) How much ethylene glycol must be added to 47.5 kg of dimethyl terephthalate to produce a linear chain structure of poly(ethylene terephthalate) according to Equation 17.5?

(b) What is the mass of the resulting polymer?

17.39 Nylon 6,6 may be formed by means of a condensation polymerization reaction in which hexamethylene diamine $[NH_2-(CH_2)_6-NH_2]$ and adipic acid react with one another with the formation of water as a by-product. What masses of hexamethylene diamine and adipic acid are necessary to yield 37.5 kg of completely linear nylon 6,6? (*Note:* the chemical equation for this reaction is the answer to Concept Check 17.8.)

Polymer Additives

17.40 What is the distinction between dye and pigment colorants?

Forming Techniques for Plastics

17.41 Cite four factors that determine what fabrication technique is used to form polymeric materials.

17.42 Contrast compression, injection, and transfer molding techniques that are used to form plastic materials.

Fabrication of Fibers and Films

17.43 Why must fiber materials that are melt-spun and then drawn be thermoplastic? Cite two reasons.

17.44 Which of the following polyethylene thin films would have the better mechanical characteristics: (1) formed by blowing, or (2) formed by extrusion and then rolled? Why?

DESIGN PROBLEMS

Heat Treatment of Steels

17.D1 A cylindrical piece of steel 20 mm in diameter is to be quenched in moderately agitated oil. Surface and center hardnesses must be at least 55 and 50 HRC, respectively. Which of the following alloys will satisfy these requirements: 1040, 5140, 4340, 4140, and 8640? Justify your choice(s).

17.D2 A cylindrical piece of steel 70 mm in diameter is to be austenitized and quenched such that a minimum hardness of 40 HRC is to be produced throughout the entire piece. Of the alloys 8660, 8640, 8630, and 8620, which will qualify if the quenching medium is **(a)** moderately agitated water and **(b)** moderately agitated oil? Justify your choice(s).

Chapter 18 Cor... ...n of Materials?

Deg...

of the environment, select a material that is relatively nonreactive, and/or protect the material from appreciable deterioration.

(a) A 1936 Deluxe Ford Sedan with a body made entirely of un...
to provide an ultimate test of the durability and corrosion resist...
of thousands of miles of everyday driving. Whereas the surface ...
car left the manufacturer's assembly line, other, nonstainless co...

Courtesy of Dan L. Greenfield, Allegheny Ludlum Corporation, Pittsburgh, PA

© EHStock/iStockphoto

(a)

the following:

6. For each of the eight forms of corrosion and hydrogen embrittlement, describe the nature of the deteriorative process and then note the proposed mechanism.
7. List five measures that are commonly used to prevent corrosion.
8. Explain why ceramic materials are, in general, very resistant to corrosion.
9. For polymeric materials, discuss (a) two degradation processes that occur when they are exposed to liquid solvents and (b) the causes and consequences of molecular chain bond rupture.

...ost materials experience some type of interaction with ...ironments. Often, such interactions impair a material's ...eterioration of its mechanical properties (e.g., ductility ...operties, or appearance. Occasionally, to the chagrin of a ...n behavior of a material for some application is ignored,

...s are different for the three material types. In metals, ...ther by dissolution (**corrosion**) or by the formation of ...dation). Ceramic materials are relatively resistant to ...occurs at elevated temperatures or in rather extreme ...requently also called corrosion. For polymers, mecha-...r from those for metals and ceramics, and the term ...ly used. Polymers may dissolve when exposed to a ...absorb the solvent and swell; also, electromagnetic ...t) and heat may cause alterations in their molecular

...of these material types is discussed in this chapter ...sm, resistance to attack by various environments, and ...degradation.

Corrosion of Metals

Corrosion is defined as the destructive and unintentional attack on a metal; it is electrochemical and ordinarily begins at the surface. The problem of metallic corrosion is significant; in economic terms, it has been estimated that approximately 5% of an industrialized nation's income is spent on corrosion prevention and the maintenance or replacement of products lost or contaminated as a result of corrosion reactions. The consequences of corrosion are all too common. Familiar examples include the rusting of automotive body panels and radiator and exhaust components.

Corrosion processes are occasionally used to advantage. For example, etching procedures, as discussed in Section 6.12, use the selective chemical reactivity of grain boundaries or various microstructural constituents.

18.2 ELECTROCHEMICAL CONSIDERATIONS

For metallic materials, the corrosion process is normally electrochemical, that is, a chemical reaction in which there is transfer of electrons from one chemical species to another. Metal atoms characteristically lose or give up electrons in what is called an **oxidation** reaction. For example, a hypothetical metal M that has a valence of n (or n valence electrons) may experience oxidation according to the reaction

oxidation

Oxidation reaction for metal M

$$M \longrightarrow M^{n+} + ne^- \tag{18.1}$$

in which M becomes an $n+$ positively charged ion and in the process loses its n valence electrons; e^- is used to symbolize an electron. Examples in which metals oxidize are

$$Fe \longrightarrow Fe^{2+} + 2e^- \tag{18.2a}$$

$$Al \longrightarrow Al^{3+} + 3e^- \tag{18.2b}$$

anode

The site at which oxidation takes place is called the **anode**; oxidation is sometimes called an anodic reaction.

reduction

The electrons generated from each metal atom that is oxidized must be transferred to and become a part of another chemical species in what is termed a **reduction** reaction. For example, some metals undergo corrosion in acid solutions, which have a high concentration of hydrogen (H^+) ions; the H^+ ions are reduced as follows:

Reduction of hydrogen ions in an acid solution

$$2H^+ + 2e^- \longrightarrow H_2 \tag{18.3}$$

and hydrogen gas (H_2) is evolved.

Other reduction reactions are possible, depending on the nature of the solution to which the metal is exposed. For an acid solution having dissolved oxygen, reduction according to

Reduction reaction in an acid solution containing dissolved oxygen

$$O_2 + 4H^+ + 4e^- \longrightarrow 2H_2O \tag{18.4}$$

will probably occur. For a neutral or basic aqueous solution in which oxygen is also dissolved,

Reduction reaction in a neutral or basic solution containing dissolved oxygen

$$O_2 + 2H_2O + 4e^- \longrightarrow 4(OH^-) \tag{18.5}$$

Any metal ions present in the solution may also be reduced; for ions that can exist in more than one valence state (multivalent ions), reduction may occur by

Reduction of a multivalent metal ion to a lower valence state

$$M^{n+} + e^- \longrightarrow M^{(n-1)+} \tag{18.6}$$

in which the metal ion decreases its valence state by accepting an electron. A metal may be totally reduced from an ionic to a neutral metallic state according to

Reduction of a metal ion to its electrically neutral atom

$$M^{n+} + ne^- \longrightarrow M \tag{18.7}$$

cathode

The location at which reduction occurs is called the **cathode**. It is possible for two or more of the preceding reduction reactions to occur simultaneously.

An overall electrochemical reaction must consist of at least one oxidation and one reduction reaction and will be their sum; often the individual oxidation and reduction reactions are termed *half-reactions*. There can be no net electrical charge accumulation from the electrons and ions—that is, the total rate of oxidation must equal the total rate of reduction, or all electrons generated through oxidation must be consumed by reduction.

For example, consider zinc metal immersed in an acid solution containing H^+ ions. At some regions on the metal surface, zinc will experience oxidation or corrosion as illustrated in Figure 18.1, according to the reaction

$$Zn \longrightarrow Zn^{2+} + 2e^- \tag{18.8}$$

Because zinc is a metal and therefore a good electrical conductor, these electrons may be transferred to an adjacent region at which the H^+ ions are reduced according to

$$2H^+ + 2e^- \longrightarrow H_2 \text{ (gas)} \tag{18.9}$$

If no other oxidation or reduction reactions occur, the total electrochemical reaction is just the sum of reactions 18.8 and 18.9, or

$$Zn \longrightarrow Zn^{2+} + 2e^-$$

$$\underline{2H^+ + 2e^- \longrightarrow H_2 \text{ (gas)}}$$

$$Zn + 2H^+ \longrightarrow Zn^{2+} + H_2 \text{ (gas)} \tag{18.10}$$

Another example is the oxidation or rusting of iron in water, which contains dissolved oxygen. This process occurs in two steps; in the first, Fe is oxidized to Fe^{2+} [as $Fe(OH)_2$],

$$Fe + \tfrac{1}{2}O_2 + H_2O \longrightarrow Fe^{2+} + 2OH^- \longrightarrow Fe(OH)_2 \tag{18.11}$$

and, in the second stage, to Fe^{3+} [as $Fe(OH)_3$] according to

$$2Fe(OH)_2 + \tfrac{1}{2}O_2 + H_2O \longrightarrow 2Fe(OH)_3 \tag{18.12}$$

The compound $Fe(OH)_3$ is the all-too-familiar rust.

As a consequence of oxidation, the metal ions may either go into the corroding solution as ions (reaction 18.8) or form an insoluble compound with nonmetallic elements as in reaction 18.12.

Concept Check 18.1 Would you expect iron to corrode in water of high purity? Why or why not?

[*The answer may be found at* www.wiley.com/college/callister (*Student Companion Site*).]

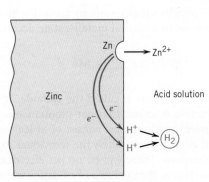

Figure 18.1 The electrochemical reactions associated with the corrosion of zinc in an acid solution. (From M. G. Fontana, *Corrosion Engineering*, 3rd edition. Copyright © 1986 by McGraw-Hill Book Company. Reproduced with permission.)

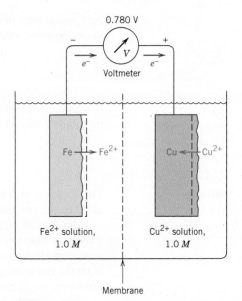

Figure 18.2 An electrochemical cell consisting of iron and copper electrodes, each of which is immersed in a 1 *M* solution of its ion. Iron corrodes while copper electrodeposits.

Electrode Potentials

Not all metallic materials oxidize to form ions with the same degree of ease. Consider the electrochemical cell shown in Figure 18.2. On the left-hand side is a piece of pure iron immersed in a solution containing Fe^{2+} ions of 1 *M* concentration.[1] The other side of the cell consists of a pure copper electrode in a 1 *M* solution of Cu^{2+} ions. The cell halves are separated by a membrane, which limits the mixing of the two solutions. If the iron and copper electrodes are connected electrically, reduction will occur for copper at the expense of the oxidation of iron, as follows:

$$Cu^{2+} + Fe \longrightarrow Cu + Fe^{2+} \tag{18.13}$$

or Cu^{2+} ions will deposit (electrodeposit) as metallic copper on the copper electrode, whereas iron dissolves (corrodes) on the other side of the cell and goes into solution as Fe^{2+} ions. Thus, the two half-cell reactions are represented by the relations

$$Fe \longrightarrow Fe^{2+} + 2e^- \tag{18.14a}$$
$$Cu^{2+} + 2e^- \longrightarrow Cu \tag{18.14b}$$

When a current passes through the external circuit, electrons generated from the oxidation of iron flow to the copper cell in order that Cu^{2+} be reduced. In addition, there will be some net ion motion from each cell to the other across the membrane. This is called a *galvanic couple*—two metals electrically connected in a liquid **electrolyte** in which one metal becomes an anode and corrodes while the other acts as a cathode.

electrolyte

molarity

---

[1]Concentration of liquid solutions is often expressed in terms of **molarity**, *M*, the number of moles of solute per liter (1000 cm³) of solution.

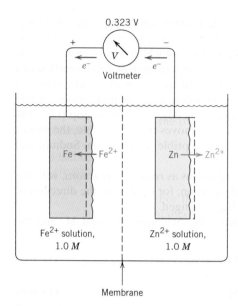

Figure 18.3 An electrochemical cell consisting of iron and zinc electrodes, each of which is immersed in a 1 M solution of its ion. The iron electrodeposits while the zinc corrodes.

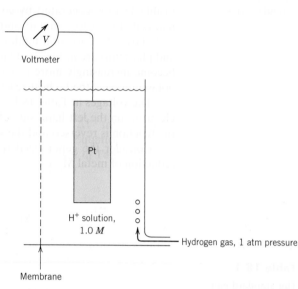

Figure 18.4 The standard hydrogen reference half-cell.

An electric potential or voltage exists between the two cell halves, and its magnitude can be determined if a voltmeter is connected in the external circuit. A potential of 0.780 V results for a copper–iron galvanic cell when the temperature is 25°C.

Now consider another galvanic couple consisting of the same iron half-cell connected to a metal zinc electrode that is immersed in a 1 M solution of Zn^{2+} ions (Figure 18.3). In this case, the zinc is the anode and corrodes, whereas the Fe becomes the cathode. The electrochemical reaction is thus

$$Fe^{2+} + Zn \longrightarrow Fe + Zn^{2+} \tag{18.15}$$

The potential associated with this cell reaction is 0.323 V.

Thus, various electrode pairs have different voltages; the magnitude of such a voltage may be thought of as representing the driving force for the electrochemical oxidation–reduction reaction. Consequently, metallic materials may be rated as to their tendency to experience oxidation when coupled to other metals in solutions of their respective ions. A half-cell similar to those just described (i.e., a pure metal electrode immersed in a 1 M solution of its ions and at 25°C) is termed a **standard half-cell**.

standard half-cell

The Standard emf Series

These measured cell voltages represent only differences in electrical potential, and thus it is convenient to establish a reference point, or reference cell, to which other cell halves may be compared. This reference cell, arbitrarily chosen, is the standard hydrogen electrode (Figure 18.4). It consists of an inert platinum electrode in a 1 M

EXAMPLE PROBLEM 18.1

Determination of Electrochemical Cell Characteristics

One-half of an electrochemical cell consists of a pure nickel electrode in a solution of Ni^{2+} ions; the other half is a cadmium electrode immersed in a Cd^{2+} solution.

(a) If the cell is a standard one, write the spontaneous overall reaction and calculate the voltage that is generated.

(b) Compute the cell potential at 25°C if the Cd^{2+} and Ni^{2+} concentrations are 0.5 and 10^{-3} M, respectively. Is the spontaneous reaction direction still the same as for the standard cell?

Solution

(a) The cadmium electrode is oxidized and nickel is reduced because cadmium is lower in the emf series; thus, the spontaneous reactions are

$$Cd \longrightarrow Cd^{2+} + 2e^-$$

$$Ni^{2+} + 2e^- \longrightarrow Ni$$

$$\overline{Ni^{2+} + Cd \longrightarrow Ni + Cd^{2+}} \qquad (18.21)$$

From Table 18.1, the half-cell potentials for cadmium and nickel are, respectively, −0.403 and −0.250 V. Therefore, from Equation 18.18,

$$\Delta V = V^0_{Ni} - V^0_{Cd} = -0.250 \text{ V} - (-0.403 \text{ V}) = +0.153 \text{ V}$$

(b) For this portion of the problem, Equation 18.20 must be used, because the half-cell solution concentrations are no longer 1 M. At this point, it is necessary to make a calculated guess as to which metal species is oxidized (or reduced). This choice will either be affirmed or refuted on the basis of the sign of ΔV at the conclusion of the computation. For the sake of argument, let us assume that in contrast to part (a), nickel is oxidized and cadmium reduced according to

$$Cd^{2+} + Ni \longrightarrow Cd + Ni^{2+} \qquad (18.22)$$

Thus,

$$\Delta V = (V^0_{Cd} - V^0_{Ni}) - \frac{RT}{n\mathscr{F}} \ln \frac{[Ni^{2+}]}{[Cd^{2+}]}$$

$$= -0.403 \text{ V} - (-0.250 \text{ V}) - \frac{0.0592}{2} \log \left(\frac{10^{-3}}{0.50} \right)$$

$$= -0.073 \text{ V}$$

Because ΔV is negative, the spontaneous reaction direction is the opposite to that of Equation 18.22, or

$$Ni^{2+} + Cd \longrightarrow Ni + Cd^{2+}$$

That is, cadmium is oxidized and nickel is reduced.

The Galvanic Series

galvanic series

Even though Table 18.1 was generated under highly idealized conditions and has limited utility, it nevertheless indicates the relative reactivities of the metals. A more realistic and practical ranking is provided by the **galvanic series**, Table 18.2. This represents the relative reactivities of a number of metals and commercial alloys in seawater. The alloys near the top are cathodic and unreactive, whereas those at the bottom are most anodic;

Table 18.2

The Galvanic Series

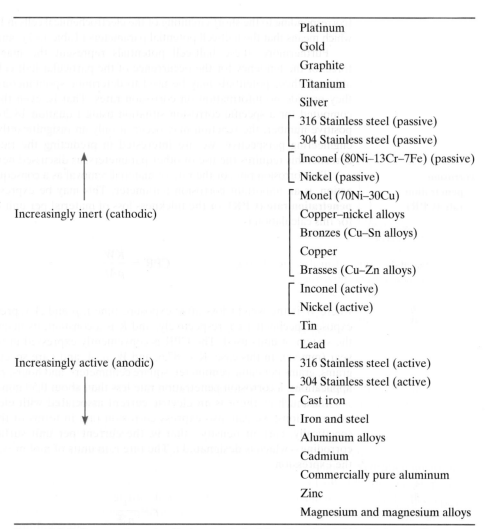

Increasingly inert (cathodic)

Increasingly active (anodic)

Platinum
Gold
Graphite
Titanium
Silver
316 Stainless steel (passive)
304 Stainless steel (passive)
Inconel (80Ni–13Cr–7Fe) (passive)
Nickel (passive)
Monel (70Ni–30Cu)
Copper–nickel alloys
Bronzes (Cu–Sn alloys)
Copper
Brasses (Cu–Zn alloys)
Inconel (active)
Nickel (active)
Tin
Lead
316 Stainless steel (active)
304 Stainless steel (active)
Cast iron
Iron and steel
Aluminum alloys
Cadmium
Commercially pure aluminum
Zinc
Magnesium and magnesium alloys

Source: M. G. Fontana, *Corrosion Engineering,* 3rd edition. Copyright 1986 by McGraw-Hill Book Company. Reprinted with permission.

no voltages are provided. Comparison of the standard emf and the galvanic series reveals a high degree of correspondence between the relative positions of the pure base metals.

Most metals and alloys are subject to oxidation or corrosion to one degree or another in a wide variety of environments—that is, they are more stable in an ionic state than as metals. In thermodynamic terms, there is a net decrease in free energy in going from metallic to oxidized states. Consequently, essentially all metals occur in nature as compounds—for example, oxides, hydroxides, carbonates, silicates, sulfides, and sulfates. Two exceptions are the noble metals gold and platinum, for which oxidation in most environments is not favorable, and, therefore, they may exist in nature in the metallic state.

18.3 CORROSION RATES

The half-cell potentials listed in Table 18.1 are thermodynamic parameters that relate to systems at equilibrium. For example, for the discussions pertaining to Figures 18.2 and 18.3, it was tacitly assumed that there was no current flow through the external circuit. Real corroding systems are not at equilibrium; there is a flow of electrons from anode to cathode where β and i_0 are constants for the particular half-cell. The parameter i_0 is termed the *exchange current density,* which deserves a brief explanation. Equilibrium for some particular half-cell reaction is really a dynamic state on the atomic level—that is, oxidation and reduction processes are occurring, but both at the same rate, so that there is no net reaction. For example, for the standard hydrogen cell (Figure 18.4), reduction of hydrogen ions in solution takes place at the surface of the platinum electrode according to

$$2H^+ + 2e^- \longrightarrow H_2$$

with a corresponding rate r_{red}. Similarly, hydrogen gas in the solution experiences oxidation as

$$H_2 \longrightarrow 2H^+ + 2e^-$$

at rate r_{oxid}. Equilibrium exists when

$$r_{red} = r_{oxid}$$

This exchange current density is just the current density from Equation 18.24 at equilibrium, or

At equilibrium, equality of rates of oxidation and reduction, and their relationship to the exchange current density

$$r_{red} = r_{oxid} = \frac{i_0}{n\mathcal{F}} \qquad (18.26)$$

Use of the term *current density* for i_0 is a little misleading inasmuch as there is no net current. Furthermore, the value for i_0 is determined experimentally and varies from system to system.

According to Equation 18.25, when overvoltage is plotted as a function of the logarithm of current density, straight-line segments result; these are shown in Figure 18.7 for the hydrogen electrode. The line segment with a slope of $+\beta$ corresponds to the oxidation half-reaction, whereas the line with a $-\beta$ slope is for reduction. Also worth noting is that both line segments originate at i_0 (H_2/H^+), the exchange current density, and at zero overvoltage, because at this point the system is at equilibrium and there is no net reaction.

Concentration Polarization

concentration
polarization

Concentration polarization exists when the reaction rate is limited by diffusion in the solution. For example, consider again the hydrogen evolution reduction reaction.

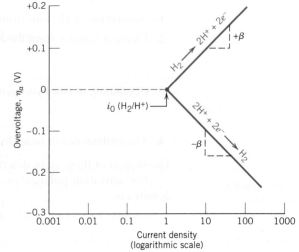

Figure 18.7 For a hydrogen electrode, plot of activation polarization overvoltage versus logarithm of current density for both oxidation and reduction reactions. (Adapted from M. G. Fontana, *Corrosion Engineering,* 3rd edition. Copyright © 1986 by McGraw-Hill Book Company. Reproduced with permission.)

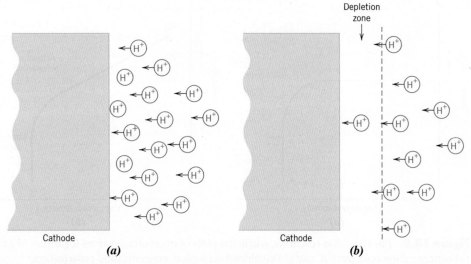

Figure 18.8 For hydrogen reduction, schematic representations of the H^+ distribution in the vicinity of the cathode for (*a*) low reaction rates and/or high concentrations and (*b*) high reaction rates and/or low concentrations, where a depletion zone is formed that gives rise to concentration polarization.
(Adapted from M. G. Fontana, *Corrosion Engineering,* 3rd edition. Copyright © 1986 by McGraw-Hill Book Company. Reproduced with permission.)

When the reaction rate is low and/or the concentration of H^+ is high, there is always an adequate supply of hydrogen ions available in the solution at the region near the electrode interface (Figure 18.8*a*). However, at high rates and/or low H^+ concentrations, a depletion zone may be formed in the vicinity of the interface because the H^+ ions are not replenished at a rate sufficient to keep up with the reaction (Figure 18.8*b*). Thus, diffusion of H^+ to the interface is rate controlling, and the system is said to be concentration polarized.

Concentration polarization data are also normally plotted as overvoltage versus the logarithm of current density; such a plot is represented schematically in Figure 18.9*a*.[2] It may be noted from this figure that overvoltage is independent of current density until i approaches i_L; at this point, η_c decreases abruptly in magnitude.

Both concentration and activation polarization are possible for reduction reactions. Under these circumstances, the total overvoltage is the sum of both overvoltage contributions. Figure 18.9*b* shows such a schematic η-versus-log i plot.

Concept Check 18.3 Briefly explain why concentration polarization is not normally rate controlling for oxidation reactions.

[*The answer may be found at* www.wiley.com/college/callister *(Student Companion Site)*.]

---

[2]The mathematical expression relating concentration polarization overvoltage η_c and current density i is

For concentration polarization, relationship between overvoltage and current density

$$\eta_c = \frac{2.3RT}{n\mathscr{F}} \log\left(1 - \frac{i}{i_L}\right) \tag{18.27}$$

where R and T are the gas constant and absolute temperature, respectively, n and $\mathscr{F}$ have the same meanings as before, and i_L is the limiting diffusion current density.

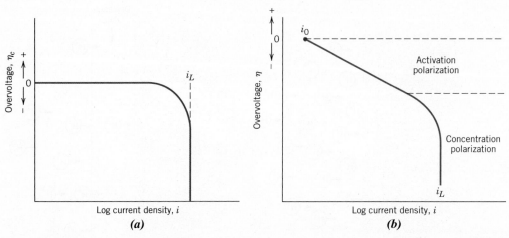

Figure 18.9 For reduction reactions, schematic plots of overvoltage versus logarithm of current density for (*a*) concentration polarization, and (*b*) combined activation–concentration polarization.

Corrosion Rates from Polarization Data

Let us now apply the concepts just developed to the determination of corrosion rates. Two types of systems are discussed. In the first case, both oxidation and reduction reactions are rate limited by activation polarization. In the second case, both concentration and activation polarization control the reduction reaction, whereas only activation polarization is important for oxidation. Case one is illustrated by considering the corrosion of zinc immersed in an acid solution (see Figure 18.1). The reduction of H^+ ions to form H_2 gas bubbles occurs at the surface of the zinc according to reaction 18.3,

$$2H^+ + 2e^- \longrightarrow H_2$$

and the zinc oxidizes as given in reaction 18.8,

$$Zn \longrightarrow Zn^{2+} + 2e^-$$

No net charge accumulation may result from these two reactions—that is, all electrons generated by reaction 18.8 must be consumed by reaction 18.3, which is to say that rates of oxidation and reduction must be equal.

Activation polarization for both reactions is expressed graphically in Figure 18.10 as cell potential referenced to the standard hydrogen electrode (not overvoltage) versus the logarithm of current density. The potentials of the uncoupled hydrogen and zinc half-cells, $V(H^+/H_2)$ and $V(Zn/Zn^{2+})$, respectively, are indicated, along with their respective exchange current densities, $i_0(H^+/H_2)$ and $i_0(Zn/Zn^{2+})$. Straight-line segments are shown for hydrogen reduction and zinc oxidation. Upon immersion, both hydrogen and zinc experience activation polarization along their respective lines. Also, oxidation and reduction rates must be equal as explained earlier, which is possible only at the intersection of the two line segments; this intersection occurs at the corrosion potential, designated V_C, and the corrosion current density i_C. The corrosion rate of zinc (which also corresponds to the rate of hydrogen evolution) may thus be computed by insertion of this i_C value into Equation 18.24.

The second corrosion case (combined activation and concentration polarization for hydrogen reduction and activation polarization for oxidation of metal M) is treated in a like manner. Figure 18.11 shows both polarization curves; as earlier, corrosion potential and corrosion current density correspond to the point at which the oxidation and reduction lines intersect.

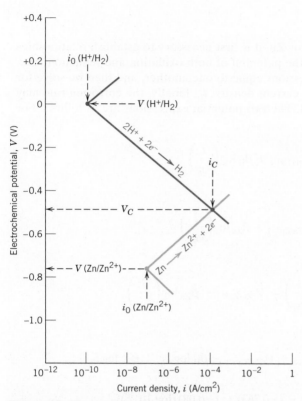

Figure 18.10 Electrode kinetic behavior of zinc in an acid solution; both oxidation and reduction reactions are rate limited by activation polarization.
(Adapted from M. G. Fontana, *Corrosion Engineering*, 3rd edition. Copyright © 1986 by McGraw-Hill Book Company. Reproduced with permission.)

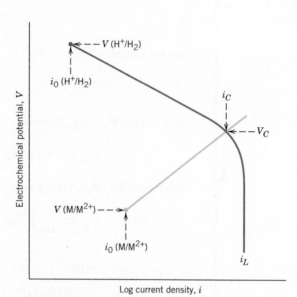

Log current density, i

Figure 18.11 Schematic electrode kinetic behavior for metal M; the reduction reaction is under combined activation–concentration polarization control.

EXAMPLE PROBLEM 18.2

Rate of Oxidation Computation

Zinc experiences corrosion in an acid solution according to the reaction

$$Zn + 2H^+ \longrightarrow Zn^{2+} + H_2$$

The rates of both oxidation and reduction half-reactions are controlled by activation polarization.

(a) Compute the rate of oxidation of Zn (in mol/cm²·s), given the following activation polarization data:

For Zn	*For Hydrogen*
$V_{(Zn/Zn^{2+})} = -0.763$ V	$V_{(H^+/H_2)} = 0$ V
$i_0 = 10^{-7}$ A/cm²	$i_0 = 10^{-10}$ A/cm²
$\beta = +0.09$	$\beta = -0.08$

(b) Compute the value of the corrosion potential.

Solution

(a) To compute the rate of oxidation for Zn, it is first necessary to establish relationships in the form of Equation 18.25 for the potential of both oxidation and reduction reactions. Next, we set these two expressions equal to one another, and then we solve for the value of i that is the corrosion current density, i_C. Finally, the corrosion rate may be calculated using Equation 18.24. The two potential expressions are as follows: For hydrogen reduction,

$$V_H = V_{(H^+/H_2)} + \beta_H \log\left(\frac{i}{i_{0_H}}\right)$$

and for Zn oxidation,

$$V_{Zn} = V_{(Zn/Zn^{2+})} + \beta_{Zn} \log\left(\frac{i}{i_{0_{Zn}}}\right)$$

Now, setting $V_H = V_{Zn}$ leads to

$$V_{(H^+/H_2)} + \beta_H \log\left(\frac{i}{i_{0_H}}\right) = V_{(Zn/Zn^{2+})} + \beta_{Zn} \log\left(\frac{i}{i_{0_{Zn}}}\right)$$

Solving for $\log i$ (i.e., $\log i_C$) leads to

$$\log i_C = \left(\frac{1}{\beta_{Zn} - \beta_H}\right)[V_{(H^+/H_2)} - V_{(Zn/Zn^{2+})} - \beta_H \log i_{0_H} + \beta_{Zn} \log i_{0_{Zn}}]$$

$$= \left[\frac{1}{0.09 - (-0.08)}\right][0 - (-0.763) - (-0.08)(\log 10^{-10})$$

$$+ (0.09)(\log 10^{-7})]$$

$$= -3.924$$

or

$$i_C = 10^{-3.924} = 1.19 \times 10^{-4} \, \text{A/cm}^2$$

From Equation 18.24,

$$r = \frac{i_C}{n\mathscr{F}}$$

$$= \frac{1.19 \times 10^{-4} \, \text{C/s·cm}^2}{(2)(96,500 \, \text{C/mol})} = 6.17 \times 10^{-10} \, \text{mol/cm}^2\text{·s}$$

(b) Now it becomes necessary to compute the value of the corrosion potential V_C. This is possible by using either of the preceding equations for V_H or V_{Zn} and substituting for i the value determined previously for i_C. Thus, using the V_H expression yields

$$V_C = V_{(H^+/H_2)} + \beta_H \log\left(\frac{i_C}{i_{0_H}}\right)$$

$$= 0 + (-0.08 \, \text{V}) \log\left(\frac{1.19 \times 10^{-4} \, \text{A/cm}^2}{10^{-10} \, \text{A/cm}^2}\right) = -0.486 \, \text{V}$$

This is the same problem that is represented and solved graphically in the voltage-versus-logarithm current density plot of Figure 18.10. It is worth noting that the i_C and V_C we have obtained by this analytical treatment are in agreement with those values occurring at the intersection of the two line segments on the plot.

18.5 PASSIVITY

passivity

Under particular environmental conditions, some normally active metals and alloys lose their chemical reactivity and become extremely inert. This phenomenon, termed **passivity**, is displayed by chromium, iron, nickel, titanium, and many of their alloys. It is believed that this passive behavior results from the formation of a highly adherent and very thin oxide film on the metal surface, which serves as a protective barrier to further corrosion. Stainless steels are highly resistant to corrosion in a rather wide variety of atmospheres as a result of passivation. They contain at least 11% chromium, which as a solid-solution alloying element in iron, minimizes the formation of rust; instead, a protective surface film forms in oxidizing atmospheres. (Stainless steels are susceptible to corrosion in some environments and therefore are not always "stainless.") Aluminum is highly corrosion resistant in many environments because it also passivates. If damaged, the protective film normally re-forms very rapidly. However, a change in the character of the environment (e.g., alteration in the concentration of the active corrosive species) may cause a passivated material to revert to an active state. Subsequent damage to a preexisting passive film could result in a substantial increase in corrosion rate, by as much as 100,000 times.

This passivation phenomenon may be explained in terms of polarization potential–log current density curves discussed in the preceding section. The polarization curve for a metal that passivates has the general shape shown in Figure 18.12. At relatively low potential values, within the "active" region the behavior is linear, as it is for normal metals. With increasing potential, the current density suddenly decreases to a very low value that remains independent of potential; this is termed the "passive" region. Finally, at even higher potential values, the current density again increases with potential in the "transpassive" region.

Figure 18.13 illustrates how a metal can experience both active and passive behavior depending on the corrosion environment. This figure shows the S-shaped oxidation polarization curve for an active–passive metal M and, in addition, reduction polarization curves for two different solutions, which are labeled 1 and 2. Curve 1 intersects the oxidation polarization curve in the active region at point A, yielding a corrosion current density $i_C(A)$. The intersection of curve 2 at point B is in the passive region and at current density $i_C(B)$.

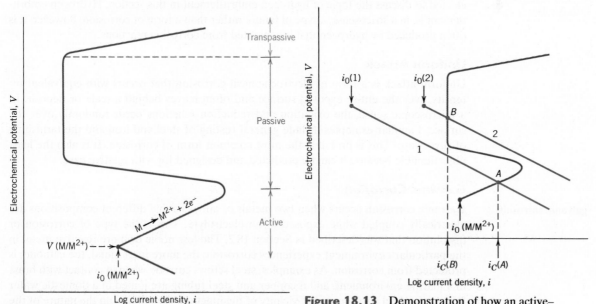

Figure 18.12 Schematic polarization curve for a metal that displays an active–passive transition.

Figure 18.13 Demonstration of how an active–passive metal can exhibit both active and passive corrosion behaviors.

The corrosion rate of metal M in solution 1 is greater than in solution 2 because $i_C(A)$ is greater than $i_C(B)$ and rate is proportional to current density according to Equation 18.24. This difference in corrosion rate between the two solutions may be significant—several orders of magnitude—given that the current density scale in Figure 18.13 is scaled logarithmically.

18.6 ENVIRONMENTAL EFFECTS

The variables in the corrosion environment, which include fluid velocity, temperature, and composition, can have a decided influence on the corrosion properties of the materials that are in contact with it. In most instances, increasing fluid velocity enhances the rate of corrosion due to erosive effects, as discussed later in the chapter. The rates of most chemical reactions rise with increasing temperature; this also holds for most corrosion situations. Increasing the concentration of the corrosive species (e.g., H^+ ions in acids) in many situations produces a more rapid rate of corrosion. However, for materials capable of passivation, raising the corrosive content may result in an active-to-passive transition, with a considerable reduction in corrosion.

Cold working or plastically deforming ductile metals is used to increase their strength; however, a cold-worked metal is more susceptible to corrosion than the same material in an annealed state. For example, deformation processes are used to shape the head and point of a nail; consequently, these positions are anodic with respect to the shank region. Thus, differential cold working on a structure should be a consideration when a corrosive environment may be encountered during service.

18.7 FORMS OF CORROSION

It is convenient to classify corrosion according to the manner in which it is manifest. Metallic corrosion is sometimes classified into eight forms: uniform, galvanic, crevice, pitting, intergranular, selective leaching, erosion–corrosion, and stress corrosion. The causes and means of prevention of each of these forms are discussed briefly. In addition, we have elected to discuss the topic of hydrogen embrittlement in this section. Hydrogen embrittlement is, in a strict sense, a type of failure rather than a form of corrosion; however, it is often produced by hydrogen that is generated from corrosion reactions.

Uniform Attack

Uniform attack is a form of electrochemical corrosion that occurs with equivalent intensity over the entire exposed surface and often leaves behind a scale or deposit. In a microscopic sense, the oxidation and reduction reactions occur randomly over the surface. Familiar examples include general rusting of steel and iron and the tarnishing of silverware. This is probably the most common form of corrosion. It is also the least objectionable because it can be predicted and designed for with relative ease.

Galvanic Corrosion

galvanic corrosion

Galvanic corrosion occurs when two metals or alloys having different compositions are electrically coupled while exposed to an electrolyte. This is the type of corrosion or dissolution that was described in Section 18.2. The less noble or more reactive metal in the particular environment experiences corrosion; the more inert metal, the cathode, is protected from corrosion. As examples, steel screws corrode when in contact with brass in a marine environment, and if copper and steel tubing are joined in a domestic water heater, the steel corrodes in the vicinity of the junction. Depending on the nature of the solution, one or more of the reduction reactions, Equations 18.3 through 18.7, occurs at the surface of the cathode material. Figure 18.14 shows galvanic corrosion.

Galvanic corrosion   Steel core

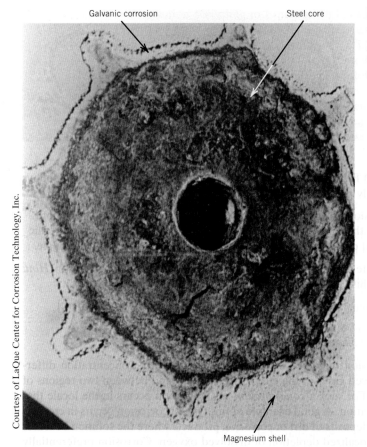

Magnesium shell

Courtesy of LaQue Center for Corrosion Technology, Inc.

Figure 18.14 Photograph showing galvanic corrosion around the inlet of a single-cycle bilge pump that is found on fishing vessels. Corrosion occurred between a magnesium shell that was cast around a steel core.

The galvanic series in Table 18.2 indicates the relative reactivities in seawater of a number of metals and alloys. When two alloys are coupled in seawater, the one lower in the series experiences corrosion. Some of the alloys in the table are grouped in brackets. Generally the base metal is the same for these bracketed alloys, and there is little danger of corrosion if alloys within a single bracket are coupled. It is also worth noting from this series that some alloys are listed twice (e.g., nickel and the stainless steels), in both active and passive states.

The rate of galvanic attack depends on the relative anode-to-cathode surface areas that are exposed to the electrolyte, and the rate is related directly to the cathode–anode area ratio—that is, for a given cathode area, a smaller anode corrodes more rapidly than a larger one because corrosion rate depends on current density (Equation 18.24)—the current per unit area of corroding surface—and not simply the current. Thus, a high current density results for the anode when its area is small relative to that of the cathode.

A number of measures may be taken to reduce the effects of galvanic corrosion significantly including the following:

1. If coupling of dissimilar metals is necessary, choose two that are close together in the galvanic series.

2. Avoid an unfavorable anode-to-cathode surface area ratio; use an anode area as large as possible.

3. Electrically insulate dissimilar metals from each other.

4. Electrically connect a third, anodic metal to the other two; this is a form of *cathodic protection*, discussed in Section 18.9.

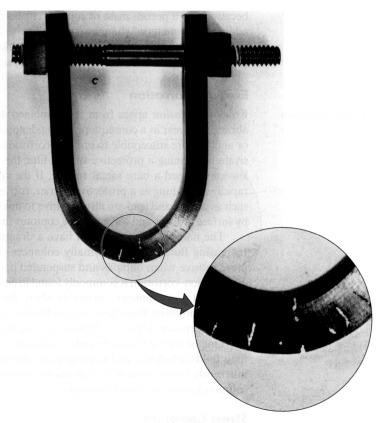

Figure 18.20 Impingement failure of an elbow that was part of a steam condensate line.
(Photograph courtesy of Mars G. Fontana. From M. G. Fontana, *Corrosion Engineering,* 3rd edition. Copyright © 1986 by McGraw-Hill Book Company. Reproduced with permission.)

Figure 18.21 A bar of steel bent into a horseshoe shape using a nut-and-bolt assembly. While immersed in seawater, stress corrosion cracks formed along the bend at those regions where the tensile stresses are the greatest.
(Photograph courtesy of F. L. LaQue. From F. L. LaQue, *Marine Corrosion, Causes and Prevention.* Copyright © 1975 by John Wiley & Sons, Inc. Reprinted by permission of John Wiley & Sons, Inc.)

hydrogen embrittlement

This phenomenon is aptly referred to as **hydrogen embrittlement**; the terms *hydrogen-induced cracking* and *hydrogen stress cracking* are sometimes also used. Strictly speaking, hydrogen embrittlement is a type of failure; in response to applied or residual tensile stresses, brittle fracture occurs catastrophically as cracks grow and rapidly propagate. Hydrogen in its atomic form (H as opposed to the molecular form, H_2) diffuses interstitially through the crystal lattice, and concentrations as low as several parts per million can lead to cracking. Furthermore, hydrogen-induced cracks are most often transgranular, although intergranular fracture is observed for some alloy systems. A number of mechanisms have been proposed to explain hydrogen embrittlement; most are based on the interference of dislocation motion by the dissolved hydrogen.

Hydrogen embrittlement is similar to stress corrosion, in that a normally ductile metal experiences brittle fracture when exposed to both a tensile stress and a corrosive atmosphere. However, these two phenomena may be distinguished on the basis of their interactions with applied electric currents. Whereas cathodic protection (Section 18.9) reduces or causes a cessation of stress corrosion, it may, however, lead to the initiation or enhancement of hydrogen embrittlement.

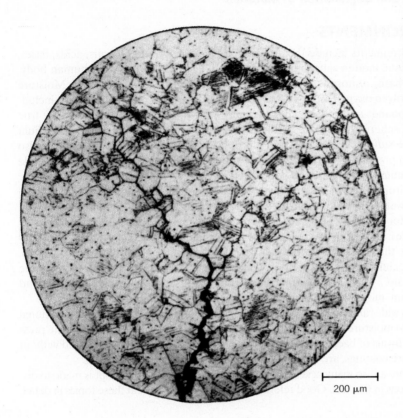

Figure 18.22 Photomicrograph showing intergranular stress corrosion cracking in brass.
(From H. H. Uhlig and R. W. Revie, *Corrosion and Corrosion Control,* 3rd edition, Fig. 5, p. 335. Copyright 1985 by John Wiley & Sons, Inc. Reprinted by permission of John Wiley & Sons, Inc.)

200 μm

For hydrogen embrittlement to occur, some source of hydrogen must be present, as well as the possibility for the formation of its atomic species. Situations in which these conditions are met include the following: pickling[3] of steels in sulfuric acid; electroplating; and the presence of hydrogen-bearing atmospheres (including water vapor) at elevated temperatures such as during welding and heat treatments. Also, the presence of what are termed *poisons* such as sulfur (i.e., H_2S) and arsenic compounds accelerates hydrogen embrittlement; these substances retard the formation of molecular hydrogen and thereby increase the residence time of atomic hydrogen on the metal surface. Hydrogen sulfide, probably the most aggressive poison, is found in petroleum fluids, natural gas, oil-well brines, and geothermal fluids.

High-strength steels are susceptible to hydrogen embrittlement, and increasing strength tends to enhance the material's susceptibility. Martensitic steels are especially vulnerable to this type of failure; bainitic, ferritic, and spheroiditic steels are more resilient. Furthermore, FCC alloys (austenitic stainless steels and alloys of copper, aluminum, and nickel) are relatively resistant to hydrogen embrittlement, mainly because of their inherently high ductilities. However, strain hardening these alloys enhances their susceptibility to embrittlement.

Techniques commonly used to reduce the likelihood of hydrogen embrittlement include reducing the tensile strength of the alloy via a heat treatment, removing the source of hydrogen, "baking" the alloy at an elevated temperature to drive out any dissolved hydrogen, and substituting a more embrittlement-resistant alloy.

---

[3] *Pickling* is a procedure used to remove surface oxide scale from steel pieces by dipping them in a vat of hot, dilute sulfuric or hydrochloric acid.

18.8 CORROSION ENVIRONMENTS

Corrosion environments include the atmosphere, aqueous solutions, soils, acids, bases, inorganic solvents, molten salts, liquid metals, and, last but not least, the human body. On a tonnage basis, atmospheric corrosion accounts for the greatest losses. Moisture containing dissolved oxygen is the primary corrosive agent, but other substances, including sulfur compounds and sodium chloride, may also contribute. This is especially true of marine atmospheres, which are highly corrosive because of the presence of sodium chloride. Dilute sulfuric acid solutions (acid rain) in industrial environments can also cause corrosion problems. Metals commonly used for atmospheric applications include alloys of aluminum and copper, and galvanized steel.

Water environments can also have a variety of compositions and corrosion characteristics. Freshwater normally contains dissolved oxygen as well as minerals, several of which account for hardness. Seawater contains approximately 3.5% salt (predominantly sodium chloride), as well as some minerals and organic matter. Seawater is generally more corrosive than freshwater, frequently producing pitting and crevice corrosion. Cast iron, steel, aluminum, copper, brass, and some stainless steels are generally suitable for freshwater use, whereas titanium, brass, some bronzes, copper–nickel alloys, and nickel–chromium–molybdenum alloys are highly corrosion resistant in seawater.

Soils have a wide range of compositions and susceptibilities to corrosion. Compositional variables include moisture, oxygen, salt content, alkalinity, and acidity, as well as the presence of various forms of bacteria. Cast iron and plain carbon steels, both with and without protective surface coatings, are economical for underground structures.

Because there are so many acids, bases, and organic solvents, no attempt is made to discuss these solutions in this text. Good references are available that treat these topics in detail.

18.9 CORROSION PREVENTION

Some corrosion prevention methods were treated relative to the eight forms of corrosion; however, only the measures specific to each of the various corrosion types were discussed. Now, some more general techniques are presented; these include material selection, environmental alteration, design, coatings, and cathodic protection.

Perhaps the most common and easiest way of preventing corrosion is through the judicious selection of materials once the corrosion environment has been characterized. Standard corrosion references are helpful in this respect. Here, cost may be a significant factor. It is not always economically feasible to employ the material that provides the optimum corrosion resistance; sometimes, either another alloy and/or some other measure must be used.

Changing the character of the environment, if possible, may also significantly influence corrosion. Lowering the fluid temperature and/or velocity usually produces a reduction in the rate at which corrosion occurs. Many times increasing or decreasing the concentration of some species in the solution will have a positive effect; for example, the metal may experience passivation.

inhibitor   **Inhibitors** are substances that, when added in relatively low concentrations to the environment, decrease its corrosiveness. The specific inhibitor depends on both the alloy and the corrosive environment. Several mechanisms may account for the effectiveness of inhibitors. Some react with and virtually eliminate a chemically active species in the solution (such as dissolved oxygen). Other inhibitor molecules attach themselves to the corroding surface and interfere with either the oxidation or the reduction reaction or form a very thin protective coating. Inhibitors are normally used in closed systems such as automobile radiators and steam boilers.

Several aspects of design consideration have already been discussed, especially with regard to galvanic and crevice corrosion and erosion–corrosion. In addition, the design should allow for complete drainage in the case of a shutdown, and easy washing. Because dissolved oxygen may enhance the corrosivity of many solutions, the design should, if possible, include provision for the exclusion of air.

Physical barriers to corrosion are applied on surfaces in the form of films and coatings. A large diversity of metallic and nonmetallic coating materials is available. It is essential that the coating maintain a high degree of surface adhesion, which undoubtedly requires some preapplication surface treatment. In most cases, the coating must be virtually nonreactive in the corrosive environment and resistant to mechanical damage that exposes the bare metal to the corrosive environment. All three material types—metals, ceramics, and polymers—are used as coatings for metals.

Cathodic Protection

cathodic protection

One of the most effective means of corrosion prevention is **cathodic protection**; it can be used for all eight different forms of corrosion as discussed earlier and may, in some situations, completely stop corrosion. Again, oxidation or corrosion of a metal M occurs by the generalized reaction 18.1,

Oxidation reaction for metal M

$$M \longrightarrow M^{n+} + ne^-$$

Cathodic protection simply involves supplying, from an external source, electrons to the metal to be protected, making it a cathode; the preceding reaction is thus forced in the reverse (or reduction) direction.

One cathodic protection technique employs a galvanic couple: the metal to be protected is electrically connected to another metal that is more reactive in the particular environment. The latter experiences oxidation and, upon giving up electrons, protects the first metal from corrosion. The oxidized metal is often called a **sacrificial anode**, and magnesium and zinc are commonly used because they lie at the anodic end of the galvanic series. This form of galvanic protection for structures buried in the ground is illustrated in Figure 18.23a.

sacrificial anode

The process of *galvanizing* is simply one in which a layer of zinc is applied to the surface of steel by hot dipping. In the atmosphere and most aqueous environments, zinc is anodic to and will thus cathodically protect the steel if there is any surface damage (Figure 18.24). Any corrosion of the zinc coating will proceed at an extremely slow rate because the ratio of the anode-to-cathode surface area is quite large.

For another method of cathodic protection, the source of electrons is an impressed current from an external dc power source, as represented in Figure 18.23b for an underground tank. The negative terminal of the power source is connected to the structure to be protected. The other terminal is joined to an inert anode (often graphite), which, in this case, is buried in the soil; high-conductivity backfill material provides good electrical contact between the anode and the surrounding soil. A current path exists between

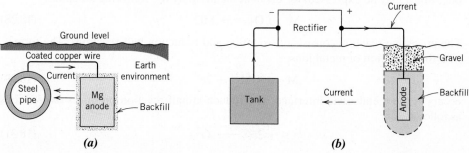

(a)   *(b)*

Figure 18.23 Cathodic protection of (*a*) an underground pipeline using a magnesium sacrificial anode and (*b*) an underground tank using an impressed current.
(From M. G. Fontana, *Corrosion Engineering,* 3rd edition. Copyright © 1986 by McGraw-Hill Book Company. Reproduced with permission.)

Figure 18.24 Galvanic protection of steel as provided by a coating of zinc.

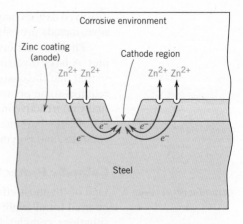

the cathode and the anode through the intervening soil, completing the electrical circuit. Cathodic protection is especially useful in preventing corrosion of water heaters, underground tanks and pipes, and marine equipment.

Concept Check 18.7 Tin cans are made of a steel, the inside of which is coated with a thin layer of tin. The tin protects the steel from corrosion by food products in the same manner as zinc protects steel from atmospheric corrosion. Briefly explain how this cathodic protection of tin cans is possible, given that tin is electrochemically less active than steel in the galvanic series (Table 18.2).

[*The answer may be found at* www.wiley.com/college/callister (*Student Companion Site*).]

18.10 OXIDATION

The discussion of Section 18.2 treated the corrosion of metallic materials in terms of electrochemical reactions that take place in aqueous solutions. In addition, oxidation of metal alloys is also possible in gaseous atmospheres, normally air, in which an oxide layer or scale forms on the surface of the metal. This phenomenon is frequently termed *scaling, tarnishing,* or *dry corrosion*. In this section, we discuss possible mechanisms for this type of corrosion, the types of oxide layers that can form, and the kinetics of oxide formation.

Mechanisms

As with aqueous corrosion, the process of oxide layer formation is an electrochemical one, which may be expressed, for divalent metal M, by the following reaction[4]:

$$M + \tfrac{1}{2}O_2 \longrightarrow MO \tag{18.28}$$

The preceding reaction consists of oxidation and reduction half-reactions. The former, with the formation of metal ions,

$$M \longrightarrow M^{2+} + 2e^- \tag{18.29}$$

occurs at the metal–scale interface. The reduction half-reaction produces oxygen ions as follows:

$$\tfrac{1}{2}O_2 + 2e^- \longrightarrow O^{2-} \tag{18.31}$$

---

[4] For other than divalent metals, this reaction may be expressed as

$$aM + \frac{b}{2}O_2 \longrightarrow M_aO_b \tag{18.30}$$

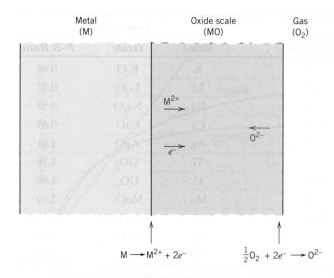

Metal (M) · Oxide scale (MO) · Gas (O₂)

M^{2+}

O^{2-}

e^-

$M \longrightarrow M^{2+} + 2e^-$ · $\frac{1}{2}O_2 + 2e^- \longrightarrow O^{2-}$

Figure 18.25 Schematic representation of processes that are involved in gaseous oxidation at a metal surface.

and takes place at the scale–gas interface. A schematic representation of this metal–scale–gas system is shown in Figure 18.25.

For the oxide layer to increase in thickness via Equation 18.28, it is necessary that electrons be conducted to the scale–gas interface, at which point the reduction reaction occurs; in addition, M^{2+} ions must diffuse away from the metal–scale interface, and/or O^{2-} ions must diffuse toward this same interface (Figure 18.25).[5] Thus, the oxide scale serves both as an electrolyte through which ions diffuse and as an electrical circuit for the passage of electrons. Furthermore, the scale may protect the metal from rapid oxidation when it acts as a barrier to ionic diffusion and/or electrical conduction; most metal oxides are highly electrically insulative.

Scale Types

Rate of oxidation (i.e., the rate of film thickness increase) and the tendency of the film to protect the metal from further oxidation are related to the relative volumes of the oxide and metal. The ratio of these volumes, termed the **Pilling–Bedworth ratio**, may be determined from the following expression[6]:

Pilling–Bedworth ratio

Pilling–Bedworth ratio for a divalent metal—dependence on the densities and atomic/formula weights of the metal and its oxide

$$\text{P–B ratio} = \frac{A_O \rho_M}{A_M \rho_O} \qquad (18.32)$$

where A_O is the molecular (or formula) weight of the oxide, A_M is the atomic weight of the metal, and ρ_O and ρ_M are the oxide and metal densities, respectively. For metals having P–B ratios less than unity, the oxide film tends to be porous and unprotective

---

[5]Alternatively, electron holes (Section 19.10) and vacancies may diffuse instead of electrons and ions.

[6]For other than divalent metals, Equation 18.32 becomes

Pilling–Bedworth ratio for a metal that is not divalent

$$\text{P–B ratio} = \frac{A_O \rho_M}{a A_M \rho_O} \qquad (18.33)$$

where a is the coefficient of the metal species for the overall oxidation reaction described by Equation 18.30.

Degradation of Polymers

Polymeric materials also experience deterioration by means of environmental interactions. However, an undesirable interaction is specified as degradation rather than corrosion because the processes are basically dissimilar. Whereas most metallic corrosion reactions are electrochemical, polymeric degradation is physiochemical; that is, it involves physical as well as chemical phenomena. Furthermore, a wide variety of reactions and adverse consequences are possible for polymer degradation. Polymers may deteriorate by swelling and dissolution. Covalent bond rupture as a result of heat energy, chemical reactions, and radiation is also possible, typically with an attendant reduction in mechanical integrity. Because of the chemical complexity of polymers, their degradation mechanisms are not well understood.

To cite a couple of examples of polymer degradation, polyethylene, if exposed to high temperatures in an oxygen atmosphere, suffers an impairment of its mechanical properties by becoming brittle, and the utility of poly(vinyl chloride) is limited because this material may become discolored when exposed to high temperatures, although such environments may not affect its mechanical characteristics.

18.11 SWELLING AND DISSOLUTION

When polymers are exposed to liquids, the main forms of degradation are swelling and dissolution. With swelling, the liquid or solute diffuses into and is absorbed within the polymer; the small solute molecules fit into and occupy positions among the polymer molecules. Thus the macromolecules are forced apart such that the specimen expands or swells. Furthermore, this increase in chain separation results in a reduction of the secondary intermolecular bonding forces; as a consequence, the material becomes softer and more ductile. The liquid solute also lowers the glass transition temperature and, if depressed below the ambient temperature, causes a once-strong material to become rubbery and weak.

Swelling may be considered a partial dissolution process in which there is only limited solubility of the polymer in the solvent. Dissolution, which occurs when the polymer is completely soluble, may be thought of as a continuation of swelling. As a rule of thumb, the greater the similarity of chemical structure between the solvent and polymer, the greater the likelihood of swelling and/or dissolution. For example, many hydrocarbon rubbers readily absorb hydrocarbon liquids such as gasoline, but absorb virtually no water. The responses of selected polymeric materials to organic solvents are given in Tables 18.4 and 18.5.

Swelling and dissolution traits also are affected by temperature, as well as by the characteristics of the molecular structure. In general, increasing molecular weight, increasing degree of crosslinking and crystallinity, and decreasing temperature result in a reduction of these deteriorative processes.

In general, polymers are much more resistant to attack by acidic and alkaline solutions than are metals. For example, hydrofluoric acid (HF) corrodes many metals as well as etch and dissolve glass, so it is stored in plastic bottles. A qualitative comparison of the behavior of various polymers in these solutions is also presented in Tables 18.4 and 18.5. Materials that exhibit outstanding resistance to attack by both solution types include polytetrafluoroethylene (and other fluorocarbons) and polyetheretherketone.

✓ *Concept Check 18.8* From a molecular perspective, explain why increasing crosslinking and crystallinity of a polymeric material will enhance its resistance to swelling and dissolution. Would you expect crosslinking or crystallinity to have the greater influence? Justify your choice. *Hint:* You may want to consult Sections 4.13 and 5.7.

[*The answer may be found at* www.wiley.com/college/callister *(Student Companion Site).*]

Table 18.4 Resistance to Degradation by Various Environments for Selected Plastic Materials[a]

Material	Nonoxidizing Acids (20% H_2SO_4)	Oxidizing Acids (10% HNO_3)	Aqueous Salt Solutions (NaCl)	Aqueous Alkalis (NaOH)	Polar Solvents (C_2H_5OH)	Nonpolar Solvents (C_6H_6)	Water
Polytetrafluoro-ethylene	S	S	S	S	S	S	S
Nylon 6,6	U	U	S	S	Q	S	S
Polycarbonate	Q	U	S	U	S	U	S
Polyester	Q	Q	S	Q	Q	U	S
Polyetherether-ketone	S	S	S	S	S	S	S
Low-density polyethylene	S	Q	S	—	S	Q	S
High-density polyethylene	S	Q	S	—	S	Q	S
Poly(ethylene terephthalate)	S	Q	S	S	S	S	S
Poly(phenylene oxide)	S	Q	S	S	S	U	S
Polypropylene	S	Q	S	S	S	Q	S
Polystyrene	S	Q	S	S	S	U	S
Polyurethane	Q	U	S	Q	U	Q	S
Epoxy	S	U	S	S	S	S	S
Silicone	Q	U	S	S	S	Q	S

[a]S = satisfactory; Q = questionable; U = unsatisfactory.

Source: Adapted from R. B. Seymour, *Polymers for Engineering Applications,* ASM International, Materials Park, OH, 1987.

Table 18.5 Resistance to Degradation by Various Environments for Selected Elastomeric Materials[a]

Material	Weather-Sunlight Aging	Oxidation	Ozone Cracking	Alkali Dilute/Concentrated	Acid Dilute/Concentrated	Chlorinated Hydrocarbons, Degreasers	Aliphatic Hydrocarbons, Kerosene, Etc.	Animal, Vegetable Oils
Polyisoprene (natural)	D	B	NR	A/C-B	A/C-B	NR	NR	D-B
Polyisoprene (synthetic)	NR	B	NR	C-B/C-B	C-B/C-B	NR	NR	D-B
Butadiene	D	B	NR	C-B/C-B	C-B/C-B	NR	NR	D-B
Styrene-butadiene	D	C	NR	C-B/C-B	C-B/C-B	NR	NR	D-B
Neoprene	B	A	A	A/A	A/A	D	C	B
Nitrile (high)	D	B	C	B/B	B/B	C-B	A	B
Silicone (polysiloxane)	A	A	A	A/A	B/C	NR	D-C	A

[a]A = excellent, B = good, C = fair, D = use with caution, NR = not recommended.

Source: *Compound Selection and Service Guide,* Seals Eastern, Inc., Red Bank, NJ, 1977.

18.12 BOND RUPTURE

scission

Polymers may also experience degradation by a process termed scission—the severance or rupture of molecular chain bonds. This causes a separation of chain segments at the point of scission and a reduction in the molecular weight. As previously discussed (Chapter 15), several properties of polymeric materials, including mechanical strength and resistance to chemical attack, depend on molecular weight. Consequently, some of the physical and chemical properties of polymers may be adversely affected by this form of degradation. Bond rupture may result from exposure to radiation or heat and from chemical reaction.

Radiation Effects

Certain types of radiation [electron beams, x-rays, β- and γ-rays, and ultraviolet (UV) radiation] possess sufficient energy to penetrate a polymer specimen and interact with the constituent atoms or their electrons. One such reaction is *ionization,* in which the radiation removes an orbital electron from a specific atom, converting that atom into a positively charged ion. As a consequence, one of the covalent bonds associated with the specific atom is broken and there is a rearrangement of atoms or groups of atoms at that point. This bond breaking leads to either scission or crosslinking at the ionization site, depending on the chemical structure of the polymer and also on the dose of radiation. Stabilizers (Section 17.13) may be added to protect polymers from radiation damage. In day-to-day use, the greatest radiation damage to polymers is caused by UV irradiation. After prolonged exposure, most polymer films become brittle, discolor, crack, and fail. For example, camping tents begin to tear, dashboards develop cracks, and plastic windows become cloudy. Radiation problems are more severe for some applications. Polymers on space vehicles must resist degradation after prolonged exposures to cosmic radiation. Similarly, polymers used in nuclear reactors must withstand high levels of nuclear radiation. Developing polymeric materials that can withstand these extreme environments is a continuing challenge.

Not all consequences of radiation exposure are deleterious. Crosslinking may be induced by irradiation to improve the mechanical behavior and degradation characteristics. For example, γ-radiation is used commercially to crosslink polyethylene to enhance its resistance to softening and flow at elevated temperatures; indeed, this process may be carried out on products that have already been fabricated.

Chemical Reaction Effects

Oxygen, ozone, and other substances can cause or accelerate chain scission as a result of chemical reaction. This effect is especially prevalent in vulcanized rubbers that have doubly bonded carbon atoms along the backbone molecular chains and that are exposed to ozone (O_3), an atmospheric pollutant. One such scission reaction may be represented by

$$-R-\underset{\underset{H}{|}}{C}=\underset{\underset{H}{|}}{C}-R'- + O_3 \longrightarrow -R-\underset{\underset{H}{|}}{C}=O + O=\underset{\underset{H}{|}}{C}-R'- + O\cdot \quad (18.37)$$

where the chain is severed at the point of the double bond; R and R′ represent groups of atoms that are unaffected during the reaction. Typically, if the rubber is in an unstressed state, an oxide film will form on the surface, protecting the bulk material from any further reaction. However, when these materials are subjected to tensile stresses, cracks and crevices form and grow in a direction perpendicular to the stress; eventually, rupture of the material may occur. This is why the sidewalls on rubber bicycle tires develop cracks as they age. Apparently these cracks result from large numbers of ozone-induced scissions. Chemical degradation is a particular problem for polymers used in areas with high levels of air pollutants such as smog and ozone. The elastomers in Table 18.5 are rated as to their resistance to degradation by exposure to ozone. Many of these chain scission reactions involve reactive groups termed *free radicals*. Stabilizers (Section 17.13) may be added to protect polymers from oxidation. The stabilizers either sacrificially react with the ozone to consume it or react with and eliminate the free radicals before the free radicals can inflict more damage.

Thermal Effects

Thermal degradation corresponds to the scission of molecular chains at elevated temperatures; as a consequence, some polymers undergo chemical reactions in which gaseous species are produced. These reactions are evidenced by a weight loss of material; a polymer's thermal stability is a measure of its resilience to this decomposition. Thermal stability is related primarily to the magnitude of the bonding energies between the various atomic constituents of the polymer: higher bonding energies result in more thermally stable materials. For example, the magnitude of the $C-F$ bond is greater than that of the $C-H$ bond, which in turn is greater than that of the $C-Cl$ bond. The fluorocarbons, having $C-F$ bonds, are among the most thermally resistant polymeric materials and may be used at relatively high temperatures. However, because of the weak $C-Cl$ bond, when poly(vinyl chloride) is heated to 200°C for even a few minutes, it discolors and gives off large amounts of HCl, which accelerates continued decomposition. Stabilizers (Section 17.13) such as ZnO can react with the HCl, providing increased thermal stability for poly(vinyl chloride).

Some of the most thermally stable polymers are the ladder polymers.[7] For example, the ladder polymer having the structure

is so thermally stable that a woven cloth of this material can be heated directly in an open flame with no degradation. Polymers of this type are used in place of asbestos for high-temperature gloves.

18.13 WEATHERING

Many polymeric materials serve in applications that require exposure to outdoor conditions. Any resultant degradation is termed *weathering*, which may be a combination of several different processes. Under these conditions, deterioration is primarily a result of oxidation, which is initiated by ultraviolet radiation from the sun. Some polymers, such as nylon and cellulose, are also susceptible to water absorption, which produces a reduction in their hardness and stiffness. Resistance to weathering among the various polymers is quite diverse. The fluorocarbons are virtually inert under these conditions; some materials, however, including poly(vinyl chloride) and polystyrene, are susceptible to weathering.

---

[7]The chain structure of a *ladder polymer* consists of two sets of covalent bonds throughout its length that are crosslinked.

Concept Check 18.9   List three differences between the corrosion of metals and each of the following:

(a) the corrosion of ceramics
(b) the degradation of polymers

[*The answer may be found at* www.wiley.com/college/callister *(Student Companion Site).*]

SUMMARY

Electrochemical Considerations

- Metallic corrosion is typically electrochemical, involving both oxidation and reduction reactions.

    Oxidation is the loss of the metal atom's valence electrons and occurs at the anode; the resulting metal ions may either go into the corroding solution or form an insoluble compound.

    During reduction (which occurs at the cathode), these electrons are transferred to at least one other chemical species. The character of the corrosion environment dictates which of several possible reduction reactions will occur.

- Not all metals oxidize with the same degree of ease, which is demonstrated with a galvanic couple.

    When in an electrolyte, one metal (the anode) will corrode, whereas a reduction reaction will occur at the other metal (the cathode).

    The magnitude of the electric potential that is established between anode and cathode is indicative of the driving force for the corrosion reaction.

- The standard emf and galvanic series are rankings of metallic materials on the basis of their tendency to corrode when coupled to other metals.

    For the standard emf series, ranking is based on the magnitude of the voltage generated when the standard cell of a metal is coupled to the standard hydrogen electrode at 25°C.

    The galvanic series consists of the relative reactivities of metals and alloys in seawater.

- The half-cell potentials in the standard emf series are thermodynamic parameters that are valid only at equilibrium; corroding systems are not in equilibrium. Furthermore, the magnitudes of these potentials provide no indication as to the rates at which corrosion reactions occur.

Corrosion Rates

- The rate of corrosion may be expressed as corrosion penetration rate, that is, the thickness loss of material per unit of time; CPR may be determined using Equation 18.23. Mils per year and millimeters per year are the common units for this parameter.

- Alternatively, rate is proportional to the current density associated with the electrochemical reaction, according to Equation 18.24.

Prediction of Corrosion Rates

- Corroding systems will experience polarization, which is the displacement of each electrode potential from its equilibrium value; the magnitude of the displacement is termed the *overvoltage*.

- The corrosion rate of a reaction is limited by polarization, of which there are two types—activation and concentration.

    Activation polarization relates to systems in which the corrosion rate is determined by the step in a series that occurs most slowly. For activation polarization, a plot of overvoltage versus logarithm of current density will appear as shown in Figure 18.7.

Concentration polarization prevails when the corrosion rate is limited by diffusion in the solution. When overvoltage versus the logarithm of current density is plotted, the resulting curve will appear as shown in Figure 18.9a.

- The corrosion rate for a particular reaction may be computed using Equation 18.24, incorporating the current density associated with the intersection point of oxidation and reduction polarization curves.

Passivity
- A number of metals and alloys passivate, or lose their chemical reactivity, under some environmental circumstances. This phenomenon is thought to involve the formation of a thin protective oxide film. Stainless steels and aluminum alloys exhibit this type of behavior.

- The active-to-passive behavior may be explained by the alloy's S-shaped electrochemical potential-versus-log current density curve (Figure 18.12). Intersections with reduction polarization curves in active and passive regions correspond, respectively, to high and low corrosion rates (Figure 18.13).

Forms of Corrosion
- Metallic corrosion is sometimes classified into nine different forms:
    Uniform attack—degree of corrosion is approximately uniform over the entire exposed surface.
    Galvanic corrosion—occurs when two different metals or alloys are electrically coupled while exposed to an electrolyte solution.
    Crevice corrosion—the situation when corrosion occurs under crevices or other areas where there is localized depletion of oxygen.
    Pitting—a type of localized corrosion in which pits or holes form from the top of horizontal surfaces.
    Intergranular corrosion—occurs preferentially along grain boundaries for specific metals/alloys (e.g., some stainless steels).
    Selective leaching—the case in which one element/constituent of an alloy is removed selectively by corrosive action.
    Erosion–corrosion—the combined action of chemical attack and mechanical wear as a consequence of fluid motion.
    Stress corrosion—the formation and propagation of cracks (and possible failure) resulting from the combined effects of corrosion and the application of a tensile stress.
    Hydrogen embrittlement—a significant reduction in ductility that accompanies the penetration of atomic hydrogen into a metal/alloy.

Corrosion Prevention
- Several measures may be taken to prevent, or at least reduce, corrosion. These include material selection, environmental alteration, the use of inhibitors, design changes, application of coatings, and cathodic protection.

- With cathodic protection, the metal to be protected is made a cathode by supplying electrons from an external source.

Oxidation
- Oxidation of metallic materials by electrochemical action is also possible in dry, gaseous atmospheres (Figure 18.25).

- An oxide film forms on the surface that may act as a barrier to further oxidation if the volumes of metal and oxide film are similar, that is, if the Pilling–Bedworth ratio (Equations 18.32 and 18.33) is near unity.

- The kinetics of film formation may follow parabolic (Equation 18.34), linear (Equation 18.35), or logarithmic (Equation 18.36) rate laws.

Corrosion of Ceramic Materials
- Ceramic materials, being inherently corrosion resistant, are frequently used at elevated temperatures and/or in extremely corrosive environments.

For Lead	For Hydrogen
$V_{(Pb/Pb^{2+})} = -0.126$ V	$V_{(H^+/H_2)} = 0$ V
$i_0 = 2 \times 10^{-9}$ A/cm^2	$i_0 = 1.0 \times 10^{-8}$ A/cm^2
$\beta = +0.12$	$\beta = -0.10$

(b) Compute the value of the corrosion potential.

18.17 The corrosion rate is to be determined for some divalent metal M in a solution containing hydrogen ions. The following corrosion data are known about the metal and solution:

For Metal M	For Hydrogen
$V_{(M/M^{2+})} = -0.47$ V	$V_{(H^+/H_2)} = 0$ V
$i_0 = 5 \times 10^{-10}$ A/cm^2	$i_0 = 2 \times 10^{-9}$ A/cm^2
$\beta = +0.15$	$\beta = -0.12$

(a) Assuming that activation polarization controls both oxidation and reduction reactions, determine the rate of corrosion of metal M (in mol/cm$^2 \cdot$ s).

(b) Compute the corrosion potential for this reaction.

18.18 The influence of increasing solution velocity on the overvoltage-versus-log-current-density behavior for a solution that experiences combined activation–concentration polarization is indicated in Figure 18.27. On the basis of this behavior, make a schematic plot of corrosion rate versus solution velocity for the oxidation of a metal; assume that the oxidation reaction is controlled by activation polarization.

Passivity

18.19 Briefly describe the phenomenon of passivity. Name two common types of alloy that passivate.

18.20 Why does chromium in stainless steels make them more corrosion resistant in many environments than plain carbon steels?

Forms of Corrosion

18.21 For each form of corrosion, other than uniform, do the following:

(a) Describe why, where, and the conditions under which the corrosion occurs.

(b) Cite three measures that may be taken to prevent or control it.

18.22 Briefly explain why cold-worked metals are more susceptible to corrosion than noncold-worked metals.

18.23 Briefly explain why, for a small anode-to-cathode area ratio, the corrosion rate will be higher than for a large ratio.

18.24 For a concentration cell, briefly explain why corrosion occurs at that region having the lower concentration.

Corrosion Prevention

18.25 (a) What are inhibitors?

(b) What possible mechanisms account for their effectiveness?

18.26 Briefly describe the two techniques that are used for galvanic protection.

Oxidation

18.27 For each of the metals listed in the following table, compute the Pilling–Bedworth ratio. Also, on the basis of this value, specify whether you would expect the oxide scale that forms on the surface to be protective, and then justify your decision. Density data for both the metal and its oxide are also tabulated.

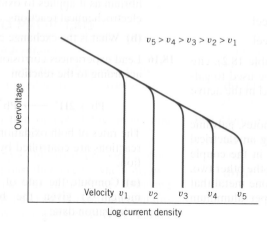

$$v_5 > v_4 > v_3 > v_2 > v_1$$

Overvoltage

Velocity v_1 $\quad v_2 \quad v_3 \quad v_4 \quad v_5$

Log current density

Figure 18.27 Plot of overvoltage versus logarithm of current density for a solution that experiences combined activation–concentration polarization at various solution velocities.

Metal	Metal Density (g/cm^3)	Metal Oxide	Oxide Density (g/cm^3)
Zr	6.51	ZrO$_2$	5.89
Sn	7.30	SnO$_2$	6.95
Bi	9.80	Bi$_2$O$_3$	8.90

18.28 According to Table 18.3, the oxide coating that forms on silver should be nonprotective, and yet Ag does not oxidize appreciably at room temperature and in air. How do you explain this apparent discrepancy?

18.29 In the following table, weight gain–time data for the oxidation of copper at an elevated temperature are given.

W (mg/cm^2)	Time (min)
0.318	20
0.525	55
0.726	105

(a) Determine whether the oxidation kinetics obey a linear, parabolic, or logarithmic rate expression.

(b) Now compute W after a time of 450 min.

18.30 In the following table, weight gain–time data for the oxidation of some metal at an elevated temperature are given.

W (mg/cm^2)	Time (min)
4.80	20
9.6	40
11.2	80

(a) Determine whether the oxidation kinetics obey a linear, parabolic, or logarithmic rate expression.

(b) Now compute W after a time of 1000 min.

18.31 In the following table, weight gain–time data for the oxidation of some metal at an elevated temperature are given.

W (mg/cm^2)	Time (min)
1.90	25
3.76	75
6.40	250

(a) Determine whether the oxidation kinetics obey a linear, parabolic, or logarithmic rate expression.

(b) Now compute W after a time of 3500 min.

Spreadsheet Problems

18.1SS Generate a spreadsheet that will determine the rate of oxidation (in mol/cm$^2 \cdot$ s) and the corrosion potential for a metal that is immersed in an acid solution. The user is allowed to input the following parameters for each of the two half-cells: the corrosion potential, the exchange current density, and the value of β.

18.2SS For the oxidation of some metal, given a set of values of weight gain and their corresponding times (at least three values), generate a spreadsheet that will allow the user to determine the following: (a) whether the oxidation kinetics obey a linear, parabolic, or logarithmic rate expression, (b) values of the constants in the appropriate rate expression, and (c) the weight gain after some other time.

DESIGN PROBLEMS

18.D1 A brine solution is used as a cooling medium in a steel heat exchanger. The brine is circulated within the heat exchanger and contains some dissolved oxygen. Suggest three methods, other than cathodic protection, for reducing corrosion of the steel by the brine. Explain the rationale for each suggestion.

18.D2 Suggest an appropriate material for each of the following applications, and, if necessary, recommend corrosion prevention measures that should be taken. Justify your suggestions.

(a) Laboratory bottles to contain relatively dilute solutions of nitric acid

(b) Barrels to contain benzene

(c) Pipe to transport hot alkaline (basic) solutions

(d) Underground tanks to store large quantities of high-purity water

(e) Architectural trim for high-rise buildings

18.D3 Each student (or group of students) is to find a real-life corrosion problem that has not been solved, conduct a thorough investigation as to the cause(s) and type(s) of corrosion, and, finally, propose possible solutions for the problem, indicating which of the solutions is best and why. Submit a report that addresses these issues.

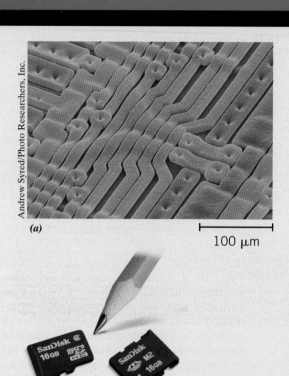

(a)

100 μm

Andrew Syred/Photo Researchers, Inc.

The functioning of modern flash memory cards (and flash drives) that are used to store digital information relies on the unique electrical properties of silicon, a semiconducting material. (Flash memory is discussed in Section 19.15.)

    (a) Scanning electron micrograph of an integrated circuit, which is composed of silicon and metallic interconnects. Integrated circuit components are used to store information in a digital format.

    (b) Three different flash memory card types.

    (c) A flash memory card being inserted into a digital camera. This memory card will be used to store photographic images (and in some cases GPS location).

Courtesy SanDisk Corporation

© GaryPhoto/iStockphoto

(b)

(c)

Consideration of the electrical properties of materials is often important when materials selection and processing decisions are being made during the design of a component or structure. For example, when we consider an integrated circuit package, the electrical behaviors of the various materials are diverse. Some need to be highly electrically conductive (e.g., connecting wires), whereas electrical insulativity is required of others (e.g., protective package encapsulation).

Learning Objectives

After studying this chapter, you should be able to do the following:

1. Describe the four possible electron band structures for solid materials.
2. Briefly describe electron excitation events that produce free electrons/holes in (a) metals, (b) semiconductors (intrinsic and extrinsic), and (c) insulators.
3. Calculate the electrical conductivities of metals, semiconductors (intrinsic and extrinsic), and insulators given their charge carrier densities and mobilities.
4. Distinguish between *intrinsic* and *extrinsic* semiconducting materials.
5. (a) On a plot of logarithm of carrier (electron, hole) concentration versus absolute temperature, draw schematic curves for both intrinsic and extrinsic semiconducting materials.

(b) On the extrinsic curve, note freeze-out, extrinsic, and intrinsic regions.
6. For a *p–n* junction, explain the rectification process in terms of electron and hole motions.
7. Calculate the capacitance of a parallel-plate capacitor.
8. Define dielectric constant in terms of permittivities.
9. Briefly explain how the charge storing capacity of a capacitor may be increased by the insertion and polarization of a dielectric material between its plates.
10. Name and describe the three types of polarization.
11. Briefly describe the phenomena of *ferroelectricity* and *piezoelectricity*.

19.1 INTRODUCTION

The prime objective of this chapter is to explore the electrical properties of materials—that is, their responses to an applied electric field. We begin with the phenomenon of electrical conduction: the parameters by which it is expressed, the mechanism of conduction by electrons, and how the electron energy band structure of a material influences its ability to conduct. These principles are extended to metals, semiconductors, and insulators. Particular attention is given to the characteristics of semiconductors and then to semiconducting devices. The dielectric characteristics of insulating materials are also treated. The final sections are devoted to the phenomena of ferroelectricity and piezoelectricity.

Electrical Conduction

19.2 OHM'S LAW

Ohm's law

One of the most important electrical characteristics of a solid material is the ease with which it transmits an electric current. **Ohm's law** relates the current I—or time rate of charge passage—to the applied voltage V as follows:

Ohm's law expression

$$V = IR \tag{19.1}$$

where R is the resistance of the material through which the current is passing. The units for V, I, and R are, respectively, volts (J/C), amperes (C/s), and ohms (V/A). The value of R is influenced by specimen configuration and for many materials is independent of

Figure 19.1 Schematic representation of the apparatus used to measure electrical resistivity.

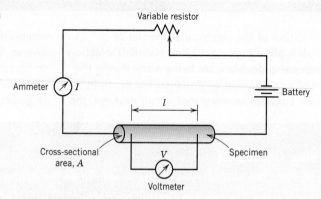

electrical resistivity

Electrical resistivity— dependence on resistance, specimen cross-sectional area, and distance between measuring points

current. The **electrical resistivity** ρ is independent of specimen geometry but related to R through the expression

$$\rho = \frac{RA}{l} \qquad (19.2)$$

where l is the distance between the two points at which the voltage is measured and A is the cross-sectional area perpendicular to the direction of the current. The units for ρ are ohm-meters ($\Omega \cdot m$). From the expression for Ohm's law and Equation 19.2,

Electrical resistivity— dependence on applied voltage, current, specimen cross-sectional area, and distance between measuring points

$$\rho = \frac{VA}{Il} \qquad (19.3)$$

Figure 19.1 is a schematic diagram of an experimental arrangement for measuring electrical resistivity.

19.3 ELECTRICAL CONDUCTIVITY

electrical conductivity

Reciprocal relationship between electrical conductivity and resistivity

Sometimes, **electrical conductivity** σ is used to specify the electrical character of a material. It is simply the reciprocal of the resistivity, or

$$\sigma = \frac{1}{\rho} \qquad (19.4)$$

and is indicative of the ease with which a material is capable of conducting an electric current. The units for σ are reciprocal ohm-meters $[(\Omega \cdot m)^{-1}]$.[1] The following discussions on electrical properties use both resistivity and conductivity.

In addition to Equation 19.1, Ohm's law may be expressed as

Ohm's law expression—in terms of current density, conductivity, and applied electric field

$$J = \sigma \mathscr{E} \qquad (19.5)$$

in which J is the current density—the current per unit of specimen area I/A—and $\mathscr{E}$ is the electric field intensity, or the voltage difference between two points divided by the distance separating them—that is,

Electric field intensity

$$\mathscr{E} = \frac{V}{l} \qquad (19.6)$$

---

[1]The SI units for electrical conductivity are siemens per meter (S/m), in which 1 S/m = 1 $(\Omega \cdot m)^{-1}$. We opted to use $(\Omega \cdot m)^{-1}$ on the basis of convention—these units are traditionally used in introductory materials science and engineering texts.

The demonstration of the equivalence of the two Ohm's law expressions (Equations 19.1 and 19.5) is left as a homework exercise.

Solid materials exhibit an amazing range of electrical conductivities, extending over 27 orders of magnitude; probably no other physical property exhibits this breadth of variation. In fact, one way of classifying solid materials is according to the ease with which they conduct an electric current; within this classification scheme there are three groupings: *conductors, semiconductors,* and *insulators.* Metals are good conductors, typically having conductivities on the order of 10^7 $(\Omega \cdot m)^{-1}$. At the other extreme are materials with very low conductivities, ranging between 10^{-10} and 10^{-20} $(\Omega \cdot m)^{-1}$; these are electrical insulators. Materials with intermediate conductivities, generally from 10^{-6} to 10^4 $(\Omega \cdot m)^{-1}$, are termed semiconductors. Electrical conductivity ranges for the various material types are compared in the bar chart of Figure 1.8.

metal

insulator

semiconductor

19.4 ELECTRONIC AND IONIC CONDUCTION

An electric current results from the motion of electrically charged particles in response to forces that act on them from an externally applied electric field. Positively charged particles are accelerated in the field direction, negatively charged particles in the direction opposite. Within most solid materials a current arises from the flow of electrons, which is termed *electronic conduction.* In addition, for ionic materials, a net motion of charged ions is possible that produces a current; this is termed ionic conduction. The present discussion deals with electronic conduction; ionic conduction is treated briefly in Section 19.16.

ionic conduction

19.5 ENERGY BAND STRUCTURES IN SOLIDS

In all conductors, semiconductors, and many insulating materials, only electronic conduction exists, and the magnitude of the electrical conductivity is strongly dependent on the number of electrons available to participate in the conduction process. However, not all electrons in every atom accelerate in the presence of an electric field. The number of electrons available for electrical conduction in a particular material is related to the arrangement of electron states or levels with respect to energy and the manner in which these states are occupied by electrons. A thorough exploration of these topics is complicated and involves principles of quantum mechanics that are beyond the scope of this book; the ensuing development omits some concepts and simplifies others.

Concepts relating to electron energy states, their occupancy, and the resulting electron configurations for isolated atoms were discussed in Section 2.3. By way of review, for each individual atom there exist discrete energy levels that may be occupied by electrons, arranged into shells and subshells. Shells are designated by integers (1, 2, 3, etc.) and subshells by letters (*s, p, d,* and *f*). For each *s, p, d,* and *f* subshell, there exist, respectively, one, three, five, and seven states. The electrons in most atoms fill only the states having the lowest energies—two electrons of opposite spin per state, in accordance with the Pauli exclusion principle. The electron configuration of an isolated atom represents the arrangement of the electrons within the allowed states.

Let us now make an extrapolation of some of these concepts to solid materials. A solid may be thought of as consisting of a large number—say, *N*—of atoms initially separated from one another that are subsequently brought together and bonded to form the ordered atomic arrangement found in the crystalline material. At relatively large separation distances, each atom is independent of all the others and has the atomic energy levels and electron configuration as if isolated. However, as the atoms come within close proximity of one another, electrons are acted upon, or *perturbed,* by the electrons and nuclei of adjacent atoms. This influence is such that each distinct atomic state may split into a series of closely spaced electron states in the solid to form what is termed an electron energy band.

electron energy band

Figure 19.5 For a metal, occupancy of electron states (a) before and (b) after an electron excitation.

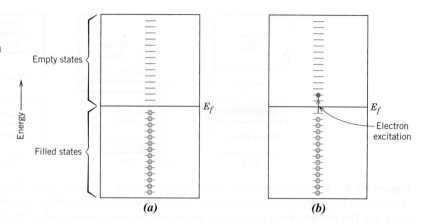

hole

entity called a **hole** is found in semiconductors and insulators. Holes have energies less than E_f and also participate in electronic conduction. The ensuing discussion shows that the electrical conductivity is a direct function of the numbers of free electrons and holes. In addition, the distinction between conductors and nonconductors (insulators and semiconductors) lies in the numbers of these free electron and hole charge carriers.

Metals

For an electron to become free, it must be excited or promoted into one of the empty and available energy states above E_f. For metals having either of the band structures shown in Figures 19.4a and 19.4b, there are vacant energy states adjacent to the highest filled state at E_f. Thus, very little energy is required to promote electrons into the low-lying empty states, as shown in Figure 19.5. Generally, the energy provided by an electric field is sufficient to excite large numbers of electrons into these conducting states.

For the metallic bonding model discussed in Section 2.6, it was assumed that all the valence electrons have freedom of motion and form an *electron gas* that is uniformly distributed throughout the lattice of ion cores. Although these electrons are not locally bound to any particular atom, they must experience some excitation to become conducting electrons that are truly free. Thus, although only a fraction are excited, this still gives rise to a relatively large number of free electrons and, consequently, a high conductivity.

Insulators and Semiconductors

For insulators and semiconductors, empty states adjacent to the top of the filled valence band are not available. To become free, therefore, electrons must be promoted across the energy band gap and into empty states at the bottom of the conduction band. This is possible only by supplying to an electron the difference in energy between these two states, which is approximately equal to the band gap energy E_g. This excitation process is demonstrated in Figure 19.6.[2] For many materials, this band gap is several electron volts wide. Most often the excitation energy is from a nonelectrical source such as heat or light, usually the former.

The number of electrons excited thermally (by heat energy) into the conduction band depends on the energy band gap width and the temperature. At a given temperature, the larger the E_g, the lower the probability that a valence electron will be promoted into an energy state within the conduction band; this results in fewer conduction electrons. In other words, the larger the band gap, the lower the electrical conductivity at a given temperature. Thus, the distinction between semiconductors and insulators lies

---

[2]The magnitudes of the band gap energy and the energies between adjacent levels in both the valence and conduction bands of Figure 19.6 are not to scale. Whereas the band gap energy is on the order of an electron volt, these levels are separated by energies on the order of 10^{-10} eV.

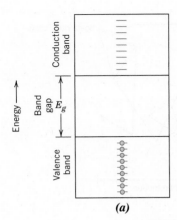
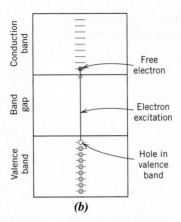

Energy ——→

(a)

(b)

Figure 19.6 For an insulator or semiconductor, occupancy of electron states (*a*) before and (*b*) after an electron excitation from the valence band into the conduction band, in which both a free electron and a hole are generated.

in the width of the band gap; for semiconductors, it is narrow, whereas for insulating materials, it is relatively wide.

Increasing the temperature of either a semiconductor or an insulator results in an increase in the thermal energy that is available for electron excitation. Thus, more electrons are promoted into the conduction band, which gives rise to an enhanced conductivity.

The conductivity of insulators and semiconductors may also be viewed from the perspective of atomic bonding models discussed in Section 2.6. For electrically insulating materials, interatomic bonding is ionic or strongly covalent. Thus, the valence electrons are tightly bound to or shared with the individual atoms. In other words, these electrons are highly localized and are not in any sense free to wander throughout the crystal. The bonding in semiconductors is covalent (or predominantly covalent) and relatively weak, which means that the valence electrons are not as strongly bound to the atoms. Consequently, these electrons are more easily removed by thermal excitation than they are for insulators.

19.7 ELECTRON MOBILITY

When an electric field is applied, a force is brought to bear on the free electrons; as a consequence, they all experience an acceleration in a direction opposite to that of the field, by virtue of their negative charge. According to quantum mechanics, there is no interaction between an accelerating electron and atoms in a perfect crystal lattice. Under such circumstances, all the free electrons should accelerate as long as the electric field is applied, which would give rise to an electric current that is continuously increasing with time. However, we know that a current reaches a constant value the instant that a field is applied, indicating that there exist what might be termed *frictional forces,* which counter this acceleration from the external field. These frictional forces result from the scattering of electrons by imperfections in the crystal lattice, including impurity atoms, vacancies, interstitial atoms, dislocations, and even the thermal vibrations of the atoms themselves. Each scattering event causes an electron to lose kinetic energy and to change its direction of motion, as represented schematically in Figure 19.7. There is, however, some net electron motion in the direction opposite to the field, and this flow of charge is the electric current.

The scattering phenomenon is manifested as a resistance to the passage of an electric current. Several parameters are used to describe the extent of this scattering; these include the *drift velocity* and the **mobility** of an electron. The drift velocity v_d represents the average electron velocity in the direction of the force imposed by the applied field. It is directly proportional to the electric field as follows:

mobility

Electron drift velocity— dependence on electron mobility and electric field intensity

$$v_d = \mu_e \mathscr{E} \tag{19.7}$$

The constant of proportionality μ_e is called the *electron mobility* and is an indication of the frequency of scattering events; its units are square meters per volt-second ($m^2/V \cdot s$).

Figure 19.7 Schematic diagram showing the path of an electron that is deflected by scattering events.

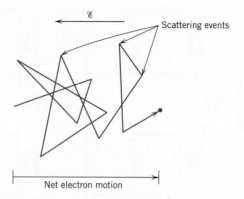

The conductivity σ of most materials may be expressed as

Electrical conductivity— dependence on electron concentration, charge, and mobility

$$\sigma = n|e|\mu_e \qquad (19.8)$$

where n is the number of free or conducting electrons per unit volume (e.g., per cubic meter) and $|e|$ is the absolute magnitude of the electrical charge on an electron (1.6×10^{-19} C). Thus, the electrical conductivity is proportional to both the number of free electrons and the electron mobility.

Concept Check 19.1   If a metallic material is cooled through its melting temperature at an extremely rapid rate, it forms a noncrystalline solid (i.e., a metallic glass). Will the electrical conductivity of the noncrystalline metal be greater or less than its crystalline counterpart? Why?

[*The answer may be found at* www.wiley.com/college/callister *(Student Companion Site).*]

19.8 ELECTRICAL RESISTIVITY OF METALS

As mentioned previously, most metals are extremely good conductors of electricity; room-temperature conductivities for several of the more common metals are given in Table 19.1. (Table B.9 in Appendix B lists the electrical resistivities of a large number of

Table 19.1

Room-Temperature Electrical Conductivities for Nine Common Metals and Alloys

Metal	*Electrical Conductivity* $[(\Omega \cdot m)^{-1}]$
Silver	6.8×10^7
Copper	6.0×10^7
Gold	4.3×10^7
Aluminum	3.8×10^7
Brass (70 Cu–30 Zn)	1.6×10^7
Iron	1.0×10^7
Platinum	0.94×10^7
Plain carbon steel	0.6×10^7
Stainless steel	0.2×10^7

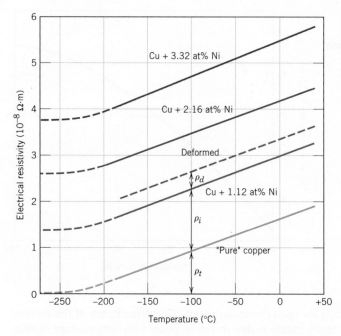

Figure 19.8 The electrical resistivity versus temperature for copper and three copper–nickel alloys, one of which has been deformed. Thermal, impurity, and deformation contributions to the resistivity are indicated at −100°C.
[Adapted from J. O. Linde, *Ann. Physik*, **5,** 219 (1932); and C. A. Wert and R. M. Thomson, *Physics of Solids,* 2nd edition, McGraw-Hill Book Company, New York, 1970.]

metals and alloys.) Again, metals have high conductivities because of the large numbers of free electrons that have been excited into empty states above the Fermi energy. Thus n has a large value in the conductivity expression, Equation 19.8.

At this point it is convenient to discuss conduction in metals in terms of the resistivity, the reciprocal of conductivity; the reason for this switch should become apparent in the ensuing discussion.

Because crystalline defects serve as scattering centers for conduction electrons in metals, increasing their number raises the resistivity (or lowers the conductivity). The concentration of these imperfections depends on temperature, composition, and the degree of cold work of a metal specimen. In fact, it has been observed experimentally that the total resistivity of a metal is the sum of the contributions from thermal vibrations, impurities, and plastic deformation—that is, the scattering mechanisms act independently of one another. This may be represented in mathematical form as follows:

Matthiessen's rule— for a metal, total electrical resistivity equals the sum of thermal, impurity, and deformation contributions

Matthiessen's rule

$$\rho_{total} = \rho_t + \rho_i + \rho_d \qquad (19.9)$$

in which ρ_t, ρ_i, and ρ_d represent the individual thermal, impurity, and deformation resistivity contributions, respectively. Equation 19.9 is sometimes known as **Matthiessen's rule.** The influence of each ρ variable on the total resistivity is demonstrated in Figure 19.8, which is a plot of resistivity versus temperature for copper and several copper–nickel alloys in annealed and deformed states. The additive nature of the individual resistivity contributions is demonstrated at −100°C.

Influence of Temperature

Dependence of thermal resistivity contribution on temperature

For the pure metal and all the copper–nickel alloys shown in Figure 19.8, the resistivity rises linearly with temperature above about −200°C. Thus,

$$\rho_t = \rho_0 + aT \qquad (19.10)$$

Figure 19.9 Room-temperature electrical resistivity versus composition for copper–nickel alloys.

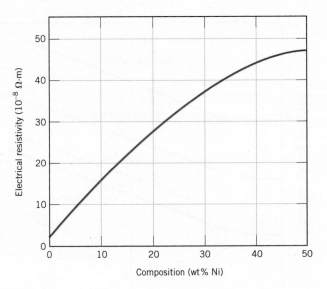

where ρ_0 and a are constants for each particular metal. This dependence of the thermal resistivity component on temperature is due to the increase with temperature in thermal vibrations and other lattice irregularities (e.g., vacancies), which serve as electron-scattering centers.

Influence of Impurities

For additions of a single impurity that forms a solid solution, the impurity resistivity ρ_i is related to the impurity concentration c_i in terms of the atom fraction (at%/100) as follows:

Impurity resistivity contribution (for solid solution)—dependence on impurity concentration (atom fraction)

$$\rho_i = Ac_i(1 - c_i) \tag{19.11}$$

where A is a composition-independent constant that is a function of both the impurity and host metals. The influence of nickel impurity additions on the room-temperature resistivity of copper is demonstrated in Figure 19.9, up to 50 wt% Ni; over this composition range nickel is completely soluble in copper (Figure 11.3a). Again, nickel atoms in copper act as scattering centers, and increasing the concentration of nickel in copper results in an enhancement of resistivity.

For a two-phase alloy consisting of α and β phases, a rule-of-mixtures expression may be used to approximate the resistivity as follows:

Impurity resistivity contribution (for two-phase alloy)—dependence on volume fractions and resistivities of two phases

$$\rho_i = \rho_\alpha V_\alpha + \rho_\beta V_\beta \tag{19.12}$$

where the Vs and ρs represent volume fractions and individual resistivities for the respective phases.

Influence of Plastic Deformation

Plastic deformation also raises the electrical resistivity as a result of increased numbers of electron-scattering dislocations. The effect of deformation on resistivity is also represented in Figure 19.8. Furthermore, its influence is much weaker than that of increasing temperature or the presence of impurities.

Concept Check 19.2   The room-temperature electrical resistivities of pure lead and pure tin are 2.06×10^{-7} and 1.11×10^{-7} $\Omega \cdot$m, respectively.

(a) Make a schematic graph of the room-temperature electrical resistivity versus composition for all compositions between pure lead and pure tin.

(b) On this same graph, schematically plot electrical resistivity versus composition at 150°C.

(c) Explain the shapes of these two curves as well as any differences between them.

Hint: You may want to consult the lead–tin phase diagram, Figure 11.7.

[*The answer may be found at* www.wiley.com/college/callister *(Student Companion Site).*]

19.9 ELECTRICAL CHARACTERISTICS OF COMMERCIAL ALLOYS

Electrical and other properties of copper render it the most widely used metallic conductor. Oxygen-free high-conductivity (OFHC) copper, having extremely low oxygen and other impurity contents, is produced for many electrical applications. Aluminum, having a conductivity only about one-half that of copper, is also frequently used as an electrical conductor. Silver has a higher conductivity than either copper or aluminum; however, its use is restricted on the basis of cost.

On occasion, it is necessary to improve the mechanical strength of a metal alloy without impairing significantly its electrical conductivity. Both solid-solution alloying (Section 9.9) and cold working (Section 9.10) improve strength at the expense of conductivity; thus, a trade-off must be made for these two properties. Most often, strength is enhanced by introducing a second phase that does not have so adverse an effect on conductivity. For example, copper–beryllium alloys are precipitation hardened (Section 17.7); even so, the conductivity is reduced by about a factor of 5 over that of high-purity copper.

For some applications, such as furnace heating elements, a high electrical resistivity is desirable. The energy loss by electrons that are scattered is dissipated as heat energy. Such materials must have not only a high resistivity, but also a resistance to oxidation at elevated temperatures and, of course, a high melting temperature. Nichrome, a nickel–chromium alloy, is commonly employed in heating elements.

MATERIALS OF IMPORTANCE

Aluminum Electrical Wires

Copper is normally used for electrical wiring in residential and commercial buildings. However, between 1965 and 1973, the price of copper increased significantly and, consequently, aluminum wiring was installed in many buildings constructed or remodeled during this period because aluminum was a less expensive electrical conductor. An inordinately high number of fires occurred in these buildings, and investigations revealed that the use of aluminum posed an increased fire hazard risk over copper wiring.

When properly installed, aluminum wiring can be just as safe as copper. These safety problems arose at connection points between the aluminum and copper; copper wiring was used for connection

(continued)

terminals on electrical equipment (circuit breakers, receptacles, switches, etc.) to which the aluminum wiring was attached.

As electrical circuits are turned on and off, the electrical wiring heats up and then cools down. This thermal cycling causes the wires to alternately expand and contract. The amounts of expansion and contraction for aluminum are greater than for copper—aluminum has a higher coefficient of thermal expansion than copper (Section 20.3).[3] Consequently, these differences in expansion and contraction between the aluminum and copper wires can cause the connections to loosen. Another factor that contributes to the loosening of copper–aluminum wire connections is creep (Section 10.12); mechanical stresses exist at these wire connections, and aluminum is more susceptible to creep deformation at or near room temperature than copper. This loosening of the connections compromises the electrical wire-to-wire contact, which increases the electrical resistance at the connection and leads to increased heating. Aluminum oxidizes more readily than copper, and this oxide coating further increases the electrical resistance at the connection. Ultimately, a connection may deteriorate to the point that electrical arcing and/ or heat buildup can ignite any combustible materials in the vicinity of the junction. Inasmuch as most receptacles, switches, and other connections are concealed,

these materials may smolder or a fire may spread undetected for an extended period of time.

Warning signs that suggest possible connection problems include warm faceplates on switches or receptacles, the smell of burning plastic in the vicinity of outlets or switches, lights that flicker or burn out quickly, unusual static on radio/television, and circuit breakers that trip for no apparent reason.

Several options are available for making buildings wired with aluminum safe.[4] The most obvious (and also most expensive) is to replace all of the aluminum wires with copper. The next-best option is to install a crimp connector repair unit at each aluminum–copper connection. With this technique, a piece of copper wire is attached to the existing aluminum wire branch using a specially designed metal sleeve and powered crimping tool; the metal sleeve is called a "COPALUM parallel splice connector." The crimping tool essentially makes a cold weld between the two wires. Finally, the connection is encased in an insulating sleeve. A schematic representation of a COPALUM device is shown in Figure 19.10. Only qualified and specially trained electricians are allowed to install these COPALUM connectors.

Two other less-desirable options are CO/ALR devices and pigtailing. A CO/ALR device is simply a switch or wall receptacle that is designed to be

Table 19.2 Compositions, Electrical Conductivities, and Coefficients of Thermal Expansion for Aluminum and Copper Alloys Used for Electrical Wiring

Alloy Name	*Alloy Designation*	*Composition* (*wt%*)	*Electrical Conductivity* [$(\Omega \cdot m)^{-1}$]	*Coefficient of Thermal Expansion* $(°C)^{-1}$
Aluminum (electrical conductor grade)	1350	99.50 Al, 0.10 Si, 0.05 Cu, 0.01 Mn, 0.01 Cr, 0.05 Zn, 0.03 Ga, 0.05 B	3.57×10^7	23.8×10^{-6}
Copper (electrolytic touch pitch)	C11000	99.90 Cu, 0.04 O	5.88×10^7	17.0×10^{-6}

[3] Coefficient of thermal expansion values, as well as compositions and other properties of the aluminum and copper alloys used for electrical wiring, are presented in Table 19.2.

[4] A discussion of the various repair options may be downloaded from the following website: http://www.cpsc.gov/cpscpub/pubs/516.pdf. (Accessed May 2013.)

used with aluminum wiring. For pigtailing, a twist-on connecting wire nut is used, which uses a grease that inhibits corrosion while maintaining a high electrical conductivity at the junction.

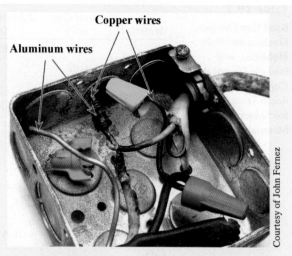

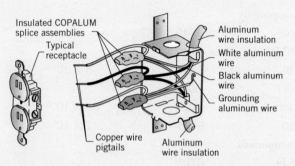

Figure 19.10 Schematic of a COPALUM connector device that is used in aluminum wire electrical circuits. (Reprinted by permission of the U.S. Consumer Product Safety Commission.)

Two copper wire–aluminum wire junctions (located in a junction box) that experienced excessive heating. The one on the right (within the yellow wire nut) failed completely.

Semiconductivity

intrinsic
   semiconductor

extrinsic
   semiconductor

The electrical conductivity of semiconducting materials is not as high as that of metals; nevertheless, they have some unique electrical characteristics that render them especially useful. The electrical properties of these materials are extremely sensitive to the presence of even minute concentrations of impurities. **Intrinsic semiconductors** are those in which the electrical behavior is based on the electronic structure inherent in the pure material. When the electrical characteristics are dictated by impurity atoms, the semiconductor is said to be **extrinsic**.

19.10 INTRINSIC SEMICONDUCTION

Intrinsic semiconductors are characterized by the electron band structure shown in Figure 19.4d: at 0 K, a completely filled valence band, separated from an empty conduction band by a relatively narrow forbidden band gap, generally less than 2 eV. The two elemental semiconductors are silicon (Si) and germanium (Ge), having band gap energies of approximately 1.1 and 0.7 eV, respectively. Both are found in Group IVA of the periodic table (Figure 2.8) and are covalently bonded.[5] In addition, a host of compound semiconducting materials also display intrinsic behavior. One such group is formed between elements of Groups IIIA and VA, for example, gallium arsenide (GaAs) and indium antimonide (InSb); these are frequently called III–V compounds. The compounds composed of elements of Groups IIB and VIA also display semiconducting behavior; these include cadmium sulfide (CdS) and zinc telluride (ZnTe). As the two elements forming these compounds become more widely separated with respect to their relative positions in the periodic table (i.e., the electronegativities become more dissimilar, Figure 2.9), the atomic bonding becomes more ionic and the magnitude of the band gap energy increases—the materials tend to become more insulative. Table 19.3 gives the band gaps for some compound semiconductors.

---

[5]The valence bands in silicon and germanium correspond to sp^3 hybrid energy levels for the isolated atom; these hybridized valence bands are completely filled at 0 K.

Table 19.3

Band Gap Energies, Electron and Hole Mobilities, and Intrinsic Electrical Conductivities at Room Temperature for Semiconducting Materials

Material	Band Gap (eV)	Electron Mobility (m^2/V·s)	Hole Mobility (m^2/V·s)	Electrical Conductivity (Intrinsic)(Ω·m)$^{-1}$
Elemental				
Ge	0.67	0.39	0.19	2.2
Si	1.11	0.145	0.050	3.4×10^{-4}
III–V Compounds				
AlP	2.42	0.006	0.045	—
AlSb	1.58	0.02	0.042	—
GaAs	1.42	0.80	0.04	3×10^{-7}
GaP	2.26	0.011	0.0075	—
InP	1.35	0.460	0.015	2.5×10^{-6}
InSb	0.17	8.00	0.125	2×10^{4}
II–VI Compounds				
CdS	2.40	0.040	0.005	—
CdTe	1.56	0.105	0.010	—
ZnS	3.66	0.060	—	—
ZnTe	2.4	0.053	0.010	—

Source: This material is reproduced with permission of John Wiley & Sons, Inc.

✓

Concept Check 19.3 Which of ZnS and CdSe has the larger band gap energy E_g? Cite reason(s) for your choice.

[*The answer may be found at* www.wiley.com/college/callister *(Student Companion Site).*]

Concept of a Hole

In intrinsic semiconductors, for every electron excited into the conduction band there is left behind a missing electron in one of the covalent bonds, or in the band scheme, a vacant electron state in the valence band, as shown in Figure 19.6b.[6] Under the influence of an electric field, the position of this missing electron within the crystalline lattice may be thought of as moving by the motion of other valence electrons that repeatedly fill in the incomplete bond (Figure 19.11). This process is expedited by treating a missing electron from the valence band as a positively charged particle called a *hole*. A hole is considered to have a charge that is of the same magnitude as that for an electron, but of opposite sign $(+1.6 \times 10^{-19}$ C). Thus, in the presence of an electric field, excited electrons and holes move in opposite directions. Furthermore, in semiconductors both electrons and holes are scattered by lattice imperfections.

Intrinsic Conductivity

Electrical conductivity for an intrinsic semiconductor— dependence on electron/hole concentrations and electron/hole mobilities

Because there are two types of charge carrier (free electrons and holes) in an intrinsic semiconductor, the expression for electrical conduction, Equation 19.8, must be modified to include a term to account for the contribution of the hole current. Therefore, we write

$$\sigma = n|e|\mu_e + p|e|\mu_h \tag{19.13}$$

---

[6]Holes (in addition to free electrons) are created in semiconductors and insulators when electron transitions occur from filled states in the valence band to empty states in the conduction band (Figure 19.6). In metals, electron transitions normally occur from empty to filled states *within the same band* (Figure 19.5), without the creation of holes.

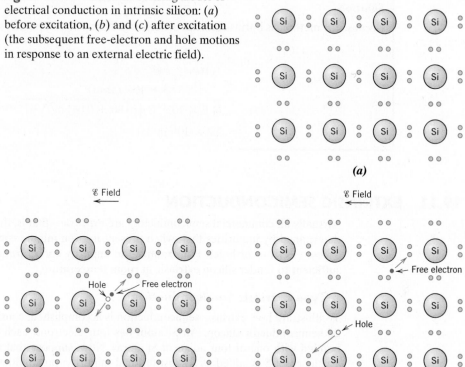

Figure 19.11 Electron-bonding model of electrical conduction in intrinsic silicon: (*a*) before excitation, (*b*) and (*c*) after excitation (the subsequent free-electron and hole motions in response to an external electric field).

where p is the number of holes per cubic meter and μ_h is the hole mobility. The magnitude of μ_h is always less than μ_e for semiconductors. For intrinsic semiconductors, every electron promoted across the band gap leaves behind a hole in the valence band; thus,

$$n = p = n_i \qquad (19.14)$$

where n_i is known as the *intrinsic carrier concentration*. Furthermore,

For an intrinsic semiconductor, conductivity in terms of intrinsic carrier concentration

$$\sigma = n|e|(\mu_e + \mu_h) = p|e|(\mu_e + \mu_h) \qquad (19.15)$$
$$= n_i|e|(\mu_e + \mu_h)$$

The room-temperature intrinsic conductivities and electron and hole mobilities for several semiconducting materials are also presented in Table 19.3.

EXAMPLE PROBLEM 19.1

Computation of the Room-Temperature Intrinsic Carrier Concentration for Gallium Arsenide

For intrinsic gallium arsenide, the room-temperature electrical conductivity is $3 \times 10^{-7}\,(\Omega \cdot \text{m})^{-1}$; the electron and hole mobilities are, respectively, 0.80 and 0.04 m²/V·s. Compute the intrinsic carrier concentration n_i at room temperature.

Solution

Because the material is intrinsic, carrier concentration may be computed using Equation 19.15 as

$$n_i = \frac{\sigma}{|e|(\mu_e + \mu_h)}$$

$$= \frac{3 \times 10^{-7}\,(\Omega \cdot m)^{-1}}{(1.6 \times 10^{-19}\,C)\big[(0.80 + 0.04)\,m^2/V \cdot s\big]}$$

$$= 2.2 \times 10^{12}\,m^{-3}$$

19.11 EXTRINSIC SEMICONDUCTION

Virtually all commercial semiconductors are *extrinsic*—that is, the electrical behavior is determined by impurities that, when present in even minute concentrations, introduce excess electrons or holes. For example, an impurity concentration of 1 atom in 10^{12} is sufficient to render silicon extrinsic at room temperature.

n-Type Extrinsic Semiconduction

To illustrate how extrinsic semiconduction is accomplished, consider again the elemental semiconductor silicon. An Si atom has four electrons, each of which is covalently bonded with one of four adjacent Si atoms. Now, suppose that an impurity atom with a valence of 5 is added as a substitutional impurity; possibilities would include atoms from the Group VA column of the periodic table (i.e., P, As, and Sb). Only four of five valence electrons of these impurity atoms can participate in the bonding because there are only four possible bonds with neighboring atoms. The extra nonbonding electron is loosely bound to the region around the impurity atom by a weak electrostatic attraction, as illustrated in Figure 19.12a. The binding energy of this electron is relatively small (on the order of 0.01 eV); thus, it is easily removed from the impurity atom, in which case it becomes a free or conducting electron (Figures 19.12b and 19.12c).

The energy state of such an electron may be viewed from the perspective of the electron band model scheme. For each of the loosely bound electrons, there exists a single energy level, or energy state, which is located within the forbidden band gap just below the bottom of the conduction band (Figure 19.13a). The electron binding energy corresponds to the energy required to excite the electron from one of these impurity states to a state within the conduction band. Each excitation event (Figure 19.13b) supplies or donates a single electron to the conduction band; an impurity of this type is aptly termed a *donor*. Because each donor electron is excited from an impurity level, no corresponding hole is created within the valence band.

At room temperature, the thermal energy available is sufficient to excite large numbers **donor state** of electrons from **donor states**; in addition, some intrinsic valence–conduction band transitions occur, as in Figure 19.6b, but to a negligible degree. Thus, the number of electrons in the conduction band far exceeds the number of holes in the valence band (or $n \gg p$), and the first term on the right-hand side of Equation 19.13 overwhelms the second—that is,

For an *n*-type extrinsic semiconductor, dependence of conductivity on concentration and mobility of electrons

$$\sigma \cong n|e|\mu_e \tag{19.16}$$

A material of this type is said to be an *n-type* extrinsic semiconductor. The electrons are *majority carriers* by virtue of their density or concentration; holes, on the other hand, are the *minority charge carriers*. For *n*-type semiconductors, the Fermi level is shifted upward in the band gap, to within the vicinity of the donor state; its exact position is a function of both temperature and donor concentration.

Figure 19.12 Extrinsic *n*-type semiconduction model (electron bonding). (*a*) An impurity atom such as phosphorus, having five valence electrons, may substitute for a silicon atom. This results in an extra bonding electron, which is bound to the impurity atom and orbits it. (*b*) Excitation to form a free electron. (*c*) The motion of this free electron in response to an electric field.

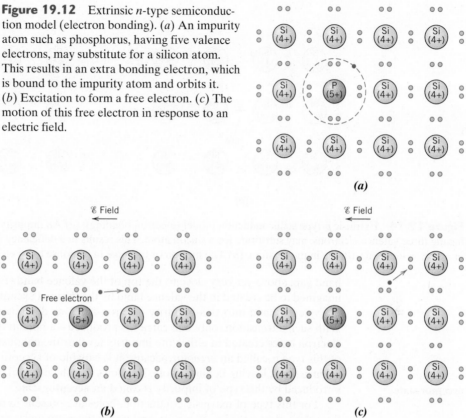

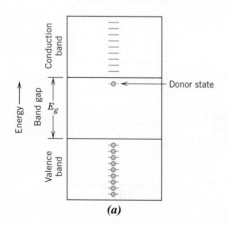 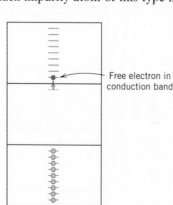

p-Type Extrinsic Semiconduction

An opposite effect is produced by the addition to silicon or germanium of trivalent substitutional impurities such as aluminum, boron, and gallium from Group IIIA of the periodic table. One of the covalent bonds around each of these atoms is deficient in an electron; such a deficiency may be viewed as a hole that is weakly bound to the impurity atom. This hole may be liberated from the impurity atom by the transfer of an electron from an adjacent bond, as illustrated in Figure 19.14. In essence, the electron and the hole exchange positions. A moving hole is considered to be in an excited state and participates in the conduction process, in a manner analogous to an excited donor electron, as described earlier.

Extrinsic excitations, in which holes are generated, may also be represented using the band model. Each impurity atom of this type introduces an energy level within the

Figure 19.13 (*a*) Electron energy band scheme for a donor impurity level located within the band gap and just below the bottom of the conduction band. (*b*) Excitation from a donor state in which a free electron is generated in the conduction band.

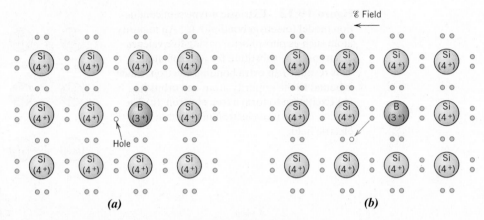

Figure 19.14 Extrinsic *p*-type semiconduction model (electron bonding). (*a*) An impurity atom such as boron, having three valence electrons, may substitute for a silicon atom. This results in a deficiency of one valence electron, or a hole associated with the impurity atom. (*b*) The motion of this hole in response to an electric field.

band gap, above yet very close to the top of the valence band (Figure 19.15*a*). A hole is imagined to be created in the valence band by the thermal excitation of an electron from the valence band into this impurity electron state, as demonstrated in Figure 19.15*b*. With such a transition, only one carrier is produced—a hole in the valence band; a free electron is *not* created in either the impurity level or the conduction band. An impurity of this type is called an *acceptor* because it is capable of accepting an electron from the valence band, leaving behind a hole. It follows that the energy level within the band gap introduced by this type of impurity is called an **acceptor state**.

acceptor state

For this type of extrinsic conduction, holes are present in much higher concentrations than electrons (i.e., $p >> n$), and under these circumstances a material is termed *p-type* because positively charged particles are primarily responsible for electrical conduction. Of course, holes are the majority carriers, and electrons are present in minority concentrations. This gives rise to a predominance of the second term on the right-hand side of Equation 19.13, or

For a *p*-type extrinsic semiconductor, dependence of conductivity on concentration and mobility of holes

$$\sigma \cong p|e|\mu_h \qquad (19.17)$$

For *p*-type semiconductors, the Fermi level is positioned within the band gap and near the acceptor level.

Extrinsic semiconductors (both *n*- and *p*-type) are produced from materials that are initially of extremely high purity, commonly having total impurity contents on the order of 10^{-7} at%. Controlled concentrations of specific donors or acceptors are then

Figure 19.15 (*a*) Energy band scheme for an acceptor impurity level located within the band gap and just above the top of the valence band. (*b*) Excitation of an electron into the acceptor level, leaving behind a hole in the valence band.

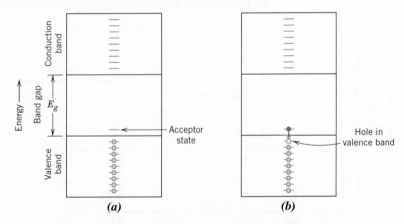

doping

intentionally added, using various techniques. Such an alloying process in semiconducting materials is termed **doping**.

In extrinsic semiconductors, large numbers of charge carriers (either electrons or holes, depending on the impurity type) are created at room temperature by the available thermal energy. As a consequence, relatively high room-temperature electrical conductivities are obtained in extrinsic semiconductors. Most of these materials are designed for use in electronic devices to be operated at ambient conditions.

Concept Check 19.4   At relatively high temperatures, both donor- and acceptor-doped semiconducting materials exhibit intrinsic behavior (Section 19.12). On the basis of discussions of Section 19.5 and this section, make a schematic plot of Fermi energy versus temperature for an *n*-type semiconductor up to a temperature at which it becomes intrinsic. Also note on this plot energy positions corresponding to the top of the valence band and the bottom of the conduction band.

Concept Check 19.5   Will Zn act as a donor or as an acceptor when added to the compound semiconductor GaAs? Why? (Assume that Zn is a substitutional impurity.)

[*The answers may be found at* www.wiley.com/college/callister *(Student Companion Site)*.]

19.12   THE TEMPERATURE DEPENDENCE OF CARRIER CONCENTRATION

Figure 19.16 plots the logarithm of the *intrinsic* carrier concentration n_i versus temperature for both silicon and germanium. A couple of features of this plot are worth noting. First, the concentrations of electrons and holes increase with temperature because, with rising temperature, more thermal energy is available to excite electrons from the valence to the conduction band (per Figure 19.6b). In addition, at all temperatures, carrier concentration in Ge is greater than in Si. This effect is due to germanium's smaller band gap (0.67 vs. 1.11 eV, Table 19.3); thus, for Ge, at any given temperature, more electrons will be excited across its band gap.

However, the carrier concentration–temperature behavior for an *extrinsic* semiconductor is much different. For example, electron concentration versus temperature for silicon that has been doped with 10^{21} m^{-3} phosphorus atoms is plotted in Figure 19.17. [For comparison, the dashed curve shown is for intrinsic Si (taken from Figure 19.16).][7] Noted on the extrinsic curve are three regions. At intermediate temperatures (between approximately 150 K and 475 K) the material is *n*-type (inasmuch as P is a donor impurity), and electron concentration is constant; this is termed the *extrinsic-temperature region*.[8] Electrons in the conduction band are excited from the phosphorus donor state (per Figure 19.13b), and because the electron concentration is approximately equal to the P content (10^{21} m^{-3}), virtually all of the phosphorus atoms have been ionized (i.e., have donated electrons). Also, intrinsic excitations across the band gap are insignificant in relation to these extrinsic donor excitations. The range of temperatures over which this extrinsic region exists depends on impurity concentration; furthermore, most solid-state devices are designed to operate within this temperature range.

At low temperatures, below about 100 K (Figure 19.17), electron concentration drops dramatically with decreasing temperature and approaches zero at 0 K. Over these temperatures, the thermal energy is insufficient to excite electrons from the P donor

---

[7]Note that the shapes of the Si curve of Figure 19.16 and the n_i curve of Figure 19.17 are not the same, even though identical parameters are plotted in both cases. This disparity is due to the scaling of the plot axes: temperature (i.e., horizontal) axes for both plots are scaled linearly; however, the carrier concentration axis of Figure 19.16 is logarithmic, whereas this same axis of Figure 19.17 is linear.

[8]For donor-doped semiconductors, this region is sometimes called the *saturation* region; for acceptor-doped materials, it is often termed the *exhaustion* region.

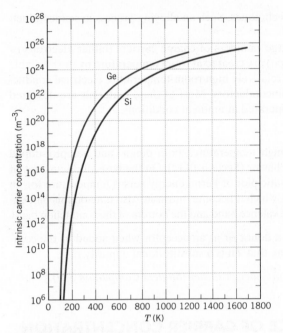

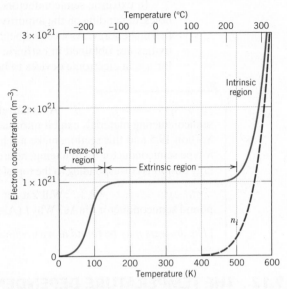

Figure 19.16 Intrinsic carrier concentration (logarithmic scale) as a function of temperature for germanium and silicon.
(From C. D. Thurmond, "The Standard Thermodynamic Functions for the Formation of Electrons and Holes in Ge, Si, GaAs, and GaP," *Journal of the Electrochemical Society,* **122**, [8], 1139 (1975). Reprinted by permission of The Electrochemical Society, Inc.)

Figure 19.17 Electron concentration versus temperature for silicon (*n*-type) that has been doped with 10^{21} m^{-3} of a donor impurity and for intrinsic silicon (dashed line). Freeze-out, extrinsic, and intrinsic temperature regimes are noted on this plot.
(From S. M. Sze, *Semiconductor Devices, Physics and Technology.* Copyright © 1985 by Bell Telephone Laboratories, Inc. Reprinted by permission of John Wiley & Sons, Inc.)

level into the conduction band. This is termed the *freeze-out temperature region* inasmuch as charged carriers (i.e., electrons) are "frozen" to the dopant atoms.

Finally, at the high end of the temperature scale of Figure 19.17, electron concentration increases above the P content and asymptotically approaches the intrinsic curve as temperature increases. This is termed the *intrinsic temperature region* because at these high temperatures the semiconductor becomes intrinsic—that is, charge carrier concentrations resulting from electron excitations across the band gap first become equal to and then completely overwhelm the donor carrier contribution with rising temperature.

Concept Check 19.6   On the basis of Figure 19.17, as dopant level is increased, would you expect the temperature at which a semiconductor becomes intrinsic to increase, to remain essentially the same, or to decrease? Why?

[*The answer may be found at* www.wiley.com/college/callister *(Student Companion Site).*]

19.13 FACTORS THAT AFFECT CARRIER MOBILITY

The conductivity (or resistivity) of a semiconducting material, in addition to being dependent on electron and/or hole concentrations, is also a function of the charge carriers' mobilities (Equation 19.13)—that is, the ease with which electrons and holes are transported through the crystal. Furthermore, magnitudes of electron and hole mobilities are influenced by the presence of those same crystalline defects that are responsible for the scattering of electrons in metals—thermal vibrations (i.e., temperature) and impurity atoms.

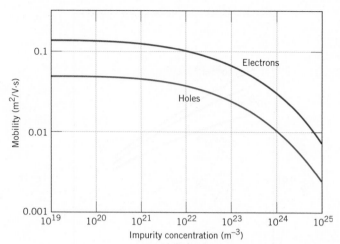

Figure 19.18 For silicon, dependence of room-temperature electron and hole mobilities (logarithmic scale) on dopant concentration (logarithmic scale).
(Adapted from W. W. Gärtner, "Temperature Dependence of Junction Transistor Parameters," *Proc. of the IRE,* **45,** 667, 1957. Copyright © 1957 IRE now IEEE.)

We now explore the manner in which dopant impurity content and temperature influence the mobilities of both electrons and holes.

Influence of Dopant Content

Figure 19.18 represents the room-temperature dependence of electron and hole mobilities in silicon as a function of the dopant (both acceptor and donor) content; note that both axes on this plot are scaled logarithmically. At dopant concentrations less than about 10^{20} m^{-3}, both carrier mobilities are at their maximum levels and independent of the doping concentration. In addition, both mobilities decrease with increasing impurity content. Also worth noting is that the mobility of electrons is always larger than the mobility of holes.

Influence of Temperature

The temperature dependences of electron and hole mobilities for silicon are presented in Figures 19.19*a* and 19.19*b*, respectively. Curves for several impurity dopant contents are shown for both carrier types; note that both sets of axes are scaled logarithmically. From these plots, note that, for dopant concentrations of 10^{24} m^{-3} and less, both electron and hole mobilities decrease in magnitude with rising temperature; again, this effect is due to enhanced thermal scattering of the carriers. For both electrons and holes and dopant levels less than 10^{20} m^{-3}, the dependence of mobility on temperature is independent of acceptor/donor concentration (i.e., is represented by a single curve). Also, for concentrations greater than 10^{20} m^{-3}, curves in both plots are shifted to progressively lower mobility values with increasing dopant level. These latter two effects are consistent with the data presented in Figure 19.18.

These previous treatments discussed the influence of temperature and dopant content on both carrier concentration and carrier mobility. Once values of n, p, μ_e, and μ_h have been determined for a specific donor/acceptor concentration and at a specified temperature (using Figures 19.16, through 19.19), computation of σ is possible using Equation 19.15, 19.16, or 19.17.

Concept Check 19.7 On the basis of the electron-concentration-versus-temperature curve for *n*-type silicon shown in Figure 19.17 and the dependence of the logarithm of electron mobility on temperature (Figure 19.19*a*), make a schematic plot of logarithm electrical conductivity versus temperature for silicon that has been doped with 10^{21} m^{-3} of a donor impurity. Now, briefly explain the shape of this curve. Recall that Equation 19.16 expresses the dependence of conductivity on electron concentration and electron mobility.

[*The answer may be found at* www.wiley.com/college/callister *(Student Companion Site).*]

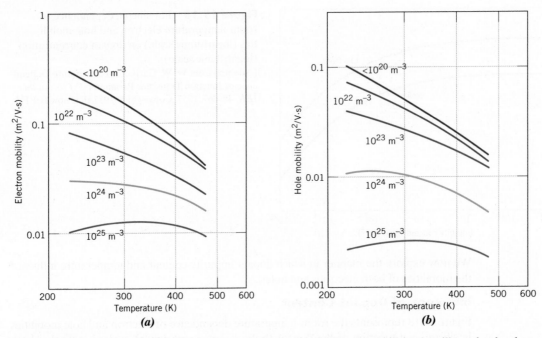

Figure 19.19 Temperature dependence of (a) electron and (b) hole mobilities for silicon that has been doped with various donor and acceptor concentrations. Both sets of axes are scaled logarithmically.
(From W. W. Gärtner, "Temperature Dependence of Junction Transistor Parameters," *Proc. of the IRE,* **45,** 667, 1957. Copyright © 1957 IRE now IEEE.)

EXAMPLE PROBLEM 19.2

Electrical Conductivity Determination for Intrinsic Silicon at 150°C

Calculate the electrical conductivity of intrinsic silicon at 150°C (423 K).

Solution

This problem may be solved using Equation 19.15, which requires specification of values for n_i, μ_e, and μ_h. From Figure 19.16, n_i for Si at 423 K is 4×10^{19} m^{-3}. Furthermore, intrinsic electron and hole mobilities are taken from the $<10^{20}$ m^{-3} curves of Figures 19.19a and 19.19b, respectively; at 423 K, $\mu_e = 0.06$ $m^2/V{\cdot}s$ and $\mu_h = 0.022$ $m^2/V{\cdot}s$ (realizing that both mobility and temperature axes are scaled logarithmically). Finally, from Equation 19.15, the conductivity is given by

$$\sigma = n_i|e|(\mu_e + \mu_h)$$
$$= (4 \times 10^{19} \text{ m}^{-3})(1.6 \times 10^{-19} \text{ C})(0.06 \text{ m}^2/\text{V}{\cdot}\text{s} + 0.022 \text{ m}^2/\text{V}{\cdot}\text{s})$$
$$= 0.52 \text{ } (\Omega{\cdot}\text{m})^{-1}$$

EXAMPLE PROBLEM 19.3

Room-Temperature and Elevated-Temperature Electrical Conductivity Calculations for Extrinsic Silicon

To high-purity silicon is added 10^{23} m^{-3} arsenic atoms.

(a) Is this material *n*-type or *p*-type?

(b) Calculate the room-temperature electrical conductivity of this material.

(c) Compute the conductivity at 100°C (373 K).

Solution

(a) Arsenic is a Group VA element (Figure 2.8) and, therefore, acts as a donor in silicon, which means that this material is *n*-type.

(b) At room temperature (298 K), we are within the extrinsic temperature region of Figure 19.17, which means that virtually all of the arsenic atoms have donated electrons (i.e., $n = 10^{23}$ m^{-3}). Furthermore, inasmuch as this material is extrinsic *n*-type, conductivity may be computed using Equation 19.16. Consequently, it is necessary to determine the electron mobility for a donor concentration of 10^{23} m^{-3}. We can do this using Figure 19.18: at 10^{23} m^{-3}, $\mu_e = 0.07$ m^2/V·s (remember that both axes of Figure 19.18 are scaled logarithmically). Thus, the conductivity is just

$$\sigma = n|e|\mu_e$$
$$= (10^{23} \text{ m}^{-3})(1.6 \times 10^{-19} \text{ C})(0.07 \text{ m}^2/\text{V·s})$$
$$= 1120 \ (\Omega\text{·m})^{-1}$$

(c) To solve for the conductivity of this material at 373 K, we again use Equation 19.16 with the electron mobility at this temperature. From the 10^{23} m^{-3} curve of Figure 19.19a, at 373 K, $\mu_e = 0.04$ m^2/V·s, which leads to

$$\sigma = n|e|\mu_e$$
$$= (10^{23} \text{ m}^{-3})(1.6 \times 10^{-19} \text{ C})(0.04 \text{ m}^2/\text{V·s})$$
$$= 640 \ (\Omega\text{·m})^{-1}$$

DESIGN EXAMPLE 19.1

Acceptor Impurity Doping in Silicon

An extrinsic *p*-type silicon material is desired having a room-temperature conductivity of 50 $(\Omega\text{·m})^{-1}$. Specify an acceptor impurity type that may be used, as well as its concentration in atom percent, to yield these electrical characteristics.

Solution

First, the elements that, when added to silicon, render it *p*-type lie one group to the left of silicon in the periodic table. These include the Group IIIA elements (Figure 2.8): boron, aluminum, gallium, and indium.

Because this material is extrinsic and *p*-type (i.e., $p \gg n$), the electrical conductivity is a function of both hole concentration and hole mobility according to Equation 19.17. In addition, it is assumed that at room temperature, all the acceptor dopant atoms have accepted electrons to form holes (i.e., that we are in the *extrinsic region* of Figure 19.17), which is to say that the number of holes is approximately equal to the number of acceptor impurities N_a.

This problem is complicated by the fact that μ_h is dependent on impurity content per Figure 19.18. Consequently, one approach to solving this problem is trial and error: assume an impurity concentration, and then compute the conductivity using this value and the corresponding hole mobility from its curve of Figure 19.18. Then, on the basis of this result, repeat the process, assuming another impurity concentration.

For example, let us select an N_a value (i.e., a *p* value) of 10^{22} m^{-3}. At this concentration, the hole mobility is approximately 0.04 m^2/V·s (Figure 19.18); these values yield a conductivity of

$$\sigma = p|e|\mu_h = (10^{22} \text{ m}^{-3})(1.6 \times 10^{-19} \text{ C})(0.04 \text{ m}^2/\text{V·s})$$
$$= 64 \ (\Omega\text{·m})^{-1}$$

which is a little on the high side. Decreasing the impurity content an order of magnitude to 10^{21} m^{-3} results in only a slight increase of μ_h to about 0.045 m^2/V·s (Figure 19.18); thus, the resulting conductivity is

$$\sigma = (10^{21}\text{ m}^{-3})(1.6 \times 10^{-19}\text{ C})(0.045\text{ m}^2/\text{V·s})$$
$$= 7.2\ (\Omega\text{·m})^{-1}$$

With some fine tuning of these numbers, a conductivity of 50 $(\Omega\text{·m})^{-1}$ is achieved when $N_a = p \cong 8 \times 10^{21}$ m^{-3}; at this N_a value, μ_h remains approximately 0.04 m^2/V·s.

It next becomes necessary to calculate the concentration of acceptor impurity in atom percent. This computation first requires the determination of the number of silicon atoms per cubic meter, N_{Si}, using Equation 6.2, which is given as follows:

$$N_{Si} = \frac{N_A \rho_{Si}}{A_{Si}}$$

$$= \frac{(6.022 \times 10^{23}\text{ atoms/mol})(2.33\text{ g/cm}^3)(10^6\text{ cm}^3/\text{m}^3)}{28.09\text{ g/mol}}$$

$$= 5 \times 10^{28}\text{ m}^{-3}$$

The concentration of acceptor impurities in atom percent (C'_a) is just the ratio of N_a and $N_a + N_{Si}$ multiplied by 100, or

$$C'_a = \frac{N_a}{N_a + N_{Si}} \times 100$$

$$= \frac{8 \times 10^{21}\text{ m}^{-3}}{(8 \times 10^{21}\text{ m}^{-3}) + (5 \times 10^{28}\text{ m}^{-3})} \times 100 = 1.60 \times 10^{-5}$$

Thus, a silicon material having a room-temperature p-type electrical conductivity of 50 $(\Omega\text{·m})^{-1}$ must contain 1.60×10^{-5} at% boron, aluminum, gallium, or indium.

19.14 THE HALL EFFECT

Hall effect

For some materials, it is occasionally desired to determine the material's majority charge carrier type, concentration, and mobility. Such determinations are not possible from a simple electrical conductivity measurement; a **Hall effect** experiment must also be conducted. This Hall effect is a result of the phenomenon by which a magnetic field applied perpendicular to the direction of motion of a charged particle exerts a force on the particle perpendicular to both the magnetic field and the particle motion directions.

To demonstrate the Hall effect, consider the specimen geometry shown in Figure 19.20—a parallelepiped specimen having one corner situated at the origin of a Cartesian coordinate system. In response to an externally applied electric field, the electrons and/or holes move in the x direction and give rise to a current I_x. When a magnetic field is imposed in the positive z direction (denoted as B_z), the resulting force on the charge carriers causes them to be deflected in the y direction—holes (positively charged carriers) to the right specimen face and electrons (negatively charged carriers) to the left face, as indicated in the figure. Thus, a voltage, termed the *Hall voltage* V_H, is established in

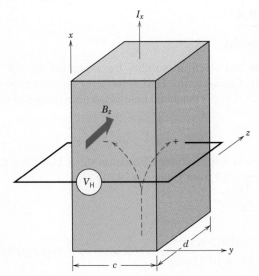

Figure 19.20 Schematic demonstration of the Hall effect. Positive and/or negative charge carriers that are part of the I_x current are deflected by the magnetic field B_z and give rise to the Hall voltage, V_H.

the y direction. The magnitude of V_H depends on I_x, B_z, and the specimen thickness d as follows:

Dependence of Hall voltage on the Hall coefficient, specimen thickness, and current and magnetic field parameters shown in Figure 19.20

$$V_H = \frac{R_H I_x B_z}{d} \qquad (19.18)$$

In this expression R_H is termed the *Hall coefficient,* which is a constant for a given material. For metals, in which conduction is by electrons, R_H is negative and is given by

Hall coefficient for metals

$$R_H = \frac{1}{n|e|} \qquad (19.19)$$

Thus, n may be determined because R_H may be found using Equation 19.18 and the magnitude of e, the charge on an electron, is known.

Furthermore, from Equation 19.8, the electron mobility μ_e is just

$$\mu_e = \frac{\sigma}{n|e|} \qquad (19.20a)$$

or, using Equation 19.19,

For metals, electron mobility in terms of the Hall coefficient and conductivity

$$\mu_e = |R_H|\sigma \qquad (19.20b)$$

Thus, the magnitude of μ_e may also be determined if the conductivity σ has also been measured.

For semiconducting materials, the determination of majority carrier type and computation of carrier concentration and mobility are more complicated and are not discussed here.

EXAMPLE PROBLEM 19.4

Hall Voltage Computation

The electrical conductivity and electron mobility for aluminum are 3.8×10^7 $(\Omega \cdot m)^{-1}$ and 0.0012 $m^2/V \cdot s$, respectively. Calculate the Hall voltage for an aluminum specimen that is 15 mm thick for a current of 25 A and a magnetic field of 0.6 tesla (imposed in a direction perpendicular to the current).

Solution

The Hall voltage V_H may be determined using Equation 19.18. However, it is first necessary to compute the Hall coefficient (R_H) from Equation 19.20b as

$$R_H = -\frac{\mu_e}{\sigma}$$

$$= -\frac{0.0012 \ m^2/V \cdot s}{3.8 \times 10^7 \ (\Omega \cdot m)^{-1}} = -3.16 \times 10^{-11} \ V \cdot m/A \cdot tesla$$

Now, use of Equation 19.18 leads to

$$V_H = \frac{R_H I_x B_z}{d}$$

$$= \frac{(-3.16 \times 10^{-11} \ V \cdot m/A \cdot tesla)(25 \ A)(0.6 \ tesla)}{15 \times 10^{-3} \ m}$$

$$= -3.16 \times 10^{-8} \ V$$

19.15 SEMICONDUCTOR DEVICES

The unique electrical properties of semiconductors permit their use in devices to perform specific electronic functions. Diodes and transistors, which have replaced old-fashioned vacuum tubes, are two familiar examples. Advantages of semiconductor devices (sometimes termed *solid-state devices*) include small size, low power consumption, and no warmup time. Vast numbers of extremely small circuits, each consisting of numerous electronic devices, may be incorporated onto a small silicon chip. The invention of semiconductor devices, which has given rise to miniaturized circuitry, is responsible for the advent and extremely rapid growth of a host of new industries in the past few years.

The *p–n* Rectifying Junction

diode

rectifying junction

A rectifier, or **diode**, is an electronic device that allows the current to flow in one direction only; for example, a rectifier transforms an alternating current into direct current. Before the advent of the *p–n* junction semiconductor rectifier, this operation was carried out using the vacuum tube diode. The *p–n* **rectifying junction** is constructed from a single piece of semiconductor that is doped so as to be *n*-type on one side and *p*-type on the other (Figure 19.21a). If pieces of *n*- and *p*-type materials are joined together, a poor rectifier results because the presence of a surface between the two sections renders the device very inefficient. Also, single crystals of semiconducting materials must be used in all devices because electronic phenomena deleterious to operation occur at grain boundaries.

Before the application of any potential across the *p–n* specimen, holes are the dominant carriers on the *p*-side, and electrons predominate in the *n*-region, as illustrated in Figure 19.21a. An external electric potential may be established across a *p–n*

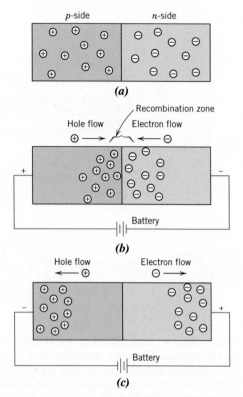

Figure 19.21 For a *p–n* rectifying junction, representations of electron and hole distributions for (*a*) no electrical potential, (*b*) forward bias, and (*c*) reverse bias.

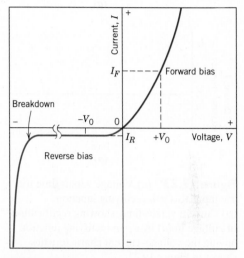

Figure 19.22 The current–voltage characteristics of a *p–n* junction for forward and reverse biases. The phenomenon of breakdown is also shown.

forward bias

reverse bias

junction with two different polarities. When a battery is used, the positive terminal may be connected to the *p*-side and the negative terminal to the *n*-side; this is referred to as a **forward bias**. The opposite polarity (minus to *p* and plus to *n*) is termed **reverse bias**.

The response of the charge carriers to the application of a forward-biased potential is demonstrated in Figure 19.21*b*. The holes on the *p*-side and the electrons on the *n*-side are attracted to the junction. As electrons and holes encounter one another near the junction, they continuously recombine and annihilate one another, according to

$$\text{electron} + \text{hole} \longrightarrow \text{energy} \tag{19.21}$$

Thus for this bias, large numbers of charge carriers flow across the semiconductor and to the junction, as evidenced by an appreciable current and a low resistivity. The current–voltage characteristics for forward bias are shown on the right-hand half of Figure 19.22.

For reverse bias (Figure 19.21*c*), both holes and electrons, as majority carriers, are rapidly drawn away from the junction; this separation of positive and negative charges (or polarization) leaves the junction region relatively free of mobile charge carriers. Recombination does not occur to any appreciable extent, so that the junction is now highly insulative. Figure 19.22 also illustrates the current–voltage behavior for reverse bias.

The rectification process in terms of input voltage and output current is demonstrated in Figure 19.23. Whereas voltage varies sinusoidally with time (Figure 19.23*a*), maximum current flow for reverse bias voltage I_R is extremely small in comparison to

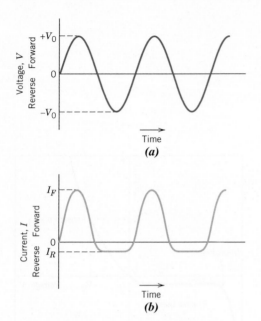

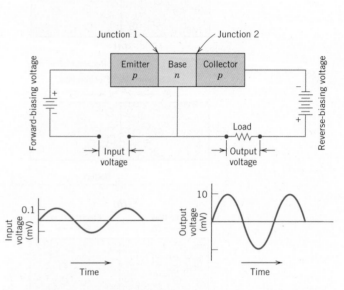

Figure 19.23 (*a*) Voltage versus time for the input to a *p–n* rectifying junction. (*b*) Current versus time, showing rectification of voltage in (*a*) by a *p–n* rectifying junction having the voltage–current characteristics shown in Figure 19.22.

Figure 19.24 Schematic diagram of a *p–n–p* junction transistor and its associated circuitry, including input and output voltage–time characteristics showing voltage amplification. (Adapted from A. G. Guy, *Essentials of Materials Science,* McGraw-Hill Book Company, New York, 1976.)

that for forward bias I_F (Figure 19.23*b*). Furthermore, correspondence between I_F and I_R and the imposed maximum voltage ($\pm V_0$) is noted in Figure 19.22.

At high reverse bias voltages—sometimes on the order of several hundred volts— large numbers of charge carriers (electrons and holes) are generated. This gives rise to a very abrupt increase in current, a phenomenon known as *breakdown,* also shown in Figure 19.22; this is discussed in more detail in Section 19.22.

The Transistor

Transistors, which are extremely important semiconducting devices in today's microelectronic circuitry, are capable of two primary types of function. First, they can perform the same operation as their vacuum-tube precursor, the triode—that is, they can amplify an electrical signal. In addition, they serve as switching devices in computers for the processing and storage of information. The two major types are the **junction** (or bimodal) **transistor** and the *metal-oxide-semiconductor field-effect transistor* (abbreviated as **MOSFET**).

junction transistor

MOSFET

Junction Transistors

The junction transistor is composed of two *p–n* junctions arranged back to back in either the *n–p–n* or the *p–n–p* configuration; the latter variety is discussed here. Figure 19.24 is a schematic representation of a *p–n–p* junction transistor along with its attendant circuitry. A very thin *n*-type *base* region is sandwiched between *p*-type *emitter* and *collector* regions. The circuit that includes the emitter–base junction (junction 1) is forward biased, whereas a reverse bias voltage is applied across the base–collector junction (junction 2).

Figure 19.25 illustrates the mechanics of operation in terms of the motion of charge carriers. Because the emitter is *p*-type and junction 1 is forward biased, large numbers of holes enter the base region. These injected holes are minority carriers in the *n*-type base,

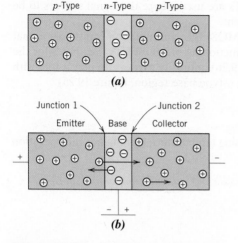

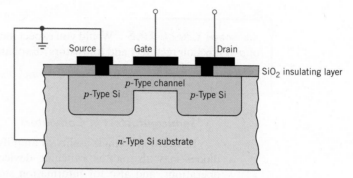

Figure 19.25 For a junction transistor (*p–n–p* type), the distributions and directions of electron and hole motion (*a*) when no potential is applied and (*b*) with appropriate bias for voltage amplification.

Figure 19.26 Schematic cross-sectional view of a MOSFET transistor.

and some combine with the majority electrons. However, if the base is extremely narrow and the semiconducting materials have been properly prepared, most of these holes will be swept through the base without recombination, then across junction 2 and into the *p*-type collector. The holes now become a part of the emitter–collector circuit. A small increase in input voltage within the emitter–base circuit produces a large increase in current across junction 2. This large increase in collector current is also reflected by a large increase in voltage across the load resistor, which is also shown in the circuit (Figure 19.24). Thus, a voltage signal that passes through a junction transistor experiences amplification; this effect is also illustrated in Figure 19.24 by the two voltage–time plots.

Similar reasoning applies to the operation of an *n–p–n* transistor, except that electrons instead of holes are injected across the base and into the collector.

The MOSFET

One variety of MOSFET[9] consists of two small islands of *p*-type semiconductor that are created within a substrate of *n*-type silicon, as shown in cross section in Figure 19.26; the islands are joined by a narrow *p*-type channel. Appropriate metal connections (source and drain) are made to these islands; an insulating layer of silicon dioxide is formed by the surface oxidation of the silicon. A final connector (gate) is then fashioned onto the surface of this insulating layer.

The conductivity of the channel is varied by the presence of an electric field imposed on the gate. For example, imposition of a positive field on the gate drives charge carriers (in this case holes) out of the channel, thereby reducing the electrical conductivity. Thus, a small alteration in the field at the gate produces a relatively large variation in current between the source and the drain. In some respects, then, the operation of a MOSFET is very similar to that described for the junction transistor. The primary difference is that the gate current is exceedingly small in comparison to the base current

---

[9]The MOSFET described here is a *depletion-mode p-type*. A *depletion-mode n-type* is also possible, wherein the *n*- and *p*-regions of Figure 19.26 are reversed.

of a junction transistor. Therefore, MOSFETs are used where the signal sources to be amplified cannot sustain an appreciable current.

Another important difference between MOSFETs and junction transistors is that although majority carriers dominate in the functioning of MOSFETs (i.e., holes for the depletion-mode *p*-type MOSFET of Figure 19.26), minority carriers do play a role with junction transistors (i.e., injected holes in the *n*-type base region, Figure 19.25).

Concept Check 19.8 Would you expect increasing temperature to influence the operation of *p–n* junction rectifiers and transistors? Explain.

[*The answer may be found at* www.wiley.com/college/callister *(Student Companion Site).*]

Semiconductors in Computers

In addition to their ability to amplify an imposed electrical signal, transistors and diodes may also act as switching devices, a feature used for arithmetic and logical operations, and also for information storage in computers. Computer numbers and functions are expressed in terms of a binary code (i.e., numbers written to the base 2). Within this framework, numbers are represented by a series of two states (sometimes designated 0 and 1). Now, transistors and diodes within a digital circuit operate as switches that also have two states—on and off, or conducting and nonconducting; "off" corresponds to one binary number state and "on" to the other. Thus, a single number may be represented by a collection of circuit elements containing transistors that are appropriately switched.

Flash (Solid-State Drive) Memory

A relatively new and rapidly evolving information storage technology that uses semiconductor devices is *flash memory*. Flash memory is programmed and erased electronically, as described in the preceding paragraph. Furthermore, this flash technology is *nonvolatile*—that is, no electrical power is needed to retain the stored information. There are no moving parts (as with magnetic hard drives and magnetic tapes; Section 21.11), which makes flash memory especially attractive for general storage and transfer of data between portable devices, such as digital cameras, laptop computers, mobile phones, digital audio players, and game consoles. In addition, flash technology is packaged as memory cards [see chapter-opening figures (*b*) and (*c*)], solid-state drives, and USB flash drives. Unlike magnetic memory, flash packages are extremely durable and are capable of withstanding relatively wide temperature extremes, as well as immersion in water. Furthermore, over time and the evolution of this flash-memory technology, storage capacity will continue to increase, physical chip size will decrease, and memory price will fall.

The mechanism of flash memory operation is relatively complicated and beyond the scope of this discussion. In essence, information is stored on a chip composed of a very large number of memory cells. Each cell consists of an array of transistors similar to the MOSFETs described earlier in this chapter; the primary difference is that flash-memory transistors have two gates instead of just one as for the MOSFETs (Figure 19.26). Flash memory is a special type of *e*lectronically *e*rasable, *p*rogrammable, *read-only m*emory (EEPROM). Data erasure is very rapid for entire blocks of cells, which makes this type of memory ideal for applications requiring frequent updates of large quantities of data (as with the applications noted in the preceding paragraph). Erasure leads to a clearing of cell contents so that it can be rewritten; this occurs by a change in electronic charge at one of the gates, which takes place very rapidly—that is, in a "flash"—hence the name.

Microelectronic Circuitry

During the past few years, the advent of microelectronic circuitry, in which millions of electronic components and circuits are incorporated into a very small space, has revolutionized the field of electronics. This revolution was precipitated, in part, by aerospace technology, which needed computers and electronic devices that were small and had low power requirements. As a result of refinement in processing and fabrication techniques, there has been an astonishing depreciation in the cost of integrated circuitry. Consequently, personal computers have become affordable to large segments of the population in many countries. Also, the use of **integrated circuits** has become infused into many other facets of our lives—calculators, communications, watches, industrial production and control, and all phases of the electronics industry.

integrated circuit

Inexpensive microelectronic circuits are mass produced by using some very ingenious fabrication techniques. The process begins with the growth of relatively large cylindrical single crystals of high-purity silicon from which thin circular wafers are cut. Many microelectronic or integrated circuits, sometimes called *chips,* are prepared on a single wafer. A chip is rectangular, typically on the order of 6 mm on a side, and contains millions of circuit elements: diodes, transistors, resistors, and capacitors. Enlarged photographs and elemental maps of a microprocessor chip are presented in Figure 19.27;

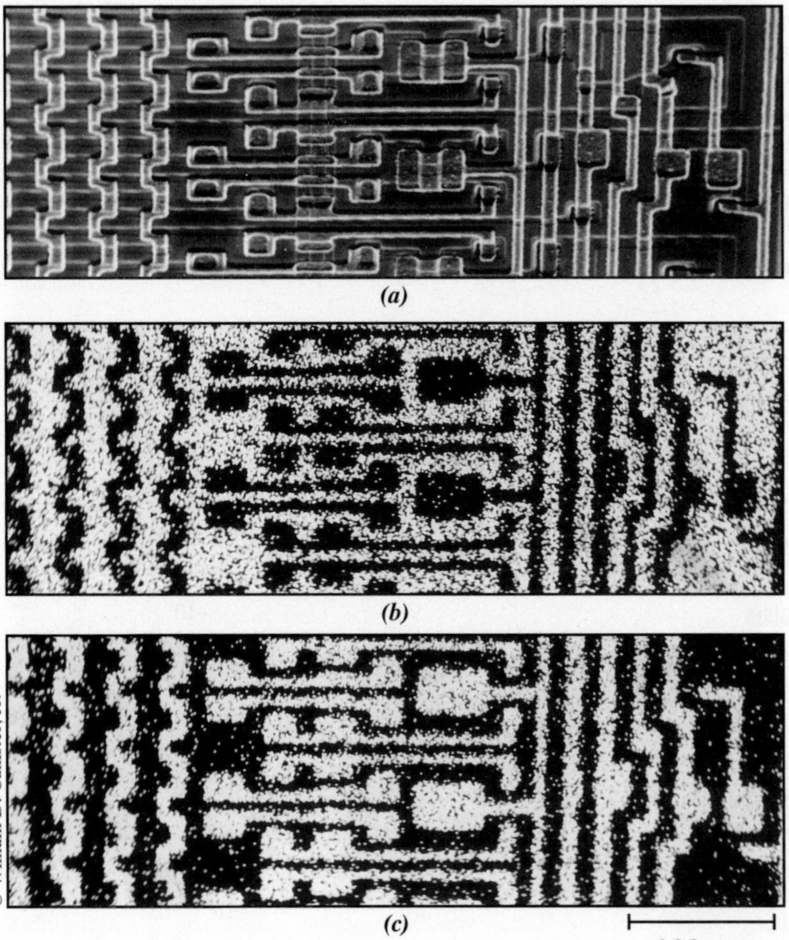

(a)

(b)

(c)

100 μm

© William D. Callister, Jr.

Figure 19.27 (*a*) Scanning electron micrograph of an integrated circuit.
(*b*) A silicon dot map for the integrated circuit above, showing regions where silicon atoms are concentrated. Doped silicon is the semiconducting material from which integrated circuit elements are made.
(*c*) An aluminum dot map. Metallic aluminum is an electrical conductor and, as such, wires the circuit elements together. Approximately 200×.
Note: The discussion of Section 6.12 mentioned that an image is generated on a scanning electron micrograph as a beam of electrons scans the surface of the specimen being examined. The electrons in this beam cause some of the specimen surface atoms to emit x-rays; the energy of an x-ray photon depends on the particular atom from which it radiates. It is possible to selectively filter out all but the x-rays emitted from one kind of atom. When projected on a cathode ray tube, small white dots are produced that indicate the locations of the particular atom type; thus, a *dot map* of the image is generated.

these micrographs reveal the intricacy of integrated circuits. At this time, microprocessor chips with densities approaching 1 billion transistors are being produced, and this number doubles about every 18 months.

Microelectronic circuits consist of many layers that lie within or are stacked on top of the silicon wafer in a precisely detailed pattern. Using photolithographic techniques, very small elements for each layer are masked in accordance with a microscopic pattern. Circuit elements are constructed by the selective introduction of specific materials [by diffusion (Section 7.6) or ion implantation] into unmasked regions to create localized n-type, p-type, high-resistivity, or conductive areas. This procedure is repeated layer by layer until the total integrated circuit has been fabricated, as illustrated in the MOSFET schematic (Figure 19.26). Elements of integrated circuits are shown in Figure 19.27 and in chapter-opening photograph (a).

Electrical Conduction in Ionic Ceramics and in Polymers

Most polymers and ionic ceramics are insulating materials at room temperature and, therefore, have electron energy band structures similar to that represented in Figure 19.4c; a filled valence band is separated from an empty conduction band by a relatively large band gap, usually greater than 2 eV. Thus, at normal temperatures, only very few electrons may be excited across the band gap by the available thermal energy, which accounts for the very small values of conductivity; Table 19.4 gives the room-temperature electrical conductivities of several of these materials. (The electrical resistivities of a large number of ceramic and polymeric materials are provided in Table B.9, Appendix B.) Many materials are used on the basis of their ability to insulate, and thus a high electrical resistivity is desirable. With rising temperature, insulating materials experience an increase in electrical conductivity, which may ultimately be greater than that for semiconductors.

Table 19.4

Typical Room-Temperature Electrical Conductivities for Thirteen Nonmetallic Materials

Material	Electrical Conductivity $[(\Omega \cdot m)^{-1}]$
Graphite	3×10^4–2×10^5
Ceramics	
Concrete (dry)	10^{-9}
Soda–lime glass	10^{-10}–10^{-11}
Porcelain	10^{-10}–10^{-12}
Borosilicate glass	$\sim 10^{-13}$
Aluminum oxide	$< 10^{-13}$
Fused silica	$< 10^{-18}$
Polymers	
Phenol-formaldehyde	10^{-9}–10^{-10}
Poly(methyl methacrylate)	$< 10^{-12}$
Nylon 6,6	10^{-12}–10^{-13}
Polystyrene	$< 10^{-14}$
Polyethylene	10^{-15}–10^{-17}
Polytetrafluoroethylene	$< 10^{-17}$

19.16 CONDUCTION IN IONIC MATERIALS

Both cations and anions in ionic materials possess an electric charge and, as a consequence, are capable of migration or diffusion when an electric field is present. Thus, an electric current results from the net movement of these charged ions, which are present in addition to current due to any electron motion. Anion and cation migrations are in opposite directions. The total conductivity of an ionic material σ_{total} is thus equal to the sum of electronic and ionic contributions, as follows:

For ionic materials, conductivity is equal to the sum of electronic and ionic contributions

$$\sigma_{total} = \sigma_{electronic} + \sigma_{ionic} \tag{19.22}$$

Either contribution may predominate, depending on the material, its purity, and temperature. A mobility μ_I may be associated with each of the ionic species as follows:

Computation of mobility for an ionic species

$$\mu_I = \frac{n_I e D_I}{kT} \tag{19.23}$$

where n_I and D_I represent, respectively, the valence and diffusion coefficient of a particular ion; e, k, and T denote the same parameters as explained earlier in the chapter. Thus, the ionic contribution to the total conductivity increases with increasing temperature, as does the electronic component. However, in spite of the two conductivity contributions, most ionic materials remain insulative, even at elevated temperatures.

19.17 ELECTRICAL PROPERTIES OF POLYMERS

Most polymeric materials are poor conductors of electricity (Table 19.4) due to the unavailability of large numbers of free electrons to participate in the conduction process; electrons in polymers are tightly bound in covalent bonds. The mechanism of electrical conduction in these materials is not well understood, but it is believed that conduction in polymers of high purity is electronic.

Conducting Polymers

Polymeric materials have been synthesized that have electrical conductivities on par with those of metallic conductors; they are appropriately termed *conducting polymers*. Conductivities as high as 1.5×10^7 $(\Omega \cdot m)^{-1}$ have been achieved in these materials; on a volume basis, this value corresponds to one-fourth of the conductivity of copper, or twice its conductivity on the basis of weight.

This phenomenon is observed in a dozen or so polymers, including polyacetylene, polyparaphenylene, polypyrrole, and polyaniline. Each of these polymers contains a system of alternating single and double bonds and/or aromatic units in the polymer chain. For example, the chain structure of polyacetylene is as follows:

The valence electrons associated with the alternating single and double chain-bonds are delocalized, which means they are shared among the backbone atoms in the polymer chain—similar to the way that electrons in a partially filled band for a metal are shared by the ion cores. In addition, the band structure of a conductive polymer is characteristic of that for an electrical insulator (Figure 19.4c)—at 0 K, a filled valence band separated from an empty conduction band by a forbidden energy band gap. In their pure forms, these polymers, which typically have band gap energies greater than 2 eV, are semiconductors or insulators. However, they become conductive when doped with appropriate impurities such as AsF_5, SbF_5, or iodine. As with semiconductors, conducting polymers may be made either n-type (i.e., free-electron dominant) or p-type (i.e., hole dominant), depending on the dopant. However, unlike semiconductors, the dopant atoms or molecules do not substitute for or replace any of the polymer atoms.

The mechanism by which large numbers of free electrons and holes are generated in these conducting polymers is complex and not well understood. In very simple terms, it appears that the dopant atoms lead to the formation of new energy bands that overlap the valence and conduction bands of the intrinsic polymer, giving rise to a partially filled band, and the production at room temperature of a high concentration of free electrons or holes. Orienting the polymer chains, either mechanically (Section 15.7) or magnetically, during synthesis results in a highly anisotropic material having a maximum conductivity along the direction of orientation.

These conducting polymers have the potential to be used in a host of applications inasmuch as they have low densities and are flexible. Rechargeable batteries and fuel cells are being manufactured that use polymer electrodes. In many respects, these batteries are superior to their metallic counterparts. Other possible applications include wiring in aircraft and aerospace components, antistatic coatings for clothing, electromagnetic screening materials, and electronic devices (e.g., transistors, diodes). Several conductive polymers display the phenomenon of *electroluminescence*—that is, light emission stimulated by an electrical current. Electroluminescent polymers are being used in applications such as solar panels and flat panel displays (see the Materials of Importance piece on light-emitting diodes in Chapter 22).

Dielectric Behavior

dielectric

electric dipole

A **dielectric** material is one that is electrically insulating (nonmetallic) and exhibits or may be made to exhibit an **electric dipole** structure—that is, there is a separation of positive and negative electrically charged entities on a molecular or atomic level. This concept of an electric dipole was introduced in Section 2.7. As a result of dipole interactions with electric fields, dielectric materials are used in capacitors.

19.18 CAPACITANCE

capacitance

When a voltage is applied across a capacitor, one plate becomes positively charged and the other negatively charged, with the corresponding electric field directed from the positive to the negative plates. The **capacitance** C is related to the quantity of charge stored on either plate Q by

Capacitance in terms of stored charge and applied voltage

$$C = \frac{Q}{V}$$

(19.24)

where V is the voltage applied across the capacitor. The units of capacitance are coulombs per volt, or farads (F).

Now, consider a parallel-plate capacitor with a vacuum in the region between the plates (Figure 19.28a). The capacitance may be computed from the relationship

Capacitance for parallel-plate capacitor, in a vacuum

$$C = \epsilon_0 \frac{A}{l}$$

(19.25)

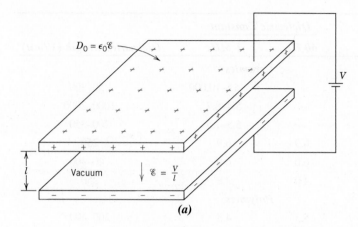

Figure 19.28 A parallel-plate capacitor (*a*) when a vacuum is present and (*b*) when a dielectric material is present. (From K. M. Ralls, T. H. Courtney, and J. Wulff, *Introduction to Materials Science and Engineering*. Copyright © 1976 by John Wiley & Sons, Inc. Reprinted by permission of John Wiley & Sons, Inc.)

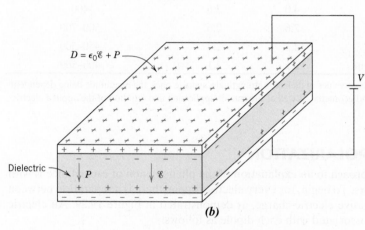

permittivity

where A represents the area of the plates and l is the distance between them. The parameter ϵ_0, called the **permittivity** of a vacuum, is a universal constant having the value of 8.85×10^{-12} F/m.

If a dielectric material is inserted into the region within the plates (Figure 19.28*b*), then

Capacitance for parallel-plate capacitor, with dielectric material

$$C = \epsilon \frac{A}{l} \qquad (19.26)$$

dielectric constant

where ϵ is the permittivity of this dielectric medium, which is greater in magnitude than ϵ_0. The relative permittivity ϵ_r, often called the **dielectric constant**, is equal to the ratio

Definition of dielectric constant

$$\epsilon_r = \frac{\epsilon}{\epsilon_0} \qquad (19.27)$$

which is greater than unity and represents the increase in charge-storing capacity upon insertion of the dielectric medium between the plates. The dielectric constant is one material property of prime consideration for capacitor design. The ϵ_r values of a number of dielectric materials are given in Table 19.5.

Table 19.5

Dielectric Constants and Strengths for Some Dielectric Materials

Material	Dielectric Constant		Dielectric Strength (V/mil)[a]
	60 Hz	**1 MHz**	
Ceramics			
Titanate ceramics	—	15–10,000	50–300
Mica	—	5.4–8.7	1000–2000
Steatite (MgO–SiO$_2$)	—	5.5–7.5	200–350
Soda–lime glass	6.9	6.9	250
Porcelain	6.0	6.0	40–400
Fused silica	4.0	3.8	250
Polymers			
Phenol-formaldehyde	5.3	4.8	300–400
Nylon 6,6	4.0	3.6	400
Polystyrene	2.6	2.6	500–700
Polyethylene	2.3	2.3	450–500
Polytetrafluoroethylene	2.1	2.1	400–500

[a]One mil = 0.025 mm. These values of dielectric strength are average ones, the magnitude being dependent on specimen thickness and geometry, as well as the rate of application and duration of the applied electric field.

19.19 FIELD VECTORS AND POLARIZATION

Perhaps the best approach to an explanation of the phenomenon of capacitance is with the aid of field vectors. To begin, for every electric dipole, there is a separation between a positive and a negative electric charge, as demonstrated in Figure 19.29. An electric dipole moment p is associated with each dipole as follows:

Electric dipole moment

$$p = qd \qquad (19.28)$$

where q is the magnitude of each dipole charge and d is the distance of separation between them. A *dipole moment* is a vector that is directed from the negative to the positive charge, as indicated in Figure 19.29. In the presence of an electric field $\mathscr{E}$, which is also a vector quantity, a force (or torque) comes to bear on an electric dipole to orient it with the applied field; this phenomenon is illustrated in Figure 19.30. The process of

polarization

dipole alignment is termed **polarization**.

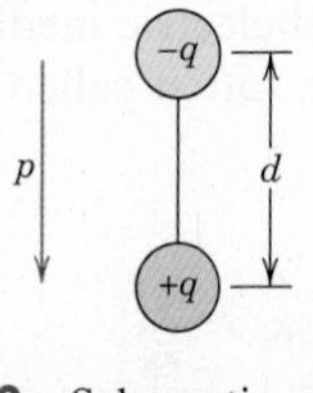

Figure 19.29 Schematic representation of an electric dipole generated by two electric charges (of magnitude q) separated by the distance d; the associated polarization vector p is also shown.

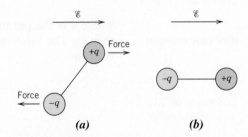

(a)   *(b)*

Figure 19.30 (*a*) Imposed forces (and torque) acting on a dipole by an electric field. (*b*) Final dipole alignment with the field.

Again, to return to the capacitor, the surface charge density D, or quantity of charge per unit area of capacitor plate (C/m^2), is proportional to the electric field. When a vacuum is present, then

Dielectric displacement (surface charge density) in a vacuum

$$D_0 = \epsilon_0 \mathcal{E} \tag{19.29}$$

where the constant of proportionality is ϵ_0. Furthermore, an analogous expression exists for the dielectric case—that is,

Dielectric displacement when a dielectric medium is present

$$D = \epsilon \mathcal{E} \tag{19.30}$$

dielectric displacement

Sometimes, D is also called the **dielectric displacement**.

The increase in capacitance, or dielectric constant, can be explained using a simplified model of polarization within a dielectric material. Consider the capacitor in Figure 19.31a—the vacuum situation—where a charge of $+Q_0$ is stored on the top plate and $-Q_0$ on the bottom plate. When a dielectric is introduced and an electric field is applied, the entire solid within the plates becomes polarized (Figure 19.31c). As a result of this polarization, there is a net accumulation of negative charge of magnitude $-Q'$ at the dielectric surface near the positively charged plate and, in a similar manner, a surplus

Figure 19.31 Schematic representations of (a) the charge stored on capacitor plates for a vacuum, (b) the dipole arrangement in an unpolarized dielectric, and (c) the increased charge-storing capacity resulting from the polarization of a dielectric material.
(Adapted from A. G. Guy, *Essentials of Materials Science*, McGraw-Hill Book Company, New York, 1976.)

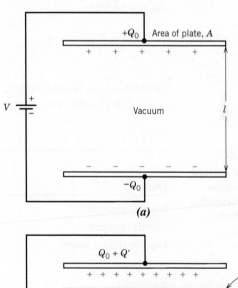
(a)

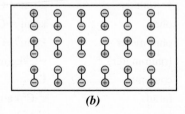
(b)

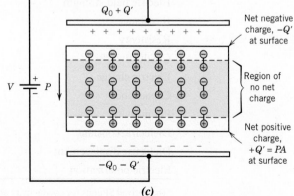
(c)

of $+Q'$ charge at the surface adjacent to the negative plate. For the region of dielectric removed from these surfaces, polarization effects are not important. Thus, if each plate and its adjacent dielectric surface are considered to be a single entity, the induced charge from the dielectric ($+Q'$ or $-Q'$) may be thought of as nullifying some of the charge that originally existed on the plate for a vacuum ($-Q_0$ or $+Q_0$). The voltage imposed across the plates is maintained at the vacuum value by increasing the charge at the negative (or bottom) plate by an amount $-Q'$ and that at the top plate by $+Q'$. Electrons are caused to flow from the positive to the negative plate by the external voltage source such that the proper voltage is reestablished. Thus, the charge on each plate is now $Q_0 + Q'$, having been increased by an amount Q'.

In the presence of a dielectric, the charge density between the plates, which is equal to the surface charge density on the plates of a capacitor, may also be represented by

Dielectric displacement—dependence on electric field intensity and polarization (of dielectric medium)

$$D = \epsilon_0 \mathscr{E} + P \qquad (19.31)$$

where P is the *polarization,* or the increase in charge density above that for a vacuum because of the presence of the dielectric; or, from Figure 19.31c, $P = Q'/A$, where A is the area of each plate. The units of P are the same as for D (C/m^2).

The polarization P may also be thought of as the total dipole moment per unit volume of the dielectric material, or as a polarization electric field within the dielectric that results from the mutual alignment of the many atomic or molecular dipoles with the externally applied field $\mathscr{E}$. For many dielectric materials, P is proportional to $\mathscr{E}$ through the relationship

Polarization of a dielectric medium—dependence on dielectric constant and electric field intensity

$$P = \epsilon_0 (\epsilon_r - 1) \mathscr{E} \qquad (19.32)$$

in which case ϵ_r is independent of the magnitude of the electric field.

Table 19.6 lists dielectric parameters along with their units.

Table 19.6

Primary and Derived Units for Various Electrical Parameters and Field Vectors

Quantity	Symbol	*SI Units* Derived	Primary
Electric potential	V	volt	kg·m^2/s^2·C
Electric current	I	ampere	C/s
Electric field strength	$\mathscr{E}$	volt/meter	kg·m/s^2·C
Resistance	R	ohm	kg·m^2/s·C^2
Resistivity	ρ	ohm-meter	kg·m^3/s·C^2
Conductivity[a]	σ	(ohm-meter)$^{-1}$	s·C^2/kg·m^3
Electric charge	Q	coulomb	C
Capacitance	C	farad	s^2·C^2/kg·m^2
Permittivity	ϵ	farad/meter	s^2·C^2/kg·m^3
Dielectric constant	ϵ_r	dimensionless	dimensionless
Dielectric displacement	D	farad-volt/m^2	C/m^2
Electric polarization	P	farad-volt/m^2	C/m^2

[a]The derived SI units for conductivity are siemens per meters (S/m).

EXAMPLE PROBLEM 19.5

Computations of Capacitor Properties

Consider a parallel-plate capacitor having an area of 6.45×10^{-4} m^2 and a plate separation of 2×10^{-3} m across which a potential of 10 V is applied. If a material having a dielectric constant of 6.0 is positioned within the region between the plates, compute the following:

(a) The capacitance

(b) The magnitude of the charge stored on each plate

(c) The dielectric displacement D

(d) The polarization

Solution

(a) Capacitance is calculated using Equation 19.26; however, the permittivity ϵ of the dielectric medium must first be determined from Equation 19.27, as follows:

$$\epsilon = \epsilon_r \epsilon_0 = (6.0)(8.85 \times 10^{-12}\,\text{F/m})$$
$$= 5.31 \times 10^{-11}\,\text{F/m}$$

Thus, the capacitance is given by

$$C = \epsilon \frac{A}{l} = (5.31 \times 10^{-11}\,\text{F/m}) \left(\frac{6.45 \times 10^{-4}\,\text{m}^{-2}}{20 \times 10^{-3}\,\text{m}} \right)$$
$$= 1.71 \times 10^{-11}\,\text{F}$$

(b) Because the capacitance has been determined, the charge stored may be computed using Equation 19.24, according to

$$Q = CV = (1.71 \times 10^{-11}\,\text{F})(10\,\text{V}) = 1.71 \times 10^{-10}\,\text{C}$$

(c) The dielectric displacement is calculated from Equation 19.30, which yields

$$D = \epsilon \mathscr{E} = \epsilon \frac{V}{l} = \frac{(5.31 \times 10^{-11}\,\text{F/m})(10\,\text{V})}{2 \times 10^{-3}\,\text{m}}$$
$$= 2.66 \times 10^{-7}\,\text{C/m}^2$$

(d) Using Equation 19.31, the polarization may be determined as follows:

$$P = D - \epsilon_0 \mathscr{E} = D - \epsilon_0 \frac{V}{l}$$
$$= 2.66 \times 10^{-7}\,\text{C/m}^2 - \frac{(8.85 \times 10^{-12}\,\text{F/m})(10\,\text{V})}{2 \times 10^{-3}\,\text{m}}$$
$$= 2.22 \times 10^{-7}\,\text{C/m}^2$$

19.20 TYPES OF POLARIZATION

Again, polarization is the alignment of permanent or induced atomic or molecular dipole moments with an externally applied electric field. There are three types or sources of polarization: electronic, ionic, and orientation. Dielectric materials typically exhibit at least one of these polarization types, depending on the material and the manner of external field application.

Electronic Polarization

electronic polarization

Electronic polarization may be induced to one degree or another in all atoms. It results from a displacement of the center of the negatively charged electron cloud relative to the positive nucleus of an atom by the electric field (Figure 19.32a). This polarization type is found in all dielectric materials and exists only while an electric field is present.

Ionic Polarization

ionic polarization

Ionic polarization occurs only in materials that are ionic. An applied field acts to displace cations in one direction and anions in the opposite direction, which gives rise to a net dipole moment. This phenomenon is illustrated in Figure 19.32b. The magnitude of the dipole moment for each ion pair p_i is equal to the product of the relative displacement d_i and the charge on each ion, or

Electric dipole moment for an ion pair

$$p_i = qd_i \tag{19.33}$$

Orientation Polarization

orientation polarization

The third type, **orientation polarization**, is found only in substances that possess permanent dipole moments. Polarization results from a rotation of the permanent moments into the direction of the applied field, as represented in Figure 19.32c. This alignment tendency is counteracted by the thermal vibrations of the atoms, such that polarization decreases with increasing temperature.

Figure 19.32 (*a*) Electronic polarization that results from the distortion of an atomic electron cloud by an electric field. (*b*) Ionic polarization that results from the relative displacements of electrically charged ions in response to an electric field. (*c*) Response of permanent electric dipoles (arrows) to an applied electric field, producing orientation polarization. (From O. H. Wyatt and D. Dew-Hughes, *Metals, Ceramics and Polymers,* Cambridge University Press, 1974. Reprinted with the permission of the Cambridge University Press.)

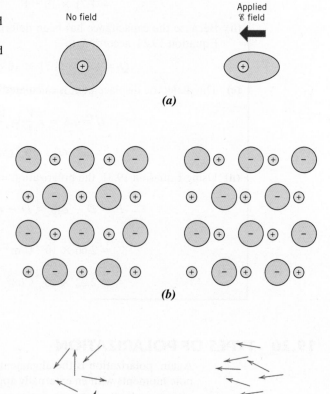

No field

Applied $\mathscr{E}$ field

(a)

(b)

(c)

The total polarization P of a substance is equal to the sum of the electronic, ionic, and orientation polarizations (P_e, P_i, and P_o, respectively), or

Total polarization of
a substance equals
the sum of electronic,
ionic, and orientation
polarizations

$$P = P_e + P_i + P_o \qquad (19.34)$$

It is possible for one or more of these contributions to the total polarization to be either absent or negligible in magnitude relative to the others. For example, ionic polarization does not exist in covalently bonded materials in which no ions are present.

Concept Check 19.9 For solid lead titanate ($PbTiO_3$), what kind(s) of polarization is (are) possible? Why? *Note:* Lead titanate has the same crystal structure as barium titanate (Figure 19.35).

[*The answer may be found at* www.wiley.com/college/callister *(Student Companion Site).*]

19.21 FREQUENCY DEPENDENCE OF THE DIELECTRIC CONSTANT

In many practical situations, the current is alternating (ac)—that is, an applied voltage or electric field changes direction with time, as indicated in Figure 19.23a. Consider a dielectric material that is subject to polarization by an ac electric field. With each direction reversal, the dipoles attempt to reorient with the field, as illustrated in Figure 19.33, in a process requiring some finite time. For each polarization type, some minimum reorientation time exists that depends on the ease with which the particular dipoles are capable of realignment.

relaxation frequency   The **relaxation frequency** is taken as the reciprocal of this minimum reorientation time.

A dipole cannot keep shifting orientation direction when the frequency of the applied electric field exceeds its relaxation frequency and, therefore, it will not make a contribution to the dielectric constant. The dependence of ϵ_r on the field frequency is represented schematically in Figure 19.34 for a dielectric medium that exhibits all three

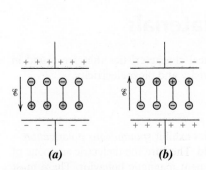

Figure 19.33 Dipole orientations for (*a*) one polarity of an alternating electric field and (*b*) the reversed polarity.
(From Richard A. Flinn and Paul K. Trojan, *Engineering Materials and Their Applications,* 4th edition. Copyright © 1990 by John Wiley & Sons, Inc. Adapted by permission of John Wiley & Sons, Inc.)

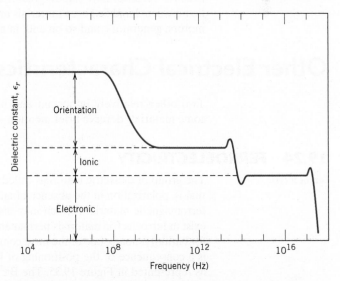

Figure 19.34 Variation of dielectric constant with frequency of an alternating electric field. Electronic, ionic, and orientation polarization contributions to the dielectric constant are indicated.

types of polarization; note that the frequency axis is scaled logarithmically. As indicated in Figure 19.34, when a polarization mechanism ceases to function, there is an abrupt drop in the dielectric constant; otherwise, ϵ_r is virtually frequency independent. Table 19.5 gave values of the dielectric constant at 60 Hz and 1 MHz; these provide an indication of this frequency dependence at the low end of the frequency spectrum.

The absorption of electrical energy by a dielectric material that is subjected to an alternating electric field is termed *dielectric loss*. This loss may be important at electric field frequencies in the vicinity of the relaxation frequency for each of the operative dipole types for a specific material. A low dielectric loss is desired at the frequency of utilization.

19.22 DIELECTRIC STRENGTH

dielectric strength

When very high electric fields are applied across dielectric materials, large numbers of electrons may suddenly be excited to energies within the conduction band. As a result, the current through the dielectric by the motion of these electrons increases dramatically; sometimes localized melting, burning, or vaporization produces irreversible degradation and perhaps even failure of the material. This phenomenon is known as *dielectric breakdown*. The **dielectric strength**, sometimes called the *breakdown strength,* represents the magnitude of an electric field necessary to produce breakdown. Table 19.5 presents dielectric strengths for several materials.

19.23 DIELECTRIC MATERIALS

A number of ceramics and polymers are used as insulators and/or in capacitors. Many of the ceramics, including glass, porcelain, steatite, and mica, have dielectric constants within the range of 6 to 10 (Table 19.5). These materials also exhibit a high degree of dimensional stability and mechanical strength. Typical applications include power line and electrical insulation, switch bases, and light receptacles. The titania (TiO_2) and titanate ceramics, such as barium titanate ($BaTiO_3$), can be made to have extremely high dielectric constants, which render them especially useful for some capacitor applications.

The magnitude of the dielectric constant for most polymers is less than for ceramics because the latter may exhibit greater dipole moments: ϵ_r values for polymers generally lie between 2 and 5. These materials are commonly used for insulation of wires, cables, motors, generators, and so on and, in addition, for some capacitors.

Other Electrical Characteristics of Materials

Two other relatively important and novel electrical characteristics that are found in some materials deserve brief mention—ferroelectricity and piezoelectricity.

19.24 FERROELECTRICITY

ferroelectric

The group of dielectric materials called **ferroelectrics** exhibit *spontaneous polarization*—that is, polarization in the absence of an electric field. They are the dielectric analogue of ferromagnetic materials, which may display permanent magnetic behavior. There must exist in ferroelectric materials permanent electric dipoles, the origin of which is explained for barium titanate, one of the most common ferroelectrics. The spontaneous polarization is a consequence of the positioning of the Ba^{2+}, Ti^{4+}, and O^{2-} ions within the unit cell, as represented in Figure 19.35. The Ba^{2+} ions are located at the corners of the unit cell, which is of *tetragonal symmetry* (a cube that has been elongated slightly in one direction). The dipole moment results from the relative displacements of the O^{2-} and Ti^{4+} ions from their symmetrical positions, as shown in the side view of the unit cell. The O^{2-} ions are located near, but slightly below, the centers of each of the six faces, whereas the Ti^{4+} ion

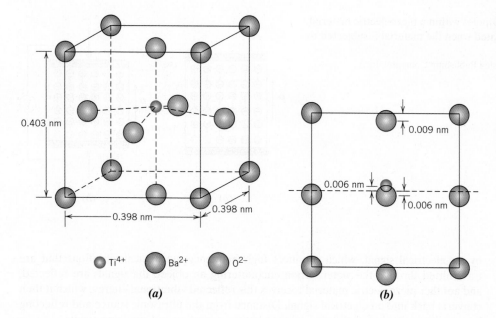

Figure 19.35 A barium titanate (BaTiO$_3$) unit cell (*a*) in an isometric projection, and (*b*) looking at one face, which shows the displacements of Ti^{4+} and O^{2-} ions from the center of the face.

is displaced upward from the unit cell center. Thus, a permanent ionic dipole moment is associated with each unit cell (Figure 19.35*b*). However, when barium titanate is heated above its *ferroelectric Curie temperature* (120°C), the unit cell becomes cubic, and all ions assume symmetric positions within the cubic unit cell; the material now has a perovskite crystal structure (Figure 4.9), and the ferroelectric behavior ceases.

Spontaneous polarization of this group of materials results as a consequence of interactions between adjacent permanent dipoles in which they mutually align, all in the same direction. For example, with barium titanate, the relative displacements of O^{2-} and Ti^{4+} ions are in the same direction for all the unit cells within some volume region of the specimen. Other materials display ferroelectricity; these include Rochelle salt (NaKC$_4$H$_4$O$_6$·4H$_2$O), potassium dihydrogen phosphate (KH$_2$PO$_4$), potassium niobate (KNbO$_3$), and lead zirconate–titanate (Pb[ZrO$_3$, TiO$_3$]). Ferroelectrics have extremely high dielectric constants at relatively low applied field frequencies; for example, at room temperature, ϵ_r for barium titanate may be as high as 5000. Consequently, capacitors made from these materials can be significantly smaller than capacitors made from other dielectric materials.

19.25 PIEZOELECTRICITY

An unusual phenomenon exhibited by a few ceramic materials (as well as some polymers) is *piezoelectricity*—literally, pressure electricity. Electric polarization (i.e., an electric field or voltage) is induced in the piezoelectric crystal as a result of a mechanical strain (dimensional change) produced from the application of an external force (Figure 19.36). Reversing the sign of the force (e.g., from tension to compression) reverses the direction of the field. The inverse piezoelectric effect is also displayed by this group of materials—that is, a mechanical strain results from the imposition of an electrical field.

piezoelectric

Piezoelectric materials may be used as transducers between electrical and mechanical energies. One of the early uses of piezoelectric ceramics was in sonar systems, in which underwater objects (e.g., submarines) are detected and their positions determined using an ultrasonic emitting and receiving system. A piezoelectric crystal is caused to oscillate

Figure 19.36 (*a*) Dipoles within a piezoelectric material. (*b*) A voltage is generated when the material is subjected to a compressive stress. (© 1989 by Addison-Wesley Publishing Company, Inc.)

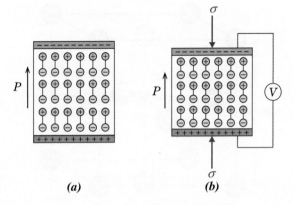

(a)   *(b)*

by an electrical signal, which produces high-frequency mechanical vibrations that are transmitted through the water. Upon encountering an object, the signals are reflected, and another piezoelectric material receives this reflected vibrational energy, which it then converts back into an electrical signal. Distance from the ultrasonic source and reflecting body is determined from the elapsed time between sending and receiving events.

More recently, the use of piezoelectric devices has grown dramatically as a consequence of increases in automation and consumer attraction to modern sophisticated gadgets. Piezoelectric devices are used in many of today's applications, including automotive—wheel balances, seat-belt buzzers, tread-wear indicators, keyless door entry, and airbag sensors; computer/electronic—microphones, speakers, microactuators for hard disks and notebook transformers; commercial/consumer—ink-jet printing heads, strain gauges, ultrasonic welders, and smoke detectors; medical—insulin pumps, ultrasonic therapy, and ultrasonic cataract-removal devices.

Piezoelectric ceramic materials include titanates of barium and lead ($BaTiO_3$ and $PbTiO_3$), lead zirconate ($PbZrO_3$), lead zirconate–titanate (PZT) [$Pb(Zr,Ti)O_3$], and potassium niobate ($KNbO_3$). This property is characteristic of materials having complicated crystal structures with a low degree of symmetry. The piezoelectric behavior of a polycrystalline specimen may be improved by heating above its Curie temperature and then cooling to room temperature in a strong electric field.

MATERIAL OF IMPORTANCE

Piezoelectric Ceramic Ink–Jet Printer Heads

Piezoelectric materials are used in one kind of ink-jet printer head that has components and a mode of operation represented in the schematic diagrams in Figure 19.37*a* through 19.37*c*. One head component is a flexible, bilayer disk that consists of a piezoelectric ceramic (orange region) bonded to a nonpiezoelectric deformable material (green region); liquid ink

and its reservoir are represented by blue areas in these diagrams. Short, horizontal arrows within the piezoelectric note the direction of the permanent dipole moment.

Printer head operation (i.e., ejection of ink droplets from the nozzle) is a result of the inverse piezoelectric effect—that is, the bilayer disk is caused to flex back and forth by the expansion and contraction of the piezoelectric layer in response to changes in bias of an applied voltage. For example, Figure 19.37a shows how the imposition of forward bias voltage causes the bilayer disk to flex in such a way as to pull (or draw) ink from the reservoir into the nozzle chamber. Reversing the voltage bias forces the bilayer disk to bend in the opposite direction—toward the nozzle—so as to eject a drop of ink (Figure 19.37b). Finally, removal of the voltage causes the disk to return to its unbent configuration (Figure 19.37c) in preparation for another ejection sequence.

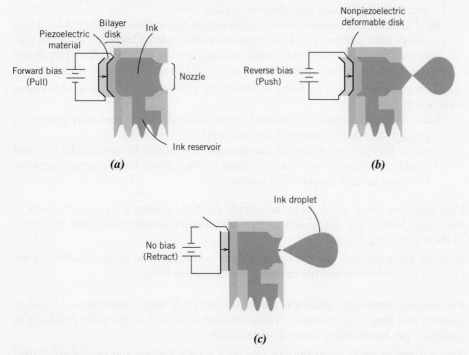

Figure 19.37 Operation sequence of a piezoelectric ceramic ink-jet printer head (schematic). (a) Imposing a forward-bias voltage draws ink into the nozzle chamber as the bilayer disk flexes in one direction. (b) Ejection of an ink drop by reversing the voltage bias and forcing the disk to flex in the opposite direction. (c) Removing the voltage retracts the bilayer disk to its unbent configuration in preparation for the next sequence.
(Images provided courtesy of Epson America, Inc.)

SUMMARY

Ohm's Law

Electrical Conductivity

- The ease with which a material is capable of transmitting an electric current is expressed in terms of electrical conductivity or its reciprocal, electrical resistivity (Equations 19.2 and 19.3).
- The relationship between applied voltage, current, and resistance is Ohm's law (Equation 19.1). An equivalent expression, Equation 19.5, relates current density, conductivity, and electric field intensity.
- On the basis of its conductivity, a solid material may be classified as a metal, a semiconductor, or an insulator.

Electronic and Ionic Conduction

- For most materials, an electric current results from the motion of free electrons, which are accelerated in response to an applied electric field.
- In ionic materials, there may also be a net motion of ions, which also makes a contribution to the conduction process.

Energy Band Structures in Solids

- The number of free electrons depends on the electron energy band structure of the material.

Conduction in Terms of Band and Atomic Bonding Models

- An electron band is a series of electron states that are closely spaced with respect to energy, and one such band may exist for each electron subshell found in the isolated atom.
- *Electron energy band structure* refers to the manner in which the outermost bands are arranged relative to one another and then filled with electrons.

  For metals, two band structure types are possible (Figures 19.4a and 19.4b)— empty electron states are adjacent to filled ones.

  Band structures for semiconductors and insulators are similar—both have a forbidden energy band gap that, at 0 K, lies between a filled valence band and an empty conduction band. The magnitude of this band gap is relatively wide (>2 eV) for insulators (Figure 19.4c) and relatively narrow (<2 eV) for semiconductors (Figure 19.4d).

- An electron becomes free by being excited from a filled state to an available empty state at a higher energy.

  Relatively small energies are required for electron excitations in metals (Figure 19.5), giving rise to large numbers of free electrons.

  Greater energies are required for electron excitations in semiconductors and insulators (Figure 19.6), which accounts for their lower free electron concentrations and smaller conductivity values.

Electron Mobility

- Free electrons being acted on by an electric field are scattered by imperfections in the crystal lattice. The magnitude of electron mobility is indicative of the frequency of these scattering events.
- In many materials, the electrical conductivity is proportional to the product of the electron concentration and the mobility (per Equation 19.8).

Electrical Resistivity of Metals

- For metallic materials, electrical resistivity increases with temperature, impurity content, and plastic deformation. The contribution of each to the total resistivity is additive—per Matthiessen's rule, Equation 19.9.
- Thermal and impurity contributions (for both solid solutions and two-phase alloys) are described by Equations 19.10, 19.11, and 19.12.

Intrinsic Semiconduction

Extrinsic Semiconduction

- Semiconductors may be either elements (Si and Ge) or covalently bonded compounds.
- With these materials, in addition to free electrons, holes (missing electrons in the valence band) may also participate in the conduction process (Figure 19.11).
- Semiconductors are classified as either intrinsic or extrinsic.

  For intrinsic behavior, the electrical properties are inherent in the pure material, and electron and hole concentrations are equal. The electrical conductivity may be computed using Equation 19.13 (or Equation 19.15).

  Electrical behavior is dictated by impurities for extrinsic semiconductors. Extrinsic semiconductors may be either *n*- or *p*-type depending on whether electrons or holes, respectively, are the predominant charge carriers.

- Donor impurities introduce excess electrons (Figures 19.12 and 19.13); acceptor impurities introduce excess holes (Figures 19.14 and 19.15).
- The electrical conductivity on an *n*-type semiconductor may be calculated using Equation 19.16; for a *p*-type semiconductor, Equation 19.17 is used.

The Temperature Dependence of Carrier Concentration

- With rising temperature, intrinsic carrier concentration increases dramatically (Figure 19.16).

Factors That Affect Carrier Mobility

- For extrinsic semiconductors, on a plot of majority carrier concentration versus temperature, carrier concentration is independent of temperature in the *extrinsic region*

(Figure 19.17). The magnitude of carrier concentration in this region is approximately equal to the impurity level.

- For extrinsic semiconductors, electron and hole mobilities (1) decrease as impurity content increases (Figure 19.18) and (2) in general, decrease with rising temperature (Figures 19.19a and 19.19b).

The Hall Effect
- Using a Hall effect experiment, it is possible to determine the charge carrier type (i.e., electron or hole), as well as carrier concentration and mobility.

Semiconductor Devices
- A number of semiconducting devices employ the unique electrical characteristics of these materials to perform specific electronic functions.
- The p–n rectifying junction (Figure 19.21) is used to transform alternating current into direct current.
- Another type of semiconductor device is the transistor, which may be used for amplification of electrical signals, as well as for switching devices in computer circuitries. Junction and MOSFET transistors (Figures 19.24, 19.25, and 19.26) are possible.

Electrical Conduction in Ionic Ceramics and in Polymers
- Most ionic ceramics and polymers are insulators at room temperature. Electrical conductivities range between about 10^{-9} and 10^{-18} $(\Omega \cdot m)^{-1}$; by way of comparison, for most metals, σ is on the order of 10^7 $(\Omega \cdot m)^{-1}$.

Dielectric Behavior
Capacitance
Field Vectors and Polarization
- A *dipole* is said to exist when there is a net spatial separation of positively and negatively charged entities on an atomic or molecular level.
- *Polarization* is the alignment of electric dipoles with an electric field.
- *Dielectric materials* are electrical insulators that may be polarized when an electric field is present.
- This polarization phenomenon accounts for the ability of the dielectrics to increase the charge-storing capability of capacitors.
- Capacitance is dependent on applied voltage and quantity of charge stored according to Equation 19.24.
- The charge-storing efficiency of a capacitor is expressed in terms of a dielectric constant or relative permittivity (Equation 19.27).
- For a parallel-plate capacitor, capacitance is a function of the permittivity of the material between the plates, as well as plate area and plate separation distance per Equation 19.26.
- The dielectric displacement within a dielectric medium depends on the applied electric field and the induced polarization according to Equation 19.31.
- For some dielectric materials, the polarization induced by an applied electric field is described by Equation 19.32.

Types of Polarization
Frequency Dependence of the Dielectric Constant
- Possible polarization types include electronic (Figure 19.32a), ionic (Figure 19.32b), and orientation (Figure 19.32c); not all types need be present in a particular dielectric.
- For alternating electric fields, whether a specific polarization type contributes to the total polarization and dielectric constant depends on frequency; each polarization mechanism ceases to function when the applied field frequency exceeds its relaxation frequency (Figure 19.34).

Other Electrical Characteristics of Materials
- Ferroelectric materials exhibit spontaneous polarization—that is, they polarize in the absence of an electric field.
- An electric field is generated when mechanical stresses are applied to a piezoelectric material.

Important Terms and Concepts

acceptor state (level)	Fermi energy	Ohm's law
capacitance	ferroelectric	permittivity
conduction band	forward bias	piezoelectric
conductivity, electrical	free electron	polarization
dielectric	Hall effect	polarization, electronic
dielectric constant	hole	polarization, ionic
dielectric displacement	insulator	polarization, orientation
dielectric strength	integrated circuit	rectifying junction
diode	intrinsic semiconductor	relaxation frequency
dipole, electric	ionic conduction	resistivity, electrical
donor state (level)	junction transistor	reverse bias
doping	Matthiessen's rule	semiconductor
electron energy band	metal	valence band
energy band gap	mobility	
extrinsic semiconductor	MOSFET	

REFERENCES

Bube, R. H., *Electrons in Solids*, 3rd edition, Academic Press, San Diego, 1992.

Hummel, R. E., *Electronic Properties of Materials*, 4th edition, Springer, New York, 2011.

Irene, E. A., *Electronic Materials Science*, Wiley, Hoboken, NJ, 2005.

Jiles, D. C., *Introduction to the Electronic Properties of Materials*, 2nd edition, CRC Press, Boca Raton, FL, 2001.

Kingery, W. D., H. K. Bowen, and D. R. Uhlmann, *Introduction to Ceramics*, 2nd edition, Wiley, New York, 1976. Chapters 17 and 18.

Kittel, C., *Introduction to Solid State Physics*, 8th edition, Wiley, Hoboken, NJ, 2005. An advanced treatment.

Livingston, J., *Electronic Properties of Engineering Materials*, Wiley, New York, 1999.

Pierret, R. F., *Semiconductor Device Fundamentals*, Addison-Wesley, Boston, 1996.

Rockett, A., *The Materials Science of Semiconductors*, Springer, New York, 2008.

Solymar, L., and D. Walsh, *Electrical Properties of Materials*, 8th edition, Oxford University Press, New York, 2010.

QUESTIONS AND PROBLEMS

Ohm's Law
Electrical Conductivity

19.1 (a) Compute the electrical conductivity of a 5.0-mm diameter cylindrical silicon specimen 50 mm long in which a current of 0.1 A passes in an axial direction. A voltage of 12.5 V is measured across two probes that are separated by 38 mm.

(b) Compute the resistance over the entire 50 mm of the specimen.

19.2 A copper wire 100 m long must experience a voltage drop of less than 1.5 V when a current of 2.5 A passes through it. Using the data in Table 19.1, compute the minimum diameter of the wire.

19.3 An aluminum wire 5 mm in diameter is to offer a resistance of no more than 2.5 Ω. Using the data in Table 19.1, compute the maximum wire length.

19.4 Demonstrate that the two Ohm's law expressions, Equations 19.1 and 19.5, are equivalent.

19.5 (a) Using the data in Table 19.1, compute the resistance of a copper wire 5 mm in diameter and 2 m long. **(b)** What would be the current flow if the potential drop across the ends of the wire is 0.05 V? **(c)** What is the current density? **(d)** What is the magnitude of the electric field across the ends of the wire?

Electronic and Ionic Conduction

19.6 What is the distinction between electronic and ionic conduction?

Energy Band Structures in Solids

19.7 How does the electron structure of an isolated atom differ from that of a solid material?

Conduction in Terms of Band and
* Atomic Bonding Models*

19.8 In terms of electron energy band structure, discuss reasons for the difference in electrical conductivity between metals, semiconductors, and insulators.

Electron Mobility

19.9 Briefly tell what is meant by the drift velocity and mobility of a free electron.

19.10 **(a)** Calculate the drift velocity of electrons in germanium at room temperature and when the magnitude of the electric field is 1000 V/m. **(b)** Under these circumstances, how long does it take an electron to traverse a 25-mm length of crystal?

19.11 At room temperature the electrical conductivity and the electron mobility for copper are 6.0×10^7 $(\Omega \cdot m)^{-1}$ and 0.0030 $m^2/V \cdot s$, respectively. **(a)** Compute the number of free electrons per cubic meter for copper at room temperature. **(b)** What is the number of free electrons per copper atom? Assume a density of 8.9 g/cm^3.

19.12 **(a)** Calculate the number of free electrons per cubic meter for gold, assuming that there are 1.5 free electrons per gold atom. The electrical conductivity and density for Au are 4.3×10^7 $(\Omega \cdot m)^{-1}$ and 19.32 g/cm^3, respectively. **(b)** Now compute the electron mobility for Au.

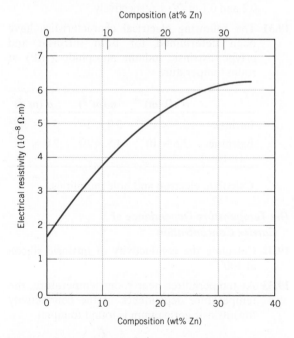

Figure 19.38 Room-temperature electrical resistivity versus composition for copper–zinc alloys.
[Adapted from *Metals Handbook: Properties and Selection: Nonferrous Alloys and Pure Metals*, Vol. 2, 9th edition, H. Baker (Managing Editor), 1979. Reproduced by permission of ASM International, Materials Park, OH.]

Electrical Resistivity of Metals

19.13 From Figure 19.38, estimate the value of A in Equation 19.11 for zinc as an impurity in copper–zinc alloys.

19.14 **(a)** Using the data in Figure 19.8, determine the values of ρ_0 and a from Equation 19.10 for pure copper. Take the temperature T to be in degrees Celsius. **(b)** Determine the value of A in Equation 19.11 for nickel as an impurity in copper, using the data in Figure 19.8. **(c)** Using the results of parts (a) and (b), estimate the electrical resistivity of copper containing 1.75 at% Ni at 100°C.

19.15 Determine the electrical conductivity of a Cu–Ni alloy that has a yield strength of 130 MPa. You will find Figure 9.16 helpful.

19.16 Tin bronze has a composition of 96 wt% Cu and 5 wt% Sn, and consists of two phases at room temperature: an α phase, which is copper containing a very small amount of tin in solid solution, and an ε phase, which consists of approximately 37 wt% Sn. Compute the room temperature conductivity of this alloy given the following data:

Phase	Electrical Resistivity ($\Omega \cdot m$)	Density (g/cm^3)
α	1.88×10^{-8}	8.94
ε	5.32×10^{-7}	8.25

19.17 A cylindrical metal wire 5 mm in diameter is required to carry a current of 12 A with a minimum of 0.03 V drop per 300 mm of wire. Which of the metals and alloys listed in Table 19.1 are possible candidates?

Intrinsic Semiconduction

19.18 **(a)** Using the data presented in Figure 19.16, determine the number of free electrons per atom for intrinsic germanium and silicon at room temperature (298 K). The densities for Ge and Si are 5.32 and 2.33 g/cm^3, respectively.

(b) Now explain the difference in these free-electron-per-atom values.

19.19 For intrinsic semiconductors, the intrinsic carrier concentration n_i depends on temperature as follows:

$$n_i \propto \exp\left(-\frac{E_g}{2kT}\right) \qquad (19.35a)$$

or, taking natural logarithms,

$$\ln n_i \propto -\frac{E_g}{2kT} \qquad (19.35b)$$

Thus, a plot of $\ln n_i$ versus $1/T$ $(K)^{-1}$ should be linear and yield a slope of $-E_g/2k$. Using this information and the data presented in Figure 19.16, determine the band gap energies for silicon and germanium, and compare these values with those given in Table 19.3.

19.20 Briefly explain the presence of the factor 2 in the denominator of Equation 19.35a.

19.21 At room temperature the electrical conductivity of PbTe is 650 $(\Omega \cdot m)^{-1}$, whereas the electron and hole mobilities are 0.16 and 0.080 $m^2/V \cdot s$, respectively. Compute the intrinsic carrier concentration for PbTe at room temperature.

19.22 Is it possible for compound semiconductors to exhibit intrinsic behavior? Explain your answer.

19.23 For each of the following pairs of semiconductors, decide which will have the smaller band gap energy, E_g, and then cite the reason for your choice. **(a)** ZnS and CdSe, **(b)** Si and C (diamond), **(c)** Al_2O_3 and ZnTe, **(d)** InSb and ZnSe, and **(e)** GaAs and AlP.

Extrinsic Semiconduction

19.24 Define the following terms as they pertain to semiconducting materials: intrinsic, extrinsic, compound, elemental. Now provide an example of each.

19.25 An *n*-type semiconductor is known to have an electron concentration of 3×10^{18} m^{-3}. If the electron drift velocity is 100 m/s in an electric field of 600 V/m, calculate the conductivity of this material.

19.26 **(a)** In your own words, explain how donor impurities in semiconductors give rise to free electrons in numbers in excess of those generated by valence band–conduction band excitations. **(b)** Also explain how acceptor impurities give rise to holes in numbers in excess of those generated by valence band–conduction band excitations.

19.27 **(a)** Explain why no hole is generated by the electron excitation involving a donor impurity atom. **(b)** Explain why no free electron is generated by the electron excitation involving an acceptor impurity atom.

19.28 Will each of the following elements act as a donor or an acceptor when added to the indicated semiconducting material? Assume that the impurity elements are substitutional.

Impurity	Semiconductor
P	Ge
S	AlP
In	CdTe
Al	Si
Cd	GaAs
Sb	ZnSe

19.29 **(a)** The room-temperature electrical conductivity of a silicon specimen is 5.93×10^{-3} $(\Omega \cdot m)^{-1}$. The hole concentration is known to be 7.0×10^{17} m^{-3}. Using the electron and hole mobilities for silicon in Table 19.3, compute the electron concentration. **(b)** On the basis of the result in part (a), is the specimen intrinsic, *n*-type extrinsic, or *p*-type extrinsic? Why?

19.30 Germanium to which 5×10^{22} m^{-3} Sb atoms have been added is an extrinsic semiconductor at room temperature, and virtually all the Sb atoms may be thought of as being ionized (i.e., one charge carrier exists for each Sb atom). **(a)** Is this material *n*-type or *p*-type? **(b)** Calculate the electrical conductivity of this material, assuming electron and hole mobilities of 0.2 and 0.1 $m^2/V \cdot s$, respectively.

19.31 The following electrical characteristics have been determined for both intrinsic and *p*-type extrinsic indium phosphide (InP) at room temperature:

	σ $(\Omega \cdot m)^{-1}$	n (m^{-3})	p (m^{-3})
Intrinsic	2.5×10^{-6}	3.0×10^{13}	3.0×10^{13}
Extrinsic (*n*-type)	3.6×10^{-5}	4.5×10^{14}	2.0×10^{12}

Calculate electron and hole mobilities.

The Temperature Dependence of Carrier Concentration

19.32 Calculate the conductivity of intrinsic silicon at 100°C.

19.33 At temperatures near room temperature, the temperature dependence of the conductivity for intrinsic germanium is found to equal

$$\sigma = CT^{-3/2} \exp\left(-\frac{E_g}{2kT}\right) \qquad (19.36)$$

where C is a temperature-independent constant and T is in Kelvins. Using Equation 19.36, calculate the intrinsic electrical conductivity of germanium at 150°C.

19.34 Using Equation 19.36 and the results of Problem 19.33, determine the temperature at which the electrical conductivity of intrinsic germanium is 22.8 $(\Omega \cdot m)^{-1}$.

19.35 Estimate the temperature at which GaAs has an electrical conductivity of 3.7×10^{-3} $(\Omega \cdot m)^{-1}$, assuming the temperature dependence for σ of Equation 19.36. The data shown in Table 19.3 may prove helpful.

19.36 Compare the temperature dependence of the conductivity for metals and intrinsic semiconductors. Briefly explain the difference in behavior.

Factors That Affect Carrier Mobility

19.37 Calculate the room-temperature electrical conductivity of silicon that has been doped with 5×10^{22} m^{-3} of boron atoms.

19.38 Calculate the room-temperature electrical conductivity of silicon that has been doped with 2×10^{23} m^{-3} of arsenic atoms.

19.39 Estimate the electrical conductivity, at 130°C, of silicon that has been doped with 10^{23} m^{-3} of aluminum atoms.

19.40 Estimate the electrical conductivity, at 85°C, of silicon that has been doped with 10^{20} m^{-3} of phosphorus atoms.

The Hall Effect

19.41 A hypothetical metal is known to have an electrical resistivity of 4×10^{-8} $(\Omega \cdot m)$. A current of 40 A is passed through a specimen of this metal that is 30 mm thick; when a magnetic field of 0.75 tesla is simultaneously imposed in a direction perpendicular to that of the current, a Hall voltage of -1.26×10^{-7} V is measured. Compute **(a)** the electron mobility for this metal and **(b)** the number of free electrons per cubic meter.

19.42 A metal alloy is known to have electrical conductivity and electron mobility values of 2.0×10^{7} $(\Omega \cdot m)^{-1}$ and 0.0025 m^2/V·s, respectively. A current of 45 A is passed through a specimen of this alloy that is 40 mm thick. What magnetic field would need to be imposed to yield a Hall voltage of -1.0×10^{-7} V?

Semiconducting Devices

19.43 Briefly describe electron and hole motions in a *p–n* junction for forward and reverse biases; then explain how these lead to rectification.

19.44 How is the energy in the reaction described by Equation 19.21 dissipated?

19.45 What are the two functions that a transistor may perform in an electronic circuit?

19.46 Cite the differences in operation and application for junction transistors and MOSFETs.

Conduction in Ionic Materials

19.47 We noted in Section 6.4 (Figure 6.4) that in FeO (wüstite), the iron ions can exist in both Fe^{2+} and Fe^{3+} states. The number of each of these ion types depends on temperature and the ambient oxygen pressure. Furthermore, we also noted that in order to retain electroneutrality, one Fe^{2+} vacancy will be created for every two Fe^{3+} ions that are formed; consequently, in order to reflect the existence of these vacancies the formula for wüstite is often represented as Fe$_{(1-x)}$O, where x is some small fraction less than unity.

In this nonstoichiometric Fe$_{(1-x)}$O material, conduction is electronic, and, in fact, behaves as a *p*-type semiconductor. That is, the Fe^{3+} ions act as electron acceptors, and it is relatively easy to excite an electron from the valence band into an Fe^{3+} acceptor state, with the formation of a hole. Determine the electrical conductivity of a specimen of wüstite that has a hole mobility of 1.0×10^{-5} m^2/V·s and for which the value of x is 0.060. Assume that the acceptor states are saturated (i.e., one hole exists for every Fe^{3+} ion). Wüstite has the sodium chloride crystal structure with a unit cell edge length of 0.437 nm.

19.48 At temperatures between 775°C (1048 K) and 1100°C (1373 K), the activation energy and preexponential for the diffusion coefficient of Fe^{2+} in FeO are 102,000 J/mol and 7.3×10^{-8} m^2/s, respectively. Compute the mobility for an Fe^{2+} ion at 1000°C (1273 K).

Capacitance

19.49 A parallel-plate capacitor using a dielectric material having an ε_r of 2.5 has a plate spacing of 1 mm. If another material having a dielectric constant of 4.0 is used and the capacitance is to be unchanged, what must be the new spacing between the plates?

19.50 A parallel-plate capacitor with dimensions of 120 mm by 30 mm and a plate separation of 3 mm must have a minimum capacitance of 38 pF (3.8×10^{-11} F) when an ac potential of 500 V is applied at a frequency of 1 MHz. Which of the materials listed in Table 19.5 are possible candidates? Why?

19.51 Consider a parallel-plate capacitor having an area of 2800 mm^2 and a plate separation of 4 mm, and with a material of dielectric constant 4.0 positioned between the plates. **(a)** What is the capacitance of this capacitor? **(b)** Compute the electric field that must be applied for 8.0×10^{-9} C to be stored on each plate.

19.52 In your own words, explain the mechanism by which charge-storing capacity is increased by the insertion of a dielectric material within the plates of a capacitor.

Field Vectors and Polarization
Types of Polarization

19.53 For NaCl, the ionic radii for Na^+ and Cl^- ions are 0.102 and 0.181 nm, respectively. If an externally applied electric field produces a 5% expansion of the lattice, compute the dipole moment for each Na^+–Cl^- pair. Assume that this material is completely unpolarized in the absence of an electric field.

19.54 The polarization P of a dielectric material positioned within a parallel-plate capacitor is to be 1.0×10^{-6} C/m^2.

(a) What must be the dielectric constant if an electric field of 5×10^4 V/m is applied?

(b) What will be the dielectric displacement D?

19.55 A charge of 4.0×10^{-11} C is to be stored on each plate of a parallel-plate capacitor having an area of 160 mm^2 and a plate separation of 3.5 mm.

(a) What voltage is required if a material having a dielectric constant of 5.0 is positioned within the plates?

(b) What voltage would be required if a vacuum were used?

(c) What are the capacitances for parts (a) and (b)?

(d) Compute the dielectric displacement for part (a).

(e) Compute the polarization for part (a).

19.56 (a) For each of the three types of polarization, briefly describe the mechanism by which dipoles are induced and/or oriented by the action of an applied electric field. **(b)** For solid lead titanate (PbTiO$_3$), gaseous neon, diamond, solid KCl, and liquid NH$_3$, what kind(s) of polarization is (are) possible? Why?

19.57 (a) Compute the magnitude of the dipole moment associated with each unit cell of BaTiO$_3$, as illustrated in Figure 19.35.

(b) Compute the maximum polarization that is possible for this material.

Frequency Dependence of the Dielectric Constant

19.58 The dielectric constant for a soda–lime glass measured at very high frequencies (on the order of 10^{15} Hz) is approximately 2.3. What fraction of the dielectric constant at relatively low frequencies (1 MHz) is attributed to ionic polarization? Neglect any orientation polarization contributions.

Ferroelectricity

19.59 Briefly explain why the ferroelectric behavior of BaTiO$_3$ ceases above its ferroelectric Curie temperature.

DESIGN PROBLEMS

Electrical Resistivity of Metals

19.D1 A 95 wt% Pt–5 wt% Ni alloy is known to have an electrical resistivity of 2.5×10^{-7} $\Omega \cdot$ m at room temperature (25°C). Calculate the composition of a platinum–nickel alloy that gives a room-temperature resistivity of 2.00×10^{-7} $\Omega \cdot$ m. The room-temperature resistivity of pure platinum may be determined from the data in Table 19.1; assume that platinum and nickel form a solid solution.

19.D2 Using information contained in Figures 19.8 and 19.38, determine the electrical conductivity of an 80 wt% Cu–20 wt% Zn alloy at −150°C (123 K).

19.D3 Is it possible to alloy copper with nickel to achieve a minimum tensile strength of 350 MPa and yet maintain an electrical conductivity of 2.5×10^6 $(\Omega \cdot$ m)$^{-1}$? If not, why? If so, what concentration of nickel is required? You may want to consult Figure 9.16a.

Extrinsic Semiconduction
Factors That Affect Carrier Mobility

19.D4 Specify an acceptor impurity type and concentration (in weight percent) that will produce a *p*-type silicon material having a room temperature electrical conductivity of 50 $(\Omega \cdot$ m)$^{-1}$.

19.D5 One integrated circuit design calls for diffusing boron into very high-purity silicon at an elevated temperature. It is necessary that at a distance 0.2 μm from the surface of the silicon wafer, the room-temperature electrical conductivity be $1.5 \times 10^3 \ (\Omega \cdot m)^{-1}$. The concentration of B at the surface of the Si is maintained at a constant level of $2.0 \times 10^{25} \ m^{-3}$; furthermore, it is assumed that the concentration of B in the original Si material is negligible, and that at room temperature the boron atoms are saturated. Specify the temperature at which this diffusion heat treatment is to take place if the treatment time is to be one hour. The diffusion coefficient for the diffusion of B in Si is a function of temperature as

$$D(m^2/s) = 2.4 \times 10^{-4} exp\left(-\frac{347 \ kJ/mol}{RT}\right)$$

Semiconductor Devices

19.D6 One of the procedures in the production of integrated circuits is the formation of a thin insulating layer of SiO_2 on the surface of chips (see Figure 19.26). This is accomplished by oxidizing the surface of the silicon by subjecting it to an oxidizing atmosphere (i.e., gaseous oxygen or water vapor) at an elevated temperature. The rate of growth of the oxide film is parabolic—that is, the thickness of the oxide layer (x) is a function of time (t) according to the following equation:

$$x^2 = Bt \qquad (19.37)$$

Here the parameter B is dependent on both temperature and the oxidizing atmosphere.

(a) For an atmosphere of O_2 at a pressure of 1 atm, the temperature dependence of B (in units of $\mu m^2/h$) is as follows:

$$B = 800 \ exp\left(-\frac{1.24 \ eV}{kT}\right) \qquad (19.38a)$$

where k is Boltzmann's constant (8.62×10^{-5} eV/atom) and T is in K. Calculate the time required to grow an oxide layer (in an atmosphere of O_2) that is 75 nm thick at both 750°C and 900°C.

(b) In an atmosphere of H_2O (1 atm pressure), the expression for B (again in units of $\mu m^2/h$) is

$$B = 215 \ exp\left(-\frac{0.70 \ eV}{kT}\right) \qquad (19.38b)$$

Now calculate the time required to grow an oxide layer that is 75 nm thick (in an atmosphere of H_2O) at both 750°C and 900°C, and compare these times with those computed in part (a).

19.D7 The base semiconducting material used in virtually all of our modern integrated circuits is silicon. However, silicon has some limitations and restrictions. Write an essay comparing the properties and applications (and/or potential applications) of silicon and gallium arsenide.

Conduction in Ionic Materials

19.D8 Problem 19.47 noted that FeO (wüstite) may behave as a semiconductor by virtue of the transformation of Fe^{2+} to Fe^{3+} and the creation of Fe^{2+} vacancies; the maintenance of electroneutrality requires that for every two Fe^{3+} ions, one vacancy is formed. The existence of these vacancies is reflected in the chemical formula of this nonstoichiometric wüstite as $Fe_{(1-x)}O$, where x is a small number having a value less than unity. The degree of nonstoichiometry (i.e., the value of x) may be varied by changing temperature and oxygen partial pressure. Compute the value of x that is required to produce an $Fe_{(1-x)}O$ material having a p-type electrical conductivity of 2500 $(\Omega \cdot m)^{-1}$; assume that the hole mobility is $2.0 \times 10^{-5} \ m^2/V \cdot s$, the crystal structure for FeO is sodium chloride (with a unit cell edge length of 0.437 nm), and the acceptor states are saturated.

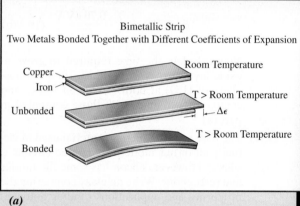

Bimetallic Strip
Two Metals Bonded Together with Different Coefficients of Expansion

Copper
Iron
Room Temperature

Unbonded
T > Room Temperature
Δε

Bonded
T > Room Temperature

(a)

Spiral Bimetal Element
Mercury Bulb

© Steven Langerman

(b)

© Kameleon007/iStockphoto

(c)

One type of *thermostat*—a device used to regulate temperature—uses the phenomenon of *thermal expansion*—the elongation of a material as it is heated. The heart of this type of thermostat is a *bimetallic strip*—strips of two metals having different coefficients of thermal expansion are bonded along their lengths. A change in temperature causes this strip to bend; upon heating, the metal having the greater expansion coefficient elongates more, producing the direction of bending shown in Figure (*a*). In the thermostat shown in Figure (*b*), the bimetallic strip is a coil or spiral; this configuration provides for a relatively long bimetallic strip, more deflection for a given temperature change, and greater accuracy. The metal having the higher expansion coefficient is located on the underside of the strip, such that, upon heating, the coil tends to unwind. Attached to the end of the coil is a *mercury switch*—a small glass bulb that contains several drops of mercury [Figure (*b*)]. This switch is mounted such that, when the temperature changes, deflections of the coil end tip the bulb one way or the other; accordingly, the blob of mercury rolls from one end of the bulb to the other. When temperature reaches the setpoint of the thermostat, electrical contact is made as the mercury rolls to one end; this switches on the heating or cooling unit (i.e., furnace or air conditioner). The unit shuts off when a limit temperature is achieved, and as the bulb tilts in the other direction, the blob of mercury rolls to the other end, and electrical contact is broken.

ASSOCIATED PRESS/© AP/Wide World Photos

(d)

Figure (*d*) shows the consequences of unseasonably high temperatures on July 24, 1978, near Asbury Park, New Jersey: rail lines buckled [which caused the derailment of a passenger car (background)] as a result of stresses from unanticipated thermal expansion.

Of the three primary material types, ceramics are the most susceptible to *thermal shock*—brittle fracture resulting from internal stresses that are established within a ceramic piece as a result of rapid changes in temperature (normally upon cooling). Thermal shock is normally an undesirable event, and the susceptibility of a ceramic material to this phenomenon is a function of its thermal and mechanical properties (coefficient of thermal expansion, thermal conductivity, modulus of elasticity, and fracture strength). From knowledge of the relationships between thermal shock parameters and these properties, it is possible (1) in some cases, to make appropriate alterations of the thermal and/or mechanical characteristics in order to render a ceramic more thermally shock resistant; and (2) for a specific ceramic material, to estimate the maximum allowable temperature change without causing fracture.

Learning Objectives

After studying this chapter, you should be able to do the following:

1. Define *heat capacity* and *specific heat*.
2. Note the primary mechanism by which thermal energy is assimilated in solid materials.
3. Determine the linear coefficient of thermal expansion, given the length alteration that accompanies a specified temperature change.
4. Briefly explain the phenomenon of thermal expansion from an atomic perspective using a potential energy–versus–interatomic separation plot.
5. Define *thermal conductivity*.
6. Note the two principal mechanisms of heat conduction in solids, and compare the relative magnitudes of these contributions for each of metals, ceramics, and polymeric materials.

20.1 INTRODUCTION

Thermal property refers to the response of a material to the application of heat. As a solid absorbs energy in the form of heat, its temperature rises and its dimensions increase. The energy may be transported to cooler regions of the specimen if temperature gradients exist, and ultimately, the specimen may melt. Heat capacity, thermal expansion, and thermal conductivity are properties that are often critical in the practical use of solids.

20.2 HEAT CAPACITY

heat capacity

A solid material, when heated, experiences an increase in temperature, signifying that some energy has been absorbed. **Heat capacity** indicates a material's ability to absorb heat from the external surroundings; it represents the amount of energy required to produce a unit temperature rise. In mathematical terms, the heat capacity C is expressed as follows:

Definition of *heat capacity*—ratio of energy change (energy gained or lost) and the resulting temperature change

$$C = \frac{dQ}{dT} \tag{20.1}$$

specific heat

where dQ is the energy required to produce a dT temperature change. Typically, heat capacity is specified per mole of material (i.e., J/mol·K, or cal/mol·K). **Specific heat** (often denoted by a lowercase c) is sometimes used; this represents the heat capacity per unit mass (J/kg·K, cal/g·K).

There are two ways in which this property may be measured, according to the environmental conditions accompanying the transfer of heat: One is the heat capacity while maintaining the specimen volume constant, C_v; the other is for constant external pressure, which is denoted C_p. The magnitude of C_p is always greater than or equal to C_v; however, any difference is very slight for most solid materials at room temperature and below.

Vibrational Heat Capacity

In most solids, the principal mode of thermal energy assimilation is by the increase in vibrational energy of the atoms. Atoms in solid materials are constantly vibrating at very high frequencies and with relatively small amplitudes. Rather than being independent of one another, the vibrations of adjacent atoms are coupled by virtue of atomic bonding. These vibrations are coordinated in such a way that traveling lattice waves are produced, a phenomenon represented in Figure 20.1. These may be thought of as elastic waves or simply sound waves, having short wavelengths and very high frequencies, which propagate through the crystal at the velocity of sound. The vibrational thermal energy for a material consists of a series of these elastic waves that have a range of distributions and frequencies. Only certain energy values are allowed (the energy is said to be *quantized*), and a single quantum of vibrational energy is called a **phonon**. (A phonon is analogous to the quantum of electromagnetic radiation, the *photon*.) On occasion, the vibrational waves themselves are termed *phonons*.

phonon

The thermal scattering of free electrons during electronic conduction (Section 19.7) is by these vibrational waves, and these elastic waves also participate in the transport of energy during thermal conduction (see Section 20.4).

Figure 20.1
Schematic represen-
tation of the genera-
tion of lattice waves
in a crystal by means
of atomic vibrations.
(Adapted from "The
Thermal Properties of
Materials" by J. Ziman.
Copyright © 1967 by
Scientific American,
Inc. All rights
reserved.)

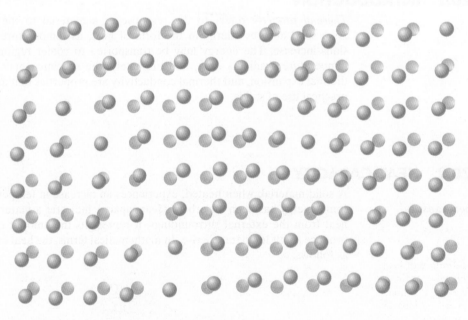

- Normal lattice positions for atoms
- Positions displaced because of vibrations

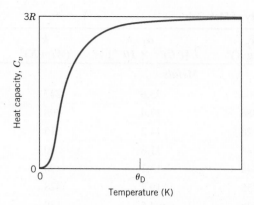

Figure 20.2 The temperature dependence of the heat capacity at constant volume; θ_D is the Debye temperature.

Temperature Dependence of the Heat Capacity

The variation with temperature of the vibrational contribution to the heat capacity at constant volume for many relatively simple crystalline solids is shown in Figure 20.2. The C_v is zero at 0 K, but it rises rapidly with temperature; this corresponds to an increased ability of the lattice waves to enhance their average energy with increasing temperature. At low temperatures, the relationship between C_v and the absolute temperature T is

Dependence of the heat capacity (at constant volume) on temperature at low temperatures (near 0 K)

$$C_v = AT^3 \tag{20.2}$$

where A is a temperature-independent constant. Above what is called the *Debye temperature* θ_D, C_v levels off and becomes essentially independent of temperature at a value of approximately $3R$, R being the gas constant. Thus, even though the total energy of the material is increasing with temperature, the quantity of energy required to produce a 1-degree temperature change is constant. The value of θ_D is less than room temperature for many solid materials, and 25 J/mol·K is a reasonable room-temperature approximation for C_v.[1] Table 20.1 presents experimental specific heats for a number of materials; c_p values for still more materials are tabulated in Table B.8 of Appendix B.

Other Heat Capacity Contributions

Other energy-absorptive mechanisms also exist that can add to the total heat capacity of a solid. In most instances, however, these are minor relative to the magnitude of the vibrational contribution. There is an electronic contribution in that electrons absorb energy by increasing their kinetic energy. However, this is possible only for free electrons—those that have been excited from filled states to empty states above the Fermi energy (Section 19.6). In metals, only electrons at states near the Fermi energy are capable of such transitions, and these represent only a very small fraction of the total number. An even smaller proportion of electrons experiences excitations in insulating and semiconducting materials. Hence, this electronic contribution is typically insignificant, except at temperatures near 0 K.

---

[1]For *solid metallic elements*, $C_v \cong 25$ J/mol·K; however, such is not the case for all solids. For example, at a temperature greater than its θ_D, the value of C_v for a ceramic material is approximately 25 joules per mole of ions—for example, the "molar" heat capacity of, say, Al_2O_3 is about $(5)(25$ J/mol·K$) = 125$ J/mol·K, given that there are five ions (two Al^{3+} ions and three O^{2-} ions) per formula (Al_2O_3) unit.

Table 20.1 Thermal Properties for a Variety of Materials

Material	c_p $(J/kg \cdot K)^a$	α_l $[(°C)^{-1} \times 10^{-6}]$	k $(W/m \cdot K)^b$	L $[\Omega \cdot W/(K)^2 \times 10^{-8}]$
Metals				
Aluminum	900	23.6	247	2.20
Copper	386	17.0	398	2.25
Gold	128	14.2	315	2.50
Iron	448	11.8	80	2.71
Nickel	443	13.3	90	2.08
Silver	235	19.7	428	2.13
Tungsten	138	4.5	178	3.20
1025 Steel	486	12.0	51.9	—
316 Stainless steel	502	16.0	15.9	—
Brass (70Cu–30Zn)	375	20.0	120	—
Kovar (54Fe–29Ni–17Co)	460	5.1	17	2.80
Invar (64Fe–36Ni)	500	1.6	10	2.75
Super Invar (63Fe–32Ni–5Co)	500	0.72	10	2.68
Ceramics				
Alumina (Al_2O_3)	775	7.6	39	—
Magnesia (MgO)	940	13.5^c	37.7	—
Spinel ($MgAl_2O_4$)	790	7.6^c	15.0^d	—
Fused silica (SiO_2)	740	0.4	1.4	—
Soda–lime glass	840	9.0	1.7	—
Borosilicate (Pyrex) glass	850	3.3	1.4	—
Polymers				
Polyethylene (high density)	1850	106–198	0.46–0.50	—
Polypropylene	1925	145–180	0.12	—
Polystyrene	1170	90–150	0.13	—
Polytetrafluoroethylene (Teflon)	1050	126–216	0.25	—
Phenol-formaldehyde, phenolic	1590–1760	122	0.15	—
Nylon 6,6	1670	144	0.24	—
Polyisoprene	—	220	0.14	—

[a]To convert to cal/g·K, multiply by 2.39×10^{-4}.
[b]To convert to cal/s·cm·K, multiply by 2.39×10^{-3}.
[c]Value measured at 100°C.
[d]Mean value taken over the temperature range 0°C to 1000°C.

Furthermore, in some materials other energy-absorptive processes occur at specific temperatures—for example, the randomization of electron spins in a ferromagnetic material as it is heated through its Curie temperature. A large spike is produced on the heat capacity–versus–temperature curve at the temperature of this transformation.

20.3 THERMAL EXPANSION

Most solid materials expand upon heating and contract when cooled. The change in length with temperature for a solid material may be expressed as follows:

For thermal expansion, dependence of fractional length change on the linear coefficient of thermal expansion and the temperature change

$$\frac{l_f - l_0}{l_0} = \alpha_l (T_f - T_0) \tag{20.3a}$$

or

$$\frac{\Delta l}{l_0} = \alpha_l \Delta T \tag{20.3b}$$

linear coefficient of thermal expansion

where l_0 and l_f represent, respectively, initial and final lengths with the temperature change from T_0 to T_f. The parameter α_l is called the **linear coefficient of thermal expansion**; it is a material property that is indicative of the extent to which a material expands upon heating and has units of reciprocal temperature $[(°C)^{-1}]$. Heating or cooling affects all the dimensions of a body, with a resultant change in volume. Volume changes with temperature may be computed from

For thermal expansion, dependence of fractional volume change on the volume coefficient of thermal expansion and the temperature change

$$\frac{\Delta V}{V_0} = \alpha_v \Delta T \tag{20.4}$$

where ΔV and V_0 are the volume change and the original volume, respectively, and α_v is the volume coefficient of thermal expansion. In many materials, the value of α_v is anisotropic; that is, it depends on the crystallographic direction along which it is measured. For materials in which the thermal expansion is isotropic, α_v is approximately $3\alpha_l$.

From an atomic perspective, thermal expansion is reflected by an increase in the average distance between the atoms. This phenomenon can best be understood by consulting the potential energy-versus-interatomic spacing curve for a solid material introduced previously (Figure 2.10b) and reproduced in Figure 20.3a. The curve is in

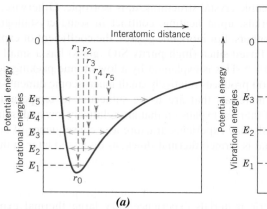
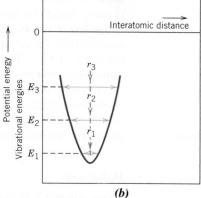

Figure 20.3 (a) Plot of potential energy versus interatomic distance, demonstrating the increase in interatomic separation with rising temperature. With heating, the interatomic separation increases from r_0 to r_1 to r_2, and so on. (b) For a symmetric potential energy–versus–interatomic distance curve, there is no increase in interatomic separation with rising temperature (i.e., $r_1 = r_2 = r_3$).

(Adapted from R. M. Rose, L. A. Shepard, and J. Wulff, *The Structure and Properties of Materials*, Vol. 4, *Electronic Properties.* Copyright © 1966 by John Wiley & Sons, New York. Reprinted by permission of John Wiley & Sons, Inc.)

Concept Check 20.1 **(a)** Explain why a brass lid ring on a glass canning jar loosens when heated.

(b) Suppose the ring is made of tungsten instead of brass. What will be the effect of heating the lid and jar? Why?

[*The answer may be found at* www.wiley.com/college/callister *(Student Companion Site).*]

20.4 THERMAL CONDUCTIVITY

thermal conductivity

For steady-state heat flow, dependence of heat flux on the thermal conductivity and the temperature gradient

Thermal conduction is the phenomenon by which heat is transported from high- to low-temperature regions of a substance. The property that characterizes the ability of a material to transfer heat is the **thermal conductivity**. It is best defined in terms of the expression

$$q = -k \frac{dT}{dx} \qquad (20.5)$$

where q denotes the *heat flux*, or heat flow, per unit time per unit area (*area* being taken as that perpendicular to the flow direction), k is the thermal conductivity, and dT/dx is the *temperature gradient* through the conducting medium.

The units of q and k are W/m^2 and $W/m\cdot K$, respectively. Equation 20.5 is valid only for steady-state heat flow—that is, for situations in which the heat flux does not change with time. The minus sign in the expression indicates that the direction of heat flow is from hot to cold, or down the temperature gradient.

Equation 20.5 is similar in form to Fick's first law (Equation 7.2) for steady-state diffusion. For these expressions, k is analogous to the diffusion coefficient D, and the temperature gradient is analogous to the concentration gradient, dC/dx.

Mechanisms of Heat Conduction

Heat is transported in solid materials by both lattice vibration waves (phonons) and free electrons. A thermal conductivity is associated with each of these mechanisms, and the total conductivity is the sum of the two contributions, or

$$k = k_l + k_e \qquad (20.6)$$

where k_l and k_e represent the lattice vibration and electron thermal conductivities, respectively; usually one or the other predominates. The thermal energy associated with phonons or lattice waves is transported in the direction of their motion. The k_l contribution results from a net movement of phonons from high- to low-temperature regions of a body across which a temperature gradient exists.

Free or conducting electrons participate in electronic thermal conduction. A gain in kinetic energy is imparted to the free electrons in a hot region of the specimen. They then migrate to colder areas, where some of this kinetic energy is transferred to the atoms (as vibrational energy) as a consequence of collisions with phonons or other imperfections in the crystal. The relative contribution of k_e to the total thermal conductivity increases with increasing free electron concentrations because more electrons are available to participate in this heat transference process.

Metals

In high-purity metals, the electron mechanism of heat transport is much more efficient than the phonon contribution because electrons are not as easily scattered as phonons and have higher velocities. Furthermore, metals are extremely good conductors of heat because relatively large numbers of free electrons exist that participate in thermal

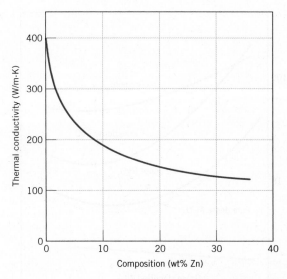

Figure 20.4 Thermal conductivity versus composition for copper–zinc alloys.
[Adapted from *Metals Handbook: Properties and Selection: Nonferrous Alloys and Pure Metals,* Vol. 2, 9th edition, H. Baker (Managing Editor), 1979. Reproduced by permission of ASM International, Materials Park, OH.]

conduction. The thermal conductivities of several common metals are given in Table 20.1; values generally range between about 20 and 400 W/m·K.

Because free electrons are responsible for both electrical and thermal conduction in pure metals, theoretical treatments suggest that the two conductivities should be related according to the *Wiedemann–Franz law*:

Wiedemann–Franz law—for metals, the ratio of thermal conductivity and the product of the electrical conductivity and temperature should be a constant

$$L = \frac{k}{\sigma T} \tag{20.7}$$

where σ is the electrical conductivity, T is the absolute temperature, and L is a constant. The theoretical value of L, 2.44×10^{-8} $\Omega \cdot W/(K)^2$, should be independent of temperature and the same for all metals if the heat energy is transported entirely by free electrons. Table 20.1 includes the experimental L values for several metals; note that the agreement between these and the theoretical value is quite reasonable (well within a factor of 2).

Alloying metals with impurities results in a reduction in the thermal conductivity, for the same reason that the electrical conductivity is decreased (Section 19.8); namely, the impurity atoms, especially if in solid solution, act as scattering centers, lowering the efficiency of electron motion. A plot of thermal conductivity versus composition for copper–zinc alloys (Figure 20.4) displays this effect.

Concept Check 20.2 The thermal conductivity of a plain carbon steel is greater than for a stainless steel. Why is this so? *Hint:* You may want to consult Section 13.2.

[*The answer may be found at* www.wiley.com/college/callister *(Student Companion Site).*]

Ceramics

Nonmetallic materials are thermal insulators inasmuch as they lack large numbers of free electrons. Thus the phonons are primarily responsible for thermal conduction: k_e is much smaller than k_l. Again, the phonons are not as effective as free electrons in the transport of heat energy as a result of the very efficient phonon scattering by lattice imperfections.

Thermal conductivity values for a number of ceramic materials are given in Table 20.1; room-temperature thermal conductivities range between approximately 2 and 50 W/m·K. Glass and other amorphous ceramics have lower conductivities than crystalline ceramics because the phonon scattering is much more effective when the atomic structure is highly disordered and irregular.

Figure 20.5 Dependence of thermal conductivity on temperature for several ceramic materials. (Adapted from W. D. Kingery, H. K. Bowen, and D. R. Uhlmann, *Introduction to Ceramics*, 2nd edition. Copyright © 1976 by John Wiley & Sons, New York. Reprinted by permission of John Wiley & Sons, Inc.)

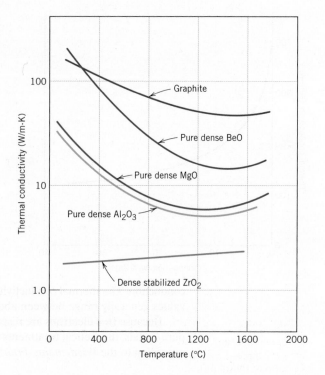

The scattering of lattice vibrations becomes more pronounced with rising temperature; hence, the thermal conductivity of most ceramic materials normally decreases with increasing temperature, at least at relatively low temperatures (Figure 20.5). As Figure 20.5 indicates, the conductivity begins to increase at higher temperatures, which is due to radiant heat transfer: significant quantities of infrared radiant heat may be transported through a transparent ceramic material. The efficiency of this process increases with temperature.

Porosity in ceramic materials may have a dramatic influence on thermal conductivity; under most circumstances, increasing the pore volume results in a reduction of the thermal conductivity. In fact, many ceramics used for thermal insulation are porous. Heat transfer across pores is typically slow and inefficient. Internal pores normally contain still air, which has an extremely low thermal conductivity—approximately 0.02 W/m·K. Furthermore, gaseous convection within the pores is also comparatively ineffective.

Concept Check 20.3   The thermal conductivity of a single-crystal ceramic specimen is slightly greater than that of a polycrystalline one of the same material. Why is this so?

[*The answer may be found at* www.wiley.com/college/callister *(Student Companion Site).*]

Polymers

As noted in Table 20.1, thermal conductivities for most polymers are on the order of 0.3 W/m·K. For these materials, energy transfer is accomplished by the vibration and rotation of the chain molecules. The magnitude of the thermal conductivity depends on the degree of crystallinity; a polymer with a highly crystalline and ordered structure has a greater conductivity than the equivalent amorphous material because of the more effective coordinated vibration of the molecular chains for the crystalline state.

Polymers are often used as thermal insulators because of their low thermal conductivities. As with ceramics, their insulative properties may be further enhanced by the introduction of small pores, which are typically introduced by foaming (Section 15.18). Foamed polystyrene is commonly used for drinking cups and insulating chests.

Concept Check 20.4   Which of a linear polyethylene ($\overline{M}_n$ = 450,000 g/mol) and a lightly branched polyethylene ($\overline{M}_n$ = 650,000 g/mol) has the higher thermal conductivity? Why? *Hint:* You may want to consult Section 4.13.

Concept Check 20.5   Explain why, on a cold day, the metal door handle of an automobile feels colder to the touch than a plastic steering wheel, even though both are at the same temperature.

[*The answers may be found at* www.wiley.com/college/callister (*Student Companion Site*).]

20.5 THERMAL STRESSES

thermal stress

Thermal stresses are stresses induced in a body as a result of changes in temperature. An understanding of the origins and nature of thermal stresses is important because these stresses can lead to fracture or undesirable plastic deformation.

Stresses Resulting from Restrained Thermal Expansion and Contraction

Let us first consider a homogeneous and isotropic solid rod that is heated or cooled uniformly—that is, no temperature gradients are imposed. For free expansion or contraction, the rod is stress-free. If, however, axial motion of the rod is restrained by rigid end supports, thermal stresses are introduced. The magnitude of the stress σ resulting from a temperature change from T_0 to T_f is

Dependence of thermal stress on elastic modulus, linear coefficient of thermal expansion, and temperature change

$$\sigma = E\alpha_l(T_0 - T_f) = E\alpha_l \Delta T \tag{20.8}$$

where E is the modulus of elasticity and α_l is the linear coefficient of thermal expansion. Upon heating ($T_f > T_0$), the stress is compressive ($\sigma < 0$) because rod expansion has been constrained. If the rod specimen is cooled ($T_f < T_0$), a tensile stress is imposed ($\sigma > 0$). Also, the stress in Equation 20.8 is the same as the stress required to elastically compress (or elongate) the rod specimen back to its original length after it had been allowed to freely expand (or contract) with the $T_0 - T_f$ temperature change.

EXAMPLE PROBLEM 20.1

Thermal Stress Created upon Heating

A brass rod is to be used in an application requiring its ends to be held rigid. If the rod is stress-free at room temperature (20°C), what is the maximum temperature to which the rod may be heated without exceeding a compressive stress of 172 MPa? Assume a modulus of elasticity of 100 GPa for brass.

Solution

Use Equation 20.8 to solve this problem, where the stress of 172 MPa is taken to be negative. Also, the initial temperature T_0 is 20°C, and the magnitude of the linear coefficient of thermal expansion from Table 20.1 is $20.0 \times 10^{-6} \, (°C)^{-1}$. Thus, solving for the final temperature T_f yields

$$T_f = T_0 - \frac{\sigma}{E\alpha_l}$$

$$= 20°C - \frac{-172 \, \text{MPa}}{(100 \times 10^3 \, \text{MPa})[20 \times 10^{-6} \, (°C)^{-1}]}$$

$$= 20°C + 86°C = 106°C$$

Stresses Resulting from Temperature Gradients

When a solid body is heated or cooled, the internal temperature distribution depends on its size and shape, the thermal conductivity of the material, and the rate of temperature change. Thermal stresses may be established as a result of temperature gradients across a body, which are frequently caused by rapid heating or cooling, in that the outside changes temperature more rapidly than the interior; differential dimensional changes restrain the free expansion or contraction of adjacent volume elements within the piece. For example, upon heating, the exterior of a specimen is hotter and, therefore, expands more than the interior regions. Hence, compressive surface stresses are induced and are balanced by tensile interior stresses. The interior–exterior stress conditions are reversed for rapid cooling such that the surface is put into a state of tension.

Thermal Shock of Brittle Materials

For ductile metals and polymers, alleviation of thermally induced stresses may be accomplished by plastic deformation. However, the nonductility of most ceramics enhances the possibility of brittle fracture from these stresses. Rapid cooling of a brittle body is more likely to inflict such thermal shock than heating because the induced surface stresses are tensile. Crack formation and propagation from surface flaws are more probable when an imposed stress is tensile (Section 14.6).

The capacity of a material to withstand this kind of failure is termed its *thermal shock resistance*. For a ceramic body that is cooled rapidly, the resistance to thermal shock depends not only on the magnitude of the temperature change, but also on the mechanical and thermal properties of the material. The thermal shock resistance is best for ceramics with high fracture strengths σ_f and high thermal conductivities, as well as low moduli of elasticity and low coefficients of thermal expansion. The resistance of many materials to this type of failure may be approximated by a thermal shock resistance parameter *TSR*:

Definition of thermal shock resistance parameter

$$TSR \cong \frac{\sigma_f k}{E\alpha_l} \tag{20.9}$$

Thermal shock may be prevented by altering the external conditions to the degree that cooling or heating rates are reduced and temperature gradients across a body are minimized. Modification of the thermal and/or mechanical characteristics in Equation 20.9 may also enhance the thermal shock resistance of a material. Of these parameters, the coefficient of thermal expansion is probably most easily changed and controlled. For example, common soda–lime glasses, which have an α_l of approximately 9×10^{-6} $(°C)^{-1}$, are particularly susceptible to thermal shock, as anyone who has baked can probably attest. Reducing the CaO and Na_2O contents while at the same time adding B_2O_3 in sufficient quantities to form borosilicate (or Pyrex) glass reduces the coefficient of expansion to about $3 \times 10^{-6} \, (°C)^{-1}$; this material is entirely suitable for kitchen oven heating and cooling cycles.[2] The introduction of some relatively large pores or a ductile second phase may also improve the thermal shock characteristics of a material; both impede the propagation of thermally induced cracks.

It is often necessary to remove thermal stresses in ceramic materials as a means of improving their mechanical strengths and optical characteristics. This may be accomplished by an annealing heat treatment, as discussed for glasses in Section 17.8.

SUMMARY

Heat Capacity
- Heat capacity represents the quantity of heat required to produce a unit rise in temperature for 1 mole of a substance; on a per-unit-mass basis, it is termed *specific heat*.
- Most of the energy assimilated by many solid materials is associated with increasing the vibrational energy of the atoms.
- Only specific vibrational energy values are allowed (the energy is said to be quantized); a single quantum of vibrational energy is called a *phonon*.
- For many crystalline solids and at temperatures within the vicinity of 0 K, the heat capacity measured at constant volume varies as the cube of the absolute temperature (Equation 20.2).
- In excess of the Debye temperature, C_v becomes temperature independent, assuming a value of approximately $3R$.

Thermal Expansion
- Solid materials expand when heated and contract when cooled. The fractional change in length is proportional to the temperature change, the constant of proportionality being the coefficient of thermal expansion (Equation 20.3).
- Thermal expansion is reflected by an increase in the average interatomic separation, which is a consequence of the asymmetric nature of the potential energy–versus–interatomic spacing curve trough (Figure 20.3a). The larger the interatomic bonding energy, the lower the coefficient of thermal expansion.
- Values of coefficients of thermal expansion for polymers are typically greater than those for metals, which in turn are greater than those for ceramic materials.

Thermal Conductivity
- The transport of thermal energy from high- to low-temperature regions of a material is termed *thermal conduction*.
- For steady-state heat transport, flux may be determined using Equation 20.5.
- For solid materials, heat is transported by free electrons and by vibrational lattice waves, or phonons.
- The high thermal conductivities for relatively pure metals are due to the large numbers of free electrons and the efficiency with which these electrons transport thermal energy. By way of contrast, ceramics and polymers are poor thermal conductors because free-electron concentrations are low and phonon conduction predominates.

Thermal Stresses
- Thermal stresses, which are introduced in a body as a consequence of temperature changes, may lead to fracture or undesirable plastic deformation.

---

[2]In the United States, some Pyrex baking glassware products are now made of less-expensive soda–lime glasses that have been tempered thermally. This glassware is not as resistant to thermal shock as a borosilicate glass. Consequently, a number of these baking dishes have shattered when subjected to reasonable temperature changes encountered during normal baking activities, sending glass shards in all directions (and in some instances causing injuries). Pyrex glassware sold in Europe is much more thermally shock resistant. A different company owns the rights to the Pyrex name in Europe, and it still uses borosilicate glass in its manufacture.

- One source of thermal stresses is the restrained thermal expansion (or contraction) of a body. Stress magnitude may be computed using Equation 20.8.

- The generation of thermal stresses resulting from the rapid heating or cooling of a body of material results from temperature gradients between the outside and interior portions and accompanying differential dimensional changes.

- *Thermal shock* is the fracture of a body resulting from thermal stresses induced by rapid temperature changes. Because ceramic materials are brittle, they are especially susceptible to this type of failure.

Important Terms and Concepts

heat capacity
linear coefficient of thermal
 expansion

phonon
specific heat
thermal conductivity

thermal shock
thermal stress

REFERENCES

Bagdade, S. D., *ASM Ready Reference: Thermal Properties of Metals,* ASM International, Materials Park, OH, 2002.

Hummel, R. E., *Electronic Properties of Materials,* 4th edition, Springer, New York, 2011.

Jiles, D. C., *Introduction to the Electronic Properties of Materials,* 2nd edition, CRC Press, Boca Raton, FL, 2001.

Kingery, W. D., H. K. Bowen, and D. R. Uhlmann, *Introduction to Ceramics,* 2nd edition, Wiley, New York, 1976. Chapters 12 and 16.

QUESTIONS AND PROBLEMS

Heat Capacity

20.1 Estimate the energy required to raise the temperature of 2 kg of the following materials from 20 to 100°C (293 to 373 K): aluminum, steel, soda–lime glass, and high-density polyethylene.

20.2 To what temperature would 11 kg of a 1025 steel specimen at 25°C (298 K) be raised if 130 KJ of heat is supplied?

20.3 **(a)** Determine the room temperature heat capacities at constant pressure for the following materials: aluminum, silver, tungsten, and 70Cu–30Zn brass. **(b)** How do these values compare with one another? How do you explain this?

20.4 For aluminum, the heat capacity at constant volume C_v at 30 K is 0.81 J/mol·K, and the Debye temperature is 375 K. Estimate the specific heat **(a)** at 50 K and **(b)** at 425 K.

20.5 The constant A in Equation 20.2 is $12\pi^4 R/5\theta_D^3$, where R is the gas constant and θ_D is the Debye temperature (K). Estimate θ_D for copper, given that the specific heat is 0.78 J/kg·K at 10 K.

20.6 **(a)** Briefly explain why C_v rises with increasing temperature at temperatures near 0 K. **(b)** Briefly explain why C_v becomes virtually independent of temperature at temperatures far removed from 0 K.

Thermal Expansion

20.7 An aluminum wire 10 m long is cooled from 38 to −1°C (311 to 272 K). How much change in length will it experience?

20.8 A 0.1 m rod of a metal elongates 0.2 mm on heating from 20 to 100°C (293 to 373 K). Determine the value of the linear coefficient of thermal expansion for this material.

20.9 Briefly explain thermal expansion using the potential-energy-versus-interatomic-spacing curve.

20.10 Compute the density for nickel at 500°C, given that its room-temperature density is 8.902 g/cm³. Assume that the volume coefficient of thermal expansion, α_v, is equal to $3\alpha_l$.

20.11 When a metal is heated its density decreases. There are two sources that give rise to this diminishment of ρ: (1) the thermal expansion of the solid and (2) the formation of vacancies (Section 6.2). Consider a specimen of copper at room temperature (20°C) that has a density of 8.940 g/cm³. **(a)** Determine its density upon heating to 1000°C when only thermal expansion is considered. **(b)** Repeat the calculation when the introduction of vacancies is taken into account. Assume that the energy of vacancy formation is 0.90 eV/atom, and that the volume coefficient of thermal expansion, α_v, is equal to $3\alpha_l$.

20.12 The difference between the specific heats at constant pressure and volume is described by the expression

$$c_p - c_v = \frac{\alpha_v^2 v_0 T}{\beta} \qquad (20.10)$$

where α_v, is the volume coefficient of thermal expansion, v_0 is the specific volume (i.e., volume per unit mass, or the reciprocal of density), β is the compressibility, and T is the absolute temperature. Compute the values of c_v at room temperature (293 K) for copper and nickel using the data in Table 20.1, assuming that $\alpha_v = 3\alpha_l$ and given that the values of β for Cu and Ni are 8.35×10^{-12} and 5.51×10^{-12} $(Pa)^{-1}$, respectively.

20.13 To what temperature must a cylindrical rod of tungsten 10.000 mm in diameter and a plate of 316 stainless steel having a circular hole 9.988 mm in diameter have to be heated for the rod to just fit into the hole? Assume that the initial temperature is 25°C.

Thermal Conductivity

20.14 **(a)** Calculate the heat flux through a sheet of steel 10 mm thick if the temperatures at the two faces are 300 and 100°C (573 and 373 K); assume steady-state heat flow. **(b)** What is the heat loss per hour if the area of the sheet is 0.25 m²? **(c)** What will be the heat loss per hour if soda–lime glass instead of steel is used? **(d)** Calculate the heat loss per hour if steel is used and the thickness is increased to 20 mm.

20.15 **(a)** Would you expect Equation 20.7 to be valid for ceramic and polymeric materials? Why or why not? **(b)** Estimate the value for the Wiedemann–Franz constant L [in $\Omega \cdot W/(K)^2$] at room temperature (293 K) for the following nonmetals: silicon (intrinsic), glass-ceramic (Pyroceram), fused silica, polycarbonate, and polytetrafluoroethylene. Consult Tables B.7 and B.9 in Appendix B.

20.16 Briefly explain why the thermal conductivities are higher for crystalline than noncrystalline ceramics.

20.17 Briefly explain why metals are typically better thermal conductors than ceramic materials.

20.18 **(a)** Briefly explain why porosity decreases the thermal conductivity of ceramic and polymeric materials, rendering them more thermally insulative. **(b)** Briefly explain how the degree of crystallinity affects the thermal conductivity of polymeric materials and why.

20.19 For some ceramic materials, why does the thermal conductivity first decrease and then increase with rising temperature?

20.20 For each of the following pairs of materials, decide which has the larger thermal conductivity. Justify your choices.

(a) Pure copper; aluminum bronze (95 wt% Cu–5 wt% Al)

(b) Fused silica; quartz

(c) Linear polyethylene; branched polyethylene

(d) Random poly(styrene-butadiene) copolymer; alternating poly(styrene-butadiene) copolymer

20.21 We might think of a porous material as being a composite wherein one of the phases is a pore phase. Estimate upper and lower limits for the room-temperature thermal conductivity of a magnesium oxide material having a volume fraction of 0.30 of pores that are filled with still air.

20.22 Nonsteady-state heat flow may be described by the following partial differential equation:

$$\frac{\partial T}{\partial t} = D_T \frac{\partial^2 T}{\partial x^2}$$

where D_T is the thermal diffusivity; this expression is the thermal equivalent of Fick's second law of diffusion (Equation 7.4b). The thermal diffusivity is defined according to

$$D_T = \frac{k}{\rho c_p}$$

In this expression, k, ρ, and c_p represent the thermal conductivity, the mass density, and the specific heat at constant pressure, respectively.

(a) What are the SI units for D_T?

(b) Determine values of D_T for aluminum, steel, aluminum oxide, soda–lime glass, polystyrene, and nylon 6,6 using the data in Table 20.1. Density values are included in Table B.1, Appendix B.

Thermal Stresses

20.23 Beginning with Equation 20.3, show that Equation 20.8 is valid.

20.24 **(a)** Briefly explain why thermal stresses may be introduced into a structure by rapid heating or cooling. **(b)** For cooling, what is the nature of the surface stresses? **(c)** For heating, what is the nature of the surface stresses?

20.25 (a) If a rod of 1025 steel 0.5 m long is heated from 20 to 80°C (293 to 353 K) while its ends are maintained rigid, determine the type and magnitude of stress that develops. Assume that at 20°C the rod is stress free. **(b)** What will be the stress magnitude if a rod 1 m long is used? **(c)** If the rod in part (a) is cooled from 20 to −10°C (293 to 263 K), what type and magnitude of stress will result?

20.26 A copper wire is stretched with a stress of 70 MPa at 20°C (293 K). If the length is held constant, to what temperature must the wire be heated to reduce the stress to 35 MPa?

20.27 If a cylindrical rod of nickel 100.00 mm long and 8.000 mm in diameter is heated from 20°C to 200°C while its ends are maintained rigid, determine its change in diameter. You may want to consult Table 8.1.

20.28 The two ends of a cylindrical rod of 1025 steel 75.00 mm long and 10.000 mm in diameter are maintained rigid. If the rod is initially at 25°C, to what temperature must it be cooled to have a 0.008-mm reduction in diameter?

20.29 What measures may be taken to reduce the likelihood of thermal shock of a ceramic piece?

DESIGN PROBLEMS

Thermal Expansion

20.D1 Railroad tracks made of 1025 steel are to be laid during the time of year when the temperature averages 10°C (283 K). If a joint space of 4.6 mm is allowed between the standard 11.9-m- long rails, what is the hottest possible temperature that can be tolerated without the introduction of thermal stresses?

Thermal Stresses

20.D2 The ends of a cylindrical rod 6.4 mm in diameter and 250 mm long are mounted between rigid supports. The rod is stress free at room temperature [20°C (293 K)]; upon cooling to −40°C (233 K), a maximum thermally induced tensile stress of 125 MPa is possible. Of which of the following metals or alloys may the rod be fabricated: aluminum, copper, brass, 1025 steel, and tungsten? Why?

20.D3 (a) What are the units for the thermal shock resistance parameter (*TSR*)? **(b)** Rank the following ceramic materials according to their thermal shock resistance: glass-ceramic (Pyroceram), partially stabilized zirconia, and borosilicate (Pyrex) glass. Appropriate data may be found in Tables B.2, B.4, B.6, and B.7 of Appendix B.

20.D4 Equation 20.9, for the thermal shock resistance of a material, is valid for relatively low rates of heat transfer. When the rate is high, then, upon cooling of a body, the maximum temperature change allowable without thermal shock, ΔT_f, is approximately

$$\Delta T_f \cong \frac{\sigma_f}{E\alpha_l}$$

where σ_f is the fracture strength. Using the data in Tables B.2, B.4, and B.6 (Appendix B), determine ΔT_f for a glass-ceramic (Pyroceram), partially stabilized zirconia, and fused silica.

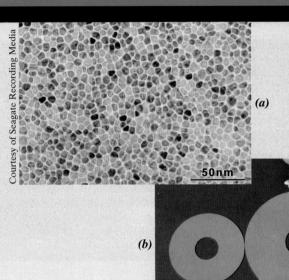

Courtesy of Seagate Recording Media

50nm

(a)

(*a*) Transmission electron micrograph showing the microstructure of the perpendicular magnetic recording medium used in hard disk drives.

(*b*) Magnetic storage hard disks used in laptop (left) and desktop (right) computers.

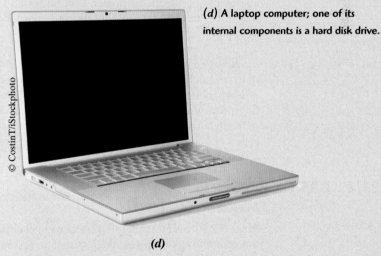

(b)

© William D. Callister, Jr.

(*c*) The inside of a hard disk drive. The circular disk will typically spin at a rotational velocity of 5400 or 7200 revolutions per minute.

(c)

Courtesy of Seagate Recording Media

(*d*) A laptop computer; one of its internal components is a hard disk drive.

© CostinT/iStockphoto

(d)

An understanding of the mechanism that explains the permanent magnetic behavior of some materials may allow us to alter and in some cases tailor the magnetic properties. For example, in Design Example 21.1 we note how the behavior of a ceramic magnetic material may be enhanced by changing its composition.

Learning Objectives

After studying this chapter, you should be able to do the following:

1. Determine the magnetization of a material given its magnetic susceptibility and the applied magnetic field strength.

2. From an electronic perspective, note and briefly explain the two sources of magnetic moments in materials.

3. Briefly explain the nature and source of (a) diamagnetism, (b) paramagnetism, and (c) ferromagnetism.

4. In terms of crystal structure, explain the source of ferrimagnetism for cubic ferrites.

5. (a) Describe magnetic hysteresis; (b) explain why ferromagnetic and ferrimagnetic materials experience magnetic hysteresis; and (c) explain why these materials may become permanent magnets.

6. Note the distinctive magnetic characteristics for both soft and hard magnetic materials.

7. Describe the phenomenon of *superconductivity*.

21.1 INTRODUCTION

Magnetism—the phenomenon by which materials exert an attractive or repulsive force or influence on other materials—has been known for thousands of years. However, the underlying principles and mechanisms that explain magnetic phenomena are complex and subtle, and their understanding has eluded scientists until relatively recent times. Many modern technological devices rely on magnetism and magnetic materials, including electrical power generators and transformers, electric motors, radio, television, telephones, computers, and components of sound and video reproduction systems.

Iron, some steels, and the naturally occurring mineral lodestone are well-known examples of materials that exhibit magnetic properties. Not so familiar, however, is the fact that all substances are influenced to one degree or another by the presence of a magnetic field. This chapter provides a brief description of the origin of magnetic fields and discusses magnetic field vectors and magnetic parameters; diamagnetism, paramagnetism, ferromagnetism, and ferrimagnetism; different magnetic materials; and superconductivity.

21.2 BASIC CONCEPTS

Magnetic Dipoles

Magnetic forces are generated by moving electrically charged particles; these magnetic forces are in addition to any electrostatic forces that may exist. Often it is convenient to think of magnetic forces in terms of fields. Imaginary lines of force may be drawn to indicate the direction of the force at positions in the vicinity of the field source. The magnetic field distributions as indicated by lines of force are shown for a current loop and a bar magnet in Figure 21.1.

Magnetic dipoles are found to exist in magnetic materials, which, in some respects, are analogous to electric dipoles (Section 19.19). Magnetic dipoles may be thought of as small bar magnets composed of north and south poles instead of positive and negative

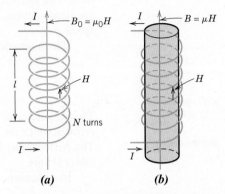

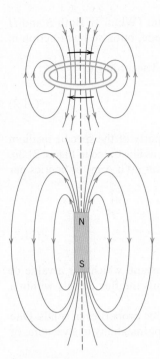

Figure 21.3 (*a*) The magnetic field *H* as generated by a cylindrical coil is dependent on the current *I*, the number of turns *N*, and the coil length *l*, according to Equation 21.1. The magnetic flux density B_0 in the presence of a vacuum is equal to $\mu_0 H$, where μ_0 is the permeability of a vacuum, $4\pi \times 10^{-7}$ H/m. (*b*) The magnetic flux density *B* within a solid material is equal to μH, where μ is the permeability of the solid material. (Adapted from A. G. Guy, *Essentials of Materials Science*, McGraw-Hill Book Company, New York, 1976.)

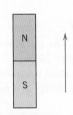

Figure 21.1 Magnetic field lines of force around a current loop and a bar magnet.

Figure 21.2 The magnetic moment as designated by an arrow.

electric charges. In our discussion, magnetic dipole moments are represented by arrows, as shown in Figure 21.2. Magnetic dipoles are influenced by magnetic fields in a manner similar to the way in which electric dipoles are affected by electric fields (Figure 19.30). Within a magnetic field, the force of the field exerts a torque that tends to orient the dipoles with the field. A familiar example is the way in which a magnetic compass needle lines up with the Earth's magnetic field.

Magnetic Field Vectors

Before discussing the origin of magnetic moments in solid materials, we describe magnetic behavior in terms of several field vectors. The externally applied magnetic field, sometimes called the **magnetic field strength**, is designated by *H*. If the magnetic field is generated by means of a cylindrical coil (or solenoid) consisting of *N* closely spaced turns having a length *l* and carrying a current of magnitude *I*, then

magnetic field strength

Magnetic field strength within a coil—dependence on number of turns, applied current, and coil length

$$H = \frac{NI}{l} \tag{21.1}$$

A schematic diagram of such an arrangement is shown in Figure 21.3*a*. The magnetic field that is generated by the current loop and the bar magnet in Figure 21.1 is an *H* field. The units of *H* are ampere-turns per meter, or just amperes per meter.

magnetic induction

magnetic flux density

The **magnetic induction**, or **magnetic flux density**, denoted by *B*, represents the magnitude of the internal field strength within a substance that is subjected to an *H* field. The units for B are *teslas* [or webers per square meter (Wb/m^2)]. Both B and H are field vectors, being characterized not only by magnitude, but also by direction in space.

The magnetic field strength and flux density are related according to

$$B = \mu H \tag{21.2}$$

The parameter μ is called the **permeability**, which is a property of the specific medium through which the H field passes and in which B is measured, as illustrated in Figure 21.3b. The permeability has dimensions of webers per ampere-meter (Wb/A·m) or henries per meter (H/m).

In a vacuum,

$$B_0 = \mu_0 H \tag{21.3}$$

where μ_0 is the *permeability of a vacuum,* a universal constant, which has a value of $4\pi \times 10^{-7}$ (1.257×10^{-6}) H/m. The parameter B_0 represents the flux density within a vacuum as demonstrated in Figure 21.3a.

Several parameters may be used to describe the magnetic properties of solids. One of these is the ratio of the permeability in a material to the permeability in a vacuum, or

$$\mu_r = \frac{\mu}{\mu_0} \tag{21.4}$$

where μ_r is called the *relative permeability,* which is unitless. The permeability or relative permeability of a material is a measure of the degree to which the material can be magnetized, or the ease with which a B field can be induced in the presence of an external H field.

Another field quantity, M, called the **magnetization** of the solid, is defined by the expression

$$B = \mu_0 H + \mu_0 M \tag{21.5}$$

In the presence of an H field, the magnetic moments within a material tend to become aligned with the field and to reinforce it by virtue of their magnetic fields; the term $\mu_0 M$ in Equation 21.5 is a measure of this contribution.

The magnitude of M is proportional to the applied field as follows:

$$M = \chi_m H \tag{21.6}$$

and χ_m is called the **magnetic susceptibility**, which is unitless.[1] The magnetic susceptibility and the relative permeability are related as follows:

$$\chi_m = \mu_r - 1 \tag{21.7}$$

---

[1]This χ_m is taken to be the volume susceptibility in SI units, which, when multiplied by H, yields the magnetization per unit volume (cubic meter) of material. Other susceptibilities are also possible; see Problem 21.3.

Table 21.1 Magnetic Units and Conversion Factors for the SI and cgs–emu Systems

| Quantity | Symbol | SI Units | | cgs–emu Unit | Conversion |
		Derived	Primary		
Magnetic induction (flux density)	B	Tesla $(Wb/m^2)^a$	$kg/s·C$	Gauss	$1\ Wb/m^2 = 10^4$ gauss
Magnetic field strength	H	Amp·turn/m	$C/m·s$	Oersted	1 amp·turn/m $= 4\pi \times 10^{-3}$ oersted
Magnetization	M (SI) I (cgs–emu)	Amp·turn/m	$C/m·s$	Maxwell/cm^2	1 amp·turn/m $= 10^{-3}$ maxwell/cm^2
Permeability of a vacuum	μ_0	Henry/m^b	$kg·m/C^2$	Unitless (emu)	$4\pi \times 10^{-7}$ henry/m $= 1$ emu
Relative permeability	μ_r (SI) μ' (cgs–emu)	Unitless	Unitless	Unitless	$\mu_r = \mu'$
Susceptibility	χ_m (SI) χ'_m (cgs–emu)	Unitless	Unitless	Unitless	$\chi_m = 4\pi \chi'_m$

^{a}Units of the weber (Wb) are volt-seconds.
^{b}Units of the henry are webers per ampere.

There is a dielectric analogue for each of the foregoing magnetic field parameters. The B and H fields are, respectively, analogous to the dielectric displacement D and the electric field $\mathscr{E}$, whereas the permeability μ is analogous to the permittivity ϵ (cf. Equations 21.2 and 19.30). Furthermore, the magnetization M and polarization P are correlates (Equations 21.5 and 19.31).

Magnetic units may be a source of confusion because there are really two systems in common use. The ones used thus far are SI [rationalized *MKS* (meter-kilogram-second)]; the others come from the *cgs–emu* (centimeter-gram-second–electromagnetic unit) system. The units for both systems as well as the appropriate conversion factors are given in Table 21.1.

Origins of Magnetic Moments

The macroscopic magnetic properties of materials are a consequence of *magnetic moments* associated with individual electrons. Some of these concepts are relatively complex and involve some quantum mechanical principles beyond the scope of this discussion; consequently, simplifications have been made and some of the details omitted. Each electron in an atom has magnetic moments that originate from two sources. One is related to its orbital motion around the nucleus; because it is a moving charge, an electron may be considered to be a small current loop, generating a very small magnetic field and having a magnetic moment along its axis of rotation, as schematically illustrated in Figure 21.4a.

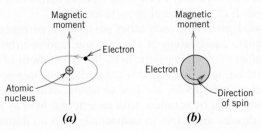

(a) **(b)**

Figure 21.4 Demonstration of the magnetic moment associated with (*a*) an orbiting electron and (*b*) a spinning electron.

Each electron may also be thought of as spinning around an axis; the other magnetic moment originates from this electron spin, which is directed along the spin axis as shown in Figure 21.4*b*. Spin magnetic moments may be only in an "up" direction or in an antiparallel "down" direction. Thus each electron in an atom may be thought of as being a small magnet having permanent orbital and spin magnetic moments.

Bohr magneton

The most fundamental magnetic moment is the **Bohr magneton** μ_B, which is of magnitude 9.27×10^{-24} A·m^2. For each electron in an atom, the spin magnetic moment is $\pm\mu_B$ (plus for spin up, minus for spin down). Furthermore, the orbital magnetic moment contribution is equal to $m_l\mu_B$, m_l being the magnetic quantum number of the electron, as mentioned in Section 2.3.

In each atom, orbital moments of some electron pairs cancel each other; this also holds true for the spin moments. For example, the spin moment of an electron with spin up cancels that of one with spin down. The net magnetic moment, then, for an atom is just the sum of the magnetic moments of each of the constituent electrons, including both orbital and spin contributions, and taking into account moment cancellation. For an atom having completely filled electron shells or subshells, when all electrons are considered, there is total cancellation of both orbital and spin moments. Thus, materials composed of atoms having completely filled electron shells are not capable of being permanently magnetized. This category includes the inert gases (He, Ne, Ar, etc.) as well as some ionic materials. The types of magnetism include diamagnetism, paramagnetism, and ferromagnetism; in addition, antiferromagnetism and ferrimagnetism are considered to be subclasses of ferromagnetism. All materials exhibit at least one of these types, and the behavior depends on the response of electron and atomic magnetic dipoles to the application of an externally applied magnetic field.

21.3 DIAMAGNETISM AND PARAMAGNETISM

diamagnetism

Diamagnetism is a very weak form of magnetism that is nonpermanent and persists only while an external field is being applied. It is induced by a change in the orbital motion of electrons due to an applied magnetic field. The magnitude of the induced magnetic moment is extremely small and in a direction opposite to that of the applied field. Thus, the relative permeability μ_r is less than unity (however, only very slightly), and the magnetic susceptibility is negative—that is, the magnitude of the B field within a diamagnetic solid is less than that in a vacuum. The volume susceptibility χ_m for diamagnetic solid materials is on the order of -10^{-5}. When placed between the poles of a strong electromagnet, diamagnetic materials are attracted toward regions where the field is weak.

Figure 21.5*a* illustrates schematically the atomic magnetic dipole configurations for a diamagnetic material with and without an external field; here, the arrows represent atomic dipole moments, whereas for the preceding discussion, arrows denoted only electron moments. The dependence of B on the external field H for a material that exhibits diamagnetic behavior is presented in Figure 21.6. Table 21.2 gives the susceptibilities of several diamagnetic materials. Diamagnetism is found in all materials, but because it is so weak, it can be observed only when other types of magnetism are totally absent. This form of magnetism is of no practical importance.

paramagnetism

For some solid materials, each atom possesses a permanent dipole moment by virtue of incomplete cancellation of electron spin and/or orbital magnetic moments. In the absence of an external magnetic field, the orientations of these atomic magnetic moments are random, such that a piece of material possesses no net macroscopic magnetization. These atomic dipoles are free to rotate, and **paramagnetism** results when they preferentially align, by rotation, with an external field as shown in Figure 21.5*b*. These magnetic dipoles are acted on individually with no mutual interaction between

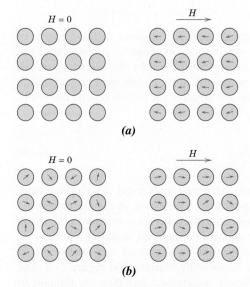

Figure 21.5 (*a*) The atomic dipole configuration for a diamagnetic material with and without a magnetic field. In the absence of an external field, no dipoles exist; in the presence of a field, dipoles are induced that are aligned opposite to the field direction. (*b*) Atomic dipole configuration with and without an external magnetic field for a paramagnetic material.

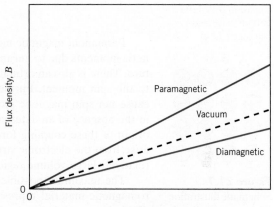

Figure 21.6 Schematic representation of the flux density B versus the magnetic field strength H for diamagnetic and paramagnetic materials.

adjacent dipoles. Inasmuch as the dipoles align with the external field, they enhance it, giving rise to a relative permeability μ_r that is greater than unity and to a relatively small but positive magnetic susceptibility. Susceptibilities for paramagnetic materials range from about 10^{-5} to 10^{-2} (Table 21.2). A schematic B-versus-H curve for a paramagnetic material is shown in Figure 21.6.

Both diamagnetic and paramagnetic materials are considered nonmagnetic because they exhibit magnetization only when in the presence of an external field. Also, for both, the flux density B within them is almost the same as it would be in a vacuum.

Table 21.2

Room-Temperature Magnetic Susceptibilities for Diamagnetic and Paramagnetic Materials

Diamagnetics		**Paramagnetics**	
Material	*Susceptibility χ_m (volume) (SI units)*	*Material*	*Susceptibility χ_m (volume) (SI units)*
Aluminum oxide	-1.81×10^{-5}	Aluminum	2.07×10^{-5}
Copper	-0.96×10^{-5}	Chromium	3.13×10^{-4}
Gold	-3.44×10^{-5}	Chromium chloride	1.51×10^{-3}
Mercury	-2.85×10^{-5}	Manganese sulfate	3.70×10^{-3}
Silicon	-0.41×10^{-5}	Molybdenum	1.19×10^{-4}
Silver	-2.38×10^{-5}	Sodium	8.48×10^{-6}
Sodium chloride	-1.41×10^{-5}	Titanium	1.81×10^{-4}
Zinc	-1.56×10^{-5}	Zirconium	1.09×10^{-4}

21.4 FERROMAGNETISM

ferromagnetism

domain

saturation magnetization

For a ferromagnetic material, relationship between magnetic flux density and magnetization

Certain metallic materials possess a permanent magnetic moment in the absence of an external field and manifest very large and permanent magnetizations. These are the characteristics of **ferromagnetism**, and they are displayed by the transition metals iron (as BCC α-ferrite), cobalt, nickel, and some rare earth metals such as gadolinium (Gd). Magnetic susceptibilities as high as 10^6 are possible for ferromagnetic materials. Consequently, $H \ll M$, and from Equation 21.5 we write

$$B \cong \mu_0 M \tag{21.8}$$

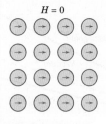

$H = 0$

Figure 21.7
Schematic illustration of the mutual alignment of atomic dipoles for a ferromagnetic material, which will exist even in the absence of an external magnetic field.

Permanent magnetic moments in ferromagnetic materials result from atomic magnetic moments due to uncanceled electron spins as a consequence of the electron structure. There is also an orbital magnetic moment contribution that is small in comparison to the spin moment. Furthermore, in a ferromagnetic material, coupling interactions cause net spin magnetic moments of adjacent atoms to align with one another, even in the absence of an external field. This is schematically illustrated in Figure 21.7. The origin of these coupling forces is not completely understood, but they are thought to arise from the electronic structure of the metal. This mutual spin alignment exists over relatively large-volume regions of the crystal called **domains** (see Section 21.7).

The maximum possible magnetization, or **saturation magnetization**, M_s, of a ferromagnetic material represents the magnetization that results when all the magnetic dipoles in a solid piece are mutually aligned with the external field; there is also a corresponding saturation flux density, B_s. The saturation magnetization is equal to the product of the net magnetic moment for each atom and the number of atoms present. For each of iron, cobalt, and nickel, the net magnetic moments per atom are 2.22, 1.72, and 0.60 Bohr magnetons, respectively.

EXAMPLE PROBLEM 21.1

Saturation Magnetization and Flux Density Computations for Nickel

Calculate **(a)** the saturation magnetization and **(b)** the saturation flux density for nickel, which has a density of 8.90 g/cm^3.

Solution

(a) The saturation magnetization is the product of the number of Bohr magnetons per atom (0.60, as given above), the magnitude of the Bohr magneton μ_B, and the number N of atoms per cubic meter, or

Saturation magnetization for nickel

$$M_s = 0.60 \,\mu_B N \tag{21.9}$$

The number of atoms per cubic meter is related to the density ρ, the atomic weight A_{Ni}, and Avogadro's number N_A, as follows:

For nickel, computation of the number of atoms per unit volume

$$N = \frac{\rho N_A}{A_{Ni}} \tag{21.10}$$

$$= \frac{(8.90 \times 10^6 \, g/m^3)(6.022 \times 10^{23} \, atoms/mol)}{58.71 \, g/mol}$$

$$= 9.13 \times 10^{28} \, atoms/m^3$$

Finally,

$$M_s = \left(\frac{0.60 \text{ Bohr magneton}}{\text{atom}}\right)\left(\frac{9.27 \times 10^{-24} \text{ A·m}^2}{\text{Bohr magneton}}\right)\left(\frac{9.13 \times 10^{28} \text{ atoms}}{\text{m}^3}\right)$$

$$= 5.1 \times 10^5 \text{ A/m}$$

(b) From Equation 21.8, the saturation flux density is

$$B_s = \mu_0 M_s$$

$$= \left(\frac{4\pi \times 10^{-7} \text{ H}}{\text{m}}\right)\left(\frac{5.1 \times 10^5 \text{ A}}{\text{m}}\right)$$

$$= 0.64 \text{ tesla}$$

21.5 ANTIFERROMAGNETISM AND FERRIMAGNETISM

Antiferromagnetism

antiferromagnetism

Magnetic moment coupling between adjacent atoms or ions also occurs in materials other than those that are ferromagnetic. In one such group, this coupling results in an antiparallel alignment; the alignment of the spin moments of neighboring atoms or ions in exactly opposite directions is termed **antiferromagnetism**. Manganese oxide (MnO) is one material that displays this behavior. Manganese oxide is a ceramic material that is ionic in character, having both Mn^{2+} and O^{2-} ions. No net magnetic moment is associated with the O^{2-} ions because there is a total cancellation of both spin and orbital moments. However, the Mn^{2+} ions possess a net magnetic moment that is predominantly of spin origin. These Mn^{2+} ions are arrayed in the crystal structure such that the moments of adjacent ions are antiparallel. This arrangement is represented schematically in Figure 21.8. The opposing magnetic moments cancel one another, and, as a consequence, the solid as a whole possesses no net magnetic moment.

Ferrimagnetism

ferrimagnetism

Some ceramics also exhibit a permanent magnetization, termed **ferrimagnetism**. The macroscopic magnetic characteristics of ferromagnets and ferrimagnets are similar; the distinction lies in the source of the net magnetic moments. The principles of ferrimagnetism are illustrated with the cubic ferrites.[2] These ionic materials may be represented by the chemical formula MFe_2O_4, in which M represents any one of several metallic elements. The prototype ferrite is Fe_3O_4—the mineral magnetite, sometimes called lodestone.

The formula for Fe_3O_4 may be written as $Fe^{2+}O^{2-}-(Fe^{3+})_2(O^{2-})_3$, in which the Fe ions exist in both +2 and +3 valence states in the ratio of 1:2. A net spin magnetic moment exists for each Fe^{2+} and Fe^{3+} ion, which corresponds to 4 and 5 Bohr magnetons, respectively, for the two ion types. Furthermore, the O^{2-} ions are magnetically neutral. There are antiparallel spin-coupling interactions between the Fe ions, similar in character to antiferromagnetism. However, the net ferrimagnetic moment arises from the incomplete cancellation of spin moments.

Cubic ferrites have the inverse spinel crystal structure, which is cubic in symmetry and is similar to the spinel structure (Sections 4.9 and 4.17). The inverse spinel crystal structure might be thought of as having been generated by the stacking of close-packed

ferrite

---

[2]Ferrite in the magnetic sense should not be confused with the ferrite α-iron discussed in Section 11.18; in the remainder of this chapter, the term **ferrite** indicates the magnetic ceramic.

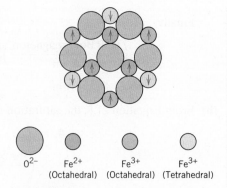

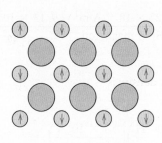

Mn²⁺   O²⁻

Figure 21.8 Schematic representation of antiparallel alignment of spin magnetic moments for antiferromagnetic manganese oxide.

O²⁻   Fe²⁺ (Octahedral)   Fe³⁺ (Octahedral)   Fe³⁺ (Tetrahedral)

Figure 21.9 Schematic diagram showing the spin magnetic moment configuration for Fe^{2+} and Fe^{3+} ions in Fe_3O_4.
(From Richard A. Flinn and Paul K. Trojan, *Engineering Materials and Their Applications,* 4th edition. Copyright © 1990 by John Wiley & Sons, Inc. Adapted by permission of John Wiley & Sons, Inc.)

WileyPLUS: VMSE
Spinel/Inverse Spinel

planes of O^{2-} ions. Again, there are two types of positions that may be occupied by the iron cations, as illustrated in Figure 4.26. For one, the coordination number is 4 (tetrahedral coordination)—that is, each Fe ion is surrounded by four oxygen nearest neighbors. For the other, the coordination number is 6 (octahedral coordination). With this inverse spinel structure, half the trivalent (Fe^{3+}) ions are situated in octahedral positions and the other half in tetrahedral positions. The divalent Fe^{2+} ions are all located in octahedral positions. The critical factor is the arrangement of the spin moments of the Fe ions, as represented in Figure 21.9 and Table 21.3. The spin moments of all the Fe^{3+} ions in the octahedral positions are aligned parallel to one another; however, they are oppositely directed to the Fe^{3+} ions disposed in the tetrahedral positions, which are also aligned. This results from the antiparallel coupling of adjacent iron ions. Thus, the spin moments of all Fe^{3+} ions cancel one another and make no net contribution to the magnetization of the solid. All the Fe^{2+} ions have their moments aligned in the same direction; this total moment is responsible for the net magnetization (see Table 21.3). Thus, the saturation magnetization of a ferrimagnetic solid may be computed from the product of the net spin magnetic moment for each Fe^{2+} ion and the number of Fe^{2+} ions; this corresponds to the mutual alignment of all the Fe^{2+} ion magnetic moments in the Fe_3O_4 specimen.

Cubic ferrites having other compositions may be produced by adding metallic ions that substitute for some of the iron in the crystal structure. Again, from the ferrite chemical formula $M^{2+}O^{2-}-(Fe^{3+})_2(O^{2-})_3$, in addition to Fe^{2+}, M^{2+} may represent

Table 21.3
The Distribution of Spin Magnetic Moments for Fe²⁺ and Fe³⁺ Ions in a Unit Cell of Fe₃O₄[a]

Cation	Octahedral Lattice Site	Tetrahedral Lattice Site	Net Magnetic Moment
Fe^{3+}	↑ ↑ ↑ ↑ ↑ ↑ ↑ ↑	↓ ↓ ↓ ↓ ↓ ↓ ↓ ↓	Complete cancellation
Fe^{2+}	↑ ↑ ↑ ↑ ↑ ↑ ↑ ↑	—	↑ ↑ ↑ ↑ ↑ ↑ ↑ ↑

[a]Each arrow represents the magnetic moment orientation for one of the cations.

Table 21.4

Net Magnetic Moments for Six Cations

Cation	Net Spin Magnetic Moment (Bohr Magnetons)
Fe^{3+}	5
Fe^{2+}	4
Mn^{2+}	5
Co^{2+}	3
Ni^{2+}	2
Cu^{2+}	1

divalent ions such as Ni^{2+}, Mn^{2+}, Co^{2+}, and Cu^{2+}, each of which possesses a net spin magnetic moment different from 4; several are listed in Table 21.4. Thus, by adjustment of composition, ferrite compounds having a range of magnetic properties may be produced. For example, nickel ferrite has the formula $NiFe_2O_4$. Other compounds may also be produced containing mixtures of two divalent metal ions such as $(Mn, Mg)Fe_2O_4$, in which the Mn^{2+}:Mg^{2+} ratio may be varied; these are called *mixed ferrites*.

Ceramic materials other than the cubic ferrites are also ferrimagnetic and include the hexagonal ferrites and garnets. Hexagonal ferrites have a crystal structure similar to the inverse spinel crystal structure, with hexagonal symmetry rather than cubic. The chemical formula for these materials may be represented by $AB_{12}O_{19}$, in which A is a divalent metal such as barium, lead, or strontium, and B is a trivalent metal such as aluminum, gallium, chromium, or iron. The two most common examples of the hexagonal ferrites are $PbFe_{12}O_{19}$ and $BaFe_{12}O_{19}$.

The garnets have a very complicated crystal structure, which may be represented by the general formula $M_3Fe_5O_{12}$; here, M represents a rare earth ion such as samarium, europium, gadolinium, or yttrium. Yttrium iron garnet ($Y_3Fe_5O_{12}$), sometimes denoted YIG, is the most common material of this type.

The saturation magnetizations for ferrimagnetic materials are not as high as for ferromagnets. However, ferrites, being ceramic materials, are good electrical insulators. For some magnetic applications, such as high-frequency transformers, a low electrical conductivity is most desirable.

Concept Check 21.1   Cite the major similarities and differences between ferromagnetic and ferrimagnetic materials.

Concept Check 21.2   What is the difference between the spinel and inverse spinel crystal structures? *Hint:* You may want to consult Sections 4.9 and 4.17.

[*The answers may be found at* www.wiley.com/college/callister (*Student Companion Site.*)]

EXAMPLE PROBLEM 21.2

Saturation Magnetization Determination for Fe_3O_4

Calculate the saturation magnetization for Fe_3O_4, given that each cubic unit cell contains 8 Fe^{2+} and 16 Fe^{3+} ions and that the unit cell edge length is 0.839 nm.

Solution

This problem is solved in a manner similar to Example Problem 21.1, except that the computational basis is per unit cell as opposed to per atom or ion.

Saturation magnetization for a ferrimagnetic material (Fe_3O_4)

The saturation magnetization is equal to the product of the number N' of Bohr magnetons per cubic meter of Fe_3O_4 and the magnetic moment per Bohr magneton μ_B,

$$M_s = N'\mu_B \qquad (21.11)$$

Now, N' is just the number of Bohr magnetons per unit cell n_B divided by the unit cell volume V_C, or

Computation of the number of Bohr magnetons per unit cell

$$N' = \frac{n_B}{V_C} \qquad (21.12)$$

Again, the net magnetization results from the Fe^{2+} ions only. Because there are 8 Fe^{2+} ions per unit cell and 4 Bohr magnetons per Fe^{2+} ion, n_B is 32. Furthermore, the unit cell is a cube, and $V_C = a^3$, a being the unit cell edge length. Therefore,

$$M_s = \frac{n_B\mu_B}{a^3} \qquad (21.13)$$

$$= \frac{(32 \text{ Bohr magnetons/unit cell})(9.27 \times 10^{-24} \text{ A·m}^2/\text{Bohr magneton})}{(0.839 \times 10^{-9} \text{ m})^3/\text{unit cell}}$$

$$= 5.0 \times 10^5 \text{ A/m}$$

DESIGN EXAMPLE 21.1

Design of a Mixed Ferrite Magnetic Material

Design a cubic mixed-ferrite magnetic material that has a saturation magnetization of 5.25×10^5 A/m.

Solution

According to Example Problem 21.2, the saturation magnetization for Fe_3O_4 is 5.0×10^5 A/m. In order to increase the magnitude of M_s, it is necessary to replace some fraction of the Fe^{2+} with a divalent metal ion that has a greater magnetic moment—for example, Mn^{2+}; from Table 21.4, note that there are 5 Bohr magnetons/Mn^{2+} ion as compared to 4 Bohr magnetons/Fe^{2+}. Let us first use Equation 21.13 to compute the number of Bohr magnetons per unit cell (n_B), assuming that the Mn^{2+} addition does not change the unit cell edge length (0.839 nm). Thus,

$$n_B = \frac{M_s a^3}{\mu_B}$$

$$= \frac{(5.25 \times 10^5 \text{ A/m})(0.839 \times 10^{-9} \text{ m})^3/\text{unit cell}}{9.27 \times 10^{-24} \text{ A·m}^2/\text{Bohr magneton}}$$

$$= 33.45 \text{ Bohr magnetons/unit cell}$$

If we let x represent the fraction of Mn^{2+} that substitute for Fe^{2+}, then the remaining unsubstituted Fe^{2+} fraction is equal to $(1 - x)$. Furthermore, inasmuch as there are 8 divalent ions per unit cell, we may write the following expression:

$$8[5x + 4(1 - x)] = 33.45$$

which leads to $x = 0.181$. Thus, if 18.1 at% of the Fe^{2+} in Fe_3O_4 are replaced with Mn^{2+}, the saturation magnetization will be increased to 5.25×10^5 A/m.

21.6 THE INFLUENCE OF TEMPERATURE ON MAGNETIC BEHAVIOR

Temperature can also influence the magnetic characteristics of materials. Recall that raising the temperature of a solid increases the magnitude of the thermal vibrations of atoms. The atomic magnetic moments are free to rotate; hence, with rising temperature, the increased thermal motion of the atoms tends to randomize the directions of any moments that may be aligned.

For ferromagnetic, antiferromagnetic, and ferrimagnetic materials, the atomic thermal motions counteract the coupling forces between the adjacent atomic dipole moments, causing some dipole misalignment, regardless of whether an external field is present. This results in a decrease in the saturation magnetization for both ferro- and ferrimagnets. The saturation magnetization is a maximum at 0 K, at which temperature the thermal vibrations are at a minimum. With increasing temperature, the saturation magnetization decreases gradually and then abruptly drops to zero at what is called the **Curie temperature** T_c. The magnetization–temperature behavior for iron and Fe_3O_4 is represented in Figure 21.10. At T_c, the mutual spin-coupling forces are completely destroyed, such that for temperatures above T_c, both ferro- and ferrimagnetic materials are paramagnetic. The magnitude of the Curie temperature varies from material to material; for example, for iron, cobalt, nickel, and Fe_3O_4, the respective values are 768°C, 1120°C, 335°C, and 585°C.

Antiferromagnetism is also affected by temperature; this behavior vanishes at what is called the *Néel temperature*. At temperatures above this point, antiferromagnetic materials also become paramagnetic.

Concept Check 21.3   Explain why repeatedly dropping a permanent magnet on the floor causes it to become demagnetized.

[*The answer may be found at* www.wiley.com/college/callister *(Student Companion Site.)*]

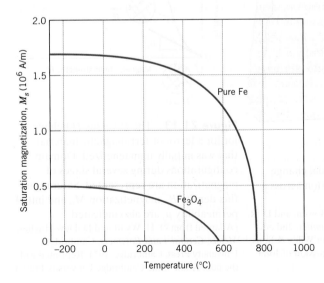

Figure 21.10 Plot of saturation magnetization as a function of temperature for iron and Fe_3O_4.
(Adapted from J. Smit and H. P. J. Wijn, *Ferrites*. Copyright © 1959 by N. V. Philips Gloeilampenfabrieken, Eindhoven (Holland). Reprinted by permission.)

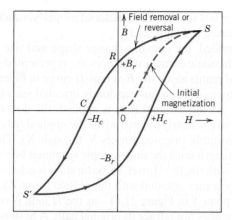

Figure 21.14 Magnetic flux density versus the magnetic field strength for a ferromagnetic material that is subjected to forward and reverse saturations (points S and S'). The hysteresis loop is represented by the solid curve; the dashed curve indicates the initial magnetization. The remanence B_r and the coercive force H_c are also shown.

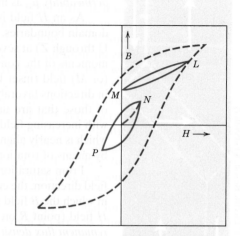

Figure 21.15 A hysteresis curve at less than saturation (curve NP) within the saturation loop for a ferromagnetic material. The B–H behavior for field reversal at other than saturation is indicated by curve LM.

✓ **Concept Check 21.4** Schematically sketch on a single plot the B-versus-H behavior for a ferromagnetic material (a) at 0 K, (b) at a temperature just below its Curie temperature, and (c) at a temperature just above its Curie temperature. Briefly explain why these curves have different shapes.

Concept Check 21.5 Schematically sketch the hysteresis behavior for a ferromagnet that is gradually demagnetized by cycling in an H field that alternates direction and decreases in magnitude.

[*The answers may be found at* www.wiley.com/college/callister (*Student Companion Site.*)]

Figure 21.16 Comparison of B-versus-H behaviors for ferromagnetic/ferrimagnetic and diamagnetic/paramagnetic materials (inset plot). Note that extremely small B fields are generated in materials that experience only diamagnetic/paramagnetic behavior, which is why they are considered nonmagnetics.

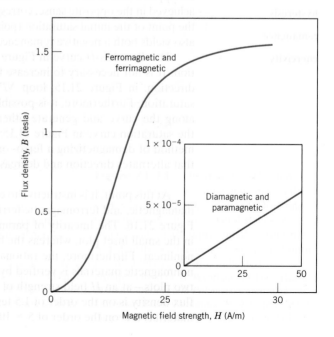

21.8 MAGNETIC ANISOTROPY

The magnetic hysteresis curves discussed in Section 21.7 have different shapes depending on various factors: (1) whether the specimen is a single crystal or polycrystalline; (2) if it is polycrystalline, whether there is any preferred orientation of the grains; (3) the presence of pores or second-phase particles; and (4) other factors such as temperature and, if a mechanical stress is applied, the stress state.

For example, the B (or M) versus H curve for a single crystal of a ferromagnetic material depends on its crystallographic orientation relative to the direction of the applied H field. This behavior is demonstrated in Figure 21.17 for single crystals of nickel (FCC) and iron (BCC), where the magnetizing field is applied in [100], [110], and [111] crystallographic directions, and in Figure 21.18 for cobalt (HCP) in [0001] and [10$\bar{1}$0]/[11$\bar{2}$0] directions. This dependence of magnetic behavior on crystallographic orientation is termed *magnetic* (or sometimes *magnetocrystalline*) *anisotropy.*

For each of these materials, there is one crystallographic direction in which magnetization is easiest—that is, saturation (of M) is achieved at the lowest H field; this is termed a direction of *easy magnetization.* For example, for Ni (Figure 21.17), this direction is [111] inasmuch as saturation occurs at point A; for [110] and [100] orientations, saturation points correspond, respectively, to points B and C. Correspondingly, easy magnetization directions for Fe and Co are [100] and [0001], respectively (Figures 21.17 and 21.18). Conversely, a *hard* crystallographic direction is that direction for which saturation magnetization is most difficult; hard directions for Ni, Fe, and Co are [100], [111], and [10$\bar{1}$0]/[11$\bar{2}$0].

As noted in Section 21.7, the insets of Figure 21.13 represent domain configurations at various stages along the B (or M) versus H curve during the magnetization of a ferromagnetic/ferrimagnetic material. Here, each of the arrows represents a domain's direction of easy magnetization; domains whose directions of easy magnetization are most closely aligned with the H field grow at the expense of the other domains, which shrink (insets V through X). Furthermore, the magnetization of the single domain in inset Y also corresponds to an easy direction. Saturation is achieved as the direction of this domain rotates away from the easy direction into the direction of the applied field (inset Z).

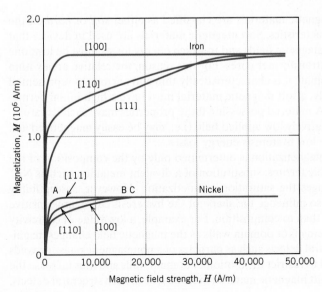

Figure 21.17 Magnetization curves for single crystals of iron and nickel. For both metals, a different curve was generated when the magnetic field was applied in each of [100], [110], and [111] crystallographic directions.
[Adapted from K. Honda and S. Kaya, "On the Magnetisation of Single Crystals of Iron," *Sci. Rep. Tohoku Univ.,* **15,** 721 (1926); and from S. Kaya, "On the Magnetisation of Single Crystals of Nickel," *Sci. Rep. Tohoku Univ.,* **17,** 639 (1928).]

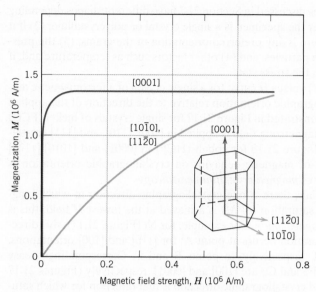

Figure 21.18 Magnetization curves for single crystals of cobalt. The curves were generated when the magnetic field was applied in [0001] and [10$\bar{1}$0]/[11$\bar{2}$0] crystallographic directions.
[Adapted from S. Kaya, "On the Magnetisation of Single Crystals of Cobalt," *Sci. Rep. Tohoku Univ.*, **17**, 1157 (1928).]

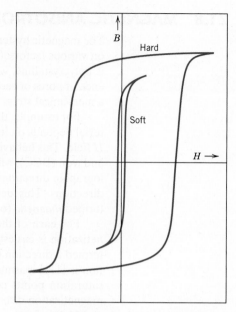

Figure 21.19 Schematic magnetization curves for soft and hard magnetic materials. (From K. M. Ralls, T. H. Courtney, and J. Wulff, *Introduction to Materials Science and Engineering.* Copyright © 1976 by John Wiley & Sons, New York. Reprinted by permission of John Wiley & Sons, Inc.)

21.9 SOFT MAGNETIC MATERIALS

soft magnetic
material

The size and shape of the hysteresis curve for ferro- and ferrimagnetic materials are of considerable practical importance. The area within a loop represents a magnetic energy loss per unit volume of material per magnetization–demagnetization cycle; this energy loss is manifested as heat that is generated within the magnetic specimen and is capable of raising its temperature.

Both ferro- and ferrimagnetic materials are classified as either *soft* or *hard* on the basis of their hysteresis characteristics. **Soft magnetic materials** are used in devices that are subjected to alternating magnetic fields and in which energy losses must be low; one familiar example consists of transformer cores. For this reason, the relative area within the hysteresis loop must be small; it is characteristically thin and narrow, as represented in Figure 21.19. Consequently, a soft magnetic material must have a high initial permeability and a low coercivity. A material possessing these properties may reach its saturation magnetization with a relatively low applied field (i.e., may be easily magnetized and demagnetized) and still have low hysteresis energy losses.

The saturation field or magnetization is determined only by the composition of the material. For example, in cubic ferrites, substitution of a divalent metal ion such as Ni^{2+} for Fe^{2+} in $FeO-Fe_2O_3$ changes the saturation magnetization. However, susceptibility and coercivity (H_c), which also influence the shape of the hysteresis curve, are sensitive to structural variables rather than to composition. For example, a low value of coercivity corresponds to the easy movement of domain walls as the magnetic field changes magnitude and/or direction. Structural defects such as particles of a nonmagnetic phase or voids in the magnetic material tend to restrict the motion of domain walls and thus increase the coercivity. Consequently, a soft magnetic material must be free of such structural defects.

M A T E R I A L S   O F   I M P O R T A N C E

An Iron–Silicon Alloy Used in Transformer Cores

As mentioned earlier in this section, transformer cores require the use of soft magnetic materials, which are easily magnetized and demagnetized (and also have relatively high electrical resistivities). One alloy commonly used for this application is the iron–silicon alloy listed in Table 21.5 (97 wt% Fe–3 wt% Si). Single crystals of this alloy are magnetically anisotropic, as are also single crystals of iron (as explained previously). Consequently, energy losses of transformers could be minimized if their cores were fabricated from single crystals such that a [100]-type direction [the direction of easy magnetization (Figure 21.17)] is oriented parallel to the direction of an applied magnetic field; this configuration for a transformer core is represented schematically in Figure 21.20. Unfortunately, single crystals are expensive to prepare, and, thus, this is an economically unpractical situation. A better, more economically attractive alternative—one used commercially—is to fabricate cores from polycrystalline sheets of this alloy that are anisotropic.

Often, the grains in polycrystalline materials are oriented randomly, with the result that their properties are isotropic (Section 3.10). However, one way of developing anisotropy in polycrystalline metals is via

plastic deformation, for example by rolling (Section 17.2, Figure 17.2b); rolling is the technique by which sheet transformer cores are fabricated. A flat sheet that has been rolled is said to have a *rolling* (or *sheet*) *texture*, or there is a preferred crystallographic orientation of the grains. For this type of texture, during the rolling operation, for most of the grains in the sheet, a specific crystallographic plane (hkl) becomes aligned parallel (or nearly parallel) to the surface of the sheet, and, in addition a direction $[uvw]$ in that plane lies parallel (or nearly parallel) to the rolling direction. Thus, a rolling texture is indicated by the plane–direction combination, $(hkl)[uvw]$. For body-centered cubic alloys (to include the iron–silicon alloy mentioned earlier), the rolling texture is (110)[001], which is represented schematically in Figure 21.21. Thus, transformer cores of this iron–silicon alloy are fabricated so that the direction in which the sheet was rolled (corresponding to a [001]-type direction for most of the grains) is aligned parallel to the direction of the magnetic field application.[3]

The magnetic characteristics of this alloy may be further improved through a series of deformation and heat-treating procedures that produce a (100)[001] texture.

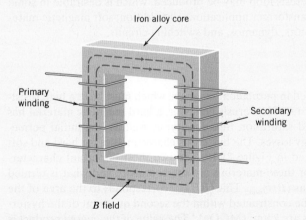

Figure 21.20 Schematic diagram of a transformer core, including the direction of the B field that is generated.

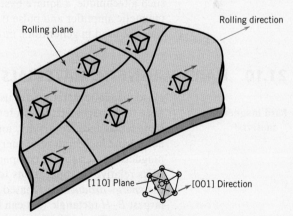

Figure 21.21 Schematic representation of the (110)[001] rolling texture for body-centered cubic iron.

---

[3]For body-centered cubic metals and alloys, [100] and [001] directions are equivalent (Section 3.6)—that is, both are directions of easy magnetization.

Table 21.5 Typical Properties for Several Soft Magnetic Materials

Material	Composition (wt%)	Initial Relative Permeability μ_i	Saturation Flux Density B_s [tesla (gauss)]	Hysteresis Loss/Cycle [J/m^3 (erg/cm^3)]	Resistivity ρ ($\Omega \cdot m$)
Commercial iron ingot	99.95 Fe	150	2.14 (21,400)	270 (2,700)	1.0×10^{-7}
Silicon–iron (oriented)	97 Fe, 3 Si	1,400	2.01 (20,100)	40 (400)	4.7×10^{-7}
45 Permalloy	55 Fe, 45 Ni	2,500	1.60 (16,000)	120 (1,200)	4.5×10^{-7}
Supermalloy	79 Ni, 15 Fe, 5 Mo, 0.5 Mn	75,000	0.80 (8,000)	—	6.0×10^{-7}
Ferroxcube A	48 MnFe$_2$O$_4$, 52 ZnFe$_2$O$_4$	1,400	0.33 (3,300)	~40 (~400)	2,000
Ferroxcube B	36 NiFe$_2$O$_4$, 64 ZnFe$_2$O$_4$	650	0.36 (3,600)	~35 (~350)	10^7

Source: Adapted from *Metals Handbook: Properties and Selection: Stainless Steels, Tool Materials and Special-Purpose Metals,* Vol. 3, 9th edition, D. Benjamin (Senior Editor), 1980. Reproduced by permission of ASM International, Materials Park, OH.

Another property consideration for soft magnetic materials is the electrical resistivity. In addition to the hysteresis energy losses described previously, energy losses may result from electrical currents that are induced in a magnetic material by a magnetic field that varies in magnitude and direction with time; these are called *eddy currents*. It is most desirable to minimize these energy losses in soft magnetic materials by increasing the electrical resistivity. This is accomplished in ferromagnetic materials by forming solid solution alloys; iron–silicon and iron–nickel alloys are examples. The ceramic ferrites are commonly used for applications requiring soft magnetic materials because they are intrinsically electrical insulators. Their applicability is somewhat limited, however, inasmuch as they have relatively small susceptibilities. The properties of some soft magnetic materials are shown in Table 21.5.

The hysteresis characteristics of soft magnetic materials may be enhanced for some applications by an appropriate heat treatment in the presence of a magnetic field. Using such a technique, a square hysteresis loop may be produced, which is desirable in some magnetic amplifier and pulse transformer applications. In addition, soft magnetic materials are used in generators, motors, dynamos, and switching circuits.

21.10 HARD MAGNETIC MATERIALS

hard magnetic material

Hard magnetic materials are used in permanent magnets, which must have a high resistance to demagnetization. In terms of hysteresis behavior, a **hard magnetic material** has high remanence, coercivity, and saturation flux density, as well as low initial permeability and high hysteresis energy losses. The hysteresis characteristics for hard and soft magnetic materials are compared in Figure 21.19. The two most important characteristics relative to applications for these materials are the coercivity and what is termed the *energy product*, designated as $(BH)_{max}$. This $(BH)_{max}$ corresponds to the area of the largest B–H rectangle that can be constructed within the second quadrant of the hysteresis curve, Figure 21.22; its units are kJ/m^3 (MGOe).[4] The value of the energy product is

---

[4]MGOe is defined as

$$1 \text{ MGOe} = 10^6 \text{ gauss-oersted}$$

Conversion from cgs–emu to SI units is accomplished by the relationship

$$1 \text{ MGOe} = 7.96 \text{ kJ/m}^3$$

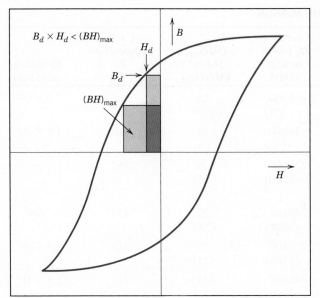

Figure 21.22 Schematic magnetization curve that displays hysteresis. Within the second quadrant are drawn two B–H energy product rectangles; the area of the rectangle labeled $(BH)_{max}$ is the largest possible, which is greater than the area defined by B_d–H_d.

representative of the energy required to demagnetize a permanent magnet—that is, the larger $(BH)_{max}$, the harder the material in terms of its magnetic characteristics.

Hysteresis behavior is related to the ease with which the magnetic domain boundaries move; by impeding domain wall motion, the coercivity and susceptibility are enhanced, such that a large external field is required for demagnetization. Furthermore, these characteristics are related to the microstructure of the material.

✔ *Concept Check 21.6*   It is possible, by various means (e.g., alteration of microstructure and impurity additions), to control the ease with which domain walls move as the magnetic field is changed for ferromagnetic and ferrimagnetic materials. Sketch a schematic B-versus-H hysteresis loop for a ferromagnetic material, and superimpose on this plot the loop alterations that would occur if domain boundary movement were hindered.

[*The answer may be found at* www.wiley.com/college/callister (*Student Companion Site.*)]

Conventional Hard Magnetic Materials

Hard magnetic materials fall within two main categories—conventional and high energy. The conventional materials have $(BH)_{max}$ values that range between about 2 and 80 kJ/m³ (0.25 and 10 MGOe). These include ferromagnetic materials—magnet steels, cunife (Cu–Ni–Fe) alloys, and alnico (Al–Ni–Co) alloys—as well as the hexagonal ferrites (BaO–6Fe₂O₃). Table 21.6 presents some critical properties of several of these hard magnetic materials.

The hard magnetic steels are normally alloyed with tungsten and/or chromium. Under the proper heat-treating conditions, these two elements readily combine with carbon in the steel to form tungsten and chromium carbide precipitate particles, which are especially effective in obstructing domain wall motion. For the other metal alloys, an appropriate heat treatment forms extremely small single-domain and strongly magnetic iron–cobalt particles within a nonmagnetic matrix phase.

Table 21.6 Typical Properties for Several Hard Magnetic Materials

Material	Composition (wt%)	Remanence B_r [tesla (gauss)]	Coercivity H_c [amp-turn/m (Oe)]	$(BH)_{max}$ [kJ/m^3 (MGOe)]	Curie Temperature T_c (°C)	Resistivity ρ (Ω·m)
Tungsten steel	92.8 Fe, 6 W, 0.5 Cr, 0.7 C	0.95 (9,500)	5,900 (74)	2.6 (0.33)	760	3.0×10^{-7}
Cunife	20 Fe, 20 Ni, 60 Cu	0.54 (5,400)	44,000 (550)	12 (1.5)	410	1.8×10^{-7}
Sintered alnico 8	34 Fe, 7 Al, 15 Ni, 35 Co, 4 Cu, 5 Ti	0.76 (7,600)	125,000 (1,550)	36 (4.5)	860	—
Sintered ferrite 3	$BaO–6Fe_2O_3$	0.32 (3,200)	240,000 (3,000)	20 (2.5)	450	$\sim 10^4$
Cobalt rare earth 1	$SmCo_5$	0.92 (9,200)	720,000 (9,000)	170 (21)	725	5.0×10^{-7}
Sintered neodymium–iron–boron	$Nd_2Fe_{14}B$	1.16 (11,600)	848,000 (10,600)	255 (32)	310	1.6×10^{-6}

Source: Adapted from *ASM Handbook*, Vol. 2, *Properties and Selection: Nonferrous Alloys and Special-Purpose Materials.* Copyright © 1990 by ASM International. Reprinted by permission of ASM International, Materials Park, OH.

High-Energy Hard Magnetic Materials

Permanent magnetic materials having energy products in excess of about 80 kJ/m^3 (10 MGOe) are considered to be of the high-energy type. These are recently developed intermetallic compounds that have a variety of compositions; the two that have found commercial exploitation are $SmCo_5$ and $Nd_2Fe_{14}B$. Their magnetic properties are also listed in Table 21.6.

Samarium–Cobalt Magnets

Samarium–cobalt, $SmCo_5$, is a member of a group of alloys that are combinations of cobalt or iron and a light rare earth element; a number of these alloys exhibit high-energy, hard magnetic behavior, but $SmCo_5$ is the only one of commercial significance. The energy product of these $SmCo_5$ materials [between 120 and 240 kJ/m^3 (15 and 30 MGOe)] are considerably higher than those of the conventional hard magnetic materials (Table 21.6); in addition, they have relatively large coercivities. Powder metallurgical techniques are used to fabricate $SmCo_5$ magnets. The appropriately alloyed material is first ground into a fine powder; the powder particles are aligned using an external magnetic field and then pressed into the desired shape. The piece is then sintered at an elevated temperature, followed by another heat treatment that improves the magnetic properties.

Neodymium–Iron–Boron Magnets

Samarium is a rare and relatively expensive material; furthermore, the price of cobalt is variable, and its sources are unreliable. Consequently, the neodymium–iron–boron, $Nd_2Fe_{14}B$, alloys have become the materials of choice for a large number and wide diversity of applications requiring hard magnetic materials. Coercivities and energy products of these materials rival those of the samarium–cobalt alloys (Table 21.6).

The magnetization–demagnetization behavior of these materials is a function of domain wall mobility, which is controlled by the final microstructure—that is, the size, shape, and orientation of the crystallites or grains, as well as the nature and distribution of any second-phase particles that are present. Microstructure depends on how the material is processed. Two different processing techniques are available for the fabrication of $Nd_2Fe_{14}B$ magnets: powder metallurgy (sintering) and rapid solidification (melt spinning). The powder metallurgical approach is similar to that used for the $SmCo_5$ materials. For rapid solidification, the alloy in molten form is quenched very rapidly such that either an amorphous or very fine-grained and thin solid ribbon is produced. This ribbon material is then pulverized, compacted into the desired shape, and subsequently heat-treated. Rapid solidification is the more involved of the two fabrication processes; nevertheless, it is continuous, whereas powder metallurgy is a batch process, which has its inherent disadvantages.

These high-energy hard magnetic materials are used in a host of different devices in a variety of technological fields. One common application is in motors. Permanent magnets are far superior to electromagnets in that their magnetic fields are maintained continuously and without the necessity to expend electrical power; furthermore, no heat is generated during operation. Motors using permanent magnets are much smaller than their electromagnetic counterparts and are used extensively in fractional horsepower units. Familiar motor applications include cordless drills and screwdrivers; in automobiles (starters; window winders, wipers, and washers; fan motors); in audio and video recorders; clocks; speakers in audio systems, lightweight earphones, and hearing aids; and computer peripherals.

21.11 MAGNETIC STORAGE

Magnetic materials are important in the area of information storage; in fact, magnetic recording[5] has become virtually the universal technology for the storage of electronic information. This is evidenced by the preponderance of disk storage media [e.g., computers (both desktop and laptop), and high-definition camcorder hard drives], credit/debit cards (magnetic stripes), and so on. Whereas in computers, semiconductor elements serve as primary memory, magnetic hard disks are normally used for secondary memory because they are capable of storing larger quantities of information and at a lower cost; however, their transfer rates are slower. Furthermore, the recording and television industries rely heavily on magnetic tapes for the storage and reproduction of audio and video sequences. In addition, tapes are used with large computer systems to back up and archive data.

In essence, computer bytes, sound, or visual images in the form of electrical signals are recorded magnetically on very small segments of the magnetic storage medium—a tape or disk. Transference to (i.e., "writing") and retrieval from (i.e., "reading") the tape or disk is accomplished by means of a recording system that consists of read and write heads. For hard drives, this head system is supported above and in close proximity to the magnetic medium by a self-generating air bearing as the medium passes beneath at relatively high rotational speeds.[6] In contrast, tapes make physical contact with the heads during read and write operations. Tape velocities run as high as 10 m/s.

As noted previously, there are two principal types of magnetic media—*hard disk drives* (*HDDs*) and *magnetic tapes*—both of which we now briefly discuss.

Hard Disk Drives

Hard disk magnetic storage hard drives consist of rigid circular disks with diameters that range between about 65 mm and 95 mm. During read and write processes, disks

---

[5]The term *magnetic recording* is often used to denote the recording and storage of audio or video and audio signals, whereas in the field of computing, *magnetic storage* is frequently preferred.

[6]It is sometimes stated that the head "flies" over the disk.

Figure 21.23 Schematic diagram of a hard disk drive that uses the perpendicular magnetic recording medium; also shown are inductive write and magnetoresistive read heads.
(Courtesy of HGST, a Western Digital Company.)

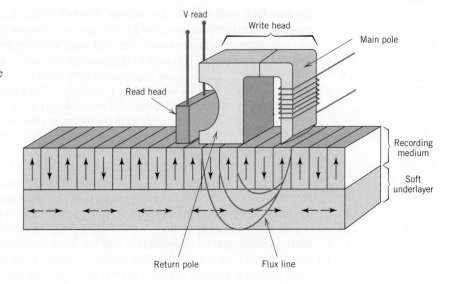

rotate at relatively high velocities—5400 and 7200 revolutions per minute are common. Rapid rates of data storage and retrieval are possible using HDDs, as are high storage densities.

For the current HDD technology, "magnetic bits" point up or down perpendicular to the plane of the disk surface; this scheme is appropriately called *perpendicular magnetic recording* (*PMR*), and is represented schematically in Figure 21.23.

Data (or bits) are introduced (written) into the storage medium using an inductive write head. For one head design, shown in Figure 21.23, a time-varying write magnetic flux is generated at the tip of the main pole—a ferromagnetic/ferrimagnetic core material around which a wire coil is wound—by an electric current (also time variable) that passes through the coil. This flux penetrates through the magnetic storage layer into a magnetically soft underlayer and then reenters the head assembly through a return pole (Figure 21.23). A very intense magnetic field is concentrated in the storage layer beneath the tip of the main pole. At this point, data are written as a very small region of the storage layer becomes magnetized. Upon removal of the field (i.e., as the disk continues its rotation), the magnetization remains—that is, the signal (i.e., data) has been stored. Digital data storage (i.e., as 1s and 0s) is in the form of minute magnetization patterns; the 1s and 0s correspond to the presence or absence of magnetic reversal directions between adjacent regions.

Data retrieval from the storage medium is accomplished using a magnetoresistive read head (Figure 21.23). During read-back, magnetic fields from the written magnetic patterns are sensed by this head; these fields produce changes in electrical resistance. The resulting signals are then processed so as to reproduce the original data.

The storage layer is composed of *granular media*—a thin film (15 to 20 nm thick) consisting of very small (~10-nm diameter) and isolated grains of an HCP cobalt–chromium alloy that are magnetically anisotropic. Other alloying elements (notably Pt and Ta) are added to enhance the magnetic anisotropy as well as to form oxide grain-boundary segregants that isolate the grains. Figure 21.24 is a transmission electron micrograph that shows the grain structure of an HDD storage layer. Each grain is a single domain that is oriented with its *c*-axis (i.e., [0001] crystallographic direction) perpendicular (or nearly perpendicular) to the disk surface. This [0001] direction is the direction of easy magnetization for Co (Figure 21.18); thus, when magnetized, the direction of magnetization of each grain has this desired perpendicular orientation. Reliable storage of data requires that each bit written on the disk

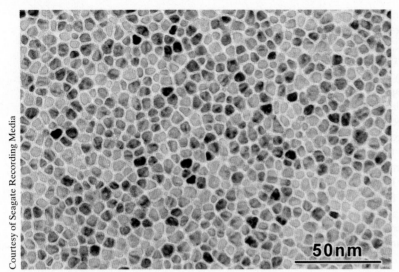

Courtesy of Seagate Recording Media

Figure 21.24 Transmission electron micrograph showing the microstructure of the perpendicular magnetic recording medium used in hard disk drives. This "granular medium" consists of small grains of a cobalt–chromium alloy (darker regions) that are isolated from one another by an oxide grain-boundary segregant (lighter regions).

encompasses approximately 100 grains. Furthermore, there is a lower limit to the grain size; for grain sizes below this limit, there is the possibility that the direction of magnetization will spontaneously reverse because of the effects of thermal agitation (Section 21.6), which causes a loss of stored data.

The current storage capacities of perpendicular HDDs are in excess of 100 Gbit/in.^{2} (10^{11} bit/in.^{2}); the ultimate goal for HDDs is a storage capacity of 1 Tbit/in.^{2} (10^{12} bit/in.^{2}).

Magnetic Tapes

The development of magnetic tape storage preceded that for the hard disk drives. Today, tape storage is less expensive than HDD; however, areal storage densities are lower for tape (by a factor of on the order of 100). Tapes (of standard 12.7-mm width) are wound onto reels and enclosed within cartridges for protection and to facilitate handling. During operation, a tape drive, using precision-synchronized motors, winds the tape from one reel onto another past a read/write head system in order to access a point of interest. Typical tape speeds are 4.8 m/s; some systems run as high as 10 m/s. Head systems for tape storage are similar to those employed for HDDs, as described previously.

For the latest tape-memory technology, storage media are particulates of magnetic materials that have dimensions on the order of tens of nanometers: ferromagnetic metal particles that are *acicular* (needle-shaped), and hexagonal and *tabular* (plate-shaped) ferrimagnetic barium–ferrite particles. Photomicrographs of both media types are shown in Figure 21.25. Tape products use one particle type or the other (not both

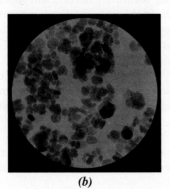

(a) *(b)*

Figure 21.25 Scanning electron micrographs showing particulate media used in tape-memory storage. (*a*) Needle-shaped ferromagnetic metal particles. (*b*) Plate-shaped ferrimagnetic barium-ferrite particles. Magnifications unknown.
(Photographs courtesy of Fujifilm, Inc., Recording Media Division.)

together), depending on application. These magnetic particles are thoroughly and uniformly dispersed in a proprietary high-molecular-weight organic binder material to form a magnetic layer approximately 50 nm thick. Beneath this layer is nonmagnetic thin-film support substrate between about 100 and 300 nm thick that is attached to the tape. Either poly(ethylene naphthalate) (PEN) or poly(ethylene terephthalate) (PET) is used for the tape.

Both particle types are magnetically *anisotropic*—that is, they have an "easy" or preferential direction along which they may be magnetized; for example, for the metal particles, this direction is parallel to their long axes. During manufacture, these particles are aligned such that this direction parallels the direction of motion of the tape past the write head. Inasmuch as each particle is a single domain that may be magnetized only in one direction or its opposite by the write head, two magnetic states are possible. These two states allow for the storage of information in digital form, as 1s and 0s.

Using the plate-shaped barium–ferrite medium, a tape-storage density of 6.7 Gbit/in.^{2} has been achieved. For the industry-standard LTO tape cartridge, this density corresponds to a storage capacity of 8 Tbytes of uncompressed data.

21.12 SUPERCONDUCTIVITY

Superconductivity is basically an electrical phenomenon; however, its discussion has been deferred to this point because there are magnetic implications relative to the superconducting state, and, in addition, superconducting materials are used primarily in magnets capable of generating high fields.

As most high-purity metals are cooled down to temperatures nearing 0 K, the electrical resistivity decreases gradually, approaching some small yet finite value characteristic of the particular metal. There are a few materials, however, for which the resistivity abruptly plunges at a very low temperature, from a finite value to one that is virtually zero and remains there upon further cooling. Materials that display this behavior are called *superconductors,* and the temperature at which they attain

superconductivity   **superconductivity** is called the critical temperature T_C.[7] The resistivity–temperature behaviors for superconductive and nonsuperconductive materials are contrasted in Figure 21.26. The critical temperature varies from superconductor to superconductor but lies between less than 1 K and approximately 20 K for metals and metal alloys. Recently, it has been demonstrated that some complex oxide ceramics have critical temperatures of greater than 100 K.

At temperatures below T_C, the superconducting state ceases upon application of a sufficiently large magnetic field, termed the *critical field* H_C, which depends on temperature and decreases with increasing temperature. The same may be said for current density—that is, a critical applied current density J_C exists below which a material is superconductive. Figure 21.27 shows schematically the boundary in temperature-magnetic field-current density space separating normal and superconducting states. The position of this boundary depends on the material. For temperature, magnetic field, and current density values lying between the origin and this boundary, the material is superconductive; outside the boundary, conduction is normal.

The superconductivity phenomenon has been satisfactorily explained by means of a rather involved theory. In essence, the superconductive state results from attractive interactions between pairs of conducting electrons; the motions of these paired electrons become coordinated such that scattering by thermal vibrations and impurity atoms is

---

[7]The symbol T_c is used to represent both the Curie temperature (Section 21.6) and the superconducting critical temperature in the scientific literature. They are different entities and should not be confused. In this discussion, they are denoted by T_c and T_C, respectively.

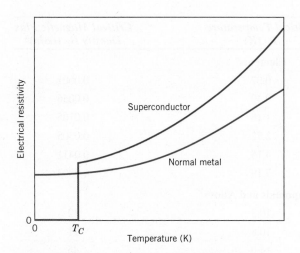

Figure 21.26 Temperature dependence of the electrical resistivity for normally conducting and superconducting materials in the vicinity of 0 K.

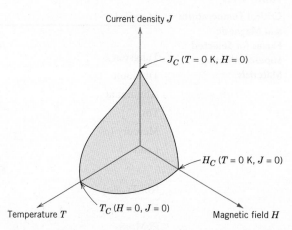

Figure 21.27 Critical temperature, current density, and magnetic field boundary separating superconducting and normal conducting states (schematic).

highly inefficient. Thus, the resistivity, being proportional to the incidence of electron scattering, is zero.

On the basis of magnetic response, superconducting materials may be divided into two classifications, type I and type II. Type I materials, while in the superconducting state, are completely diamagnetic—that is, all of an applied magnetic field is excluded from the body of material, a phenomenon known as the *Meissner effect,* which is il- lustrated in Figure 21.28. As H is increased, the material remains diamagnetic until the critical magnetic field H_C is reached. At this point, conduction becomes normal, and complete magnetic flux penetration takes place. Several metallic elements, including aluminum, lead, tin, and mercury, belong to the type I group.

Type II superconductors are completely diamagnetic at low applied fields, and field exclusion is total. However, the transition from the superconducting state to the normal state is gradual and occurs between lower critical and upper critical fields, designated H_{C1} and H_{C2}, respectively. The magnetic flux lines begin to penetrate into the body of material at H_{C1}, and with increasing applied magnetic field, this penetration continues; at H_{C2}, field penetration is complete. For fields between H_{C1} and H_{C2}, the material exists in what is termed a *mixed state*—both normal and superconducting regions are present.

Type II superconductors are preferred over type I for most practical applications by virtue of their higher critical temperatures and critical magnetic fields. The three

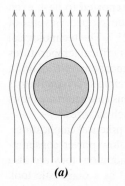

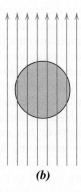

(a)   *(b)*

Figure 21.28 Representation of the Meissner effect. (*a*) While in the superconducting state, a body of material (circle) excludes a magnetic field (arrows) from its interior. (*b*) The magnetic field penetrates the same body of material once it becomes normally conductive.

Table 21.7

Critical Temperatures and Magnetic Fluxes for Selected Superconducting Materials

Material	Critical Temperature T_C (K)	Critical Magnetic Flux Density B_C (tesla)[a]
Elements[b]		
Tungsten	0.02	0.0001
Titanium	0.40	0.0056
Aluminum	1.18	0.0105
Tin	3.72	0.0305
Mercury (α)	4.15	0.0411
Lead	7.19	0.0803
Compounds and Alloys[b]		
Nb–Ti alloy	10.2	12
Nb–Zr alloy	10.8	11
$PbMo_6S_8$	14.0	45
V_3Ga	16.5	22
Nb_3Sn	18.3	22
Nb_3Al	18.9	32
Nb_3Ge	23.0	30
Ceramic Compounds		
$YBa_2Cu_3O_7$	92	—
$Bi_2Sr_2Ca_2Cu_3O_{10}$	110	—
$Tl_2Ba_2Ca_2Cu_3O_{10}$	125	—
$HgBa_2Ca_2Cu_2O_8$	153	—

[a]The critical magnetic flux density ($\mu_0 H_C$) for the elements was measured at 0 K. For alloys and compounds, the flux is taken as $\mu_0 H_{C2}$ (in teslas), measured at 0 K.

[b]**Source:** Adapted from *Materials at Low Temperatures,* R. P. Reed and A. F. Clark (Editors), 1983. Reproduced by permission of ASM International, Materials Park, OH.

most commonly used superconductors are niobium–zirconium (Nb–Zr) and niobium–titanium (Nb–Ti) alloys and the niobium–tin intermetallic compound Nb_3Sn. Table 21.7 lists several type I and II superconductors, their critical temperatures, and their critical magnetic flux densities.

Recently, a family of ceramic materials that are normally electrically insulative have been found to be superconductors with inordinately high critical temperatures. Initial research has centered on yttrium barium copper oxide, $YBa_2Cu_3O_7$, which has a critical temperature of about 92 K. This material has a complex perovskite-type crystal structure (Section 4.9). New superconducting ceramic materials reported to have even higher critical temperatures have been and are currently being developed. Several of these materials and their critical temperatures are listed in Table 21.7. The technological potential of these materials is extremely promising because as their critical temperatures are above 77 K, which permits the use of liquid nitrogen, a very inexpensive coolant in comparison to liquid hydrogen and liquid helium. These new ceramic superconductors are not without drawbacks, chief of which is their brittle nature. This characteristic limits the ability of these materials to be fabricated into useful forms such as wires.

The phenomenon of superconductivity has many important practical implications. Superconducting magnets capable of generating high fields with low power consumption are being used in scientific test and research equipment. In addition, they are also used for magnetic resonance imaging (MRI) in the medical field as a diagnostic tool. Abnormalities in body tissues and organs can be detected on the basis of the production of cross-sectional images. Chemical analysis of body tissues is also possible using magnetic resonance spectroscopy (MRS). Numerous other potential applications of superconducting materials also exist. Some of the areas being explored include (1) electrical power transmission through superconducting materials—power losses would be extremely low, and the equipment would operate at low voltage levels; (2) magnets for high-energy particle accelerators; (3) higher-speed switching and signal transmission for computers; and (4) high-speed magnetically levitated trains, for which the levitation results from magnetic field repulsion. The chief deterrent to the widespread application of these superconducting materials is the difficulty in attaining and maintaining extremely low temperatures. It can be hoped that this problem will be overcome with the development of the new generation of superconductors with reasonably high critical temperatures.

SUMMARY

Basic Concepts

- The macroscopic magnetic properties of a material are a consequence of interactions between an external magnetic field and the magnetic dipole moments of the constituent atoms.

- The magnetic field strength (H) within a coil of wire is proportional to the number of wire turns and the magnitude of the current and inversely proportional to the coil length (Equation 21.1).

- Magnetic flux density and magnetic field strength are proportional to one another.

  In a vacuum, the constant of proportionality is the permeability of a vacuum (Equation 21.3).

  When some material is present, this constant is the permeability of the material (Equation 21.2).

- Associated with each individual electron are both orbital and spin magnetic moments.

  The magnitude of an electron's orbital magnetic moment is equal to the product of the value of the Bohr magneton and the electron's magnetic quantum number.

  An electron's spin magnetic moment is plus or minus the value of the Bohr magneton (plus for spin up and minus for spin down).

- The net magnetic moment for an atom is the sum of the contributions of each of its electrons, in which there is spin and orbital moment cancellation of electron pairs. When cancellation is complete, the atom possesses no magnetic moment.

Diamagnetism and Paramagnetism

- *Diamagnetism* results from changes in electron orbital motion that are induced by an external field. The effect is extremely small (with susceptibilities on the order of -10^{-5}) and in opposition to the applied field. All materials are diamagnetic.

- *Paramagnetic materials* are those having permanent atomic dipoles, which are acted on individually and aligned in the direction of an external field.

- Diamagnetic and paramagnetic materials are considered nonmagnetic because the magnetizations are relatively small and persist only while an applied field is present.

Ferromagnetism

- Large and permanent magnetizations may be established within the ferromagnetic metals (Fe, Co, Ni).

- Atomic magnetic dipole moments are of spin origin, which are coupled and mutually aligned with moments of adjacent atoms.

Antiferromagnetism and Ferrimagnetism

- Antiparallel coupling of adjacent cation spin moments is found for some ionic materials. Those in which there is total cancellation of spin moments are termed *antiferromagnetic*.

- With ferrimagnetism, permanent magnetization is possible because spin moment cancellation is incomplete.
- For cubic ferrites, the net magnetization results from the divalent ions (e.g., Fe^{2+}) that reside on octahedral lattice sites, the spin moments of which are all mutually aligned.

The Influence of Temperature on Magnetic Behavior

- With rising temperature, increased thermal vibrations tend to counteract the dipole coupling forces in ferromagnetic and ferrimagnetic materials. Consequently, the saturation magnetization gradually decreases with temperature, up to the Curie temperature, at which point it drops to near zero (Figure 21.10).
- Above T_c, ferromagnetic and ferrimagnetic materials are paramagnetic.

Domains and Hysteresis

- Below its Curie temperature, a ferromagnetic or ferrimagnetic material is composed of *domains*—small-volume regions in which all net dipole moments are mutually aligned and the magnetization is saturated (Figure 21.11).
- The total magnetization of the solid is just the appropriately weighted vector sum of the magnetizations of all these domains.
- As an external magnetic field is applied, domains having magnetization vectors oriented in the direction of the field grow at the expense of domains that have unfavorable magnetization orientations (Figure 21.13).
- At total saturation, the entire solid is a single domain and the magnetization is aligned with the field direction.
- The change in domain structure with increase or reversal of a magnetic field is accomplished by the motion of domain walls. Both hysteresis (the lag of the B field behind the applied H field) and permanent magnetization (or *remanence*) result from the resistance to movement of these domain walls.
- From a complete hysteresis curve for a ferromagnetic/ferrimagnetic material, the following may be determined:
  Remanence—value of the B field when $H = 0$ (B_r, Figure 21.14)
  Coercivity—value of the H field when $B = 0$ (H_c, Figure 21.14)

Magnetic Anisotropy

- The M (or B) versus H behavior for a ferromagnetic single crystal is *anisotropic*—that is, it depends on the crystallographic direction along which the magnetic field is applied.
- The crystallographic direction for which M_s is achieved at the lowest H field is an easy magnetization direction.
- For Fe, Ni, and Co, easy magnetization directions are, respectively, [100], [111], and [0001].
- Energy losses in transformer cores made of magnetic ferrous alloys may be minimized by taking advantage of anisotropic magnetic behavior.

Soft Magnetic Materials

- For soft magnetic materials, domain wall movement is easy during magnetization and demagnetization. Consequently, they have small hysteresis loops and low energy losses.

Hard Magnetic Materials

- Domain wall motion is much more difficult for the hard magnetic materials, which results in larger hysteresis loops; because greater fields are required to demagnetize these materials, the magnetization is more permanent.

Magnetic Storage

- Information storage is accomplished by using magnetic materials; the two principal types of magnetic media are hard disk drives and magnetic tapes.
- The storage medium for hard disk drives is composed of nanometer-size grains of an HCP cobalt–chromium alloy. These grains are oriented such that their direction of easy magnetization (i.e., [0001]) is perpendicular to the plane of the disk.

- For tape-memory storage, either needle-shape ferromagnetic metal particles or plate-shape ferromagnetic barium–ferrite particles are used. Particle size is on the order of tens of nanometers.

Superconductivity
- Superconductivity has been observed in a number of materials; upon cooling and in the vicinity of absolute zero temperature, the electrical resistivity vanishes (Figure 21.26).
- The superconducting state ceases to exist if temperature, magnetic field, or current density exceeds a critical value.
- For type I superconductors, magnetic field exclusion is complete below a critical field, and field penetration is complete once H_C is exceeded. This penetration is gradual with increasing magnetic field for type II materials.
- New complex oxide ceramics are being developed with relatively high critical temperatures that allow inexpensive liquid nitrogen to be used as a coolant.

Important Terms and Concepts

antiferromagnetism	ferromagnetism	magnetization
Bohr magneton	hard magnetic material	paramagnetism
coercivity	hysteresis	permeability
Curie temperature	magnetic field strength	remanence
diamagnetism	magnetic flux density	saturation magnetization
domain	magnetic induction	soft magnetic material
ferrimagnetism	magnetic susceptibility	superconductivity
ferrite (ceramic)		

REFERENCES

Brockman, F. G., "Magnetic Ceramics—A Review and Status Report," *American Ceramic Society Bulletin,* Vol. 47, No. 2, February 1968, pp. 186–194.

Coey, J. M. D., *Magnetism and Magnetic Materials*, Cambridge University Press, Cambridge, 2009.

Cullity, B. D., and C. D. Graham, *Introduction to Magnetic Materials*, 2nd edition, Wiley, Hoboken, NJ, 2009.

Hilzinger, R., and W. Rodewald, *Magnetic Materials: Fundamentals, Products, Properties, Applications*, Wiley, Hoboken, NJ, 2013.

Jiles, D., *Introduction to Magnetism and Magnetic Materials,* 2nd edition, CRC Press, Boca Raton, FL, 1998.

Spaldin, N. A., *Magnetic Materials: Fundamentals and Device Applications,* 2nd edition, Cambridge University Press, Cambridge, 2011.

QUESTIONS AND PROBLEMS

Basic Concepts

21.1 A coil of wire 0.30 m long and having 200 turns carries a current of 15 A.

(a) What is the magnitude of the magnetic field strength H?

(b) Compute the flux density B if the coil is in a vacuum.

(c) Compute the flux density inside a bar of titanium that is positioned within the coil.

The susceptibility for titanium is found in Table 21.2.

(d) Compute the magnitude of the magnetization M.

21.2 Demonstrate that the relative permeability and the magnetic susceptibility are related according to Equation 21.7.

21.3 It is possible to express the magnetic susceptibility χ_m in several different units. For the discussion of this chapter, χ_m was used to designate the volume susceptibility in SI units, that is, the quantity that gives the magnetization per unit volume (m^3) of material when multiplied by H. The mass susceptibility χ_m (kg) yields the magnetic moment (or magnetization) per kilogram of material when multiplied by H; similarly, the atomic susceptibility $\chi_m(a)$ gives the magnetization per kilogram-mole. The latter two quantities are related to χ_m through the relationships

$$\chi_m = \chi_m(\text{kg}) \times \text{mass density (in kg/m}^3)$$
$$\chi_m(a) = \chi_m(\text{kg}) \times \text{atomic weight (in kg)}$$

From Table 21.2, χ_m for silver is -2.38×10^{-5}; convert this value into the other five susceptibilities.

21.4 **(a)** Explain the two sources of magnetic moments for electrons.

(b) Do all electrons have a net magnetic moment? Why or why not?

(c) Do all atoms have a net magnetic moment? Why or why not?

Diamagnetism and Paramagnetism
Ferromagnetism

21.5 The magnetic flux density within a bar of some material is 0.438 tesla at an H field of 3.44×10^5 A/m. Compute the following for this material: **(a)** the magnetic permeability and **(b)** the magnetic susceptibility. **(c)** What type(s) of magnetism would you suggest is (are) being displayed by this material? Why?

21.6 The magnetization within a bar of some metal alloy is 3.6×10^5 A/m at an H field of 60 A/m. Compute the following: **(a)** the magnetic susceptibility, **(b)** the permeability, and **(c)** the magnetic flux density within this material. **(d)** What type(s) of magnetism would you suggest as being displayed by this material? Why?

21.7 Compute **(a)** the saturation magnetization and **(b)** the saturation flux density for cobalt, which has a net magnetic moment per atom of 1.72 Bohr magnetons and a density of 8.90 g/cm^3.

21.8 Confirm that there are 2.2 Bohr magnetons associated with each iron atom, given that the saturation magnetization is 1.75×10^6 A/m, that iron has a BCC crystal structure, and that the unit cell edge length is 0.2866 nm.

21.9 Assume there exists some hypothetical metal that exhibits ferromagnetic behavior and that has (1) a simple cubic crystal structure (Figure 4.2), (2) an atomic radius of 0.153 nm,

and (3) a saturation flux density of 0.76 tesla. Determine the number of Bohr magnetons per atom for this material.

21.10 There is associated with each atom in paramagnetic and ferromagnetic materials a net magnetic moment. Explain why ferromagnetic materials can be permanently magnetized whereas paramagnetic ones cannot.

Antiferromagnetism and Ferrimagnetism

21.11 Consult another reference in which Hund's rule is outlined, and on its basis explain the net magnetic moments for each of the cations listed in Table 21.4.

21.12 Estimate **(a)** the saturation magnetization and **(b)** the saturation flux density of nickel ferrite [$(NiFe_2O_4)_8$], which has a unit cell edge length of 0.8337 nm.

21.13 The chemical formula for manganese ferrite may be written as $(MnFe_2O_4)_8$ because there are eight formula units per unit cell. If this material has a saturation magnetization of 5.6×10^5 A/m and a density of 5.00 g/cm^3, estimate the number of Bohr magnetons associated with each Mn^{2+} ion.

21.14 The formula for yttrium iron garnet ($Y_3Fe_5O_{12}$) may be written in the form $Y_3^c Fe_2^a Fe_3^d O_{12}$, where the superscripts a, c, and d represent different sites on which the Y^{3+} and Fe^{3+} ions are located. The spin magnetic moments for the Y^{3+} and Fe^{3+} ions positioned in the a and c sites are oriented parallel to one another and antiparallel to the Fe^{3+} ions in d sites. Compute the number of Bohr magnetons associated with each Y^{3+} ion, given the following information: (1) each unit cell consists of eight formula ($Y_3Fe_5O_{12}$) units; (2) the unit cell is cubic with an edge length of 1.2376 nm; (3) the saturation magnetization for this material is 1.0×10^4 A/m; and (4) there are five Bohr magnetons associated with each Fe^{3+} ion.

The Influence of Temperature on
* Magnetic Behavior*

21.15 Briefly explain why the magnitude of the saturation magnetization decreases with increasing temperature for ferromagnetic materials, and why ferromagnetic behavior ceases above the Curie temperature.

Domains and Hysteresis

21.16 Briefly describe the phenomenon of magnetic hysteresis, and why it occurs for ferromagnetic and ferrimagnetic materials.

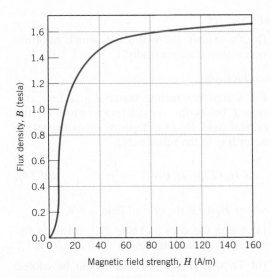

Figure 21.29 Initial magnetization *B*-versus-*H* curve for an iron–silicon alloy.

21.17 A coil of wire 0.5 m long and having 20 turns carries a current of 2.0 A.

(a) Compute the flux density if the coil is within a vacuum.

(b) A bar of an iron–silicon alloy, the *B–H* behavior for which is shown in Figure 21.29, is positioned within the coil. What is the flux density within this bar?

(c) Suppose that a bar of molybdenum is now situated within the coil. What current must be used to produce the same *B* field in the Mo as was produced in the iron–silicon alloy [part (b)] using 1.0 A?

21.18 A ferromagnetic material has a remanence of 1.30 teslas and a coercivity of 50,000 A/m. Saturation is achieved at a magnetic field intensity of 100,000 A/m, at which the flux density is 1.50 teslas. Using these data, sketch the entire hysteresis curve in the range $H = -100,000$ to $+ 100,000$ A/m. Be sure to scale and label both coordinate axes.

21.19 The following data are for a transformer steel:

H (A/m)	*B* (teslas)	*H* (A/m)	*B* (teslas)
0	0	200	1.04
10	0.03	400	1.28
20	0.07	600	1.36
50	0.23	800	1.39
100	0.70	1000	1.41
150	0.92		

(a) Construct a graph of *B* versus *H*.

(b) What are the values of the initial permeability and initial relative permeability?

(c) What is the value of the maximum permeability?

(d) At about what *H* field does this maximum permeability occur?

(e) To what magnetic susceptibility does this maximum permeability correspond?

21.20 An iron bar magnet having a coercivity of 4500 A/m is to be demagnetized. If the bar is inserted within a cylindrical wire coil 0.18 m long and having 100 turns, what electric current is required to generate the necessary magnetic field?

21.21 A bar of an iron–silicon alloy having the *B–H* behavior shown in Figure 21.29 is inserted within a coil of wire 0.25 m long and having 80 turns, through which passes a current of 0.1 A.

(a) What is the *B* field within this bar?

(b) At this magnetic field,

  (i) What is the permeability?

  (ii) What is the relative permeability?

  (iii) What is the susceptibility?

  (iv) What is the magnetization?

Magnetic Anisotropy

21.22 Estimate saturation values of *H* for single-crystal iron in [100], [110], and [111] directions.

21.23 The energy (per unit volume) required to magnetize a ferromagnetic material to saturation (E_s) is defined by the following equation:

$$E_s = \int_0^{M_s} \mu_0 H \, dM$$

That is, E_s is equal to the product of μ_0 and the area under an *M* versus *H* curve, to the point of saturation referenced to the ordinate (or *M*) axis—for example, in Figure 21.17 the area between the vertical axis and the magnetization curve to M_s. Estimate E_s values (in J/m^3) for single-crystal nickel in [100], [110], and [111] directions.

Soft Magnetic Materials
Hard Magnetic Materials

21.24 Cite the differences between hard and soft magnetic materials in terms of both hysteresis behavior and typical applications.

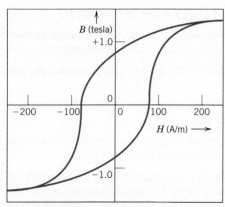

Figure 21.30 Complete magnetic hysteresis loop for a steel alloy.

21.25 Assume that the commercial iron (99.95 wt% Fe) in Table 21.5 just reaches the point of saturation when inserted within the coil in Problem 21.1. Compute the saturation magnetization.

21.26 Figure 21.30 shows the B-versus-H curve for a steel alloy.

   (a) What is the saturation flux density?

   (b) What is the saturation magnetization?

   (c) What is the remanence?

   (d) What is the coercivity?

   (e) On the basis of data in Tables 21.5 and 21.6, would you classify this material as a soft or hard magnetic material? Why?

Magnetic Storage

21.27 Briefly explain the manner in which information is stored magnetically.

Superconductivity

21.28 For a superconducting material at a temperature T below the critical temperature T_C, the critical field $H_C(T)$, depends on temperature according to the relationship

$$H_C(T) = H_C(0)\left(1 - \frac{T^2}{T_C^2}\right) \qquad (21.14)$$

where $H_C(0)$ is the critical field at 0 K.

   (a) Using the data in Table 21.7, calculate the critical magnetic fields for tin at 2.0 and 3.0 K.

   (b) To what temperature must tin be cooled in a magnetic field of 22,000 A/m for it to be superconductive?

21.29 Using Equation 21.14, determine which of the superconducting elements in Table 21.7 are superconducting at 3 K and in a magnetic field of 15,000 A/m.

21.30 Cite the differences between type I and type II superconductors.

21.31 Briefly describe the Meissner effect.

21.32 Cite the primary limitation of the new superconducting materials that have relatively high critical temperatures.

DESIGN PROBLEMS

Ferromagnetism

21.D1 A cobalt–nickel alloy is desired that has a saturation magnetization of 1.3×10^6 A/m. Specify its composition in weight percent nickel. Cobalt has an HCP crystal structure with c/a ratio of 1.623, whereas the maximum solubility of Ni in Co at room temperature is approximately 35 wt%. Assume that the unit cell volume for this alloy is the same as for pure Co.

Ferrimagnetism

21.D2 Design a cubic mixed-ferrite magnetic material that has a saturation magnetization of 4.8×10^5 A/m.

(*a*) Schematic diagram illustrating the operation of a photovoltaic solar cell. The cell is made of polycrystalline silicon that has been fabricated to form a *p–n* junction (see Sections 19.11 and 19.15). Photons that originate as light from the sun excite electrons into the conduction band on the *n* side of the junction and create holes on the *p* side. These electrons and holes are drawn away from the junction in opposite directions and become part of an external current.

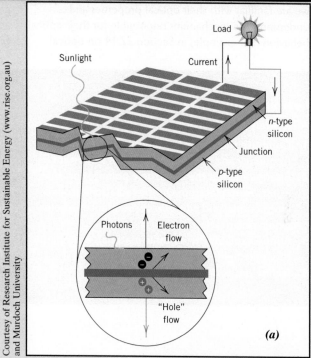

Courtesy of Research Institute for Sustainable Energy (www.rise.org.au) and Murdoch University

(*a*)

(*c*) A home with a number of solar panels.

© Brainstorm1962/iStockphoto

(*c*)

(*b*) An array of polycrystalline silicon photovoltaic cells.

© Gabor Izso/iStockphoto

(*b*)

When materials are exposed to electromagnetic radiation, it is sometimes important to be able to predict and alter their responses. This is possible when we are familiar with their optical properties and understand the mechanisms responsible for their optical behaviors. For example, in Section 22.14 on optical fiber materials in communications, we note that the performance of optical fibers is increased by introducing a gradual variation of the index of refraction (i.e., a graded index) at the outer surface of the fiber. This is accomplished by the addition of specific impurities in controlled concentrations.

Learning Objectives

After studying this chapter, you should be able to do the following:

1. Compute the energy of a photon given its frequency and the value of Planck's constant.
2. Briefly describe the electronic polarization that results from electromagnetic radiation–atomic interactions, and cite two consequences of electronic polarization.
3. Briefly explain why metallic materials are opaque to visible light.
4. Define *index of refraction*.
5. Describe the mechanism of photon absorption for (a) high-purity insulators and semiconductors and (b) insulators and semiconductors that contain electrically active defects.
6. For inherently transparent dielectric materials, note three sources of internal scattering that can lead to translucency and opacity.
7. Briefly describe the construction and operation of ruby and semiconductor lasers.

22.1 INTRODUCTION

Optical property refers to a material's response to exposure to electromagnetic radiation and, in particular, to visible light. This chapter first discusses some of the basic principles and concepts relating to the nature of electromagnetic radiation and its possible interactions with solid materials. Then it explores the optical behaviors of metallic and nonmetallic materials in terms of their absorption, reflection, and transmission characteristics. The final sections outline luminescence, photoconductivity, and light amplification by stimulated emission of radiation (laser), the practical use of these phenomena, and the use of optical fibers in communications.

Basic Concepts

22.2 ELECTROMAGNETIC RADIATION

In the classical sense, electromagnetic radiation is considered to be wavelike, consisting of electric and magnetic field components that are perpendicular to each other and also to the direction of propagation (Figure 22.1). Light, heat (or radiant energy), radar, radio waves, and x-rays are all forms of electromagnetic radiation. Each is characterized primarily by a specific range of wavelengths and also according to the technique by which it is generated. The *electromagnetic spectrum* of radiation spans the wide range from γ-rays (emitted by radioactive materials) having wavelengths on the order of 10^{-12} m (10^{-3} nm) through x-rays, ultraviolet, visible, infrared, and finally radio waves with wavelengths as long as 10^5 m. This spectrum is shown on a logarithmic scale in Figure 22.2.

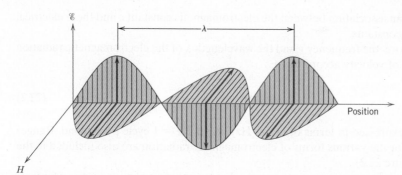

Figure 22.1 An electromagnetic wave showing electric field $\mathcal{E}$ and magnetic field H components and the wavelength λ.

Visible light lies within a very narrow region of the spectrum, with wavelengths ranging between about 0.4 μm (4×10^{-7} m) and 0.7 μm. The perceived color is determined by wavelength; for example, radiation having a wavelength of approximately 0.4 μm appears violet, whereas green and red occur at about 0.5 and 0.65 μm, respectively. The spectral ranges for several colors are included in Figure 22.2. White light is simply a mixture of all colors. The ensuing discussion is concerned primarily with this visible radiation, by definition the only radiation to which the eye is sensitive.

All electromagnetic radiation traverses a vacuum at the same velocity, that of light—namely, 3×10^8 m/s. This velocity, c, is related to the electric permittivity of a vacuum ϵ_0 and the magnetic permeability of a vacuum μ_0 through

For a vacuum, dependence of the velocity of light on electric permittivity and magnetic permeability

$$c = \frac{1}{\sqrt{\epsilon_0 \mu_0}}$$

(22.1)

Energy (eV)   Wavelength (m)

Frequency (Hz)

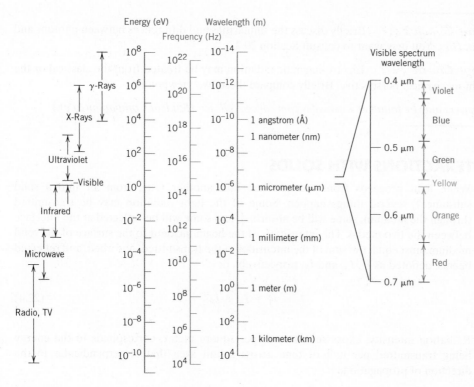

Figure 22.2 The spectrum of electromagnetic radiation, including wavelength ranges for the various colors in the visible spectrum.

Thus, there is an association between the electromagnetic constant c and these electrical and magnetic constants.

Furthermore, the frequency v and the wavelength λ of the electromagnetic radiation are a function of velocity according to

For electromagnetic radiation, relationship among velocity, wavelength, and frequency

$$c = \lambda v \qquad (22.2)$$

Frequency is expressed in terms of hertz (Hz), and 1 Hz = 1 cycle per second. Ranges of frequency for the various forms of electromagnetic radiation are also included in the spectrum (Figure 22.2).

Sometimes it is more convenient to view electromagnetic radiation from a quantum-mechanical perspective, in which the radiation, rather than consisting of waves, is composed of groups or packets of energy called **photons**. The energy E of a photon is said to be *quantized,* or can only have specific values, defined by the relationship

photon

For a photon of electromagnetic radiation, dependence of energy on frequency, and also velocity and wavelength

$$E = hv = \frac{hc}{\lambda} \qquad (22.3)$$

Planck's constant

where h is a universal constant called **Planck's constant**, which has a value of 6.63×10^{-34} J·s. Thus, photon energy is proportional to the frequency of the radiation, or inversely proportional to the wavelength. Photon energies are also included in the electromagnetic spectrum (Figure 22.2).

When describing optical phenomena involving the interactions between radiation and matter, an explanation is often facilitated if light is treated in terms of photons. On other occasions, a wave treatment is preferred; both approaches are used in this discussion, as appropriate.

Concept Check 22.1 Briefly discuss the similarities and differences between photons and phonons. *Hint:* You may want to consult Section 20.2.

Concept Check 22.2 Electromagnetic radiation may be treated from the classical or the quantum-mechanical perspective. Briefly compare these two viewpoints.

[*The answers may be found at* www.wiley.com/college/callister *(Student Companion Site).*]

22.3 LIGHT INTERACTIONS WITH SOLIDS

When light proceeds from one medium into another (e.g., from air into a solid substance), several things happen. Some of the light radiation may be transmitted through the medium, some will be absorbed, and some will be reflected at the interface between the two media. The intensity I_0 of the beam incident to the surface of the solid medium must equal the sum of the intensities of the transmitted, absorbed, and reflected beams, denoted as I_T, I_A, and I_R, respectively, or

Intensity of an incident beam at an interface is equal to the sum of the intensities of transmitted, absorbed, and reflected beams

$$I_0 = I_T + I_A + I_R \qquad (22.4)$$

Radiation intensity, expressed in watts per square meter, corresponds to the energy being transmitted per unit of time across a unit area that is perpendicular to the direction of propagation.

An alternate form of Equation 22.4 is

$$T + A + R = 1 \qquad (22.5)$$

where T, A, and R represent, respectively, the transmissivity (I_T/I_0), absorptivity (I_A/I_0), and reflectivity (I_R/I_0), or the fractions of incident light that are transmitted, absorbed, and reflected by a material; their sum must equal unity because all the incident light is transmitted, absorbed, or reflected.

transparent

translucent

opaque

Materials that are capable of transmitting light with relatively little absorption and reflection are **transparent**—one can see through them. **Translucent** materials are those through which light is transmitted diffusely; that is, light is scattered within the interior to the degree that objects are not clearly distinguishable when viewed through a specimen of the material. Materials that are impervious to the transmission of visible light are termed **opaque**.

Bulk metals are opaque throughout the entire visible spectrum—that is, all light radiation is either absorbed or reflected. However, electrically insulating materials can be made to be transparent. Furthermore, some semiconducting materials are transparent, whereas others are opaque.

22.4 ATOMIC AND ELECTRONIC INTERACTIONS

The optical phenomena that occur within solid materials involve interactions between the electromagnetic radiation and atoms, ions, and/or electrons. Two of the most important of these interactions are electronic polarization and electron energy transitions.

Electronic Polarization

One component of an electromagnetic wave is simply a rapidly fluctuating electric field (Figure 22.1). For the visible range of frequencies, this electric field interacts with the electron cloud surrounding each atom within its path in such a way as to induce electronic polarization or to shift the electron cloud relative to the nucleus of the atom with each change in direction of electric field component, as demonstrated in Figure 19.32*a*. Two consequences of this polarization are as follows: (1) some of the radiation energy may be absorbed, and (2) light waves are decreased in velocity as they pass through the medium. The second consequence is manifested as refraction, a phenomenon to be discussed in Section 22.5.

Electron Transitions

The absorption and emission of electromagnetic radiation may involve electron transitions from one energy state to another. For the sake of this discussion, consider an isolated atom, the electron energy diagram for which is represented in Figure 22.3. An electron may be excited from an occupied state at energy E_2 to a vacant and higher-lying one, denoted E_4, by the absorption of a photon of energy. The change in energy experienced by the electron, ΔE, depends on the radiation frequency as follows:

For an electron transition, change in energy equals the product of Planck's constant and the frequency of radiation absorbed (or emitted)

$$\Delta E = hv \qquad (22.6)$$

where, again, h is Planck's constant. At this point, it is important to understand several concepts: First, because the energy states for the atom are discrete, only specific ΔEs exist between the energy levels; thus, only photons of frequencies corresponding to the possible ΔEs for the atom can be absorbed by electron transitions. Furthermore, all of a photon's energy is absorbed in each excitation event.

Figure 22.3 For an isolated atom, a schematic illustration of photon absorption by the excitation of an electron from one energy state to another. The energy of the photon $(h\nu_{42})$ must be exactly equal to the difference in energy between the two states $(E_4 - E_2)$.

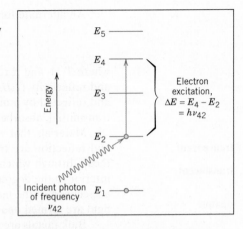

A second important concept is that a stimulated electron cannot remain in an **excited state** indefinitely; after a short time, it falls or decays back into its **ground state**, or unexcited level, with reemission of electromagnetic radiation. Several decay paths are possible, and these are discussed later. In any case, there must be a conservation of energy for absorption and emission electron transitions.

As the ensuing discussions show, the optical characteristics of solid materials that relate to absorption and emission of electromagnetic radiation are explained in terms of the electron band structure of the material (possible band structures are discussed in Section 19.5) and the principles relating to electron transitions, as outlined in the preceding two paragraphs.

excited state

ground state

Optical Properties of Metals

Consider the electron energy band schemes for metals as illustrated in Figures 19.4*a* and 19.4*b*; in both cases, a high-energy band is only partially filled with electrons. Metals are opaque because the incident radiation having frequencies within the visible range excites electrons into unoccupied energy states above the Fermi energy, as demonstrated in Figure 22.4*a*; as a consequence, the incident radiation is absorbed, in accordance with Equation 22.6. Total absorption is within a very thin outer layer, usually less than 0.1 μm; thus only metallic films thinner than 0.1 μm are capable of transmitting visible light.

All frequencies of visible light are absorbed by metals because of the continuously available empty electron states, which permit electron transitions as in Figure 22.4*a*. In fact, metals are opaque to all electromagnetic radiation on the low end of the frequency

Figure 22.4 (*a*) Schematic representation of the mechanism of photon absorption for metallic materials in which an electron is excited into a higher-energy unoccupied state. The change in energy of the electron ΔE is equal to the energy of the photon. (*b*) Reemission of a photon of light by the direct transition of an electron from a high to a low energy state.

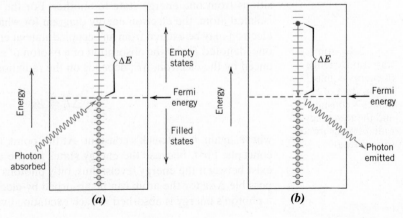

spectrum, from radio waves, through infrared and the visible, and into about the middle of the ultraviolet radiation. Metals are transparent to high-frequency (x- and γ-ray) radiation.

Most of the absorbed radiation is reemitted from the surface in the form of visible light of the same wavelength, which appears as reflected light; an electron transition accompanying reradiation is shown in Figure 22.4b. The reflectivity for most metals is between 0.90 and 0.95; some small fraction of the energy from electron decay processes is dissipated as heat.

Because metals are opaque and highly reflective, the perceived color is determined by the wavelength distribution of the radiation that is reflected and not absorbed. A bright silvery appearance when exposed to white light indicates that the metal is highly reflective over the entire range of the visible spectrum. In other words, for the reflected beam, the composition of these reemitted photons, in terms of frequency and number, is approximately the same as for the incident beam. Aluminum and silver are two metals that exhibit this reflective behavior. Copper and gold appear red-orange and yellow, respectively, because some of the energy associated with light photons having short wavelengths is not reemitted as visible light.

Concept Check 22.3   Why are metals transparent to high-frequency x-ray and γ-ray radiation?

[*The answer may be found at* www.wiley.com/college/callister *(Student Companion Site).*]

Optical Properties of Nonmetals

By virtue of their electron energy band structures, nonmetallic materials may be transparent to visible light. Therefore, in addition to reflection and absorption, refraction and transmission phenomena must also be considered.

22.5 REFRACTION

refraction

index of refraction

Definition of *index of refraction*—the ratio of light velocities in a vacuum and in the medium of interest

Light that is transmitted into the interior of transparent materials experiences a decrease in velocity, and, as a result, is bent at the interface; this phenomenon is termed **refraction**. The **index of refraction** n of a material is defined as the ratio of the velocity in a vacuum c to the velocity in the medium v, or

$$n = \frac{c}{v} \tag{22.7}$$

The magnitude of n (or the degree of bending) depends on the wavelength of the light. This effect is graphically demonstrated by the familiar dispersion or separation of a beam of white light into its component colors by a glass prism (as shown in the margin photograph on the next page). Each color is deflected by a different amount as it passes into and out of the glass, which results in the separation of the colors. Not only does the index of refraction affect the optical path of light, but also, as explained shortly, it influences the fraction of incident light reflected at the surface.

Velocity of light in a medium, in terms of the medium's electric permittivity and magnetic permeability

Just as Equation 22.1 defines the magnitude of c, an equivalent expression gives the velocity of light v in a medium as

$$v = \frac{1}{\sqrt{\epsilon\mu}} \tag{22.8}$$

Index of refraction of
a medium—in terms
of the medium's
dielectric constant
and relative magnetic
permeability

where ϵ and μ are, respectively, the permittivity and permeability of the particular substance. From Equation 22.7, we have

$$n = \frac{c}{v} = \frac{\sqrt{\epsilon\mu}}{\sqrt{\epsilon_0\mu_0}} = \sqrt{\epsilon_r\mu_r} \qquad (22.9)$$

where ϵ_r and μ_r are the dielectric constant and the relative magnetic permeability, respectively. Because most substances are only slightly magnetic, $\mu_r \cong 1$, and

Relationship
between index
of refraction and
dielectric constant
for a nonmagnetic
material

$$n \cong \sqrt{\epsilon_r} \qquad (22.10)$$

The dispersion of
white light as it
passes through a
prism.
(© PhotoDisc/Getty
Images.)

Thus, for transparent materials, there is a relation between the index of refraction and the dielectric constant. As already mentioned, the phenomenon of refraction is related to electronic polarization (Section 22.4) at the relatively high frequencies for visible light; thus, the electronic component of the dielectric constant may be determined from index of refraction measurements using Equation 22.10.

Because the retardation of electromagnetic radiation in a medium results from electronic polarization, the size of the constituent atoms or ions has considerable influence on the magnitude of this effect—generally, the larger an atom or ion, the greater the electronic polarization, the slower the velocity, and the greater the index of refraction. The index of refraction for a typical soda–lime glass is approximately 1.5. Additions of large barium and lead ions (as BaO and PbO) to a glass increases n significantly. For example, highly leaded glasses containing 90 wt% PbO have an index of refraction of approximately 2.1.

For crystalline ceramics with cubic crystal structures and for glasses, the index of refraction is independent of crystallographic direction (i.e., it is isotropic). Noncubic crystals, however, have an anisotropic n—that is, the index is greatest along the directions that have the highest density of ions. Table 22.1 gives refractive indices for several

Table 22.1

Refractive Indices for
Some Transparent
Materials

Material	Average Index of Refraction
Ceramics	
Silica glass	1.458
Borosilicate (Pyrex) glass	1.47
Soda–lime glass	1.51
Quartz (SiO_2)	1.55
Dense optical flint glass	1.65
Spinel ($MgAl_2O_4$)	1.72
Periclase (MgO)	1.74
Corundum (Al_2O_3)	1.76
Polymers	
Polytetrafluoroethylene	1.35
Poly(methyl methacrylate)	1.49
Polypropylene	1.49
Polyethylene	1.51
Polystyrene	1.60

glasses, transparent ceramics, and polymers. Average values are provided for the crystalline ceramics in which n is anisotropic.

Concept Check 22.4   Which of the following oxide materials, when added to fused silica (SiO_2), increases its index of refraction: Al_2O_3, TiO_2, NiO, MgO? Why? You may find Table 4.4 helpful.

[*The answer may be found at* www.wiley.com/college/callister (*Student Companion Site*).]

22.6   REFLECTION

When light radiation passes from one medium into another having a different index of refraction, some of the light is scattered at the interface between the two media, even if both are transparent. The reflectivity R represents the fraction of the incident light that is reflected at the interface, or

Definition of *reflectivity*—in terms of intensities of reflected and incident beams

$$R = \frac{I_R}{I_0} \qquad (22.11)$$

where I_0 and I_R are the intensities of the incident and reflected beams, respectively. If the light is normal (or perpendicular) to the interface, then

Reflectivity (for normal incidence) at the interface between two media having indices of refraction of n_1 and n_2

$$R = \left(\frac{n_2 - n_1}{n_2 + n_1} \right)^2 \qquad (22.12)$$

where n_1 and n_2 are the indices of refraction of the two media. If the incident light is not normal to the interface, R depends on the angle of incidence. When light is transmitted from a vacuum or air into a solid s, then

$$R = \left(\frac{n_s - 1}{n_s + 1} \right)^2 \qquad (22.13)$$

because the index of refraction of air is very nearly unity. Thus, the higher the index of refraction of the solid, the greater the reflectivity. For typical silicate glasses, the reflectivity is approximately 0.05. Just as the index of refraction of a solid depends on the wavelength of the incident light, so does the reflectivity vary with wavelength. Reflection losses for lenses and other optical instruments are minimized significantly by coating the reflecting surface with very thin layers of dielectric materials such as magnesium fluoride (MgF_2).

22.7   ABSORPTION

Nonmetallic materials may be opaque or transparent to visible light; if transparent, they often appear colored. In principle, light radiation is absorbed in this group of materials by two basic mechanisms that also influence the transmission characteristics of these nonmetals. One of these is electronic polarization (Section 22.4). Absorption by electronic polarization is important only at light frequencies in the vicinity of the relaxation frequency of the constituent atoms. The other mechanism involves valence

Figure 22.5 (a) Mechanism of photon absorption for nonmetallic materials in which an electron is excited across the band gap, leaving behind a hole in the valence band. The energy of the photon absorbed is ΔE, which is necessarily greater than the band gap energy E_g. (b) Emission of a photon of light by a direct electron transition across the band gap.

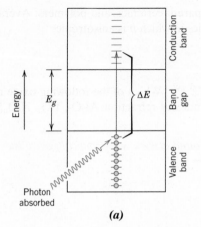
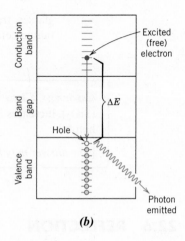

(a) *(b)*

band–conduction band electron transitions, which depend on the electron energy band structure of the material; band structures for semiconductors and insulators are discussed in Section 19.5.

Absorption of a photon of light may occur by the promotion or excitation of an electron from the nearly filled valence band, across the band gap, and into an empty state within the conduction band, as demonstrated in Figure 22.5a; a free electron in the conduction band and a hole in the valence band are created. Again, the energy of excitation ΔE is related to the absorbed photon frequency through Equation 22.6. These excitations with the accompanying absorption can take place only if the photon energy is greater than that of the band gap E_g—that is, if

For a nonmetallic material, condition for absorption of a photon (of radiation) by an electron transition in terms of radiation frequency

$$hv > E_g \tag{22.14}$$

or, in terms of wavelength,

For a nonmetallic material, condition for absorption of a photon (of radiation) by an electron transition in terms of radiation wavelength

$$\frac{hc}{\lambda} > E_g \tag{22.15}$$

The minimum wavelength for visible light, $\lambda(\min)$, is about 0.4 μm, and because $c = 3 \times 10^8$ m/s and $h = 4.13 \times 10^{-15}$ eV·s, the maximum band gap energy $E_g(\max)$ for which absorption of visible light is possible is

Maximum possible band gap energy for absorption of visible light by valence band–to–conduction band electron transitions

$$E_g(\max) = \frac{hc}{\lambda(\min)}$$

$$= \frac{(4.13 \times 10^{-15} \text{ eV·s})(3 \times 10^8 \text{ m/s})}{4 \times 10^{-7} \text{ m}} \tag{22.16a}$$

$$= 3.1 \text{ eV}$$

In other words, no visible light is absorbed by nonmetallic materials having band gap energies greater than about 3.1 eV; these materials, if of high purity, appear transparent and colorless.

However, the maximum wavelength for visible light, $\lambda(max)$, is about 0.7 µm; computation of the minimum band gap energy $E_g(min)$ for which there is absorption of visible light gives

Minimum possible band gap energy for absorption of visible light by valence band–to–conduction band electron transitions

$$E_g(min) = \frac{hc}{\lambda(max)}$$

$$= \frac{(4.13 \times 10^{-15} \text{ eV·s})(3 \times 10^8 \text{ m/s})}{7 \times 10^{-7} \text{ m}} = 1.8 \text{ eV} \qquad (22.16b)$$

This result means that all visible light is absorbed by valence band–to–conduction band electron transitions for semiconducting materials that have band gap energies less than about 1.8 eV; thus, these materials are opaque. Only a portion of the visible spectrum is absorbed by materials having band gap energies between 1.8 and 3.1 eV; consequently, these materials appear colored.

Every nonmetallic material becomes opaque at some wavelength, which depends on the magnitude of its E_g. For example, diamond, having a band gap of 5.6 eV, is opaque to radiation having wavelengths less than about 0.22 µm.

Interactions with light radiation can also occur in dielectric solids having wide band gaps, involving other than valence band–conduction band electron transitions. If impurities or other electrically active defects are present, electron levels within the band gap may be introduced, such as the donor and acceptor levels (Section 19.11), except that they lie closer to the center of the band gap. Light radiation of specific wavelengths may be emitted as a result of electron transitions involving these levels within the band gap. For example, consider Figure 22.6a, which shows the valence band–conduction band electron excitation for a material that has one such impurity level. Again, the electromagnetic energy that is absorbed by this electron excitation must be dissipated in some manner; several mechanisms are possible. For one, this dissipation may occur via direct electron and hole recombination according to the reaction

Reaction describing electron-hole recombination with the generation of energy

$$\text{electron} + \text{hole} \longrightarrow \text{energy} (\Delta E) \qquad (22.17)$$

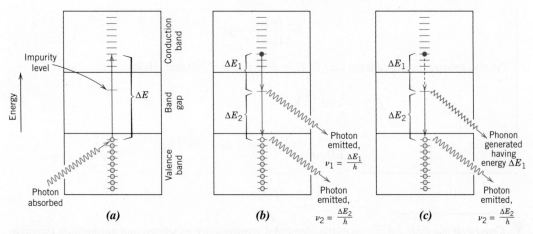

Figure 22.6 (*a*) Photon absorption via a valence band–conduction band electron excitation for a material that has an impurity level that lies within the band gap. (*b*) Emission of two photons involving electron decay first into an impurity state and finally to the ground state. (*c*) Generation of both a phonon and a photon as an excited electron falls first into an impurity level and finally back to its ground state.

which is represented schematically in Figure 22.5b. In addition, multiple-step electron transitions may occur that involve impurity levels lying within the band gap. One possibility, as indicated in Figure 22.6b, is the emission of two photons; one is emitted as the electron drops from a state in the conduction band to the impurity level, the other as it decays back into the valence band. Alternatively, one of the transitions may involve the generation of a phonon (Figure 22.6c), in which the associated energy is dissipated in the form of heat.

The intensity of the net absorbed radiation is dependent on the character of the medium and the path length within. The intensity of transmitted or nonabsorbed radiation I'_T continuously decreases with the distance x that the light traverses:

Intensity of nonabsorbed radiation—dependence on the absorption coefficient and the distance light traverses through the absorbing medium

$$I'_T = I'_0 e^{-\beta x} \qquad (22.18)$$

where I'_0 is the intensity of the nonreflected incident radiation and β, the *absorption coefficient* (in mm^{-1}), is characteristic of the particular material; β varies with the wavelength of the incident radiation. The distance parameter x is measured from the incident surface into the material. Materials with large β values are considered highly absorptive.

EXAMPLE PROBLEM 22.1

Computation of the Absorption Coefficient for Glass

The fraction of nonreflected light that is transmitted through a 200-mm thickness of glass is 0.98. Calculate the absorption coefficient of this material.

Solution

This problem calls for us to solve for β in Equation 22.18. We first rearrange this expression as

$$\frac{I'_T}{I'_0} = e^{-\beta x}$$

Taking logarithms of both sides of the preceding equation leads to

$$\ln\left(\frac{I'_T}{I'_0}\right) = -\beta x$$

Finally, solving for β, realizing that $I'_T/I'_0 = 0.98$ and $x = 200$ mm, yields

$$\beta = -\frac{1}{x}\ln\left(\frac{I'_T}{I'_0}\right)$$

$$= -\frac{1}{200\,\text{mm}}\ln(0.98) = 1.01 \times 10^{-4}\,\text{mm}^{-1}$$

Concept Check 22.5   Are the elemental semiconductors silicon and germanium transparent to visible light? Why or why not? *Hint:* You may want to consult Table 19.3.

[*The answer may be found at* www.wiley.com/college/callister (*Student Companion Site*).]

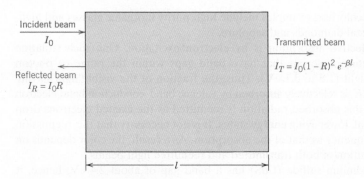

Figure 22.7 The transmission of light through a transparent medium for which there is reflection at the front and back faces, as well as absorption within the medium. (Adapted from R. M. Rose, L. A. Shepard, and J. Wulff, *The Structure and Properties of Materials,* Vol. 4, *Electronic Properties.* Copyright © 1966 by John Wiley & Sons, New York. Reprinted by permission of John Wiley & Sons, Inc.)

22.8 TRANSMISSION

The phenomena of absorption, reflection, and transmission may be applied to the passage of light through a transparent solid, as shown in Figure 22.7. For an incident beam of intensity I_0 that impinges on the front surface of a specimen of thickness l and absorption coefficient β, the transmitted intensity at the back face I_T is

Intensity of radiation transmitted through a specimen of thickness l, accounting for all absorption and reflection losses

$$I_T = I_0(1 - R)^2 e^{-\beta l} \tag{22.19}$$

where R is the reflectance; for this expression, it is assumed that the same medium exists outside both front and back faces. The derivation of Equation 22.19 is left as a homework problem.

Thus, the fraction of incident light that is transmitted through a transparent material depends on the losses that are incurred by absorption and reflection. Again, the sum of the reflectivity R, absorptivity A, and transmissivity T, is unity according to Equation 22.5. Also, each of the variables R, A, and T depends on light wavelength. This is demonstrated over the visible region of the spectrum for a green glass in Figure 22.8. For example, for light having a wavelength of 0.4 μm, the fractions transmitted, absorbed, and reflected are approximately 0.90, 0.05, and 0.05, respectively. However, at 0.55 μm, the respective fractions shift to about 0.50, 0.48, and 0.02.

22.9 COLOR

color

Transparent materials appear colored as a consequence of specific wavelength ranges of light that are selectively absorbed; the **color** discerned is a result of the combination of wavelengths that are transmitted. If absorption is uniform for all visible wavelengths,

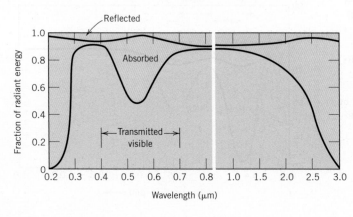

Figure 22.8 The variation with wavelength of the fractions of incident light transmitted, absorbed, and reflected through a green glass. (From W. D. Kingery, H. K. Bowen, and D. R. Uhlmann, *Introduction to Ceramics,* 2nd edition. Copyright © 1976 by John Wiley & Sons, New York. Reprinted by permission of John Wiley & Sons, Inc.)

the material appears colorless; examples include high-purity inorganic glasses and high-purity and single-crystal diamonds and sapphire.

Usually, any selective absorption is by electron excitation. One such situation involves semiconducting materials that have band gaps within the range of photon energies for visible light (1.8 to 3.1 eV). Thus, the fraction of the visible light having energies greater than E_g is selectively absorbed by valence band–conduction band electron transitions. Some of this absorbed radiation is reemitted as the excited electrons drop back into their original, lower-lying energy states. It is not necessary that this reemission occur at the same frequency as that of the absorption. As a result, the color depends on the frequency distribution of both transmitted and reemitted light beams.

For example, cadmium sulfide (CdS) has a band gap of about 2.4 eV; hence, it absorbs photons having energies greater than about 2.4 eV, which correspond to the blue and violet portions of the visible spectrum; some of this energy is reradiated as light having other wavelengths. Nonabsorbed visible light consists of photons having energies between about 1.8 and 2.4 eV. Cadmium sulfide takes on a yellow–orange color because of the composition of the transmitted beam.

With insulator ceramics, specific impurities also introduce electron levels within the forbidden band gap, as discussed previously. Photons having energies less than the band gap may be emitted as a consequence of electron decay processes involving impurity atoms or ions, as demonstrated in Figures 22.6b and 22.6c. Again, the color of the material is a function of the distribution of wavelengths in the transmitted beam.

For example, high-purity and single-crystal aluminum oxide or sapphire is colorless. Ruby, which has a brilliant red color, is sapphire to which has been added 0.5% to 2% chromium oxide (Cr_2O_3). The Cr^{3+} ion substitutes for the Al^{3+} ion in the Al_2O_3 crystal structure and introduces impurity levels within the wide energy band gap of the sapphire. Light radiation is absorbed by valence band–conduction band electron transitions, some of which is then reemitted at specific wavelengths as a consequence of electron transitions to and from these impurity levels. The transmittance as a function of wavelength for sapphire and ruby is presented in Figure 22.9. For the sapphire, transmittance is relatively constant with wavelength over the visible spectrum, which accounts for the colorlessness of this material. However, strong absorption peaks (or minima) occur for the ruby—one in the blue–violet region (at about 0.4 μm) and the other for yellow–green light (at about 0.6 μm). That nonabsorbed or transmitted light mixed with reemitted light imparts to ruby its deep-red color.

Inorganic glasses are colored by incorporating transition or rare earth ions while the glass is in the molten state. Representative color–ion pairs include Cu^{2+}, blue–green; Co^{2+}, blue–violet; Cr^{3+}, green; Mn^{2+}, yellow; and Mn^{3+}, purple. These colored glasses are also used as glazes and decorative coatings on ceramic ware.

Figure 22.9 Transmission of light radiation as a function of wavelength for sapphire (single-crystal aluminum oxide) and ruby (aluminum oxide containing some chromium oxide). The sapphire appears colorless, whereas the ruby has a red tint due to selective absorption over specific wavelength ranges.
(Adapted from "The Optical Properties of Materials," by A. Javan. Copyright © 1967 by Scientific American, Inc. All rights reserved.)

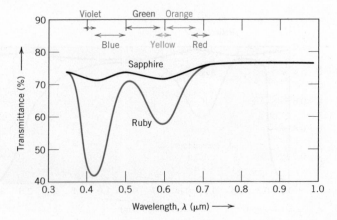

Concept Check 22.6 Compare the factors that determine the characteristic colors of metals and transparent nonmetals.

[*The answer may be found at* www.wiley.com/college/callister (*Student Companion Site*).]

22.10 OPACITY AND TRANSLUCENCY IN INSULATORS

The extent of translucency and opacity for inherently transparent dielectric materials depends to a great degree on their internal reflectance and transmittance characteristics. Many dielectric materials that are intrinsically transparent may be made translucent or even opaque because of interior reflection and refraction. A transmitted light beam is deflected in direction and appears diffuse as a result of multiple scattering events. Opacity results when the scattering is so extensive that virtually none of the incident beam is transmitted undeflected to the back surface.

This internal scattering may result from several different sources. Polycrystalline specimens in which the index of refraction is anisotropic normally appear translucent. Both reflection and refraction occur at grain boundaries, which causes a diversion in the incident beam. This results from a slight difference in index of refraction n between adjacent grains that do not have the same crystallographic orientation.

Scattering of light also occurs in two-phase materials in which one phase is finely dispersed within the other. Again, the beam dispersion occurs across phase boundaries when there is a difference in the refractive index for the two phases; the greater this difference, the more efficient the scattering. Glass-ceramics (Section 14.11), which may consist of both crystalline and residual glass phases, appear highly transparent if the sizes of the crystallites are smaller than the wavelength of visible light and when the indices of refraction of the two phases are nearly identical (which is possible by adjustment of composition).

As a consequence of fabrication or processing, many ceramic pieces contain some residual porosity in the form of finely dispersed pores. These pores also effectively scatter light radiation.

Figure 22.10 demonstrates the difference in optical transmission characteristics of single-crystal, fully dense polycrystalline, and porous (~5% porosity) aluminum oxide specimens. Whereas the single crystal is totally transparent, polycrystalline and porous materials are, respectively, translucent and opaque.

For intrinsic polymers (without additives and impurities), the degree of translucency is influenced primarily by the extent of crystallinity. Some scattering of visible light

Figure 22.10 The light transmittance of three aluminum oxide specimens. From left to right; single-crystal material (sapphire), which is transparent; a polycrystalline and fully dense (nonporous) material, which is translucent; and a polycrystalline material that contains approximately 5% porosity, which is opaque. (Specimen preparation, P. A. Lessing.)

occurs at the boundaries between crystalline and amorphous regions, again as a result of different indices of refraction. For highly crystalline specimens, this degree of scattering is extensive, which leads to translucency and, in some instances, even opacity. Highly amorphous polymers are completely transparent.

Applications of Optical Phenomena

22.11 LUMINESCENCE

luminescence

Some materials are capable of absorbing energy and then reemitting visible light in a phenomenon called **luminescence**. Photons of emitted light are generated from electron transitions in the solid. Energy is absorbed when an electron is promoted to an excited energy state; visible light is emitted when it falls back to a lower energy state if $1.8 \text{ eV} < h\nu < 3.1 \text{ eV}$. The absorbed energy may be supplied as higher-energy electromagnetic radiation (causing valence band–conduction band transitions, Figure 22.6a) such as ultraviolet light; other sources such as high-energy electrons; or heat, mechanical, or chemical energy. Furthermore, luminescence is classified according to the magnitude of the delay time between absorption and reemission events. If reemission occurs for times much less than 1 s, the phenomenon is termed **fluorescence**; for longer times, it is called **phosphorescence**. A number of materials can be made to fluoresce or phosphoresce, including in some sulfides, oxides, tungstates, and a few organic materials. Typically, pure materials do not display these phenomena, and to induce them, impurities in controlled concentrations must be added.

fluorescence
phosphorescence

Luminescence has a number of commercial applications. For example, fluorescent lamps consist of a glass housing coated on the inside with specially prepared tungstates or silicates. Ultraviolet light is generated within the tube from a mercury glow discharge, which causes the coating to fluoresce and emit white light. The new *compact fluorescence* lights (or lamps), or *CLF*s, are replacing general-service incandescent lights. These CFL bulbs are constructed from a tube that is curved or folded so as to fit into the space formerly occupied by the incandescent bulb and also to mount in their fixtures. Compact fluorescents emit the same amount of visible light, consume between one-fifth to one-third of the electrical power, and have much longer lifetimes than incandescent bulbs. However, they are more expensive and disposal is more complicated because they contain mercury.

22.12 PHOTOCONDUCTIVITY

photoconductivity

The conductivity of semiconducting materials depends on the number of free electrons in the conduction band and the number of holes in the valence band, according to Equation 19.13. Thermal energy associated with lattice vibrations can promote electron excitations in which free electrons and/or holes are created, as described in Section 19.6. Additional charge carriers may be generated as a consequence of photon-induced electron transitions in which light is absorbed; the attendant increase in conductivity is called **photoconductivity**. Thus, when a specimen of a photoconductive material is illuminated, the conductivity increases.

This phenomenon is used in photographic light meters. A photoinduced current is measured, and its magnitude is a direct function of the intensity of the incident light radiation, or the rate at which the photons of light strike the photoconductive material. Visible light radiation must induce electronic transitions in the photoconductive material; cadmium sulfide is commonly used in light meters.

Sunlight may be directly converted into electrical energy in solar cells, which also use semiconductors. The operation of these devices is, in a sense, the reverse of that for the light-emitting diode. A *p–n* junction is used in which photoexcited electrons and holes are drawn away from the junction, in opposite directions and become part of an external current, as illustrated in chapter-opening diagram (a).

MATERIALS OF IMPORTANCE

Light–Emitting Diodes

I n Section 19.15 we discussed semiconductor p–n junctions, and how they may be used as diodes or as rectifiers of an electric current.[1] In some situations, when a forward-biased potential of relatively high magnitude is applied across a p–n junction diode, visible light (or infrared radiation) is emitted. This conversion of electrical energy into light energy is termed **electroluminescence**, and the device that produces it is termed a **light-emitting diode (LED)**. The forward-biased potential attracts electrons on the n-side toward the junction, where some of them pass into (or are "injected" into) the p-side (Figure 22.11a). Here, the electrons are minority charge carriers and therefore "recombine" with, or are annihilated by, the holes in the region

electro-luminescence

light-emitting diode (LED)

near the junction, according to Equation 22.17, where the energy is in the form of photons of light (Figure 22.11b). An analogous process occurs on the p-side—holes travel to the junction and recombine with the majority electrons on the n-side.

The elemental semiconductors silicon and germanium are not suitable for LEDs due to the detailed nature of their band gap structures. Rather, some of the III-V semiconducting compounds such as gallium arsenide (GaAs), and indium phosphide (InP), and alloys composed of these materials (e.g., $GaAs_xP_{1-x}$, where x is a small number less than unity) are frequently used. The wavelength (i.e., color) of the emitted radiation is related to the band gap of the semiconductor (which is normally the same for both n- and p-sides of the diode). For example, red, orange, and yellow colors are possible for the GaAs–InP system. Blue and green LEDs have also been developed using (Ga,In)N semiconducting alloys. Thus, with this complement of colors, full-color displays are possible using LEDs.

Important applications for semiconductor LEDs include digital clocks and illuminated watch displays, optical mice (computer input devices), and film scanners. Electronic remote controls (for televisions, DVD players, etc.) also use LEDs that emit an infrared beam; this beam transmits coded signals that are picked up by detectors in the receiving devices. In addition, LEDs are being used for light sources. They are more energy efficient than incandescent lights, generate very little heat, and have much longer lifetimes (because there is no filament that can burn out). Most new traffic control signals use LEDs instead of incandescent lights.

We noted in Section 19.17 that some polymeric materials may be semiconductors (both n- and p-type). As a consequence, light-emitting diodes made of polymers are possible, of which there are two types: (1) *organic light-emitting diodes* (or *OLEDs*), which have relatively low molecular weights; and (2) high-molecular-weight *polymer light-emitting diodes* (or *PLEDs*). For these LED types, amorphous

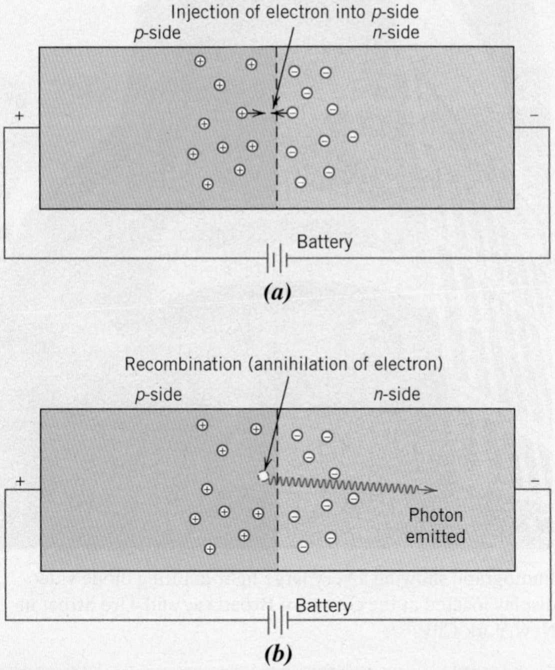

Figure 22.11 Schematic diagram of a forward-biased semiconductor p–n junction showing (a) the injection of an electron from the n-side into the p-side, and (b) the emission of a photon of light as this electron recombines with a hole.

[1] Schematic diagrams showing electron and hole distributions on both sides of the junction with no applied electric potential, as well as for both forward and reverse biases are presented in Figure 19.21. In addition, Figure 19.22 shows the current-versus-voltage behavior for a p–n junction.

(continued)

polymers are used in the form of thin layers that are sandwiched together with electrical contacts (anodes and cathodes). In order for the light to be emitted from the LED, one of the contacts must be transparent. Figure 22.12 is a schematic illustration that shows the components and configuration of an OLED. A wide variety of colors is possible using OLEDs and PLEDs, and more than a single color may be produced from each device (which is not possible with semiconductor LEDs)—thus, combining colors makes it possible to generate white light.

Although the semiconductor LEDs currently have longer lifetimes than these organic emitters, OLEDs/PLEDs have distinct advantages. In addition to generating multiple colors, they are easier to fabricate (by "printing" onto their substrates with an ink-jet printer), are relatively inexpensive, have slimmer profiles, and can be patterned to give high-resolution and full-color images. OLED displays are being marketed for use on digital cameras, cell phones, and car audio components. Potential applications include larger displays for televisions, computers, and

billboards. In addition, with the right combination of materials, these displays can also be flexible. Imagine having a computer monitor or television that can be rolled up like a projection screen, or a lighting fixture that is wrapped around an architectural column or is mounted on a room wall to make ever-changing wallpaper.

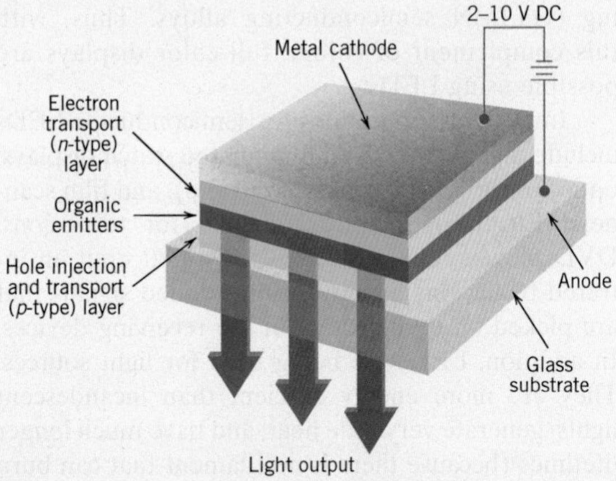

Photograph showing a very large light-emitting diode video display located at the corner of Broadway and 43rd Street in New York City.

© Sean Pavone Photo/Shutterstock.com

Figure 22.12 Schematic diagram showing the components and configuration of an organic light-emitting diode (OLED). (Reproduced by arrangement with *Silicon Chip* magazine.)

Labels in figure:
- 2–10 V DC
- Metal cathode
- Electron transport (*n*-type) layer
- Organic emitters
- Hole injection and transport (*p*-type) layer
- Anode
- Glass substrate
- Light output

✓ *Concept Check 22.7* Is the semiconductor zinc selenide (ZnSe), which has a band gap of 2.58 eV, photoconductive when exposed to visible light radiation? Why or why not?

[*The answer may be found at* www.wiley.com/college/callister *(Student Companion Site).*]

22.13 LASERS

laser

All the radiative electron transitions heretofore discussed are *spontaneous*—that is, an electron falls from a high-energy state to a lower-energy state without any external provocation. These transition events occur independently of one another and at random times, producing radiation that is *incoherent*—that is, the light waves are out of phase with one another. With lasers, however, coherent light is generated by electron transitions initiated by an external stimulus—**laser** is just the acronym for *l*ight *a*mplification by *s*timulated *e*mission of *r*adiation.

Although there are several different varieties of laser, we explain the principles of operation using the solid-state ruby laser. Ruby is a single crystal of Al_2O_3 (sapphire) to which has been added on the order of 0.05% Cr^{3+} ions. As previously explained (Section 22.9), these ions impart to ruby its characteristic red color; more important, they provide electron states that are essential for the laser to function. The ruby laser is in the form of a rod, the ends of which are flat, parallel, and highly polished. Both ends are silvered such that one is totally reflecting and the other partially transmitting.

The ruby is illuminated with light from a xenon flash lamp (Figure 22.13). Before this exposure, virtually all the Cr^{3+} ions are in their ground states; that is, electrons fill the lowest-energy levels, as represented schematically in Figure 22.14. However, photons of wavelength 0.56 μm from the xenon lamp excite electrons from the Cr^{3+} ions into higher-energy states. These electrons can decay back into their ground state by two different paths. Some fall back directly; associated photon emissions are not part of the laser beam. Other electrons decay into a metastable intermediate state (path *EM*, Figure 22.14), where they may reside for up to 3 ms (milliseconds) before spontaneous emission (path *MG*). In terms of electronic processes, 3 ms is a relatively long time, which means that a large number of these metastable states may become occupied. This situation is indicated in Figure 22.15*b*.

The initial spontaneous photon emission by a few of these electrons is the stimulus that triggers an avalanche of emissions from the remaining electrons in the metastable state (Figure 22.15*c*). Of the photons directed parallel to the long axis of the ruby rod, some are transmitted through the partially silvered end; others, incident to the totally

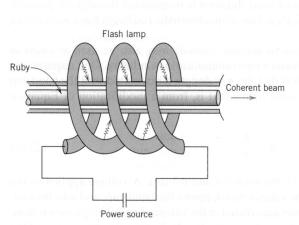

Figure 22.13 Schematic diagram of the ruby laser and xenon flash lamp.
(From R. M. Rose, L. A. Shepard, and J. Wulff, *The Structure and Properties of Materials*, Vol. 4, *Electronic Properties*. Copyright © 1966 by John Wiley & Sons, New York. Reprinted by permission of John Wiley & Sons, Inc.)

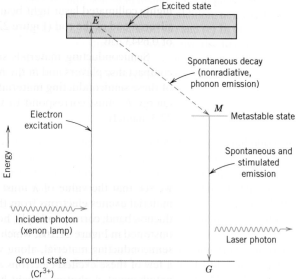

Figure 22.14 Schematic energy diagram for the ruby laser, showing electron excitation and decay paths.

Figure 22.17 Schematic diagram showing the layered cross section of a GaAs semiconducting laser. Holes, excited electrons, and the laser beam are confined to the GaAs layer by the adjacent *n*- and *p*-type GaAlAs layers.
(Adapted from "Photonic Materials," by J. M. Rowell. Copyright © 1986 by Scientific American, Inc. All rights reserved.)

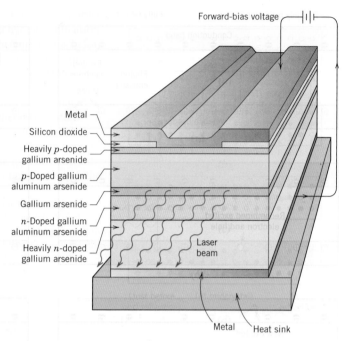

Forward-bias voltage

Metal
Silicon dioxide
Heavily *p*-doped gallium arsenide
p-Doped gallium aluminum arsenide
Gallium arsenide
n-Doped gallium aluminum arsenide
Heavily *n*-doped gallium arsenide

Laser beam

Metal   Heat sink

A variety of other substances may be used for lasers, including some gases and glasses. Table 22.2 lists several common lasers and their characteristics. Laser applications are diverse. Because laser beams may be focused to produce localized heating, they are used in surgical procedures and for cutting, welding, and machining metals. Lasers are also used as light sources for optical communication systems. Furthermore, because the beam is highly coherent, lasers may be used for making very precise distance measurements.

Table 22.2 Characteristics and Applications of Several Types of Lasers

Laser	*Wavelength (μm)*	*Average Power Range*	*Applications*
Carbon dioxide	10.6	Milliwatts to tens of kilowatts	Heat treating, welding, cutting, scribing, marking
Nd:YAG	1.06 0.532	Millwatts to hundreds of watts Milliwatts to watts	Welding, hole piercing, cutting
Nd:glass	1.05	Watts[a]	Pulse welding, hole piercing
Diodes	Visible and infrared	Milliwatts to kilowatts	Bar-code reading, CDs and DVDs, optical communications
Argon-ion	0.5415 0.488	Milliwatts to tens of watts Milliwatts to watts	Surgery, distance measurements, holography
Fiber	Infrared	Watts to kilowatts	Telecommunications, spectroscopy, directed energy weapons
Excimer	Ultraviolet	Watts to hundreds of watts[b]	Eye surgery, micromachining, microlithography

[a]Although glass lasers produce relatively low average powers, they almost always run in pulsed mode, where their peak powers can reach the gigawatt levels.
[b]Excimers are also pulsed lasers, and are capable of peak powers in the tens of megawatts.

Source: Adapted from C. Breck, J. J. Ewing, and J. Hecht, *Introduction to Laser Technology,* 4th edition. Copyright © 2012 by John Wiley & Sons, Inc., Hoboken, NJ. Reprinted with permission of John Wiley & Sons, Inc.

22.14 OPTICAL FIBERS IN COMMUNICATIONS

The communications field recently experienced a revolution with the development of optical fiber technology; virtually all telecommunications are transmitted via this medium rather than through copper wires. Signal transmission through a metallic wire conductor is electronic (i.e., by electrons), whereas using optically transparent fibers, signal transmission is *photonic,* meaning that it uses photons of electromagnetic or light radiation. Use of fiber-optic systems has improved speed of transmission, information density, and transmission distance, with a reduction in error rate; furthermore, there is no electromagnetic interference with fibers. The bandwidth (i.e., data transfer rate) of optical fibers is amazing; in 1 s, an optical fiber can transmit 15.5 terabits of data over a distance of 7000 km; at this rate it would take approximately 30 s to transmit the entire iTunes catalog from New York to London. Furthermore, a single fiber is capable of transmitting 250 million phone conversations every second. It would require 30,000 kg of copper to transmit the same amount of information as only 0.1 kg of optical fiber.

The present treatment centers on the characteristics of optical fibers; however, it is worthwhile to first briefly discuss the components and operation of the transmission system. A schematic diagram showing these components is presented in Figure 22.18. The information (e.g., a telephone conversation) in electronic form must first be digitized into bits—that is, 1s and 0s; this is accomplished in the encoder. It is next necessary to convert this electrical signal into an optical (photonic) one, which takes place in the electrical-to-optical converter (Figure 22.18). This converter is normally a semiconductor laser, as described in the previous section, which emits monochromatic and coherent light. The wavelength normally lies between 0.78 and 1.6 μm, which is in the infrared region of the electromagnetic spectrum; absorption losses are low within this range of wavelengths. The output from this laser converter is in the form of pulses of light; a binary 1 is represented by a high-power pulse (Figure 22.19a), whereas a 0 corresponds to a low-power pulse (or the absence of one), (Figure 22.19b). These photonic pulse signals are then fed into and carried through the fiber-optic cable (sometimes called a *waveguide*) to the receiving end. For long transmissions, *repeaters* may be required; these are devices that amplify and regenerate the signal. Finally, at the receiving end the photonic signal is reconverted to an electronic one and then decoded (undigitized).

The heart of this communication system is the optical fiber. It must guide these light pulses over long distances without significant signal power loss (i.e., attenuation) and pulse distortion. Fiber components are the core, cladding, and coating; these are represented in the cross-section profile shown in Figure 22.20. The signal passes through the core, whereas the surrounding cladding constrains the light rays to travel within the core; the outer coating protects core and cladding from damage that might result from abrasion and external pressures.

High-purity silica glass is used as the fiber material; fiber diameters normally range between about 5 and 100 μm. The fibers are relatively flaw-free and, thus, remarkably strong; during production the continuous fibers are tested to ensure that they meet minimum strength standards.

Containment of the light to within the fiber core is made possible by total internal reflection—that is, any light rays traveling at oblique angles to the fiber axis are reflected

Figure 22.18 Schematic diagram showing the components of an optical-fiber communications system.

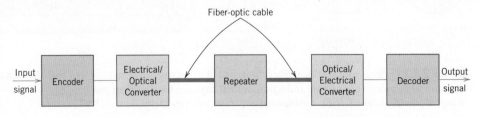

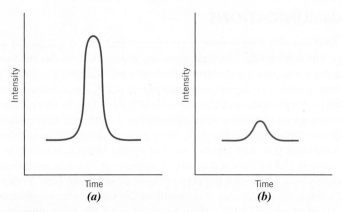

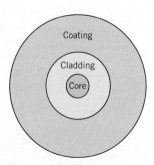

Figure 22.19 Digital encoding scheme for optical communications. (*a*) A high-power pulse of photons corresponds to a 1 in the binary format. (*b*) A low-power photon pulse represents a 0.

Figure 22.20 Schematic cross section of an optical fiber.

back into the core. Internal reflection is accomplished by varying the index of refraction of the core and cladding glass materials. In this regard, two design types are used. With one type (termed *step-index*), the index of refraction of the cladding is slightly lower than that of the core. The index profile and the manner of internal reflection are shown in Figures 22.21*b* and 22.21*d*. For this design, the output pulse is broader than the input one (Figures 22.21*c* and 22.21*e*), a phenomenon that is undesirable because it limits the rate of transmission. Pulse broadening results because various light rays, although injected at approximately the same instant, arrive at the output at different times; they traverse different trajectories and, thus, have a variety of path lengths.

Pulse broadening is largely avoided by use of the *graded-index* design. Here, impurities such as boron oxide (B_2O_3) or germanium dioxide (GeO_2) are added to the silica glass such that the index of refraction is made to vary parabolically across the cross section (Figure 22.22*b*). Thus, the velocity of light within the core varies with radial position, being greater at the periphery than at the center. Consequently, light rays that traverse longer path lengths through the outer periphery of the core travel faster in this lower-index material and arrive at the output at approximately the same time as undeviated rays that pass through the center portion of the core.

Exceptionally pure and high-quality fibers are fabricated using advanced and sophisticated processing techniques, which are not discussed here. Impurities and other defects that absorb, scatter, and thus attenuate the light beam must be eliminated. The

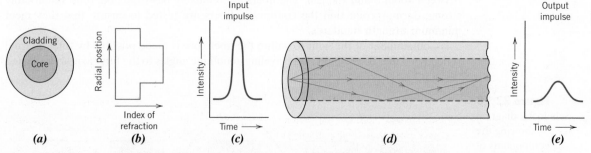

Figure 22.21 Step-index optical fiber design. (*a*) Fiber cross section. (*b*) Fiber radial index of refraction profile. (*c*) Input light pulse. (*d*) Internal reflection of light rays. (*e*) Output light pulse.
(Adapted from S. R. Nagel, *IEEE Communications Magazine,* Vol. 25, No. 4, p. 34, 1987.)

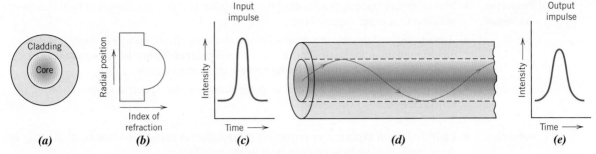

Figure 22.22 Graded-index optical-fiber design. (*a*) Fiber cross section. (*b*) Fiber radial index of refraction profile. (*c*) Input light pulse. (*d*) Internal reflection of a light ray. (*e*) Output light pulse. (Adapted from S. R. Nagel, *IEEE Communications Magazine*, Vol. 25, No. 4, p. 34, 1987.)

presence of copper, iron, and vanadium is especially detrimental; their concentrations are reduced to on the order of several parts per billion. Likewise, water and hydroxyl contaminant contents are extremely low. Uniformity of fiber cross-sectional dimensions and core roundness is critical; tolerances of these parameters to within 1 μm over 1 km of length are possible. In addition, bubbles within the glass and surface defects have been virtually eliminated. The attenuation of light in this glass material is imperceptibly small. For example, the power loss through a 16-km thickness of optical-fiber glass is equivalent to the power loss through a 25-mm thickness of ordinary window glass!

SUMMARY

Electromagnetic Radiation

- The optical behavior of a solid material is a function of its interactions with electromagnetic radiation having wavelengths within the visible region of the spectrum (about 0.4 μm to 0.7 μm).

- From a quantum-mechanical perspective, electromagnetic radiation may be considered to be composed of photons—groups or packets of energy that are *quantized*—that is, they can have only specific values of energy.

- Photon energy is equal to the product of Planck's constant and radiation frequency (Equation 22.3).

Light Interactions with Solids

- Possible interactive phenomena that may occur as light radiation passes from one medium to another are refraction, reflection, absorption, and transmission.

- Regarding degree of light transmissivity, materials are classified as follows:
  Transparent—light is transmitted through the material with very little absorption and reflection.
  Translucent—light is transmitted diffusely; there is some scattering within the interior of the material.
  Opaque—virtually all light is scattered or reflected such that none is transmitted through the material.

Atomic and Electronic Interactions

- One possible interaction between electromagnetic radiation and matter is electronic polarization—the electric field component of a light wave induces a shift of the electron cloud around an atom relative to its nucleus (Figure 19.32*a*).

- Two consequences of electronic polarization are absorption and refraction of light.

- Electromagnetic radiation may be absorbed by causing the excitation of electrons from one energy state to a higher state (Figure 22.3).

Optical Properties of Metals

- Metals appear opaque as a result of the absorption and then reemission of light radiation within a thin outer surface layer.

- Absorption occurs via the excitation of electrons from occupied energy states to unoccupied ones above the Fermi energy level (Figure 22.4*a*). Reemission takes place by decay electron transitions in the reverse direction (Figure 22.4*b*).

- The perceived color of a metal is determined by the spectral composition of the reflected light.

Refraction

- Light radiation experiences refraction in transparent materials—that is, its velocity is decreased, and the light beam is bent at the interface.

- The phenomenon of refraction is a consequence of electronic polarization of the atoms or ions. The larger an atom or ion, the greater the index of refraction.

Reflection

- When light passes from one transparent medium to another having a different index of refraction, some of it is reflected at the interface.

- The degree of the reflectance depends on the indices of refraction of both media, as well as the angle of incidence. For normal incidence, reflectivity may be calculated using Equation 22.12.

Absorption

- Pure nonmetallic materials are either intrinsically transparent or opaque.

  Opacity results in relatively narrow-band-gap materials ($E_g < 1.8$ eV) as a result of absorption whereby a photon's energy is sufficient to promote valence band–conduction band electron transitions (Figure 22.5).

  Transparent nonmetals have band gaps greater than 3.1 eV.

  For nonmetallic materials that have band gaps between 1.8 and 3.1 eV, only a portion of the visible spectrum is absorbed; these materials appear colored.

- Some light absorption occurs in even transparent materials as a consequence of electronic polarization.

- For wide-band-gap insulators that contain impurities, decay processes involving excited electrons to states within the band gap are possible with the emission of photons having energies less than the band gap energy (Figure 22.6).

Color

- Transparent materials appear colored as a consequence of specific wavelength ranges of light that are selectively absorbed (usually by electron excitations).

- The color discerned is a result of the distribution of wavelength ranges in the transmitted beam.

Opacity and Translucency in Insulators

- Normally transparent materials may be made translucent or even opaque if the incident light beam experiences interior reflection and/or refraction.

- Translucency and opacity as a result of internal scattering may occur as follows:

  (1) in polycrystalline materials that have anisotropic indices of refraction

  (2) in two-phase materials

  (3) in materials containing small pores

  (4) in highly crystalline polymers

Luminescence

- With luminescence, energy is absorbed as a consequence of electron excitations, which is subsequently reemitted as visible light.

When light is reemitted less than 1 s after excitation, the phenomenon is called *fluorescence*.

For longer reemission times, the term *phosphorescence* is used.

- *Electroluminescence* is the phenomenon by which light is emitted as a result of electron–hole recombination events that are induced in a forward-biased diode (Figure 22.11).
- The device that experiences electroluminescence is the light-emitting diode (LED).

Photoconductivity

- *Photoconductivity* is the phenomenon by which the electrical conductivity of some semiconductors may be enhanced by photo-induced electron transitions, by which additional free electrons and holes are generated.

Lasers

- Coherent and high-intensity light beams are produced in lasers by stimulated electron transitions.
- With the ruby laser, a beam is generated by electrons that decay back into their ground Cr^{3+} states from metastable excited states.
- The beam from a semiconducting laser results from the recombination of excited electrons in the conduction band with valence band holes.

Optical Fibers in Communications

- Use of fiber-optic technology in modern telecommunications provides for the transmission of information that is interference-free, rapid, and intense.
- An optical fiber is composed of the following elements:

  A core through which the pulses of light propagate

  The cladding, which provides for total internal reflection and containment of the light beam within the core

  The coating, which protects the core and cladding from damage

Important Terms and Concepts

absorption	laser	Planck's constant
color	light-emitting diode (LED)	reflection
electroluminescence	luminescence	refraction
excited state	opaque	translucent
fluorescence	phosphorescence	transmission
ground state	photoconductivity	transparent
index of refraction	photon	

REFERENCES

Fox, M., *Optical Properties of Solids,* 2nd edition, Oxford University Press, Oxford, 2010.

Gupta, M. C., and J. Ballato, *The Handbook of Photonics,* 2nd edition, CRC Press, Boca Raton, FL, 2007.

Hecht, J., *Understanding Lasers: An Entry-Level Guide,* 3rd edition, Wiley-IEEE Press, Hoboken/Piscataway, NJ, 2008.

Kingery, W. D., H. K. Bowen, and D. R. Uhlmann, *Introduction to Ceramics,* 2nd edition, Wiley, New York, 1976, Chapter 13.

Rogers, A., *Essentials of Photonics,* 2nd edition, CRC Press, Boca Raton, FL, 2008.

Saleh, B. E. A., and M. C. Teich, *Fundamentals of Photonics,* 2nd Edition, Wiley, Hoboken, NJ, 2007.

Svelto, O., *Principles of Lasers,* 5th edition, Springer, New York, 2010.

QUESTIONS AND PROBLEMS

Electromagnetic Radiation

22.1 Visible light having a wavelength of 6×10^{-7} m appears orange. Compute the frequency and energy of a photon of this light.

Light Interactions with Solids

22.2 Distinguish between materials that are opaque, translucent, and transparent in terms of their appearance and light transmittance.

Atomic and Electronic Interactions

22.3 **(a)** Briefly describe the phenomenon of electronic polarization by electromagnetic radiation.

(b) What are two consequences of electronic polarization in transparent materials?

Optical Properties of Metals

22.4 Briefly explain why metals are opaque to electromagnetic radiation having photon energies within the visible region of the spectrum.

Refraction

22.5 In ionic materials, how does the size of the component ions affect the extent of electronic polarization?

22.6 Can a material have an index of refraction less than unity? Why or why not?

22.7 Compute the velocity of light in calcium fluoride (CaF_2), which has a dielectric constant ϵ_r of 2.056 (at frequencies within the visible range) and a magnetic susceptibility of -1.43×10^{-5}.

22.8 The indices of refraction of fused silica and a soda–lime glass within the visible spectrum are 1.460 and 1.53, respectively. For each of these materials determine the fraction of the relative dielectric constant at 60 Hz that is due to electronic polarization, using the data of Table 19.5. Neglect any orientation polarization effects.

22.9 Using the data in Table 22.1, estimate the dielectric constants for borosilicate glass, periclase (MgO), poly(methyl methacrylate), and polypropylene, and compare these values with those cited in the following table. Briefly explain any discrepancies.

Material	Dielectric Constant (1 MHz)
Borosilicate glass	4.7
Periclase	9.7
Poly(methyl methacrylate)	2.8
Polypropylene	2.35

22.10 Briefly describe the phenomenon of dispersion in a transparent medium.

Reflection

22.11 It is desired that the reflectivity of light at normal incidence to the surface of a transparent medium be less than 6.0%. Which of the following materials in Table 22.1 are likely candidates: silica glass, Pyrex glass, corundum, spinel, polystyrene, and polytetrafluoroethylene? Justify your selection(s).

22.12 Briefly explain how reflection losses of transparent materials are minimized by thin surface coatings.

22.13 The index of refraction of corundum (Al_2O_3) is anisotropic. Suppose that visible light is passing from one grain to another of different crystallographic orientation and at normal incidence to the grain boundary. Calculate the reflectivity at the boundary if the indices of refraction for the two grains are 10.12 and 8.25 cm in the direction of light propagation.

Absorption

22.14 Zinc telluride has a band gap of 2.23 eV. Over what range of wavelengths of visible light is it transparent?

22.15 Briefly explain why the magnitude of the absorption coefficient (β in Equation 22.18) depends on the radiation wavelength.

22.16 The fraction of nonreflected radiation that is transmitted through a 10-mm thickness of a transparent material is 0.90. If the thickness is increased to 20 mm, what fraction of light will be transmitted?

Transmission

22.17 Derive Equation 22.19, starting from other expressions given in the chapter.

22.18 The transmissivity T of a transparent material 20 mm thick to normally incident light is 0.85. If the index of refraction of this material is 1.6, compute the thickness of material that will yield a transmissivity of 0.75. All reflection losses should be considered.

Color

22.19 Briefly explain what determines the characteristic color of **(a)** a metal and **(b)** a transparent nonmetal.

22.20 Briefly explain why some transparent materials appear colored whereas others are colorless.

Opacity and Translucency in Insulators

22.21 Briefly describe the three absorption mechanisms in nonmetallic materials.

22.22 Briefly explain why amorphous polymers are transparent, while predominantly crystalline polymers appear opaque or, at best, translucent.

Luminescence
Photoconductivity
Lasers

22.23 **(a)** In your own words, briefly describe the phenomenon of luminescence.

(b) What is the distinction between fluorescence and phosphorescence?

22.24 In your own words, briefly describe the phenomenon of photoconductivity.

22.25 Briefly explain the operation of a photographic lightmeter.

22.26 In your own words, describe how a ruby laser operates.

22.27 Compute the difference in energy between metastable and ground electron states for the ruby laser.

Optical Fibers in Communications

22.28 At the end of Section 22.14 it was noted that the intensity of light absorbed while passing through a 16-kilometer length of optical fiber glass is equivalent to the light intensity absorbed through for a 25-mm thickness of ordinary window glass. Calculate the absorption coefficient β of the optical fiber glass if the value of β for the window glass is 5×10^{-4} mm^{-1}.

DESIGN PROBLEM

Atomic and Electronic Interactions

21.D1 Gallium arsenide (GaAs) and gallium phosphide (GaP) are compound semiconductors that have room-temperature band gap energies of 1.42 and 2.25 eV, respectively, and form solid solutions in all proportions. Furthermore, the band gap of the alloy increases approximately linearly with GaP additions (in mol%). Alloys of these two materials are used for light-emitting diodes wherein light is generated by conduction band-to-valence band electron transitions. Determine the composition of a GaAs–GaP alloy that will emit orange light having a wavelength of 0.60 μm.

Chapter 23 Economic, Environmental, and Societal Issues in Materials Science and Engineering

(a) Beverage cans made of an aluminum alloy (left) and a steel alloy (right). The steel beverage can has corroded significantly and, therefore, is biodegradable and nonrecyclable. In contrast, the aluminum can is nonbiodegradable and recyclable because it experienced very little corrosion.

© William D. Callister, Jr.

(a)

(b) A fork made of the biodegradable polymer poly(lactic acid) at various stages of degradation. As noted, the total degradation process took about 45 days.

Day 0   Day 12   Day 33   Day 45

© Roger Ressmeyer/Corbis

(b)

Courtesy of Jennifer Welter

(c)

(c) Familiar picnic items, some of which are recyclable and/or possibly nonbiodegradable (one of them is edible).

It is essential for the engineer to know about and understand economic issues because the company or institution for which he or she works must realize a profit from the products it manufactures. Materials engineering decisions have economic consequences with regard to both material and production costs.

An awareness of environmental and societal issues is important for the engineer because, over time, greater demands are being made on the world's natural resources. Furthermore, levels of pollution are ever increasing. Materials engineering decisions have impacts on the consumption of raw materials and energy, on the contamination of our water and atmosphere, on human health, on global climate change, and on the ability of the consumer to recycle or dispose of spent products. The quality of life for this and future generations depends, to some degree, on how these issues are addressed by the global engineering community.

Learning Objectives

After studying this chapter, you should be able to do the following:

1. List and briefly discuss three factors over which an engineer has control that affect the cost of a product.
2. Diagram the total materials cycle, and briefly discuss relevant issues that pertain to each stage of this cycle.
3. List the two inputs and five outputs for the life cycle analysis/assessment scheme.
4. Cite issues that are relevant to the "green design" philosophy of product design.
5. Discuss recyclability/disposability issues relative to (a) metals, (b) glass, (c) plastics and rubber, and (d) composite materials.

23.1 INTRODUCTION

In previous chapters, we dealt with a variety of materials science and materials engineering issues to include criteria that may be used in the materials selection process. Many of these selection criteria relate to material properties or property combinations—mechanical, electrical, thermal, corrosion, and so on; the performance of some component depends on the properties of the material from which it is made. Processability or ease of fabrication of the component may also play a role in the selection process. Virtually all of this book, in one way or another, has addressed these property and fabrication issues.

In engineering practice, other important criteria must be considered in the development of a marketable product. Some of these are economic in nature, which, to some degree, are unrelated to scientific principles and engineering practice and yet are significant if a product is to be competitive in the commercial marketplace. Other criteria that should be addressed involve environmental and societal issues such as pollution, disposal, recycling, toxicity, and energy. This final chapter offers relatively brief overviews of economic, environmental, and societal considerations that are important in engineering practice.

Economic Considerations

Engineering practice involves using scientific principles to design components and systems that perform reliably and satisfactorily. Another critical driving force in engineering practice is that of economics; simply stated, a company or an institution must realize a profit from the products that it manufactures and sells. The engineer may design the perfect component; however, as manufactured, it must be offered for sale at a price that is attractive to the consumer and return a suitable profit to the company.

Furthermore, in today's world and the global marketplace, economics does not always mean just the final cost of a product. Many countries have specific regulations regarding the chemicals used, CO_2 emissions, and end-of-life procedures. Companies must consider a myriad of such factors. For example, in some instances, deleting toxic chemicals (which are regulated) in a product results in a cheaper fabrication process.

Only a brief overview of important economic considerations as they apply to the materials engineer are provided. The student may want to consult references provided at the end of this chapter that address engineering economics in detail.

The materials engineer has control over three factors that affect the cost of a product: (1) component design, (2) the material(s) used, and (3) the manufacturing technique(s). These factors are interrelated, in that component design may affect which material is used, and both component design and the material used influence the choice of manufacturing technique(s). Economic considerations for each of these factors are now briefly discussed.

23.2 COMPONENT DESIGN

Some fraction of the cost of a component is associated with its design. In this context, component design is the specification of size, shape, and configuration, which affects in-service component performance. For example, if mechanical forces are present, then stress analyses may be required. Detailed drawings of the component must be prepared; computers are normally employed, using software that has been generated for this specific function.

A single component is often part of a complex device or system consisting of a large number of components (a television, an automobile, a DVD player/recorder, etc.). Thus, design must take into consideration each component's contribution to the efficient operation of the complete system.

The approximate cost of a product is determined by this up-front design, even before the product has been manufactured. Thus, a creative design and the selection of appropriate materials can have a significant impact later on.

Component design is a highly iterative process that involves many compromises and trade-offs. The engineer should keep in mind that an optimal component design may not be possible because of system constraints.

23.3 MATERIALS

In terms of economics, we should select the material or materials with the appropriate combination(s) of properties that are the least expensive, which may also include consideration of availability. Once a family of materials has been selected that satisfy the design constraints, cost comparisons of the various candidate materials may be made on the basis of cost per part. Material price is usually quoted per unit mass. The part volume may be determined from its dimensions and geometry, which is then converted into mass using the density of the material. In addition, during manufacturing, there is typically some unavoidable material waste that should also be taken into account in these computations. Current prices for a wide variety of engineering materials are given in Appendix C.

23.4 MANUFACTURING TECHNIQUES

As already stated, the choice of manufacturing process is influenced by both the material selected and part design. The entire manufacturing process normally consists of primary and secondary operations. *Primary operations* are those that convert the raw material into a recognizable part (casting, plastic forming, powder compaction, molding, etc.), whereas *secondary operations* are those subsequently used to produce the finished part (e.g., heat treatments, welding, grinding, drilling, painting, decorating). The major cost considerations for these processes include capital equipment, tooling, labor, repairs, machine downtime, and waste. In this cost analysis, rate of production is an important consideration. If this particular part is one component of a system, then assembly costs must also be addressed. Finally, there are costs associated with inspection, packaging, and transportation of the final product.

As a sidelight, there are also other factors not directly related to design, material, or manufacturing that figure into the product selling price. These factors include labor fringe benefits, supervisory and management labor, research and development, property and rent, insurance, profit, taxes, and so on.

Environmental and Societal Considerations

Modern technologies and the manufacturing of their associated products affect society in a variety of ways—some are positive, others are adverse. Furthermore, these impacts are economic and environmental in type and international in scope inasmuch as (1) the resources required for a new technology often come from many different countries, (2) the economic prosperity resulting from technological development is global in extent, and (3) environmental impacts may extend beyond the boundaries of a single country.

Materials play a crucial role in this technology–economy–environment scheme. A material that is used in some end product and then discarded passes through several stages or phases; these stages are represented in Figure 23.1, which is sometimes termed the *total materials cycle* or just *materials cycle* and represents the "cradle-to-grave" life circuit of a material. Beginning on the far left side of Figure 23.1, raw materials are extracted from their natural earthly habitats by mining, drilling, harvesting, and so on. These raw materials are then purified, refined, and converted into bulk forms such as

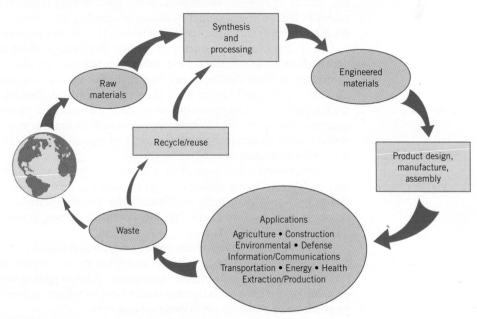

Figure 23.1 Schematic representation of the total materials cycle.
(Adapted from M. Cohen, *Advanced Materials & Processes,* Vol. 147, No. 3, p. 70, 1995. Copyright © 1995 by ASM International. Reprinted by permission of ASM International, Materials Park, OH.)

metals, cements, petroleum, rubber, and fibers. Further synthesis and processing results in products that are what may be termed *engineered materials,* such as metal alloys, ceramic powders, glass, plastics, composites, semiconductors, and elastomers. Next, these engineered materials are further shaped, treated, and assembled into products, devices, and appliances that are ready for the consumer—this constitutes the "product design, manufacture, assembly" stage of Figure 23.1. The consumer purchases these products and uses them (the "applications" stage) until they wear out or become obsolete and are discarded. At this time, the product constituents may either be recycled/reused (by which they reenter the materials cycle) or disposed of as waste, normally being either incinerated or dumped as solid waste in municipal landfills—and so they return to the Earth and complete the materials cycle.

It has been estimated that worldwide, about 15 billion tons of raw materials are extracted from the Earth every year; some of these are renewable and some are not. Over time, it is becoming more apparent that the Earth is virtually a closed system relative to its constituent materials and that its resources are finite. In addition, as societies mature and populations increase, the available resources become scarcer, and greater attention must be paid to more effective use of these resources relative to the materials cycle.

Furthermore, energy must be supplied at each cycle stage; in the United States it has been estimated that approximately one-half of the energy consumed by manufacturing industries goes to produce and fabricate materials. Energy is a resource that, to some degree, is limited in supply, and measures must be taken to conserve and use it more effectively in the production, application, and disposal of materials.

Finally, there are interactions with and impacts on the natural environment at all stages of the materials cycle. The condition of the Earth's atmosphere, water, and land depends to a large extent on how carefully we traverse the materials cycle. Some ecological damage and landscape spoilage undoubtedly result during the extraction of raw materials. Pollutants may be generated that are expelled into the air and water during synthesis and processing; in addition, any toxic chemicals that are produced must be disposed of or discarded. The final product, device, or appliance should be designed so that during its lifetime, any impact on the environment is minimal; furthermore, at the end of its life, provision should be made for recycling its component materials, or at least for their disposal with little ecological impact (i.e., it should be biodegradable).

Recycling of used products rather than disposing of them as waste is a desirable approach for several reasons. First, using recycled material obviates the need to extract raw materials from the Earth and thus conserves natural resources and eliminates any associated ecological impact from the extraction phase. Second, energy requirements for the refinement and processing of recycled materials are normally less than for their natural counterparts; for example, approximately 28 times as much energy is required to refine natural aluminum ores as to recycle aluminum beverage can scrap. Finally, there is no need to dispose of recycled materials.

Thus, the materials cycle (Figure 23.1) is really a system that involves interactions and exchanges among materials, energy, and the environment. Furthermore, future engineers, worldwide, must understand the interrelationships among these various stages so as to use the Earth's resources effectively and minimize adverse ecological affects on our environment.

In many countries, environmental problems and issues are being addressed by the establishment of standards that are mandated by governmental regulatory agencies (e.g., the use of lead in electronic components is being phased out). Furthermore, from an industrial perspective, it becomes incumbent on engineers to propose viable solutions to existing and potential environmental concerns.

Correcting any environmental problems associated with manufacturing influences product price. A common misconception is that a more environmentally friendly product or process is inherently more costly than one that is environmentally unfriendly. Engineers

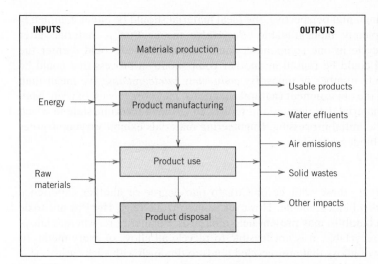

INPUTS   OUTPUTS

Materials production

Energy →   Product manufacturing   → Usable products

                                                       → Water effluents

                                                       → Air emissions

Raw   Product use   → Solid wastes
materials

                                                       → Other impacts

                       Product disposal

Figure 23.2 Schematic representation of an input/output inventory for the life-cycle assessment of a product.
(Adapted from J. L. Sullivan and S. B. Young, *Advanced Materials & Processes,* Vol. 147, No. 2, p. 38, 1995. Copyright © 1995 by ASM International. Reprinted by permission of ASM International, Materials Park, OH.)

who use "out-of-the-box" thinking can generate better and cheaper products/processes. Another consideration relates to how one defines *cost;* in this regard, it is essential to look at the entire life cycle and take into account all relevant factors (including disposal and environmental impact issues).

One approach being implemented by industry to improve the environmental performance of products is termed *life cycle analysis/assessment.* With this approach to product design, consideration is given to the cradle-to-grave environmental assessment of the product, from material extraction to product manufacture to product use and, finally, to recycling and disposal; sometimes this approach is also labeled *green design.* One important phase of this approach is to quantify the various inputs (i.e., materials and energy) and outputs (i.e., wastes) for each phase of the life cycle; this is represented schematically in Figure 23.2. In addition, an assessment is conducted relative to the impact on both global and local environments in terms of the effects on ecology, human health, and resource reserves.

One of the current environmental/economic/societal buzzwords is *sustainability.* In this context, sustainability represents the ability to maintain an acceptable lifestyle at the present level and into the indefinite future while preserving the environment. This means that over time and as populations increase in size, the Earth's resources must be used at a rate such that they can be replenished naturally and that emission levels of pollutants are maintained at acceptable levels. For engineers, this concept of sustainability translates into being responsible for the development of sustainable products. An internationally accepted standard, ISO 14001, has been established to help organizations comply with applicable laws and regulations and address the delicate balance between being profitable and reducing impacts on the environment.[1]

23.5 RECYCLING ISSUES IN MATERIALS SCIENCE AND ENGINEERING

Important stages in the materials cycle where materials science and engineering plays a significant role are recycling and disposal. The issues of recyclability and disposability are important when new materials are being designed and synthesized. Furthermore, during the materials selection process, the ultimate disposition of the materials used should be an important criterion. We conclude this section by discussing briefly several of these recyclability/disposability issues.

----

[1]The *International Organization for Standardization,* also known as *ISO,* is a worldwide body composed of representatives from various national standards organizations that establishes and disseminates industrial and commercial standards.

From an environmental perspective, the ideal material should be either completely recyclable or completely biodegradable. *Recyclable* means that a material, after completing its life cycle in one component, could be reprocessed, could reenter the materials cycle, and could be reused in another component—a process that could be repeated an indefinite number of times. By *completely biodegradable*, we mean that, by interactions with the environment (natural chemicals, microorganisms, oxygen, heat, sunlight, etc.), the material deteriorates and returns to virtually the same state in which it existed prior to the initial processing. Engineering materials exhibit varying degrees of recyclability and biodegradability.

Metals

Most metal alloys (e.g., those with Fe or Cu), to one degree or another experience corrosion and are also biodegradable. However, some metals (e.g., Hg, Pb) are toxic and, when placed in landfills, may present health hazards. Furthermore, whereas alloys of most metals are recyclable, it is not feasible to recycle all alloys of every metal. In addition, the quality of alloys that are recycled tends to diminish with each cycle.

Product designs should allow for the dismantling of components composed of different alloys. Another problem with recycling involves separation of various alloy types (e.g., aluminum from ferrous alloys) after dismantling and shredding; in this regard, some rather ingenious separation techniques have been devised (e.g., magnetic, gravity-driven). Joining of dissimilar alloys presents contamination problems; for example, if two similar alloys are to be joined, welding is preferred over bolting or riveting. Coatings (paints, anodized layers, claddings, etc.) may also act as contaminants and render the material nonrecyclable. These examples illustrate why it is so important to consider the entire life cycle of a product at the beginning stages of its design.

Aluminum alloys are very corrosion resistant and, therefore, nonbiodegradable. Fortunately, however, they may be recycled; in fact, aluminum is the most important recyclable nonferrous metal. Because aluminum is not easily corroded, it may be totally reclaimed. A low ratio of energy is required to refine recycled aluminum relative to its primary production. In addition, a large number of commercially available alloys have been designed to accommodate impurity contamination. The primary sources of recycled aluminum are used beverage cans and scrapped automobiles.

Glass

The ceramic material consumed by the general public in the greatest quantities is glass, in the form of containers. Glass is a relatively inert material and, as such, does not decompose; thus, it is not biodegradable. A significant proportion of municipal landfills consists of waste glass; this is also true of incinerator residue.

In addition, there is no significant economic driving force for recycling glass. Its basic raw materials (sand, soda ash, and limestone) are inexpensive and readily available. Furthermore, salvaged glass (also called *cullet*) must be sorted by color (e.g., clear, amber, green), by type (plate vs. container), and by composition (lime, lead, and borosilicate [or Pyrex]); these sorting procedures are time-consuming and expensive. Therefore, scrap glass has a low market value, which reduces its recyclability. Advantages of using recycled glass include more rapid and increased production rates and a reduction in pollutant emissions.

Plastics and Rubber

One reason that synthetic polymers (including rubber) are so popular as engineering materials is their chemical and biological inertness. On the down side, this characteristic is a liability when it comes to waste disposal. Most polymers are not biodegradable and, therefore, do not biodegrade in landfills; major sources of waste are from packaging,

Table 23.1 Recycle Codes, Uses of Virgin Material, and Recycled Products for Several Commercial Polymers

Recycle Code	Polymer Name	Uses of Virgin Material	Recycled Products
1	Poly(ethylene terephthalate) (PET or PETE)	Plastic beverage containers; mouthwash jars; peanut butter and salad dressing bottles	Liquid-soap bottles, strapping, fiberfill for winter coats, surfboards, paint brushes, fuzz on tennis balls, soft-drink bottles, film, egg cartons, skis, carpets, boats
2	High-density polyethylene (HDPE)	Milk, water, and juice containers; grocery bags; toys; liquid detergent bottles	Soft-drink bottle base caps, flower pots, drain pipes, signs, stadium seats, trash cans, recycling bins, traffic-barrier cones, golf-bag liners, detergent bottles, toys
3	Poly(vinyl chloride) or vinyl (V)	Clear food packaging; shampoo bottles	Floor mats, pipes, hoses, mud flaps
4	Low-density polyethylene (LDPE)	Bread bags; frozen-food bags; grocery bags	Garbage can liners, grocery bags, multipurpose bags
5	Polypropylene (PP)	Ketchup bottles; yogurt containers; margarine tubs, medicine bottles, carpet fibers	Manhole steps, paint buckets, ice scrapers, fast-food trays, lawnmower wheels, automobile battery parts
6	Polystyrene (PS)	Compact disc jackets; coffee cups; knives, spoons, and forks; cafeteria trays; grocery store meat trays; fast-food sandwich containers	License-plate holders, golf-course and septic-tank drainage systems, desktop accessories, hanging files, food-service trays, flowerpots, trash cans

Source: Reproduced with permission from SPI: The Plastics Industry Trade Association.

junked automobiles, automobile tires, and domestic durable goods. Biodegradable polymers have been synthesized, but they are relatively expensive to produce (see the Materials of Importance box that follows). However, because some polymers are combustible and do not yield appreciable toxic or polluting emissions, they may be disposed of by incineration.

Thermoplastic polymers, specifically poly(ethylene terephthalate), polyethylene, and polypropylene, are those most amenable to reclamation and recycling because they may be re-formed upon heating. Sorting by type and color is necessary. In some countries, type sorting of packaging materials is facilitated by using a number identification code; for example, a 1 denotes poly(ethylene terephthalate) (PET or PETE). Table 23.1 presents these recycling code numbers and their associated materials. Also included in the table are uses of virgin and recycled materials. Plastics recycling is complicated by the presence of fillers (Section 17.13) that were added to modify the original properties. The recycled plastic is less costly than the original material, and quality and appearance are generally degraded with each recycle. Typical applications for recycled plastics include shoe soles, tool handles, and industrial products such as pallets.

The recycling of thermoset resins is much more difficult because these materials are not easily remolded or reshaped because of their crosslinked or network structures. Some thermosets are ground up and added to the virgin molding material prior to processing; therefore, they are recycled as filler materials.

MATERIALS OF IMPORTANCE

Biodegradable and Biorenewable Polymers/Plastics

Most polymers manufactured today are synthetic and petroleum based. These synthetic materials (e.g., polyethylene, polystyrene) are extremely stable and resistant to degradation, particularly in moist environments. In the 1970s and 1980s it was feared that the large volume of plastic waste being generated would contribute to the filling of all available landfill capacity. Thus, the resistance to degradation of polymers was viewed as a liability rather than as an asset. The introduction of biodegradable polymers was perceived as a means to eliminate some of this landfill waste, and the response of the polymer industry was to start developing biodegradable materials.

Biodegradable polymers are those that degrade naturally in the environment, normally by microbial action. With regard to degradation mechanism, microbes sever polymer chain bonds, which leads to a decrease in molecular size; these smaller molecules may then be ingested by microbes in a process that is similar to the composting of plants. Of course, natural polymers such as wool, cotton, and wood are biodegradable inasmuch as microbes can readily digest these materials.

The first generation of these degradable materials was based on common polymers such as polyethylene. Compounds were added to make these materials decompose in sunlight (i.e., photodegrade), to oxidize by reacting with oxygen in the air, and/or to degrade biologically. Unfortunately, this first generation did not measure up to expectations. They degraded slowly (if at all) and the anticipated reduction in landfill waste was not realized. These initial disappointments gave degradable polymers a bad reputation that hindered their development. By way of response, the polymer industry instituted standards that measure degradation rate accurately and characterize the mode of degradation. These developments led to a renewed interest in biodegradable polymers.

Development of the current generation of biodegradable polymers is frequently directed to niche applications that take advantage of their short lifetimes. For example, biodegradable leaf and yard waste bags can be used to contain compostable matter, which eliminates the need to debag the material.

Another important application of biodegradable plastics is as mulch films for farming (Figure 23.3).

Figure 23.3 Biodegradable plastic mulch films that have been laid out on cultivated farmland.

Courtesy of Dubois Agrinovation

In colder regions of the world, covering seedbeds with plastic sheets can extend the growing season so as to increase crop yields and, in addition, reduce costs. The plastic sheets absorb heat, raise the ground temperature, and increase moisture retention. Traditionally, black polyethylene (nonbiodegradable) sheets were used. However, at the end of the growing season, these sheets had to be gathered from the field and disposed of manually because they did not decompose/biodegrade. More recently, biodegradable plastics have been developed for use as mulch films. After the crops have been harvested, these films are simply plowed into and enrich the soil as they decompose.

Other potential opportunities for this group of materials exist in the fast-food industry. For example, if all plates, cups, packaging, and so on were based on biodegradable materials, they could be commingled with food waste and then composted in large-scale operations. Not only would these measures reduce the amount of material entering landfills, but if the polymers were derived from renewable materials, a reduction in greenhouse gas emissions would result.

In order to reduce our dependence on petroleum and the emissions of greenhouse gases, there has been a major effort to develop biodegradable polymers that are also *biorenewable*—based on plant-derived materials (*biomass*[2]). These new materials must be cost-competitive with existing polymers and capable of being processed using conventional techniques (extrusion, injection molding, etc.).

Over the past 30 or so years, a number of biorenewable polymers have been synthesized that have properties comparable to those of petroleum-derived materials; some are biodegradable, and others are not. Perhaps the best known of these bioderived polymers is *poly(l-lactic acid)* (abbreviated *PLA*), which has the following repeat unit structure:

$$\left[\begin{array}{c} \overset{\displaystyle O}{\underset{\displaystyle CH_3}{\overset{\|}{C}}} \overset{\displaystyle O}{\underset{}{\overset{\|}{C}}} - O \end{array} \right]_n$$

Commercially, PLA is derived from lactic acid; however, the raw materials for its manufacture are starch-rich renewable products such as corn, sugar beets, and wheat. Mechanically, the modulus of elasticity and tensile strength of PLA are comparable to poly(ethylene terephthalate), and copolymerization with other biodegradable polymers [e.g., poly(glycolic acid) (PGA)] promotes property alterations to allow the use of conventional manufacturing processes such as injection molding, extrusion, blow molding, and fiber forming. Other properties make PLA desirable as a packaging material, especially for beverages and food products—transparency, resistance to attack by moisture and grease, odorlessness, and odor-barrier characteristics. PLA is also *bioresorbable*, meaning that it is assimilated (or absorbed) in biological systems—for example the human body. Hence, it has been used in a variety of biomedical applications, including resorbable sutures, implants, and controlled release of drugs.

The primary obstacle to the widespread use of PLA and other biodegradable polymers has been that of high cost, a common problem associated with the introduction of new materials. However, the development of more efficient and economical synthesis and processing techniques has resulted in a significant reduction in the cost of this class of materials, making them more competitive with conventional petroleum-based polymers.

Although PLA is biodegradable, it degrades only under carefully controlled circumstances—that is, at elevated temperatures generated in commercial composting facilities. At room temperature and normal ambient conditions, it is stable indefinitely. The degradation products consist of water, carbon dioxide, and organic matter. The initial stages of the degradation process in which a high-molecular-weight polymer is broken into smaller pieces is not truly one of "biodegradation" as described earlier; rather, it involves hydrolytic cleavage of the polymer backbone chain, and there is little or no evidence of microbial action. However, the subsequent degradation of these lower-molecular-weight fragments is microbial.

Poly(lactic acid) is also recyclable—with the right equipment, it can be converted back into the original monomer and then resynthesized to form PLA.

A number of other characteristics of PLA make it an especially attractive material, in particular for textile applications. For example, it may be spun into fibers using conventional melt-spinning processes (Section 17.16). In addition, PLA has excellent crimp and crimp retention, is resistant to degradation when exposed to ultraviolet light (i.e., resists fading), and is relatively non-flammable. Other potential applications for this material include household furnishings such as drapes, upholstery, and awnings, as well as diapers and industrial wipes.

Courtesy of NatureWorks LLC and International Paper, Inc.

Examples of applications for biodegradable/biorenewable poly(lactic acid): films, packaging, and fabrics.

---

[2]*Biomass* refers to biological material such as the stems, leaves, and seeds of plants that can be used as fuel or as an industrial feedstock.

Rubber materials present disposal and recycling challenges. When vulcanized, they are thermoset materials, which makes chemical recycling difficult. In addition, they may also contain a variety of fillers. The major source of rubber scrap in the United States is discarded automobile tires, which are highly nonbiodegradable. Scrap tires have been used as a fuel for some industrial applications (e.g., cement plants) but yield dirty emissions. Recycled rubber tires that have been split and reshaped are used in a variety of applications, such as automotive bumper guards, mud flaps, doormats, and conveyor rollers; of course, used tires may also be recapped. In addition, rubber tires may be ground into small chunks that are then recombined into the desired shape using some type of adhesive; the resulting material may be used in a number of nondemanding applications such as placemats and rubber toys.

The most viable recyclable alternatives to traditional rubber materials are thermoplastic elastomers (Section 15.19). Being thermoplastic in nature, they are not chemically crosslinked and, thus, are easily reshaped. Furthermore, production energy requirements are lower than for the thermoset rubbers because a vulcanization step is not required in their manufacture.

Composite Materials

Composites are inherently difficult to recycle because they are multiphase. The two or more phases/materials that constitute the composite are normally intermixed on a very fine scale, and trying to separate them complicates the recycling process. However, some techniques have been developed, with modest success, for recycling polymer-matrix composites. Recycling technologies differ only slightly for thermoset-matrix and thermoplastic-matrix composite materials.

The first step in recycling both thermoset- and thermoplastic-matrix composites is shredding/grinding, in which the components are reduced in size to relatively small particles. In some instances, these ground particles are used as filler materials that are blended with a polymer (and perhaps other fillers) before fabrication (usually using some type of molding technique) into postconsumer products. Other recycling processes allow for separating the fibers and/or matrix materials. With some techniques, the matrix is volatilized; with others, it is recovered as a monomer. The recovered fibers have short lengths as a result of the shredding/grinding process. In addition, fibers experience a reduction of mechanical strength, the degree of which depends on the specific recovery process, as well as fiber type.

SUMMARY

Economic Considerations	• To minimize product cost, materials engineers must take into account component design, the materials used, and manufacturing processes.
	• Other significant economic factors include fringe benefits, labor, insurance, and profit.
Environmental and Societal Considerations	• Environmental and societal impacts of production are becoming significant engineering issues. In this regard, the material cradle-to-grave life cycle is an important consideration.
	• The materials cycle consists of extraction, synthesis/processing, product design/manufacture, application, and disposal stages (Figure 23.1).
	• Efficient operation of the materials cycle is facilitated using an input/output inventory for the life-cycle assessment of a product. Materials and energy are the input parameters, whereas outputs include usable products, water effluents, air emissions, and solid wastes (Figure 23.2).

- The Earth is a closed system, in that its materials resources are finite; to some degree, the same may be said of energy resources. Environmental issues involve ecological damage, pollution, and waste disposal.

- Recycling of used products and the use of green design obviate some of these environmental problems.

Recycling Issues in Materials Science and Engineering

- Recyclability and disposability issues are important in the context of materials science and engineering. Ideally, a material should be at best recyclable, and at least biodegradable or disposable.

- With regard to the recyclability/disposability of the various material types:

  Among metal alloys, there are varying degrees of recyclability and biodegradability (i.e., susceptibility to corrosion). Some metals are toxic and, therefore, not disposable.

  Glass is the most common commercial ceramic. It is not biodegradable; furthermore, at present, there is no economic incentive to recycle glass.

  Most plastics and rubber are nonbiodegradable. Thermoplastic polymers are recyclable, whereas thermosetting materials are (for the most part) nonrecyclable.

  Composite materials are very difficult to recycle because they are composed of two or more phases that are normally intermixed on a fine scale.

REFERENCES

Engineering Economics

Newnan, D. G., T. G. Eschenbach, and J. P. Lavelle, *Engineering Economic Analysis,* 11th edition, Oxford University Press, New York, 2011.

Park, C. S., *Fundamentals of Engineering Economics,* 3rd edition, Prentice Hall, Upper Saddle River, NJ, 2012.

White, J. A., K. E. Case, and D. B. Pratt, *Principles of Engineering Economics Analysis,* 6th edition, Wiley, Hoboken, NJ, 2012.

Societal

Cohen, M., "Societal Issues in Materials Science and Technology," *Materials Research Society Bulletin,* September, 1994, pp. 3–8.

Environmental

Ackerman, F., *Why Do We Recycle? Markets, Values, and Public Policy,* Island Press, Washington, DC, 1997.

Ashby, M. F., *Materials and the Environment: Eco-Informed Material Choice,* 2nd edition, Butterworth-Heinemann/Elsevier, Oxford, 2012.

Azapagic, A., A. Emsley, and I. Hamerton, *Polymers, the Environment and Sustainable Development,* Wiley, West Sussex, UK, 2003.

McDonough, W., and M. Braungart, *Cradle to Cradle: Remaking the Way We Make Things,* North Point Press, New York, 2002.

Nemerow, N. L., F. J. Agardy, and J. A. Salvato (Editors), *Environmental Engineering,* 6th edition, Wiley, Hoboken, NJ, 2009. Three volumes.

Porter, R. C., *The Economics of Waste,* Resources for the Future Press, Washington, DC, 2002.

Young, G. C., *Municipal Solid Waste to Energy Conversion Processes: Economic, Technical and Renewable Comparisons,* Wiley, Hoboken, NJ, 2010.

DESIGN QUESTIONS

23.D1 Glass, aluminum, and various plastic materials are used for containers (see the chapter-opening photographs for Chapter 1 and the photographs accompanying the Materials of Importance box for this chapter). Make a list of the advantages and disadvantages of using each of these three material types; include such factors as cost, recyclability, and energy consumption for container production.

23.D2 Discuss why it is important to consider the entire life cycle, rather than just the first stage.

23.D3 Discuss how materials engineering can play a role in "green design."

23.D4 Suggest other consumer actions for minimal environmental impact than just recycling.

Appendix A The International System of Units (SI)

Units in the *International System of Units* fall into two classifications: base and derived. Base units are fundamental and not reducible. Table A.1 lists the base units of interest in the discipline of materials science and engineering.

Derived units are expressed in terms of the base units, using mathematical signs for multiplication and division. For example, the SI units for density are kilogram per cubic meter (kg/m^3). For some derived units, special names and symbols exist; for example, N is used to denote the newton—the unit of force—which is equivalent to $1 \ kg \cdot m/s^2$. Table A.2 lists a number of important derived units.

It is sometimes necessary, or convenient, to form names and symbols that are decimal multiples or submultiples of SI units. Only one prefix is used when a multiple of an SI unit is formed, which should be in the numerator. These prefixes and their approved symbols are given in Table A.3. Symbols for all units used in this book, SI or otherwise, are given inside the front cover.

Table A.1

The SI Base Units

Quantity	Name	Symbol
Length	meter, metre	m
Mass	kilogram	kg
Time	second	s
Electric current	ampere	A
Thermodynamic temperature	kelvin	K
Amount of substance	mole	mol

Table A.2

Some SI-Derived Units

Quantity	Name	Formula	Special Symbol
Area	square meter	m^2	—
Volume	cubic meter	m^3	—
Velocity	meter per second	m/s	—
Density	kilogram per cubic meter	kg/m^3	—
Concentration	moles per cubic meter	mol/m^3	—
Force	newton	$kg \cdot m/s^2$	N
Energy	joule	$kg \cdot m^2/s^2$, N·m	J
Stress	pascal	$kg/m \cdot s^2$, N/m^2	Pa
Strain	—	m/m	—
Power, radiant flux	watt	$kg \cdot m^2/s^3$, J/s	W
Viscosity	pascal-second	kg/m·s	Pa·s
Frequency (of a periodic phenomenon)	hertz	s^{-1}	Hz
Electric charge	coulomb	A·s	C
Electric potential	volt	$kg \cdot m^2/s^2 \cdot C$	V
Capacitance	farad	$s^2 \cdot C^2/kg \cdot m^2$	F
Electric resistance	ohm	$kg \cdot m^2/s \cdot C^2$	Ω
Magnetic flux	weber	$kg \cdot m^2/s \cdot C$	Wb
Magnetic flux density	tesla	$kg/s \cdot C$, Wb/m^2	$(T)^a$

[a]T is a special symbol approved for SI but not used in this text; here, the name *tesla* is used instead of the symbol.

Table A.3

SI Multiple and Submultiple Prefixes

Factor by Which Multiplied	Prefix	Symbol
10^9	giga	G
10^6	mega	M
10^3	kilo	k
10^{-2}	centi[a]	c
10^{-3}	milli	m
10^{-6}	micro	μ
10^{-9}	nano	n
10^{-12}	pico	p

[a]Avoided when possible.

This appendix compiles important properties for approximately 100 common engineering materials. Each table gives data values of one particular property for this chosen set of materials; also included is a tabulation of the compositions of the various metal alloys that are considered (Table B.10). Data are tabulated by material type (metals and metal alloys; graphite, ceramics, and semiconducting materials; polymers; fiber materials; and composites). Within each classification, the materials are listed alphabetically.

Note that data entries are expressed either as ranges of values or as single values that are typically measured. Also, on occasion, (*min*) is associated with an entry, indicating that the value cited is a minimum one.

Table B.1

Room-Temperature Density Values for Various Engineering Materials

Material	Density	
	g/cm³	*lbₘ/in.³*
METALS AND METAL ALLOYS		
Plain Carbon and Low-Alloy Steels		
Steel alloy A36	7.85	0.283
Steel alloy 1020	7.85	0.283
Steel alloy 1040	7.85	0.283
Steel alloy 4140	7.85	0.283
Steel alloy 4340	7.85	0.283
Stainless Steels		
Stainless alloy 304	8.00	0.289
Stainless alloy 316	8.00	0.289
Stainless alloy 405	7.80	0.282
Stainless alloy 440A	7.80	0.282
Stainless alloy 17-7PH	7.65	0.276

Table B.1
(Continued)

Material	Density	
	g/cm³	*lbₘ/in.³*

Cast Irons		
Gray irons		
• Grade G1800	7.30	0.264
• Grade G3000	7.30	0.264
• Grade G4000	7.30	0.264
Ductile irons		
• Grade 60-40-18	7.10	0.256
• Grade 80-55-06	7.10	0.256
• Grade 120-90-02	7.10	0.256
Aluminum Alloys		
Alloy 1100	2.71	0.0978
Alloy 2024	2.77	0.100
Alloy 6061	2.70	0.0975
Alloy 7075	2.80	0.101
Alloy 356.0	2.69	0.0971
Copper Alloys		
C11000 (electrolytic tough pitch)	8.89	0.321
C17200 (beryllium–copper)	8.25	0.298
C26000 (cartridge brass)	8.53	0.308
C36000 (free-cutting brass)	8.50	0.307
C71500 (copper–nickel, 30%)	8.94	0.323
C93200 (bearing bronze)	8.93	0.322
Magnesium Alloys		
Alloy AZ31B	1.77	0.0639
Alloy AZ91D	1.81	0.0653
Titanium Alloys		
Commercially pure (ASTM grade 1)	4.51	0.163
Alloy Ti–5Al–2.5Sn	4.48	0.162
Alloy Ti–6Al–4V	4.43	0.160
Precious Metals		
Gold (commercially pure)	19.32	0.697
Platinum (commercially pure)	21.45	0.774
Silver (commercially pure)	10.49	0.379
Refractory Metals		
Molybdenum (commercially pure)	10.22	0.369
Tantalum (commercially pure)	16.6	0.599
Tungsten (commercially pure)	19.3	0.697
Miscellaneous Nonferrous Alloys		
Nickel 200	8.89	0.321
Inconel 625	8.44	0.305
Monel 400	8.80	0.318

Table B.1
(Continued)

Material	Density	
	g/cm³	*lbₘ/in.³*

Material	Density g/cm^3	Density $lb_m/in.^3$
Haynes alloy 25	9.13	0.330
Invar	8.05	0.291
Super invar	8.10	0.292
Kovar	8.36	0.302
Chemical lead	11.34	0.409
Antimonial lead (6%)	10.88	0.393
Tin (commercially pure)	7.17	0.259
Lead–tin solder (60Sn–40Pb)	8.52	0.308
Zinc (commercially pure)	7.14	0.258
Zirconium, reactor grade 702	6.51	0.235

**GRAPHITE, CERAMICS, AND
SEMICONDUCTING MATERIALS**

Aluminum oxide		
• 99.9% pure	3.98	0.144
• 96% pure	3.72	0.134
• 90% pure	3.60	0.130
Concrete	2.4	0.087
Diamond		
• Natural	3.51	0.127
• Synthetic	3.20–3.52	0.116–0.127
Gallium arsenide	5.32	0.192
Glass, borosilicate (Pyrex)	2.23	0.0805
Glass, soda–lime	2.5	0.0903
Glass-ceramic (Pyroceram)	2.60	0.0939
Graphite		
• Extruded	1.71	0.0616
• Isostatically molded	1.78	0.0643
Silica, fused	2.2	0.079
Silicon	2.33	0.0841
Silicon carbide		
• Hot-pressed	3.3	0.119
• Sintered	3.2	0.116
Silicon nitride		
• Hot-pressed	3.3	0.119
• Reaction-bonded	2.7	0.0975
• Sintered	3.3	0.119
Zirconia, 3 mol% Y_2O_3, sintered	6.0	0.217

POLYMERS

Elastomers		
• Butadiene–acrylonitrile (nitrile)	0.98	0.0354
• Styrene–butadiene (SBR)	0.94	0.0339
• Silicone	1.1–1.6	0.040–0.058
Epoxy	1.11–1.40	0.0401–0.0505

Table B.1

(Continued)

Material	Density	
	g/cm^3	**$lb_m/in.^3$**
Nylon 6,6	1.14	0.0412
Phenolic	1.28	0.0462
Poly(butylene terephthalate) (PBT)	1.34	0.0484
Polycarbonate (PC)	1.20	0.0433
Polyester (thermoset)	1.04–1.46	0.038–0.053
Polyetheretherketone (PEEK)	1.31	0.0473
Polyethylene		
• Low density (LDPE)	0.925	0.0334
• High density (HDPE)	0.959	0.0346
• Ultrahigh molecular weight (UHMWPE)	0.94	0.0339
Poly(ethylene terephthalate) (PET)	1.35	0.0487
Poly(methyl methacrylate) (PMMA)	1.19	0.0430
Polypropylene (PP)	0.905	0.0327
Polystyrene (PS)	1.05	0.0379
Polytetrafluoroethylene (PTFE)	2.17	0.0783
Poly(vinyl chloride) (PVC)	1.30–1.58	0.047–0.057
FIBER MATERIALS		
Aramid (Kevlar 49)	1.44	0.0520
Carbon		
• Standard modulus (PAN precursor)	1.78	0.0643
• Intermediate modulus (PAN precursor)	1.78	0.0643
• High modulus (PAN precursor)	1.81	0.0653
• Ultrahigh modulus (pitch precursor)	2.12–2.19	0.077–0.079
E-glass	2.58	0.0931
COMPOSITE MATERIALS		
Aramid fibers–epoxy matrix ($V_f = 0.60$)	1.4	0.050
High-modulus carbon fibers–epoxy matrix ($V_f = 0.60$)	1.7	0.061
E-glass fibers–epoxy matrix ($V_f = 0.60$)	2.1	0.075
Wood		
• Douglas fir (12% moisture)	0.46–0.50	0.017–0.018
• Red oak (12% moisture)	0.61–0.67	0.022–0.024

Sources: *ASM Handbooks,* Volumes 1 and 2, *Engineered Materials Handbook,* Volume 4, *Metals Handbook: Properties and Selection: Nonferrous Alloys and Pure Metals,* Vol. 2, 9th edition, and *Advanced Materials & Processes,* Vol. 146, No. 4, ASM International, Materials Park, OH; *Modern Plastics Encyclopedia '96,* The McGraw-Hill Companies, New York, NY; R. F. Floral and S. T. Peters, "Composite Structures and Technologies," tutorial notes, 1989; and manufacturers' technical data sheets.

Table B.2

Room-Temperature Modulus of Elasticity Values for Various Engineering Materials

Material	Modulus of Elasticity	
	GPa	**10^6 psi**
METALS AND METAL ALLOYS		
Plain Carbon and Low-Alloy Steels		
Steel alloy A36	207	30
Steel alloy 1020	207	30
Steel alloy 1040	207	30

Table B.2
(Continued)

	Modulus of Elasticity	
Material	*GPa*	*10^6 psi*
Steel alloy 4140	207	30
Steel alloy 4340	207	30
Stainless Steels		
Stainless alloy 304	193	28
Stainless alloy 316	193	28
Stainless alloy 405	200	29
Stainless alloy 440A	200	29
Stainless alloy 17-7PH	204	29.5
Cast Irons		
Gray irons		
• Grade G1800	66–97[a]	9.6–14[a]
• Grade G3000	90–113[a]	13.0–16.4[a]
• Grade G4000	110–138[a]	16–20[a]
Ductile irons		
• Grade 60-40-18	169	24.5
• Grade 80-55-06	168	24.4
• Grade 120-90-02	164	23.8
Aluminum Alloys		
Alloy 1100	69	10
Alloy 2024	72.4	10.5
Alloy 6061	69	10
Alloy 7075	71	10.3
Alloy 356.0	72.4	10.5
Copper Alloys		
C11000 (electrolytic tough pitch)	115	16.7
C17200 (beryllium–copper)	128	18.6
C26000 (cartridge brass)	110	16
C36000 (free-cutting brass)	97	14
C71500 (copper–nickel, 30%)	150	21.8
C93200 (bearing bronze)	100	14.5
Magnesium Alloys		
Alloy AZ31B	45	6.5
Alloy AZ91D	45	6.5
Titanium Alloys		
Commercially pure (ASTM grade 1)	103	14.9
Alloy Ti–5Al–2.5Sn	110	16
Alloy Ti–6Al–4V	114	16.5
Precious Metals		
Gold (commercially pure)	77	11.2
Platinum (commercially pure)	171	24.8
Silver (commercially pure)	74	10.7

Table B.2
(Continued)

Material	Modulus of Elasticity	
	GPa	10^6 psi
Refractory Metals		
Molybdenum (commercially pure)	320	46.4
Tantalum (commercially pure)	185	27
Tungsten (commercially pure)	400	58
Miscellaneous Nonferrous Alloys		
Nickel 200	204	29.6
Inconel 625	207	30
Monel 400	180	26
Haynes alloy 25	236	34.2
Invar	141	20.5
Super invar	144	21
Kovar	207	30
Chemical lead	13.5	2
Tin (commercially pure)	44.3	6.4
Lead–tin solder (60Sn–40Pb)	30	4.4
Zinc (commercially pure)	104.5	15.2
Zirconium, reactor grade 702	99.3	14.4
GRAPHITE, CERAMICS, AND SEMICONDUCTING MATERIALS		
Aluminum oxide		
• 99.9% pure	380	55
• 96% pure	303	44
• 90% pure	275	40
Concrete	25.4–36.6[a]	3.7–5.3[a]
Diamond		
• Natural	700–1200	102–174
• Synthetic	800–925	116–134
Gallium arsenide, single crystal		
• In the $\langle 100 \rangle$ direction	85	12.3
• In the $\langle 110 \rangle$ direction	122	17.7
• In the $\langle 111 \rangle$ direction	142	20.6
Glass, borosilicate (Pyrex)	70	10.1
Glass, soda–lime	69	10
Glass-ceramic (Pyroceram)	120	17.4
Graphite		
• Extruded	11	1.6
• Isostatically molded	11.7	1.7
Silica, fused	73	10.6
Silicon, single crystal		
• In the $\langle 100 \rangle$ direction	129	18.7
• In the $\langle 110 \rangle$ direction	168	24.4
• In the $\langle 111 \rangle$ direction	187	27.1
Silicon carbide		
• Hot-pressed	207–483	30–70
• Sintered	207–483	30–70

Table B.2
(Continued)

	Modulus of Elasticity	
Material	*GPa*	*10^6 psi*
Silicon nitride		
• Hot-pressed	304	44.1
• Reaction-bonded	304	44.1
• Sintered	304	44.1
Zirconia, 3 mol% Y_2O_3	205	30
POLYMERS		
Elastomers		
• Butadiene acrylonitrile (nitrile)	0.0034[b]	0.00049[b]
• Styrene-butadiene (SBR)	0.002–0.010[b]	0.0003–0.0015[b]
Epoxy	2.41	0.35
Nylon 6,6	1.59–3.79	0.230–0.550
Phenolic	2.76–4.83	0.40–0.70
Poly(butylene terephthalate) (PBT)	1.93–3.00	0.280–0.435
Polycarbonate (PC)	2.38	0.345
Polyester (thermoset)	2.06–4.41	0.30–0.64
Polyetheretherketone (PEEK)	1.10	0.16
Polyethylene		
• Low density (LDPE)	0.172–0.282	0.025–0.041
• High density (HDPE)	1.08	0.157
• Ultrahigh molecular weight (UHMWPE)	0.69	0.100
Poly(ethylene terephthalate) (PET)	2.76–4.14	0.40–0.60
Poly(methyl methacrylate) (PMMA)	2.24–3.24	0.325–0.470
Polypropylene (PP)	1.14–1.55	0.165–0.225
Polystyrene (PS)	2.28–3.28	0.330–0.475
Polytetrafluoroethylene (PTFE)	0.40–0.55	0.058–0.080
Poly(vinyl chloride) (PVC)	2.41–4.14	0.35–0.60
FIBER MATERIALS		
Aramid (Kevlar 49)	131	19
Carbon		
• Standard modulus (PAN precursor)	230	33.4
• Intermediate modulus (PAN precursor)	285	41.3
• High modulus (PAN precursor)	400	58
• Ultrahigh modulus (pitch precursor)	520–940	75–136
E-glass	72.5	10.5
COMPOSITE MATERIALS		
Aramid fibers–epoxy matrix ($V_f = 0.60$)		
Longitudinal	76	11
Transverse	5.5	0.8
High-modulus carbon fibers–epoxy matrix ($V_f = 0.60$)		
Longitudinal	220	32
Transverse	6.9	1.0
E-glass fibers–epoxy matrix ($V_f = 0.60$)		
Longitudinal	45	6.5
Transverse	12	1.8

Table B.2
(Continued)

Material	Modulus of Elasticity	
	GPa	*10^6 psi*
Wood		
• Douglas fir (12% moisture)		
Parallel to grain	10.8–13.6[c]	1.57–1.97[c]
Perpendicular to grain	0.54–0.68[c]	0.078–0.10[c]
• Red oak (12% moisture)		
Parallel to grain	11.0–14.1[c]	1.60–2.04[c]
Perpendicular to grain	0.55–0.71[c]	0.08–0.10[c]

[a] Secant modulus taken at 25% of ultimate strength.
[b] Modulus taken at 100% elongation.
[c] Measured in bending.

Sources: *ASM Handbooks,* Volumes 1 and 2, *Engineered Materials Handbooks,* Volumes 1 and 4, *Metals Handbook: Properties and Selection: Nonferrous Alloys and Pure Metals,* Vol. 2, 9th edition, and *Advanced Materials & Processes,* Vol. 146, No. 4, ASM International, Materials Park, OH; *Modern Plastics Encyclopedia '96,* The McGraw-Hill Companies, New York, NY; R. F. Floral and S. T. Peters, "Composite Structures and Technologies," tutorial notes, 1989; and manufacturers' technical data sheets.

Table B.3 Room-Temperature Poisson's Ratio Values for Various Engineering Materials

Material	Poisson's Ratio	Material	Poisson's Ratio
METALS AND METAL ALLOYS		**Copper Alloys**	
Plain Carbon and Low-Alloy Steels		C11000 (electrolytic tough pitch)	0.33
Steel alloy A36	0.30	C17200 (beryllium–copper)	0.30
Steel alloy 1020	0.30	C26000 (cartridge brass)	0.35
Steel alloy 1040	0.30	C36000 (free-cutting brass)	0.34
Steel alloy 4140	0.30	C71500 (copper–nickel, 30%)	0.34
Steel alloy 4340	0.30	C93200 (bearing bronze)	0.34
Stainless Steels		**Magnesium Alloys**	
Stainless alloy 304	0.30	Alloy AZ31B	0.35
Stainless alloy 316	0.30	Alloy AZ91D	0.35
Stainless alloy 405	0.30	**Titanium Alloys**	
Stainless alloy 440A	0.30	Commercially pure (ASTM grade 1)	0.34
Stainless alloy 17-7PH	0.30	Alloy Ti–5Al–2.5Sn	0.34
Cast Irons		Alloy Ti–6Al–4V	0.34
Gray irons		**Precious Metals**	
• Grade G1800	0.26	Gold (commercially pure)	0.42
• Grade G3000	0.26	Platinum (commercially pure)	0.39
• Grade G4000	0.26	Silver (commercially pure)	0.37
Ductile irons		**Refractory Metals**	
• Grade 60-40-18	0.29	Molybdenum (commercially pure)	0.32
• Grade 80-55-06	0.31	Tantalum (commercially pure)	0.35
• Grade 120-90-02	0.28	Tungsten (commercially pure)	0.28
Aluminum Alloys		**Miscellaneous Nonferrous Alloys**	
Alloy 1100	0.33	Nickel 200	0.31
Alloy 2024	0.33	Inconel 625	0.31
Alloy 6061	0.33	Monel 400	0.32
Alloy 7075	0.33		
Alloy 356.0	0.33		

Table B.3 (*Continued*)

Material	Poisson's Ratio	Material	Poisson's Ratio
Chemical lead	0.44	Silicon nitride	
Tin (commercially pure)	0.33	• Hot-pressed	0.30
Zinc (commercially pure)	0.25	• Reaction-bonded	0.22
Zirconium, reactor grade 702	0.35	• Sintered	0.28
		Zirconia, 3 mol% Y_2O_3	0.31

GRAPHITE, CERAMICS, AND SEMICONDUCTING MATERIALS (left column)

POLYMERS (right column)

Material	Poisson's Ratio	Material	Poisson's Ratio
Aluminum oxide		Nylon 6,6	0.39
• 99.9% pure	0.22	Polycarbonate (PC)	0.36
• 96% pure	0.21	Polyethylene	
• 90% pure	0.22	• Low density (LDPE)	0.33–0.40
Concrete	0.20	• High density (HDPE)	0.46
Diamond		Poly(ethylene terephthalate) (PET)	0.33
• Natural	0.10–0.30	Poly(methyl methacrylate) (PMMA)	0.37–0.44
• Synthetic	0.20	Polypropylene (PP)	0.40
Gallium arsenide		Polystyrene (PS)	0.33
• ⟨100⟩ orientation	0.30	Polytetrafluoroethylene (PTFE)	0.46
Glass, borosilicate (Pyrex)	0.20	Poly(vinyl chloride) (PVC)	0.38
Glass, soda–lime	0.23	**FIBER MATERIALS**	
Glass-ceramic (Pyroceram)	0.25	E–glass	0.22
Silica, fused	0.17	**COMPOSITE MATERIALS**	
Silicon		Aramid fibers–epoxy matrix ($V_f = 0.6$)	0.34
• ⟨100⟩ orientation	0.28	High-modulus carbon fibers–epoxy matrix ($V_f = 0.6$)	0.25
• ⟨111⟩ orientation	0.36		
Silicon carbide		E-glass fibers–epoxy matrix ($V_f = 0.6$)	0.19
• Hot-pressed	0.17		
• Sintered	0.16		

Sources: *ASM Handbooks,* Volumes 1 and 2, and *Engineered Materials Handbooks,* Volumes 1 and 4, ASM International, Materials Park, OH; R. F. Floral and S. T. Peters, "Composite Structures and Technologies," tutorial notes, 1989; and manufacturers' technical data sheets.

Table B.4 Typical Room-Temperature Yield Strength, Tensile Strength, and Ductility (Percent Elongation) Values for Various Engineering Materials

Material/Condition	Yield Strength (MPa [ksi])	Tensile Strength (MPa [ksi])	Percent Elongation
METALS AND METAL ALLOYS **Plain Carbon and Low-Alloy Steels**			
Steel alloy A36			
• Hot-rolled	220–250 (32–36)	400–500 (58–72.5)	23
Steel alloy 1020			
• Hot-rolled	210 (30) (min)	380 (55) (min)	25 (min)
• Cold-drawn	350 (51) (min)	420 (61) (min)	15 (min)
• Annealed (@ 870°C)	295 (42.8)	395 (57.3)	36.5
• Normalized (@ 925°C)	345 (50.3)	440 (64)	38.5
Steel alloy 1040			
• Hot-rolled	290 (42) (min)	520 (76) (min)	18 (min)
• Cold-drawn	490 (71) (min)	590 (85) (min)	12 (min)
• Annealed (@ 785°C)	355 (51.3)	520 (75.3)	30.2
• Normalized (@ 900°C)	375 (54.3)	590 (85)	28.0

Table B.4 *(Continued)*

Material/Condition	Yield Strength (MPa [ksi])	Tensile Strength (MPa [ksi])	Percent Elongation
Steel alloy 4140			
• Annealed (@ 815°C)	417 (60.5)	655 (95)	25.7
• Normalized (@ 870°C)	655 (95)	1020 (148)	17.7
• Oil-quenched and tempered (@ 315°C)	1570 (228)	1720 (250)	11.5
Steel alloy 4340			
• Annealed (@ 810°C)	472 (68.5)	745 (108)	22
• Normalized (@ 870°C)	862 (125)	1280 (185.5)	12.2
• Oil-quenched and tempered (@ 315°C)	1620 (235)	1760 (255)	12
Stainless Steels			
Stainless alloy 304			
• Hot-finished and annealed	205 (30) (min)	515 (75) (min)	40 (min)
• Cold-worked ($\frac{1}{4}$ hard)	515 (75) (min)	860 (125) (min)	10 (min)
Stainless alloy 316			
• Hot-finished and annealed	205 (30) (min)	515 (75) (min)	40 (min)
• Cold-drawn and annealed	310 (45) (min)	620 (90) (min)	30 (min)
Stainless alloy 405			
• Annealed	170 (25)	415 (60)	20
Stainless alloy 440A			
• Annealed	415 (60)	725 (105)	20
• Tempered (@ 315°C)	1650 (240)	1790 (260)	5
Stainless alloy 17-7PH			
• Cold-rolled	1210 (175) (min)	1380 (200) (min)	1 (min)
• Precipitation-hardened (@ 510°C)	1310 (190) (min)	1450 (210) (min)	3.5 (min)
Cast Irons			
Gray irons			
• Grade G1800 (as cast)	—	124 (18) (min)	—
• Grade G3000 (as cast)	—	207 (30) (min)	—
• Grade G4000 (as cast)	—	276 (40) (min)	—
Ductile irons			
• Grade 60-40-18 (annealed)	276 (40) (min)	414 (60) (min)	18 (min)
• Grade 80-55-06 (as cast)	379 (55) (min)	552 (80) (min)	6 (min)
• Grade 120-90-02 (oil-quenched and tempered)	621 (90) (min)	827 (120) (min)	2 (min)
Aluminum Alloys			
Alloy 1100			
• Annealed (O temper)	34 (5)	90 (13)	40
• Strain-hardened (H14 temper)	117 (17)	124 (18)	15
Alloy 2024			
• Annealed (O temper)	75 (11)	185 (27)	20
• Heat-treated and aged (T3 temper)	345 (50)	485 (70)	18
• Heat-treated and aged (T351 temper)	325 (47)	470 (68)	20
Alloy 6061			
• Annealed (O temper)	55 (8)	124 (18)	30
• Heat-treated and aged (T6 and T651 tempers)	276 (40)	310 (45)	17
Alloy 7075			
• Annealed (O temper)	103 (15)	228 (33)	17
• Heat-treated and aged (T6 temper)	505 (73)	572 (83)	11

Table B.4 *(Continued)*

Material/Condition	Yield Strength (MPa [ksi])	Tensile Strength (MPa [ksi])	Percent Elongation
Alloy 356.0			
• As cast	124 (18)	164 (24)	6
• Heat-treated and aged (T6 temper)	164 (24)	228 (33)	3.5
Copper Alloys			
C11000 (electrolytic tough pitch)			
• Hot-rolled	69 (10)	220 (32)	45
• Cold-worked (H04 temper)	310 (45)	345 (50)	12
C17200 (beryllium–copper)			
• Solution heat-treated	195–380 (28–55)	415–540 (60–78)	35–60
• Solution heat-treated and aged (@ 330°C)	965–1205 (140–175)	1140–1310 (165–190)	4–10
C26000 (cartridge brass)			
• Annealed	75–150 (11–22)	300–365 (43.5–53.0)	54–68
• Cold-worked (H04 temper)	435 (63)	525 (76)	8
C36000 (free-cutting brass)			
• Annealed	125 (18)	340 (49)	53
• Cold-worked (H02 temper)	310 (45)	400 (58)	25
C71500 (copper–nickel, 30%)			
• Hot-rolled	140 (20)	380 (55)	45
• Cold-worked (H80 temper)	545 (79)	580 (84)	3
C93200 (bearing bronze)			
• Sand cast	125 (18)	240 (35)	20
Magnesium Alloys			
Alloy AZ31B			
• Rolled	220 (32)	290 (42)	15
• Extruded	200 (29)	262 (38)	15
Alloy AZ91D			
• As cast	97–150 (14–22)	165–230 (24–33)	3
Titanium Alloys			
Commercially pure (ASTM grade 1)			
• Annealed	170 (25) (min)	240 (35) (min)	24
Alloy Ti–5Al–2.5Sn			
• Annealed	760 (110) (min)	790 (115) (min)	16
Alloy Ti–6Al–4V			
• Annealed	830 (120) (min)	900 (130) (min)	14
• Solution heat-treated and aged	1103 (160)	1172 (170)	10
Precious Metals			
Gold (commercially pure)			
• Annealed	nil	130 (19)	45
• Cold-worked (60% reduction)	205 (30)	220 (32)	4
Platinum (commercially pure)			
• Annealed	<13.8 (2)	125–165 (18–24)	30–40
• Cold-worked (50%)	—	205–240 (30–35)	1–3
Silver (commercially pure)			
• Annealed	—	170 (24.6)	44
• Cold-worked (50%)	—	296 (43)	3.5

Table B.4 (*Continued*)

Material/Condition	Yield Strength (MPa [ksi])	Tensile Strength (MPa [ksi])	Percent Elongation
Refractory Metals			
Molybdenum (commercially pure)	500 (72.5)	630 (91)	25
Tantalum (commercially pure)	165 (24)	205 (30)	40
Tungsten (commercially pure)	760 (110)	960 (139)	2
Miscellaneous Nonferrous Alloys			
Nickel 200 (annealed)	148 (21.5)	462 (67)	47
Inconel 625 (annealed)	517 (75)	930 (135)	42.5
Monel 400 (annealed)	240 (35)	550 (80)	40
Haynes alloy 25	445 (65)	970 (141)	62
Invar (annealed)	276 (40)	517 (75)	30
Super invar (annealed)	276 (40)	483 (70)	30
Kovar (annealed)	276 (40)	517 (75)	30
Chemical lead	6–8 (0.9–1.2)	16–19 (2.3–2.7)	30–60
Antimonial lead (6%) (chill cast)	—	47.2 (6.8)	24
Tin (commercially pure)	11 (1.6)	—	57
Lead–tin solder (60Sn–40Pb)	—	52.5 (7.6)	30–60
Zinc (commercially pure)			
• Hot-rolled (anisotropic)	—	134–159 (19.4–23.0)	50–65
• Cold-rolled (anisotropic)	—	145–186 (21–27)	40–50
Zirconium, reactor grade 702			
• Cold-worked and annealed	207 (30) (min)	379 (55) (min)	16 (min)
GRAPHITE, CERAMICS, AND SEMICONDUCTING MATERIALS[a]			
Aluminum oxide			
• 99.9% pure	—	282–551 (41–80)	—
• 96% pure	—	358 (52)	—
• 90% pure	—	337 (49)	—
Concrete[b]	—	37.3–41.3 (5.4–6.0)	—
Diamond			
• Natural	—	1050 (152)	—
• Synthetic	—	800–1400 (116–203)	—
Gallium arsenide			
• {100} orientation, polished surface	—	66 (9.6)[c]	—
• {100} orientation, as-cut surface	—	57 (8.3)[c]	—
Glass, borosilicate (Pyrex)	—	69 (10)	—
Glass, soda–lime	—	69 (10)	—
Glass-ceramic (Pyroceram)	—	123–370 (18–54)	—
Graphite			
• Extruded (with the grain direction)	—	13.8–34.5 (2.0–5.0)	—
• Isostatically molded	—	31–69 (4.5–10)	—
Silica, fused	—	104 (15)	—
Silicon			
• {100} orientation, as-cut surface	—	130 (18.9)	—
• {100} orientation, laser scribed	—	81.8 (11.9)	—

Table B.4 *(Continued)*

Material/Condition	*Yield Strength (MPa [ksi])*	*Tensile Strength (MPa [ksi])*	*Percent Elongation*
Silicon carbide			
• Hot-pressed	—	230–825 (33–120)	—
• Sintered	—	96–520 (14–75)	—
Silicon nitride			
• Hot-pressed	—	700–1000 (100–150)	—
• Reaction-bonded	—	250–345 (36–50)	—
• Sintered	—	414–650 (60–94)	—
Zirconia, 3 mol% Y_2O_3 (sintered)	—	800–1500 (116–218)	—
POLYMERS			
Elastomers			
• Butadiene–acrylonitrile (nitrile)	—	6.9–24.1 (1.0–3.5)	400–600
• Styrene–butadiene (SBR)	—	12.4–20.7 (1.8–3.0)	450–500
• Silicone	—	10.3 (1.5)	100–800
Epoxy		27.6–90.0 (4.0–13)	3–6
Nylon 6,6			
• Dry, as molded	55.1–82.8 (8–12)	94.5 (13.7)	15–80
• 50% relative humidity	44.8–58.6 (6.5–8.5)	75.9 (11)	150–300
Phenolic	—	34.5–62.1 (5.0–9.0)	1.5–2.0
Poly(butylene terephthalate) (PBT)	56.6–60.0 (8.2–8.7)	56.6–60.0 (8.2–8.7)	50–300
Polycarbonate (PC)	62.1 (9)	62.8–72.4 (9.1–10.5)	110–150
Polyester (thermoset)	—	41.4–89.7 (6.0–13.0)	<2.6
Polyetheretherketone (PEEK)	91 (13.2)	70.3–103 (10.2–15.0)	30–150
Polyethylene			
• Low density (LDPE)	9.0–14.5 (1.3–2.1)	8.3–31.4 (1.2–4.55)	100–650
• High density (HDPE)	26.2–33.1 (3.8–4.8)	22.1–31.0 (3.2–4.5)	10–1200
• Ultrahigh molecular weight (UHMWPE)	21.4–27.6 (3.1–4.0)	38.6–48.3 (5.6–7.0)	350–525
Poly(ethylene terephthalate) (PET)	59.3 (8.6)	48.3–72.4 (7.0–10.5)	30–300
Poly(methyl methacrylate) (PMMA)	53.8–73.1 (7.8–10.6)	48.3–72.4 (7.0–10.5)	2.0–5.5
Polypropylene (PP)	31.0–37.2 (4.5–5.4)	31.0–41.4 (4.5–6.0)	100–600
Polystyrene (PS)	25.0–69.0 (3.63–10.0)	35.9–51.7 (5.2–7.5)	1.2–2.5
Polytetrafluoroethylene (PTFE)	13.8–15.2 (2.0–2.2)	20.7–34.5 (3.0–5.0)	200–400
Poly(vinyl chloride) (PVC)	40.7–44.8 (5.9–6.5)	40.7–51.7 (5.9–7.5)	40–80
FIBER MATERIALS			
Aramid (Kevlar 49)	—	3600–4100 (525–600)	2.8
Carbon			
• Standard modulus (longitudinal) (PAN precursor)	—	3800–4200 (550–610)	2
• Intermediate modulus (longitudinal) (PAN precursor)	—	4650–6350 (675–920)	1.8
• High modulus (longitudinal) (PAN precursor)	—	2500–4500 (360–650)	0.6
• Ultrahigh modulus (longitudinal) (pitch precursor)	—	2620–3630 (380–526)	0.30–0.66
E-glass	—	3450 (500)	4.3

Table B.4 *(Continued)*

Material/Condition	Yield Strength (MPa [ksi])	Tensile Strength (MPa [ksi])	Percent Elongation
COMPOSITE MATERIALS			
Aramid fibers–epoxy matrix (aligned, V_f = 0.6)			
• Longitudinal direction	—	1380 (200)	1.8
• Transverse direction	—	30 (4.3)	0.5
High-modulus carbon fibers–epoxy matrix (aligned, V_f = 0.6)			
• Longitudinal direction	—	760 (110)	0.3
• Transverse direction	—	28 (4)	0.4
E-glass fibers–epoxy matrix (aligned, V_f = 0.6)			
• Longitudinal direction	—	1020 (150)	2.3
• Transverse direction	—	40 (5.8)	0.4
Wood			
• Douglas fir (12% moisture)			
Parallel to grain	—	108 (15.6)	—
Perpendicular to grain	—	2.4 (0.35)	—
• Red oak (12% moisture)			
Parallel to grain	—	112 (16.3)	—
Perpendicular to grain	—	7.2 (1.05)	—

[a]The strength of graphite, ceramics, and semiconducting materials is taken as flexural strength.
[b]The strength of concrete is measured in compression.
[c]Flexural strength value at 50% fracture probability.

Sources: *ASM Handbooks,* Volumes 1 and 2, *Engineered Materials Handbooks,* Volumes 1 and 4, *Metals Handbook: Properties and Selection: Nonferrous Alloys and Pure Metals,* Vol. 2, 9th edition, *Advanced Materials & Processes,* Vol. 146, No. 4, and *Materials & Processing Databook (1985),* ASM International, Materials Park, OH; *Modern Plastics Encyclopedia '96,* The McGraw-Hill Companies, New York, NY; R. F. Floral and S. T. Peters, "Composite Structures and Technologies," tutorial notes, 1989; and manufacturers' technical data sheets.

Table B.5

Room-Temperature Plane Strain Fracture Toughness and Strength Values for Various Engineering Materials

Material	Fracture Toughness		Strength[a] (MPa)
	MPa$\sqrt{m}$	ksi$\sqrt{in.}$	
METALS AND METAL ALLOYS			
Plain Carbon and Low-Alloy Steels			
Steel alloy 1040	54.0	49.0	260
Steel alloy 4140			
• Tempered @ 370°C	55–65	50–59	1375–1585
• Tempered @ 482°C	75–93	68.3–84.6	1100–1200
Steel alloy 4340			
• Tempered @ 260°C	50.0	45.8	1640
• Tempered @ 425°C	87.4	80.0	1420
Stainless Steels			
Stainless alloy 17-7PH			
• Precipitation hardened @ 510°C	76	69	1310
Aluminum Alloys			
Alloy 2024-T3	44	40	345
Alloy 7075-T651	24	22	495

Table B.5
(Continued)

Material	Fracture Toughness		Strength[a] (MPa)
	MPa$\sqrt{m}$	ksi$\sqrt{in.}$	
Magnesium Alloys			
Alloy AZ31B			
• Extruded	28.0	25.5	200
Titanium Alloys			
Alloy Ti–5Al–2.5Sn			
• Air-cooled	71.4	65.0	876
Alloy Ti–6Al–4V			
• Equiaxed grains	44–66	40–60	910
GRAPHITE, CERAMICS, AND SEMICONDUCTING MATERIALS			
Aluminum oxide			
• 99.9% pure	4.2–5.9	3.8–5.4	282–551
• 96% pure	3.85–3.95	3.5–3.6	358
Concrete	0.2–1.4	0.18–1.27	—
Diamond			
• Natural	3.4	3.1	1050
• Synthetic	6.0–10.7	5.5–9.7	800–1400
Gallium arsenide			
• In the {100} orientation	0.43	0.39	66
• In the {110} orientation	0.31	0.28	—
• In the {111} orientation	0.45	0.41	—
Glass, borosilicate (Pyrex)	0.77	0.70	69
Glass, soda–lime	0.75	0.68	69
Glass-ceramic (Pyroceram)	1.6–2.1	1.5–1.9	123–370
Silica, fused	0.79	0.72	104
Silicon			
• In the {100} orientation	0.95	0.86	—
• In the {110} orientation	0.90	0.82	—
• In the {111} orientation	0.82	0.75	—
Silicon carbide			
• Hot-pressed	4.8–6.1	4.4–5.6	230–825
• Sintered	4.8	4.4	96–520
Silicon nitride			
• Hot-pressed	4.1–6.0	3.7–5.5	700–1000
• Reaction-bonded	3.6	3.3	250–345
• Sintered	5.3	4.8	414–650
Zirconia, 3 mol% Y_2O_3	7.0–12.0	6.4–10.9	800–1500
POLYMERS			
Epoxy	0.6	0.55	—
Nylon 6,6	2.5–3.0	2.3–2.7	44.8–58.6
Polycarbonate (PC)	2.2	2.0	62.1
Polyester (thermoset)	0.6	0.55	—
Poly(ethylene terephthalate) (PET)	5.0	4.6	59.3
Poly(methyl methacrylate) (PMMA)	0.7–1.6	0.6–1.5	53.8–73.1
Polypropylene (PP)	3.0–4.5	2.7–4.1	31.0–37.2
Polystyrene (PS)	0.7–1.1	0.6–1.0	—
Poly(vinyl chloride) (PVC)	2.0–4.0	1.8–3.6	40.7–44.8

[a]For metal alloys and polymers, strength is taken as yield strength; for ceramic materials, flexural strength is used.

Sources: *ASM Handbooks,* Volumes 1 and 19, *Engineered Materials Handbooks,* Volumes 2 and 4, and *Advanced Materials & Processes,* Vol. 137, No. 6, ASM International, Materials Park, OH.

Table B.6

Room-Temperature Linear Coefficient of Thermal Expansion Values for Various Engineering Materials

Material	Coefficient of Thermal Expansion	
	$10^{-6}\ (°C)^{-1}$	$10^{-6}\ (°F)^{-1}$
METALS AND METAL ALLOYS		
Plain Carbon and Low-Alloy Steels		
Steel alloy A36	11.7	6.5
Steel alloy 1020	11.7	6.5
Steel alloy 1040	11.3	6.3
Steel alloy 4140	12.3	6.8
Steel alloy 4340	12.3	6.8
Stainless Steels		
Stainless alloy 304	17.2	9.6
Stainless alloy 316	16.0	8.9
Stainless alloy 405	10.8	6.0
Stainless alloy 440A	10.2	5.7
Stainless alloy 17-7PH	11.0	6.1
Cast Irons		
Gray irons		
• Grade G1800	11.4	6.3
• Grade G3000	11.4	6.3
• Grade G4000	11.4	6.3
Ductile irons		
• Grade 60-40-18	11.2	6.2
• Grade 80-55-06	10.6	5.9
Aluminum Alloys		
Alloy 1100	23.6	13.1
Alloy 2024	22.9	12.7
Alloy 6061	23.6	13.1
Alloy 7075	23.4	13.0
Alloy 356.0	21.5	11.9
Copper Alloys		
C11000 (electrolytic tough pitch)	17.0	9.4
C17200 (beryllium–copper)	16.7	9.3
C26000 (cartridge brass)	19.9	11.1
C36000 (free-cutting brass)	20.5	11.4
C71500 (copper–nickel, 30%)	16.2	9.0
C93200 (bearing bronze)	18.0	10.0
Magnesium Alloys		
Alloy AZ31B	26.0	14.4
Alloy AZ91D	26.0	14.4
Titanium Alloys		
Commercially pure (ASTM grade 1)	8.6	4.8
Alloy Ti–5Al–2.5Sn	9.4	5.2
Alloy Ti–6Al–4V	8.6	4.8

Material	Coefficient of Thermal Expansion	
	10^{-6} $(°C)^{-1}$	10^{-6} $(°F)^{-1}$
Precious Metals		
Gold (commercially pure)	14.2	7.9
Platinum (commercially pure)	9.1	5.1
Silver (commercially pure)	19.7	10.9
Refractory Metals		
Molybdenum (commercially pure)	4.9	2.7
Tantalum (commercially pure)	6.5	3.6
Tungsten (commercially pure)	4.5	2.5
Miscellaneous Nonferrous Alloys		
Nickel 200	13.3	7.4
Inconel 625	12.8	7.1
Monel 400	13.9	7.7
Haynes alloy 25	12.3	6.8
Invar	1.6	0.9
Super invar	0.72	0.40
Kovar	5.1	2.8
Chemical lead	29.3	16.3
Antimonial lead (6%)	27.2	15.1
Tin (commercially pure)	23.8	13.2
Lead–tin solder (60Sn–40Pb)	24.0	13.3
Zinc (commercially pure)	23.0–32.5	12.7–18.1
Zirconium, reactor grade 702	5.9	3.3
GRAPHITE, CERAMICS, AND SEMICONDUCTING MATERIALS		
Aluminum oxide		
• 99.9% pure	7.4	4.1
• 96% pure	7.4	4.1
• 90% pure	7.0	3.9
Concrete	10.0–13.6	5.6–7.6
Diamond (natural)	0.11–1.23	0.06–0.68
Gallium arsenide	5.9	3.3
Glass, borosilicate (Pyrex)	3.3	1.8
Glass, soda–lime	9.0	5.0
Glass-ceramic (Pyroceram)	6.5	3.6
Graphite		
• Extruded	2.0–2.7	1.1–1.5
• Isostatically molded	2.2–6.0	1.2–3.3
Silica, fused	0.4	0.22
Silicon	2.5	1.4
Silicon carbide		
• Hot-pressed	4.6	2.6
• Sintered	4.1	2.3

Table B.6
(Continued)

Material	Coefficient of Thermal Expansion	
	$10^{-6}\,(°C)^{-1}$	$10^{-6}\,(°F)^{-1}$
Silicon nitride		
• Hot-pressed	2.7	1.5
• Reaction-bonded	3.1	1.7
• Sintered	3.1	1.7
Zirconia, 3 mol% Y_2O_3	9.6	5.3
POLYMERS		
Elastomers		
• Butadiene–acrylonitrile (nitrile)	235	130
• Styrene–butadiene (SBR)	220	125
• Silicone	270	150
Epoxy	81–117	45–65
Nylon 6,6	144	80
Phenolic	122	68
Poly(butylene terephthalate) (PBT)	108–171	60–95
Polycarbonate (PC)	122	68
Polyester (thermoset)	100–180	55–100
Polyetheretherketone (PEEK)	72–85	40–47
Polyethylene		
• Low density (LDPE)	180–400	100–220
• High density (HDPE)	106–198	59–110
• Ultrahigh molecular weight (UHMWPE)	234–360	130–200
Poly(ethylene terephthalate) (PET)	117	65
Poly(methyl methacrylate) (PMMA)	90–162	50–90
Polypropylene (PP)	146–180	81–100
Polystyrene (PS)	90–150	50–83
Polytetrafluoroethylene (PTFE)	126–216	70–120
Poly(vinyl chloride) (PVC)	90–180	50–100
FIBER MATERIALS		
Aramid (Kevlar 49)		
• Longitudinal direction	−2.0	−1.1
• Transverse direction	60	33
Carbon		
• Standard modulus (PAN precursor)		
Longitudinal direction	−0.6	−0.3
Transverse direction	10.0	5.6
• Intermediate modulus (PAN precursor)		
Longitudinal direction	−0.6	−0.3
• High modulus (PAN precursor)		
Longitudinal direction	−0.5	−0.28
Transverse direction	7.0	3.9
• Ultrahigh modulus (pitch precursor)		
Longitudinal direction	−1.6	−0.9
Transverse direction	15.0	8.3
E-glass	5.0	2.8

Table B.6

(Continued)

Sources: *ASM Handbooks,* Volumes 1 and 2, *Engineered Materials Handbooks,* Volumes 1 and 4, *Metals Handbook: Properties and Selection: Nonferrous Alloys and Pure Metals,* Vol. 2, 9th edition, and *Advanced Materials & Processes,* Vol. 146, No. 4, ASM International, Materials Park, OH; *Modern Plastics Encyclopedia '96,* The McGraw-Hill Companies, New York, NY; R. F. Floral and S. T. Peters, "Composite Structures and Technologies," tutorial notes, 1989; and manufacturers' technical data sheets.

| | Coefficient of Thermal Expansion | |
Material	10^{-6} $(°C)^{-1}$	10^{-6} $(°F)^{-1}$
COMPOSITE MATERIALS		
Aramid fibers–epoxy matrix ($V_f = 0.6$)		
• Longitudinal direction	−4.0	−2.2
• Transverse direction	70	40
High-modulus carbon fibers–epoxy matrix ($V_f = 0.6$)		
• Longitudinal direction	−0.5	−0.3
• Transverse direction	32	18
E-glass fibers–epoxy matrix ($V_f = 0.6$)		
• Longitudinal direction	6.6	3.7
• Transverse direction	30	16.7
Wood		
• Douglas fir (12% moisture)		
Parallel to grain	3.8–5.1	2.2–2.8
Perpendicular to grain	25.4–33.8	14.1–18.8
• Red oak (12% moisture)		
Parallel to grain	4.6–5.9	2.6–3.3
Perpendicular to grain	30.6–39.1	17.0–21.7

Table B.7

Room-Temperature Thermal Conductivity Values for Various Engineering Materials

| | Thermal Conductivity | |
Material	W/m·K	Btu/ft·h·°F
METALS AND METAL ALLOYS **Plain Carbon and Low-Alloy Steels**		
Steel alloy A36	51.9	30
Steel alloy 1020	51.9	30
Steel alloy 1040	51.9	30
Stainless Steels		
Stainless alloy 304 (annealed)	16.2	9.4
Stainless alloy 316 (annealed)	15.9	9.2
Stainless alloy 405 (annealed)	27.0	15.6
Stainless alloy 440A (annealed)	24.2	14.0
Stainless alloy 17-7PH (annealed)	16.4	9.5
Cast Irons		
Gray irons		
• Grade G1800	46.0	26.6
• Grade G3000	46.0	26.6
• Grade G4000	46.0	26.6
Ductile irons		
• Grade 60-40-18	36.0	20.8
• Grade 80-55-06	36.0	20.8
• Grade 120-90-02	36.0	20.8

Table B.7
(Continued)

Material	Thermal Conductivity	
	W/m·K	Btu/ft·h·°F
Aluminum Alloys		
Alloy 1100 (annealed)	222	128
Alloy 2024 (annealed)	190	110
Alloy 6061 (annealed)	180	104
Alloy 7075-T6	130	75
Alloy 356.0-T6	151	87
Copper Alloys		
C11000 (electrolytic tough pitch)	388	224
C17200 (beryllium–copper)	105–130	60–75
C26000 (cartridge brass)	120	70
C36000 (free-cutting brass)	115	67
C71500 (copper–nickel, 30%)	29	16.8
C93200 (bearing bronze)	59	34
Magnesium Alloys		
Alloy AZ31B	96[a]	55[a]
Alloy AZ91D	72[a]	43[a]
Titanium Alloys		
Commercially pure (ASTM grade 1)	16	9.2
Alloy Ti–5Al–2.5Sn	7.6	4.4
Alloy Ti–6Al–4V	6.7	3.9
Precious Metals		
Gold (commercially pure)	315	182
Platinum (commercially pure)	71[b]	41[b]
Silver (commercially pure)	428	247
Refractory Metals		
Molybdenum (commercially pure)	142	82
Tantalum (commercially pure)	54.4	31.4
Tungsten (commercially pure)	155	89.4
Miscellaneous Nonferrous Alloys		
Nickel 200	70	40.5
Inconel 625	9.8	5.7
Monel 400	21.8	12.6
Haynes alloy 25	9.8	5.7
Invar	10	5.8
Super invar	10	5.8
Kovar	17	9.8
Chemical lead	35	20.2
Antimonial lead (6%)	29	16.8
Tin (commercially pure)	60.7	35.1
Lead–tin solder (60Sn–40Pb)	50	28.9
Zinc (commercially pure)	108	62
Zirconium, reactor grade 702	22	12.7

Material	Thermal Conductivity	
	W/m·K	*Btu/ft·h·°F*
GRAPHITE, CERAMICS, AND SEMICONDUCTING MATERIALS		
Aluminum oxide		
• 99.9% pure	39	22.5
• 96% pure	35	20
• 90% pure	16	9.2
Concrete	1.25–1.75	0.72–1.0
Diamond		
• Natural	1450–4650	840–2700
• Synthetic	3150	1820
Gallium arsenide	45.5	26.3
Glass, borosilicate (Pyrex)	1.4	0.81
Glass, soda–lime	1.7	1.0
Glass-ceramic (Pyroceram)	3.3	1.9
Graphite		
• Extruded	130–190	75–110
• Isostatically molded	104–130	60–75
Silica, fused	1.4	0.81
Silicon	141	82
Silicon carbide		
• Hot-pressed	80	46.2
• Sintered	71	41
Silicon nitride		
• Hot-pressed	29	17
• Reaction-bonded	10	6
• Sintered	33	19.1
Zirconia, 3 mol% Y_2O_3	2.0–3.3	1.2–1.9
POLYMERS		
Elastomers		
• Butadiene-acrylonitrile (nitrile)	0.25	0.14
• Styrene-butadiene (SBR)	0.25	0.14
• Silicone	0.23	0.13
Epoxy	0.19	0.11
Nylon 6,6	0.24	0.14
Phenolic	0.15	0.087
Poly(butylene terephthalate) (PBT)	0.18–0.29	0.10–0.17
Polycarbonate (PC)	0.20	0.12
Polyester (thermoset)	0.17	0.10
Polyethylene		
• Low density (LDPE)	0.33	0.19
• High density (HDPE)	0.48	0.28
• Ultrahigh molecular weight (UHMWPE)	0.33	0.19
Poly(ethylene terephthalate) (PET)	0.15	0.087
Poly(methyl methacrylate) (PMMA)	0.17–0.25	0.10–0.15
Polypropylene (PP)	0.12	0.069
Polystyrene (PS)	0.13	0.075

Table B.7

(Continued)

Sources: *ASM Handbooks,* Volumes 1 and 2, *Engineered Materials Handbooks,* Volumes 1 and 4, *Metals Handbook: Properties and Selection: Nonferrous Alloys and Pure Metals,* Vol. 2, 9th edition, and *Advanced Materials & Processes,* Vol. 146, No. 4, ASM International, Materials Park, OH; *Modern Plastics Encyclopedia '96* and *Modern Plastics Encyclopedia 1977–1978,* The McGraw-Hill Companies, New York, NY; and manufacturers' technical data sheets.

	Thermal Conductivity	
Material	*W/m·K*	*Btu/ft·h·°F*
Polytetrafluoroethylene (PTFE)	0.25	0.14
Poly(vinyl chloride) (PVC)	0.15–0.21	0.08–0.12
FIBER MATERIALS		
Carbon (longitudinal)		
• Standard modulus (PAN precursor)	11	6.4
• Intermediate modulus (PAN precursor)	15	8.7
• High modulus (PAN precursor)	70	40
• Ultrahigh modulus (pitch precursor)	320–600	180–340
E-glass	1.3	0.75
COMPOSITE MATERIALS		
Wood		
• Douglas fir (12% moisture)		
Perpendicular to grain	0.14	0.08
• Red oak (12% moisture)		
Perpendicular to grain	0.18	0.11

[a]At 100°C.
[b]At 0°C.

Table B.8

Room-Temperature Specific Heat Values for Various Engineering Materials

	Specific Heat	
Material	*J/kg·K*	10^{-2} *Btu/lb$_m$·°F*
METALS AND METAL ALLOYS		
Plain Carbon and Low-Alloy Steels		
Steel alloy A36	486[a]	11.6[a]
Steel alloy 1020	486[a]	11.6[a]
Steel alloy 1040	486[a]	11.6[a]
Stainless Steels		
Stainless alloy 304	500	12.0
Stainless alloy 316	502	12.1
Stainless alloy 405	460	11.0
Stainless alloy 440A	460	11.0
Stainless alloy 17-7PH	460	11.0
Cast Irons		
Gray irons		
• Grade G1800	544	13
• Grade G3000	544	13
• Grade G4000	544	13
Ductile irons		
• Grade 60-40-18	544	13
• Grade 80-55-06	544	13
• Grade 120-90-02	544	13
Aluminum Alloys		
Alloy 1100	904	21.6
Alloy 2024	875	20.9
Alloy 6061	896	21.4

Table B.8
(*Continued*)

Material	Specific Heat	
	J/kg·K	*10^{-2} Btu/lb$_m$·°F*
Alloy 7075	960[b]	23.0[b]
Alloy 356.0	963[b]	23.0[b]
Copper Alloys		
C11000 (electrolytic tough pitch)	385	9.2
C17200 (beryllium–copper)	420	10.0
C26000 (cartridge brass)	375	9.0
C36000 (free-cutting brass)	380	9.1
C71500 (copper–nickel, 30%)	380	9.1
C93200 (bearing bronze)	376	9.0
Magnesium Alloys		
Alloy AZ31B	1024	24.5
Alloy AZ91D	1050	25.1
Titanium Alloys		
Commercially pure (ASTM grade 1)	528[c]	12.6[c]
Alloy Ti–5Al–2.5Sn	470[c]	11.2[c]
Alloy Ti–6Al–4V	610[c]	14.6[c]
Precious Metals		
Gold (commercially pure)	128	3.1
Platinum (commercially pure)	132[d]	3.2[d]
Silver (commercially pure)	235	5.6
Refractory Metals		
Molybdenum (commercially pure)	276	6.6
Tantalum (commercially pure)	139	3.3
Tungsten (commercially pure)	138	3.3
Miscellaneous Nonferrous Alloys		
Nickel 200	456	10.9
Inconel 625	410	9.8
Monel 400	427	10.2
Haynes alloy 25	377	9.0
Invar	500	12.0
Super invar	500	12.0
Kovar	460	11.0
Chemical lead	129	3.1
Antimonial lead (6%)	135	3.2
Tin (commercially pure)	222	5.3
Lead–tin solder (60Sn–40Pb)	150	3.6
Zinc (commercially pure)	395	9.4
Zirconium, reactor grade 702	285	6.8

Table B.8
(Continued)

Material	Specific Heat	
	J/kg·K	10^{-2} Btu/lb$_m$·°F
GRAPHITE, CERAMICS, AND SEMICONDUCTING MATERIALS		
Aluminum oxide		
• 99.9% pure	775	18.5
• 96% pure	775	18.5
• 90% pure	775	18.5
Concrete	850–1150	20.3–27.5
Diamond (natural)	520	12.4
Gallium arsenide	350	8.4
Glass, borosilicate (Pyrex)	850	20.3
Glass, soda–lime	840	20.0
Glass-ceramic (Pyroceram)	975	23.3
Graphite		
• Extruded	830	19.8
• Isostatically molded	830	19.8
Silica, fused	740	17.7
Silicon	700	16.7
Silicon carbide		
• Hot-pressed	670	16.0
• Sintered	590	14.1
Silicon nitride		
• Hot-pressed	750	17.9
• Reaction-bonded	870	20.7
• Sintered	1100	26.3
Zirconia, 3 mol% Y_2O_3	481	11.5
POLYMERS		
Epoxy	1050	25
Nylon 6,6	1670	40
Phenolic	1590–1760	38–42
Poly(butylene terephthalate) (PBT)	1170–2300	28–55
Polycarbonate (PC)	840	20
Polyester (thermoset)	710–920	17–22
Polyethylene		
• Low density (LDPE)	2300	55
• High density (HDPE)	1850	44.2
Poly(ethylene terephthalate) (PET)	1170	28
Poly(methyl methacrylate) (PMMA)	1460	35
Polypropylene (PP)	1925	46
Polystyrene (PS)	1170	28
Polytetrafluoroethylene (PTFE)	1050	25
Poly(vinyl chloride) (PVC)	1050–1460	25–35
FIBER MATERIALS		
Aramid (Kevlar 49)	1300	31
E-glass	810	19.3

Table B.8
(Continued)

Material	Specific Heat	
	J/kg·K	*10^{-2} Btu/lb$_m$·°F*
COMPOSITE MATERIALS		
Wood		
• Douglas fir (12% moisture)	2900	69.3
• Red oak (12% moisture)	2900	69.3

[a]At temperatures between 50°C and 100°C.
[b]At 100°C.
[c]At 50°C.
[d]At 0°C.

Sources: *ASM Handbooks,* Volumes 1 and 2, *Engineered Materials Handbooks,* Volumes 1, 2, and 4, *Metals Handbook: Properties and Selection: Nonferrous Alloys and Pure Metals,* Vol. 2, 9th edition, and *Advanced Materials & Processes,* Vol. 146, No. 4, ASM International, Materials Park, OH; *Modern Plastics Encyclopedia 1977–1978,* The McGraw-Hill Companies, New York, NY; and manufacturers' technical data sheets.

Table B.9

Room-Temperature
Electrical Resistivity
Values for Various
Engineering Materials

Material	Electrical Resistivity, $\Omega \cdot m$
METALS AND METAL ALLOYS	
Plain Carbon and Low-Alloy Steels	
Steel alloy A36[a]	1.60×10^{-7}
Steel alloy 1020 (annealed)[a]	1.60×10^{-7}
Steel alloy 1040 (annealed)[a]	1.60×10^{-7}
Steel alloy 4140 (quenched and tempered)	2.20×10^{-7}
Steel alloy 4340 (quenched and tempered)	2.48×10^{-7}
Stainless Steels	
Stainless alloy 304 (annealed)	7.2×10^{-7}
Stainless alloy 316 (annealed)	7.4×10^{-7}
Stainless alloy 405 (annealed)	6.0×10^{-7}
Stainless alloy 440A (annealed)	6.0×10^{-7}
Stainless alloy 17-7PH (annealed)	8.3×10^{-7}
Cast Irons	
Gray irons	
• Grade G1800	15.0×10^{-7}
• Grade G3000	9.5×10^{-7}
• Grade G4000	8.5×10^{-7}
Ductile irons	
• Grade 60-40-18	5.5×10^{-7}
• Grade 80-55-06	6.2×10^{-7}
• Grade 120-90-02	6.2×10^{-7}
Aluminum Alloys	
Alloy 1100 (annealed)	2.9×10^{-8}
Alloy 2024 (annealed)	3.4×10^{-8}
Alloy 6061 (annealed)	3.7×10^{-8}
Alloy 7075 (T6 treatment)	5.22×10^{-8}
Alloy 356.0 (T6 treatment)	4.42×10^{-8}

Table B.9
(Continued)

Material	Electrical Resistivity, $\Omega \cdot m$
Copper Alloys	
C11000 (electrolytic tough pitch, annealed)	1.72×10^{-8}
C17200 (beryllium–copper)	5.7×10^{-8}–1.15×10^{-7}
C26000 (cartridge brass)	6.2×10^{-8}
C36000 (free-cutting brass)	6.6×10^{-8}
C71500 (copper–nickel, 30%)	37.5×10^{-8}
C93200 (bearing bronze)	14.4×10^{-8}
Magnesium Alloys	
Alloy AZ31B	9.2×10^{-8}
Alloy AZ91D	17.0×10^{-8}
Titanium Alloys	
Commercially pure (ASTM grade 1)	4.2×10^{-7}–5.2×10^{-7}
Alloy Ti–5Al–2.5Sn	15.7×10^{-7}
Alloy Ti–6Al–4V	17.1×10^{-7}
Precious Metals	
Gold (commercially pure)	2.35×10^{-8}
Platinum (commercially pure)	10.60×10^{-8}
Silver (commercially pure)	1.47×10^{-8}
Refractory Metals	
Molybdenum (commercially pure)	5.2×10^{-8}
Tantalum (commercially pure)	13.5×10^{-8}
Tungsten (commercially pure)	5.3×10^{-8}
Miscellaneous Nonferrous Alloys	
Nickel 200	0.95×10^{-7}
Inconel 625	12.90×10^{-7}
Monel 400	5.47×10^{-7}
Haynes alloy 25	8.9×10^{-7}
Invar	8.2×10^{-7}
Super invar	8.0×10^{-7}
Kovar	4.9×10^{-7}
Chemical lead	2.06×10^{-7}
Antimonial lead (6%)	2.53×10^{-7}
Tin (commercially pure)	1.11×10^{-7}
Lead–tin solder (60Sn–40Pb)	1.50×10^{-7}
Zinc (commercially pure)	62.0×10^{-7}
Zirconium, reactor grade 702	3.97×10^{-7}
GRAPHITE, CERAMICS, AND SEMICONDUCTING MATERIALS	
Aluminum oxide	
• 99.9% pure	$>10^{13}$
• 96% pure	$>10^{12}$
• 90% pure	$>10^{12}$

Material	Electrical Resistivity, $\Omega \cdot m$
Concrete (dry)	10^9
Diamond	
• Natural	10–10^{14}
• Synthetic	1.5×10^{-2}
Gallium arsenide (intrinsic)	10^6
Glass, borosilicate (Pyrex)	~10^{13}
Glass, soda–lime	10^{10}–10^{11}
Glass-ceramic (Pyroceram)	2×10^{14}
Graphite	
• Extruded (with grain direction)	7×10^{-6}–20×10^{-6}
• Isostatically molded	10×10^{-6}–18×10^{-6}
Silica, fused	$>10^{18}$
Silicon (intrinsic)	2500
Silicon carbide	
• Hot-pressed	1.0–10^9
• Sintered	1.0–10^9
Silicon nitride	
• Hot isostatic pressed	$>10^{12}$
• Reaction-bonded	$>10^{12}$
• Sintered	$>10^{12}$
Zirconia, 3 mol% Y_2O_3	10^{10}
POLYMERS	
Elastomers	
• Butadiene–acrylonitrile (nitrile)	3.5×10^8
• Styrene–butadiene (SBR)	6×10^{11}
• Silicone	10^{13}
Epoxy	10^{10}–10^{13}
Nylon 6,6	10^{12}–10^{13}
Phenolic	10^9–10^{10}
Poly(butylene terephthalate) (PBT)	4×10^{14}
Polycarbonate (PC)	2×10^{14}
Polyester (thermoset)	10^{13}
Polyetheretherketone (PEEK)	6×10^{14}
Polyethylene	
• Low density (LDPE)	10^{15}–5×10^{16}
• High density (HDPE)	10^{15}–5×10^{16}
• Ultrahigh molecular weight (UHMWPE)	$>5 \times 10^{14}$
Poly(ethylene terephthalate) (PET)	10^{12}
Poly(methyl methacrylate) (PMMA)	$>10^{12}$
Polypropylene (PP)	$>10^{14}$
Polystyrene (PS)	$>10^{14}$
Polytetrafluoroethylene (PTFE)	10^{17}
Poly(vinyl chloride) (PVC)	$>10^{14}$

Table B.9
(Continued)

Material	Electrical Resistivity, $\Omega \cdot m$
FIBER MATERIALS	
Carbon	
• Standard modulus (PAN precursor)	17×10^{-6}
• Intermediate modulus (PAN precursor)	15×10^{-6}
• High modulus (PAN precursor)	9.5×10^{-6}
• Ultrahigh modulus (pitch precursor)	$1.35 \times 10^{-6} – 5 \times 10^{-6}$
E-glass	4×10^{14}
COMPOSITE MATERIALS	
Wood	
• Douglas fir (oven dry)	
Parallel to grain	$10^{14} – 10^{16}$
Perpendicular to grain	$10^{14} – 10^{16}$
• Red oak (oven dry)	
Parallel to grain	$10^{14} – 10^{16}$
Perpendicular to grain	$10^{14} – 10^{16}$

[a]At 0°C.

Sources: *ASM Handbooks,* Volumes 1 and 2, *Engineered Materials Handbooks,* Volumes 1, 2, and 4, *Metals Handbook: Properties and Selection: Nonferrous Alloys and Pure Metals,* Vol. 2, 9th edition, and *Advanced Materials & Processes,* Vol. 146, No. 4, ASM International, Materials Park, OH; *Modern Plastics Encyclopedia 1977–1978,* The McGraw-Hill Companies, New York, NY; and manufacturers' technical data sheets.

Table B.10 Compositions of Metal Alloys for Which Data Are Included in Tables B.1 through B.9

Alloy (UNS Designation)	Composition (wt%)
PLAIN CARBON AND LOW-ALLOY STEELS	
A36 (ASTM A36)	98.0 Fe (min), 0.29 C, 1.0 Mn, 0.28 Si
1020 (G10200)	99.1 Fe (min), 0.20 C, 0.45 Mn
1040 (G10400)	98.6 Fe (min), 0.40 C, 0.75 Mn
4140 (G41400)	96.8 Fe (min), 0.40 C, 0.90 Cr, 0.20 Mo, 0.9 Mn
4340 (G43400)	95.2 Fe (min), 0.40 C, 1.8 Ni, 0.80 Cr, 0.25 Mo, 0.7 Mn
STAINLESS STEELS	
304 (S30400)	66.4 Fe (min), 0.08 C, 19.0 Cr, 9.25 Ni, 2.0 Mn
316 (S31600)	61.9 Fe (min), 0.08 C, 17.0 Cr, 12.0 Ni, 2.5 Mo, 2.0 Mn
405 (S40500)	83.1 Fe (min), 0.08 C, 13.0 Cr, 0.20 Al, 1.0 Mn
440A (S44002)	78.4 Fe (min), 0.70 C, 17.0 Cr, 0.75 Mo, 1.0 Mn
17-7PH (S17700)	70.6 Fe (min), 0.09 C, 17.0 Cr, 7.1 Ni, 1.1 Al, 1.0 Mn
CAST IRONS	
Grade G1800 (F10004)	Fe (bal), 3.4–3.7 C, 2.8–2.3 Si, 0.65 Mn, 0.15 P, 0.15 S
Grade G3000 (F10006)	Fe (bal), 3.1–3.4 C, 2.3–1.9 Si, 0.75 Mn, 0.10 P, 0.15 S
Grade G4000 (F10008)	Fe (bal), 3.0–3.3 C, 2.1–1.8 Si, 0.85 Mn, 0.07 P, 0.15 S
Grade 60-40-18 (F32800)	Fe (bal), 3.4–4.0 C, 2.0–2.8 Si, 0–1.0 Ni, 0.05 Mg
Grade 80-55-06 (F33800)	Fe (bal), 3.3–3.8 C, 2.0–3.0 Si, 0–1.0 Ni, 0.05 Mg
Grade 120-90-02 (F36200)	Fe (bal), 3.4–3.8 C, 2.0–2.8 Si, 0–2.5 Ni, 0–1.0 Mo, 0.05 Mg

Table B.10 *(Continued)*

Alloy (*UNS Designation*)	Composition (*wt%*)
ALUMINUM ALLOYS	
1100 (A91100)	99.00 Al (min), 0.20 Cu (max)
2024 (A92024)	90.75 Al (min), 4.4 Cu, 0.6 Mn, 1.5 Mg
6061 (A96061)	95.85 Al (min), 1.0 Mg, 0.6 Si, 0.30 Cu, 0.20 Cr
7075 (A97075)	87.2 Al (min), 5.6 Zn, 2.5 Mg, 1.6 Cu, 0.23 Cr
356.0 (A03560)	90.1 Al (min), 7.0 Si, 0.3 Mg
COPPER ALLOYS	
(C11000)	99.90 Cu (min), 0.04 O (max)
(C17200)	96.7 Cu (min), 1.9 Be, 0.20 Co
(C26000)	Zn (bal), 70 Cu, 0.07 Pb, 0.05 Fe (max)
(C36000)	60.0 Cu (min), 35.5 Zn, 3.0 Pb
(C71500)	63.75 Cu (min), 30.0 Ni
(C93200)	81.0 Cu (min), 7.0 Sn, 7.0 Pb, 3.0 Zn
MAGNESIUM ALLOYS	
AZ31B (M11311)	94.4 Mg (min), 3.0 Al, 0.20 Mn (min), 1.0 Zn, 0.1 Si (max)
AZ91D (M11916)	89.0 Mg (min), 9.0 Al, 0.13 Mn (min), 0.7 Zn, 0.1 Si (max)
TITANIUM ALLOYS	
Commercial, grade 1 (R50250)	99.5 Ti (min)
Ti–5Al–2.5Sn (R54520)	90.2 Ti (min), 5.0 Al, 2.5 Sn
Ti–6Al–4V (R56400)	87.7 Ti (min), 6.0 Al, 4.0 V
MISCELLANEOUS ALLOYS	
Nickel 200	99.0 Ni (min)
Inconel 625	58.0 Ni (min), 21.5 Cr, 9.0 Mo, 5.0 Fe, 3.65 Nb + Ta, 1.0 Co
Monel 400	63.0 Ni (min), 31.0 Cu, 2.5 Fe, 0.2 Mn, 0.3 C, 0.5 Si
Haynes alloy 25	49.4 Co (min), 20 Cr, 15 W, 10 Ni, 3 Fe (max), 0.10 C, 1.5 Mn
Invar (K93601)	64 Fe, 36 Ni
Super invar	63 Fe, 32 Ni, 5 Co
Kovar	54 Fe, 29 Ni, 17 Co
Chemical lead (L51120)	99.90 Pb (min)
Antimonial lead, 6% (L53105)	94 Pb, 6 Sb
Tin (commercially pure) (ASTM B339A)	98.85 Pb (min)
Lead–tin solder (60Sn–40Pb) (ASTM B32 grade 60)	60 Sn, 40 Pb
Zinc (commercially pure) (Z21210)	99.9 Zn (min), 0.10 Pb (max)
Zirconium, reactor grade 702 (R60702)	99.2 Zr + Hf (min), 4.5 Hf (max), 0.2 Fe + Cr

Sources: *ASM Handbooks,* Volumes 1 and 2, ASM International, Materials Park, OH.

This appendix contains price information for the set of materials for which Appendix B gives the properties. The collection of valid cost data for materials is an extremely difficult task, which explains the dearth of materials pricing information in the literature. One reason for this is that there are three pricing tiers: manufacturer, distributor, and retail. Under most circumstances, we cite distributor prices. For some materials (e.g., specialized ceramics, such as silicon carbide and silicon nitride), it is necessary to use manufacturers' prices. In addition, there may be significant variation in the cost for a specific material. There are several reasons for this. First, each vendor has its own pricing scheme. Furthermore, cost depends on quantity of material purchased and, in addition, how it was processed or treated. We endeavored to collect data for relatively large orders—that is, quantities on the order of 900 kg (2000 lb$_m$) for materials that are typically sold in bulk lots—and also for common shapes/treatments. When possible, we obtained price quotes from at least three distributors/manufacturers.

This pricing information was collected in January 2007. Cost data are in U.S. dollars per kilogram; in addition, these data are expressed as both price ranges and single-price values. The absence of a price range (i.e., when a single value is cited) means either that the variation is small or that, on the basis of limited data, it is not possible to identify a range of prices. Furthermore, inasmuch as material prices change over time, it was decided to use a relative cost index; this index represents the per-unit-mass cost (or average per-unit-mass cost) of a material divided by the average per-unit-mass cost of a common engineering material—A36 plain carbon steel. Although the price of a specific material varies over time, the price ratio between that material and another will, most likely, change more slowly.

Material/Condition	Cost ($US/kg)	Relative Cost
PLAIN CARBON AND LOW-ALLOY STEELS		
Steel alloy A36		
• Plate, hot-rolled	0.90–1.50	1.00
• Angle bar, hot-rolled	1.00–1.65	1.0
Steel alloy 1020		
• Plate, hot-rolled	0.90–1.65	1.0
• Plate, cold-rolled	0.85–1.40	0.9
Steel alloy 1040		
• Plate, hot-rolled	0.90–0.95	0.7
• Plate, cold-rolled	2.20	1.7
Steel alloy 4140		
• Bar, normalized	1.50–2.60	1.6
• H grade (round), normalized	5.00	3.9
Steel alloy 4340		
• Bar, annealed	2.55	2.0
• Bar, normalized	3.60	2.8

Material/Condition	Cost ($US/kg)	Relative Cost
STAINLESS STEELS		
Stainless alloy 304	6.20–9.20	6.0
Stainless alloy 316	6.20–11.70	7.3
Stainless alloy 17-7PH	9.20	7.1
CAST IRONS		
Gray irons (all grades)	1.75–2.40	1.7
Ductile irons (all grades)	2.00–3.20	2.0
ALUMINUM ALLOYS		
Aluminum (unalloyed)	2.65–2.75	2.1
Alloy 1100		
• Sheet, annealed	5.30–5.50	4.2
Alloy 2024		
• Sheet, T3 temper	12.50–19.50	12.9
• Bar, T351 temper	11.00–21.00	13.4
Alloy 5052		
• Sheet, H32 temper	4.85–5.10	3.9
Alloy 6061		
• Sheet, T6 temper	6.60–8.50	5.7
• Bar, T651 temper	5.10–7.50	5.0
Alloy 7075		
• Sheet, T6 temper	11.30–14.70	10.0
Alloy 356.0		
• As cast, high production	2.70–3.35	2.4
• As cast, custom pieces	17.50	13.6
• T6 temper, custom pieces	18.90	14.7
COPPER ALLOYS		
Copper (unalloyed)	5.60–7.00	4.8
Alloy C11000 (electrolytic tough pitch), sheet	7.60–11.60	7.4
Alloy C17200 (beryllium–copper), sheet	9.00–36.00	17.5
Alloy C26000 (cartridge brass), sheet	7.10–12.80	7.5
Alloy C36000 (free-cutting brass), sheet, rod	7.20–10.90	7.0
Alloy C71500 (copper–nickel, 30%), sheet	27.00	21.0
Alloy C93200 (bearing bronze)		
• Bar	9.70	7.5
• As cast, custom piece	23.00	17.9
MAGNESIUM ALLOYS		
Magnesium (unalloyed)	3.00–3.30	2.4
Alloy AZ31B		
• Sheet (rolled)	17.60–46.00	23.4
• Extruded	9.90–14.30	9.4
Alloy AZ91D (as cast)	3.40	2.6
TITANIUM ALLOYS		
Commercially pure		
• ASTM grade 1, annealed	100.00–120.00	85.6
• ASTM grade 2, annealed	90.00–160.00	95.9

Material/Condition	Cost ($US/kg)	Relative Cost
Alloy Ti–5Al–2.5Sn	110.00–120.00	89.3
Alloy Ti–6Al–4V	66.00–154.00	94.2
PRECIOUS METALS		
Gold, bullion	18,600–20,900	15,300
Platinum, bullion	32,100–40,000	28,400
Silver, bullion	350–450	313
REFRACTORY METALS		
Molybdenum, commercial purity	180–300	161
Tantalum, commercial purity	400–420	318
Tungsten, commercial purity	225	175
MISCELLANEOUS NONFERROUS ALLOYS		
Nickel, commercial purity	25.00–34.50	23.7
Nickel 200	35.00–74.00	46.8
Inconel 625	59.00–88.00	55.5
Monel 400	15.00–33.00	16.8
Haynes alloy 25	143.00–165.00	120
Invar	44.00–54.00	37.2
Super invar	44.00	34.2
Kovar	50.00–66.00	44.3
Chemical lead		
• Ingot	1.50–2.00	1.4
• Plate	2.15–4.40	2.5
Antimonial lead (6%)		
• Ingot	2.30–3.90	2.4
• Plate	3.10–6.10	3.4
Tin, commercial purity	9.75–10.75	8.0
Solder (60Sn–40Pb), bar	8.10–16.50	9.4
Zinc, commercial purity, ingot or anode	2.00–4.65	2.8
Zirconium, reactor grade 702, plate	46.00–88.00	52.2
GRAPHITE, CERAMICS, AND SEMICONDUCTING MATERIALS		
Aluminum oxide		
• Calcined powder, 99.8% pure, particle size between 0.4 and 5 μm	1.85–2.80	1.8
• Ball grinding media, 99% pure, $\frac{1}{4}$ in. dia.	39.00–52.00	35.1
• Ball grinding media, 96% pure, $\frac{1}{4}$ in. dia.	33.00	25.6
• Ball grinding media, 90% pure, $\frac{1}{4}$ in. dia.	16.00	12.4
Concrete, mixed	0.05	0.04
Diamond		
• Synthetic, 30–40 mesh, industrial grade	7700	6000
• Natural, powder, 45 μm, polishing abrasive	2300	1800
• Natural, industrial, $\frac{1}{3}$ carat	50,000–85,000	52,400
Gallium arsenide		
• Mechanical grade, 75-mm-diameter wafers, ~625 μm thick	3900	3000
• Prime grade, 75-mm-diameter wafers, ~625 μm thick	6500	5000
Glass, borosilicate (Pyrex), plate	9.20–11.30	7.9

Material/Condition	Cost ($US/kg)	Relative Cost
Glass, soda–lime, plate	0.56–1.35	0.7
Glass-ceramic (Pyroceram), plate	12.65–16.55	11.3
Graphite		
• Powder, synthetic, 99+% pure, particle size ~10 μm	1.80–7.00	3.1
• Isostatically pressed parts, high purity, particle size ~20 μm	50.00–125.00	65.3
Silica, fused, plate	1200–1700	1100
Silicon		
• Test grade, undoped, 100-mm-diameter wafers, ~425 μm thick	5100–9000	5500
• Prime grade, undoped, 100-mm-diameter wafers, ~425 μm thick	8000–14,000	8800
Silicon carbide		
• α-phase ball grinding media, $\frac{1}{4}$ in. diameter, sintered	250.00	194
Silicon nitride		
• Powder, submicron particle size	100–200	100
• Balls, finished ground, 0.25–0.50 in. diameter, hot isostatically pressed	1000–4000	1600
Zirconia (5 mol% Y_2O_3), 15-mm-diameter ball grinding media	50–200	97.1
POLYMERS		
Butadiene-acrylonitrile (nitrile) rubber		
• Raw and unprocessed	4.00	3.1
• Extruded sheet ($\frac{1}{4}$–$\frac{1}{8}$ in. thick)	8.25	6.4
• Calendered sheet ($\frac{1}{4}$–$\frac{1}{8}$ in. thick)	5.25–7.40	4.9
Styrene-butadiene (SBR) rubber		
• Raw and unprocessed	1.70	1.3
• Extruded sheet ($\frac{1}{4}$–$\frac{1}{8}$ in. thick)	5.05	3.9
• Calendered sheet ($\frac{1}{4}$–$\frac{1}{8}$ in. thick)	3.25–3.75	2.7
Silicone rubber		
• Raw and unprocessed	9.90–14.00	9.5
• Extruded sheet ($\frac{1}{4}$–$\frac{1}{8}$ in. thick)	28.00–29.50	22.4
• Calendered sheet ($\frac{1}{4}$–$\frac{1}{8}$ in. thick)	7.75–12.00	7.7
Epoxy resin, raw form	2.20–2.80	1.9
Nylon 6,6		
• Raw form	3.20–4.00	2.8
• Extruded	12.80	9.9
Phenolic resin, raw form	1.65–1.90	1.4
Poly(butylene terephthalate) (PBT)		
• Raw form	4.00–7.00	4.3
• Sheet	40.00–100.00	54.3
Polycarbonate (PC)		
• Raw form	3.00–4.70	2.9
• Sheet	10.50	8.2
Polyester (thermoset), raw form	3.10–4.30	2.7
Polyetheretherketone (PEEK), raw form	90.00–105.00	76.0

Material/Condition	Cost ($US/kg)	Relative Cost
Polyethylene		
• Low density (LDPE), raw form	1.60–1.85	1.3
• High density (HDPE), raw form	1.20–1.75	1.2
• Ultrahigh molecular weight (UHMWPE), raw form	2.20–3.00	2.1
Poly(ethylene terephthalate) (PET)		
• Raw form	1.50–1.75	1.3
• Sheet	3.30–5.40	3.4
Poly(methyl methacrylate) (PMMA)		3.1
• Raw form	2.60–5.40	3.1
• Extruded sheet ($\frac{1}{4}$ in. thick)	4.65–6.05	4.1
Polypropylene (PP), raw form	1.05–1.70	1.2
Polystyrene (PS), raw form	1.55–1.95	1.4
Polytetrafluoroethylene (PTFE)		
• Raw form	14.80–16.90	11.9
• Rod	21.00	16.3
Poly(vinyl chloride) (PVC), raw form	1.10–1.85	1.2
FIBER MATERIALS		
Aramid (Kevlar 49), continuous	35.00–100.00	38.8
Carbon (PAN precursor), continuous		
• Standard modulus	40.00–80.00	48.1
• Intermediate modulus	60.00–130.00	69.1
• High modulus	220.00–275.00	193
• Ultrahigh modulus	1750–2650	1700
E-glass, continuous	1.55–2.65	1.6
COMPOSITE MATERIALS		
Aramid (Kevlar 49) continuous-fiber, epoxy prepreg	75.00–100.00	66.8
Carbon continuous-fiber, epoxy prepreg		
• Standard modulus	49.00–66.00	43.1
• Intermediate modulus	75.00–240.00	123
• High modulus	120.00–725.00	330
E-glass continuous-fiber, epoxy prepreg	24.00–50.00	28.3
Woods		
• Douglas fir	0.61–0.97	0.6
• Ponderosa pine	1.15–1.50	1.0
• Red oak	3.35–3.75	2.8

Chemical Name	Repeat Unit Structure
Epoxy (diglycidyl ether of bisphenol A, DGEPA)	
Melamine-formaldehyde (melamine)	
Phenol-formaldehyde (phenolic)	
Polyacrylonitrile (PAN)	
Poly(amide-imide) (PAI)	

Chemical Name	*Repeat Unit Structure*
Polybutadiene	
Poly(butylene terephthalate) (PBT)	
Polycarbonate (PC)	
Polychloroprene	
Polychlorotrifluoroethylene	
Poly(dimethyl siloxane) (silicone rubber)	
Polyetheretherketone (PEEK)	
Polyethylene (PE)	
Poly(ethylene terephthalate) (PET)	
Poly(hexamethylene adipamide) (nylon 6,6)	

Chemical Name	Repeat Unit Structure
Polyimide	
Polyisobutylene	
cis-Polyisoprene (natural rubber)	
Poly(methyl methacrylate) (PMMA)	
Poly(phenylene oxide) (PPO)	
Poly(phenylene sulfide) (PPS)	
Poly(paraphenylene terephthalamide) (aramid)	
Polypropylene (PP)	

Chemical Name	*Repeat Unit Structure*
Polystyrene (PS)	$\begin{bmatrix} \overset{\displaystyle H}{\underset{\displaystyle H}{-C}} - \overset{\displaystyle H}{\underset{\displaystyle \bigcirc}{C}} - \end{bmatrix}$
Polytetrafluoroethylene (PTFE)	$\begin{bmatrix} \overset{\displaystyle F}{\underset{\displaystyle F}{-C}} - \overset{\displaystyle F}{\underset{\displaystyle F}{C}} - \end{bmatrix}$
Poly(vinyl acetate) (PVAc)	$\begin{bmatrix} & & \overset{O}{\diagup}C\overset{CH_3}{\diagdown} \\ & H & O \\ -C & - & C- \\ & H & H \end{bmatrix}$
Poly(vinyl alcohol) (PVA)	$\begin{bmatrix} \overset{\displaystyle H}{\underset{\displaystyle H}{-C}} - \overset{\displaystyle H}{\underset{\displaystyle OH}{C}} - \end{bmatrix}$
Poly(vinyl chloride) (PVC)	$\begin{bmatrix} \overset{\displaystyle H}{\underset{\displaystyle H}{-C}} - \overset{\displaystyle H}{\underset{\displaystyle Cl}{C}} - \end{bmatrix}$
Poly(vinyl fluoride) (PVF)	$\begin{bmatrix} \overset{\displaystyle H}{\underset{\displaystyle H}{-C}} - \overset{\displaystyle H}{\underset{\displaystyle F}{C}} - \end{bmatrix}$
Poly(vinylidene chloride) (PVDC)	$\begin{bmatrix} \overset{\displaystyle H}{\underset{\displaystyle H}{-C}} - \overset{\displaystyle Cl}{\underset{\displaystyle Cl}{C}} - \end{bmatrix}$
Poly(vinylidene fluoride) (PVDF)	$\begin{bmatrix} \overset{\displaystyle H}{\underset{\displaystyle H}{-C}} - \overset{\displaystyle F}{\underset{\displaystyle F}{C}} - \end{bmatrix}$

Polymer	Glass Transition Temperature [°C (°F)]	Melting Temperature [°C (°F)]
Aramid	375 (705)	~640 (~1185)
Polyimide (thermoplastic)	280–330 (535–625)	a
Poly(amide-imide)	277–289 (530–550)	a
Polycarbonate	150 (300)	265 (510)
Polyetheretherketone	143 (290)	334 (635)
Polyacrylonitrile	104 (220)	317 (600)
Polystyrene		
• Atactic	100 (212)	a
• Isotactic	100 (212)	240 (465)
Poly(butylene terephthalate)	—	220–267 (428–513)
Poly(vinyl chloride)	87 (190)	212 (415)
Poly(phenylene sulfide)	85 (185)	285 (545)
Poly(ethylene terephthalate)	69 (155)	265 (510)
Nylon 6,6	57 (135)	265 (510)
Poly(methyl methacrylate)		
• Syndiotactic	3 (35)	105 (220)
• Isotactic	3 (35)	45 (115)
Polypropylene		
• Isotactic	−10 (15)	175 (347)
• Atactic	−18 (0)	175 (347)
Poly(vinylidene chloride)		
• Atactic	−18 (0)	175 (347)
Poly(vinyl fluoride)	−20 (−5)	200 (390)
Poly(vinylidene fluoride)	−35 (−30)	—
Polychloroprene (chloroprene rubber or neoprene)	−50 (−60)	80 (175)
Polyisobutylene	−70 (−95)	128 (260)
cis-Polyisoprene	−73 (−100)	28 (80)
Polybutadiene		
• Syndiotactic	−90 (−130)	154 (310)
• Isotactic	−90 (−130)	120 (250)
High-density polyethylene	−90 (−130)	137 (279)
Polytetrafluoroethylene	−97 (−140)	327 (620)
Low-density polyethylene	−110 (−165)	115 (240)
Poly(dimethyl siloxane) (silicone rubber)	−123 (−190)	−54 (−65)

[a]These polymers normally exist as at least 95% noncrystalline.

Glossary

A

abrasive. A hard and wear-resistant material (commonly a ceramic) that is used to wear, grind, or cut away other material.

absorption. The optical phenomenon by which the energy of a photon of light is assimilated within a substance, normally by electronic polarization or by an electron excitation event.

acceptor state (level). For a semiconductor or insulator, an energy level lying within yet near the bottom of the energy band gap that may accept electrons from the valence band, leaving behind holes. The level is normally introduced by an impurity atom.

activation energy (Q). The energy required to initiate a reaction, such as diffusion.

activation polarization. The condition in which the rate of an electrochemical reaction is controlled by the slowest step in a sequence of steps that occur in series.

addition (or chain reaction) polymerization. The process by which monomer units are attached one at a time, in chainlike fashion, to form a linear polymer macromolecule.

adhesive. A substance that bonds together the surfaces of two other materials (termed *adherends*).

age hardening. See **precipitation hardening**.

allotropy. The possibility of the existence of two or more different crystal structures for a substance (generally an elemental solid).

alloy. A metallic substance that is composed of two or more elements.

alloy steel. A ferrous (or iron-based) alloy that contains appreciable concentrations of alloying elements (other than C and residual amounts of Mn, Si, S, and P). These alloying elements are usually added to improve mechanical and corrosion-resistance properties.

alternating copolymer. A copolymer in which two different repeat units alternate positions along the molecular chain.

amorphous. Having a noncrystalline structure.

anelastic deformation. Time-dependent elastic (nonpermanent) deformation.

anion. A negatively charged ion.

anisotropic. Exhibiting different values of a property in different crystallographic directions.

annealing. A generic term used to denote a heat treatment in which the microstructure and, consequently, the properties of a material are altered. *Annealing* frequently refers to a heat treatment whereby a previously cold-worked metal is softened by allowing it to recrystallize.

annealing point (glass). The temperature at which residual stresses in a glass are eliminated within about 15 min; this corresponds to a glass viscosity of about 10^{12} Pa·s.

anode. The electrode in an electrochemical cell or galvanic couple that experiences oxidation, or gives up electrons.

antiferromagnetism. A phenomenon observed in some materials (e.g., MnO): complete magnetic moment cancellation occurs as a result of antiparallel coupling of adjacent atoms or ions. The macroscopic solid possesses no net magnetic moment.

artificial aging. For precipitation hardening, aging above room temperature.

atactic. A type of polymer chain configuration (stereoisomer) in which side groups are randomly positioned on one side of the chain or the other.

athermal transformation. A reaction that is not thermally activated, and usually diffusionless, as with the martensitic transformation. Normally, the transformation takes place with great speed (i.e., is independent of time), and the extent of reaction depends on temperature.

atomic mass unit (amu). A measure of atomic mass; 1/12 of the mass of an atom of C^{12}.

atomic number (Z). For a chemical element, the number of protons within the atomic nucleus.

atomic packing factor (APF). The fraction of the volume of a unit cell that is occupied by *hard-sphere* atoms or ions.

atomic vibration. The vibration of an atom about its normal position in a substance.

atomic weight (A). The weighted average of the atomic masses of an atom's naturally occurring isotopes. It may be expressed in terms of atomic mass units (on an atomic basis), or the mass per mole of atoms.

atom percent (at%). A concentration specification on the basis of the number of moles (or atoms) of a particular element relative to the total number of moles (or atoms) of all elements within an alloy.

austenite. Face-centered cubic iron; also iron and steel alloys that have the FCC crystal structure.

austenitizing. Forming austenite by heating a ferrous alloy above its upper critical temperature—to within the austenite phase region from the phase diagram.

B

bainite. An austenitic transformation product found in some steels and cast irons. It forms at temperatures between those at which pearlite and martensite transformations occur. The microstructure consists of α-ferrite and a fine dispersion of cementite.

band gap energy (E_g). For semiconductors and insulators, the energies that lie between the valence and conduction bands; for intrinsic materials, electrons are forbidden to have energies within this range.

bifunctional. Designates monomers that may react to form two covalent bonds with other monomers to create a two-dimensional chainlike molecular structure.

block copolymer. A linear copolymer in which identical repeat units are clustered in blocks along the molecular chain.

body-centered cubic (BCC). A common crystal structure found in some elemental metals. Within the cubic unit cell, atoms are located at corner and cell center positions.

Bohr atomic model. An early atomic model in which electrons are assumed to revolve around the nucleus in discrete orbitals.

Bohr magneton (μ_B). The most fundamental magnetic moment, of magnitude 9.27×10^{-24} A·m^2.

Boltzmann's constant (k). A thermal energy constant having the value of 1.38×10^{-23} J/atom·K. See also **gas constant (R)**.

bonding energy. The energy required to separate two atoms that are chemically bonded to each other. It may be expressed on a per-atom basis or per mole of atoms.

Bragg's law. A relationship (Equation 4.16) that stipulates the condition for diffraction by a set of crystallographic planes.

branched polymer. A polymer having a molecular structure of secondary chains that extend from the primary main chains.

brass. A copper-rich copper–zinc alloy.

brazing. A metal-joining technique that uses a molten filler metal alloy having a melting temperature greater than about 425°C.

brittle fracture. Fracture that occurs by rapid crack propagation and without appreciable macroscopic deformation.

bronze. A copper-rich copper–tin alloy; aluminum, silicon, and nickel bronzes are also possible.

Burgers vector (b). A vector that denotes the magnitude and direction of lattice distortion associated with a dislocation.

C

calcination. A high-temperature reaction by which one solid material dissociates to form a gas and another solid. It is one step in the production of cement.

capacitance (C). The charge-storing ability of a capacitor, defined as the magnitude of charge stored on either plate divided by the applied voltage.

carbon–carbon composite. A composite composed of continuous fibers of carbon that are embedded in a carbon matrix. The matrix was originally a polymer resin that was subsequently pyrolyzed to form carbon.

carburizing. The process by which the surface carbon concentration of a ferrous alloy is increased by diffusion from the surrounding environment.

case hardening. Hardening of the outer surface (or *case*) of a steel component by a carburizing or nitriding process; used to improve wear and fatigue resistance.

cast iron. Generically, a ferrous alloy, the carbon content of which is greater than the maximum solubility in austenite at the eutectic temperature. Most commercial cast irons contain between 3.0 and 4.5 wt% C and between 1 and 3 wt% Si.

cathode. The electrode in an electrochemical cell or galvanic couple at which a reduction reaction occurs; thus the electrode that receives electrons from an external circuit.

cathodic protection. A means of corrosion prevention by which electrons are supplied to the structure to be protected from an external source such as another, more reactive metal or a dc power supply.

cation. A positively charged ion.

cement. A substance (often a ceramic) that by chemical reaction binds particulate aggregates into a cohesive structure. With hydraulic cements the chemical reaction is one of hydration, involving water.

cementite. Iron carbide (Fe$_3$C).

ceramic. A compound of metallic and nonmetallic elements, for which the interatomic bonding is predominantly ionic.

ceramic-matrix composite (CMC). A composite for which both matrix and dispersed phases are ceramic materials. The dispersed phase is normally added to improve fracture toughness.

cermet. A composite material consisting of a combination of ceramic and metallic materials. The most common cermets are the cemented carbides, composed of an extremely hard ceramic (e.g., WC, TiC), bonded together by a ductile metal such as cobalt or nickel.

chain-folded model. For crystalline polymers, a model that describes the structure of platelet crystallites. Molecular alignment is accomplished by chain folding that occurs at the crystallite faces.

Charpy test. One of two tests (see also **Izod test**) that may be used to measure the impact energy or notch toughness of a standard notched specimen. An impact blow is imparted to the specimen by means of a weighted pendulum.

cis. For polymers, a prefix denoting a type of molecular structure. For some unsaturated carbon chain atoms within a repeat unit, a side atom or group may be situated on one side of the double bond or directly opposite at a 180° rotation position. In a cis structure, two such side groups within the same repeat unit reside on the same side (e.g., *cis*-isoprene).

coarse pearlite. Pearlite for which the alternating ferrite and cementite layers are relatively thick.

coercivity (or coercive field, H_c). The applied magnetic field necessary to reduce to zero the magnetic flux density of a magnetized ferromagnetic or ferrimagnetic material.

cold working. The plastic deformation of a metal at a temperature below that at which it recrystallizes.

color. Visual perception stimulated by the combination of wavelengths of light that are transmitted to the eye.

colorant. An additive that imparts a specific color to a polymer.

compacted graphite iron. A cast iron alloyed with silicon and a small amount of magnesium, cerium, or other additives, in which the graphite exists as wormlike particles.

component. A chemical constituent (element or compound) of an alloy that may be used to specify its composition.

composition (C_i). The relative content of a particular element or constituent (i) within an alloy, usually expressed in weight percent or atom percent.

concentration. See **composition**.

concentration gradient (dC/dx). The slope of the concentration profile at a specific position.

concentration polarization. The condition in which the rate of an electrochemical reaction is limited by the rate of diffusion in the solution.

concentration profile. The curve that results when the concentration of a chemical species is plotted versus position in a material.

concrete. A composite material consisting of aggregate particles bound together in a solid body by a cement.

condensation (or step reaction) polymerization. The formation of polymer macromolecules by an intermolecular reaction, usually with the production of a by-product of low molecular weight, such as water.

conduction band. For electrical insulators and semiconductors, the lowest-lying electron energy band that is empty of electrons at 0 K. Conduction electrons are those that have been excited to states within this band.

conductivity, electrical (σ). The proportionality constant between current density and applied electric field; also, a measure of the ease with which a material is capable of conducting an electric current.

congruent transformation. A transformation of one phase to another of the same composition.

continuous-cooling-transformation (CCT) diagram. A plot of temperature versus the logarithm of time for a steel alloy of definite composition. Used to indicate when transformations occur as the initially austenitized material is continuously cooled at a specified rate; in addition, the final microstructure and mechanical characteristics may be predicted.

coordination number. The number of atomic or ionic nearest neighbors.

copolymer. A polymer that consists of two or more dissimilar repeat units in combination along its molecular chains.

corrosion. Deteriorative loss of a metal as a result of dissolution environmental reactions.

corrosion fatigue. A type of failure that results from the simultaneous action of a cyclic stress and chemical attack.

corrosion penetration rate (CPR). Thickness loss of material per unit of time as a result of corrosion; usually expressed in terms of mils per year or millimeters per year.

coulombic force. A force between charged particles such as ions; the force is attractive when the particles are of opposite charge.

covalent bond. A primary interatomic bond that is formed by the sharing of electrons between neighboring atoms.

creep. The time-dependent permanent deformation that occurs under stress; for most materials it is important only at elevated temperatures.

crevice corrosion. A form of corrosion that occurs within narrow crevices and under deposits of dirt or corrosion products (i.e., in regions of localized depletion of oxygen in the solution).

critical resolved shear stress (τ_{crss}). The shear stress, resolved within a slip plane and direction, required to initiate slip.

crosslinked polymer. A polymer in which adjacent linear molecular chains are joined at various positions by covalent bonds.

crystalline. The state of a solid material characterized by a periodic and repeating three-dimensional array of atoms, ions, or molecules.

crystallinity. For polymers, the state in which a periodic and repeating atomic arrangement is achieved by molecular chain alignment.

crystallite. A region within a crystalline polymer in which all the molecular chains are ordered and aligned.

crystallization (glass-ceramics). The process in which a glass (noncrystalline or vitreous solid) transforms into a crystalline solid.

crystal structure. For crystalline materials, the manner in which atoms or ions are arrayed in space. It is defined in terms of the unit cell geometry and the atom positions within the unit cell.

crystal system. A scheme by which crystal structures are classified according to unit cell geometry. This geometry is specified in terms of the relationships between edge lengths and interaxial angles. There are seven different crystal systems.

Curie temperature (T_c). The temperature above which a ferromagnetic or ferrimagnetic material becomes paramagnetic.

D

defect structure. Relating to the kinds and concentrations of vacancies and interstitials in a ceramic compound.

degradation. Used to denote the deteriorative processes that occur with polymeric materials, including swelling, dissolution, and chain scission.

degree of polymerization (DP). The average number of repeat units per polymer chain molecule.

design stress (σ_d). Product of the calculated stress level (on the basis of estimated maximum load) and a design factor (which has a value greater than unity). Used to protect against unanticipated failure.

diamagnetism. A weak form of induced or nonpermanent magnetism for which the magnetic susceptibility is negative.

die. An individual integrated circuit chip with a thickness on the order of 0.4 mm and with a square or rectangular geometry, each side measuring on the order of 6 mm.

dielectric. Any material that is electrically insulating.

dielectric constant (ϵ_r). The ratio of the permittivity of a medium to that of a vacuum. Often called the *relative dielectric constant* or *relative permittivity*.

dielectric displacement (D). The magnitude of charge per unit area of capacitor plate.

dielectric (breakdown) strength. The magnitude of an electric field necessary to cause significant current passage through a dielectric material.

diffraction (x-ray). Constructive interference of x-ray beams scattered by atoms of a crystal.

diffusion. Mass transport by atomic motion.

diffusion coefficient (D). The constant of proportionality between the diffusion flux and the concentration gradient in Fick's first law. Its magnitude is indicative of the rate of atomic diffusion.

diffusion flux (J). The quantity of mass diffusing through and perpendicular to a unit cross-sectional area of material per unit time.

diode. An electronic device that rectifies an electrical current—that is, allows current flow in one direction only.

dipole (electric). A pair of equal and opposite electrical charges separated by a small distance.

dislocation. A linear crystalline defect around which there is atomic misalignment. Plastic deformation corresponds to the motion of dislocations in response to an applied shear stress. Edge, screw, and mixed dislocations are possible.

dislocation density. The total dislocation length per unit volume of material; alternatively, the number of dislocations that intersect a unit area of a random surface section.

dislocation line. The line that extends along the end of the extra half-plane of atoms for an edge dislocation and along the center of the spiral of a screw dislocation.

dispersed phase. For composites and some two-phase alloys, the discontinuous phase surrounded by the matrix phase.

dispersion strengthening. A means of strengthening materials in which very small particles (usually <0.1 μm) of a hard, inert phase are uniformly dispersed within a loadbearing matrix phase.

domain. A volume region of a ferromagnetic or ferrimagnetic material in which all atomic or ionic magnetic moments are aligned in the same direction.

donor state (level). For a semiconductor or insulator, an energy level lying within and near the top of the energy band gap and from which electrons may be excited into the conduction band. It is normally introduced by an impurity atom.

doping. The intentional alloying of semiconducting materials with controlled concentrations of donor or acceptor impurities.

drawing (metals). A forming technique used to fabricate metal wire and tubing. Deformation is accomplished by pulling the material through a die by means of a tensile force applied on the exit side.

drawing (polymers). A deformation technique in which polymer fibers are strengthened by elongation.

driving force. The impetus behind a reaction, such as diffusion, grain growth, or a phase transformation. Usually attendant to the reaction is a reduction in some type of energy (e.g., free energy).

ductile fracture. A mode of fracture attended by extensive gross plastic deformation.

ductile iron. A cast iron alloyed with silicon and a small concentration of magnesium and/or cerium and in which the free graphite exists in nodular form. Sometimes called *nodular iron*.

ductile-to-brittle transition. The transition from ductile to brittle behavior with a decrease in temperature exhibited by some low-strength steel (BCC) alloys; the temperature range over which the transition occurs is determined by Charpy and Izod impact tests.

ductility. A measure of a material's ability to undergo appreciable plastic deformation before fracture; it may be expressed as percent elongation (%EL) or percent reduction in area (%RA) from a tensile test.

E

edge dislocation. A linear crystalline defect associated with the lattice distortion produced in the vicinity of the end of an extra half-plane of atoms within a crystal. The Burgers vector is perpendicular to the dislocation line.

elastic deformation. Deformation that is nonpermanent—that is, totally recovered upon release of an applied stress.

elastic recovery. Nonpermanent deformation recovered or regained upon release of a mechanical stress.

elastomer. A polymeric material that may experience large and reversible elastic deformations.

electrical conductivity. See **conductivity, electrical (σ)**.

electric dipole. See **dipole (electric)**.

electric field ($\mathcal{E}$). The gradient of voltage.

electroluminescence. The emission of visible light by a *p–n* junction across which a forward-biased voltage is applied.

electrolyte. A solution through which an electric current may be carried by the motion of ions.

electromotive force (emf) series. A ranking of metallic elements according to their standard electrochemical cell potentials.

electron configuration. For an atom, the manner in which possible electron states are filled with electrons.

electronegative. For an atom, having a tendency to accept valence electrons. Also used to describe nonmetallic elements.

electron energy band. A series of electron energy states that are very closely spaced with respect to energy.

electroneutrality. The state of having exactly the same numbers of positive and negative electrical charges (ionic and electronic)—that is, of being electrically neutral.

electron state (level). One of a set of discrete, quantized energies that are allowed for electrons. In the atomic case, each state is specified by four quantum numbers.

electron volt (eV). A convenient unit of energy for atomic and subatomic systems. It is equivalent to the energy acquired by an electron when it falls through an electric potential of 1 V.

electropositive. For an atom, having a tendency to release valence electrons. Also used to describe metallic elements.

endurance limit. See **fatigue limit**.

energy band gap. See **band gap energy (E_g)**.

engineering strain. See **strain, engineering (ϵ)**.

engineering stress. See **stress, engineering (σ)**.

equilibrium (phase). The state of a system in which the phase characteristics remain constant over indefinite time periods. At equilibrium the free energy is a minimum.

erosion–corrosion. A form of corrosion that arises from the combined action of chemical attack and mechanical wear.

eutectic phase. One of the two phases found in the eutectic structure.

eutectic reaction. A reaction in which, upon cooling, a liquid phase transforms isothermally and reversibly into two intimately mixed solid phases.

eutectic structure. A two-phase microstructure resulting from the solidification of a liquid having the eutectic composition; the phases exist as lamellae that alternate with one another.

eutectoid reaction. A reaction in which, upon cooling, one solid phase transforms isothermally and reversibly into two new solid phases that are intimately mixed.

excited state. An electron energy state, not normally occupied, to which an electron may be promoted (from a lower-energy state) by the absorption of some type of energy (e.g., heat, radiative).

extrinsic semiconductor. A semiconducting material for which the electrical behavior is determined by impurities.

extrusion. A forming technique by which a material is forced, by compression, through a die orifice.

F

face-centered cubic (FCC). A crystal structure found in some common elemental metals. Within the cubic unit cell, atoms are located at all corner and face-centered positions.

fatigue. Failure, at relatively low stress levels, of structures that are subjected to fluctuating and cyclic stresses.

fatigue life (N_f). The total number of stress cycles that cause a fatigue failure at some specified stress amplitude.

fatigue limit. For fatigue, the maximum stress amplitude level below which a material can endure an essentially infinite number of stress cycles and not fail.

fatigue strength. The maximum stress level that a material can sustain without failing, for some specified number of cycles.

Fermi energy (E_f). For a metal, the energy corresponding to the highest filled electron state at 0 K.

ferrimagnetism. Permanent and large magnetizations found in some ceramic materials resulting from antiparallel spin coupling and incomplete magnetic moment cancellation.

ferrite (ceramic). Ceramic oxide materials composed of both divalent and trivalent cations (e.g., Fe^{2+} and Fe^{3+}), some of which are ferrimagnetic.

ferrite (iron). Body-centered cubic iron; also iron and steel alloys that have the BCC crystal structure.

ferroelectric. A dielectric material that may exhibit polarization in the absence of an electric field.

ferromagnetism. Permanent and large magnetizations found in some metals (e.g., Fe, Ni, and Co) resulting from the parallel alignment of neighboring magnetic moments.

ferrous alloy. A metal alloy for which iron is the prime constituent.

fiber. Any polymer, metal, or ceramic that has been drawn into a long and thin filament.

fiber-reinforced composite. A composite in which the dispersed phase is in the form of a fiber (i.e., a filament that has a large length-to-diameter ratio).

fiber reinforcement. Strengthening or reinforcement of a relatively weak material by embedding a strong fiber phase within the weak matrix material.

Fick's first law. The diffusion flux is proportional to the concentration gradient. This relationship is used for steady-state diffusion situations.

Fick's second law. The time rate of change of concentration is proportional to the second derivative of concentration. This relationship is used in nonsteady-state diffusion situations.

filler. An inert foreign substance added to a polymer to improve or modify its properties.

fine pearlite. Pearlite in which the alternating ferrite and cementite layers are relatively thin.

firing. A high-temperature heat treatment that increases the density and strength of a ceramic piece.

flame retardant. A polymer additive that increases flammability resistance.

flexural strength (σ_{fs}). Stress at fracture from a bend (or flexure) test.

fluorescence. Luminescence that occurs for times much less than 1 s after an electron excitation event.

foam. A polymer that has been made porous (or spongelike) by the incorporation of gas bubbles.

forging. Mechanical forming of a metal by heating and hammering.

forward bias. The conducting bias for a p–n junction rectifier such that electron flow is to the n side of the junction.

fracture mechanics. A technique of fracture analysis used to determine the stress level at which preexisting cracks of known size will propagate, leading to fracture.

fracture toughness (K_c). The measure of a material's resistance to fracture when a crack is present.

free electron. An electron that has been excited into an energy state above the Fermi energy (or into the conduction band for semiconductors and insulators) and may participate in the electrical conduction process.

free energy. A thermodynamic quantity that is a function of both the internal energy and entropy (or randomness) of a system. At equilibrium, the free energy is at a minimum.

Frenkel defect. In an ionic solid, a cation–vacancy and cation–interstitial pair.

full annealing. For ferrous alloys, austenitizing, followed by cooling slowly to room temperature.

functionality. The number of covalent bonds a monomer can form when reacting with other monomers.

G

galvanic corrosion. The preferential corrosion of the more chemically active of two metals that are electrically coupled and exposed to an electrolyte.

galvanic series. A ranking of metals and alloys as to their relative electrochemical reactivity in seawater.

gas constant (R). Boltzmann's constant per mole of atoms. $R = 8.31$ J/mol·K.

Gibbs phase rule. For a system at equilibrium, an equation (Equation 11.16) that expresses the relationship between the number of phases present and the number of externally controllable variables.

glass-ceramic. A fine-grained crystalline ceramic material formed as a glass and subsequently crystallized.

glass transition temperature (T_g). The temperature at which, upon cooling, a noncrystalline ceramic or polymer transforms from a supercooled liquid into a rigid glass.

graft copolymer. A copolymer in which homopolymer side branches of one monomer type are grafted to homopolymer main chains of a different monomer type.

grain. An individual crystal in a polycrystalline metal or ceramic.

grain boundary. The interface separating two adjoining grains having different crystallographic orientations.

grain growth. The increase in average grain size of a polycrystalline material; for most materials, an elevated-temperature heat treatment is necessary.

grain size. The average grain diameter as determined from a random cross section.

gray cast iron. A cast iron alloyed with silicon in which the graphite exists in the form of flakes. A fractured surface appears gray.

green ceramic body. A ceramic piece, formed as a particulate aggregate, that has been dried but not fired.

ground state. A normally filled electron energy state from which an electron excitation may occur.

growth (particle). During a phase transformation and subsequent to nucleation, the increase in size of a particle of a new phase.

H

Hall effect. The phenomenon by which a force is brought to bear on a moving electron or hole by a magnetic field applied perpendicular to the direction of motion. The force direction is perpendicular to both the magnetic field and the particle motion directions.

hardenability. A measure of the depth to which a specific ferrous alloy may be hardened by the formation of martensite upon quenching from a temperature above the upper critical temperature.

hard magnetic material. A ferrimagnetic or ferromagnetic material that has large coercive field and remanence values, normally used in permanent magnet applications.

hardness. The measure of a material's resistance to deformation by surface indentation or by abrasion.

heat capacity (C_p, C_v). The quantity of heat required to produce a unit temperature rise per mole of material.

hexagonal close-packed (HCP). A crystal structure found for some metals. The HCP unit cell is of hexagonal geometry and is generated by the stacking of close-packed planes of atoms.

high polymer. A solid polymeric material having a molecular weight greater than about 10,000 g/mol.

high-strength, low-alloy (HSLA) steels. Relatively strong, low-carbon steels, with less than about 10 wt% total of alloying elements.

hole (electron). For semiconductors and insulators, a vacant electron state in the valence band that behaves as a positive charge carrier in an electric field.

homopolymer. A polymer having a chain structure in which all repeat units are of the same type.

hot working. Any metal-forming operation performed above a metal's recrystallization temperature.

hybrid composite. A composite that is fiber reinforced by two or more types of fibers (e.g., glass and carbon).

hydrogen bond. A strong secondary interatomic bond that exists between a bound hydrogen atom (its unscreened proton) and the electrons of adjacent atoms.

hydrogen embrittlement. The loss or reduction of ductility of a metal alloy (often steel) as a result of the diffusion of atomic hydrogen into the material.

hydroplastic forming. The molding or shaping of clay-based ceramics that have been made plastic and pliable by adding water.

hypereutectoid alloy. For an alloy system displaying a eutectoid, an alloy for which the concentration of solute is greater than the eutectoid composition.

hypoeutectoid alloy. For an alloy system displaying a eutectoid, an alloy for which the concentration of solute is less than the eutectoid composition.

hysteresis (magnetic). The irreversible magnetic flux density–versus–magnetic field strength (B-versus-H) behavior found for ferromagnetic and ferrimagnetic materials; a closed B–H loop is formed upon field reversal.

I

impact energy (notch toughness). A measure of the energy absorbed during the fracture of a specimen of standard dimensions and geometry when subjected to very rapid (impact) loading. Charpy and Izod impact tests are used to measure this parameter, which is important in assessing the ductile-to-brittle transition behavior of a material.

imperfection. A deviation from perfection; normally applied to crystalline materials in which there is a deviation from atomic/molecular order and/or continuity.

index of refraction (n). The ratio of the velocity of light in a vacuum to the velocity in some medium.

inhibitor. A chemical substance that, when added in relatively low concentrations, retards a chemical reaction.

insulator (electrical). A nonmetallic material that has a filled valence band at 0 K and a relatively wide energy band gap. Consequently, the room-temperature electrical conductivity is very low, less than about 10^{-10} $(\Omega \cdot m)^{-1}$.

integrated circuit. Millions of electronic circuit elements (transistors, diodes, resistors, capacitors, etc.) incorporated on a very small silicon chip.

interdiffusion. Diffusion of atoms of one metal into another metal.

intergranular corrosion. Preferential corrosion along grain-boundary regions of polycrystalline materials.

intergranular fracture. Fracture of polycrystalline materials by crack propagation along grain boundaries.

intermediate solid solution. A solid solution or phase having a composition range that does not extend to either of the pure components of the system.

intermetallic compound. A compound of two metals that has a distinct chemical formula. On a phase diagram it appears as an intermediate phase that exists over a very narrow range of compositions.

interstitial diffusion. A diffusion mechanism by which atomic motion is from interstitial site to interstitial site.

interstitial solid solution. A solid solution in which relatively small solute atoms occupy interstitial positions between the solvent or host atoms.

intrinsic semiconductor. A semiconductor material for which the electrical behavior is characteristic of the pure material—that is, in which electrical conductivity depends only on temperature and the band gap energy.

ionic bond. A coulombic interatomic bond that exists between two adjacent and oppositely charged ions.

isomerism. The phenomenon by which two or more polymer molecules or repeat units have the same composition but different structural arrangements and properties.

isomorphous. Having the same structure. In the phase diagram sense, *isomorphicity* means having the same crystal structure or complete solid solubility for all compositions (see Figure 11.3a).

isotactic. A type of polymer chain configuration (stereoisomer) in which all side groups are positioned on the same side of the chain molecule.

isothermal. At a constant temperature.

isothermal transformation (T–T–T) diagram. A plot of temperature versus the logarithm of time for a steel alloy of definite composition. Used to determine when transformations begin and end for an isothermal (constant-temperature) heat treatment of a previously austenitized alloy.

isotopes. Atoms of the same element that have different atomic masses.

isotropic. Having identical values of a property in all crystallographic directions.

Izod test. One of two tests (see also **Charpy test**) that may be used to measure the impact energy of a standard notched specimen. An impact blow is imparted to the specimen by a weighted pendulum.

J

Jominy end-quench test. A standardized laboratory test used to assess the hardenability of ferrous alloys.

junction transistor. A semiconducting device composed of appropriately biased n–p–n or p–n–p junctions, used to amplify an electrical signal.

K

kinetics. The study of reaction rates and the factors that affect them.

L

laminar composite. A series of two-dimensional sheets, each having a preferred high-strength direction, fastened one on top of the other at different orientations; strength in the plane of the laminate is highly isotropic.

large-particle composite. A type of particle-reinforced composite in which particle-matrix interactions cannot be treated on an atomic level; the particles reinforce the matrix phase.

laser. Acronym for *l*ight *a*mplification by *s*timulated *e*mission of *r*adiation—a source of light that is coherent.

lattice. The regular geometrical arrangement of points in crystal space.

lattice parameters. The combination of unit cell edge lengths and interaxial angles that defines the unit cell geometry.

lattice strains. Slight displacements of atoms relative to their normal lattice positions, normally imposed by crystalline defects such as dislocations, and interstitial and impurity atoms.

lever rule. A mathematical expression, such as Equation 11.1b or Equation 11.2b, by which the relative phase amounts in a two-phase alloy at equilibrium may be computed.

light-emitting diode (LED). A diode composed of a semiconducting material that is *p*-type on one side and *n*-type on the other side. When a forward-biased potential is applied across the junction between the two sides, recombination of electrons and holes occurs, with the emission of light radiation.

linear coefficient of thermal expansion. See **thermal expansion coefficient, linear (α_l)**.

linear polymer. A polymer produced from bifunctional monomers in which each polymer molecule consists of repeat units joined end to end in a single chain.

liquid crystal polymer (LCP). A group of polymeric materials having extended and rod-shape molecules that, structurally, do not fall within traditional liquid, amorphous, crystalline, or semicrystalline classifications. In the molten (or liquid) state, they can become aligned in highly ordered (crystal-like) conformations. They are used in digital displays and a variety of applications in electronics and medical equipment industries.

liquidus line. On a binary phase diagram, the line or boundary separating liquid- and liquid + solid–phase regions. For an alloy, the *liquidus temperature* is the temperature at which a solid phase first forms under conditions of equilibrium cooling.

longitudinal direction. The lengthwise dimension. For a rod or fiber, in the direction of the long axis.

lower critical temperature. For a steel alloy, the temperature below which, under equilibrium conditions, all austenite has transformed into ferrite and cementite phases.

luminescence. The emission of visible light as a result of electron decay from an excited state.

M

macromolecule. A huge molecule made up of thousands of atoms.

magnetic field strength (H). The intensity of an externally applied magnetic field.

magnetic flux density (B). The magnetic field produced in a substance by an external magnetic field.

magnetic induction (B). See **magnetic flux density (B)**.

magnetic susceptibility (χ_m). The proportionality constant between the magnetization M and the magnetic field strength H.

magnetization (M). The total magnetic moment per unit volume of material. Also, a measure of the contribution to the magnetic flux by some material within an H field.

malleable cast iron. White cast iron that has been heat-treated to convert the cementite into graphite clusters; a relatively ductile cast iron.

martensite. A metastable iron phase supersaturated in carbon that is the product of a diffusionless (athermal) transformation from austenite.

matrix phase. The phase in a composite or two-phase alloy microstructure that is continuous or completely surrounds the other (or dispersed) phase.

Matthiessen's rule. The total electrical resistivity of a metal is equal to the sum of temperature-, impurity-, and cold-work-dependent contributions.

melting point (glass). The temperature at which the viscosity of a glass material is 10 Pa·s.

melting temperature. The temperature at which, upon heating, a solid (and crystalline) phase transforms into a liquid.

metal. The electropositive elements and the alloys based on these elements. The electron band structure of metals is characterized by a partially filled electron band.

metallic bond. A primary interatomic bond involving the nondirectional sharing of nonlocalized valence electrons ("sea of electrons") that are mutually shared by all the atoms in the metallic solid.

metal-matrix composite (MMC). A composite material that has a metal or metal alloy as the matrix phase. The dispersed phase may be particulates, fibers, or whiskers, which normally are stiffer, stronger, and/or harder than the matrix.

metastable. A nonequilibrium state that may persist for a very long time.

microconstituent. An element of the microstructure that has an identifiable and characteristic structure. It may consist of more than one phase, such as with pearlite.

microelectromechanical system (MEMS). A large number of miniature mechanical devices that are integrated with electrical elements on a silicon substrate. Mechanical components act as microsensors and microactuators and are in the form of beams, gears, motors, and membranes. In response to microsensor stimuli, the electrical elements render decisions that direct responses to the microactuator devices.

microscopy. The investigation of microstructural elements using some type of microscope.

microstructure. The structural features of an alloy (e.g., grain and phase structure) subject to observation under a microscope.

Miller indices. A set of three integers (four for hexagonal) that designate crystallographic planes, as determined from reciprocals of fractional axial intercepts.

mixed dislocation. A dislocation that has both edge and screw components.

mobility (electron, μ_e, and hole, μ_h). The proportionality constant between the carrier drift velocity and applied electric field; also, a measure of the ease of charge carrier motion.

modulus of elasticity (E). The ratio of stress to strain when deformation is totally elastic; also a measure of the stiffness of a material.

molarity (M). Concentration in a liquid solution, in terms of the number of moles of a solute dissolved in 1 L (10^3 cm^3) of solution.

molding (plastics). Shaping a plastic material by forcing it, under pressure and at an elevated temperature, into a mold cavity.

mole. The quantity of a substance corresponding to 6.022×10^{23} atoms or molecules.

molecular chemistry (polymer). With regard only to composition, not the structure of a repeat unit.

molecular structure (polymer). With regard to atomic arrangements within and interconnections between polymer molecules.

molecular weight. The sum of the atomic weights of all the atoms in a molecule.

monomer. A stable molecule from which a polymer is synthesized.

MOSFET. Metal-oxide-semiconductor field-effect transistor, an integrated circuit element.

N

nanocarbon. A particle having a size of less than about 100 nm composed of carbon atoms that are bonded together with sp^2 hybridized electron orbitals. Three nanocarbon types are fullerenes, carbon nanotubes, and graphene.

nanocomposite. A composite composed of nanosize particles (i.e., *nanoparticles*) embedded in matrix material. Nanoparticle types include nanocarbons, nanoclays, and nanocrystals. The most common matrix materials are polymers.

natural aging. For precipitation hardening, aging at room temperature.

network polymer. A polymer produced from multifunctional monomers having three or more active covalent bonds, resulting in the formation of three-dimensional molecules.

nodular iron. See **ductile iron**.

noncrystalline. The solid state in which there is no long-range atomic order. Sometimes the terms *amorphous, glassy,* and *vitreous* are used synonymously.

nonferrous alloy. A metal alloy of which iron is *not* the prime constituent.

nonsteady-state diffusion. The diffusion condition for which there is some net accumulation or depletion of diffusing species. The diffusion flux is dependent on time.

normalizing. For ferrous alloys, austenitizing above the upper critical temperature, then cooling in air. The objective of this heat treatment is to enhance toughness by refining the grain size.

n-type semiconductor. A semiconductor for which the predominant charge carriers responsible for electrical conduction are electrons. Normally, donor impurity atoms give rise to the excess electrons.

nucleation. The initial stage in a phase transformation. It is evidenced by the formation of small particles (nuclei) of the new phase that are capable of growing.

O

octahedral position. The void space among close-packed, hard-sphere atoms or ions for which there are six nearest neighbors. An octahedron (double pyramid) is circumscribed by lines constructed from centers of adjacent spheres.

Ohm's law. The applied voltage is equal to the product of the current and resistance; equivalently, the current density is equal to the product of the conductivity and electric field intensity.

opaque. Being impervious to the transmission of light as a result of absorption, reflection, and/or scattering of incident light.

optical fiber. A thin (5- to 100-μm diameter), ultra-high-purity silica fiber through which information may be transmitted via photonic (light radiation) signals.

overaging. During precipitation hardening, aging beyond the point at which strength and hardness are at their maxima.

oxidation. The removal of one or more electrons from an atom, ion, or molecule.

P

paramagnetism. A relatively weak form of magnetism that results from the independent alignment of atomic dipoles (magnetic) with an applied magnetic field.

particle-reinforced composite. A composite for which the dispersed phase is equiaxed.

passivity. The loss of chemical reactivity, under particular environmental conditions, by some active metals and alloys, often due to the formation of a protective film.

Pauli exclusion principle. The postulate that for an individual atom, at most two electrons, which necessarily have opposite spins, can occupy the same state.

pearlite. A two-phase microstructure found in some steels and cast irons; it results from the transformation of austenite of eutectoid composition and consists of alternating layers (or *lamellae*) of α-ferrite and cementite.

periodic table. The arrangement of the chemical elements with increasing atomic number according to the periodic variation in electron structure. Nonmetallic elements are positioned at the far right-hand side of the table.

peritectic reaction. A reaction in which, upon cooling, a solid and a liquid phase transform isothermally and reversibly to a solid phase having a different composition.

permeability (magnetic, μ). The proportionality constant between B and H fields. The value of the permeability of a vacuum (μ_0) is 1.257×10^{-6} H/m.

permittivity (ϵ). The proportionality constant between the dielectric displacement D and the electric field $\mathscr{E}$. The value of the permittivity ϵ_0 for a vacuum is 8.85×10^{-12} F/m.

phase. A homogeneous portion of a system that has uniform physical and chemical characteristics.

phase diagram. A graphical representation of the relationships among environmental constraints (e.g., temperature and sometimes pressure), composition, and regions of phase stability, typically under conditions of equilibrium.

phase equilibrium. See **equilibrium (phase)**.

phase transformation. A change in the number and/or character of the phases that constitute the microstructure of an alloy.

phonon. A single quantum of vibrational or elastic energy.

phosphorescence. Luminescence that occurs at times greater than on the order of 1 s after an electron excitation event.

photoconductivity. Electrical conductivity that results from photon-induced electron excitations in which light is absorbed.

photomicrograph. A photograph made with a microscope that records a microstructural image.

photon. A quantum unit of electromagnetic energy.

piezoelectric. A dielectric material in which polarization is induced by the application of external forces.

Pilling–Bedworth ratio (P–B ratio). The ratio of metal oxide volume to metal volume; used to predict whether a scale that forms will protect a metal from further oxidation.

pitting. A form of very localized corrosion in which small pits or holes form, usually in a vertical direction.

plain carbon steel. A ferrous alloy in which carbon is the prime alloying element.

Planck's constant (h). A universal constant that has a value of 6.63×10^{-34} J·s. The energy of a photon of electromagnetic radiation is the product of h and the radiation frequency.

plane strain. The condition, important in fracture mechanical analyses, in which, for tensile loading, there is zero strain in a direction perpendicular to both the stress axis and the direction of crack propagation; this condition is found in thick plates, and the zero-strain direction is perpendicular to the plate surface.

plane strain fracture toughness (K_{Ic}). For the condition of plane strain, the measure of a material's resistance to fracture when a crack is present.

plastic. A solid organic polymer of high molecular weight that has some structural rigidity under load and is used in general-purpose applications. It may also contain additives such as fillers, plasticizers, and flame retardants.

plastic deformation. Deformation that is permanent or nonrecoverable after release of the applied load. It is accompanied by permanent atomic displacements.

plasticizer. A low-molecular-weight polymer additive that enhances flexibility and workability and reduces stiffness and brittleness, resulting in a decrease in the glass transition temperature T_g.

point defect. A crystalline defect associated with one or, at most, several atomic sites.

Poisson's ratio (ν). For elastic deformation, the negative ratio of lateral and axial strains that result from an applied axial stress.

polarization (P). The total electric dipole moment per unit volume of dielectric material. Also, a measure of the contribution to the total dielectric displacement by a dielectric material.

polarization (corrosion). The displacement of an electrode potential from its equilibrium value as a result of current flow.

polarization (electronic). For an atom, the displacement of the center of the negatively charged electron cloud relative to the positive nucleus induced by an electric field.

polarization (ionic). Polarization as a result of the displacement of anions and cations in opposite directions.

polarization (orientation). Polarization resulting from the alignment (by rotation) of permanent electric dipole moments with an applied electric field.

polar molecule. A molecule in which there exists a permanent electric dipole moment by virtue of the asymmetrical distribution of positively and negatively charged regions.

polycrystalline. Crystalline materials composed of more than one crystal or grain.

polymer. A compound of high molecular weight (normally organic), the structure of which is composed of chains of small repeat units.

polymer-matrix composite (PMC). A composite material for which the matrix is a polymer resin and fibers (normally glass, carbon, or aramid) are the dispersed phase.

polymorphism. The ability of a solid material to exist in more than one form or crystal structure.

powder metallurgy (P/M). The fabrication of metal pieces having intricate and precise shapes by the compaction of metal powders, followed by a densification heat treatment.

precipitation hardening. Hardening and strengthening of a metal alloy by extremely small and uniformly dispersed particles that precipitate from a supersaturated solid solution; sometimes also called *age hardening*.

precipitation heat treatment. A heat treatment used to precipitate a new phase from a supersaturated solid solution. For precipitation hardening, it is termed *artificial aging*.

prepreg. Continuous fiber reinforcement preimpregnated with a polymer resin that is then partially cured.

prestressed concrete. Concrete into which compressive stresses have been introduced using steel wires or rods.

primary bonds. Interatomic bonds that are relatively strong and for which bonding energies are relatively large. Primary bonding types are ionic, covalent, and metallic.

primary phase. A phase that exists in addition to the eutectic structure.

principle of combined action. The supposition, often valid, that new properties, better properties, better property combinations, and/or a higher level of properties can be fashioned by the judicious combination of two or more distinct materials.

process annealing. Annealing of previously cold-worked products (commonly steel alloys in sheet or wire form) below the lower critical (eutectoid) temperature.

proeutectoid cementite. Primary cementite that exists in addition to pearlite for hypereutectoid steels.

proeutectoid ferrite. Primary ferrite that exists in addition to pearlite for hypoeutectoid steels.

property. A material trait expressed in terms of the measured response to a specific imposed stimulus.

proportional limit. The point on a stress–strain curve at which the straight-line proportionality between stress and strain ceases.

p-type semiconductor. A semiconductor for which the predominant charge carriers responsible for electrical conduction are holes. Normally, acceptor impurity atoms give rise to the excess holes.

Q

quantum mechanics. A branch of physics that deals with atomic and subatomic systems; it allows only discrete values of energy. By contrast, for classical mechanics, continuous energy values are permissible.

quantum numbers. A set of four numbers, the values of which are used to label possible electron states. Three of the quantum numbers are integers that specify the size, shape, and spatial orientation of an electron's probability density; the fourth number designates spin orientation.

R

random copolymer. A polymer in which two different repeat units are randomly distributed along the molecular chain.

recovery. The relief of some of the internal strain energy of a previously cold-worked metal, usually by heat treatment.

recrystallization. The formation of a new set of strain-free grains within a previously cold-worked material; normally, an annealing heat treatment is necessary.

recrystallization temperature. For a particular alloy, the minimum temperature at which complete recrystallization occurs within approximately 1 h.

rectifying junction. A semiconductor p–n junction conductive for a current flow in one direction and highly resistive for the opposite direction.

reduction. The addition of one or more electrons to an atom, ion, or molecule.

reflection. Deflection of a light beam at the interface between two media.

refraction. Bending of a light beam upon passing from one medium into another; the velocity of light differs in the two media.

refractory. A metal or ceramic that may be exposed to extremely high temperatures without deteriorating rapidly or without melting.

reinforced concrete. Concrete that is reinforced (or strengthened in tension) by the incorporation of steel rods, wires, or mesh.

relative magnetic permeability (μ_r). The ratio of the magnetic permeability of some medium to that of a vacuum.

relaxation frequency. The reciprocal of the minimum reorientation time for an electric dipole within an alternating electric field.

relaxation modulus [$E_r(t)$]. For viscoelastic polymers, the time-dependent modulus of elasticity. It is determined from stress relaxation measurements as the ratio of stress (taken at some time after the load application—normally 10 s) to strain.

remanence (remanent induction, B_r). For a ferromagnetic or ferrimagnetic material, the magnitude of residual flux density that remains when a magnetic field is removed.

repeat unit. The most fundamental structural unit in a polymer chain. A polymer molecule is composed of a large number of repeat units linked together.

residual stress. A stress that persists in a material free of external forces or temperature gradients.

resilience. The capacity of a material to absorb energy when it is elastically deformed.

resistivity (ρ). The reciprocal of electrical conductivity; a measure of a material's resistance to the passage of electric current.

resolved shear stress. An applied tensile or compressive stress resolved into a shear component along a specific plane and direction within that plane.

reverse bias. The insulating bias for a p–n junction rectifier; electrons flow into the p side of the junction.

rolling. A metal-forming operation that reduces the thickness of sheet stock; elongated shapes may be fashioned using grooved circular rolls.

rule of mixtures. The properties of a multiphase alloy or composite material are a weighted average (usually on the basis of volume) of the properties of the individual constituents.

rupture. Failure accompanied by significant plastic deformation; often associated with creep failure.

S

sacrificial anode. An active metal or alloy that preferentially corrodes and protects another metal or alloy to which it is electrically coupled.

safe stress (σ_w). A stress used for design purposes; for ductile metals, it is the yield strength divided by a factor of safety.

sandwich panel. A type of structural composite consisting of two stiff and strong outer faces that are separated by a lightweight core material.

saturated. A carbon atom that participates in only single covalent bonds with four other atoms.

saturation magnetization, flux density (M_s, B_s). The maximum magnetization (or flux density) for a ferromagnetic or ferrimagnetic material.

scanning electron microscope (SEM). A microscope that produces an image by using an electron beam that scans the surface of a specimen; an image is produced by reflected electron beams. Examination of surface and/or microstructural features at high magnifications is possible.

scanning probe microscope (SPM). A microscope that does not produce an image using light radiation. Rather, a very small and sharp probe raster scans across the specimen surface; out-of-surface plane deflections in response to electronic or other interactions with the probe are monitored, from which a topographical map of the specimen surface (on a nanometer scale) is produced.

Schottky defect. In an ionic solid, a defect consisting of a cation–vacancy and anion–vacancy pair.

scission. A polymer degradation process by which molecular chain bonds are ruptured by chemical reactions or by exposure to radiation or heat.

screw dislocation. A linear crystalline defect associated with the lattice distortion created when normally parallel planes are joined together to form a helical ramp. The Burgers vector is parallel to the dislocation line.

secondary bonds. Interatomic and intermolecular bonds that are relatively weak and for which bonding energies are relatively small. Normally, atomic or molecular dipoles are involved. Examples of secondary bonding types are van der Waals forces and hydrogen bonding.

selective leaching. A form of corrosion in which one element or constituent of an alloy is dissolved preferentially.

self-diffusion. Atomic migration in pure metals.

self-interstitial. A host atom or ion positioned on an interstitial lattice site.

semiconductor. A nonmetallic material that has a filled valence band at 0 K and a relatively narrow energy band gap. The room-temperature electrical conductivity ranges between about 10^{-6} and 10^4 $(\Omega \cdot m)^{-1}$.

shear. A force applied so as to cause or tend to cause two adjacent parts of the same body to slide relative to each other in a direction parallel to their plane of contact.

shear strain (γ). The tangent of the shear angle that results from an applied shear load.

shear stress (τ). The instantaneous applied shear load divided by the original cross-sectional area across which it is applied.

single crystal. A crystalline solid for which the periodic and repeated atomic pattern extends throughout its entirety without interruption.

sintering. Particle coalescence of a powdered aggregate by diffusion that is accomplished by firing at an elevated temperature.

slip. Plastic deformation as the result of dislocation motion; also, the shear displacement of two adjacent planes of atoms.

slip casting. A forming technique used for some ceramic materials. A slip, or suspension of solid particles in water, is poured into a porous mold. A solid layer forms on the inside wall as water is absorbed by the mold, leaving a shell (or ultimately a solid piece) having the shape of the mold.

slip system. The combination of a crystallographic plane and, within that plane, a crystallographic direction along which slip (i.e., dislocation motion) occurs.

softening point (glass). The maximum temperature at which a glass piece may be handled without permanent deformation; this corresponds to a viscosity of approximately 4×10^6 Pa·s.

soft magnetic material. A ferromagnetic or ferrimagnetic material having a small B-versus-H hysteresis loop; it may be magnetized and demagnetized with relative ease.

soldering. A technique for joining metals using a filler metal alloy that has a melting temperature less than about 425°C.

solid solution. A homogeneous crystalline phase that contains two or more chemical species. Both substitutional and interstitial solid solutions are possible.

solid-solution strengthening. Hardening and strengthening of metals that result from alloying in which a solid solution is formed. The presence of impurity atoms restricts dislocation mobility.

solidus line. On a phase diagram, the locus of points at which solidification is complete upon equilibrium cooling, or at which melting begins upon equilibrium heating.

solubility limit. The maximum concentration of solute that may be added without forming a new phase.

solute. One component or element of a solution present in a minor concentration. It is dissolved in the solvent.

solution heat treatment. The process used to form a solid solution by dissolving precipitate particles. Often, the solid solution is supersaturated and metastable at ambient conditions as a result of rapid cooling from an elevated temperature.

solvent. The component of a solution present in the greatest amount. It is the component that dissolves a solute.

solvus line. The locus of points on a phase diagram representing the limit of solid solubility as a function of temperature.

specific heat (c_p, c_v)**.** The heat capacity per unit mass of material.

specific modulus (specific stiffness). The ratio of elastic modulus to specific gravity for a material.

specific strength. The ratio of tensile strength to specific gravity for a material.

spheroidite. Microstructure found in steel alloys consisting of spherelike cementite particles within an α-ferrite matrix. It is produced by an appropriate elevated-temperature heat treatment of pearlite, bainite, or martensite, and is relatively soft.

spheroidizing. For steels, a heat treatment normally carried out at a temperature just below the eutectoid in which the spheroidite microstructure is produced.

spherulite. An aggregate of ribbon-like polymer crystallites (lamellae) radiating from a common central nucleation site; the crystallites are separated by amorphous regions.

spinning. The process by which fibers are formed. A multitude of fibers are spun as molten or dissolved material is forced through many small orifices.

stabilizer. A polymer additive that counteracts deteriorative processes.

stainless steel. A steel alloy that is highly resistant to corrosion in a variety of environments. The predominant alloying element is chromium, which must be present in a concentration of at least 11 wt%; other alloy additions, to include nickel and molybdenum, are also possible.

standard half-cell. An electrochemical cell consisting of a pure metal immersed in a 1 M aqueous solution of its ions, which is electrically coupled to the standard hydrogen electrode.

steady-state diffusion. The diffusion condition for which there is no net accumulation or depletion of diffusing species. The diffusion flux is independent of time.

stereoisomerism. Polymer isomerism in which side groups within repeat units are bonded along the molecular chain in the same order but in different spatial arrangements.

stoichiometry. For ionic compounds, the state of having exactly the ratio of cations to anions specified by the chemical formula.

strain, engineering (ϵ)**.** The change in gauge length of a specimen (in the direction of an applied stress) divided by its original gauge length.

strain hardening. The increase in hardness and strength of a ductile metal as it is plastically deformed below its recrystallization temperature.

strain point (glass). The maximum temperature at which glass fractures without plastic deformation; this corresponds to a viscosity of about 3×10^{13} Pa·s.

strain, true. See **true strain** (ϵ_T)**.**

stress concentration. The concentration or amplification of an applied stress at the tip of a notch or small crack.

stress corrosion (cracking). A form of failure that results from the combined action of a tensile stress and a corrosion environment; it occurs at lower stress levels than are required when the corrosion environment is absent.

stress, engineering (σ)**.** The instantaneous load applied to a specimen divided by its cross-sectional area before any deformation.

stress raiser. A small flaw (internal or surface) or a structural discontinuity at which an applied tensile stress will be amplified and from which cracks may propagate.

stress relief. A heat treatment for the removal of residual stresses.

stress, true. See **true stress** (σ_T)**.**

structural clay products. Ceramic products made principally of clay and used in applications in which structural integrity is important (e.g., bricks, tiles, pipes).

structural composite. A composite whose properties depend on the geometrical design of the structural elements. Laminar composites and sandwich panels are two subclasses.

structure. The arrangement of the internal components of matter: electron structure (on a subatomic level), crystal structure (on an atomic level), and microstructure (on a microscopic level).

substitutional solid solution. A solid solution in which the solute atoms replace or substitute for the host atoms.

superconductivity. A phenomenon observed in some materials: the disappearance of the electrical resistivity at temperatures approaching 0 K.

supercooling. Cooling to below a phase transition temperature without the occurrence of the transformation.

superheating. Heating to above a phase transition temperature without the occurrence of the transformation.

syndiotactic. A type of polymer chain configuration (stereoisomer) in which side groups regularly alternate positions on opposite sides of the chain.

system. Two meanings are possible: (1) a specific body of material being considered, and (2) a series of possible alloys consisting of the same components.

T

temper designation. A letter–digit code used to designate the mechanical and/or thermal treatment to which a metal alloy has been subjected.

tempered martensite. The microstructural product resulting from a tempering heat treatment of a martensitic steel. The microstructure consists of extremely small and uniformly dispersed cementite particles embedded within a continuous α-ferrite matrix. Toughness and ductility are enhanced significantly by tempering.

tempering (glass). See **thermal tempering**.

tensile strength (*TS*)**.** The maximum engineering stress, in tension, that may be sustained without fracture. Often termed *ultimate (tensile) strength*.

terminal solid solution. A solid solution that exists over a composition range extending to either composition extreme of a binary phase diagram.

tetrahedral position. The void space among close-packed, hard-sphere atoms or ions for which there are four nearest neighbors.

thermal conductivity (k). For steady-state heat flow, the proportionality constant between the heat flux and the temperature gradient. Also, a parameter characterizing the ability of a material to conduct heat.

thermal expansion coefficient, linear (α_l). The fractional change in length divided by the change in temperature.

thermal fatigue. A type of fatigue failure in which the cyclic stresses are introduced by fluctuating thermal stresses.

thermally activated transformation. A reaction that depends on atomic thermal fluctuations; the atoms having energies greater than an activation energy spontaneously react or transform.

thermal shock. The fracture of a brittle material as a result of stresses introduced by a rapid temperature change.

thermal stress. A residual stress introduced within a body resulting from a change in temperature.

thermal tempering. Increasing the strength of a glass piece by the introduction of residual compressive stresses within the outer surface using an appropriate heat treatment.

thermoplastic elastomer (TPE). A copolymeric material that exhibits elastomeric behavior yet is thermoplastic in nature. At the ambient temperature, domains of one repeat unit type form at molecular chain ends that crystallize to act as physical crosslinks.

thermoplastic (polymer). A semicrystalline polymeric material that softens when heated and hardens upon cooling. While in the softened state, articles may be formed by molding or extrusion.

thermosetting (polymer). A polymeric material that, once having been cured (or hardened) by a chemical reaction, will not soften or melt when subsequently heated.

tie line. A horizontal line constructed across a two-phase region of a binary phase diagram; its intersections with the phase boundaries on either end represent the equilibrium compositions of the respective phases at the temperature in question.

time–temperature–transformation (T–T–T) diagram. See **isothermal transformation (T–T–T) diagram**.

toughness. A mechanical characteristic that may be expressed in three contexts: (1) the measure of a material's resistance to fracture when a crack (or other stress-concentrating defect) is present; (2) the ability of a material to absorb energy and plastically deform before fracturing; and (3) the total area under the material's tensile engineering stress–strain curve taken to fracture.

trans. For polymers, a prefix denoting a type of molecular structure. For some unsaturated carbon chain atoms within a repeat unit, a single side atom or group may be situated on one side of the double bond, or directly opposite at a 180° rotation position. In a trans structure, two such side groups within the same repeat unit reside on opposite sides (e.g., *trans*-isoprene).

transformation rate. The reciprocal of the time necessary for a reaction to proceed halfway to its completion.

transgranular fracture. Fracture of polycrystalline materials by crack propagation through the grains.

translucent. Having the property of transmitting light only diffusely; objects viewed through a translucent medium are not clearly distinguishable.

transmission electron microscope (TEM). A microscope that produces an image by using electron beams that are *transmitted* (pass through) the specimen. Examination of internal features at high magnifications is possible.

transparent. Having the property of transmitting light with relatively little absorption, reflection, and scattering, so that objects viewed through a transparent medium can be readily distinguished.

transverse direction. A direction that crosses (usually perpendicularly) the longitudinal or lengthwise direction.

trifunctional. Designating monomers that may react to form three covalent bonds with other monomers.

true strain (ϵ_T). The natural logarithm of the ratio of instantaneous gauge length to original gauge length of a specimen being deformed by a uniaxial force.

true stress (σ_T). The instantaneous applied load divided by the instantaneous cross-sectional area of a specimen.

U

ultimate (tensile) strength. See **tensile strength (TS)**.

ultra-high-molecular-weight polyethylene (UHMWPE). A polyethylene polymer that has an extremely high molecular weight (approximately 4×10^6 g/mol). Distinctive characteristics of this material include high impact and abrasion resistance and low coefficient of friction.

unit cell. The basic structural unit of a crystal structure. It is generally defined in terms of atom (or ion) positions within a parallelepiped volume.

unsaturated. Describes carbon atoms that participate in double or triple covalent bonds and, therefore, do not bond to a maximum of four other atoms.

upper critical temperature. For a steel alloy, the minimum temperature above which, under equilibrium conditions, only austenite is present.

V

vacancy. A normally occupied lattice site from which an atom or ion is missing.

vacancy diffusion. The diffusion mechanism in which net atomic migration is from a lattice site to an adjacent vacancy.

valence band. For solid materials, the electron energy band that contains the valence electrons.

valence electrons. The electrons in the outermost occupied electron shell, which participate in interatomic bonding.

van der Waals bond. A secondary interatomic bond between adjacent molecular dipoles that may be permanent or induced.

viscoelasticity. A type of deformation exhibiting the mechanical characteristics of viscous flow and elastic deformation.

viscosity (η). The ratio of the magnitude of an applied shear stress to the velocity gradient that it produces—that is, a measure of a noncrystalline material's resistance to permanent deformation.

vitrification. During firing of a ceramic body, the formation of a liquid phase that, upon cooling, becomes a glass-bonding matrix.

vulcanization. A nonreversible chemical reaction involving sulfur or another suitable agent in which crosslinks are formed between molecular chains in rubber materials. The rubber's modulus of elasticity and strength are enhanced.

W

wave mechanical model. An atomic model in which electrons are treated as being wavelike.

weight percent (wt%). A concentration specification on the basis of weight (or mass) of a particular element relative to the total alloy weight (or mass).

weld decay. Intergranular corrosion that occurs in some welded stainless steels at regions adjacent to the weld.

welding. A technique for joining metals in which actual melting of the pieces to be joined occurs in the vicinity of the bond. A filler metal may be used to facilitate the process.

whisker. A very thin, single crystal of high perfection that has an extremely large length-to-diameter ratio. Whiskers are used as the reinforcing phase in some composites.

white cast iron. A low-silicon and very brittle cast iron in which the carbon is in combined form as cementite; a fractured surface appears white.

whiteware. A clay-based ceramic product that becomes white after high-temperature firing; whitewares include porcelain, china, and plumbing sanitary ware.

working point (glass). The temperature at which a glass is easily deformed, which corresponds to a viscosity of 10^3 Pa·s.

wrought alloy. A metal alloy that is relatively ductile and amenable to hot working or cold working during fabrication.

Y

yielding. The onset of plastic deformation.

yield strength (σ_y). The stress required to produce a very slight yet specified amount of plastic strain; a strain offset of 0.002 is commonly used.

Young's modulus. See **modulus of elasticity (E)**.

Answers to Selected Problems

Chapter 2

2.3 **(a)** 1.66×10^{-24} g/amu;
(b) 6.022×10^{23} atoms/g-mol

2.14

$$r_0 = \left(\frac{A}{nB}\right)^{1/(1-n)}$$

$$E_0 = -\frac{A}{\left(\dfrac{A}{nB}\right)^{1/(1-n)}} + \frac{B}{\left(\dfrac{A}{nB}\right)^{n/(1-n)}}$$

2.15 **(c)** $r_0 = 0.279$ nm, $E_0 = -4.57$ eV

2.22 70.2% for TiO_2; 2.2% for InSb

Chapter 3

3.4 $000, 100, 110, 010, 001, 101, 111, 011, \frac{1}{2}\frac{1}{2}0,$
$\frac{1}{2}\frac{1}{2}1, 1\frac{1}{2}\frac{1}{2}, 0\frac{1}{2}\frac{1}{2}, \frac{1}{2}0\frac{1}{2}$, and $\frac{1}{2}1\frac{1}{2}$

3.11 Direction 1: $[012]$

3.13 Direction A: $[0\bar{1}\bar{1}]$;
Direction C: $[112]$

3.14 Direction B: $[2\bar{3}2]$;
Direction D: $[13\bar{6}]$

3.15 **(b)** $[\bar{1}\bar{1}0], [\bar{1}10]$, and $[1\bar{1}0]$

3.17 Direction A: $[10\bar{1}1]$

3.23 Plane B: $(\bar{1}\bar{1}2)$ or $(11\bar{2})$

3.24 Plane A: $(32\bar{2})$

3.25 Plane B: (221)

3.26 **(c)** $[010]$ or $[0\bar{1}0]$

3.28 **(b)** $(10\bar{1}0)$

Chapter 4

4.1 $V_C = 6.62 \times 10^{-29}$ m^3

4.7 $R = 0.136$ nm

4.10 **(a)** $V_C = 1.40 \times 10^{-28}$ m^3;
(b) $a = 0.323$ nm, $c = 0.515$ nm

4.13 Metal B: face-centered cubic

4.15 **(a)** $n = 8.0$ atoms/unit cell;
(b) $\rho = 4.96$ g/cm^3

4.18 $V_C = 8.08 \times 10^{-2}$ nm^3

4.25 APF = 0.73

4.26 **(a)** FCC; **(b)** tetrahedral; **(c)** one-half

4.27 **(a)** tetrahedral; **(b)** all

4.32 **(a)** $a = 0.421$ nm; **(b)** $a = 0.424$ nm

4.34 ρ(calculated) = 3.80 g/cm^3;
ρ(measured) = 3.80 g/cm^3

4.36 **(a)** $\rho = 4.21$ g/cm^3

4.38 Cesium chloride

4.40 APF = 0.755

4.42 APF = 0.684

4.48 **(a)** $\rho_a = 2.000$ g/cm^3, $\rho_c = 2.301$ g/cm^3;
(b) % crystallinity = 87.9%

4.51 **(a)** (010) and $(\bar{1}00)$

4.55 **(a)** $LD_{100} = \dfrac{1}{2R\sqrt{2}}$

4.56 **(b)** $LD_{111}(W) = 3.65 \times 10^9$ m^{-1}

4.57 **(a)** $PD_{111} = \dfrac{1}{2R^2\sqrt{3}}$

4.58 **(b)** $PD_{110}(V) = 1.522 \times 10^{19}$ m^{-2}

4.61 $2\theta = 81.24°$

4.62 $d_{110} = 0.2862$ nm

4.64 **(a)** $d_{321} = 0.1520$ nm;
(b) $R = 0.2463$ nm

4.66 $d_{110} = 0.2012$ nm, $a = 0.2845$ nm

Chapter 5

5.3 $DP = 23{,}760$

5.5 **(a)** $\overline{M}_n = 33{,}040$ g/mol; **(c)** $DP = 785$

5.8 **(a)** $C_{Cl} = 29.0$ wt%

5.9 $L = 1254$ nm; $r = 15.4$ nm

5.16 8530 of both styrene and butadiene repeat units

5.18 Propylene

5.21 f(isoprene) = 0.88, f(isobutylene) = 0.12

Chapter 6

6.1 $N_v/N = 4.3 \times 10^{-5}$

6.3 $Q_v = 0.75$ eV/atom

6.5 $N_s/N = 4.03 \times 10^{-6}$

6.10 For FCC, $r = 0.41R$

6.12 **(a)** O^{2-} vacancy; one O^{2-} vacancy for every two Li^+ added

6.15 $C'_{Zn} = 29.4$ at%, $C'_{Cu} = 70.6$ at%

6.16 $C_{Pb} = 10.0$ wt%, $C_{Sn} = 90.0$ wt%

6.18 $C'_{Sn} = 71.9$ at%, $C'_{Pb} = 28.1$ at%

6.22 $N_{Al} = 6.05 \times 10^{28}$ atoms/m^3

6.25 $a = 0.289$ nm

6.28 $N_{Au} = 3.36 \times 10^{21}$ atoms/cm^3

6.32 $C_{Nb} = 35.2$ wt%

6.40 **(a)** $d \cong 0.066$ mm

6.42 **(b)** $n_1 = 198,300$ grains/cm^2

6.D1 $C_{Li} = 2.06$ wt%

Chapter 7

7.6 $M = 3.2 \times 10^{-3}$ kg/h

7.8 $D = 4.23 \times 10^{-11}$ m^2/s

7.11 $t = 19.7$ h

7.15 $t = 40$ h

7.18 $T = 1152$ K (879°C)

7.21 **(a)** $Q_d = 252.4$ kJ/mol,
$D_0 = 2.2 \times 10^{-5}$ m^2/s;
(b) $D = 5.4 \times 10^{-15}$ m^2/s

7.24 $T = 1051$ K (778°C)

7.29 $x = 1.52$ mm

7.33 $t_p = 47.4$ min

7.D1 Not possible

Chapter 8

8.4 $l_0 = 257$ mm

8.7 **(a)** $F = 91,000$ N
(b) $l = 120.29$ mm

8.10 $\Delta l = 0.0525$ mm

8.13
$$\left(\frac{dF}{dr}\right)_{r_0} = -\frac{2A}{\left(\dfrac{A}{nB}\right)^{3/(1-n)}} + \frac{(n)(n+1)B}{\left(\dfrac{A}{nB}\right)^{(n+2)/(1-n)}}$$

8.15 **(a)** $\Delta l = 0.47$ mm
(b) $\Delta d = -1.5 \times 10^{-2}$ mm
decrease

8.16 $F = 32,500$ N

8.17 $\nu = 0.280$

8.19 $E = 170.5$ GPa

8.22 **(a)** $\Delta l = 0.075$ mm
(b) $\Delta d = -3 \times 10^{-3}$ mm

8.24 Steel

8.27 **(a)** Both elastic and plastic;
(b) $\Delta l = 0.39$ mm

8.29 **(b)** $E = 62.5$ GPa
(c) $\sigma_y = 285$ MPa
(d) $TS = 370$ MPa
(e) %EL = 16%;
(f) $U_r = 6.5 \times 10^5$ J/m^3

8.36 Figure 8.12: $U_r = 3.32 \times 10^5$ J/m^3
Figure 8.22: $U_r = 4.0 \times 10^5$ J/m^3

8.38 $\sigma_y = 394$ MPa

8.42 $\epsilon_T = 0.237$

8.44 $\sigma_T = 440$ MPa

8.46 Toughness = 3.65×10^9 J/m^3

8.48 $n = 0.136$

8.50 **(a)** ϵ(elastic) $\cong 0.00226$,
ϵ(plastic) $\cong 0.00774$;
(b) $l_i = 463.6$ mm

8.52 **(a)** 125 HB (70 HRB)

8.57 Figure 8.12: $\sigma_w = 125$ MPa,
Figure 8.22: $\sigma_w = 200$ MPa

8.D2 **(a)** $\Delta x = 1.9$ mm; **(b)** $\sigma = 26.44$ MPa

Chapter 9

9.9 Cu: $|\mathbf{b}| = 0.2555$ nm, Mo: $|\mathbf{b}| = 0.2725$ nm;

9.11 $\cos \lambda \cos \phi = 0.408$

9.13 **(b)** $\tau_{crss} = 0.74$ MPa

9.14 $\tau_{crss} = 0.57$ MPa

9.15 For $(111)-[10\bar{1}]$: $\sigma_y = 5.39$ MPa

9.24 $d = 1.5 \times 10^{-2}$ mm

9.25 $d = 6.9 \times 10^{-3}$ mm

9.28 $r_d = 9.17$ mm

9.30 $r_0 = 10.6$ mm

9.32 $\tau_{crss} = 20.2$ MPa

9.37 **(b)** $t \cong 10$ min

9.38 **(b)** $d = 0.085$ mm

9.D1 Is possible

9.D6 Cold work to between 21 and 23% CW [to
$d_0' \cong 12.8$ mm], anneal, then cold work to give
a final diameter of 11.3 mm.

Chapter 10

10.1 $\sigma_m = 2452$ MPa

10.3 $\sigma_c = 16.2$ MPa

10.6 Fracture will not occur

10.8 $a_c = 24$ mm

10.10 Is not subject to detection because $a < 4.1$ mm

10.12 **(b)** -105°C (168 K); **(c)** -95°C (178 K)

10.14 **(a)** $\sigma_{max} = 275$ MPa, $\sigma_{min} = -175$ MPa;
(b) $R = -0.64$;
(c) $\sigma_r = 450$ MPa

10.16 $N_f \cong 1 \times 10^5$ cycles

10.18 **(b)** $S = 250$ MPa; **(c)** $N_f \cong 2 \times 10^6$ cycles

10.19 **(a)** $\tau = 130$ MPa; **(c)** $\tau = 195$ MPa

10.21 **(a)** $t = 120$ min; **(c)** $t = 222$ h

10.27 $\Delta\epsilon/\Delta t = 7.0 \times 10^{-3}$ min^{-1}

10.28 $\Delta l = 22.4$ mm

10.30 $t_r = 600$ h

10.32 650°C (923 K): $n = 11.2$, 730°C (1003 K):
$n = 11.2$, 815°C (1088 K): $n = 8.7$, 925°C
(1198 K): $n = 7.8$

10.33 **(a)** $Q_c = 480,000$ J/mol

10.35 $\dot{\epsilon}_s = 0.118$ s^{-1}

10.D4 $T = 991$ K (718°C)

10.D6 For 5 years: $\sigma = 260$ MPa

Chapter 11

11.1 **(a)** $m_s = 5022$ g;

(b) $C_L = 64$ wt% sugar;

(c) $m_s = 2355$ g

11.5 **(a)** The pressure must be raised to approximately 570 atm (57741 KPa).

11.8 **(a)** $\epsilon + \eta$; $C_\epsilon = 87$ wt% Zn–13 wt% Cu, $C_\eta = 97$ wt% Zn–3 wt% Cu;

(c) Liquid; $C_L = 55$ wt% Ag–45 wt% Cu;

(e) $\beta + \gamma$, $C_\beta = 49$ wt% Zn–51 wt% Cu, $C_\gamma = 58$ wt% Zn–42 wt% Cu;

(g) α; $C_\alpha = 63.8$ wt% Ni–36.2 wt% Cu

11.9 Is not possible

11.12 **(a)** $T = 560°C$ (833 K).

(b) $C_\alpha = 21$ wt% Pb–79 wt% Mg;

(c) $T = 465°C$ (738 K).

(d) $C_L = 67$ wt% Pb–33 wt% Mg

11.14 **(a)** $W_\epsilon = 0.70$, $W_\eta = 0.30$;

(c) $W_L = 1.0$;

(e) $W_\beta = 0.56$, $W_\gamma = 0.44$;

(g) $W_\alpha = 1.0$

11.15 **(a)** $T = 295°C$ (568 K)

11.18 **(a)** $T \cong 270°C$ (543 K)

(b) $C_\alpha = 15$ wt% Sn; $C_L = 43$ wt% Sn

11.19 $C_\alpha = 90$ wt% A–10 wt% B; $C_\beta = 20.2$ wt% A–79.8 wt% B

11.21 Not possible

11.24 **(a)** $V_\epsilon = 0.70$, $V_\eta = 0.30$

11.30 Is possible

11.33 $C_0 = 81.6$ wt% Sn–18.4 wt% Pb

11.35 Schematic sketches of the microstructures called for are shown here.

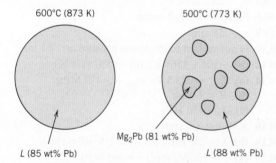

600°C (873 K)   500°C (773 K)

Mg₂Pb (81 wt% Pb)

L (85 wt% Pb)   L (88 wt% Pb)

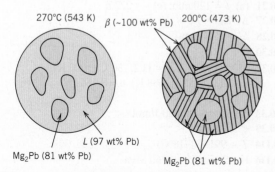

270°C (543 K)   β (~100 wt% Pb) 200°C (473 K)

L (97 wt% Pb)

Mg₂Pb (81 wt% Pb)   Mg₂Pb (81 wt% Pb)

11.42 Eutectics: (1) 12 wt% Nd, 632°C (905 K), $L \to Al + Al_{11}Nd_3$; (2) 97 wt% Nd, 635°C (908 K), $L \to AlNd_3 + Nd$;

Congruent melting point: 73 wt% Nd, 1460°C (1733 K), $L \to Al_2Nd$

Peritectics: (1) 59 wt% Nd, 1235°C (1508 K), $L + Al_2Nd \to Al_{11}Nd_3$; (2) 84 wt% Nd, 940°C (613 K), $L + Al_2Nd \to AlNd$; (3) 91 wt% Nd, 795°C (1068 K), $L + AlNd \to AlNd_2$; (4) 94 wt% Nd, 675°C (948 K), $L + AlNd_2 \to AlNd_3$. No eutectoids are present.

11.45 For point B, $F = 2$

11.48 $C_0' = 0.42$ wt% C

11.51 **(a)** α-ferrite;

(b) 2.27 kg of ferrite, 0.23 kg of Fe_3C;

(c) 0.38 kg of proeutectoid ferrite, 2.12 kg of pearlite

11.53 $C_0' = 0.55$ wt% C

11.55 $C_1' = 0.61$ wt% C

11.58 Is not possible

11.61 Two answers are possible: $C_0 = 1.2$ wt% C and 0.70 wt% C

11.64 HB (alloy) = 128

11.66 **(a)** T(eutectoid) = 650°C

(b) ferrite;

(c) $W_{\alpha'} = 0.65$, $W_p = 0.35$

Chapter 12

12.3 $r^* = 1.31$ nm

12.6 $t = 305$ s

12.8 $r = 4.96 \times 10^{-3}$ min^{-1}

12.10 $y = 0.52$

12.11 **(c)** $t \cong 103$ days

12.15 **(b)** 265 HB (27 HRC)

12.18 **(a)** 50% coarse pearlite and 50% martensite;

(d) 100% martensite;

(e) 40% bainite and 60% martensite;

(g) 100% fine pearlite

12.20 **(a)** martensite;

(c) bainite;

(e) ferrite, medium pearlite, bainite, and martensite;

(g) proeutectoid ferrite, pearlite, and martensite

12.23 **(a)** martensite

12.27 **(a)** martensite;

(c) martensite, proeutectoid ferrite, and bainite

12.36 **(b)** 180 HB (87 HRB); **(g)** 265 HB (27 HRC)

12.38 **(c)** $TS = 915$ MPa

12.39 **(a)** Rapidly cool to about 675°C (948 K) hold for at least 200 s, then cool to room temperature.

12.D1 Not possible

12.D5 Temper at between 400 and 450°C (673 and 723 K) for 1 h

Chapter 13

13.4 $V_{Gr} = 10.4$ vol%

Chapter 14

14.2 **(a)** 8.1% of Mg^{2+} vacancies

14.3 **(a)** $C = 45.9$ wt% Al_2O_3–54.1 wt% SiO_2

14.5 $\rho_t = 0.39$ nm

14.8 $R = 4.26$ mm

14.9 $F_f = 9533$ N

14.12 **(a)** $E_0 = 342$ GPa

(b) $E = 280$ GPa

14.14 **(b)** $P = 0.19$

14.19 **(a)** $T = 2000°C$ (2273 K)

14.21 **(a)** $W_L = 0.71$; **(c)** $W_L = 0.73$

14.22 **(b)** $T \cong 2800°C$ (3073 K); pure MgO

Chapter 15

15.6 $E_r(10) = 4.3$ MPa

15.17 $TS = 41$ MPa

15.25 Fraction sites vulcanized = 0.187

15.27 Fraction of repeat unit sites crosslinked = 0.470

Chapter 16

16.2 $k_{max} = 31.2$ W/m · K; $k_{min} = 28.75$ W/m · K

16.6 $\tau_c = 28.75$ MPa

16.9 Is possible

16.10 $E_f = 70.4$ GPa
$E_m = 2.79$ GPa

16.13 **(a)** $F_f/F_m = 23.4$;

(b) $F_f = 42{,}773$ N
$F_m = 1827$ N

(c) $\sigma_f = 438$ MPa
$\sigma_m = 8.03$ MPa

(d) $\epsilon = 3.35 \times 10^{-3}$

16.15 $\sigma_{cl}^* = 633$ MPa

16.17 $\sigma_{cd}^* = 1340$ MPa

16.26 **(b)** $E_{cl} = 69.1$ GPa

16.D2 Carbon (PAN standard-modulus) and aramid

16.D3 Not possible

Chapter 17

17.9 **(a)** At least 885°C (1158 K)

17.10 **(b)** At least 815°C (1088 K)

17.24 **(b)** $Q_{vis} = 364{,}000$ J/mol

17.38 **(a)** dimethyl terephthalate = 15.183 kg

(b) poly(ethylene terephthalate) = 47.008 kg

17.D5 Maximum diameter = 83 mm

17.D6 Maximum diameter = 75 mm

17.D10 Heat for between 3 and 10 h at 149°C (422 K),
or between about 35 and 500 h at 121°C (394 K)

Chapter 18

18.4 **(a)** $\Delta V = -0.034$ V;

(b) $Fe^{2+} + Cd \rightarrow Fe + Cd^{2+}$

18.6 $[Pb^{2+}] = 2.3 \times 10^{-2}$ M

18.11 $t = 11$ yr

18.16 **(a)** $r = 8.03 \times 10^{-14}$ mol/cm^2·s;

(b) $V_C = -0.019$ V

18.27 Sn: P–B ratio = 1.33; protective

Chapter 19

19.2 $d = 1.88$ mm

19.5 **(a)** $R = 1.7 \times 10^{-3}$ Ω; **(b)** $I = 29.4$ A;

(c) $J = 1.5 \times 10^6$ A/m^2;

(d) $\mathscr{E} = 2.5 \times 10^{-2}$ V/m

19.11 **(a)** $n = 1.25 \times 10^{29}$ m^{-3};

(b) 1.48 free electrons/atom

19.14 **(a)** $\rho_0 = 1.58 \times 10^{-8}$ Ω·m,
$a = 6.5 \times 10^{-11}$ (Ω·m)/°C;

(b) $A = 1.18 \times 10^{-6}$ Ω·m;

(c) $\rho = 4.25 \times 10^{-8}$ Ω·m

19.16 $\sigma = 10.91 \times 10^6$ (Ω·m)$^{-1}$

19.18 **(a)** For Si, 1.40×10^{-12}; for Ge,
1.13×10^{-9}

19.25 $\sigma = 0.082$ (Ω·m)$^{-1}$

19.29 **(a)** $n = 1.44 \times 10^{16}$ m^{-3};

(b) p-type extrinsic

19.31 $\mu_e = 0.50$ m^2/V·s; $\mu_h = 0.02$ m^2/V·s

19.33 $\sigma = 61.6$ (Ω·m)$^{-1}$

19.37 $\sigma = 224$ (Ω·m)$^{-1}$

19.39 $\sigma = 240$ (Ω·m)$^{-1}$

19.42 $B_z = 0.53$ tesla

19.49 $l = 1.6$ mm

19.53 $p_i = 2.26 \times 10^{-30}$ C·m

19.55 **(a)** $V = 19.8$ V;

(b) $V = 98.9$ V;

(e) $P = 2.00 \times 10^{-7}$ C/m^2

19.58 Fraction of ϵ_r due to $P_i = 0.67$

19.D2 $\sigma = 2.44 \times 10^7$ (Ω·m)$^{-1}$

19.D3 Is possible; 23 wt% $< C_{Ni} <$ 32.5 wt%

Chapter 20

20.2 $T_f = 49°C$ (322 K)

20.4 **(a)** $c_v = 139$ J/kg · K;

(b) $c_v = 923$ J/kg · K

20.7 $\Delta l = -9.2$ mm

20.13 $T_f = 129.5°C$ (402.5 K)

20.14 **(b)** $dQ/dt = 9.3 \times 10^8$ J/h

20.21 k(upper) = 26.4 W/m · K

20.25 **(a)** $\sigma = -150$ MPa; compression

20.26 $T_f = 39°C$ (312 K)

20.27 $\Delta d = 0.0251$ mm

20.D1 $T_f = 42.2°C$ (315.2 K)

20.D4 Glass-ceramic: $\Delta T_f = 317°C$ (590 K)

Chapter 21

21.1 **(a)** $H = 10{,}000$ A-turns/m;

(b) $B_0 = 1.257 \times 10^{-2}$ tesla;

(c) $B \cong 1.257 \times 10^{-2}$ tesla;

(d) $M = 1.81$ A/m

21.5 (a) $\mu = 1.2732 \times 10^{-6}$ H/m;
(b) $\chi_m = 12 \times 10^{-3}$
21.7 (a) $M_s = 1.45 \times 10^6$ A/m
21.13 4.6 Bohr magnetons/Mn^{2+} ion
21.19 (b) $\mu_i \cong 3 \times 10^{-3}$ H/m, $\mu_{ri} = 2387$;
(c) $\mu(max) \cong 8.70 \times 10^{-3}$ H/m
21.21 (b) (i) $\mu = 1.10 \times 10^{-2}$ H/m,
(iii) $\chi_m \cong 8750$
21.25 $M_s = 1.69 \times 10^6$ A/m
21.28 (a) 3.0 K: 1.9×10^4 A/m;
(b) 1.14 K

Chapter 22

22.7 $v = 2.09 \times 10^8$ m/s
22.8 Fused silica: 2.13; soda–lime glass: 2.34
22.9 Borosilicate glass: $\epsilon_r = 2.16$; polypropylene:
$\epsilon_r = 2.22$
22.16 $I_T'/I_0' = 0.81$
22.18 $l = 67.3$ mm
22.27 $\Delta E = 1.78$ eV

Index

Power

1 W = 0.239 cal/s	1 cal/s = 4.184 W
1 W = 3.414 Btu/h	1 Btu/h = 0.293 W
1 cal/s = 14.29 Btu/h	1 Btu/h = 0.070 cal/s

Viscosity

1 Pa-s = 10 P	1 P = 0.1 Pa-s

Temperature, T

$$T(K) = 273 + T(°C)$$
$$T(K) = \tfrac{5}{9}[T(°F) - 32] + 273$$
$$T(°C) = \tfrac{5}{9}[T(°F) - 32]$$

$$T(°C) = T(K) - 273$$
$$T(°F) = \tfrac{9}{5}[T(K) - 273] + 32$$
$$T(°F) = \tfrac{9}{5}[T(°C)] + 32$$

Specific Heat

$1 \text{ J/kg} \cdot \text{K} = 2.39 \times 10^{-4} \text{ cal/g} \cdot \text{K}$	$1 \text{ cal/g} \cdot °\text{C} = 4184 \text{ J/kg} \cdot \text{K}$
$1 \text{ J/kg} \cdot \text{K} = 2.39 \times 10^{-4} \text{ Btu/lb}_m \cdot °\text{F}$	$1 \text{ Btu/lb}_m \cdot °\text{F} = 4184 \text{ J/kg} \cdot \text{K}$
$1 \text{ cal/g} \cdot °\text{C} = 1.0 \text{ Btu/lb}_m \cdot °\text{F}$	$1 \text{ Btu/lb}_m \cdot °\text{F} = 1.0 \text{ cal/g} \cdot \text{K}$

Thermal Conductivity

$1 \text{ W/m} \cdot \text{K} = 2.39 \times 10^{-3} \text{ cal/cm} \cdot \text{s} \cdot \text{K}$	$1 \text{ cal/cm} \cdot \text{s} \cdot \text{K} = 418.4 \text{ W/m} \cdot \text{K}$
$1 \text{ W/m} \cdot \text{K} = 0.578 \text{ Btu/ft} \cdot \text{h} \cdot °\text{F}$	$1 \text{ Btu/ft} \cdot \text{h} \cdot °\text{F} = 1.730 \text{ W/m} \cdot \text{K}$
$1 \text{ cal/cm} \cdot \text{s} \cdot \text{K} = 241.8 \text{ Btu/ft} \cdot \text{h} \cdot °\text{F}$	$1 \text{ Btu/ft} \cdot \text{h} \cdot °\text{F} = 4.136 \times 10^{-3} \text{ cal/cm} \cdot \text{s} \cdot \text{K}$

Periodic Table of the Elements

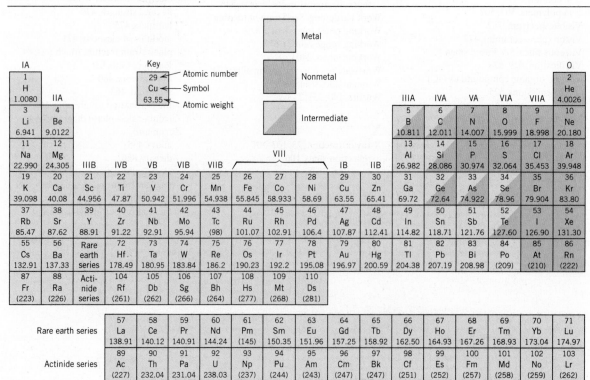

Unit Conversion Factors

Length

$1 \text{ m} = 10^{10} \text{ Å}$	$1 \text{ Å} = 10^{-10} \text{ m}$
$1 \text{ m} = 10^{9} \text{ nm}$	$1 \text{ nm} = 10^{-9} \text{ m}$
$1 \text{ m} = 10^{6} \text{ μm}$	$1 \text{ μm} = 10^{-6} \text{ m}$
$1 \text{ m} = 10^{3} \text{ mm}$	$1 \text{ mm} = 10^{-3} \text{ m}$
$1 \text{ m} = 10^{2} \text{ cm}$	$1 \text{ cm} = 10^{-2} \text{ m}$
$1 \text{ mm} = 0.0394 \text{ in.}$	$1 \text{ in.} = 25.4 \text{ mm}$
$1 \text{ cm} = 0.394 \text{ in.}$	$1 \text{ in.} = 2.54 \text{ cm}$
$1 \text{ m} = 3.28 \text{ ft}$	$1 \text{ ft} = 0.3048 \text{ m}$

Area

$1 \text{ m}^2 = 10^{4} \text{ cm}^2$	$1 \text{ cm}^2 = 10^{-4} \text{ m}^2$
$1 \text{ mm}^2 = 10^{-2} \text{ cm}^2$	$1 \text{ cm}^2 = 10^{2} \text{ mm}^2$
$1 \text{ m}^2 = 10.76 \text{ ft}^2$	$1 \text{ ft}^2 = 0.093 \text{ m}^2$
$1 \text{ cm}^2 = 0.1550 \text{ in.}^2$	$1 \text{ in.}^2 = 6.452 \text{ cm}^2$

Volume

$1 \text{ m}^3 = 10^{6} \text{ cm}^3$	$1 \text{ cm}^3 = 10^{-6} \text{ m}^3$
$1 \text{ mm}^3 = 10^{-3} \text{ cm}^3$	$1 \text{ cm}^3 = 10^{3} \text{ mm}^3$
$1 \text{ m}^3 = 35.32 \text{ ft}^3$	$1 \text{ ft}^3 = 0.0283 \text{ m}^3$
$1 \text{ cm}^3 = 0.0610 \text{ in.}^3$	$1 \text{ in.}^3 = 16.39 \text{ cm}^3$

Mass

$1 \text{ Mg} = 10^{3} \text{ kg}$	$1 \text{ kg} = 10^{-3} \text{ Mg}$
$1 \text{ kg} = 10^{3} \text{ g}$	$1 \text{ g} = 10^{-3} \text{ kg}$
$1 \text{ kg} = 2.205 \text{ lb}_m$	$1 \text{ lb}_m = 0.4536 \text{ kg}$
$1 \text{ g} = 2.205 \times 10^{-3} \text{ lb}_m$	$1 \text{ lb}_m = 453.6 \text{ g}$

Density

$1 \text{ kg/m}^3 = 10^{-3} \text{ g/cm}^3$	$1 \text{ g/cm}^3 = 10^{3} \text{ kg/m}^3$
$1 \text{ Mg/m}^3 = 1 \text{ g/cm}^3$	$1 \text{ g/cm}^3 = 1 \text{ Mg/m}^3$
$1 \text{ kg/m}^3 = 0.0624 \text{ lb}_m/\text{ft}^3$	$1 \text{ lb}_m/\text{ft}^3 = 16.02 \text{ kg/m}^3$
$1 \text{ g/cm}^3 = 62.4 \text{ lb}_m/\text{ft}^3$	$1 \text{ lb}_m/\text{ft}^3 = 1.602 \times 10^{-2} \text{ g/cm}^3$
$1 \text{ g/cm}^3 = 0.0361 \text{ lb}_m/\text{in.}^3$	$1 \text{ lb}_m/\text{in.}^3 = 27.7 \text{ g/cm}^3$

Force

$1 \text{ N} = 10^{5} \text{ dynes}$	$1 \text{ dyne} = 10^{-5} \text{ N}$
$1 \text{ N} = 0.2248 \text{ lb}_f$	$1 \text{ lb}_f = 4.448 \text{ N}$

Stress

$1 \text{ MPa} = 145 \text{ psi}$	$1 \text{ psi} = 6.90 \times 10^{-3} \text{ MPa}$
$1 \text{ MPa} = 0.102 \text{ kg/mm}^2$	$1 \text{ kg/mm}^2 = 9.806 \text{ MPa}$
$1 \text{ Pa} = 10 \text{ dynes/cm}^2$	$1 \text{ dyne/cm}^2 = 0.10 \text{ Pa}$
$1 \text{ kg/mm}^2 = 1422 \text{ psi}$	$1 \text{ psi} = 7.03 \times 10^{-4} \text{ kg/mm}^2$

Fracture Toughness

$1 \text{ psi}\sqrt{\text{in.}} = 1.099 \times 10^{-3} \text{ MPa}\sqrt{\text{m}}$	$1 \text{ MPa}\sqrt{\text{m}} = 910 \text{ psi}\sqrt{\text{in.}}$

Energy

$1 \text{ J} = 10^{7} \text{ ergs}$	$1 \text{ erg} = 10^{-7} \text{ J}$
$1 \text{ J} = 6.24 \times 10^{18} \text{ eV}$	$1 \text{ eV} = 1.602 \times 10^{-19} \text{ J}$
$1 \text{ J} = 0.239 \text{ cal}$	$1 \text{ cal} = 4.184 \text{ J}$
$1 \text{ J} = 9.48 \times 10^{-4} \text{ Btu}$	$1 \text{ Btu} = 1054 \text{ J}$
$1 \text{ J} = 0.738 \text{ ft} \cdot \text{lb}_f$	$1 \text{ ft} \cdot \text{lb}_f = 1.356 \text{ J}$
$1 \text{ eV} = 3.83 \times 10^{-20} \text{ cal}$	$1 \text{ cal} = 2.61 \times 10^{19} \text{ eV}$
$1 \text{ cal} = 3.97 \times 10^{-3} \text{ Btu}$	$1 \text{ Btu} = 252.0 \text{ cal}$